Optics in Our Time

Mohammad D. Al-Amri
Mohamed M. El-Gomati
M. Suhail Zubairy

Optics in Our Time

Editors
Mohammad D. Al-Amri
National Center for Applied Physics
King Abdulaziz City for Science
   and Technology
Riyadh, Saudi Arabia

Mohamed M. El-Gomati
Department of Electronics
University of York
Heslington, UK

M. Suhail Zubairy
Department of Physics and Astronomy
Texas A&M University
College Station, TX, USA

ISBN 978-3-319-31902-5   ISBN 978-3-319-31903-2   (eBook)
DOI 10.1007/978-3-319-31903-2

Library of Congress Control Number: 2016959602

Printed on acid-free paper

This Springer imprint is published by Springer Nature
The registered company is Springer International Publishing AG
The registered company address is: Gewerbestrasse 11, 6330 Cham, Switzerland

Foreword

Light has occupied a position of importance in our attempt to understand the world around us. The earliest studies going back to the dawn of civilization related to our attempts to understand vision and the properties of optical materials. The modern era in optics is rooted in the great work of Ibn al-Haytham whose work on the nature of light and its applications had a long-lasting impact. In our generation, the discovery of laser has opened up not only new areas of research but also had great impact on a number of technologies. Lasers have revolutionized the fields of communication, medicine, and biotechnology. It has influenced the art, architecture, and printing. It is therefore befitting that United Nations has declared 2015 as the International Year of Light and Light-Based Technologies to celebrate these great achievements. Saudi Arabia is one of the sponsors of this initiative that has ignited a number of activities all around the world. This volume, which covers the history of light and its applications to many diverse branches of science, contains articles by some of the leading scientists who have played a key role in advancing the frontiers in our own times.

Turki S.M. Al-Saud
King Abdulaziz City for Science and Technology, Riyadh, Saudi Arabia

Preface

Light and light-based technologies have played an important role in transforming our lives via scientific contributions spanned over thousands of years. In this book, we present a vast collection of articles on various aspects of light and its applications in the contemporary world at a popular or semi-popular level. These chapters are written by the world authorities in their respective fields. This is therefore a rare volume where the world experts have come together to present the developments in this most important field of science in an almost pedagogical manner.

This volume covers five aspects related to light. The first presents two articles, one on the history of the nature of light and the other on the scientific achievements of Ibn al-Haitham (Alhazen), who is broadly considered the father of modern optics. These are then followed by an article on ultrafast phenomena and the invisible world. The third part includes papers on specific sources of light, the discoveries of which have revolutionized optical technologies in our lifetime. They discuss the nature and the characteristics of lasers, solid-state lighting based on the light emitting diode (LED) technology, and finally modern electron optics and its relationship to the Muslim golden age in science. The book's fourth part discusses various applications of optics and light in today's world, including biophotonics, art, optical communication, nanotechnology, the eye as an optical instrument, remote sensing, and optics in medicine. In turn, the last part focuses on quantum optics, a modern field that grew out of the interaction of light and matter. Topics addressed include atom optics, slow, stored and stationary light, optical tests of the foundation of physics, quantum mechanical properties of light fields carrying orbital angular momentum, quantum communication, and wave-particle dualism in action.

We are grateful to many individuals and organizations whose contributions and cooperation were invaluable in compiling this book. First and foremost, we are grateful to all the authors who took their time in writing these articles for the general audience. We are very grateful to the leadership and the staff at *King Abdulaziz City for Science and Technology* (KACST) for their generous support in the completion of this project. Khalid Al Zahrani ought to be thanked with whom the idea of this book was triggered over a cup of tea.

We have however one deep regret: one of the authors, Nobel Laureate Ahmed Zewail, who enthusiastically supported this volume and contributed an important chapter passed away on August 2, 2016, before the publication of this book.

Mohammad D. Al-Amri
Riyadh, Saudi Arabia

Mohamed M. El-Gomati
York, UK

M. Suhail Zubairy
College Station, TX, USA

Contents

V   Quantum Optics

About the Editors and Authors

Mohammad D. Al-Amri

received his MSc in physics with distinction from Sussex University, UK, in 2001. He went on and got his PhD from York University and was the recipient of the Stott Prize in Physics for the best PhD thesis in 2004. He then joined the National Center for Applied Physics (NCAP) at King Abdulaziz City for Science and Technology, where he has been working as a professor.

He was the recipient of the CO/ICTP Gallieno Denardo Award Winner (2013) and the senior membership of the Optical Society of America (2012). He has been working on different areas of research related to quantum optics and quantum informatics, where the focus is on quantum optical lithography and microscopy, weak measurement, and direct quantum communication. He has published around 55 journal papers, within about 35 published in the Journal of PRL and PRA, and has got 5 patents. Some of Dr. AlAmri's research work has been highlighted in semipopular press. He has given many lectures, seminars, and invited talks at universities and international conferences.

Mohamed M. El-Gomati

is professor of electronics at the University of York, UK. His research interests are in the fields of surface science and electron optics with particular emphasis on the development of novel instrumentation for nanostructure analysis. He is the author and co-author of more than 200 articles and patents in these fields. He is a fellow of the Institute of Physics (IoP) and the Royal Microscopical Society (RMS). His interests extend to history of physics with particular emphasis on history of optics within Muslim civilization. He is the chairman of the Foundation for Science, Technology and Civilisation (UK) and a trustee of the educational charity, Curriculum Enrichment for the Future, and is an advisor to a number of UK and overseas universities. He was awarded the UKESCA Award (1993), the Cosslett Award by the Microbeam Society of America (2008), the Fazlur Rahman Prize for Science and Engineering (2009), and the British Muslims Award for Science (2013). In 2012, Professor El-Gomati was awarded an OBE for his services to science.

M. Suhail Zubairy

is a university distinguished professor of physics and the holder of the Munnerlyn-Heep Chair in Quantum Optics at the Texas A&M University. He received his PhD from the University of Rochester in 1978. He served as professor of electronics and the founding chairman of the Department of Electronics at the Quaid-i-Azam University before joining Texas A&M University in 2000. Prof. Zubairy's research interests include quantum optics and laser physics. He has published over 300 research papers on topics such as precision microscopy and lithography, quantum computing, noise-free amplification, and atomic coherence effects. He is the co-author of two books, one on quantum optics and the other on quantum computing devices. He has received many honors including the Willis E. Lamb Award for Laser Science and Quantum Optics, Alexander von Humboldt Research Prize, the Outstanding Physicist Award from the Organization of Islamic Countries, the Abdus Salam Prize in Physics, the International Khwarizmi Award from the president of Iran, the Orders of Hilal-e-Imtiaz and Sitara-e-Imtiaz from the president of Pakistan, and the George H. W. Bush Award for Excellence in International Research. He is an elected member of the Pakistan Academy of Sciences and a fellow of the American Physical Society and the Optical Society of America.

Govind P. Agrawal

received the MS and PhD degrees from the Indian Institute of Technology, New Delhi, in 1971 and 1974, respectively. After holding positions at the Ecole Polytechnique, France, the City University of New York, and AT&T Bell Laboratories, Dr. Agrawal joined in 1989 the faculty of the Institute of Optics at University of Rochester, where he is currently James C. Wyant Professor of Optics. His research interests focus on optical communications, nonlinear photonics, and laser physics. He is an author or co-author of more than 400 research papers and eight books. His books on nonlinear fiber optics (Academic Press, 5th ed., 2013) and fiber-optic communication systems (Wiley, 4th ed., 2010) are used worldwide for research and teaching. Since 2014, he is serving as editor-in-chief of the journal *Advances in Optics and Photonics*.

Prof. Agrawal is a fellow of the IEEE and OSA (the Optical Society) and a life fellow of the Optical Society of India. In 2012, the IEEE Photonics Society honored him with its prestigious Quantum Electronics Award. He received in 2013 the Riker University Award for Excellence in Graduate Teaching. More recently, he was awarded the 2015 Esther Hoffman Beller Medal of the Optical Society.

Pablo Artal

is a full professor of optics at the University of Murcia, Spain. He spent several periods doing collaborative research in laboratories in Europe, Australia, and the USA. He is a fellow member of the OSA, ARVO, and EOS. He received the prestigious 2013 Edwin H. Land Medal Award, and he is the recipient of the exclusive "ERC advanced grant" in 2013. He received the "Rey Jaime I" Award for Applied Research in 2015. He has published more than 180 reviewed papers that received 7000 citations (h-index: 43) and presented more than 150 invited talks in international meetings and is also a co-inventor of 20 international patents. He has pioneered a number of highly innovative advances in the methods for studying the optics of the eye and has contributed substantially to our understanding of the factors that limit human visual resolution. Dr. Artal is the founder of Voptica SL, a spin-off company developing the concept he invented of adaptive optics vision analyzers. He has been the mentor of many graduate and postdoctoral students. His personal science blog is followed by readers, mostly graduate students and fellow researchers, from around the world. He has been editor of the *Journal of the Optical Society of America A* and the *Journal of Vision*.

Robert W. Boyd

was born in Buffalo, New York. He received the BS degree in physics from MIT and the PhD degree in physics from the University of California at Berkeley. His PhD thesis was supervised by Charles Townes and involves the use of nonlinear optical techniques in infrared detection for astronomy. Professor Boyd joined the faculty of the University of Rochester in 1977 and in 2001 became the M. Parker Givens Professor of Optics and Professor of Physics. In 2010, he became professor of physics and Canada Excellence Research Chair in Quantum Nonlinear Optics at the University of Ottawa. His research interests include studies of "slow" and "fast" light propagation, quantum imaging techniques, nonlinear optical interactions, studies of the nonlinear optical properties of materials, and the development of photonic devices including photonic biosensors. Professor Boyd has written two books, co-edited two anthologies, published over 400 research papers ($\approx$29,000 citations, Google h-index 71), and been awarded nine patents. He is the 2009 recipient of the Willis E. Lamb Award for Laser Science and Quantum Optics, the 2010 recipient of a Humboldt Research Prize, and the 2014 recipient of the Quantum Electronics Award of the IEEE Photonics Society.

Charles M. Falco

has joint appointments as professor of optical sciences and professor of physics at the University of Arizona where he holds the UA Chair of Condensed Matter Physics. He is a fellow of four professional societies (the American Physical Society, the Institute of Electrical and Electronics Engineers (IEEE), the Optical Society of America, and the Society of Photo-optical Instrumentation Engineers (SPIE)), has published more than 275 scientific manuscripts, co-edited two books, has seven US patents, and has given over 400 invited talks at conferences, research institutions, and cultural organizations in 33 countries. In addition to his scientific research, he was co-curator of the Solomon R. Guggenheim museum's "The Art of the Motorcycle" which, with over 2 million visitors in New York, Chicago, Bilbao, and the Guggenheim Las Vegas, was by far the most successful exhibition of industrial design ever assembled. More recently, he and the world-renowned artist David Hockney found artists of such repute as van Eyck, Bellini, and Caravaggio who used optical projections in creating portions of their work. Three international conferences have been organized around these discoveries, and recognition for them includes the 2008 Ziegfeld Lecture Award from the National Art Education Association.

Michael Fleischhauer

is professor of theoretical physics at the University of Kaiserslautern, Germany. He made his PhD in physics at the University of Friedrich-Schiller University Jena on the theory of nonclassical light and his Habilitation on Electromagnetically Induced Transparency (EIT) and its applications at the Ludwig Maximilian University (LMU) of Munich, both in Germany. His interests are in quantum optics and many-body physics of ultracold quantum gases. He is an expert in numerical methods for strongly interacting systems in low dimensions, and his research interests include topological systems. Among other things, he has developed the method of light storage using EIT, which is by now one of the main techniques for quantum memories for light and key for photon-based quantum networks. He served as department head of the selection committee of the Alexander von Humboldt Foundation and at the boards of several physics journals and is currently a member of the editorial board of *Physical Review Letters*. Michael Fleischhauer is member of the executive board of the German Physical Society where he has been the spokesperson of the division of quantum optics and photonics. He is a member of the senate of the German Research Foundation and has been elected to the Academy of Sciences and Literature in Mainz.

Edward S. Fry

distinguished professor and former head of the Department of Physics and Astronomy at Texas A&M University, holds the George P. Mitchell Chair in Experimental Physics and has been with Texas A&M since 1969. He is past chair of the Texas Section of the American Physical Society and is a fellow of both the American Physical Society and the Optical Society of America. He received the Association of Former Students Distinguished Faculty Teaching Award (1993), the Texas A&M Distinguished Scientist Award of Sigma Xi (2001), and the Association of Former Students Distinguished Faculty Achievement Award (2012).

Dr. Fry's research interests cover the gamut from basic research to applied research. Some notable achievements include (i) one of the first, and definitive, Bell inequality tests of the foundations of quantum mechanics, addressing questions first raised by Einstein; (ii) the first observations of lasing without (population) inversion (LWI); (iii) a new integrating cavity technique for the measurement of optical absorption in the presence of even severe scattering, leading to what are now widely considered the standard reference data for pure water absorption; and (iv) a new diffuse reflector whose reflectivity is so high that ring-down spectroscopy in an integrating cavity is now possible.

Colin J. Humphreys

is professor of materials science and director of research in the Department of Materials Science and Metallurgy, University of Cambridge, and a fellow of Selwyn College, Cambridge. He is a fellow of the Royal Society and a fellow of the Royal Academy of Engineering. He founded and directs the Cambridge Centre for Gallium Nitride (GaN). He founded two spin-off companies to exploit the research of his group on low-cost LEDs for home and office lighting. The companies were acquired in February 2012 by Plessey, which is now manufacturing LEDs based on this technology at their factory in Plymouth, UK. He also founded and directs the Cambridge/Rolls-Royce Centre for Advanced Materials for Aerospace.

Gediminas Juzeliūnas

is a director of the Institute of Theoretical Physics and Astronomy of Vilnius University, Lithuania. He is also a principal researcher (research professor) at the institute. Dr. Juzeliunas completed a PhD in 1986 in theoretical condensed matter physics at Vilnius University, studying optical properties of excitons in confined geometries. Subsequently he held a two-year postdoctoral appointment at the University of East Anglia, England, shifting his research area toward quantum optics. Dr. Juzeliunas was a Humboldt research fellow at the University of Ulm, Germany (1997–1998), and a Fulbright scholar at the University of Oregon in the USA (2000–2001). Dr. Juzeliunas received a National State Prize for Science of Lithuania in 2008, a Vilnius University Rector's Award in 2010, and a Jucys Prize for Theoretical Physics of the Lithuanian Academy of Science in 2013. His current research focuses on ultracold atomic gases, slow light, and metamaterials. In particular, this includes a pioneering theoretical work on light-induced gauge potential for ultracold atoms.

Olga Kocharovskaya

is the distinguished professor in the Department of Physics and Astronomy, Texas A&M University. She joined the Texas A&M faculty in 1998 after 12 years at the Institute of Applied Physics of the Russian Academy of Sciences. She made a number of contributions to laser science and quantum optics.

These include the predictions of the phenomena of electromagnetically induced transparency and lasing without inversion as well as suggestion and experimental realization of the various schemes for coherent control of gamma-ray nuclear transitions. A fellow of both the American Physical Society and Optical Society of America, she has earned the Willis E. Lamb Medal for Laser Physics and Quantum Electronics, the Sigma Xi Distinguished Scientist Award, the Texas A&M Association of Former Students Distinguished Achievement Award in Research, and the Texas A&M University Distinguished Professor Award.

Mario Krenn

is a PhD student since 2012 in the group of Anton Zeilinger at the University of Vienna and Institute for Quantum Optics and Quantum Information (IQOQI) of the Austrian Academy of Sciences. He is investigating quantum properties of spatial structures of photons and high-dimensional Hilbert spaces. This involves the certification and long-distance transmission of spatial mode entanglement. In connection with complex high-dimensional multipartite quantum states, he is interested in automated computer-designed quantum experiments. In his free time, he enjoys playing table soccer and fantasizing about artificial general intelligence.

Mehul Malik

is a Marie Curie postdoctoral fellow in the group of Professor Anton Zeilinger at the Institute for Quantum Optics and Quantum Information in Vienna, Austria. Originally from New Delhi, India, Mehul received his PhD in optics in 2013 from the University of Rochester under the supervision of Professor Robert Boyd and a bachelor of arts in 2006 from Colgate University. He is currently working on creating the first multi-photon entangled states in high dimensions using the orbital angular momentum property of light. His broader research interests lie in the fields of fundamental quantum optics, quantum imaging, and quantum information. Mehul has published 19 papers in internationally recognized journals such as *Reviews of Modern Physics*, *Nature Communications*, and *Physical Review Letters*. He has delivered talks on his research in 11 different countries, including 5 invited talks at international conferences. Outside the laboratory, Mehul is an avid skier, loves to cook, and enjoys dancing salsa.

Alexis Méndez

received a PhD degree in electrical engineering from Brown University, USA, in 1992. He is president of MCH Engineering LLC—a consulting firm specializing in optical fiber sensing technology. Dr. Mendez was the former group leader of the Fiber Optic Sensors Lab within ABB Corporate Research (USA) where he led R&D activities for the development of fiber sensors for use in industrial plant, oil and gas, and high-voltage electric power applications. He has written 60 technical publications, taught several short courses on fiber sensors, holds 5 US patents, and is a recipient of an R&D100 award.

Dr. Mendez is a member of the OFS International Steering Committee, is a fellow of SPIE, and was past chairman of the 2006 International Optical Fiber Sensors Conference (OFS-18) and past technical chair of the 2nd Workshop on Specialty Optical Fibers and Their Applications (WSOF-2). He is also a member of the International Society for Health Monitoring of Intelligent Infrastructure (ISHMII) Committee. He is co-editor of the *Specialty Optical Fibers Handbook* and co-author of SPIE's *Fiber Optical Sensors Book: Fundamentals and Applications, 4th Ed.*

Pierre Meystre

obtained his physics diploma and PhD from the Swiss Federal Institute of Technology in Lausanne, Switzerland, and the Habilitation in Theoretical Physics from the University of Munich, Germany. He is a Regents Professor of Physics and Optical Sciences at the University of Arizona and since 2013 serves as lead editor of *Physical Review Letters*. His research interests include theoretical quantum optics, atomic physics, ultracold science, and quantum optomechanics. He has published well over 300 refereed papers and is the author of the text *Elements of Quantum Optics*, together with Murray Sargent III, and of the monograph "Atom Optics." He is a recipient of the Humboldt Foundation Research Prize for Senior US Scientists, the R.W. Wood Prize of the Optical Society of America, and the Willis E. Lamb Award for Laser Science and Quantum Optics. He is a fellow of the American Physical Society, the Optical Society of America, and the American Association for the Advancement of Science.

Munir H. Nayfeh

nanotechnology pioneer, Islamic Academy of Science fellow, and physics professor at the University of Illinois, received his BSc from the American University of Beirut and PhD from Stanford University in atomic laser spectroscopy. He was a researcher at Oak Ridge National Laboratory, lecturer at Yale University, and consultant at Argonne National Laboratory. He is chairman of the advisory board at center of excellence in nanotechnology (King Fahd University), ex-vice-chairman of the International Science Council at King Abdullah Institute for Nanotechnology, and advisor at the Nanotechnology Center at An-Najah University and Saigon Hi-Tech Park, Vietnam. He served on panels for King Abdullah University (KAUST) and

Royal Commission, Jubail. Nayfeh co-authored *Electricity and Magnetism* (translated into Farsi) and co-edited three laser books. He presents nanotechnology science fiction using the trademark "Dr. Nano." He holds the largest number of patents in nanosilicon worldwide (22 US-European, 18 issued) and is a founder of nanotechnology companies: NanoSi Advanced Technologies, Nano Silicon Solar, and Parasat-Nanosi (Kazakhstan). Professor Nayfeh developed a process for creating highly luminescent ultrasmall silicon nanoparticles with electronics, photonics, and biomedicine applications, a process for combining lasers with high electric fields of scanning electron microscopes to pin/write atoms on surfaces with atomic resolution, and a process using strong laser light to detect single atoms.

Roshdi Rashed

is emeritus research director (distinguished class) at the Centre National de la Recherche Scientifique (CNRS, Paris) and the former director of the Center for History of Arabic and Medieval Sciences and Philosophy at Paris VII—Denis Diderot University. He is honorary professor at the University of Tokyo and emeritus professor at the University of Mansoura. From 1965 on, his field of research was the history of mathematics and mathematical sciences and their applications. He is the author of about fifty books, mainly in French and in English, including six substantial volumes devoted to Ibn al-Haytham's mathematics, optics, and astronomy. These volumes, written in French—with a critical edition of Ibn al-Haytham's texts—have been translated into English and Arabic.

Bahaa Saleh

has been dean of CREOL, the College of Optics and Photonics at the University of Central Florida, since 2009. He was born in Cairo, Egypt, and received the PhD degree from Johns Hopkins University in 1971. He held positions at the University of Santa Catarina in Brazil; Kuwait University; Max Planck Institute in Germany; University of California-Berkeley; European Molecular Biology Laboratory; Columbia University; University of Wisconsin-Madison, where he was chair of electrical and computer engineering (1990–1994); and Boston University, where he was chair of electrical and computer engineering (1994–2008). He has made significant contributions to coherence and statistical optics, nonlinear optics, quantum optics, and image science, and his publications include more than 600 journal and conference papers and 3 books: *Photoelectron Statistics* (Springer, 1978), *Fundamentals of Photonics* (Wiley, 2007, with M. C. Teich), and *Subsurface Imaging* (Cambridge, 2011). He served as editor-in-chief of the *Journal of the Optical Society of America A* (1991–1997) and founding editor of OSA's *Advances in Optics and Photonics* (2008–2013). He is a fellow of the IEEE, OSA, SPIE, APS, and Guggenheim Foundation and a recipient of the OSA Beller Award, OSA Meese Medal, OSA Distinguished Service Award, SPIE BACUS award, and Kuwait Prize.

Thomas Scheidl

is a senior scientist at the Institute for Quantum Optics and Quantum Information of the Austrian Academy of Sciences and a team leader within the group of Prof. Anton Zeilinger. He studied experimental physics at the University of Vienna and received a doctor's degree in 2009. His main field of research is the experimental investigation of fundamental questions in quantum physics as well as the development of prototypes for secure quantum communications and quantum information, mainly for free-space systems. He was experimentally active in many international collaborations and responsible for a number of projects funded by the European Space Agency. He has published more than 12 papers in scientific journals such as in *Nature*, *Nature Physics*, and *Physical Review Letters*, and in 2012 and 2013, he received fellowships from the Energie AG (Innovation Fund) and from the Sohmen Far East Foundation for his contribution to the field of long-distance quantum communication.

Wolfgang P. Schleich

is engaged in research on quantum optics ranging from the foundations of quantum physics via general relativity to number theory. He was educated at the Ludwig Maximilian University (LMU) of Munich and studied with Marlan O. Scully at the University of New Mexico, Albuquerque, and the Max Planck Institute for Quantum Optics, Garching. Moreover, he was also a postdoctoral fellow with John Archibald Wheeler at the University of Texas at Austin. Professor Schleich is a member of several national and international academies and has received numerous prizes and honors for his scientific work such as the Gottfried Wilhelm Leibniz Prize, the Max Planck Research Award, and the Willis E. Lamb Award for Laser Science and Quantum Optics. He is also a distinguished adjunct professor at the University of North Texas and a faculty fellow at Texas A&M University Institute for Advanced Study. His textbook, *Quantum Optics in Phase Space*, has been translated into Russian, and a Chinese edition was published in 2010.

Marlan O. Scully

(Baylor, Princeton, and Texas A&M) has worked on a variety of problems in laser physics and quantum optics including the first quantum theory of the laser with Lamb, the laser phase transition analogy and its applications to the Bose condensate, experimental demonstrations of lasing without inversion, and ultraslow light in hot gases via quantum coherence. His introduction of entanglement interferometry to quantum optics has shed light on the foundations of quantum mechanics, e.g., the quantum eraser. Recently, he and his colleagues have applied quantum coherence to remote sensing of anthrax and probing through turbid medium such as skin and plant tissue. Scully is currently a distinguished university professor at Texas A&M University and also holds positions at Princeton and Baylor Universities. He has been elected to the US National Academy of Sciences and the Max Planck Society. He has recently been awarded the OSA Frederic Ives Medal/Quinn Prize, the DPG/OSA Herbert Walther Award, and the Commemorative Silver Medal of the Senate of the Czech Republic (by K. Chapin).

Yanhua H. Shih

professor of physics at the University of Maryland, Baltimore Campus (UMBC), received his PhD in 1987 from the University of Maryland at College Park, USA. He started the Quantum Optics Laboratory at UMBC in the fall of 1989. His laboratory has been recognized as one of the leading groups in the field of quantum optics that attempts to probe the foundations of quantum theory. In the past 10 years, he published more than 100 papers in leading refereed journals and given more than 100 invited presentations in national and international professional conferences and workshops. His book *An Introduction to Quantum Optics: Photon and Biphoton Physics*, published in 2011, is a good summary of his theoretical and experimental research. Yanhua Shih is a winner of the 2002 Willis E. Lamb Medal for pioneering contributions to quantum electronics and especially the study of spatial coherence effects of multi-photon entangled states.

Alexei V. Sokolov

obtained a master's diploma from Moscow Institute of Physics and Technology (1994) and a physics PhD from Stanford University (2001). Currently at Texas A&M University, Sokolov holds a professor position in physics and astronomy and a Stephen Harris Professorship in Quantum Optics. His overall expertise is in the field of laser physics, nonlinear optics, ultrafast science, and spectroscopy. His research interests center around applications of molecular coherence to quantum optics, ultrafast laser science and technology including generation of sub-cycle optical pulses with prescribed temporal shapes, and studies of ultrafast atomic, molecular, and nuclear processes, as well as applications of quantum coherence in biological and defense-oriented areas. Sokolov is an OSA fellow; his awards include the Lomb Medal (OSA, 2003), the Hyer Award (TX section APS, 2007), and the Treat Award (Texas A&M Research Foundation, 2011).

Rupert Ursin

is a group leader and senior scientist at the IQOQI. His main field of research is to develop quantum communication and quantum information processing technologies, mainly for free-space but also for fiber-based systems. The scope of his work ranges from near-term engineering solutions for secure key sharing (quantum cryptography) to more speculative research (de-coherence of entangled states in gravitational fields). Experiments on quantum communication and teleportation using entangled photon pairs are also among his interests, with the long-term goal of a future global quantum network based on quantum repeaters. He has published more than 49 papers in peer-reviewed scientific journals such as *Science and Nature*. He has been experimentally active in numerous international collaborations in Germany, Italy, Spain, and the USA, as well as in Japan. To date, several of his publications were selected as yearly highlights by the British PhysicsWeb and others. In 2008, he received the Award for the Telecommunications Advancement Research Fellowship (National Institute of Information and Communications Technology (NICT), Tokyo, Japan) and in 2010, the Christian Doppler Prize. He has presented invited talks on original scientific results at more than 60 prestigious international conferences. He currently holds a guest professorship at the University of Science and Technology (USTC) in Shanghai, China.

Dmitri V. Voronine

is a research assistant professor at the Institute for Quantum Science and Engineering, Texas A&M University, and a visiting scholar at Baylor University. His research interests are in the experimental and theoretical spectroscopy, quantum optics, and ultrafast nano-photonics with applications to the investigation, imaging, and control of complex systems. He has made contributions in coherent control of nano-optical excitations and multidimensional spectroscopy, time-resolved surface-enhanced coherent Raman spectroscopy, and quantum biophotonics.

Thomas Walther

studied physics at the Ludwig Maximilian University (LMU) of Munich. After receiving his PhD from the University of Zürich in physical chemistry in 1994, he joined the Physics Department of Texas A&M University, College Station, TX, as a research assistant scientist, where he became an assistant professor in 1998. In 2002, he moved to the TU Darmstadt, Germany, as a full professor. His research interests include basic research in quantum optics as well as lasers and their applications.

Vladislav V. Yakovlev

is currently professor in the Department of Biomedical Engineering at Texas A&M University. He also holds a joint appointment in the Department of Physics and Astronomy. He got his BS/MS/PhD degrees in physics/quantum electronics from Moscow State University, Russia. He completed his postdoctoral training in the Department of Chemistry and Biochemistry, University of California-San Diego, where he first developed the concept of optical parametric amplification of white-light continuum, which is nowadays employed as a standard tool in ultrafast optical spectroscopy, and then performed a series of groundbreaking experiments demonstrating coherent control of molecular systems. Since joining the Physics Department at the University of Wisconsin-Milwaukee, Dr. Yakovlev has been involved in instrumentation development for biomedical research. He was the first to demonstrate broadband coherent anti-Stokes Raman microscopy, Raman photoacoustic imaging, and deep-tissue coherent Raman imaging, including anthrax detection in the mail. In 2012, he moved to Texas A&M University where he continuously developed new spectroscopic tools for biological, medical, and environmental sensing and imaging. He has published over 200 research publications and gave over 150 scientific talks. Dr. Yakovlev is a fellow of the Optical Society of America, the American Institute of Medical and Biological Engineering, and the International Society for

Optics and Photonics. His current research interests include biomechanics on a microscale level, nanoscopic optical imaging of molecular and cellular structures, protein spectroscopy and structural dynamics, bioanalytical applications of optical technology and spectroscopy, environmental sensing, and deep-tissue imaging and sensing.

Anton Zeilinger

(PhD 1971, Vienna University) is professor of physics at the University of Vienna and president of the Austrian Academy of Sciences. He has performed many groundbreaking experiments in quantum mechanics, from important fundamental tests all the way to innovative applications. Most of his research concerns the fundamental aspects and applications of quantum entanglement. He is one of the pioneers of the new field of quantum information science. The most important stations of his career include M.I.T., the Collége de France, the Technical Universities of Munich and Vienna, Oxford University, and the University of Innsbruck. Among his many awards and prizes are the German Order Pour le Mérite, Wolf Prize, the Inaugural Isaac Newton Medal of the Institute of Physics, the Medaille du Collége de France, the Great Cross of Merit with Star of the Federal Republic of Germany, and the King Faisal Prize. He is a member of the Austrian, Berlin-Brandenburg, Slovak, Belarus, and Serbian Academies of Science, the German National Academy Leopoldina, the Académie des Sciences Paris, the US National Academy of Sciences, and the European Academy of Sciences and a fellow of the American Association for the Advancement of Science (AAAS), the American Physical Society (APS), and the World Academy of Sciences (TWAS).

Ahmed H. Zewail

is the Linus Pauling Chair Professor of Chemistry and Physics and director of the Center for Physical Biology at Caltech. He is the sole recipient of the 1999 Nobel Prize for the development of the field of *femtochemistry*. In the post-Nobel era, he developed *4D electron microscopy* for the direct visualization of matter in space and time. Dr. Zewail's other honors include fifty honorary degrees, orders of merits, postage stamps, and more than a hundred international awards.

He has published some 600 articles and 14 books and is known for his effective public lectures and writings, not only on science but also in global affairs. For his leadership role in these world affairs, he received, among others, the "Top American Leaders Award" from *The Washington Post* and Harvard University. In 2009, President Barack Obama appointed him to the Council of Advisors on Science and Technology, and in the same year, he was named the first US Science Envoy to the Middle East. Subsequently, the Secretary General of the United Nations Ban Ki-moon invited Dr. Zewail to join the UN Scientific Advisory Board. In Egypt, he serves in the Council of Advisors to the President. Following the 2011 Egyptian revolution, the government established "Zewail City of Science and Technology" as the national project for scientific renaissance, and Dr. Zewail became its first chairman of the board of trustees.

Aleksey Zheltikov

received his PhD degree from M.V. Lomonosov Moscow State University in 1990. He received his doctor of science degree from the same university in 1999. He has been a professor at M.V. Lomonosov Moscow State University since 2000, the head of Neurophotonics Laboratory at Kurchatov Institute Russian Research Center since 2010, and the head of Advanced Photonics International Laboratory at the Russian Quantum Center since 2012. Since 2010, he is a professor at Texas A&M University. He is the author of more than 600 peer-reviewed scientific publications. He is the winner of the Russian Federation State Prize for young researchers (1997), Lamb Award for achievements in quantum electronics (2010), Shuvalov Prize for research at Moscow State University (2001), and Kurchatov Prize for achievements in neurophotonics (2014).

Dandan Zhu

received the BE and MS degrees in materials science from Tsinghua University, Beijing, China, in 2001 and 2003, respectively, and the PhD degree in materials science from the University of Cambridge, UK, in 2007.

Her PhD project was on metal-organic vapor phase epitaxy growth and characterization of near-UV emitting quantum well and light-emitting diode (LED) structures on sapphire substrates. After graduation, she worked as a research associate with the Cambridge Centre for Gallium Nitride, focusing on the development of crack-free LED structures on large area Si (111) substrates. She subsequently joined Plessey Semiconductors as chief materials engineer, working on technology transfer and R&D on GaN-on-Si technology for mass production of LEDs. In 2015, she returned to academia to continue her research interest in nitride materials. She has over 30 publications and 2 patents in the field of nitride-based LEDs.

History

Contents

A *Very* Brief History of Light

M. Suhail Zubairy

M.S. Zubairy (✉)
Institute for Quantum Science and Engineering (IQSE) and Department of Physics and Astronomy, Texas A&M
University, College Station, TX 77843-4242, USA
e-mail: zubairy@tamu.edu

© The Author(s) 2016
M.D. Al-Amri et al. (eds.), *Optics in Our Time*, DOI 10.1007/978-3-319-31903-2_1

1.1   Introduction

While tracing the history of ideas that shaped our understanding of nature and the properties of light, it is quite remarkable to see how one can almost neatly divide the geographical regions where human thoughts progressed during a certain time period followed by a decay and setting of the dark ages. We can divide the history of light into four distinct eras. The first era, with its center initially in Athens and then Alexandria, belonged to the Greeks. This era extended from about 800 BC till around 200 AD. It seems that hardly anything of significance in our understanding of light was contributed between 200 AD till around 750 AD when Muslims burst onto the scene. The second era belongs to the Islamic civilization, with its centers in Baghdad and Cordoba. It had its golden age till around middle to late thirteenth century when Mongol invasion destroyed the eastern center in Baghdad in 1258 and the decay set in the Western Center of Cordoba. The third era started in Europe around the fourteenth century when medieval Europe that had slipped into a dark age after the fall of Roman Empire started to emerge out of it. The crusades (1095–1272) and the conquest of Islamic Spain made the Muslim scholarship and the Greek traditions accessible to the Europeans, helping to initiate the glorious era of scientific revolution in the West. The last era started with the dawn of twentieth century that opened not only with new and revolutionary theories of Physics but also with a revolution in communication technology. This has helped to make science, and optics, a global preoccupation.

1.2   Greeks and Antiquity

The Greek civilization flourished in the eastern Mediterranean area, extending from Athens in Greece to Anatolia, Syria, and Egypt from Archaic period in about eighth century BC till about 200 AD. This civilization produced the highest level of intellect in many branches of human thought such as mathematics, philosophy, ethics, and astronomy. Through the galaxy of thinkers, such as Archimedes, Socrates, Plato, Aristotle, Euclid, Ptolemy, and Galen, they left a lasting imprint on the human civilization. Their most lasting legacy is not the theories that these giants of history presented as most of them have either been overturned or replaced during the evolution of human thought. Their lasting legacy to the mankind lies in placing the rational thinking at the apex of creation that has reverberated through millennia, long after the Greek civilization disappeared.

Light in Antiquity   The earliest studies concerning light had to do with understanding vision. For example, the ancient Egyptians believed that light was the activity of their god Ra seeing. When Ra's eye (the Sun) was open, it was day. When it was closed, night fell. The earliest studies on the nature of light and vision can be attributed to the Greek and Hellenistic traditions. The Greek period, extending from the Archaic period till around 320 BC and centered in Athens, produced many earliest ideas about vision through the works of Democritus, Epicurus, Plato, and Aristotle. After the death of Alexander, the center shifted to Alexandria where Ptolemy I, a general in the Alexander's army, established a new dynasty that lasted till the Roman conquest of Egypt in the first century BC. In this Hellenistic period, the glorious traditions of Greek scholarship in the field of light and vision continued through the works of Euclid, Hero of Alexandria, Ptolemy, and Galen.

The theory of vision attempts to explain how objects, near and far, their shape, size, and color, are perceived by us. The earliest systematic studies of vision are attributed to atomists who reduced every sensation, including vision, to the impact of atoms from the observed object on the organ of observation. There were different schools of thought among atomists. For example, Democritus (460 BC–370 BC) believed that the visual image did not arise directly in the eye, but the air between the object and the eye is contracted and stamped by the object seen and the observing eye. The pressed air contains the details of the object and this information is transferred to the eye. Epicurus (341 BC–270 BC), on the other hand, proposed that atoms flow continuously from the body of the object into the eye. However the body does not shrink because other particles replace and fill in the empty space.

An alternate theory of vision due to Plato (428 BC–328 BC) and his followers advocated that light consisted of rays emitted by the eyes. The striking of the rays on the object allows the viewer to perceive things such as the color, shape, and size of the object. Our vision was initiated by our eyes reaching out to "touch" or feel something at a distance. This is the essence of extramission theory of light that would be influential for almost a 1000 years until Alhazen would conclusively prove it to be wrong.

Hellenistic Era, Euclid, Hero of Alexandria, and Ptolemy   Euclid (b. 300 BC) is the father of Geometry. His book *Elements* laid down the foundation of axiomatic approach to geometry and is one of the most influential books ever written. Little original references are available about Euclid and what we know about him was written centuries after he lived by Proclus (c. 450 AD) and Pappus of Alexandria (c. 320 AD). His work in optics follows the same methodology as *Elements* and gives a geometrical treatment of the subject. Euclid believed in extramission and his theory of vision is founded in the following postulates:

1. Rectilinear rays proceeding from the eye diverge indefinitely;
2. The figure contained by the set of visual rays is a cone of which the vertex is at the eye and the base at the surface of the object seen;
3. Those things are seen upon which visual rays fall and those things are not seen upon which visual rays do not fall;
4. Things seen under a larger angle appear larger, those under a smaller angle appear smaller, and those under equal angles appear equal;
5. Things seen by higher visual rays appear higher, and things seen by lower visual rays appear lower;
6. Similarly, things seen by rays further to the right appear further to the right, and things seen by the rays further to the left appear further to the left;
7. Things seen under more angles are seen more clearly.

Euclid did not define the physical nature of these visual rays. However, using the principles of geometry, he discussed the effects of perspective and the rounding of things seen at a distance.

Euclid had restricted his analysis to vision. Hero of Alexandria (10–70), who also believed in the extramission theory of Euclid, extended the principles of geometrical optics to consider the problems of catoptrics, particularly, reflection from smooth surfaces. Hero derived the law of reflection by invoking the principle of least distance. According to him, light from a point A to another point B follows a path that is shortest. On this basis, he showed that when light reflects from a surface, angle of incidence is equal to the angle of reflection. Specifically, the image appears to be as far behind the mirror as the object is in front of the mirror. Hero's principle of least distance would be replaced by the principle of least time by Pierre Fermat more than 1500 years later to derive the law of refraction.

The most influential and perhaps last important figure in optics of the Greek–Egyptian era was Claudius Ptolemy (90–168). He is most well known for championing the geocentric model for the movement of planets, a view that would survive for almost 1400 years until it was replaced by a heliocentric model through the work of Nicholas Copernicus in 1543. His book on the subject *Almagest* was very influential in shaping the thinking on astronomy and, along with *Elements* by Euclid, was the longest read book in the history of science. Ptolemy wrote *Optics* in which he discussed the theory of vision, reflection, refraction, and optical illusions. Like Euclid and Hero, Ptolemy championed the extramission theory of vision. He considered visual rays as propagating from the eye to the object seen. However, instead of considering visual rays as discrete lines as postulated by Euclid, he considered them forming a continuous cone. Ptolemy carried out careful experiments on refraction and concluded that, for light propagating from one medium to another, the ratio of the angle of incidence to the angle of refraction was constant and depended on the properties of the two media. He thus derived the small angle approximation of the law of refraction. The formulation of theory based on experimental results, frequently supported by the construction of special apparatus, is the most striking feature of Ptolemy's *Optics*.

1.3   Islamic Period

Islam has its roots in Mekkah, a city that was on the cross road of trade route from Syria to Yemen in the sixth and seventh century. The founder of the religion, Prophet Muhammad (PBUH), was born there in 570 and claimed to have received his first revelations from God in 610. Under severe opposition from his kinsmen to the new religion, he migrated to the northern city of Madinah in 622. This marked the beginning of the Islamic era. When Muhammad (PBUH) died in 632, Islam had spread throughout the Arabian Peninsula. He was followed by four caliphs, Abu Bakar, Umar, Usman, and Ali, in the leadership of the Islamic community. Under Umar, the conquests of Persia, Syria, and Egypt expanded the Islamic writ to a major part of the Middle East. These caliphs were followed by the Ummayad dynasty (660–750) when North Africa, Spain, Western China (Xinjiang), and Western India (modern Pakistan) came under the Muslim rule. The capital of Ummayads was Damascus. In 750, the Ummayads were replaced by Abbasids (descendants of an uncle of the Prophet named Abbas). They continued to rule till 1258 when Mongols attacked and conquered their capital Baghdad. The Ummayad's rule in the Iberian Peninsula continued till 1492 (the year when Columbus landed in the new world). The tenth and eleventh century Egypt was ruled by Fatimids, a dynasty founded by the descendants of Fatima, daughter of Prophet Muhammad (PBUH).

Contrary to some modern claims and perceptions, Islam was an enlightened religion in its beginning, deeply rooted in the search of knowledge. The Islamic holy book, Qur'an, that was believed by Muslims to be the direct word of God as revealed to Prophet Muhammad (PBUH) exhorted human beings to contemplate and seek knowledge through words such as "And say, Lord increase my knowledge" (Qur'an 20:114) and "He (God) has subjected to you, as from Him, all that is in the heavens and on earth: behold in that are Signs indeed for those who reflect" (Qur'an 45:13). Similarly the sayings, such as "the ink of a scholar is more holy than the blood of a martyr" and "The most learned of men is the one who gathers knowledge from others on his own; the most worthy of men is the most knowing and the meanest is the most ignorant", attributed to Prophet Muhammad (PBUH) emphasize the importance of the pursuit of knowledge. These and other similar injunctions in the Qur'an and the Prophetic traditions helped to develop an attitude in the Muslim community that supported the quest of knowledge and promoted an environment where open discussion was encouraged. Science was not seen to be contrary to the faith. Rather it was considered to be a religious duty to seek knowledge and understanding. As the Islamic Empire increased in size so did the thirst for more knowledge in all fields.

Armed with this attitude, Muslims built a civilization in the Eighth century that would last for several centuries and contributed to almost all aspects of human knowledge. It is unprecedented in the annals of history that an empire would bring with it a great civilization as well. The Muslims built a body of knowledge by first learning and then expanding on the older traditions, particularly Greek. Their own contributions would, in turn, provide a foundation for the emergence of the modern Western civilization.

Bayt-al-Hikmah (House of Wisdom)   Traditionally the beginning of the Islamic Golden Age of science is attributed to the Abbasid caliph Harun al-Rashid (763–809) who ruled an empire stretching from modern Pakistan to North Africa to the shores of Atlantic Ocean from 786 till 809. However the age may have started earlier, even in the Ummayad period, when the foundations of Islamic Jurisprudence were being laid down through discussion and reason. This tradition crossed into the secular body of knowledge, leading first to assimilating what old sages from Greek, Indian, and other civilizations had contributed and then building their own contributions in fields ranging from philosophy and medicine to mathematics and physical sciences.

Harun al-Rashid and his court is fantasized in the book *One Thousand and One Nights*. He laid the foundation of *Bayt-al-Hikma* (House of Wisdom) in the newly built capital city of Baghdad. However it was formally completed in 830 in the era of his equally brilliant son, al-Mamun (786–833) who ruled from 813 till 833. Originally the House of Wisdom was a scientific academy and a public library where books from all parts of the empire were brought and translated in Arabic. These included old texts from India, Greece, and Persia in the fields of philosophy, mathematics, astronomy, medicine, and optics. By 850, House of Wisdom had the largest repository of manuscripts of its time. Gradually this center turned into a center of research and many famous names of Islamic Golden era were associated with it. These included, among others, Jabir bin Hayyan (721–815) who is regarded as father of chemistry, Al-Khwarizmi (780–850) who is credited with inventing algebra, Al-Kindi (800–873) who is regarded as the first Muslim philosopher, Hunayn ibn-Ishaq (809–873) whose contributions in medicine were influential till the modern era, and Alhazen (865–1040) who is regarded as the father of optics.

Al-Mamun himself took great interest in the progress of the House of Wisdom and is reputed to have intellectual discussions with the scholars that had started coming from distant lands. Al-Mamun supported organized research in areas such as developing detailed maps of the world, measurement of the circumference of the Earth, and the confirmation of data from Ptolemy's *Almagest*. This is the first known example of the state sponsored research. Gradually other institutions of higher learning, such as Al-Azhar University (970) in Cairo and Al-Nizamiyya (1095) in Baghdad, developed in and outside Baghdad.

Al-Kindi and Optics   Abu Yusef Yaqoub ibn Ishaq Al-Kindi (800–873) was the first great philosopher of the Islamic era. He synthesized, adopted, and promoted Greek philosophy in the Islamic world. He worked with a group of translators at the House of Wisdom who rendered works of Aristotle, Plato, Euclid, and other Greek mathematicians and scientists into Arabic. Al-Kindi's main authority in philosophical matters was Aristotle. His philosophical treatises include *On First Philosophy*, in which he argues that the world is not eternal and that God is a simple *One*. Al-Kindī tried to demonstrate that philosophy is compatible with Islamic traditions and had a great influence on later Muslim philosophers Abdullah ibn-Sina (known in the West as Avicenna) and Ibn-Rushd (known in the West as Averroes).

Al-Kindi was also the first to undertake serious studies in optics and the theory of vision. His work on optics, *De Aspectibus* in Latin translation, exerted a strong impact on Islamic and Western optics throughout the middle ages. In optics, Al-Kindi followed the traditions of Euclid, and carried on by Ptolemy and others in which geometrical constructions were used to explain phenomena such as vision, reflection, refraction, shadows, and burning mirrors. Whereas Euclid considered the straightness of a visual ray as an axiom, Al-Kindi proved it experimentally by considering the shadows projected by different opaque objects. He treated the geometry of visual cone, rejecting the discreteness of Euclid's rays and replacing them with a cone of continuous beam of radiation similar to Ptolemy.

Al-Kindi's work was followed in tenth century by Al-Razi and Al-Farbi who started objecting to the extramission theory of light. A series of strong arguments against the notion of visual fire were put forward around 1000 by the great ibn-Sina. He argued that the visual fire cannot reach remote objects as it will have to fill an enormous space each time we opened our eyes. He also argued that Euclid's discrete rays may leave large areas of a distant object unobservable.

Ibn-Sahl and Snell's Law 600 Years Before Snell   Snell's law of refraction is an important law relating to the propagation of light between two media with different refractive indices. Refractive index of a medium is inversely proportional to the speed of light in that medium. The law of refraction thus forms the basis of understanding the bending of light rays from various kinds of lenses. The credit of the discovery of the law of refraction is given to Willebrord Snellius (1580–1626) who derived it using trigonometric methods in 1621. However recent studies indicate that this law was discovered more than 600 hundred years earlier during the Islamic Golden Age of Baghdad by a scientist named Abu Sad Al Alla Ibn Sahl (940–1000). Ibn Sahl excelled in optics and wrote a treatise *On Burning Mirrors and Lenses* in 984 in which he discussed the focusing properties of the parabolic and elliptical burning mirrors. He also presented an analysis of how hyperbolic glass lenses bend and focus light. As a lemma in his derivation of the focusing property of light by a plano-convex hyperbolic lens, he presented a geometric argument based on the sine law of refraction. This appears to be a major achievement and shows how far Muslims had advanced in pure and applied mathematics as well as optics by the end of tenth century. A question, however, remains about how such a major discovery could remain ignored for so many centuries. A plausible explanation is that Ibn Sahl did not state the law of refraction explicitly. Instead it was hidden as a sort of lemma and his emphasis was on the focusing property of lenses.

Alhazen, Father of Modern Optics   Abu Ali al-Hasan ibn al- Hasan ibn al-Haytham (965–1040), known in the west as Alhazen, is a central figure in science. He is often described as the greatest physicist between Archimedes and Newton. He was the first person to follow the scientific method, the systematic observation of physical phenomena and their relation to theory, thus earning the title First Scientist from many. His most important contribution in optics is his book *Kitab-al Manzir* (Book of Optics) which was completed around 1027. This book, comprising seven volumes, was the first comprehensive treatment of optics and covered subjects such as the nature of light, the physiological treatment of eye, and the bending and focusing properties of lenses and mirrors. This book was most influential in the transition from the Greek ideas about light and vision to the modern day optics. Alhazen's Book of Optics was translated in Latin at the end of twelfth century under the title *De Aspectibus* and would remain the most influential book in optics till Newton's *Opticks* published in 1704.

Alhazen proved the long held theory of Euclid, Hero, and Ptolemy that light originated from the eye to be wrong and showed that light originated from the light sources. He did this by carrying out a simple experiment in a dark room where light was sent through a hole by two lanterns held at different heights outside the room. He could then see two spots on the wall corresponding to the light rays that originated from each lantern passing through the hole onto the wall. When he covered one lantern, the bright spot corresponding to that lantern disappeared. He thus concluded that light does not emanate from the human eye, but is emitted by objects such as lanterns and travels from these objects in straight lines. Based on these experiments, he invented the first pinhole camera (that Kepler would use and call *camera obscura* in the seventeenth century) and explained why the image in a pinhole camera was upside down.

Alhazen's theory of vision was not limited to the description of light rays originating from the objects and entering the eye. He also understood that an explanation of vision must also take into account the anatomical and psychological factors. He proved that the perception of an image occurs not in the eyes but in the brain and that the location of an image is largely determined by psychological factors.

Alhazen did not invent the telescope but he explained how a lens worked as a magnifier. He contended that magnification was due to the bending, or refraction, of light rays at the glass-to-air boundary and not, as was thought, to something in the glass. He correctly deduced that the curvature of the glass, or lens, produced the magnification. He concluded that the magnification takes place at the surface of the lens, and not within it.

His work on catoptrics in Book V of the Book of Optics dealt with problems of reflection from spherical and parabolic mirrors. It also contains a discussion of what is now known as Alhazen's problem. The problem was first formulated by Ptolemy in 150 AD. Draw lines from two points in the plane of a circle such that they meet at a point on the circumference, making equal angles with the normal at that point. The problem was to locate this point. This problem is equivalent to the Billiard table problem: On a circular table there are two balls; at what point along the circumference must one be aimed at in order for it to strike the other after rebounding off the edge. Alhazen's interest in this problem stemmed from the following formulation of the problem: When light is sent from a source towards a spherical mirror, find the point on the mirror where the light will be reflected to the eye of an observer. The problem is insoluble using a compass and a ruler because the solution requires solution of an equation of fourth degree. Alhazen solved this problem geometrically by the aid of a hyperbola intersecting a circle. This problem remained unsolved using algebraic methods for almost a thousand years until it was finally solved in 1997 by the Oxford mathematician Peter M. Neumann.

1.4 Scientific Revolution

The publication in 1543 of Nicholaus Copernicus's *De Revolutionibus Orbium Coelestium* (On the Revolutions of the Heavenly Spheres) marks the beginning of the scientific revolution. He proposed a heliocentric model of the solar system, a system in which Sun was held at rest and all the planets including Earth circled around it, replacing the long held Ptolemaic geocentric model in which earth was at rest. Without the benefit of the knowledge of the law of gravitation, it was hard to believe how earth could be moving around the sun still maintaining the stability of all objects including the humans on its surface. The hostility to a model that took away the centrality of earth in a solar system was so great (particularly in the Church) that Copernicus could not publish his heliocentric theory till the end of his life. According to a legend, Copernicus received the copy of his book *De Revolutionibus* on the very last day of his life, thus dying without knowing that his work heralded a new era of human history. There were, however, birth pangs of this new world of science, most famous being Galileo's heresy conviction in 1633 for his support of Copernicus.

Kepler   The most important figure to follow Copernicus was the German astronomer, Johannes Kepler (1571–1630), whose laws on planetary motion would prove pivotal for Newton's law of gravitation. Kepler is a key figure in the history of light and vision as well. His interest in the subject appears to have originated in his observation of a solar eclipse on July 10, 1600 by means of *camera obscura*. Several years ago, Tycho Brahe (1546–1601), the greatest naked-eye astronomer of the time, had observed that the angular diameter of the moon appeared to be larger during a solar eclipse when observed through the pin-hole camera than when observed directly. Kepler understood that this anomaly could not be explained without a full understanding of the optical instruments, in this case, the *camera obscura*. He noted that the finite diameter of the pinhole should be responsible for this anomaly. He discovered the solution by an experimental technique where he stretched a thread through an aperture from a simulated luminous source to the surface on which the image was formed. He traced out the image cast by each point on the luminous body seeing, in the process, the geometry of radiation in three-dimensional terms. In this way, Kepler was able to formulate a satisfactory theory of radiation through apertures based on the rectilinear propagation of light rays.

Kepler did not stop at explaining Tycho Brahe's problem of seemingly variable lunar diameter. In 1601, he noted that eye itself possesses an aperture and should be treated in the same way as the aperture in a pinhole camera. Kepler published his theory of vision in 1604.

Descartes   Until Kepler, the main motivation of studying the nature of light came from a desire to understand vision. René Descartes (1590–1650) appears to be the first person to concern himself with the intrinsic nature of light and the laws of optics. Descartes was a French philosopher and mathematician who had a great impact on western philosophy. He is heralded as the Father of Modern Philosophy. His mathematical contributions included a connection between geometry and algebra that allowed for the solving of geometrical problems using algebraic equations. Descartes promoted the accounting of physical phenomena by way of mechanical explanations.

Descartes main contribution to optics is his book *Dioptrics* that was published in 1637. It deals with many topics relating to the nature of light and the laws of optics. He compares light to a stick that allows a blind person to discern his environment through touch.

Descartes used a tennis ball analogy to derive the laws of reflection and refraction of light. The credit of the discovery of the law of refraction is given to Willebrord Snellius (1580–1626) who derived it using trigonometric methods in

1621. However Snell did not publish his work in his lifetime. Descartes published the law of refraction 16 years after Snell's death, as Descarte's law of refraction.

Fermat and Principle of Least Time Together with Descartes, Pierre de Fermat (1601–1675) was one of the two greatest French mathematicians of the first half of the seventeenth century. A lawyer by profession, Fermat made a number of important contributions in analytical geometry, probability, and number theory. He is most well known for the Fermat's Last Theorem (no three positive integers a, b, c can satisfy the relation $a^n + b^n = c^n$ for any integer n that is larger than 2) that he conjectured in 1621 but could not be proved till 1994.

Fermat's major contribution in optics relates to his derivation of Snell's law using the principle of least time. Just as Hero of Alexandria had derived the law of reflection on the basis of the principle of least distance 1400 years ago, Fermat argued that light rays going from a point located in a region where it propagates with a particular speed to a point in another region where it propagates with a different speed, it would follow a path that takes the shortest time. This yielded the correct Snell's law.

Newton Sir Isaac Newton (1642–1727) is definitely the defining figure in the history of science. His *Principia* laid down the foundation of classical mechanics. His law of gravitation is a bright example of the nature of scientific law–law that applies equally well for all objects, big and small. His contributions in mathematics, particularly his co-discovery of calculus with Wilhelm Leibniz, provided tools that would be so vital for almost all the subsequent major discoveries in physics and other branches of science. They played key role in shaping the physics of the coming centuries.

It is, however, interesting to note that the most important experimental contributions to physics made by Newton are all in the field of optics. He was the first to show that the color is the property of light and not of the medium. Through ingenious experiments he could show that the light generated by sun consisted of all the colors. For example, when sun light passes through a prism, it is dispersed in a rainbow of colors. The red color bends the least and the violet color bends the most. This ability of glass prisms to generate multiple colors was known since antiquity but it was not attributed to light. Instead color was considered as a characteristic of the material. What Newton showed was that when a particular color passed through the prism, no such dispersion took place. In a relatively complicated setup, when these colors were combined together and passed through the prism again, Newton recovered white light, proving that white light consisted of all the colors.

The other major contribution of Newton towards optics is his design of reflecting telescope. All the telescopes through his time were unwieldy refracting telescopes that suffered from chromatic aberrations. The earliest refracting telescope, built in 1608, is credited to Hans Lippershey who got the patent for the design. These refracting telescopes consisted of a convex objective lens and a concave eyepiece. Galileo used this design in 1609. In 1611, Kepler described how a telescope could be made with a convex objective lens and a convex eyepiece lens. Newton designed a reflecting telescope where incoming light is reflected by a concave mirror onto a plane mirror that reflected the light to the observer. This design was simple and less susceptible to chromatic aberrations. All the major telescopes that exist today are improved versions of Newton's reflecting telescope.

Newton was also concerned with the nature of light and advocated corpuscular theory of light. According to him, light is made up of extremely small corpuscles, whereas ordinary matter was made of grosser corpuscles. He speculated that through a kind of alchemical transmutation they change into one another. According to him, "Are not gross Bodies and Light convertible into one another,..... And may not Bodies receive much of their Activity from the Particles of Light which enter their Composition?" It is surprising that Newton advocated corpuscular theory of light when there was evidence that supported the wave behavior. For example, Francesco Grimaldi (1618–1663) made the first observation of the phenomenon that he called diffraction of light. He showed through experimentation that when light passed through a hole, it did not follow a rectilinear path as would be expected if it consisted of particles but took on the shape of a cone. Newton explained that the phenomenon of diffraction was only a special case of refraction that was caused somehow by the ethereal atmosphere near the surface of the bodies. Newton could explain the phenomenon of reflection with his theory. However he could explain refraction by incorrectly assuming that light accelerated upon entering a denser medium because the gravitational pull was stronger.

Huygens, Wave Nature of Light   When Newton was expounding a corpuscular nature of light, his contemporary, Christian Huygens (1629–1695) suggested a wave picture of light. Huygens published his results in his *Traite de la lumie*re (Treatise on light) in 1690. Crucial to his wave theory was the result recently obtained by Olaus Romer (1679) that the speed of light is finite. He considered light waves propagating through the *ether* just as sound waves propagate through air. He explained the high but finite speed of light by the elastic collisions of a succession of spheres that made the *ether*. The light waves, according to Huygens, were thus longitudinal waves as opposed to the later studies by Fresnel and Maxwell that showed light to consist of transverse waves.

Huygens formulated a principle (that now bears his name) which describes wave propagation as the interference of secondary wavelets arising from point sources on the existing wavefront. In propagation each ether particle collides with all the surrounding particles so that "... around each particle there is made a wave of which that particle is the center."

Young and Double-Slit Experiment   Till the beginning of nineteenth century, Newton's status was so great particularly in the British Isles that few dared to challenge his corpuscular theory of light. It was, however, Thomas Young (1773–1829) who, in 1802, conclusively demonstrated the wave nature of light through his double-slit experiment. He described his experiment in these words in *The Course of Lectures on Natural Philosophy and the Mechanical Arts* (1807): "..... when a beam of homogeneous light falls on a screen in which there are two very small holes or slits, which may be considered as centres of divergence, from whence the light is diffracted in every direction. In this case, when the two newly formed beams are received on a surface placed so as to intercept them, their light is divided by dark stripes into portions nearly equal, but becoming wider as the surface is more remote from the apertures, so as to subtend very nearly equal angles from the apertures at all distances, and wider also in the same proportion as the apertures are closer to each other. The middle of the two portions is always light, and the bright stripes on each side are at such distances, that the light coming to them from one of the apertures, must have passed through a longer space than that which comes from the other, by an interval which is equal to the breadth of one, two, three, or more of the supposed undulations, while the intervening dark spaces correspond to a difference of half a supposed undulation, of one and a half, of two and a half, or more." With this he firmly established the wave nature of light.

By repeating his experiment, Young could relate color to wavelength and was able to calculate approximately the wavelengths of the seven colors recognized by Newton that composed white light. According to him "…. it appears that the breadth of the undulations constituting the extreme red light must be supposed to be, in air, about one 36 thousandth of an inch, and those of the extreme violet about one 60 thousandth."

The Young's double-slit experiment was not only decisive in debunking Newton's corpuscular theory of light, but it also continued to play a crucial role in our understanding of the nature of light and matter even in the twentieth century. For example, in 2002, *Physics World* published the results of a survey on the all-time Ten Most Beautiful Experiments in Physics. Young's double-slit experiment made not one but two appearances on this prestigious list—at number 1 was the double-slit experiment applied to the interference of electrons and, at number 5 was the original experiment by Young.

Young's double-slit experiment was, however, regarded highly controversial and counterintuitive in his own time. How can a screen uniformly illuminated by a single aperture develop dark fringes with the introduction of a second aperture? And how could the addition of *more* light result in *less* illumination? Young's theory would eventually find broad acceptance, particularly through the works of Fresnel in France.

Fresnel, Theory of Diffraction, and Polarization of Light   Augustin Jean Fresnel (1788–1827), a contemporary of Young, championed the wave nature of light based on his work on diffraction. Fresnel began by undertaking experiments with diffraction. He noted that when light passed through a diffractor, one could see a series of dark and bright bands behind the diffractor. However when he blocked one edge of the diffractor, the bright bands within the shadow vanished and the bright bands remained only on the unblocked side of the diffractor. From this, he concluded that the bright bands in the shadow were produced by light coming from both edges and the bright bands, when one edge was blocked, resulted from the reflection of light from one edge of the diffractor. He was able to develop a mathematical theory for these observations based on a wave theory of light and could predict the position of bright and dark lines based on where the vibrations were in phase and out of phase. He published his first paper on wave theory of diffraction in 1815.

An episode indicates the stunning success of the wave nature of light as formulated by Fresnel. In 1819, Fresnel presented his work on wave theory of diffraction in a competition by the French Academy of Sciences. The committee of judges, headed by Francois Arago, included Jean-Baptiste Biot, Pierre-Simon Laplace, and Simeon-Denis Poisson. They were all prominent advocates of Newton's corpuscular theory and were not well disposed to the wave theory of light. Poisson was, however, impressed by Fresnel's submission and extended his calculations to come up with an interesting consequence: "Let parallel light impinge on an opaque disk, the surrounding being perfectly transparent. The disk casts a shadow - of course - but the very center of the shadow will be bright. Succinctly, there is no darkness anywhere along the central perpendicular behind an opaque disk (except immediately behind the disk)". According to the corpuscular theory, there could be no bright spot behind the disk. As Chair of the Committee, Arago asked Fresnel to verify Poisson's prediction and amazingly Fresnel found the bright spot as predicted. This discovery was an impressive vindication of the wave theory and Fresnel won the competition. This spot is now known as "Poisson spot."

Despite the triumph of the wave theory of light, the properties of the polarized light still provided a strong argument in favor of the corpuscular theory, since no explanation from a wave theory had ever been made. Following the success of the wave theory in explaining the interference and diffraction phenomena, Fresnel and Arago embarked upon explaining the properties of the polarized light based on Fresnel's theory. In 1817, Fresnel became the first person to obtain what was later called circularly polarized light. The only hypothesis that could explain the experimental results was that light is a transverse wave. In 1821, Fresnel published a paper in which he claimed that light is a transverse wave. Young had independently reached the same conclusion. The assertion that light is a transverse wave was not readily accepted by many, including Arago. Again Fresnel was vindicated when he could explain the double refraction from the transverse wave hypothesis. This helped to seal the status of light as a transverse wave.

Maxwell and Electromagnetic Waves It was left to James Clerk Maxwell (1831–1879) to complete the classical picture of light as consisting of electric and magnetic waves. This was a truly remarkable outcome of his efforts to unify the two known forces of nature: electric force and magnetic force. It was known through the work of Michael Faraday that a time rate of change of magnetic field yielded electric force. The insight due to Maxwell was that if electricity and magnetism were the two sides of the same coin then a change of electric field should similarly result in a magnetic field. This motivated him to add a term in the Ampere's law that corresponded to a time rate of change of the electric field. This addition immediately yielded a wave equation for an electromagnetic wave propagating at the same velocity as known for light, 3×10^8 m/s. The picture of light that emerged was thus that of undulations of mutually perpendicular electric and magnetic fields propagating. The direction of propagation was perpendicular to both the electric and magnetic fields. Maxwell's results were published in 1865. Thus the light waves were shown to be transverse waves in line with Young and Fresnel as opposed to the picture adopted by Huygens where light was seen as a longitudinal wave propagating through the medium *ether*. This description of light as an electromagnetic wave was experimentally demonstrated by Heinrich Hertz (1857–1894) in 1888.

1.5   Light in Twentieth Century

According to a quote, attributed to Lord Kelvin in an address to the British Association for the Advancement of Science in 1900, "There is nothing new to be discovered in physics now. All that remains is more and more precise measurement." The classical theories of mechanics, electromagnetics, thermodynamic, and, of course, light were firmly in place and it was a justified feeling to believe that the basic laws of nature were fully understood.

There were, however, two "clouds" on the horizon of physics at the dawn of the twentieth century. Interestingly enough, both of these involved light. The first cloud, the Rayleigh–Jeans ultraviolet (UV) catastrophe and the nature of blackbody radiation, led to the advent of quantum mechanics, which of course was a radical change in physical thought up to that point. The second cloud, namely the null result of the Michelson–Morley experiment, led to special relativity, which is the epitome of classical mechanics, and the logical capstone of classical physics. These theories, quantum mechanics and the theory of relativity, were major departures from the classical theory as first formally introduced by Newton.

They would shape the physics of the twentieth century. They also dramatically revised our understanding of the nature of light.

Black-Body Radiation, Kirchhoff and Planck The concept of a black body was introduced by Gustav Kirchhoff (1824–1887) in 1860. Kirchhoff knew from looking at the spectral lines from the sun that there was heat energy in empty space, and postulated equilibrium radiation. However the knowledge of what it consisted of was still primitive. By 1860, Maxwell's equations had not yet been postulated, and the electromagnetic nature of both heat and light rays had not yet been established. Nor had the existence of atoms in the walls of a cavity, nor that an oscillator radiates and absorbs electromagnetic energy, or that such energy carries momentum. Thus it is rather amazing that Kirchhoff should have established on the basis of relatively simple arguments that within a cavity at equilibrium, this radiation should be independent of the substance of the walls of the cavity, and that at a fixed temperature a good emitter of radiation should be a good absorber. A perfect absorber should then radiate an energy equivalent to everything that falls upon it within the cavity at equilibrium, independently at each frequency. He called the radiation emitted by such a perfect absorber the black-body radiation, and postulated that there should be a universal function $u(\nu,T)$ that describes the radiation density in equilibrium with the walls, that on average gets both absorbed and reemitted, at any particular frequency ν and temperature T. The challenge was to find the explicit form of the function $u(\nu,T)$. This search would eventually lead to the birth of quantum mechanics in early twentieth century.

In 1888 Hertz showed the reality of Maxwell waves. In 1893 Wien applied the laws of thermodynamics and electromagnetism to the problem of black-body radiation and succeeded in reducing Kirchhoff's universal function to a function of one variable. That is as far as one can go in classical physics. Wien tackled the problem of including the frequency in the black-body law by considering an adiabatic motion of a wall of the cavity. This induced a Doppler shift on the radiation, while at the same time the wall did work on the radiation.

The Rayleigh–Jeans formula gave results in agreement with the experimental observations at low frequencies; however, it failed miserably at high frequencies. The radiancy, according to Rayleigh and Jeans, is inversely proportional to the fourth power of the frequency, which indicates that at high frequencies the radiancy will approach infinity, thus leading to unphysical results in the ultraviolet region of the spectrum. By 1900, this failure, known as the Rayleigh–Jeans ultraviolet catastrophe, had caused people to question the basic concepts of classical physics and thermodynamics.

It was, however, Max Planck (1858–1947) who would eventually present the radiation formula that matched the experimentally observed black-body radiation spectra for the entire range of frequency spectrum. Planck presented his results that would eventually revolutionize our understanding of the laws of nature, literally at the close of the nineteenth century, on December 15, 1900, at a meeting of the German Physical Society.

When Planck addressed the problem of black-body radiation, he realized that since the results were independent of the nature of the material in the cavity, one could use a simple model for the cavity. So he chose to consider a damped harmonic oscillator as a model for the material in the walls. Planck's derivation consisted of three steps. In the electromagnetic step, he calculated the equilibrium energy of these harmonic oscillators of frequency ν driven by the periodic electric field of frequency ω. In the thermodynamic step, he calculated the entropy of the linear oscillators that gave the correct value for the function $u(\nu,T)$. In the third and crucial statistical step, he calculated the entropy of the linear oscillators and showed that the expression for entropy in the thermodynamic step could only be recovered if he assumed the total energy of the oscillators was made up of finite energy elements, and each element had an energy ε that is equal to $n h\nu$. Here n is an integer and h is a constant that eventually carried Planck's name and is called Planck's constant. This last step was a departure from a classical description and Planck would later describe it as "an act of desperation" to get the correct expression for the Kirchhoff function that agreed with experiments. It is important to realize that the Planck relation $\epsilon = n h\nu$, for integer values of n, is a significant departure from classical thought in two ways. First, it postulates that energy is proportional to frequency, not amplitude, as would be expected for a classical oscillator. Second, for a given frequency ν, the energy is quantized, i.e., it comes in units of $h\nu$.

Planck's derivation for the black-body spectrum was based on the quantization of the material of the cavity and not the radiation itself. However it would have far reaching consequences for the ultimate description of the nature of light through the work of Albert Einstein and others.

Einstein and the Notion of Photon   The revival of the particle theory of light, and the beginning of the modern concept of the photon, is due to Albert Einstein (1879–1955). Einstein is a giant in the history of science. He is the founding figure of both quantum mechanics and the theory of relativity. His impact on our understanding of the nature of light is immense.

In his 1905 paper on the photoelectric effect, the emission of electrons from a metallic surface irradiated by UV rays, Einstein was forced to postulate that light comes in discrete bundles, or quanta of energy, borrowing Planck's hypothesis: $\epsilon = n h\nu$. This re-introduced the particulate nature of light into physical discourse, not as localization in space in the manner of Newton's corpuscles, but as discreteness in energy. This gave the Planck hypothesis a new and bold meaning.

There were three issues associated with the photoelectric effect: When light of frequency ν falls on a photoemissive surface, energy of the ejected electron T_e obeys $h\nu = \Phi + T_e$; rate of emission is proportional to the intensity of incident light; and there is no time delay between the time in which the field begins falling on the photoactive surface and the instance of photoelectron emission. The first two of these phenomena can, in contrast to what we read in most textbooks, be explained fully by simply quantizing the atoms associated with the photodetector. However, the third point, namely the lack of a delay is a bit more subtle. It may be reasonably argued that quantum mechanics teaches us that the rate of ejection is finite even for small times, i.e., times involving a few optical cycles of the radiation field. Nevertheless, one may argue that the concept of the photon is really explicit here in the sense that conservation of energy is at stake. That is, if we have only a short period of time τ elapsing between the instants that the radiation field E begins to interact with the photoemitting atoms and the emission of the photoelectron, the amount of energy which has fallen on the surface would be given by $\epsilon_0 E^2 A \tau$, where A is the cross-section of the incident beam. For sufficiently short times, the energy which has fallen on the photodetector may not exceed $h\nu$. This clearly shows that we are not able to conserve energy if we take a semiclassical point of view. However, the photon concept in which the ejection of the photo-electron implies that a photon is annihilated gets around this problem completely. This is one of the triumphs of the quantum field theory. In any case, it is a tribute to Einstein's deep understanding of physics that he was able to introduce the photon concept from such limited, and in some ways, misleading information.

This was a difficult situation. On the one hand, were the interference and diffraction experiments that required a wave nature of light for their explanation and, on the other, was the photoelectric effect that could be understood by invoking a particle type picture. A complete resolution and the formal theory that would rigorously explain all these phenomena would have to wait almost a quarter century—till the birth of quantum mechanics in the summer of 1925.

Before discussing these developments, we briefly discuss the other "cloud" at the end of the nineteenth century—the null result of the Michelson–Morley experiment—and the birth of the theory of relativity.

Michelson–Morley Experiment and the Birth of the Theory of Relativity   Towards the end of nineteenth century, the concept of ether was firmly ingrained within the physics community. For example, Maxwell stated in an article entitled *Ether* for the *Encyclopedia Britannica* (1878): "There can be no doubt that the interplanetary and intersteller spaces are not empty but are occupied by a material substance or body, which is certainly the largest, and probably the most uniform, body of which we have any knowledge." He himself attempted unsuccessfully to measure the influence of ether's drag on the motion of the earth. It was, however, Albert Michelson (1852–1931) and Edward Morley (1838–1923), who carried out an experiment in 1887 to decisively establish the existence of Ether. They sent white light, through a half-silvered mirror into an interferometer, now called Michelson interferometer. The light beam was split into two beams, one of them traveling straight to a mirror in one arm and the other propagating at right angles to another mirror, with both beams recombining at the beam splitter after traveling equal distances They thus produced a pattern of constructive and destructive interference whose transverse displacement would depend on the relative time the light took to traverse the paths in the two arms. If the earth moves through the ether medium, the beam traveling along ether would take a longer time than the beam traveling in the perpendicular direction. Michelson and Morley expected a fringe shift equal to 0.4 fringes. What they measured was the maximum displacement of 0.02 and an average shift much less than 0.01. They thus concluded that the hypothesis concerning the existence of Ether medium is false. This null result—the most famous null result in the history of physics—was initially a major disappointment.

A resolution of the null result of the Michelson–Morley experiment came in 1889 by an Irish physicist, George FitzGerald. He postulated that the results of Michelson–Morley experiment could be explained using the hypothesis of the contraction of moving bodies in the direction of motion, the amount of contraction being just the right amount to give the same time difference as to explain the null result. According to him: "I would suggest that almost the only hypothesis that can reconcile. . . is that the length of the material bodies changes, according as they are moving through the ether or across it, by an amount depending on the square of the ratio of their velocities to that of light." In 1892, unaware of FitzGerald's hypothesis, Lorentz came to the same conclusion. Today we call it FitzGerald–Lorentz contraction. This was, however, an ad-hoc solution to the null result of the Michelson–Morley experiment with no basis in theory. This set up Lorentz (1853–1928) on the road to the derivation of Lorentz transformation and Einstein to his theory of relativity.

In 1905, Einstein formulated the theory of special relativity, based on the two postulates: The principle of relativity, i.e., the laws of physics do not change, even for objects moving in inertial (constant speed) frames of reference and the principle of the speed of light, i.e., the speed of light c is the same for all observers, regardless of their motion relative to the light source. Based on these postulates, Einstein could derive the Lorentz transformation and the length contraction. This represented a major departure from the Newton's notion of absolute space and

1

time. A most celebrated consequence of the theory of relativity was the equivalence of energy E and mass m via the relation $E = mc^2$. The notion of stationary ether that had been the ever existing background in all the theories since antiquity played no role in the theory of relativity.

Einstein also went on to develop a general theory of relativity that would provide a geometric theory of gravitation. This theory is based on the equivalence principle under which the states of accelerated motion and being at rest in a gravitational field are physically identical. In 1915, Einstein published a paper in which he described gravity as a geometric property of space and time. In particular, the curvature of spacetime changes in the vicinity of a massive object. The predictions of this theory were at variance with Newton's theory of gravitation and motivated one of the most dramatic experiments in the history of Physics—an experiment that would pit the two giants of science, Isaac Newton and Albert Einstein, and their conflicting theories of gravitation against each other. The experiment was the bending of light by a massive object.

Bending of Light, Newton, Einstein, and Eddington   We recall that Newton championed a corpuscular nature for light. Newton had also noted, while formulating his theory of gravitation, that any material particle moving at a finite speed would experience a force while passing in the vicinity of a massive object. This pull by gravity should bend the trajectory of the particle and the bending angle should be independent of the mass of the particle. Thus if light is composed of small particles, they should also experience such deflection. Newton himself did not calculate this deflection as, in his time, the finite speed of light was not well established. However he postulated this deflection and, towards the end of his treatise *Opticks* (1704), he noted "Do not Bodies act upon Light at a distance, and by their action bend its Rays, and is not this action strongest at the least distance?". The finite speed of light was well established by early nineteenth century and a German astronomer, Johann Georg von Soldner (1804), presented calculations based on Newton's corpuscular theory that light weighs and bends like high speed projectiles in a gravitational field. He produced a value of 0.87 arc sec bending angle for light grazing the Sun.

In 1911, more than hundred years later, Einstein calculated the bending of light by combining the equivalence principle with special theory of relativity to predict a deflection of light from the sun by the angle of 0.87 arc second. This is the same value that Newtonian theory predicted. He obtained this result before he formulated the general theory of relativity and the associated curved space time. When he included the effects of general theory of relativity, the predicted value for the bending of light doubled to 1.83 arc second. This result was published on November 18, 1915. Thus the predictions of Newton and Einstein were at odds with each other and an experimental activity followed soon to decide who was right. The bending of light by a massive object also became the first test of the esoteric Einstein's general theory of relativity.

After the First World War was over, Sir Arthur Eddington (1882–1944) organized an expedition to the island of Principe near Africa to watch the solar eclipse on May 29, 1919 and to measure the observed curving of light from distant stars by the gravitational pull of the sun. While the expedition was being planned, Eddington wrote: "The present eclipse expeditions may for the first time demonstrate the weight of light (i.e. Newton's value) and they may also confirm the added effect of Einstein's weird theory of non-Euclidean space, or they may lead to a result of yet more far reaching consequences of no deflection". When the results were announced, they agreed with Einstein's predicted value. Einstein became an overnight international celebrity and an iconic figure.

Birth of Quantum Mechanics   The first quarter of twentieth century was perhaps the most remarkable period in the history of Physics. Through the discoveries of the quantum theory of Planck, Einstein, and Niels Bohr and the Einstein's theories of special and general relativity, the outlook on conventional or classical Physics had completely transformed. Newtonian Physics was unable to explain effects that happened at sub-atomic level or at high speeds, speeds comparable to the speed of light. The capstone of these developments was the birth of quantum mechanics that took place in the summer and winter of 1925 through the works of Werner Heisenberg, Max Born, Pascual Jordan, and Paul Dirac, on one hand, and Erwin Schrodinger, on the other. This new theory, that replaced Newton's and Maxwell's theories, would have revolutionary consequences in our story on the nature of light. An important underlying feature of the new theory was the notion of complementarity, namely two observables are *complementary* if precise knowledge of one of them implies that all possible outcomes of measuring the other one are equally probable. This injected the notion of wave-particle duality in the discourse on the nature of both light and matter.

Dirac, Quantum Theory of Light   With the advent of quantum mechanics, the dual nature of light was apparent. There were phenomena such as interference and diffraction that could be explained based on the wave nature of light. Then there were phenomena such as excitation of an atom by absorbing a photon that required a particle nature of light. It was Paul Adrien Dirac (1902–1982) who, in a seminal paper published in 1927, synthesized the wave and particle natures of light in a single theory. According to the Maxwell's theory, the light consisted of electromagnetic waves of different frequencies. The oscillating waves could be looked upon as a sort of simple harmonic oscillators. Central to Dirac's quantum theory of radiation was the notion that each mode of the electromagnetic field could be identified as a quantized simple harmonic oscillator. Both satisfy the same commutation relation $[\hat{q}, \hat{p}] = i\hbar$, although q and p represent different things in the two cases. In the case of harmonic oscillator, they represent the position and momentum of the oscillating particle, while in the case of electromagnetic case, they represent the electric (E) and magnetic (B) fields of the light in a given wavevector and polarization mode k. Thus, the quantum electromagnetic field consists of an *infinite* product of such generalized harmonic oscillators, one for each mode of the field. A Heisenberg-type uncertainty relation applies to the Maxwell fields: $\Delta E \Delta B \geq \hbar/2 \times$ constant, i.e., the electric and the magnetic fields associated with light cannot be measured arbitrarily precisely. Such field fluctuations are an intrinsic feature of the quantized theory. The uncertainty relation can also be formulated in terms of the in-phase and in-quadrature components of the electric field. To introduce the notion of a photon, it is convenient to recast the above quantization of the field in terms of the annihilation ($\hat{a}$) and creation ($\hat{a}^{\dagger}$) operators of a harmonic oscillator. These correspond to the positive and negative frequency parts of the electric field operator, respectively.

By analogy to the theory of the harmonic oscillator, the application of $\hat{a}$ produces a state with one *less* quantum of energy, and the application of $\hat{a}^{\dagger}$ produces a state with one *more* quantum of energy. This naturally leads to discrete energies for the oscillator in each mode: $n_k = 0, 1, 2, \ldots$. In the absence of any medium, the modes $E_k(\mathbf{r})$ are just the plane-wave solutions to the Maxwell equations. Alternately, we can define a localized "pulse" basis for the photon by summing over many wave vectors and frequencies, just as for classical waves. Thus, quantum electrodynamics permits both wave and particle perspectives on light. The wave perspective is exemplified by the picture of a stochastic electro-magnetic field. The particle perspective follows from the language of annihilation and creation operators which are subject to the appropriate commutation relation. Combining these perspectives, one can adopt a rigorous definition of the photon as follows: A photon corresponds to a single excitation of a particular mode k of the electromagnetic field in a suitably defined cavity, such that the annihilation and the creation operators for the field mode satisfy a Boson commutation relation. The wave-particle picture of light embedded in the Dirac's theory of light had novel and important consequences. The most important was the reshaping of our concept of vacuum.

Quantum Vacuum   Before the advent of quantum field theory, the vacuum was perceived as *nothing*—a place where no light existed, nothing moved, and there was no energy present. The quantum mechanical picture of vacuum turned out to be dramatically different. According to Dirac's theory of light, the quantum harmonic oscillator associated with each electromagnetic wave of frequency ν has an energy equal to $h\nu/2$ in vacuum. There are infinite number of mode in the universe, each associated with a frequency ν. Thus the total energy in the universe can be calculated by adding this vacuum energy for each mode and the result is an infinite amount of energy. In addition, as noted above, there are quantum mechanical fluctuations as a result of Heisenberg's uncertainty relation that cannot be neglected, even in vacuum. This forbids a classical description of absolutely zero electric and magnetic fields in vacuum. Instead we have fluctuations—randomly nonzero fields at any time. We thus have a revolutionary new way of thinking about light that field quantization introduced into the scientific discourse, namely that the electromagnetic field, when quantized, has the ability to exist in a state of pure nothingness—the so-called vacuum state—and yet have observable consequences in the material world.

Spontaneous Emission   An important consequence of the fluctuations in the vacuum field is the phenomenon of spontaneous emission by an atom. A photon is created in response to these fluctuations. Thus, even in the absence of an applied field, an atom in the excited state can decay to the ground state and spontaneously emit a photon. Since the direction and time of emission are random, this process represents a fundamental source of quantum noise, and a limitation to any coherent process (such as lasing). The excited atomic level acquires a finite bandwidth which is the inverse of the emission lifetime. We can use quantum theory to calculate the spatio-temporal profile of the emitted photon as detected by a photodetector.

Lamb Shift   Perhaps the greatest triumph of field quantization is the explana-tion of the Lamb shift between, for example, the $2s_{1/2}$ and $2p_{1/2}$ levels in a hydrogenic atom. Relativistic quantum mechanics predicts that these levels should be at the same energy. Willis Lamb (1913–2008), however, experimen-tally observed in 1947 a frequency splitting of about 1 GHz in contradiction to the theoretical prediction. We can understand the shift intuitively by pictur-ing the electron forced to fluctuate about its first-quantized position in the atom due to random kicks from the surrounding, fluctuating vacuum field. Its average displacement $< \Delta \mathbf{r} >$ is zero, but the squared displacement $< (\Delta \mathbf{r})^2 >$ is slightly nonzero, with the result that the electron "senses" a slightly different Coulomb pull from the positively charged nucleus than it normally would. The effect is more prominent nearer the nucleus where the Coulomb potential falls off more steeply, thus the s orbital is affected more than the p orbital. This is manifested as the Lamb shift between the levels.

Casimir Force In 1947, Hendrick Casimir (1909–2000) predicted that if two conducting plates separated by a distance a are placed in vacuum, and no external force is acting on them, they would attract each other with a force equal to $\hbar c\pi^2/240a^4$. This Casimir force was experimentally observed in 1958. Casimir explained this force arising purely as a consequence of the quantized modes of the radiation in vacuum. When the two conducting plates are inserted in the vacuum, the space is divided into three regions, the two infinite regions outside the plates and another region inside the two plates. The regions outside the plates have continuum of frequencies, i.e., all possible frequencies, resulting in an infinite amount of energy when we add the contributions of all the modes. The region inside the plates, however, allows only discrete number of modes satisfied by the resonance condition $a = \pi nc/\nu_n$, where ν_n is the frequency of the nth mode. These are also infinite number of modes, one for each value of n. The total amount of the vacuum field energy between the plates is also infinite. Thus we have an infinite amount of energy outside the plates and an infinite amount of energy between the plates. The truly dramatic result is that when we subtract these two infinities, the outcome is finite. As the system tends to evolve to a state with minimum energy there is a resulting force and this force is attractive. This is a highly counterintuitive result. Julian Schwinger called it "One of the least intuitive consequences of quantum electrodynamics" and according to Bryce DeWitt: "What startled me, in addition to the crazy idea that a pair of electrically neutral conductors should attract one another, was the way in which Casimir said the force could be computed, namely, by examining the effect on the zero-point energy of the electromagnetic vacuum caused by the mere presence of the plates. I had always been taught that the zero-point energy of a quantized field was unphysical."

Laser: A Coherent Light Source All the studies on light from the antiquity until the middle of nineteenth century were based on incoherent light sources such as the sun, candle light, sodium lamp, or light bulb. In 1950s a new coherent source of light was invented, first in the microwave region and then in the optical region. This new kind of light source, laser, is one of the greatest inventions of the second part of the twentieth century. It has helped to revolutionize many branches of science and technology ranging from bio-technology and precision measurements to communication and remote sensing. The physical process behind conventional light sources is spontaneous emission and they operate in thermal equilibrium. Initially majority of atoms and molecules are in their ground state. When energy is supplied to the atoms or molecules, some of them go to the excited states and then radiate via spontaneous emission. As discussed above, the spontaneous emission process is due to the ubiquitous vacuum fluctuations and each atom radiates independently of each other. The resulting light is a white light sent in all directions and is incoherent. On the other hand, the dominant emission process in a laser is stimulated emission. By a clever design, the radiated photons by the atoms or molecules are able to stimulate other atoms to radiate with the same frequency and same direction. The resulting radiation is coherent, monochromatic, and highly directional.

In 1954, Gordon, Zeiger, and Charles Townes (1915–2015) showed that coherent electromagnetic radiation can be generated in the radio frequency range by the so-called maser (microwave amplification by stimulated emission of

1

radiation). The first maser action was observed in ammonia. The maser principle was extended by Arthur Schawlow (1921–1999) and Townes, and also by Basov and Prokhorov, to the optical domain, thus obtaining a LASER (light amplification by stimulated emission of radiation). A laser consists of a set of atoms interacting with an electromagnetic field inside a cavity. The cavity supports only a specific set of modes corresponding to a discrete sequence of frequencies. The active atoms, i.e., the ones that are pumped to the upper level of the laser transition, are in resonance with one of these frequencies of the cavity. A resonant electromagnetic field gives rise to stimulated emission, and the atoms transfer their excitation energy to the radiation field. The emitted radiation is still at resonance. If the upper level is sufficiently populated, this radiation gives rise to further transitions in other atoms. In this way all the excitation energy of the atoms is transferred to a single mode of the radiation field.

The first pulsed laser operation was demonstrated by Theodore Maiman (1927–2007) in ruby in 1960. The first continuous wave (cw) laser, a He–Ne gas laser, was built by Ali Javan later in the same year. Since then, a large variety of systems have been demonstrated to exhibit lasing action; generating coherent light over a frequency domain ranging from infrared to ultraviolet. These include dye lasers, chemical lasers, and semiconductor lasers.

The Birth of Quantum Optics   The advent of laser required a careful description of the various sources of light. The question was: What is the fundamental difference between the conventional light sources, such as the sun, and the newly discovered laser light? An answer to this question led to a new field of study in Physics: Quantum Optics. Roy Glauber, in a series of seminal papers in 1963, at first controversial, differentiated between laser (coherent) light and normal (blackbody) light in terms of the photon statistics. This work had far reaching consequences as it showed that there could be all kinds of light sources that have to be distinguished by their quantum states and the corresponding statistical properties. These sources could range from a photon number state, in which light quanta could behave like particles, to a coherent state, which is as close to a classical Maxwellian description of light as electromagnetic wave as the quantum mechanics allows.

Quantum Interference and Delayed Choice Quantum Eraser   Before closing this brief story of light, we observe that the paradigm of quantum interference, the interference of probability amplitudes associated with different paths taken by a photon, defines our present understanding on the nature of light. In some ways, this is a culmination of the centuries-old debate on the nature of light reviewed through this article. The modern quantum perspective on this debate is that light is neither wave nor particle, but an elusive, intermediate entity that obeys the superposition principle. The quintessential experiment that demonstrates wave-particle duality is the Young's two-slit interference experiment. When a single photon goes through the slits, it registers as a point-like event on the screen (measured say by a CCD array). An accumulation of such events over repeated trials builds up a probabilistic fringe pattern that is characteristic of wave interference. However, if we arrange to measure which slit the photon goes through, the interference always disappears.

This picture is, however, not so simple. The counterintuitive aspect of complementarity is epitomized in the problem of quantum eraser, as was shown by Marlan Scully in 1982. The inability to discern which-path information, or the indistinguishability of interfering pathways, in the double-slit experiment is the key to preserving the wave properties of the photon and the appearance of fringes on the screen. What if, rather than subject the photon to a classical measurement, we can have it interact *quantum mechanically* with a localized marker particle (such as an atom) and leave behind a record of its path? The interference pattern then survives or not depends on the marker states, which carry the tell-tale information about which path the photon took to the detector. The coherence is destroyed as soon as we have the which-path information. One then wonders whether it might not be possible to retrieve the coherence, and the fringes, by destroying the which-path information contained in the marker—long after the photon is detected on the screen. This is the essence of the quantum eraser idea. An experimental demonstration of quantum eraser elicits a response of incredulity as the following quote by Brian Greene in his beautiful book *The Fabric of the Cosmos* indicates: "These experiments are a magnificent affront to our conventional notions of space and time. . . . . For a few days after I learned of these experiments, I remember feeling elated. I felt I'd been given a glimpse into a veiled side of reality."

1.6 Epilogue

We have come a long way from the earliest studies on light, trying to understand vision as light emanating from our eyes, to the description of light as rays, then as particles, and then waves, and finally exhibiting both particle and wave natures. We can only speculate how our present understanding of light will be perceived decades or centuries from now. Will our picture of light quanta as both waves and particles survive or will something more intuitive replace this incomprehensible picture? It is an irony that the greatest strides taken in the scientific understanding have come in our time, yet we feel least certain of our understanding of what light is, what photon is. In spite of the great success of the mathematical theory to describe light and its amazing agreement with experiment, the question "What is Light?" can ignite a heated discussion. To quote Albert Einstein (1954), "All the fifty years of conscious brooding have brought me no closer to the answer to the question: What are light quanta? Of course today every rascal thinks he knows the answer, but he is deluding himself."

Bibliography

Lindberg DC (1976) Theories of vision: from Al-Kindi to Kepler. University of Chicago Press, Chicago
Pais A (1982) Subtle is the lord: the science and the life of Albert Einstein. Oxford University Press, New York

Scully MO, Suhail Zubairy M (1997) Quantum optics. Cambridge University Press, Cambridge

Aharonov Y, Suhail Zubairy M (2005) Time and the quantum: erasing the past and impacting the future. Science 307:875

Muthukrishnan A, Scully MO, Suhail Zubairy M (2003) The concept of the photon-revisited? Opt Photonics News 14(10):18–27

Greenberger DM, Erez N, Scully MO, Svidzinsky AA, Suhail Zubairy M (2007) Planck, photon statistics, and Bose-Einstein condensate. In: Wolf E (ed) Progress in optics, vol 50. Elsevier, Amsterdam, p 275

Lyons J (2009) The house of wisdom: how the Arabs transformed western civilization. Bloomsbury Press, New York

Darrigol O (2012) A history of light: from Greek antiquity to the nineteenth century. Oxford University Press, Oxford

Mark Smith A (2015) From sight to light: the passage from ancient to modern optics. University of Chicago Press, Chicago

Weinberg S (2015) To explain the world: the discovery of modern science. HarperCollins, New York

Ibn al-Haytham's Scientific Research Programme

Roshdi Rashed

R. Rashed (✉)
Université Paris Diderot, Sorbonne Paris Cité, SPHERE, UMR 7219, CNRS, 5 rue Thomas Mann, Bâtiment Condorcet, Case 7093, 75205 Paris Cedex 13, France
e-mail: rashed@paris7.jussieu.fr

© The Author(s) 2016
M.D. Al-Amri et al. (eds.), *Optics in Our Time*, DOI 10.1007/978-3-319-31903-2_2

2

2.1   Introduction

For the vast majority of historians, and, more generally, of laymen, Ibn al-Haytham's major contribution concerns the vision in all its aspects (physical, physiological and psychological) and, namely, the causes of perceptual and cognitive effects. The reform of Ibn al-Haytham, according to them, was mainly to abandon the traditional theory of vision, to a new one. Henceforth he belongs to ancient and mediaeval traditions, in spite of this reform, in so far that he was concerned with vision and sight.

I will argue here that this reform was a minor consequence of a more general and more fundamental research programme, and even his conception of the science of optics is quite different as so far that his main task was about light, its fundamental properties and how they determine its physical behaviour, as reflection, refraction, focalization, etc.

Some historians of optics consider that, up to the seventeenth century in Europe, the science in optics before Kepler was aimed primarily at explaining vision. The merest glance at the optical works of Ibn al-Haytham leaves no doubt that this global judgement is far from being correct. Indeed, this statement is correct as far as it concerns the history of optics before the shift done by Ibn al-Haytham and the reform he accomplished. Successor of Ptolemy, al-Kindī and Ibn Sahl, to mention only a few, he unified the different branches of optics: optics, dioptrics, anaclastics, meteorological optics, etc. This unification was possible only for a mathematician who focused on light, and not on vision. Nobody, as far as I know, before Ibn al-Haytham, wrote such books titled: *On Light*; *On the Light of the Moon*; *On the Light of the Stars*; *On the Shadows*, among others, in which nothing concerns sight. At the same time, three books from his famous *Book of Optics* are devoted strictly to the theory of light. None of the authors before him, who were mainly interested in vision, wrote a very important contribution on physical optics such as the one on *The Burning Sphere*.

I begin by quoting the expression which Ibn al-Haytham repeated more than once in his different writings on optics. At the beginning of this famous *Book of Optics*, he writes:

>   Our subject is obscure and the way leading to knowledge of its nature difficult, moreover our inquiry requires a combination of the natural and mathematical sciences.[1]

But such a combination in optics, for instance, requires one to examine the entire foundations and to invent the means and the procedures to apply mathematics on the ideas of natural phenomena. For Ibn al-Haytham, it was the only way to obtain a rigorous body of knowledge.

Why this particular turn, at that time? Let me remind that Ibn al-Haytham lived in the turn of the first millennium. He was the heir of two centuries of scientific research and scientific translations, in mathematics, in astronomy, in statics, in optics, etc. His time was of intense research in all these fields. He himself wrote more in mathematics and in astronomy than in optics per se. According to early bio-bibliographers, Ibn al-Haytham wrote 25 astronomical works: twice as many works on the subject as he did in optics. The number of his writings alone indicates the huge size of the task accomplished by him and the importance of astronomy in his life work. In all branches of mathematics, he wrote more than all his writings in astronomy and in optics put together. If he wrote in optics the famous huge book, *Kitāb al-Manāẓir—The Book of Optics*, in astronomy likewise

---

1   Ibn al-Haytham [2], p. 4.

he wrote a huge book entitled *The Configuration of the Motions of each of the Seven Wandering Stars.*

Before coming back in some details to these contributions, let me characterize Ibn al-Haytham's research programme.

1. It is a new one, concerning the relationships between mathematics and natural phenomena, never conceived before. His aim is to mathematize every empirical science. This application of mathematics can take different forms, not only given to the different disciplines, but also in one and the same discipline.

2. It does not concern only optics, but every natural science, i.e., for the epoch, astronomy and statics.

3. Its success depends on the means—mathematical, linguistic and technical—by which mathematics control the semantic and syntactical structures of natural phenomena.

2.2   Between Ptolemy and Kepler: Ibn al-Haytham's Celestial Kinematics

To put the facts right, I will turn at first, quite briefly, to Ibn al-Haytham's astronomy. He wrote at least three books criticizing the astronomical theory of Ptolemy:

1. *The Doubts concerning Ptolemy*
2. *Corrections to the* Almagest
3. *The Resolution of Doubts concerning the* Almagest

In the *Doubts*, Ibn al-Haytham comes to the conclusion that "the configuration Ptolemy assumes for the motions of the five planets is a false one".[2] A few lines further on, he continues: "The order in which Ptolemy had placed the motions of the five planets conflicts with the theory <that he had proposed>".[3] A little later, he states: "The configurations that Ptolemy assumed for the <motions of> the five planets are false ones. He decided on them knowing they were false, because he was unable <to propose> other ones.[4]" After such comments, and many others like them in several places of his writings, Ibn al-Haytham had no option but to construct a planetary theory of his own, on a solid mathematical basis, and free from the internal contradictions found in Ptolemy's *Almagest*. For this purpose, he conceived the idea of writing his monumental and fundamental book *The Configuration of the Motions of the Seven Wandering Stars*. If we wish to characterize the irreducible inconsistencies that, according to Ibn al-Haytham, vitiate Ptolemy's astronomy, we may say that they arise from the poor fit between a mathematical theory of the planets and a cosmology; that is, the combination between mathematics and physics. Ibn al-Haytham was familiar with similar, though of course not identical, situations when, in optics, as we shall see, he encountered the inconsistency between geometrical optics and physical optics as understood not only by Euclid and Ptolemy, but also by Aristotle and the philosophers.

In *The Configuration of the Motions* he deals with the apparent motions of the planets, without ever raising the question of the physical explanation of these motions in terms of dynamics. It is not the causes of celestial motions that interest Ibn al-Haytham, but only the motions themselves observed in space and time. Thus, to proceed with the systematic mathematical treatment, and to avoid the

---

2   See Rashed [8], p. 13.

3   See footnote 2.

4   See footnote 2.

2

obstacles that Ptolemy had encountered, he first needed to break away from any kind of cosmology. Thus the purpose of Ibn al-Haytham's *Configuration of the Motions* is clear: instead of constructing, as his predecessors, a cosmology, or a kind of dynamics, he constructs the first geometrical kinematics.

A close examination of the way he organizes his exposition of planetary theory shows that Ibn al-Haytham begins by omitting physical spheres and by proposing simple—in effect, descriptive—models of the motions of each of the seven planets. As the exposition progresses, he makes the models more complicated and increasingly subordinates them to the discipline of mathematics. This growing mathematization leads him to regroup the motions of several planets under a single model. This step obviously has the effect of privileging a property that is common to several motions. In this way Ibn al-Haytham opens up the way to achieving his principal objective: to establish a system of celestial kinematics. He does so without as yet formulating the concept of instantaneous speed, but by using the concept of mean speed, represented by a ratio of arcs.

In the course of his research, which I analysed elsewhere,[5] we encounter a concept of astronomy that is new in several respects. Ibn al-Haytham sets himself the task of describing the motions of the planets exactly in accordance with the paths they draw on the celestial sphere. He is neither trying 'to save the phenomena', like Ptolemy, that is, to explain the irregularities in the assumed motion by means of artifices such as the equant; nor trying to account for the observed motions by appealing to underlying mechanisms or hidden natures. He wants to give a rigorously exact description of the observed motions in terms of mathematics. Thus his theory for the motion of the planets calls upon no more than observation and conceptual constructs susceptible of explaining the data, such as the eccentric circle and in some cases the epicycle. However, this theory does not aim to describe anything beyond observation and these concepts, and in no way is it concerned to propose a causal explanation of the motions.

The new astronomy no longer aims at constructing a model of the universe, as in the *Almagest*, but only at describing the apparent motion of each planet, a motion composed of elementary motions, and, for the inferior planets, also of an epicycle. Ibn al-Haytham considers various properties of this apparent motion: localization and the kinematic properties of the variations in speed.

In this new astronomy, as in the old one, every observed motion is circular and uniform, or composed of circular and uniform motions. To find these motions, Ibn al-Haytham uses various systems of spherical coordinates: equatorial coordinates (the required time and its proper inclination); horizon coordinates (altitude and azimuth) and ecliptic coordinates. The use of equatorial coordinates as a primary system of reference marks a break with Hellenistic astronomy. In the latter, the motion of the orbs was measured against the ecliptic, and all coordinates were ecliptic ones (latitude and longitude). Thus, basing the analysis of the planets' motion on their apparent motions drives a change in the reference system for the data; we are now dealing with right ascension and declination. Ibn al-Haytham's book thus transports us into a different system of analysis.

To sum up, in the *The Configuration of the Motions*, Ibn al-Haytham's purpose is purely kinematics; more precisely, he wanted to lay the foundations of a completely geometrical kinematics tradition. But carrying out such a project involves first of all developing some branches of geometry, as also of plane and spherical trigonometry. In both fields, Ibn al-Haytham obtained new and important results.

In astronomy, properly, there are two major processes that are jointly involved in carrying through this project: freeing celestial kinematics from cosmological

---

5   See Rashed [8].

connections, that is, from all considerations of dynamics, in the ancient sense of the term; and to reduce physical entities to geometrical ones. The centres of the motions are geometrical points without physical significance; the centres to which speeds are referred are also geometrical points without physical significance; even more radically, all that remains of physical time is the 'required time', that is, a geometrical magnitude. In short, in this new kinematics, we are concerned with nothing that identifies celestial bodies as physical bodies. All in all, though it is not yet that of Kepler, this new kinematics is no longer that of Ptolemy nor of any of Ibn al-Haytham's predecessors; it is *sui generis*, half way between Ptolemy and Kepler. It shares two important ideas with ancient kinematics: every celestial motion is composed of elementary uniform circular motions, and the centre of observation is the same as the centre of the Universe. On the other hand, it has in common with modern kinematics the fact that the physical centres of motions and speeds are replaced by geometrical centres.

In fact, once Ibn al-Haytham had engaged upon mathematizing astronomy and had noted not only the internal contradictions in Ptolemy, but doubtless also the difficulty of constructing a self-consistent mathematical theory of material spheres using an Aristotelian physics, he conceived the project of giving a completely geometrized kinematic account.

Ibn al-Haytham had the same experience in optics. In astronomy, kinematics and cosmology are entirely separated to effect a reform of the discipline, just as in optics, work on light and its propagation is entirely separated from work on vision to effect a reform of optics; in the one case as in the other, we shall see, Ibn al-Haytham arrived at a new idea of the science concerned.

2.3   Ibn al-Haytham's Reform of Optics

It is now time to come to Ibn al-Haytham's optics. As we have said above, Ibn al-Haytham was preceded by two centuries of translation into Arabic of the main Greek optical writings, as well of inventive research. Among his Arabic predecessors, al-Kindī, Qusṭā ibn Lūqā, Aḥmad ibn 'Īsā 'Uṭārid, etc. During these two centuries, the interest shown in the study of burning mirrors is an essential part of the comprehension of the development of catoptrics, anaclastics and dioptrics, as the book produced between 983 and 985 by the mathematician al-'Alā' ibn Sahl testifies. Before this contribution of Ibn Sahl, the catoptricians like Diocles, Anthemius of Tralles, al-Kindī etc.[6] asked themselves about geometrical properties of mirrors and about light they reflect at a given distance. Ibn Sahl modifies the question by considering not only mirrors but also burning instruments, i.e. those which are susceptible to light not only by reflection, but also by refraction; and how in each case the focalization of light is obtained. Ibn Sahl studies then, according to the distance of the source (finite or infinite) and the type of lighting (reflection or refraction) the parabolic mirror, the ellipsoidal mirror, the plano-convex lens and the biconvex lens. In each of these, he proceeds to a mathematical study of the curve, and, then, expounds a mechanical continuous drawing of it. For the plano-convex lens, for instance, he starts by studying the hyperbola as a conic section, in order then to take up again a study of the tangent plane to the surface engendered by the rotation of the arc of hyperbola around a fixed straight line, and, finally, the curve as an anaclastic curve, and the laws of refraction.

These studies which focused on light and its physical behaviour were instrumental in the discovery by Ibn Sahl of the concept of a constant ratio, characteristic

---

6   See Rashed [5, 6].

2

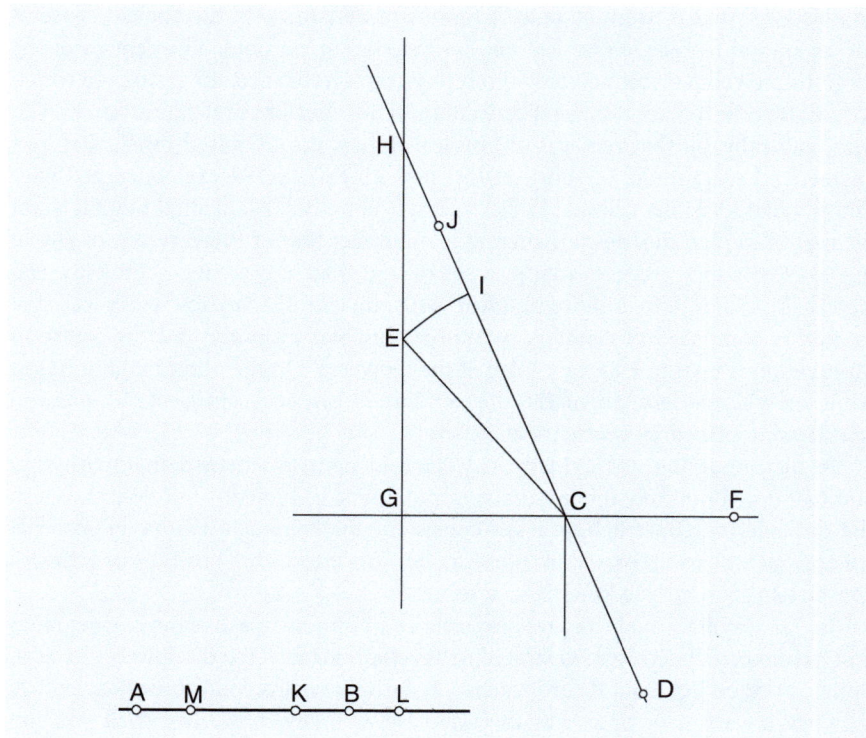

□ Fig. 2.1 Ibn Sahl illustration of a ray of light (*DC*) refracted as it crosses the boundary (*GF*) of two media of different refractive indices (see text for more details)

of the medium, which is a masterpiece in his study of refraction in lenses, as well as his discovery of the so-called Snellius' law.

At the beginning of his study, Ibn Sahl considers a plane surface *GF* surrounding a piece of transparent and homogeneous crystal. He next considers the straight line *CD* along which the light propagates in the crystal, the straight line *CE* along which it refracts itself in the air, and the normal at *G* on the surface *GF* which intersects the straight line *CD* at *H* and the ray refracted at *E*.

Obviously, Ibn Sahl is here applying the known law of Ptolemy according to which the ray *CD* in the crystal, the ray *CE* in the air and the normal *GE* to the plane surface of the crystal are found in the same plane (□ Fig. 2.1). He writes then, in a brief way, and, according to his habit, with no conceptual commentary:

» Straight line *CE* is therefore smaller than straightline *CH*. From straight line *CH*, we separate the straight line *CI* equal to straight line *CE*; we divide *HI* into two halves at point *J*; we make the ratio of straight line *AK* to straight line *AB* equal to the ratio of straight line *CI* to straight line *CJ*. We draw the line *BL* on the prolongation of straight line *AB* and we make it equal to straight line *BK*.[7]

In these few phrases, Ibn Sahl draws the conclusion first that *CE/CH* < 1, which he will use throughout his research into lenses made in the *same* crystal. In effect he does not fail to give this same ratio again, nor to reproduce this same figure, each time that he discusses refraction in this crystal (□ Fig. 2.2).

---

7   See Rashed [7], p. 106.

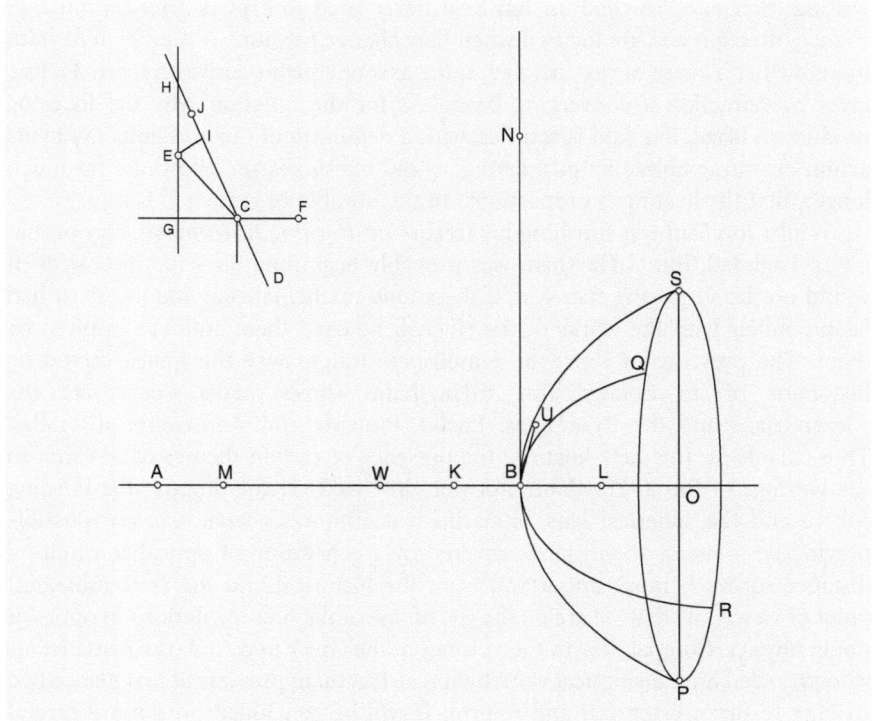

Fig. 2.2 Ibn Sahls diagram depicting refraction with plano-convex lenses (see text for more details)

But the ratio is nothing other than the inverse of the index of refraction in this crystal in relation to the air. Considering the i_1 and i_2 as the angles formed, respectively, by CD and by CE with the normal GH, we have

$$\frac{1}{n} = \frac{\sin i_1}{\sin i_2} = \frac{CG}{CH} \cdot \frac{CE}{CG} = \frac{CE}{CH}.$$

Ibn Sahl takes on the segment CH a point I such that $CI=CE$, and a point J at the midpoint of IH. This gives

$$\frac{CI}{CH} = \frac{1}{n}.$$

The division $CIJH$ characterizes this crystal for all refraction.

Ibn Sahl shows, moreover, in the course of his research into the plano-convex lens and the biconvex lens, that the choice of hyperbola to fashion the lens depends on the nature of the crystal, since the eccentricity of the hyperbola is $e = 1/n$.

Thus, Ibn Sahl had conceived and put together an area of research into burning instruments and, also, anaclastics. But, obliged to think about conical figures other than the parabola and the ellipse—the hyperbola for example—as anaclastic curves, he was quite naturally led to the discovery of the law of Snellius. We understand therefore that dioptrics, when it was developed by Ibn Sahl, only dealt with matters involving the propagation of light, independently of problems of vision. The eye did not have its place within the area of burning instruments, nor did the rest of the subject of vision. It is thus an objective point of view which is deliberately adopted in the analysis of luminous phenomena. Rich in technical material, this new discipline is in fact very poor on physical content: it is evanescent and reduces a few energy considerations. By way of example, at least in his writings that have reached us, Ibn Sahl never tried to explain why certain rays change direction and are focused when they change medium: it is enough for him to know that a beam of rays parallel to the axis of a plano-convex hyperbolic lens gives by refraction a converging beam. As for the question why the focusing produces a blaze, Ibn Sahl is satisfied with a definition of the luminous ray by its action of setting ablaze by postulating, as did his successors elsewhere for much longer, that the heating is proportional to the number of rays.

Whilst Ibn Sahl was finishing his treatise on *Burning Instruments* very probably in Baghdad, Ibn al-Haytham was probably beginning his scientific career. It would not be surprising therefore if the young mathematician and physicist had been familiar with the works of the elder, if he cited them and was inspired by them. The presence of Ibn Sahl demolishes straightaway the image carved by historians of an isolated Ibn al-Haytham whose predecessors were the Alexandrians and the Byzantines: Euclid, Ptolemy and Anthemius of Tralles. Thus, thanks to this new filiation, the presence of certain themes of research in the writings of Ibn al-Haytham, not only his work on the dioptre, the burning sphere and the spherical lens, is clarified; it authorizes what was not possible previously: to assess the distance covered by a generation of optical research—a distance so much more important, from the historical and the epistemological point of view, now that we are on the eve of one of the first revolutions in optics, if not in physics. Compared with the writings of the Greek and Arab mathematicians who preceded him, the optical work by Ibn al-Haytham presents at first glance two striking features: extension and reform. It will be concluded on a more careful examination that the first trait is the material trace of the second. In fact no one before Ibn al-Haytham had embraced so many domains in his research, collecting together fairly independent traditions: mathematical, philosophical, medical. The titles of his books serve moreover to illustrate this large spectrum: *The Light of the Moon, The Light of the Stars, The Rainbow and the Halo, Spherical Burning Mirrors, Parabolical Burning Mirrors, The Burning Sphere, The Shape of the Eclipse, The Formation of Shadow*s, *On Light*, as well as his *Book of Optics* translated into Latin in the twelfth century and studied and commented on in Arabic and Latin until the seventeenth century. Ibn al-Haytham therefore studied not only the traditional themes of optical research but also other new ones to cover finally the following areas: optics, catoptrics, dioptrics, physical optics, meteorological optics, burning mirrors, the burning sphere.

A more meticulous look reveals that, in the majority of these writings, Ibn al-Haytham pursued the realization of his programme to reform the discipline, which brought clearly to take up each different problem in turn. The founding action of this reform consisted in making clear the distinction, for the first time in the history of optics, between the conditions of propagation of light and the conditions of vision of objects. It led, on one hand, to providing physical support for the rules of propagation—it concerns a mathematically guaranteed analogy between a mechanical model of the movement of a solid ball thrown against an obstacle, and that of the light—and, on the other hand, to proceeding everywhere geometrically and by observation and experimentation. It led also to the definition of the concept of light ray and light bundle as a set of straight lines on which light propagates, rays independent from each other which propagate in a homogeneous region of space. These rays are not modified by other rays which propagate in the same region. Thanks to the concept of light bundle, Ibn al-Haytham was able to study the propagation and diffusion of light mathematically and experimentally. Optics no longer has the meaning that is assumed formerly: a geometry of perception. It includes henceforth two parts: a theory of vision, with which is also associated a physiology of the eye and a psychology of perception, and a theory of light, to which are linked geometrical optics and physical optics. Without doubt traces of the ancient optics are still detected: the survival of ancient terms, or a tendency to pose the problem in relation to the subject of vision without that being really necessary. But these relics do not have to deceive: their effect is no longer the same, nor is their meaning. The organization of his *Book of Optics* reflects already the new situation. In it are books devoted in full to propagation—the third chapter of the first book and Books IV to VII; others deal with vision and related problems. This reform led, amongst other things, to the emergence of new problems, never previously posed, such as the famous "problem of Alhazen" on catoptrics, the examination of the spherical lens and the spherical dioptre, not only as burning instruments but also as optical instruments, in dioptrics; and to experimental control as a practice of investigation as well as the norm for proofs in optics and more generally in physics.

Let us follow now the realization of his reform in the *Book of Optics* and in other treatises. This book opens with a rejection and a reformulation. Ibn al-Haytham rejects straightaway all the variants of the doctrine on the visual ray, to ally himself with philosophers who defended an intromissionist doctrine on the form of visible objects. A fundamental difference remains nevertheless between him and the philosophers, such as his contemporary Avicenna: Ibn al-Haytham did not consider the forms perceived by the eyes as "totalities" which radiate from the visible object under the effect of light, but as reducible to their elements: from every point of the visible object radiate a ray towards the eye. The latter has become without soul, without πνεῦμα ὀπτικόν, a simple optical instrument. The whole problem was then to explain how the eye perceives the visible object with the aid of these rays emitted from every visible point.

After a short introductory chapter, Ibn al-Haytham devotes two successive chapters—the second and the third books of his *Book of Optics*—to the foundations of the new structure. In one, he defines the conditions for the possibility of vision, while the other is about the conditions for the possibility of light and its propagation. These conditions, which Ibn al-Haytham presents in the two cases as empirical notions, i.e. as resulting from an ordered observation or a controlled experiment, are effectively constraints on the elaboration of the theory of vision, and in this way on the new style of optics. The conditions for vision detailed by Ibn al-Haytham are six: the visible object must be luminous by itself or illuminated by another; it must be opposite to the eye, i.e. one can draw a straight line to the eye from each of its points; the medium that separates it from the eye must be transparent, without being cut into by any opaque obstacle; the visible object must be more opaque than this medium; it must be of a certain volume, in relation to the visual sharpness. These are the notions, writes Ibn al-Haytham, "without which vision cannot take place". These conditions, one cannot fail to notice, do not refer, as in the ancient optics, to those of light or its propagation. Of these, the most important, established by Ibn al-Haytham, are the following: light exists independently of vision and exterior to it; it moves with great speed and not instantaneously; it loses intensity as it moves away from the source; the light from a luminous source—substantial—and that from an illuminated object—second or accidental—propagate onto bodies which surround them, penetrate transparent media, and light up opaque bodies which in turn emit light; the light propagates from every point of the luminous or illuminated object in straight lines in transparent media and in all directions; these virtual straight lines along which light propagates form with it "the rays"; these lines can be parallel or cross one another, but the light does not mix in either case; the reflected or refracted light propagates along straight lines in particular directions. As can be noted, none of these notions relate to vision. Ibn al-Haytham completes them with other notions relative to colour. According to him, the colours exist independently from the light in opaque bodies, and as a consequence only light emitted by these bodies—second or accidental light—accompanies the colours which propagate then according to the same principles and laws as the light. As we have explained elsewhere, it is this

2

doctrine on colours which imposed on Ibn al-Haytham concessions to the philosophical tradition, obliging him to keep the language of "forms", already devoid of content when he only deals with light.[8]

A theory of vision must henceforth answer not only the six conditions of vision, but also the conditions of light and its propagation. Ibn al-Haytham devotes the rest of the first book of his *Book of Optics* and the two following books to the elaboration of this theory, where he takes up again the physiology of the eye and a psychology of perception as an integral part of this new intromissionist theory.

Three books of the *Book of Optics*—the fourth to the sixth—deal with catoptrics. This area, as ancient as the discipline itself, amply studied by Ptolemy in his *Optics*, has never been the object of so extensive a study as that by Ibn al-Haytham. Besides the three voluminous books of his *Book of Optics*, Ibn al-Haytham devotes other essays to it which complete them, on the subject of connected problems such as that of burning mirrors. Research into catoptrics by Ibn al-Haytham distinguishes itself, among other traits, by the introduction of physical ideas, both to explain the known ideas and to grasp new phenomena. It is in the course of this study that Ibn al-Haytham poses himself new questions, such as the problem that bears his name.

Let us consider some aspect of this research into catoptrics by Ibn al-Haytham. He restates the law of reflection, and explains it with the help of the mechanical model already mentioned. Then he studies this law for different mirrors: plane, spherical, cylindrical and conical. In each case, he applies himself above all to the determination of the tangent plane to the surface of the mirror at the point of incidence, in order to determine the plane perpendicular to this last plane, which includes the incident ray, the reflected ray and the normal to the point of incidence. Here as in his other studies, to prove these results experimentally, he conceives and builds an apparatus inspired by the one that Ptolemy constructed to study reflection, but more complicated and adaptable to every case. Ibn al-Haytham also studies the image of an object and its position in the different mirrors. He applies himself to a whole class of problems: the determination of the incidence of a given reflection in the different mirrors and conversely. He also poses for the different mirrors the problem which his name is associated with: given any two points in front of a mirror, how does one determine on the surface of the mirror a point such that the straight line which joins the point to one of the two given points is the incident ray, whilst the straight line that joins this point to the other given point is the reflected ray. This problem, which rapidly becomes more complicated, has been solved by Ibn al-Haytham.

Ibn al-Haytham pursues this catoptric research in other essays, some of which are later than the *Book of Optics*, such as *Spherical Burning Mirrors*.[9] It is in this essay of a particular interest that Ibn al-Haytham discovers the longitudinal spherical aberration; it is also in this text that he proves the following proposition:

On a sphere of centre E let there be a zone surrounded by two circles of axis EB; let IJ be the generator arc of this zone, and D its midpoint. Ibn al-Haytham has shown in two previous propositions that to each of the two circles is associated a point of the axis towards which the incident rays parallel to the axis reflect on this circle. He shows here that all the rays reflected on the zone meet the segment thus defined: if GD is the medium ray of the zone, the point H is associated with D, and the segment is on either side of H. The length of this segment depends on the arc IJ (◘ Fig. 2.3).

8   See Rashed [4], pp. 271–298.
9   Ibn al-Haytham [1].

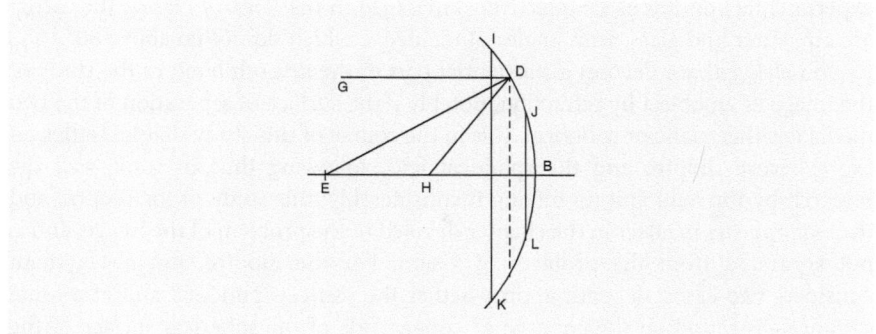

□ **Fig. 2.3** Ibn al-Hayhtam illustration of the longitudinal spherical aberration

The seventh and last book of the *Book of Optics* by Ibn al-Haytham is devoted to dioptrics. In the same way as he did for catoptrics, Ibn al-Haytham inserts in this book the elements of a physical—mechanical—explanation of refraction. Moreover, his book is completed by his essays, such as his treatise on the *Burning Sphere* or his *Discourse on Light*, where he comes back to the notion about the medium, following Ibn Sahl.

In this seventh book of the *Book of Optics*, Ibn al-Haytham starts by taking on the two qualitative laws of refraction, and several quantitative rules, all controlled experimentally with the help of an apparatus that he conceives and builds as in the previous case. The two quantitative laws known by his predecessors, Ptolemy and Ibn Sahl, can be expressed as follows: (1) the incident ray, the normal at the point of refraction and the refracted ray are in the same plane; the refracted ray approaches (or moves away from) the normal if the light passes from a less (respectively more) refractive medium to a more (respectively less) refractive medium; (2) the principle of the inverse return.

But, instead of following the way opened by Ibn Sahl through his discovery of the law of Snellius, Ibn al-Haytham returns to the ratios of angles and establishes his quantitative rules.

1. The angles of deviation vary in direct proportion to the angles of incidence: if in medium n_1 one takes $i' > i$, one will have, in medium n_2, $d' > d$ (i is the angle of incidence, r the angle of refraction and d the angle of deviation; $d = |i - r|$).
2. If the angle of incidence increases by a certain amount, the angle of deviation increases by a smaller quantity: if $i' > i - I$ and $d' > d$, one will have $d' - d < i' - i$.
3. The angle of refraction increases in proportion to the angle of incidence: if $i' > i$, one will have $r' > r$.
4. If the light penetrates from a less refractive medium into a more refractive medium, $n_1 < n_2$, one has $d < \frac{1}{2}i$; in the opposite path, one has $d < \frac{(i+d)}{2}$, and one will have $2i > r$.
5. Ibn al-Haytham takes up again the rules stated by Ibn Sahl in his book on *The Celestial Sphere*; he affirms that, if the light penetrates from a medium n_1, with the same angle of incidence, into two different media n_2 and n_3, then the angle of deviation is different for each of these media because of the difference in opaqueness. If, for example, n_3 is more opaque than n_2, then the angle of deviation will be larger in n_3 than in n_2. Conversely, if n_1 is more opaque than n_2, and n_2 more opaque than n_3, the angle of deviation will be larger in n_3 than in n_2.

Contrary to what Ibn al-Haytham believes, these quantitative rules are not all valid in a general sense. But to his credit all are provable within the limits of the experimental conditions he effectively envisaged in his *Book of Optics*; the media are air, water and glass, with angles of incidence which do not go above 80°.[10]

Ibn al-Haytham devotes a substantial part of the seventh book to the study of the image of an object by refraction, notably if the surface of separation of the two media is either plane or spherical. It is in the course of this study that he settles on the spherical dioptre and the spherical lens, following thus in some way the research by Ibn Sahl, but modifying it considerably; this study of the dioptre and the lens appears in effect in the chapter devoted to the problem of the image, and is not separated from the problem of vision. For the dioptre, Ibn al-Haytham considers two cases, depending on whether the source—punctual and at a finite distance—is found on the concave or convex side of the spherical surface of the dioptre.

Ibn al-Haytham studies the spherical lens, giving particular attention to the image that it gives of an object. He restricts himself nevertheless to the examination of only one case, when the eye and the object are on the same diameter. Put another way, he studies the image through a spherical lens of an object placed in a particular position on the diameter passing through the eye. His procedure is not without similarities to that of Ibn Sahl when he studied the biconvex hyperbolic lens. Ibn al-Haytham considers two dioptres separately, and applies the results obtained previously. It is in the course of his study of the spherical lens that Ibn al-Haytham returns to the spherical aberration of a point at a finite distance in the case of the dioptre, in order to study the image of a segment which is a portion of the segment defined by the spherical aberration.

In his treatise on the *Burning Sphere*, Ibn al-Haytham explains and refines certain results on the spherical lens which he had already obtained in his *Book of Optics*. However, he returns to the question of the burning by means of that lens. It is in this treatise that we encounter the first deliberate study of spherical aberration for parallel rays falling on a glass sphere and undergoing two refractions. In the course of this study, Ibn al-Haytham uses numerical data given in the *Optics* by Ptolemy for the two angles of incidence 40° and 50°, and, to explain this phenomenon of focusing of light propagated along trajectories parallel to the diameter of the sphere, he returns to angular values instead of applying what is called the law of Snellius.

In this treatise on the *Burning Sphere*, as in the seventh book of his *Book of Optics* or in other writings on dioptrics, Ibn al-Haytham exposes his research in a somewhat paradoxical way: while he takes a lot of care to invent, fashion and describe some experimental devices that are advanced for this age, allowing the determination of numerical values, in most cases he avoids giving these values. When he does give them, as in the treatise on the *Burning Sphere*, it is with economy and circumspection. For this attitude, already noted, at least two reasons can perhaps be found. The first is in the style of the scientific practice itself: quantitative description does not yet seem to be a compelling norm. The second is no doubt linked: the experimental devices can only give approximate values. It is for this reason that Ibn al-Haytham took into account the values which he had borrowed from the *Optics* by Ptolemy.

This book on the *Burning Sphere* is undoubtedly one of the summits of research in classical optics. Kamāl al-Dīn al-Fārisī (d. 1319) was able to put this book to work in order to explain for the first time the rainbow and the hallo. In this book, Ibn al-Haytham returns to the problem of combustion with the help of a spherical lens. Here then, is a text that enables us to follow the evolution of Ibn al-Haytham's thought on spherical lenses, by examining how he takes up the problem raised by his predecessor Ibn Sahl: to cause combustion by refraction,

---

10   See Rashed [3].

with the help of a lens. For Ibn al-Haytham, this research is an integral part of optics.

He begins this book by proving several propositions two of which are particularly important:

1. $\frac{i}{4} < d < \frac{i}{2}$ (i, angle of incidence in the glass; d, angle of deviation)
2. Let α and β be two arcs of a circle; we suppose that $\alpha > \beta$; $\alpha = \alpha_1 + \alpha_2$ and $\beta = \beta_1 + \beta_2$, such that

$$\frac{\alpha_2}{\alpha_1} = \frac{\beta_2}{\beta_1} = k < 1$$

and

$$\alpha_1 < \frac{\pi}{2}$$

Then $\dfrac{\sin \beta_1}{\sin \beta_2} > \dfrac{\sin \alpha_1}{\sin \alpha_2}$.

With the help of these two propositions, as well as his rules of refraction, Ibn al-Haytham studies the propagation of a bundle of parallel rays falling upon a glass or crystal sphere. Let us sketch how he proceeds.

In a first proposition, he shows that all parallel rays falling on the sphere with one and the same angle of incidence converge, after two refractions, towards one and the same point of the diameter which is parallel to the ray. This point is the focus associated with incidence i.

Thus, he considers a ray (HN) parallel to the diameter AC, falling upon the sphere at M. The refracted ray corresponding to it meets the sphere at B, and meets AC at point S. Point S is the focus associated with incidence i, and it belongs to the segment $[CK]$ where K is the intersection of MB with AC (◨ Fig. 2.4).

In a second proposition, he proves that the total deviation is twice each of the deviations: $D = 2d$.

He proves then that a given point S, beyond C on the diameter, can be obtained only from a single point M; that is to say, S corresponds to a single incidence.

In a third proposition, he proves that the two incidences i and i', correspond two distinct points S and S'.

In a fourth proposition, he proves: for $i > i'$, we have S and S' such that $CS' > CS$. Therefore, when i increases, CS decreases. To a given point S, therefore, there corresponds one single incidence i.

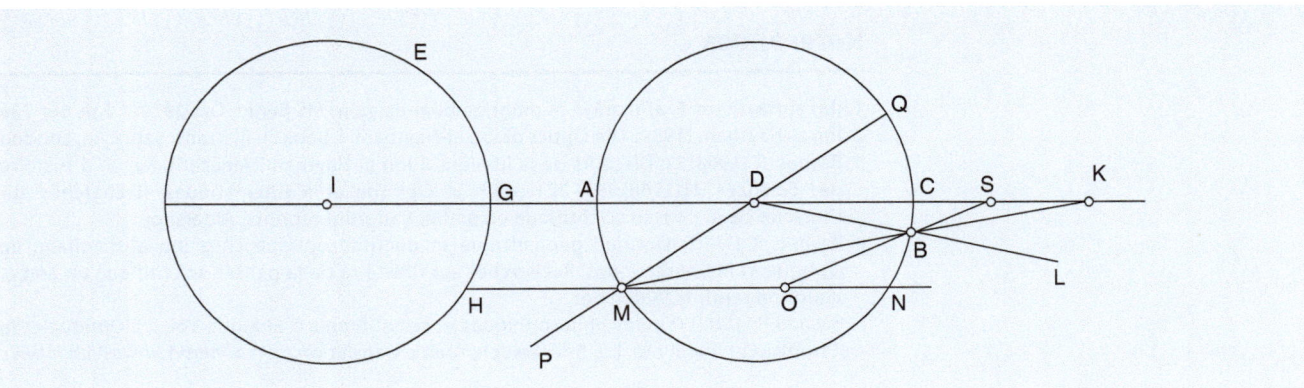

◨ **Fig. 2.4** Illustration of spherical aberration in glass (crystal) spheres

Ibn al-Haytham proposes to determine the extremities of the segment on which the points S are located. With this in view, he studies the positions of B—the point of the second refraction—when the angle of incidence varies. As far as we know, this is the first deliberate study of spherical aberration for parallel rays which fall on a glass sphere and undergo two refractions.[11]

2.4   Conclusion

Let us stop at this point on spherical aberration, to conclude.

With Ibn al-Haytham, one result has been definitively obtained: the half century which separates him from Ibn Sahl should be counted among the distinctive moments in the history of optics: dioptrics appears to have extended its domain of validity and, by its very progress, to have changed its orientation. With Ibn al-Haytham, the conception of dioptrics as a geometry of lenses has become outdated. Here again, in his own words, we must combine mathematics and physics in order to study dioptres and lenses, whether burning or not. The mathematization could only be achieved with Ibn al-Haytham because he separated the study of the natural phenomenon of light from vision and sight. The step taken suggests already that the domain carved out by Ibn Sahl was not long-lived and wound up, 50 years later, exploding under the assault of the mathematician and physicist Ibn al-Haytham. In optics as in astronomy the research programme of Ibn al-Haytham is the same: mathematize the discipline and combine this mathematization with the ideas of the natural phenomena.

References

1. Ibn al-Haytham. Fī al-marāyā al-muḥriqa bi-al-dawā'ir, MS Berlin, Oct 2970/7, fols 66r-73v
2. Ibn al-Haytham (1989) The Optics of Ibn al-Haytham, I, Books I–III (trans: Sabra AI). London
3. Rashed R (1968) Le Discours de la lumière d'Ibn al-Haytham (Alhazen). Revue d'Histoire des Sciences 21(1968):197–224, repr. in Optique et Mathématiques: Recherches sur l'histoire de la pensée scientifique en arabe, Variorum reprints, Aldershot
4. Rashed R (1992) Optique géométrique et doctrine optique chez Ibn al-Haytham. In: Optique et Mathématiques: Recherches sur l'histoire de la pensée scientifique en arabe. Variorum reprints, Aldershot
5. Rashed R (1997) Œuvres philosophiques et scientifiques d'al-Kindī. Vol. I: L'Optique et la Catoptrique d'al-Kindī. E.J. Brill, Leiden; Arabic translation: 'Ilm al-manāẓir wa-'ilm in'ikās

---

11   See Rashed [7], p. 164.

al-ḍaw', Silsilat Tārīkh al-'ulūm 'inda al-'Arab 6, Beirut, Markaz Dirāsat al-Waḥda al-'Arabiyya

6. Rashed R (2000) Les Catoptriciens grecs. I: Les Miroirs ardents, édition, traduction et commentaire, Collection des Universités de France, publiée sous le patronage de l'Association Guillaume Budé. Les Belles Lettres, Paris
7. Rashed R (2005) Geometry and Dioptrics in Classical Islam. Al-Furqān, London
8. Rashed R (2014) Ibn al-Haytham. New Spherical Geometry and Astronomy. A History of Arabic Sciences and Mathematics, vol. 4, Culture and Civilization in the Middle East, London, Centre for Arab Unity Studies, Routledge

Ultrafast Phenomena and the Invisible World

Contents

Ultrafast Light and Electrons: Imaging the Invisible

Ahmed H. Zewail

A.H. Zewail (✉)
Physical Biology Center for Ultrafast Science and Technology, Arthur Amos Noyes Laboratory for Chemical Physics, California Institute of Technology, Pasadena, CA 91125, USA
e-mail: zewail@caltech.edu

© The Author(s) 2016
M.D. Al-Amri et al. (eds.), *Optics in Our Time*, DOI 10.1007/978-3-319-31903-2_3

3

3.1   Origins

The ever-increasing progress made by humans in making the very small and the very distant visible and tangible is truly remarkable. The human eye has spatial and temporal resolutions that are limited to about 100 μm and a fraction of a second, respectively. Today we are aided by tools that enable the visualization of objects that are below a nanometer in size and that move in femtoseconds ([98] and the references therein). This chapter, which is based on review articles by the author [94, 99–101], some commentaries [83–85, 87], and a book [103], provides a road map for the evolutionary and revolutionary developments in the fields of ultrafast light and electrons used to image the invisibles of matter.

How did it all begin? Surely, the power of light for observation has been with humans since their creation. Stretching back over six millennia, one finds its connection to the science of time clocking [95] (first in calendars) and to the mighty monotheistic faiths and rituals (◘ Fig. 3.1). Naturally, the philosophers of the past must have been baffled by the question: what is light and what gives rise to the associated optical phenomena?

◘ **Fig. 3.1** The significance of the light–life interaction as perceived more than three millennia ago, at the time of Akhenaton and Nefertiti. Note the light's "ray diagram" from a spherical source, the Sun. Adapted from Zewail [96]

Fig. 3.2 The concept of the *camera obscura* as perceived a 1000 years ago by Alhazen (Ibn al-Haytham), who coined the term and experimented with the light rays (see Sec. 3.1). Note the formation of the inverted image through a ray diagram. Adapted from Al-Hassani et al. [1]

A leading contribution to this endeavor was made by the Arab polymath Alhazen (Ibn al-Haytham; AD 965–1040) nearly a millennium ago. He is recognized for his quantitative experimentation and thoughts on light reflection and refraction, and is also credited with explaining correctly the mechanism of vision, prior to the contributions of Kepler, Descartes, Da Vinci, Snell, and Newton. But of relevance to our topic is his conceptual analysis of the *camera obscura*, the "dark chamber," which aroused the photographic interests of J. W. Strutt (later known as Lord Rayleigh) in the 1890s [81]. Alhazen's idea that light must travel along straight lines and that the object is inverted in the image plane is no different from the modern picture of ray diagrams taught in optics today (■ Fig. 3.2). His brilliant work was published in the *Book of Optics* or, in Arabic, *Kitab al-Manazir*.

3.2 Optical Microscopy and the Phenomenon of Interference

In the fourteenth and fifteenth centuries, the art of grinding lenses was perfected in Europe, and the idea of optical microscopy was developed. In 1665, Robert Hooke (the scientist who coined the word "cell") published his studies in *Micrographia* ([39]; ■ Fig. 3.3), and among these studies was a description of plants, feathers, as well as cork cells and their ability to float in water. Contemporaneously, Anton van Leeuwenhoek used a simple, one-lens microscope to examine blood, insects, and other objects, and was the first to visualize bacteria, among other microscopic objects.

More than a 100 years later, an experiment by the physicist, physician, and Egyptologist, Thomas Young, demonstrated the interference of light, an experiment that revolutionized our view of the nature of light. His double-slit experiment of 1801 performed at the Royal Institution of Great Britain led to the rethinking of Newton's corpuscular theory of light. Of relevance here is the phenomenon of diffraction due to interferences of waves (coherence). Much later, such diffraction

3

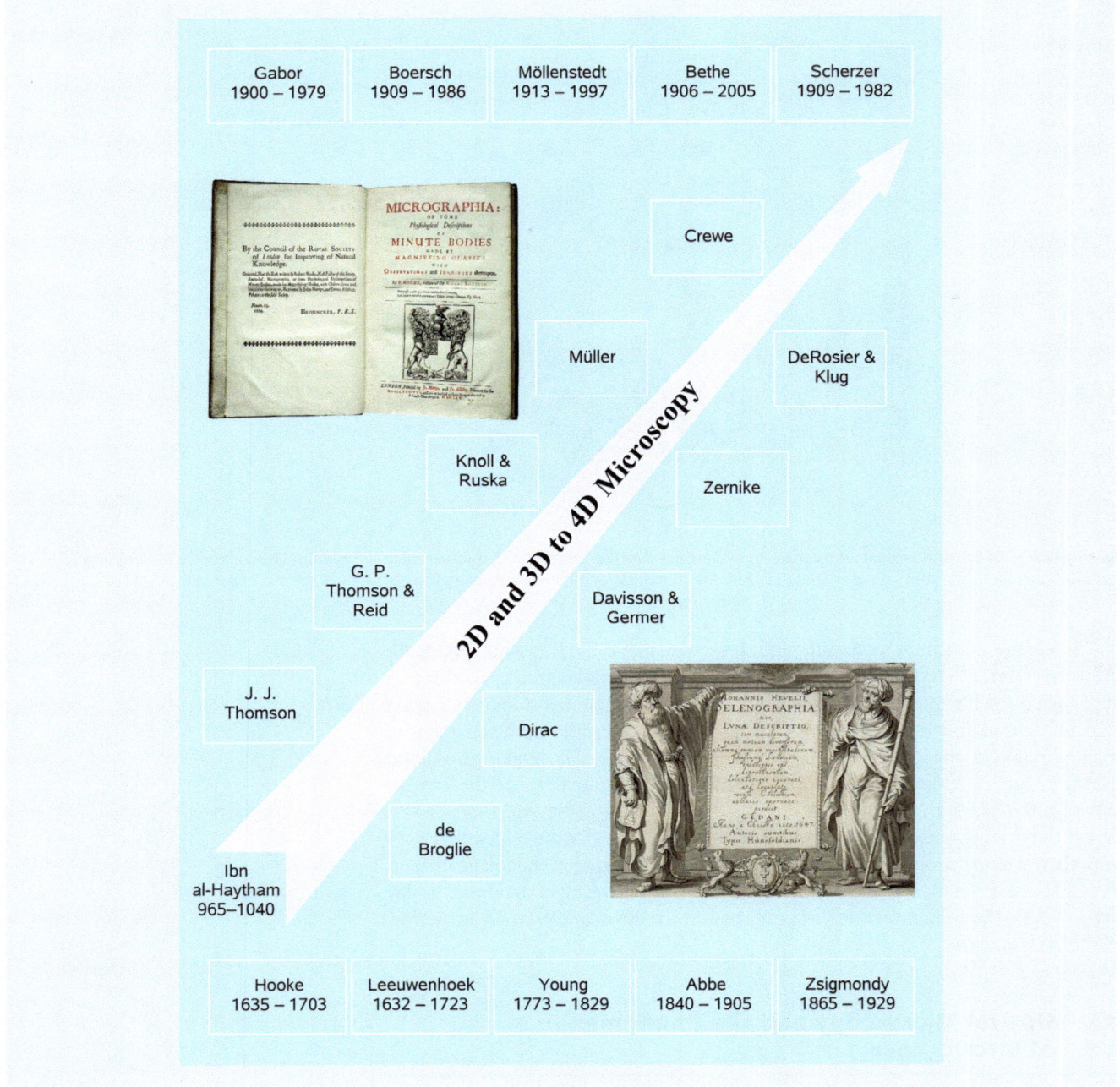

Fig. 3.3 Microscopy time line, from *camera obscura* to 3D electron microscopes. 4D ultrafast electron microscopy and diffraction were developed a decade ago (see Sec. 3.7). The top inset is the frontispiece of Hooke's [39] *Micrographia* published by the Royal Society of London. In the frontispiece to Hevelius's *Selenographia* (bottom inset), Ibn al-Haytham represents *Ratione* (the use of reason) with his geometrical proof and Galileo represents *Sensu* (the use of the senses) with his telescope. The two scientists hold the book's title page between them, suggesting a harmony between the methods [72, 80]. Adapted from Zewail and Thomas [98]

was found to yield the (microscopic) interatomic distances characteristic of molecular and crystal structures, as discovered in 1912 by von Laue and elucidated later that year by W. L. Bragg.

Resolution in microscopic imaging was brought to a whole new level by two major developments in optical microscopy. In 1878, Ernst Abbe formulated a mathematical theory correlating resolution to the wavelength of light (beyond what we now designate the empirical Rayleigh criterion for incoherent sources), and hence the optimum parameters for achieving higher resolution. At the beginning of the twentieth century, Richard Zsigmondy, by extending the work of Faraday and Tyndall, developed the "ultramicroscope" to study colloidal particles; for this work, he received the Nobel Prize in Chemistry in 1925. Then came the penetrating developments in the 1930s by Frits Zernike, who introduced the phase-contrast concept in microscopy; he, too, received the Nobel Prize, in Physics, in 1953. It was understood that the spatial resolution of optical microscopes was limited by the wavelength of the visible light used. Recently, optical techniques have led to considerable improvement in the spatial resolution, and the 2014 Nobel Prize in Chemistry was awarded to Eric Betzig, Stefan Hell, and William Moerner for reaching the spatial resolution beyond the diffraction limit.

3.3   The Temporal Resolution: From Visible to Invisible Objects

In 1872 railroad magnate Leland Stanford wagered $25,000 that a galloping horse, at some point in stride, lifts all four hooves off the ground. To prove it, Stanford employed English photographer Eadweard Muybridge. After many attempts, Muybridge developed a camera shutter that opened and closed for only two thousandths of a second, enabling him to capture on film a horse flying through the air (◘ Fig. 3.4). During the past century, all scientific disciplines from astrophysics to zoology have exploited "high-speed" photography to understand motion of objects and animals that are quicker than the eye can follow.

The time resolution, or shutter speed, needed to photograph the ultrafast motions of atoms and molecules is beyond any conventional scale. When a molecule breaks apart into fragments or when it combines with another to form a new molecule, the chemical bonds between atoms break or form in less than a trillionth of a second, or one picosecond. Scientists have hoped to observe molecular motions in real time and to witness the birth of molecules: the instant at which the fate of the molecular reaction is decided and the final products are determined. Like Muybridge, they needed to develop a shutter, but it had to work ten billion times faster than that of Muybridge.

In the 1980s, our research group at the California Institute of Technology developed new techniques to observe the dynamics of molecules in real time [95]. We integrated our system of advanced lasers with molecular beams, rays of isolated molecules, to a point at which we were able to record the motions of molecules as they form and break chemical bonds on their energy landscapes (◘ Fig. 3.4). The chemical reaction can now be seen as it proceeds from reactants through transition states, the ephemeral structures between reactants and products, and finally to products—chemistry as it happens!

Because transition states exist for less than a trillionth of a second, the time resolution should be shorter—a few quadrillionths of a second, or a few femtoseconds (1 fs is equal to 10^{-15} s). A femtosecond is to a second what a second is to 32 million years. Furthermore, whereas in 1 s light travels nearly 300,000 km—almost the distance between the Earth and the Moon—in 1 fs light travels 0.3 μm, about the diameter of the smallest bacterium.

Over half a century ago, techniques were introduced to study the so-called chemical intermediates using fast kinetics. Ronald Norrish and George Porter of the University of Cambridge and Manfred Eigen of the Max Planck Institute for Physical Chemistry were able to resolve chemical events that lasted about a thousandth of a second, and in some cases a millionth of a second. This time scale was ideal for the intermediates studied, but too long for capturing the transition states—by orders of magnitude for what was needed. Over the millisecond—or even the nano/picosecond—time scale, the transition states are not in the picture, and hence not much is known about the reaction scenery.

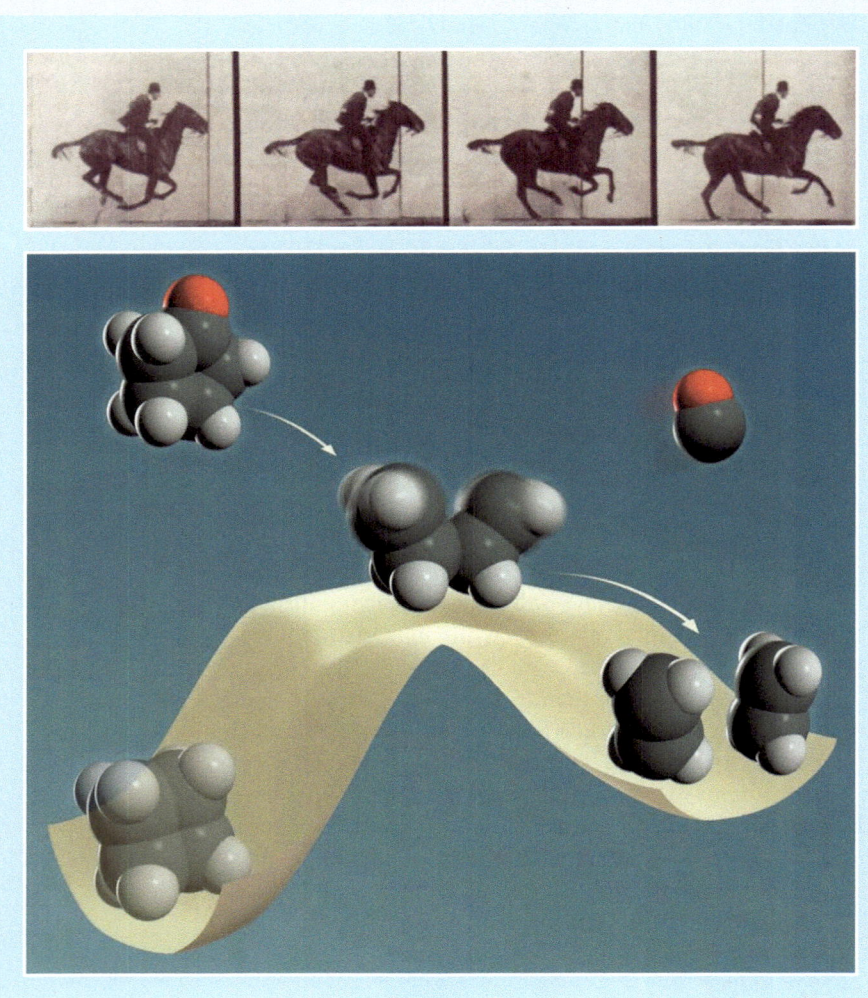

Fig. 3.4 Ultrafast and fast motions. The snapshots of a galloping horse in the upper panel were taken in 1887 by Eadweard Muybridge. He photographed a single horse and the shutter speed was 0.002 s per frame. In the molecular case (*femtochemistry*), a ten-orders-of-magnitude improvement in the temporal resolution was required, and methods for synchronization of billions of molecules had to be developed. The yellow surface in the lower panel represents the energy landscape for the transitory journey of the chemical reaction. Adapted from Zewail [94, 95]

Chemist Sture Forsén of Lund University came up with an insightful analogy that illustrates the importance of understanding transitory stages in the dynamics. He compared researchers to a theater audience watching a drastically shortened version of a classical drama. The audience is shown only the opening scenes of, say, *Hamlet* and its finale. Forsén writes, "*The main characters are introduced, then the curtain falls for change of scenery and, as it rises again, we see on the scene floor a considerable number of 'dead' bodies and a few survivors. Not an easy task for the inexperienced to unravel what actually took place in between*" [25].

The principles involved in the ultrafast molecular "camera" [42] have some similarity to those applied by Muybridge. The key to his work was a special camera shutter that exposed a film for only 0.002 s. To set up the experiment, Muybridge spaced 12 of these cameras half a meter apart alongside a horse track. For each camera he stretched a string across the track to a mechanism that would trigger the shutter when a horse broke through the string. With this system, Muybridge attained a resolution in each picture of about two centimeters, assuming the horse was galloping at a speed of about 10 m/s. (The resolution, or definition, is

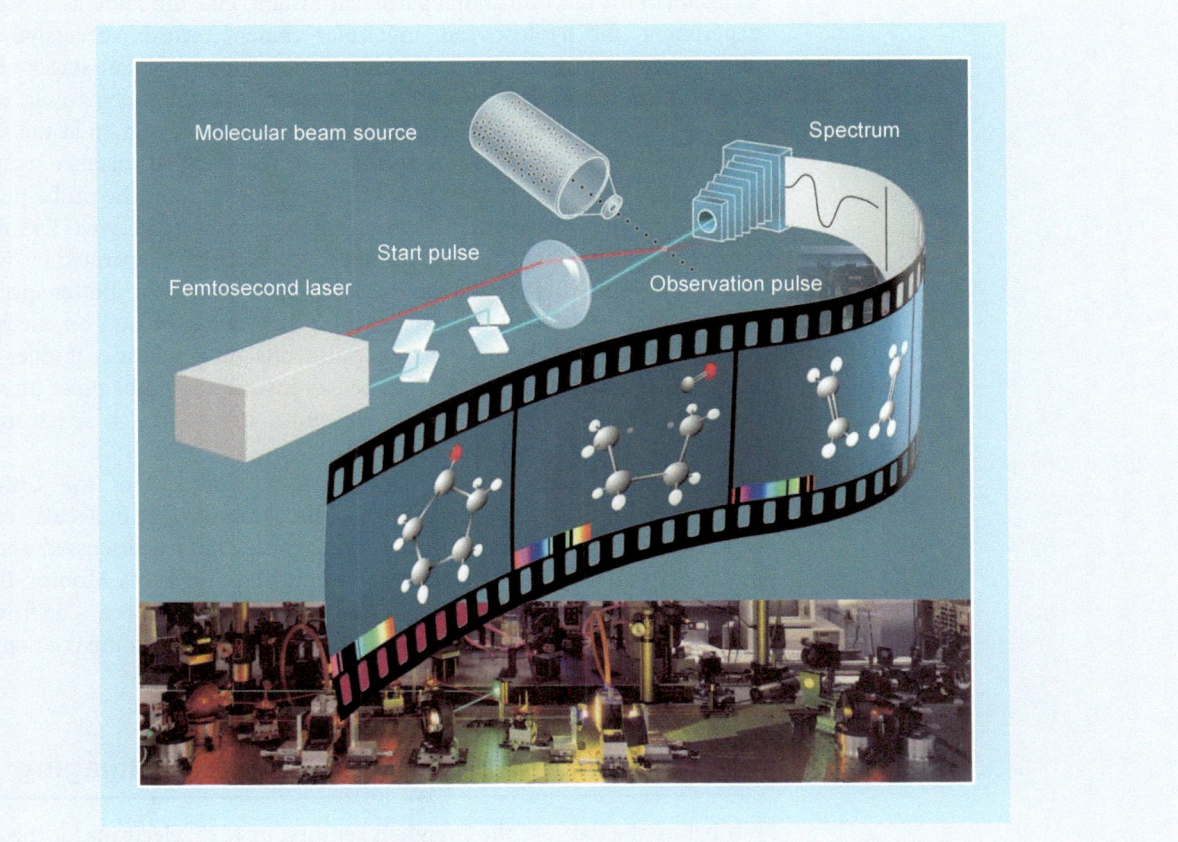

Fig. 3.5 "The fastest camera in the world" [42] records what happens during a molecular transformation by initiating the reaction with a femtosecond laser pulse ("start pulse," *red*). A short time later a second pulse ("observation pulse," *blue*) takes a "picture" of the reacting molecule(s). By successively delaying the observation pulse in relation to the start pulse a "film" is obtained of the course of the reaction. With this first ultrafast "camera" built at Caltech, the ephemeral *transition states* were identified and characterized. The bottom inset shows part of the camera. It is a complex array of lasers, mirrors, lenses, prisms, molecular beams, detection equipment, and more. Adapted from Zewail [95]

simply the velocity of the motion multiplied by the exposure time.) The speed of the motion divided by the distance between cameras equals the number of frames per second—20 in this case. The motion within a picture becomes sharper as the shutter speed increases. The resolution of the motion improves as the distance between the cameras decreases.

Two aspects are relevant to the femtosecond, molecular camera (Fig. 3.5). First, a continuous motion is broken up into a series of snapshots or frames. Thus, one can slow down a fast motion as much as one likes so that the eye can see it. Second, both methods must produce enough frames in rapid succession so that the frames can be reassembled to give the illusion of a continuous motion. The change in position of an object from one frame to the next should be gradual, and at least 30 frames should be taken to provide 1 s of the animation. In our case—the femtosecond, molecular camera—the definition of the frame and the number of frames per second must be adjusted to resolve the elementary nuclear motions of reactions and, most importantly, the ephemeral transition states. The frame definition must be shorter than 0.1 nm. Because the speed of the molecular motion is typically 1 km/s, the shutter resolution must be in a time range of better than 100 fs.

Experimentally, we utilize the femtosecond pulses in the configuration shown in Fig. 3.5. The first femtosecond pulse, called the pump pulse, hits the molecule to initialize the reaction and set the experimental clock "at zero." The second laser pulse, called the probe pulse, arrives several femtoseconds later and records a

3

snapshot of the reaction at that particular instant. Like the cameras in Muybridge's experiment, the femtosecond, molecular camera records successive images at different times in order to obtain information about different stages of the reaction. To produce time delays between the pump and the probe pulses, we initially tune the optical system so that both pulses reach the specimen at the same time. We then divert the probe pulse so that it travels a longer distance than does the pump pulse before it reaches the specimen (◘ Fig. 3.5). If the probe pulse travels one micrometer farther than the pump pulse, it will be delayed 3.33 fs, because light travels at 300,000 km/s. Accordingly, pulses that are separated by distances of 1–100 μm resolve the motion during 3.33–333 fs periods. A shutter speed of a few femtoseconds is beyond the capability of any camera based on mechanical or electrical devices. When the probe pulse hits the molecule, it does not then transmit an image to a detector; see below. Instead the probe pulse interacts with the molecule, and then the molecule emits, or absorbs, a spectrum of light (◘ Fig. 3.6) or changes its mass.

When my colleague and friend Richard Bernstein of the University of California at Los Angeles learned about the femtosecond, molecular camera, he was very enthusiastic about the development, and we discussed the exciting possibilities created by the technique. At his house in Santa Monica, the idea of naming the emerging field of research *femtochemistry* was born. The field has now matured in many laboratories around the globe, and it encompasses applications in chemistry, biology, and materials science.

3.4   Electron Microscopy: Time-Averaged Imaging

Just before the dawn of the twentieth century, in 1897, electrons, or the *corpuscles* of J. J. Thomson, were discovered, but they were not conceived as imaging rays until Louis de Broglie formulated the concept of particle–wave duality in 1924. The duality character of an electron, which is quantified in the relationship $\lambda_{\text{deBroglie}} = h/p$, where h is Planck's constant and p is the momentum of the particle, suggested the possibility of achieving waves of picometer wavelength, which became essential to the understanding of diffraction and imaging. The first experimental evidence of the wave character of the electron was established in 1927 by Davisson and Germer (diffraction from a nickel surface) and, independently, by G. P. Thomson (the son of J. J. Thomson), who, with Reid, observed diffraction of electrons penetrating a thin foil. Around the same time, in 1923, Dirac postulated the concept of "single-particle interference."

With these concepts in the background, Knoll and Ruska [46] invented the electron microscope (EM), in the transmission mode—TEM, using accelerated electrons. Initially the resolution was close to that of optical microscopy, but, as discussed below, it now reaches the atomic scale! Many contributions (◘ Fig. 3.3) to this field have laid the foundation ([56] and the references therein, [77, 79]) for advances of the fundamentals of microscopy and for recent studies of electron interferometry [88, 89]. A comprehensive overview is given by Zewail and Thomas [98].

3.5   2D Imaging and Visualization of Atoms

The first images of individual atoms were obtained in 1951 by Müller [64, 86, 90], who introduced the technique of field-ion microscopy to visualize them at fine tips of metals and alloys, and to detect vacancies and atomic steps and kinks at surfaces. With the invention of field-emission sources and scanning TEM, pioneered in 1970 by Crewe, isolated heavy atoms became readily visible [15, 82]. (The scanning

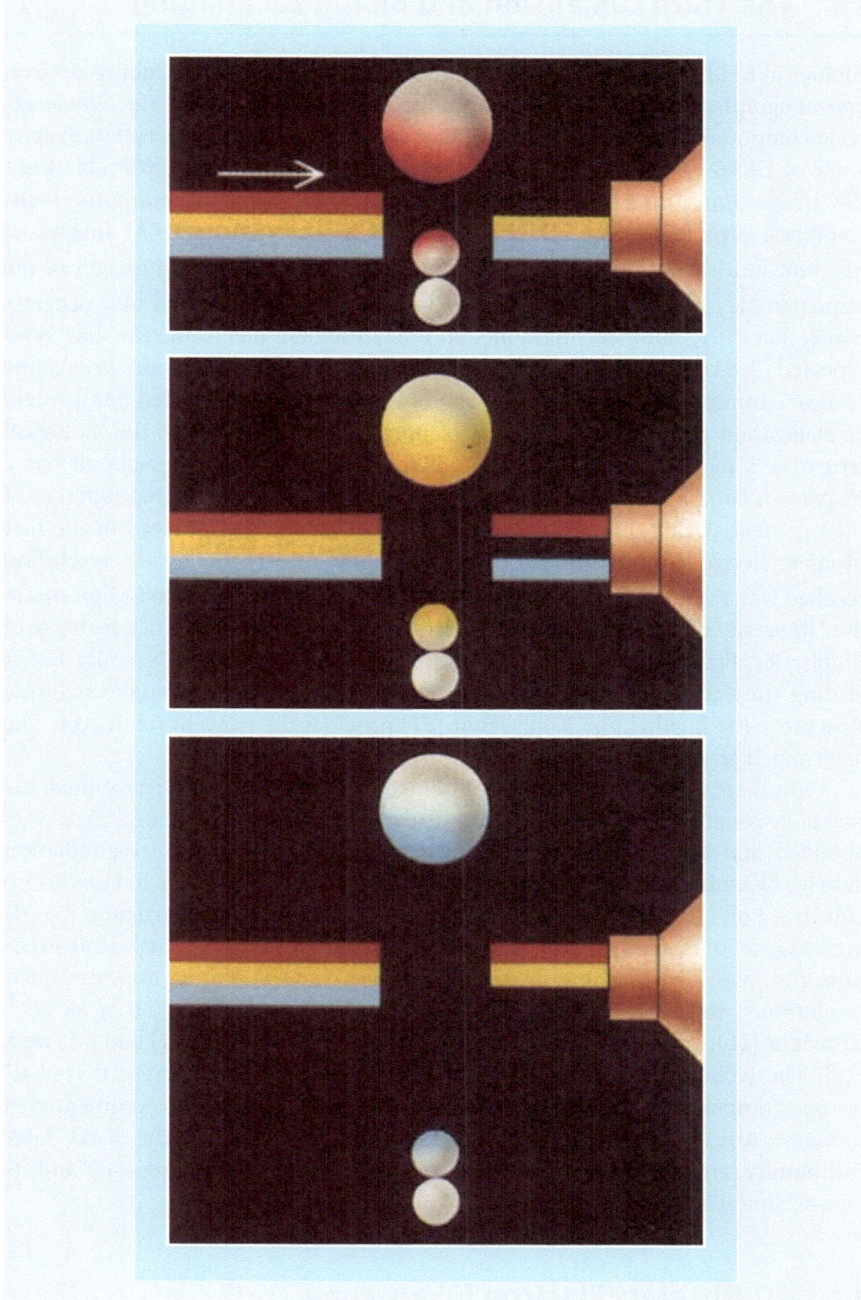

Fig. 3.6 Femtosecond spectroscopy of transient species. A given state of a molecular system can be identified by the light that the molecule absorbs. When atoms in a molecule are relatively close together, they tend to absorb long wavelengths of light (*red*, for example). When the atoms are farther apart, they tend to absorb short wavelengths of light (*blue*, for instance). The change in the spectrum is the fingerprint of the atoms in motion. Adapted from Zewail [94]

tunneling microscope was developed in the 1980s and made possible atomic-scale images of conducting surfaces.) Today, with aberration-corrected microscopes, imaging has reached a resolution of less than an ångström [65]. This history would be incomplete if I did not mention that the totality of technical developments and applications in the investigations of inorganic and organic materials have benefited enormously from the contributions of many other scientists, and for more details I refer the reader to the books by Cowley [14], Humphreys [41], Gai and Boyes [28], Spence [78], and Hawkes and Spence [34], and the most recent papers by Hawkes [33] and Howie [40].

3

3.6   The Third Dimension and Biological Imaging

Biological EM has been transformed by several major advances, including electron crystallography, single-particle tomography, and cryo-microscopy, aided by large-scale computational processing. Beginning with the 1968 electron crystallography work of DeRosier and Klug, see [45], 3D density maps became retrievable from EM images and diffraction patterns. Landmark experiments revealing the high-resolution structure from 2D crystals, single-particle 3D cryo-EM images of different but identical particles (6 Å resolution), and 3D cryo-EM images of the same particle (tomography with 6 Å resolution) represent the impressive progress made. Recently, another milestone in EM structural determination has been reported [2, 51]. Cryo-EM is giving us the first glimpse of mitochondrial ribosome at near-atomic resolution, and, as importantly, without the need for protein crystallization or extensive protocols of purification. In this case, the biological structure is massive, being three megadalton in content, and the subunit has a 39 protein complex which is clearly critical for the energy-producing function of the organelle. Using direct electron detection, a method we involved in the first ultrafast electron diffraction (UED) experiments [18, 91], the spatial resolution reached was 3.2 Å. With the recent structural advances made by the EM groups at the University of California, San Francisco (ion channels), Max Planck Institute of Biophysics, Frankfurt am Main (hydrogenase), and others, it is clear that EM is leading the way in the determination of macromolecular (and noncrystalline!) structures; the highlight by Kühlbrandt [51] provides the relevant references. The determined structures, however, represent an average over time.

With these methods, the first membrane protein structure was determined, the first high-resolution density maps for the protein shell of an icosahedral virus were obtained, and the imaging of whole cells was accomplished. Minimizing radiation damage by embedding the biological macromolecules and machines in vitreous ice affords a non-invasive, high-resolution imaging technique for visualizing the 3D organization of eukaryotic cells, with their dynamic organelles, cytoskeletal structure, and molecular machines in an unperturbed context, with an unprecedented resolution. I refer the reader to the papers by Henderson [35], Sali et al. [73], Crowther [16], and Glaeser [30], and the books by Glaeser et al. [31] and by Frank [27]. The Nobel paper by Roger Kornberg [48] on RNA polymerase II (pol II) transcription machinery is a must for reading. Recently, Henderson commented in a *Nature* article on the overzealous claims made by some in the X-ray laser community, emphasizing the unique advantages of electron microscopy and its cryo-techniques for biological imaging [36].

3.7   4D Ultrafast Electron Microscopy

Whereas in all of the above methods the processes of imaging, diffraction, and chemical analysis have been conducted in a *static* (time-averaged) manner, with the advent of femtosecond light pulses it has now become possible to unite the temporal domain with the (3D) spatial one, thereby creating 4D electron micrographs (◘ Fig. 3.7; [99, 102]); the new approach is termed *4D ultrafast electron microscopy* or—for short—4D UEM. This development owes its success to the advancement of the concept of coherent *single-electron imaging* [99], with the electron packets being liberated from a photocathode using femtosecond optical pulses. In such a mode of electron imaging, the repulsion between electrons is negligible, and thus atomic-scale spatiotemporal resolution can be achieved. Atomic motions, phase transitions, mechanical movements, and the nature of fields at interfaces are examples of phenomena that can be charted in

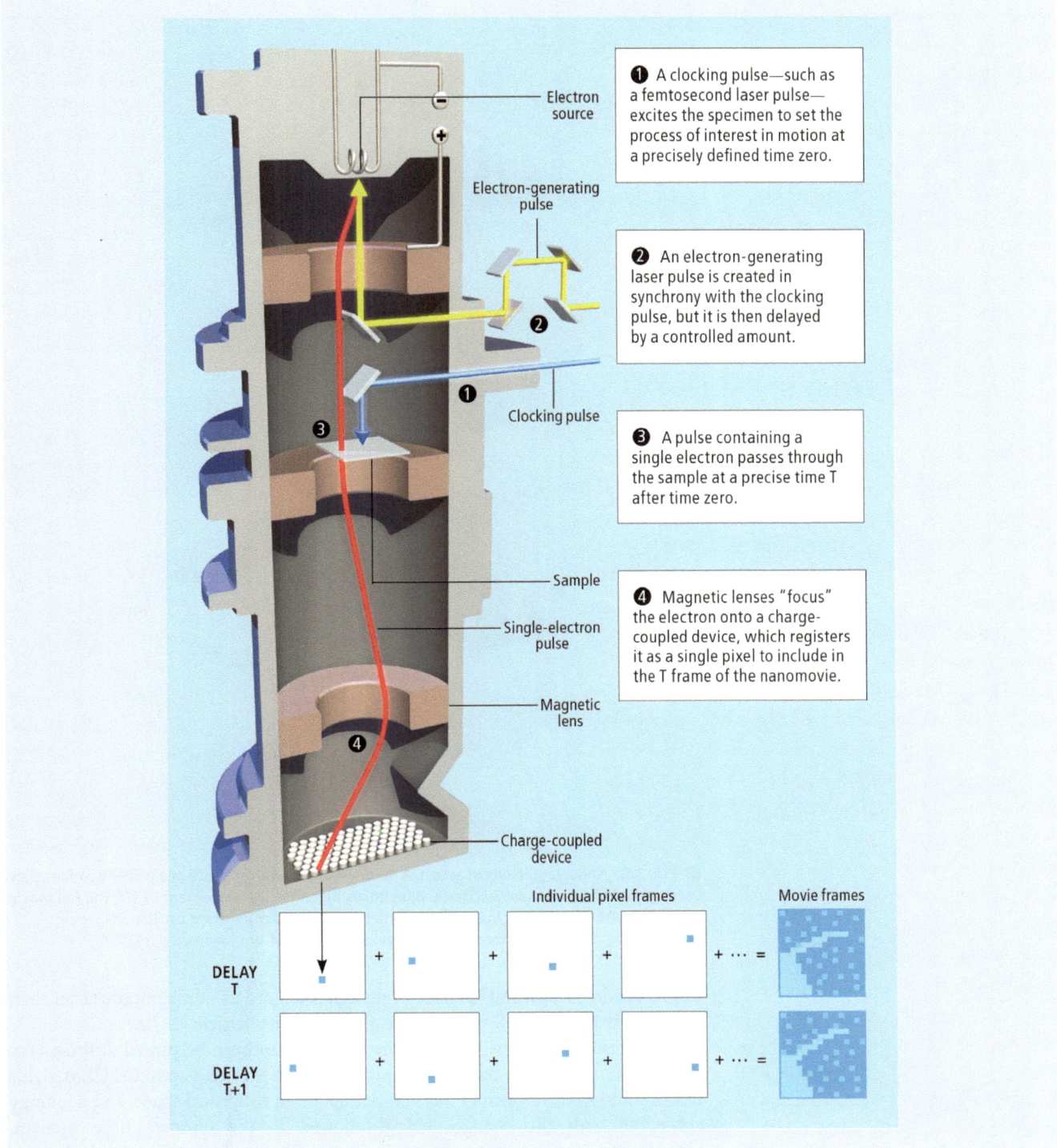

1 A clocking pulse—such as a femtosecond laser pulse—excites the specimen to set the process of interest in motion at a precisely defined time zero.

2 An electron-generating laser pulse is created in synchrony with the clocking pulse, but it is then delayed by a controlled amount.

3 A pulse containing a single electron passes through the sample at a precise time T after time zero.

4 Magnetic lenses "focus" the electron onto a charge-coupled device, which registers it as a single pixel to include in the T frame of the nanomovie.

Electron source

Electron-generating pulse

Clocking pulse

Sample

Single-electron pulse

Magnetic lens

Charge-coupled device

Individual pixel frames

Movie frames

DELAY T

DELAY T+1

Fig. 3.7 Concept of single-electron 4D UEM. A standard electron microscope records still images of a nanoscopic sample by sending a beam of electrons through the sample and focusing it onto a detector. By employing single-electron pulses, a 4D electron microscope produces movie frames representing time steps as short as femtoseconds (10^{-15} s). Each frame of the nanomovie is built up by repeating this process thousands of times with the same delay and combining all the pixels from the individual shots. The microscope may also be used in other modes, e.g., with one many-electron pulse per frame, depending on the kind of movie to be obtained. The single-electron mode produces the finest spatial resolution and captures the shortest time spans in each frame. Adapted from Zewail [101]

3

Fig. 3.8 Nano-cantilevers in action. A 50-nm-wide cantilever made of a nickel-titanium alloy oscillates after a laser pulse excites it. Blue boxes highlight the movement in 3D. The full movie has one frame every 10 ns. Material properties determined from these oscillations would influence the design of nanomechanical devices. Adapted from Kwon et al. [52]

unprecedented structural detail at a rate that is ten orders of magnitude faster than hitherto (◘ Figs. 3.4, 3.8, and 3.9; see also the review article [24]).

Furthermore, because electrons are focusable and can be pulsed at these very high rates, and because they have appreciable inelastic cross sections, UEM yields information in three distinct ways: in real space, in reciprocal space, and in energy space, all with the changes being followed in the ultrafast time domain. Convergent-beam imaging was shown to provide nanoscale diffraction of heterogeneous ensembles [92], and the power of tomography was also demonstrated for a complex structure [53]. Perhaps the most significant discovery in UEM was the photon induced near-field electron microscopy (PINEM; [3]), which uncovered the nature of the electromagnetic field in nanostructures; shown in ◘ Fig. 3.9 are images for PINEM of a single carbon nanotube and the coherent interaction between two particles at nanoscale separations. For biological PINEM imaging [23], see the upcoming ◘ Sect. 3.9. Thus, besides structural imaging, the energy landscapes of macromolecules, chemical compositions, and valence and core-energy states can be studied. The 3D structures (from tomography) can also be visualized.

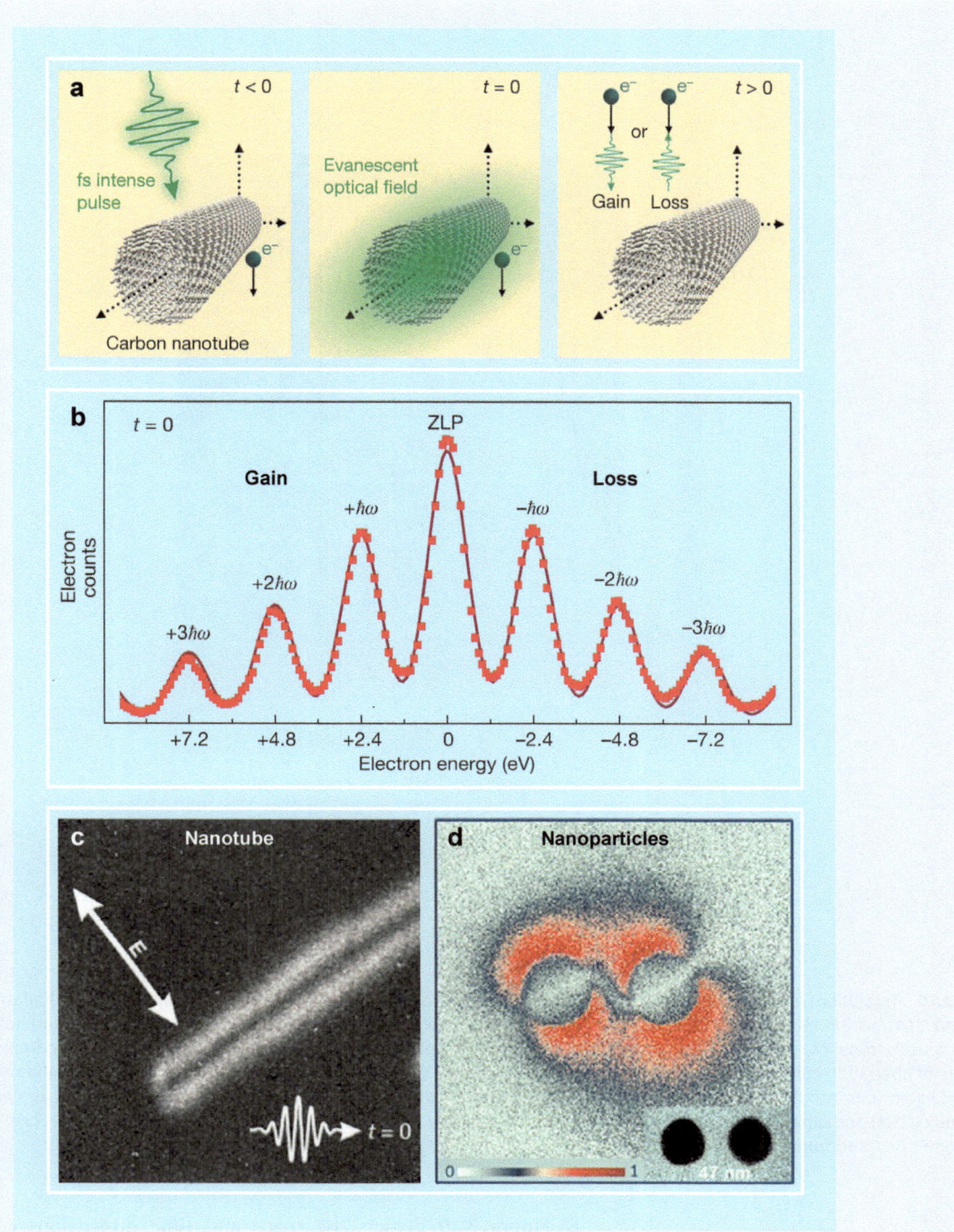

■ **Fig. 3.9** Principles of PINEM and experimental examples. (**a**) *Left frame* shows the moment of arrival of the electron packet at the nanotube prior to the femtosecond laser pulse excitation ($t < 0$); no spatiotemporal overlap has yet occurred. *Middle frame* shows the precise moment at $t = 0$ when the electron packet, femtosecond laser pulse, and evanescent field are at maximum overlap at the carbon nanotube. *Right frame* depicts the process during and immediately after the interaction ($t > 0$) when the electron gains/loses energy equal to integer multiples of femtosecond laser photons. (**b**) PINEM electron energy spectrum obtained at $t = 0$. The spectrum is given in reference to the loss/gain of photon quanta by the electrons with respect to the zero loss peak (ZLP). (**c**) Image taken with the E-field polarization of the femtosecond laser pulse being perpendicular to the long-axis of the carbon nanotube at $t = 0$, when the interaction between electrons, photons, and the evanescent field is at a maximum. (**d**) Near-fields of a nanoparticle pair with an edge-to-edge distance of 47 nm with false-color mapping. When the separation between the particles is reduced from 47 to 32 nm, a "channel" is formed between them. Adapted from Barwick et al. [3], Yurtsever et al. [93]

3

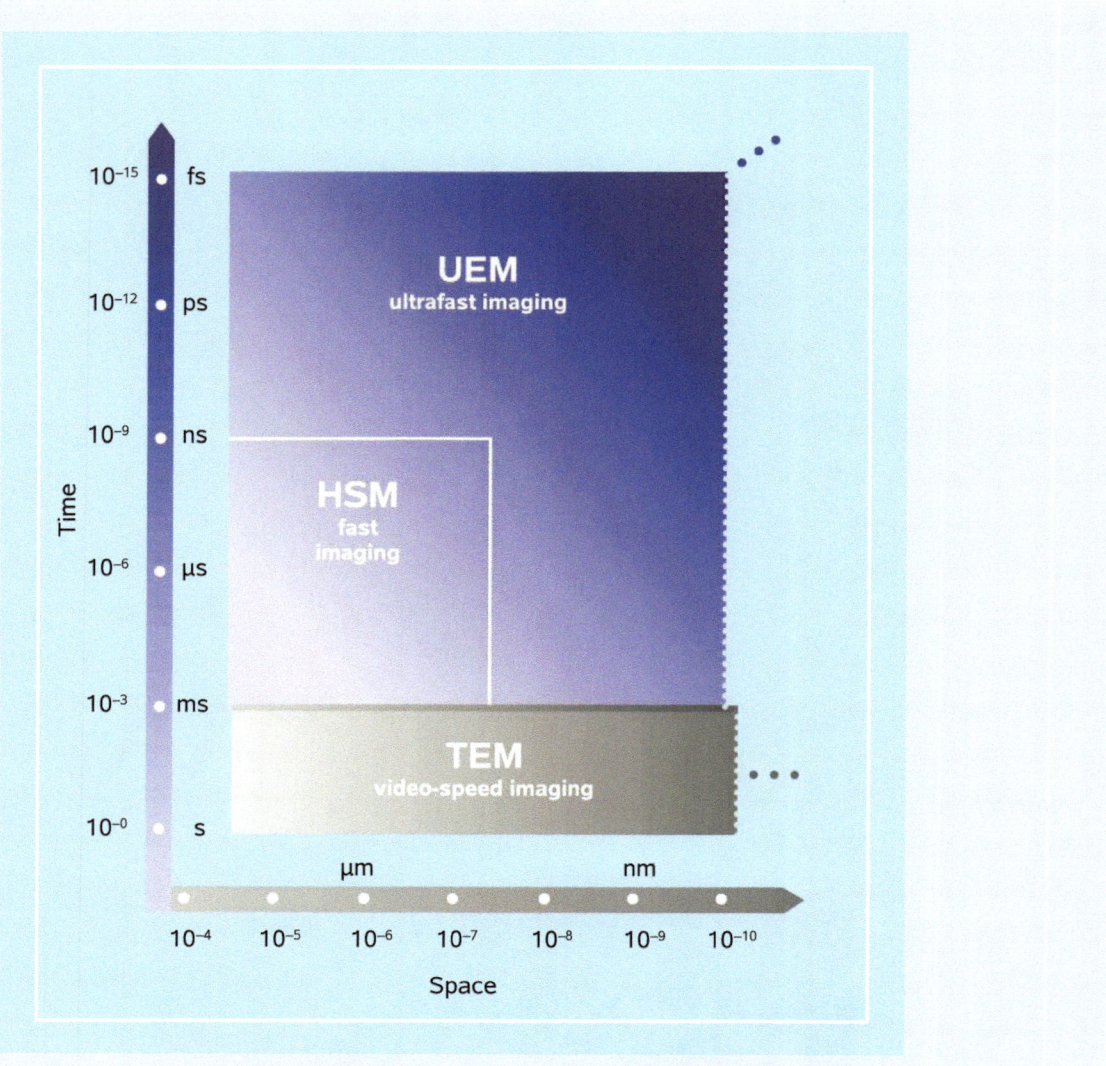

 Fig. 3.10 Resolutions in space and time achieved in electron microscopy. The focus here is on the comparison of ultrafast electron microscopy (UEM) and transmission electron microscopy (TEM), but other variants of the techniques (scanning EM, tomography and holography, as well as electron spectroscopy) can similarly be considered. The horizontal dimension represents the spatial resolution achieved from the early years of EM to the era of aberration-corrected instruments. The vertical axis depicts the temporal resolution scale achieved up to the present time and the projected extensions into the near future. The domains of "fast" and "ultrafast" temporal resolutions are indicated by the areas of high-speed microscopy (HSM) and ultrafast electron microscopy (UEM), respectively [98]. Care should be taken in not naming the HSM "ultrafast electron microscopy". Adapted from Zewail [99]

■ Figure 3.10 depicts the space and time dimensions of UEM, and—for comparison—those of TEM. The boundaries of the time resolution are representative of the transition from the millisecond video speed used in TEM imaging, to the fast, or high-speed nanosecond-to-microsecond imaging, and on to the ultrafast femtosecond-to-attosecond imaging regime. The spatial resolution in the high-speed, nanosecond domain indicated in the Figure is limited by electron–electron (space–charge) repulsion in the nanosecond pulses of electrons. The UEM landscape is that of single-electron imaging, which, owing to the absence of inter-electron repulsion, reaches the spatial resolution of the TEM, but with the temporal resolution being ultrafast. Examples of time-averaged EM and of UEM studies can be found in [98]. The key concepts pertinent to UEM are outlined in ■ Fig. 3.7.

3.8   Coherent Single-Electrons in Ultrafast Electron Microscopy

The concept of single-electron imaging is based on the premise that the trajectories of coherent and timed, single-electron packets can provide an image equivalent to that obtained using many electrons in conventional microscopes. Unlike the random electron distribution of conventional microscopes, in UEM the packets are timed with femtosecond precision, and each electron has a unique coherence volume. As such, each electron of finite de Broglie wavelength is (transversely) coherent over the object length scale to be imaged, with a longitudinal coherence length that depends on its velocity. On the detector, the electron produces a "click" behaving as a classical particle, and when a sufficient number of such clicks are accumulated stroboscopically, the whole image emerges (◻ Fig. 3.7). This was the idea realized in electron microscopy for the first time at Caltech. Putting it in Dirac's famous dictum: *each electron interferes only with itself*. In the microscope, this "stop-motion imaging" yields a real-time movie of the process, and the methodology being used is similar to that described in ◻ Sect. 3.3. We note that, in contrast with Muybridge's experiments, which deal with a single object (the horse), here we have to synchronize the motion of numerous independent atoms or molecules so that all of them have reached a similar point in the course of their structural evolution; to achieve such synchronization for millions—or billions—of the studied objects, the relative timing of the clocking and probe pulses must be of femtosecond precision, and the launch configuration must be defined to sub-ångström resolution.

Unlike with photons, in imaging with electrons we must also consider the consequences of the Pauli exclusion principle. The maximum number of electrons that can be packed into a state (or a cell of phase space) is two, one for each spin; in contrast, billions of photons can be condensed in a state of the laser radiation. This characteristic of electrons represents a fundamental difference in what is termed the "degeneracy," or the mean number of electrons per cell in phase space. Typically it is about 10^{-4} to 10^{-6} but it is possible in UEM to increase the degeneracy by orders of magnitude, a feature that could be exploited for studies in quantum electron optics [98]. I note here that the definition of "single-electron packet" is reserved for the case when each timed packet contains one or a small number of electrons such that coulombic repulsion is effectively absent.

At Caltech, three UEM microscopes operate at 30, 120, and 200 keV. Upon the initiation of the structural change by heating of the specimen, or through electronic excitation, using the ultrashort clocking pulse, a series of frames for real space images, and similarly for diffraction patterns or electron-energy-loss spectra (EELS), is obtained. In the single-electron mode of operation, which affords studies of reversible processes or repeatable exposures, the train of strobing electron pulses is used to build up the image. By contrast, in the single-pulse mode, each recorded frame is created with a single pulse that contains 10^4–10^6 electrons. One has the freedom to operate the apparatus in either single-electron or single-pulse mode.

It is known from the Rayleigh criterion that, with a wavelength λ of the probing beam, the smallest distance that can be resolved is given by approximately 0.5 λ. Thus, in conventional optical microscopy, green light cannot resolve distances smaller than approximately 3000 Å (300 nm). Special variants of optical microscopy can nowadays resolve small objects of sub-hundreds of nanometers in size, which is below the diffraction limit ([37] and the references therein). When, however, an electron is accelerated in a 100 kV microscope, its wavelength approaches 4×10^{-2} Å, i.e., the picometer scale, a distance far shorter than that separating the atoms in a solid or molecule. The resolution of an electron microscope can, in principle, reach the sub-ångström limit [66]. One important advantage in optical microscopy is the ability to study objects with attached chromophores in water. Advances in environmental EM [28] for the studies of catalysis have been achieved, and liquid-cell EM ([17] and the references therein) has been successful in the studies of nanomaterials and cells.

Of the three kinds of primary beams (neutrons, X-rays, and electrons) suitable for structural imaging, the most powerful are coherent electrons, which are readily produced from field-emission guns [98]. The source brightness, as well as the temporal and spatial coherence of such electrons, significantly exceeds the values achievable for neutrons and X-rays: moreover, the minimum probe diameter of an electron beam is as small as 1 Å, and its elastic mean free path is approximately 100 Å (for carbon), much less than for neutrons and X-rays [35]. For larger samples and for those studied in liquids, X-ray absorption spectroscopy, when time-resolved, provides unprecedented details of energy pathways and electronic structural changes [8, 11, 43]. It is significant to note that in large samples the precision is high but it represents an average over the micrometer-scale specimens.

As a result of these developments and inventions, new fields of research continue to emerge. First, by combining energy-filtered electron imaging with electron tomography, chemical compositions of sub-attogram (less than 10^{-18} g) quantities located at the interior of microscopic or mesoscopic objects may be retrieved non-destructively. Second, transmission electron microscopes fitted with field-emission guns to provide coherent electron rays can be readily adapted for electron holography to record the magnetic fields within and surrounding nanoparticles or metal clusters, thereby yielding the lines of force of, for example, a nanoferromagnet encapsulated within a multi-walled carbon nanotube. Third, advances in the design of aberration-corrected high-resolution EMs have greatly enhanced the quality of structural information pertaining to nanoparticle metals, binary semiconductors, ceramics, and complex oxides. Moreover, electron tomography sheds light on the shape, size, and composition of materials. Finally, with convergent-beam and near-field 4D UEM [3, 92], the structural dynamics and plasmonics of a nanoscale single site (particle), and of nanoscale interface fields, can be visualized, reaching unprecedented resolutions in space and time [98].

3.9 Visualization and Complexity

Realization of the importance of visualization and observation is evident in the exploration of natural phenomena, from the very small to the very large. A century ago, the atom appeared complex, a "raisin or plum pie of no structure," until it was visualized on the appropriate length and time scales. Similarly, with telescopic observations, a central dogma of the cosmos was changed and complexity yielded to the simplicity of the heliocentric structure and motion in the entire Solar System. From the atom to the universe, the length and time scales span extremes of powers of 10. The electron in the first orbital of a hydrogen atom has a "period" of sub-femtoseconds, and the size of atoms is on the nanometer scale or less. The lifetime of our universe is approximately 13 billion years and, considering the light year (approx. 10^{16} m), its length scale is of the order of 10^{26} m. In between these scales lies the world of life processes, with scales varying from nanometers to centimeters and from femtoseconds to seconds.

In the early days of DNA structural determination (1950s), a cardinal concept, in vogue at that time, was encapsulated in Francis Crick's statement: *If you want to know the function, determine the structure*. This view dominated the thinking at the time, and it was what drove Max Perutz and John Kendrew earlier in their studies of proteins. But as we learn more about complexity, it becomes clear that the so-called structure–function correlation is insufficient to establish the mechanisms that determine the behavior of complex systems [98]. For example, the structures of many proteins have been determined, but we still do not understand how they fold, how they selectively recognize other molecules, how the matrix water assists folding and the role it plays in directionality, selectivity, and recognition; see, e.g., [61] for protein behavior in water (hydrophobic effect) and [62] for complexity, even in isolated systems. The proteins hemoglobin and myoglobin (a subunit of hemoglobin) have unique functions: the former is responsible for transporting oxygen in the blood of vertebrates, while the latter carries and stores oxygen in muscle cells. The three-dimensional structures of the two proteins have been determined (by Perutz and Kendrew), but we still do not understand the differences in behavior in the oxygen uptake by these two related proteins, the role of hydration, and the exact nature of the forces that control the dynamics of oxygen binding and liberation from the haem group. Visualization of the changing structures during the course of their functional operation is what is needed (see, e.g., [12, 75]).

In biological transformations, the energy landscape involves very complicated pathways, including those that lead to a multitude of conformations, with some that are "active" and others that are "inactive" in the biological function. Moreover, the landscapes define "good" and "bad" regions, the latter being descriptive of the origin of molecular diseases. It is remarkable that the robustness and function of these "molecules of life" are the result of a balance of weak forces—hydrogen bonding, electrostatic forces, dispersion, and hydrophobic interactions—all of energy of the order of a few kcal·mol^{-1}, or approximately 0.1 eV or less. Determination of time-averaged molecular structures is important and has led to an impressive list of achievements, for which more than ten Nobel Prizes have been awarded, but the structures relevant to function are those that exist in the non-equilibrium state. Understanding their behavior requires an integration of the trilogy: structure, dynamics, and function.

■ Figure 3.11 depicts the experimental PINEM field of *Escherichia coli* bacterium which decays on the femtosecond time scale; in the same Figure, we display the conceptual framework of cryo-UEM for the study of folding/unfolding in proteins. Time-resolved cryo-EM has been successfully introduced in the studies of amyloids [21, 22]. In parallel, theoretical efforts ([57–60, 98] and the references therein) have been launched at Caltech to explore the areas of research pertaining to biological structures, dynamics, and the energy landscapes, with focus on the elementary processes involved.

Large-scale complexity is also evident in correlated physical systems exhibiting, e.g., superconductivity, phase transitions, or self-assembly, and in biological systems with emergent behavior [96]. For materials, an assembly of atoms in a lattice can undergo a change, which leads to a new structure with properties different from the original ones. In other materials, the structural transformation leads to a whole new material phase, as in the case of metal–insulator phase transitions. Questions of fundamental importance pertain to the time and length scales involved and to the elementary pathways that describe the mechanism. Recently, a number of such questions have been addressed by means of 4D electron imaging. Of significance are two regimes of structural transformation: the first one involves an initial (coherent) bond dilation that triggers unit-cell expansion and phase growth [5, 9], and the second one involves phase transformations in a diffusionless (collective) process that emerges from an initial random motion of atoms [67]. The elementary processes taking place in superconducting materials are now being examined [26] by direct probing of the "ultrafast phonons," which are critically involved, and—in this laboratory—by studying the effect of optical excitation [10, 29] in ultrafast electron crystallography (UEC).

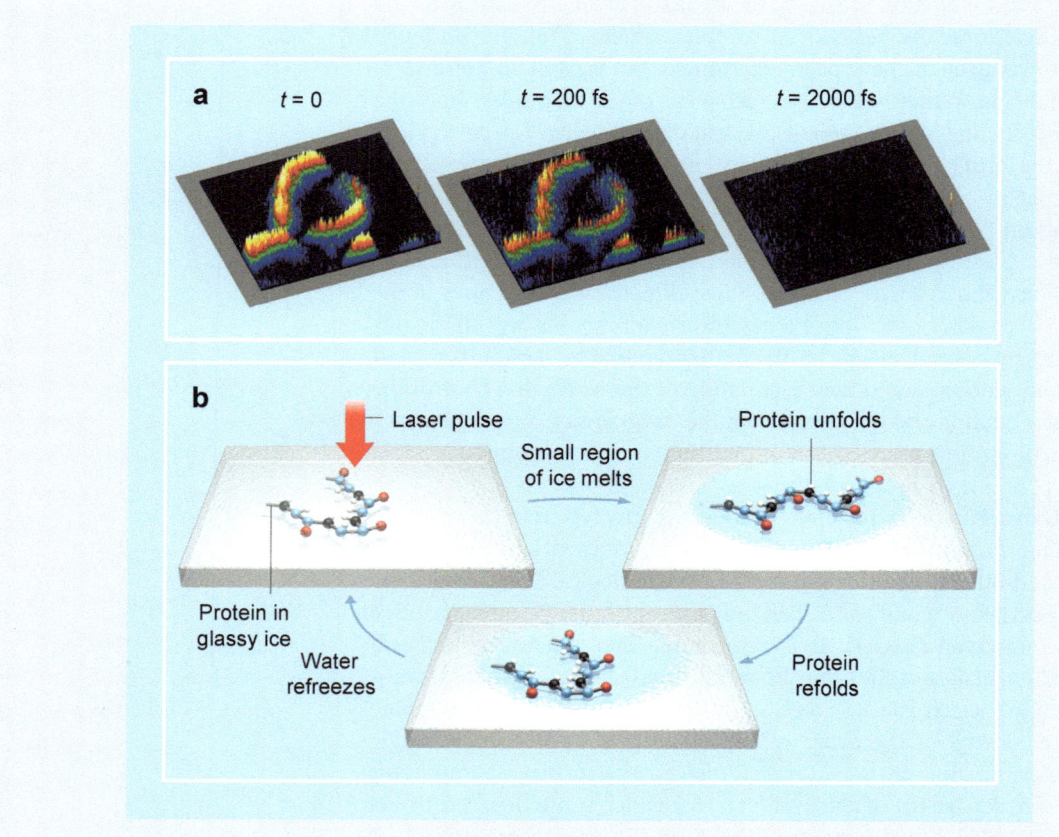

□ Fig. 3.11 Biological UEM. (**a**) An *Escherichia coli* bacterium imaged with PINEM. A femtosecond laser pulse generates an evanescent electromagnetic field in the cell's membrane at time zero. By collecting only the imaging electrons that gained energy from this field, the technique produces high-contrast, relatively high-spatial resolution snapshot of the membrane. The false-color contour plot depicts the intensity recorded. The method can capture events occurring on very short timescales, as is evinced by the field's significant decay after 200 fs. The field vanishes by 2000 fs. (**b**) By adapting the technique called cryo-imaging, we proposed the use of 4D UEM for the observation of biological processes such as protein folding. A glassy (noncrystalline) ice holds the protein. For each shot of the movie, a laser pulse melts the ice around the sample, causing the protein to unfold in the warm water. The movie records the protein refolding before the water cools and refreezes. The protein could be anchored to the substrate to keep it in the same position for each shot. Adapted from Flannigan et al. [23], Zewail [101]

3.10 Attosecond Pulse Generation

For photons, attosecond pulse generation has already been demonstrated [49], and with a unique model that describes the processes involved [13]. Several review articles have detailed the history shaped by many involved in the field and the potential for applications; I recommend the reviews by Corkum [44], Krausz and Stockman [50], and Vrakking [55], and the critique by Leone et al. [54]. It is true that the pulse width can approach the 100 attosecond duration or less (see □ Fig. 3.12 for the electric field pattern), yielding a few-cycles-long pulse. However, there is a price to pay; the band width in the energy domain becomes a challenge in designing experiments. A 20-attosecond pulse has an associated energy bandwidth of $\Delta E \approx 30$ eV. In the femtochemistry domain, the bandwidth permits the mapping of dynamics on a given potential energy surface, and with selectivity for atomic motions. With electrons, processes of ionization and electron density change can be examined, but not with the selectivity mentioned above, as the pulse energy width covers, in case of molecules, numerous energy states. Experimental success has so far been reported for ionized atoms, Auger processes, and the direct measurement of the current produced by valence–conduction-band excitation in SiO_2.

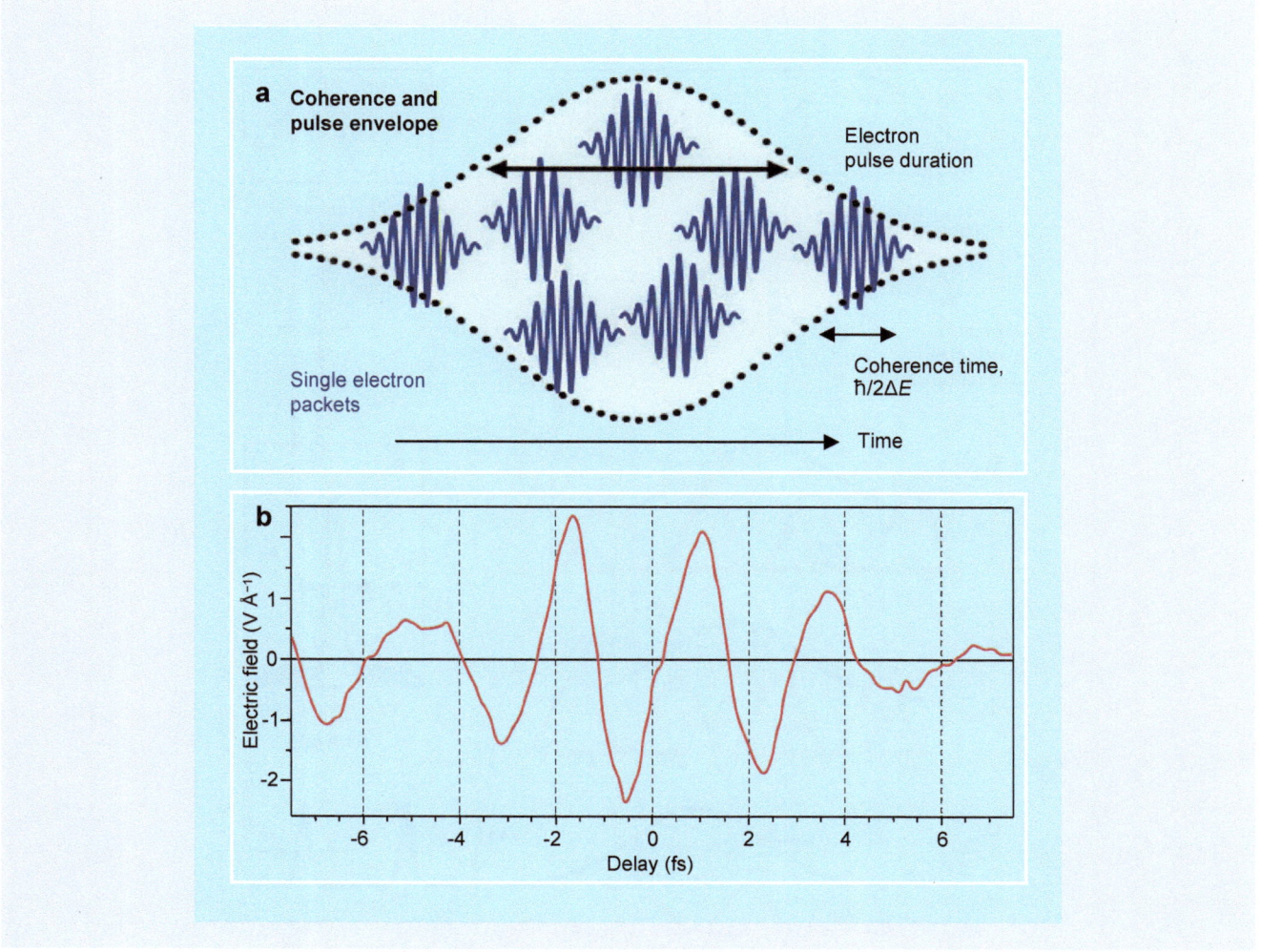

Fig. 3.12 Electron pulse coherence and its packets, together with measured attosecond pulse envelope. (**a**) Single-electron packets and electron pulses. Shown are the effective pulse parameters and the coherence time involved. Each single-electron (*blue*) is a coherent packet consisting of many cycles of the de Broglie wave and has different timing due to the statistics of generation. On average, multiple single-electron packets form an effective electron pulse (dotted envelope). (**b**) Electric field of a few-cycle laser pulse impinging on an SiO_2 sample. Adapted from Baum and Zewail [7], Krausz and Stockman [50]

On the other hand, for electrons in UEM, the challenge is to push the limit of the temporal width into the femtosecond domain and on to the attosecond regime. Several schemes have been proposed and discussed in the review article by Baum and Zewail [7]. In ☐ Fig. 3.12 we display a schematic for a single-electron packet and pulse. The electron trajectories obtained for the temporal optical-grating, tilted-pulses, and temporal-lens methodology are given in ☐ Fig. 3.13. One well-known technique is that of microwave compression [20] of electron pulses, which has recently been applied in femtosecond electron diffraction setups [63].

A more promising method for compression directly to the attosecond domain involves the creation of "temporal lenses" made by ultrashort laser pulses [6, 38]. The technique relies on the ponderomotive force (or ponderomotive potential) that influences electrons when they encounter an intense electromagnetic field. To create trains of attosecond electron pulses, appropriate optical intensity patterns have to be synchronized with the electron pulse. This is done by using counter-propagating laser pulses to create a standing optical wave that must be both spatially and temporally overlapped with the femtosecond electron pulse to get the desired compression. To make the standing wave in the rest frame

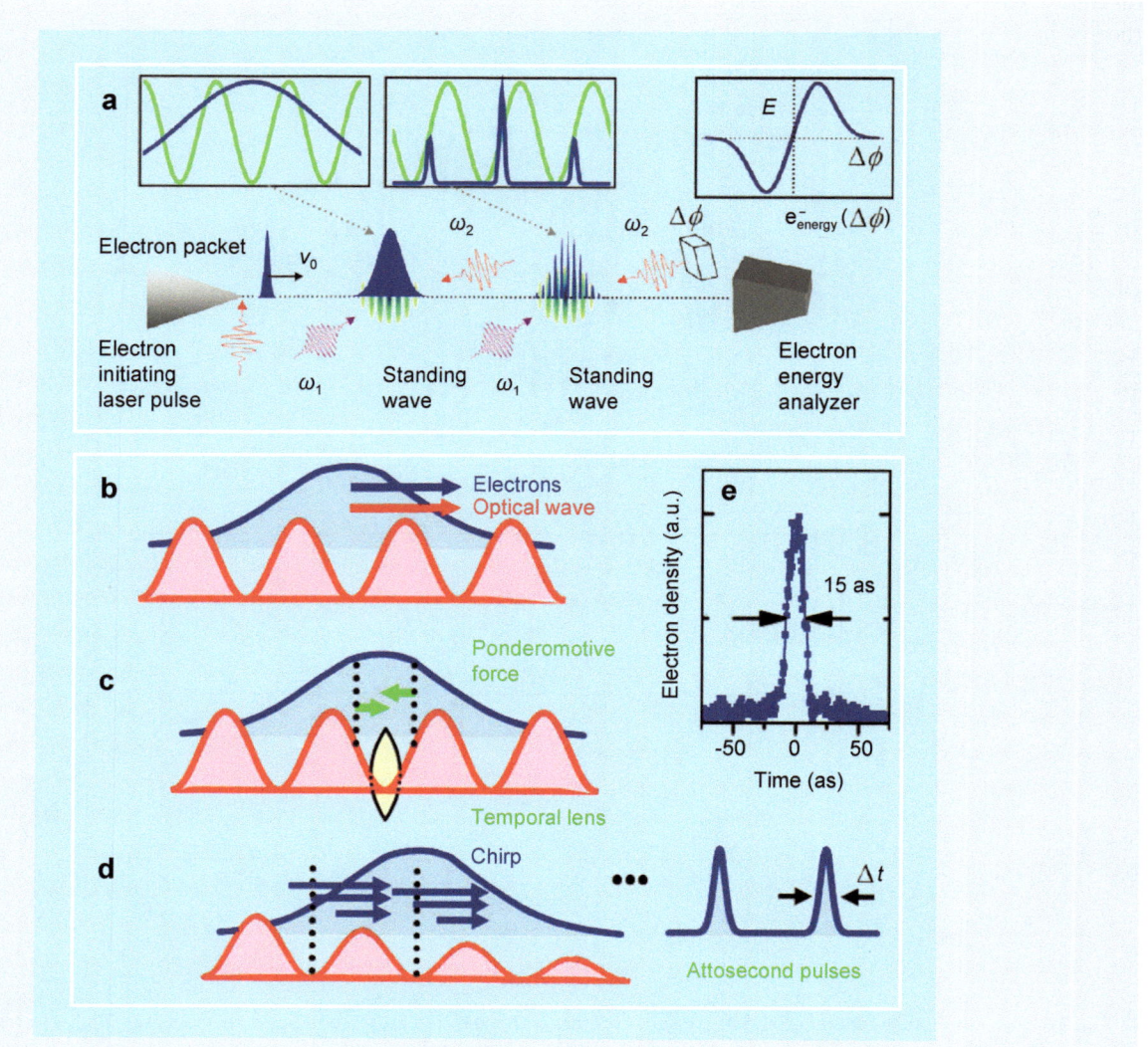

Fig. 3.13 Schemes for creating and measuring attosecond compressed electron packets. (**a**) Temporal lensing. To measure the duration of the attosecond pulses, a second co-propagating standing wave is made to coincide with the electron pulse at the focal position. Instead of using a temporal delay, a phase shift, $\Delta\varphi$, is introduced into one of the laser pulses that creates the probing standing wave. By varying this phase shift, the nodes of the standing wave shift position. The average electron energy can thus be plotted vs. this phase shift. As the electron pulses become shorter than the period of the standing wave, the change in the average energy will increase. To use the attosecond electron pulse train as a probing beam in a UEM, the specimen would need to be positioned where the second standing wave appears in the figure; see Sec. 3.10. (**b–e**) Temporal optical gratings for the generation of free attosecond electron pulses for use in diffraction. (**b**) A femtosecond electron packet (*blue*) is made to co-propagate with a moving optical intensity grating (*red*). (**c**) The ponderomotive force pushes electrons toward the minima and thus creates a temporal lens. (**d**) The induced electron chirp leads to compression to the attosecond duration at a later time. (**e**) The electron pulse duration from 10^5 trajectories reaches into the domain of few attoseconds. Adapted from Barwick and Zewail [4], Baum and Zewail [7]

of the electron pulse, the two counter-propagating electromagnetic waves must have different frequencies [6] or be angled appropriately [38]; see Fig. 3.13.

The standing wave that appears in the rest frame of the traveling electron pulse introduces a series of high and low intensity regions, and in this periodic potential each of the individual ponderomotive potential "wells" causes a compression of the local portion of the electron pulse that encounters it [38]. After interaction with the optical potential well, the electrons that have encountered steep intensity gradients get sped up or slowed down, depending on their position in the potential. After additional propagation, the electron pulse self-compresses into a train of attosecond pulses, with the pulse train spacing being equal to the periodicity of the optical standing wave. By placing the compression potential at an appropriate distance before the specimen, the pulse maximally compresses when encountering the system under study.

3.11 Optical Gating of Electrons and Attosecond Electron Microscopy

As discussed in this chapter, PINEM can be used to generate attosecond pulses [7]. Theoretically, it was clear that spatial and temporal coherences of partial waves (the so-called Bessel functions) can produce attosecond pulse trains, whereas incoherent waves cannot [68–70]. With this in mind, such pulse trains have been generated [19, 47], and the coherent interference of a plasmonic near-field was visualized [71]. Very recently at Caltech, we have developed a new variant of PINEM, which constitutes a breakthrough in electron pulse slicing and imaging.

In all the previous experiments conducted in 4D ultrafast electron microscopy, only "*one optical pulse*" is used to initiate the change in the nanostructure. In a recent report [32], based on the conceptual framework given by Park and Zewail [70], we have used "*two optical pulses*" for the excitation and "*one electron pulse*" for probing. The implementation of this pulse sequence led to the concept of "photon gating" of electron pulses, as shown in ■ Fig. 3.14. The sequence gives rise

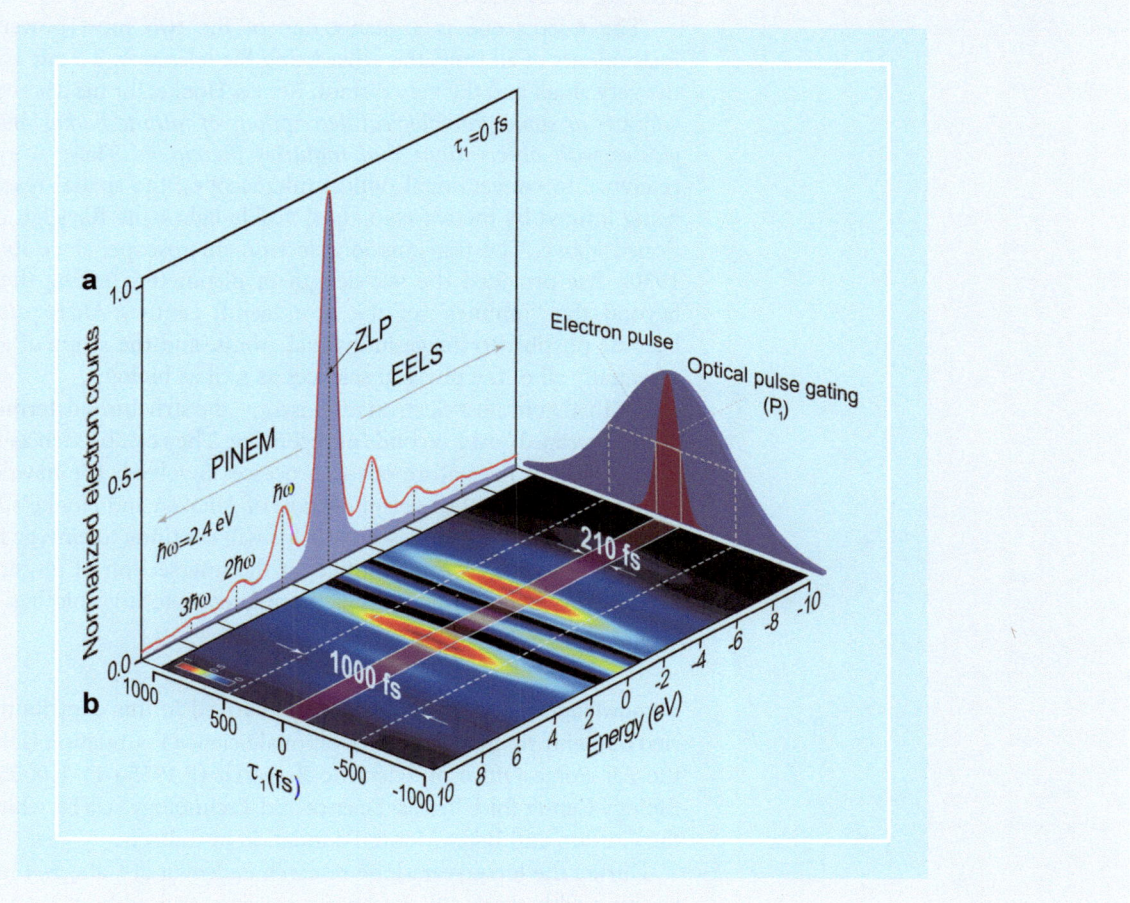

■ **Fig. 3.14** Ultrafast "optical gating" of electrons using three-pulse sequence in PINEM. (**a**) PINEM spectrum at $\tau_1 = 0$ fs, which consists of discrete peaks on the higher and lower energy sides of the zero loss peak (ZLP) separated by multiple photon-energy quanta (2.4 eV). The shaded curve presents the normalized ZLP measured at $\tau_1 = 1000$ fs. (**b**) PINEM spectrogram of photon–electron coupling of the first optical and electron pulse as a function of the first optical pulse delay (τ_1). The ZLP area between −1.5 eV and 1.5 eV has been reduced for visualization of the adjacent discrete peaks. Optical gating is clearly manifested in the narrow strip corresponding to the width of the optical pulse (210 ± 35 fs) shown in red in the vertical plane at right, which is superimposed on the ultrafast electron pulse (1000 fs) in blue. The material studied is vanadium dioxide nanoparticles which undergo metal-to-insulator phase transition when appropriately excited. Adapted from Hassan et al. [32]

to an electron pulse width limited only by the optical-gate pulse width. A picosecond electron pulse was shown to compress into the femtosecond width of the exciting optical pulse. This is a very important advance with the potential for reaching the attosecond time domain with many applications in 4D materials visualization.

3.12 Conclusion

The generation of femtosecond light has its origins in several advances: mode locking by Anthony DeMaria, dye lasers by Peter Sorokin and Fritz Schäfer, colliding pulses in dye lasers by Erich Ippen and Chuck Shank (see the books edited by Schäfer and Shapiro [74, 76]); chirped pulse amplification by Gérard Mourou; white-light continuum by Bob Alfano and colleagues; and the streaking methods by Dan Bradley, and others, have all contributed to the advances made. Prior to the development of femtochemistry, there were significant contributions made in picosecond chemistry and physics (by Peter Rentzepis, Ken Eisenthal, the late Robin Hochstrasser, and Wolfgang Kaiser). In our laboratories at Caltech, the development of *femtochemistry* in the 1980s involved the use of spectroscopy, photoelectron detection, and mass spectrometry. But nothing can be compared with the capability we developed from the year 2000 and until now for microscopy imaging in 4D UEM.

The microscope is arguably one of the two most powerful human-made instruments of all time, the other being the telescope. To our vision they brought the very small and the very distant. Robert Hooke, for his *Micrographia*, chose the subtitle: *or some physiological descriptions of minute bodies made by magnifying glasses with observations and inquiries thereupon.* These words were made in reference to conventional optical microscopes; the spatial resolution of them is being limited by the wavelength of visible light—the Rayleigh criterion, as mentioned above. The transmission electron microscope, since its invention in the 1930s, has provided the wavelength of picometers, taking the field of imaging beyond the "minutes" of the seventeenth century *Micrographia*—it has now become possible to image individual atoms, and the scope of applications spans essentially all of the physical sciences as well as biology.

With 4D ultrafast electron microscopy, the structures determined are no longer "time-averaged" over seconds of recording. They can be seen as frames of a movie that elucidates the nature of the processes involved. We have come a long way from the epochs of the *camera obscura* of Alhazen and Hooke's *Micrographia*, but I am confident that new research fields will continue to emerge in the twenty-first century, especially within frontiers at the intersection of physical, chemical, and biological sciences [97]. Indeed, the microscopic invisible has become visible—thanks to ultrafast photons and electrons.

Acknowledgements The research summarized in this contribution had been carried out with support from the National Science Foundation (DMR-0964886) and the Air Force Office of Scientific Research (FA9550-11-1-0055) in the Physical Biology Center for Ultrafast Science and Technology (UST), which is supported by the Gordon and Betty Moore Foundation at Caltech.

During the forty-years-long research endeavor at Caltech, I had the pleasure of working with some 400 research associates, and without their efforts the above story would not have been told. References are included here to highlight selected contributions, but the work in its totality could not be covered because of the limited space and the article focus.

I am especially grateful to Dr. Dmitry Shorokhov, not only for his technical support but also for the intellectual discussions of the science involved and possible future projects. Dr. Dmitry Shorokhov, together with Dr. Milo Lin, has made major contributions to biological dynamics in the isolated phase (see Sec. 3.9).

References

1. Al-Hassani STS, Woodcock E, Saoud R (eds) (2006) 1001 inventions: Muslim heritage in our world. Foundation for Science, Technology and Civilization, Manchester
2. Amunts A, Brown A, Bai X et al (2014) Structure of the yeast mitochondrial large ribosomal subunit. Science 343:1485–1489
3. Barwick B, Flannigan DJ, Zewail AH (2009) Photon induced near-field electron microscopy. Nature 462:902–906
4. Barwick B, Zewail AH (2015) Photonics and plasmonics in 4D ultrafast electron microscopy. ACS Photonics 2:1391–1402
5. Baum P, Yang DS, Zewail AH (2007) 4D visualization of transitional structures in phase transformations by electron diffraction. Science 318:788–792
6. Baum P, Zewail AH (2007) Attosecond electron pulses for 4D diffraction and microscopy. Proc Natl Acad Sci U S A 104:18409–18414
7. Baum P, Zewail AH (2009) 4D attosecond imaging with free electrons: diffraction methods and potential applications. Chem Phys 366:2–8
8. Bressler C, Milne C, Pham VT et al (2009) Femtosecond XANES study of the light-induced spin crossover dynamics in an iron(II) complex. Science 323:489–492
9. Cavalleri A (2007) All at once. Science 318:755–756
10. Carbone F, Yang DS, Giannini E et al (2008) Direct role of structural dynamics in electron-lattice coupling of superconducting cuprates. Proc Natl Acad Sci USA 105:20161–20166
11. Chergui M, Zewail AH (2009) Electron and X-ray methods of ultrafast structural dynamics: advances and applications. Chem Phys Chem 10:28–43
12. Cho HS, Schotte F, Dashdorj N et al (2013) Probing anisotropic structure changes in proteins with picosecond time-resolved small-angle X-ray scattering. J Phys Chem B 117:15825–15832
13. Corkum PB, Krausz F (2007) Attosecond science. Nat Phys 3:381–387
14. Cowley JM (1995) Diffraction physics, 3rd edn. Elsevier, Amsterdam
15. Crewe AV, Wall J, Langmore J (1970) Visibility of single atoms. Science 168:1338–1340
16. Crowther RA (2008) The Leeuwenhoek lecture 2006. Microscopy goes cold: frozen viruses reveal their structural secrets. Philos Trans R Soc B 363:2441–2451
17. de Jonge N, Peckys DB, Kremers GJ et al (2009) Electron microscopy of whole cells in liquid with nanometer resolution. Proc Natl Acad Sci U S A 106:2159–2164
18. Dantus M, Kim SB, Williamson JC et al (1994) Ultrafast electron diffraction V: experimental time resolution and applications. J Phys Chem 98:2782–2796
19. Feist A, Echternkamp KE, Schauss J et al (2015) Quantum coherent optical phase modulation in an ultrafast transmission electron microscope. Nature 521:200–203

3

20. Fill E, Veisz L, Apolonski A et al (2006) Sub-fs electron pulses for ultrafast electron diffraction. New J Phys 8:272
21. Fitzpatrick AWP, Lorenz UJ, Vanacore GM et al (2013) 4D Cryo-electron microscopy of proteins. J Am Chem Soc 135:19123–19126
22. Fitzpatrick AWP, Park ST, Zewail AH et al (2013) Exceptional rigidity and biomechanics of amyloid revealed by 4D electron microscopy. Proc Natl Acad Sci USA 110:10976–10981
23. Flannigan DJ, Barwick B, Zewail AH (2010) Biological imaging with 4D ultrafast electron microscopy. Proc Natl Acad Sci U S A 107:9933–9937
24. Flannigan DJ, Zewail AH (2012) 4D electron microscopy: principles and applications. Acc Chem Res 45:1828–1839
25. Forsén S (1992) The Nobel prize for chemistry. In: Frängsmyr T, Malmström BG (eds) Nobel lectures in chemistry, 1981–1990. World Scientific, Singapore, p 257, transcript of the presentation made during the 1986 Nobel Prize in chemistry award ceremony
26. Först M, Mankowsky R, Cavalleri A (2015) Mode-selective control of the crystal lattice. Acc Chem Res 48:380–387
27. Frank J (2006) Three-dimensional electron microscopy of macromolecular assemblies: visualization of biological molecules in their native state. Oxford University Press, New York
28. Gai PL, Boyes ED (2003) Electron microscopy in heterogeneous catalysis. Series in microscopy in materials science. IOP Publishing, Bristol
29. Gedik N, Yang DS, Logvenov G et al (2007) Nonequilibrium phase transitions in cuprates observed by ultrafast electron crystallography. Science 316:425–429
30. Glaeser RM (2008) Macromolecular structures without crystals. Proc Natl Acad Sci U S A 105:1779–1780
31. Glaeser RM, Downing K, DeRosier D et al (2007) Electron crystallography of biological macromolecules. Oxford University Press, New York
32. Hassan MT, Liu H, Baskin JS et al (2015) Photon gating in four-dimensional ultrafast electron microscopy. Proc Natl Acad Sci U S A 112:12944–12949
33. Hawkes PW (2009) Aberration correction: past and present. Phil Trans R Soc A 367:3637–3664
34. Hawkes PW, Spence JCH (eds) (2007) Science of microscopy. Springer, New York
35. Henderson R (1995) The potential and limitations of neutrons, electrons and X-rays for atomic resolution microscopy of unstained biological molecules. Q Rev Biophys 28:171–193
36. Henderson R (2002) Excitement over X-ray lasers is excessive. Nature 415:833
37. Hell SW (2015) Nanoscopy with focused light. Angew Chem Int Ed 54:8054–8066, transcript of the Nobel lecture given in 2014
38. Hilbert SA, Uiterwaal C, Barwick B et al (2009) Temporal lenses for attosecond and femtosecond electron pulses. Proc Natl Acad Sci U S A 106:10558–10563
39. Hooke R (1665) Micrographia: or some physiological descriptions of minute bodies made by magnifying glasses with observations and inquiries thereupon. Royal Society, London
40. Howie A (2009) Aberration correction: zooming out to overview. Phil Trans R Soc A 367:3859–3870
41. Humphreys CJ (ed) (2002) Understanding materials: a festschrift for Sir Peter Hirsch. Maney, London
42. Jain VK (1995) The world's fastest camera. The World and I 10:156–163
43. Kim TK, Lee JH, Wulff M et al (2009) Spatiotemporal kinetics in solution studied by time-resolved X-ray liquidography (solution scattering). Chem Phys Chem 10:1958–1980
44. Kim KT, Villeneuve DM, Corkum PB (2014) Manipulating quantum paths for novel attosecond measurement methods. Nat Photonics 8:187–194
45. Klug A (1983) From macromolecules to biological assemblies. Angew Chem Int Ed 22:565–636, transcript of the Nobel lecture given in 1982
46. Knoll M, Ruska E (1932) Das Elektronenmikroskop. Z Phys 78:318–339
47. Kociak M (2015) Microscopy: quantum control of free electrons. Nature 521:166–167
48. Kornberg R (2007) The molecular basis of eukaryotic transcription. Angew Chem Int Ed 46:6956–6965, transcript of the Nobel lecture given in 2006
49. Krausz F, Ivanov M (2009) Attosecond physics. Rev Mod Phys 81:163–234
50. Krausz F, Stockman MI (2014) Attosecond metrology: from electron capture to future signal processing. Nat Photonics 8:205–213
51. Kühlbrandt W (2014) The resolution revolution. Science 343:1443–1444
52. Kwon OH, Park HS, Baskin JS et al (2010) Nonchaotic, nonlinear motion visualized in complex nanostructures by stereographic 4D electron microscopy. Nano Lett 10:3190–3198
53. Kwon OH, Zewail AH (2010) 4D electron tomography. Science 328:1668–1673

54. Leone SR, McCurdy CW, Burgdörfer J et al (2014) What will it take to observe processes in 'real time'? Nat Photonics 8:162–166

55. Lépine F, Ivanov MY, Vrakking MJJ (2014) Attosecond molecular dynamics: fact or fiction? Nat Photonics 8:195–204

56. Lichte H (2002) Electron interference: mystery and reality. Philos Trans R Soc Lond A 360:897–920

57. Lin MM, Shorokhov D, Zewail AH (2006) Helix-to-coil transitions in proteins: helicity resonance in ultrafast electron diffraction. Chem Phys Lett 420:1–7

58. Lin MM, Meinhold L, Shorokhov D et al (2008) Unfolding and melting of DNA (RNA) hairpins: the concept of structure-specific 2D dynamic landscapes. Phys Chem Chem Phys 10:4227–4239

59. Lin MM, Shorokhov D, Zewail AH (2009) Structural ultrafast dynamics of macromolecules: diffraction of free DNA and effect of hydration. Phys Chem Chem Phys 11:10619–10632

60. Lin MM, Shorokhov D, Zewail AH (2009) Conformations and coherences in structure determination by ultrafast electron diffraction. J Phys Chem A 113:4075–4093

61. Lin MM, Zewail AH (2012) Protein folding: simplicity in complexity. Ann Phys 524:379–391

62. Lin MM, Shorokhov D, Zewail AH (2014) Dominance of misfolded intermediates in the dynamics of α-helix folding. Proc Natl Acad Sci U S A 111:14424–14429

63. Mancini GF, Mansart B, Pagano S et al (2012) Design and implementation of a flexible beamline for fs electron diffraction experiments. Nucl Instrum Methods Phys Res A 691:113–122

64. Müller EW (1951) Das Feldionenmikroskop. Z Phys 131:136–142

65. Nellist PD, Chisholm MF, Dellby N et al (2004) Direct sub-ångström imaging of a crystal lattice. Science 305:1741

66. O'Keefe MA (2008) Seeing atoms with aberration-corrected sub-ångström electron microscopy. Ultramicroscopy 108:196–209

67. Park HS, Kwon OH, Baskin JS et al (2009) Direct observation of martensitic phase-transformation dynamics in iron by 4D single-pulse electron microscopy. Nano Lett 9:3954–3962

68. Park ST, Lin MM, Zewail AH (2010) Photon induced near-field electron microscopy (PINEM): theoretical and experimental. New J Phys 12:123028

69. Park ST, Kwon OH, Zewail AH (2012) Chirped imaging pulses in four-dimensional electron microscopy: femtosecond pulsed hole burning. New J Phys 14:053046

70. Park ST, Zewail AH (2012) Enhancing image contrast and slicing electron pulses in 4D near-field electron microscopy. Chem Phys Lett 521:1–6

71. Piazza L, Lummen TTA, Quiñonez E et al (2015) Simultaneous observation of the quantization and the interference pattern of a plasmonic near-field. Nat Commun 6:6407

72. Sabra AI (2003) Ibn al-Haytham. Harvard Mag 9–10:54–55

73. Sali A, Glaeser RM, Earnest T et al (2003) From words to literature in structural proteomics. Nature 422:216–225

74. Schäfer FP (ed) (1973) Dye lasers (Topics in applied physics, vol. 1) Springer, Heidelberg

75. Schotte F, Cho HS, Kaila VRI et al (2012) Watching a signaling protein function in real time via 100-ps time-resolved Laue crystallography. Proc Natl Acad Sci USA 109:19256–19261

76. Shapiro SL (ed) (1977) Ultrashort light pulses, picosecond techniques and applications (Topics in applied physics, vol. 18) Springer, Heidelberg

77. Silverman MP, Strange W, Spence JCH (1995) The brightest beam in science: new directions in electron microscopy and interferometry. Am J Phys 63:800–813

78. Spence JCH (2003) High-resolution electron microscopy, (Monographs on the physics and chemistry of materials, vol. 60), 3rd edn. Oxford University Press, New York

79. Spence JCH (2009) Electron interferometry. In: Greenberger D, Hentschel K, Weinert F (eds) Compendium of quantum physics: concepts, experiments, history and philosophy. Springer, Berlin, pp 188–195

80. Steffens B (2007) Ibn al-Haytham: first scientist. Morgan Reynolds, Greensboro

81. Strutt JW (1891) On pin-hole photography. Philos Mag 31:87–99

82. Thomas JM (1979) Direct imaging of atoms. Nature 281:523–524

83. Thomas JM (1991) Femtosecond diffraction. Nature 351:694–695

84. Thomas JM (2004) Ultrafast electron crystallography: the dawn of a new era. Angew Chem Int Ed 43:2606–2610

85. Thomas JM (2005) A revolution in electron microscopy. Angew Chem Int Ed 44:5563–5566

86. Thomas JM (2008) Revolutionary developments from atomic to extended structural imaging. In: Zewail AH (ed) Physical biology: from atoms to medicine. Imperial College Press, London, pp 51–114

87. Thomas JM (2009) The renaissance and promise of electron energy-loss spectroscopy. Angew Chem Int Ed 48:8824–8826

88. Tonomura A (1998) The quantum world unveiled by electron waves. World Scientific, Singapore

89. Tonomura A (1999) Electron holography, 2nd edn. Springer, Berlin

90. Tsong TT (2006) Fifty years of seeing atoms. Phys Today 59:31–37

91. Williamson JC, Cao J, Ihee H et al (1997) Clocking transient chemical changes by ultrafast electron diffraction. Nature 386:159–162

92. Yurtsever A, Zewail AH (2009) 4D nanoscale diffraction observed by convergent-beam ultrafast electron microscopy. Science 326:708–712

93. Yurtsever A, Baskin JS, Zewail AH (2012) Entangled nanoparticles: discovery by visualization in 4D electron microscopy. Nano Lett 12:5027–5032

94. Zewail AH (1990) The birth of molecules. Sci Am 263:76–82

95. Zewail AH (2000) Femtochemistry: atomic-scale dynamics of the chemical bond using ultrafast lasers. In: Frängsmyr T (ed) Les prix nobel: the nobel prizes 1999. Almqvist & Wiksell, Stockholm, pp 110–203, Also published in Angew Chem Int Ed 39:2587–2631; transcript of the Nobel lecture given in 1999

96. Zewail AH (2008) Physical biology: 4D visualization of complexity. In: Zewail AH (ed) Physical biology: from atoms to medicine. Imperial College Press, London, pp 23–49

97. Zewail AH (2009) Chemistry at a historic crossroads. Chem Phys Chem 10:23

98. Zewail AH, Thomas JM (2009) 4D electron microscopy: imaging in space and time. Imperial College Press, London

99. Zewail AH (2010) 4D electron microscopy. Science 328:187–193

100. Zewail AH (2010) Micrographia of the 21st century: from camera obscura to 4D microscopy. Philos Trans R Soc Lond A 368:1191–1204

101. Zewail AH (2010) Filming the invisible in 4D. Sci Am 303:74–81

102. Zewail AH (2012) 4D imaging in an ultrafast electron microscope. US Patent 8,203,120

103. Zewail AH (2014) 4D visualization of matter: recent collected works. Imperial College Press, London

Optical Sources

Contents

The Laser

Bahaa Saleh

B. Saleh (✉)
CREOL, The College of Optics and Photonics, University of Central Florida, Orlando, FL, USA
e-mail: besaleh@creol.ucf.edu

© The Author(s) 2016
M.D. Al-Amri et al. (eds.), *Optics in Our Time*, DOI 10.1007/978-3-319-31903-2_4

4

One of the greatest inventions of the twentieth century, if not of all times, the laser is regarded as humankind's most versatile light source, a truly new kind of light with remarkable properties unlike anything that existed before—a light fantastic! Since it was first used to generate light in 1960 by Theodore Maiman at Hughes Research Laboratories with the help of a ruby crystal, the laser has been at the core of most light-based technologies and has garnered applications in all facets of life. It is a marvelous tool that has enabled many scientific discoveries. Numerous books, textbooks, and articles have been written about laser physics and engineering. This article is a brief tutorial introducing some of the basic principles underlying the development of the laser and highlighting some of its remarkable characteristics.

4.1 Introduction: A Laser in the Hands of Ibn al-Haytham

As the millennium-old contributions of al-Hasan ibn al-Haytham (ca. 965–ca. 1040) to light and vision are being celebrated in this *International Year of Light*, one might wonder if the laser could have been invented in al-Hasan's time. It is even tempting to imagine him being handed a red light beam from a laser, shining through the smoke created by incense somewhere in Fatimid Cairo, and to speculate on his possible reaction. As the reigning expert of optics and vision of his time, would he have found laser light to be truly remarkable? Would he have used it to corroborate his original observations pertaining to reflection, refraction, and the focusing of light?

Seeing the laser beam, al-Hasan would probably have surmised that it could be sunlight shaped into a thin beam by the use of some ingenious contraption of mirrors, a tiny version of the mirror systems said to have been used by Archimedes to destroy the Roman fleet in 212 BC. What about the red color of the laser beam? Simple: it's sunlight transmitted through a piece of red glass, much like those made and traded by the Phoenician glassmakers. Puzzled about the unusual thinness of the beam, its very limited divergence, and its exceptional brightness, al-Hasan would have waited for the Sun to set; seeing that the light beam still shined brilliantly through the smoke, he might then have concluded that the contraption uses a miniaturized version of a red lantern, similar to those he saw as a child in Basra. He would then have again contemplated the exceptional brightness of the light. And as to the narrowness of the beam, he might have speculated that it is collimated light passed through tiny holes much like those used in the then-known camera obscura.

al-Hasan Ibn al-Haytham
(Alhazen) ca. 965 – ca. 1040

Known for his methodical reliance on experimentation and controlled testing, al-Hasan would have conducted an experiment using the brightest red lantern available, along with mirrors, magnifying lenses, and pinholes, to produce a similar beam of light. He probably would have succeeded in creating a dim red beam of light, a miniaturized version of the beam produced at the Pharos (lighthouse) of Alexandria, which used a mirror to reflect sunlight during the day and a fire lit at night. But no matter what he might have done with the oil-based light sources of the day, it would not have been possible for him to come close to the brightness and narrowness of the laser beam.

Frustrated with his failure, al-Hasan would probably have engaged in a set of experiments benefitting from the available "magic" light source to confirm his findings about reflection and refraction from his stock of mirrors and lenses. He would not have known that it took centuries to discover the wave nature of light (propagation, diffraction, interference, and coherence), its electromagnetic nature (polarization and propagation through anisotropic media), its quantum nature (photons and light–matter interaction), and to develop concepts such as thermal equilibrium, oscillation, modes, spectral analysis, and transient dynamics—all of which are necessary to truly understand, design, and use lasers.

4.2 The Laser: An Optical Oscillator

The laser is simply an oscillator of light, and the phenomenon of *oscillation* is one of its underlying foundational principles.

4.2.1 Oscillators

An oscillator is a device, system, or structure that produces oscillation at some frequency, with little or no excitation at that frequency. An example of a mechanical oscillator producing sound is the familiar tuning fork. When struck, it vibrates, or oscillates, creating sound at its characteristic (resonance) frequency (◘ Fig. 4.1). The oscillation eventually decays since there is no energy source to sustain it. A musical instrument is made of mechanical structures (chords or pipes) that oscillate at distinct frequencies, which may be altered by changing their dimensions or shapes, as the instrument is played. Another example is the undesirable acousto-electrical oscillation often encountered when a microphone is connected to an audio amplifier feeding a close-by loudspeaker (◘ Fig. 4.2). A small disturbance sensed by the microphone is amplified, and if the sound produced by the loudspeaker reaches the microphone, it gets amplified once more, and the cycle is repeated. Circulation through this feedback loop results in

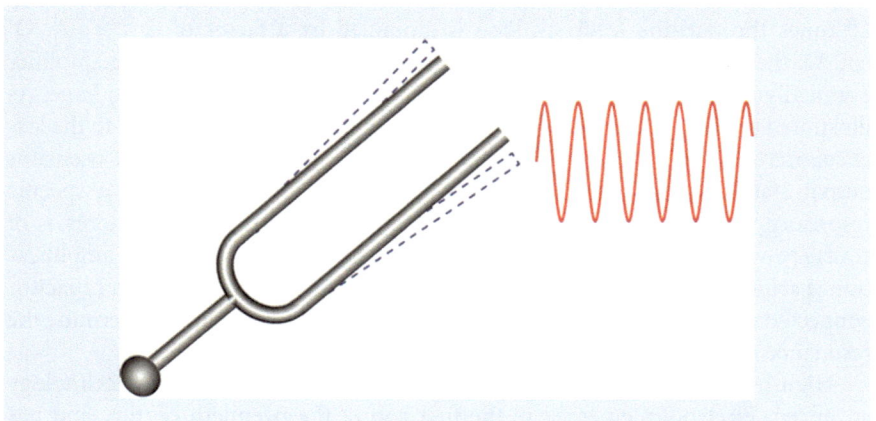

◘ **Fig. 4.1** When struck, a tuning fork vibrates at its characteristic resonance frequency

4

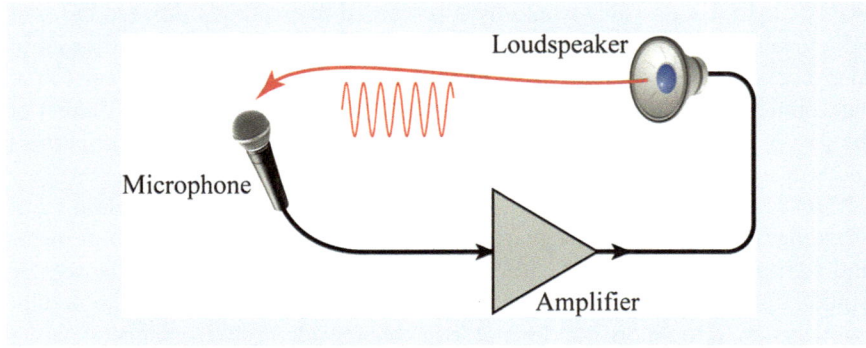

Fig. 4.2 Undesirable oscillation is created when the loudspeaker sound reaches the microphone

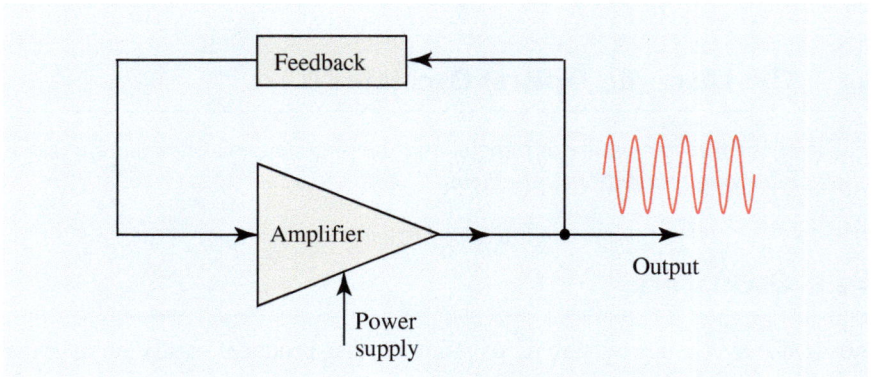

Fig. 4.3 An electronic oscillator using an amplifier and a feedback loop

oscillation—the familiar whistle sound. The frequency of oscillation is character-istic of the overall system.

The oscillator is a basic building block of virtually all electronic systems, analog or digital. An electronic oscillator comprises a resonant electronic amplifier along with a feedback system that directs the output of the amplifier back to its input, in the form of a feedback loop, as illustrated in ▣ Fig. 4.3. It is essential that the feedback be positive, i.e., the feedback signal re-entering the amplifier must be in phase with the original signal. Oscillation is initiated by noise, which contains a broad spectrum of all frequencies. The resonant amplifier amplifies a selected frequency component, and the feedback circuit brings the output back to the input for further amplification. For example, with gain of 2, a tiny input traveling 20 times through the feedback loop is amplified by a factor of $2^{20} \approx 10^6$. Of course, the output of the device cannot grow without bound, since the amplifier eventually saturates, i.e., its gain is reduced when its input becomes too large. As illustrated in ▣ Fig. 4.4, when the reduced gain eventually becomes equal to the loss encountered in the loop, growth of the circulating signal ceases, and the oscillator output stabilizes. The ultimate result is the generation of energy at a specific resonance frequency, with little initial excitation at that frequency. Energy is of course provided by the amplifier power supply, e.g., a battery. Resonant amplifica-tion is achieved by means of a resonant element, usually in the form of a capacitor connected to an inductor in the domain of electronics, whose values determine the resonance frequency.

High-frequency electronic oscillators were developed as electronics technology advanced. Electronics emerged in the first half of the twentieth century and has advanced steadily with efforts to achieve miniaturization and greater switching

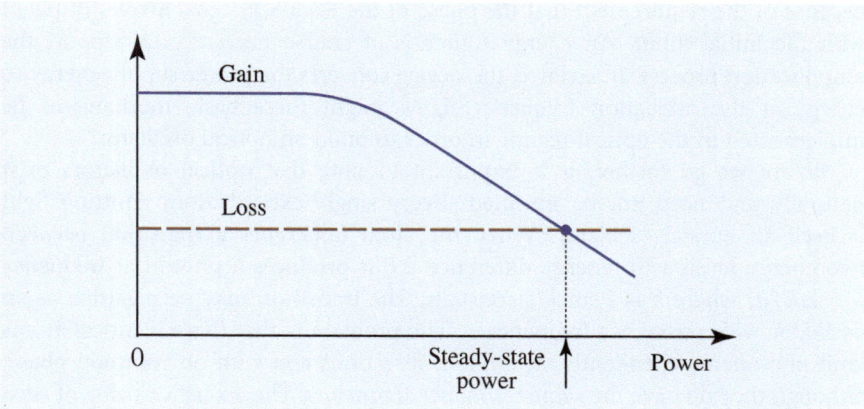

□ Fig. 4.4 Buildup of oscillation. As the power increases, the amplifier gain is reduced; when it equals the loss in the feedback loop, a steady-state power is reached

speeds—and this continues to this day. The quest to build electronic oscillators that operate at higher and higher frequencies was fueled by the desire to make use of wider and wider bands of the electromagnetic spectrum. The earliest electronic oscillators operated at audio frequencies (AF) and radio frequencies (RF), in the kHz and MHz ranges. They used transistor-based amplifiers and inductor–capacitor resonant circuits. Early microwave (MW) oscillators in the GHz range employed special vacuum-tube amplifiers, such as magnetrons and klystrons, which were based on ingenious mechanisms for forging interactions of the microwave field with electron beams, along with microwave cavity resonators that provided the prerequisite feedback. These oscillators were used for decades following their introduction in the 1940s, but were ultimately replaced by solid-state devices such as Gunn and IMPATT diodes.

The development of AF, RF, and MW oscillators was motivated by the needs of the electronics and telecommunications industries, and the development of oscillators at frequencies in the GHz range was fueled by the need for radar systems during the Second World War. This evolutionary process was bound to lead naturally to the development of oscillators in the THz frequency range and beyond, including optical frequencies (the frequency of visible light lies in the 400–770 THz range, and ultraviolet light and X-rays have even greater frequencies), but this natural evolution took several decades. This may be because the need for a new kind of light source was not pressing since many conventional light sources already existed. Gas-discharge lamps emitting over a narrow band of frequencies, filtered both temporally and spatially, met the needs of most scientific applications envisioned at the time.

Perhaps it was the development of the maser, the microwave predecessor of the laser, that stimulated the extrapolation to optical frequencies; both the laser and the maser are based on the same amplification principle: *stimulated emission*. In fact, after the laser was invented, it was known as an *optical maser* in the early technical literature. The term "maser" is an acronym for Microwave Amplification by Stimulated Emission of Radiation.

4.2.2 The Optical Oscillator

As mentioned earlier, two basic mechanisms are necessary to build an oscillator: resonant amplification and positive feedback. The frequency of oscillation is dictated not only by the resonant amplifier, but also by the feedback system, because of the requirement that the phase of the feedback signal arrives in phase with the initial input. An energy source is of course necessary to support the amplification process. In essence, the device converts the power supply energy to energy at the oscillation frequency. How might these basic mechanisms be implemented in the optical regime in order to build an optical oscillator?

Before we go further, it is important to note that optical oscillators exist naturally and need not be invented! Every single excited atom emitting light is itself an optical oscillator. When the atom undergoes a transition between two energy levels with energy difference ΔE it produces a photon at frequency $\nu = \Delta E/h$, where h is Planck's constant. The transition may be regarded as an oscillator with resonance frequency ν. The problem is that these identical atoms emit photons independently, in random directions and with no common phase, although they do have the same resonance frequency. The acoustic analog of such an incoherent source is a large set of independently struck tuning forks, or a large orchestra playing without a conductor. The outcome, which is a superposition of light emitted by say 10^{23} independent atoms, is an optical field that oscillates randomly in both time and space. This is light that lacks temporal and spatial coherence, i.e., it is basically optical noise. What is needed instead is a *single* optical oscillator producing a coherent optical field, rather than an extensive set of independent oscillators, each with miniscule power. The laser does just that!

The laser is an optical oscillator that employs a *coherent* optical amplifier—a medium, which when illuminated by incoming light, produces more light in the same direction and with the same properties as the incoming light. Gordon Gould, one of the early pioneers of the laser, is credited with calling this mechanism Light Amplification by Stimulated Emission of Radiation, thereby introducing the acronym LASER. Since the laser is an oscillator, rather than an amplifier, a more appropriate name would perhaps be Light Oscillation by Stimulated Emission of Radiation, but the associated acronym would have been imprudent.

To construct a system that acts as a single oscillator, i.e., a coherent light source, feedback is necessary. With adequate feedback, the stimulated emission mechanism synchronizes the individual independent atomic oscillators to act collectively as a single optical oscillator. Feedback can be readily implemented by means of an arrangement of mirrors or reflective surfaces that form an optical resonator within which light is repeatedly passed through the amplifying (or gain) medium. The simplest optical resonator takes the form of two parallel planar or spherical mirrors between which the light circulates back and forth, as illustrated in ◘ Fig. 4.5. This structure, known in the optics community as a Fabry–Pérot interferometer, was not readily recognized as a resonator until the laser was invented. Amplification occurs at each passage since the gain medium amplifies in both directions. Useful light is extracted by making one of the mirrors partially transmitting.

Every oscillator has a built-in mechanism for stabilization that governs the steady output power (see ◘ Fig. 4.4). In lasers, if the pump power is sufficiently high such that the gain in the medium is greater than the resonator loss, then

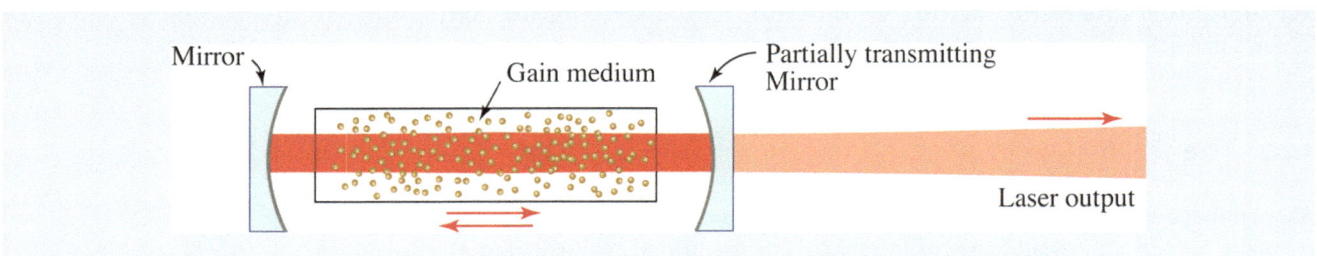

◘ Fig. 4.5 The laser is an optical oscillator that makes use of an amplifying (gain) medium placed inside a resonator

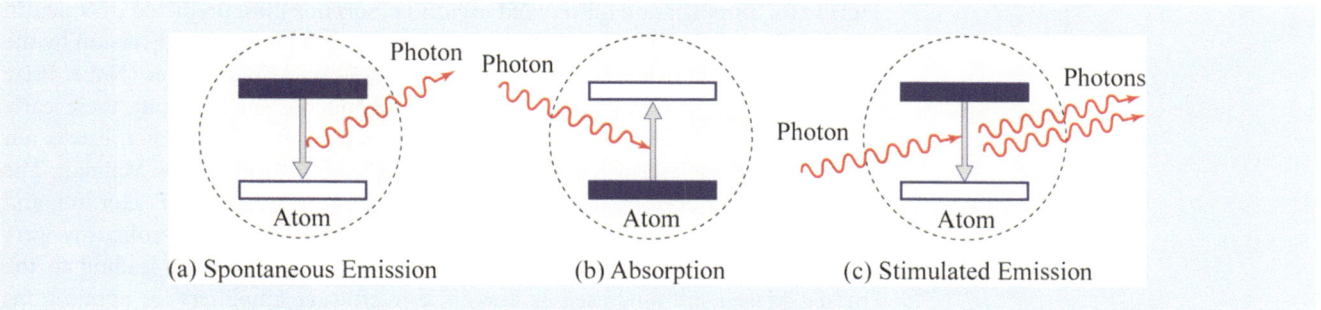

○ Fig. 4.6 Interaction of light with matter by transitions between two energy levels. (**a**) Spontaneous emission of a photon by an excited atom. (**b**) Absorption of a photon by an unexcited atom. (**c**) A photon stimulates an excited atom to emit a second photon

lasing is initiated and the optical power increases exponentially. However, since atoms undergoing stimulated emission become de-excited in the process, the population inversion is thereby reduced so that further growth of the optical power is suppressed. When the decreased gain eventually equals the loss, the system reaches equilibrium and a steady laser power is delivered.

4.2.3 Optical Amplification by Stimulated Emission

There are three processes inherent in the interaction between photons and atoms (see ○ Fig. 4.6). The first is *spontaneous emission*, which does not generate coherent light. The second is *absorption*, which causes the medium to become an attenuator rather than an amplifier. When unexcited atoms absorb light, they are excited to the higher energy level and subsequently decay back spontaneously, either by emitting light or via some other non-radiative means. *Stimulated emission* is the third component of the interaction. The relation among these processes was set forth by Albert Einstein in 1917 when he revisited Max Planck's law of radiation (the spectral distribution of blackbody radiation in thermal equilibrium).

Einstein showed that while an atom in an unexcited state (lower energy level) might absorb a photon at a rate proportional to the incoming photon flux density, an atom in an excited state (higher energy level) is just as likely to be stimulated by the incoming photons and to emit a photon, so that the flux of photons increases. An important recognition was that the photon emitted via stimulated emission is in the same direction and has the same properties as the stimulating photon. These three mechanisms—absorption, and stimulated and spontaneous emission—govern the interaction of photons with atoms and underlie the law of radiation in thermal equilibrium as well as the generation of laser light.

For a medium with an excess of unexcited atoms, absorption exceeds stimulated emission and the medium provides net attenuation. This is the situation when the medium is in thermal equilibrium. If this equilibrium is somehow disturbed to create an excess of excited atoms, stimulated emission dominates absorption and the medium provides net gain. The medium then serves as a coherent optical amplifier. This non-equilibrium state, called *population inversion*, may be achieved via a process known as *pumping*. In fact, the pumping process supplies the medium with the energy needed to realize the desired gain. Since transitions occur only when the optical frequency matches the resonance frequency of the atomic transition (i.e., the energy of the photon $h\nu$ matches that of the atomic transition energy ΔE), the amplifier is a resonant amplifier. It provides gain only within a narrow spectral band dictated by the atomic transition.

Following Einstein's conception in 1917, the fact that stimulated emission could provide optical gain was confirmed in 1928 by Rudolf W. Ladenburg.

Its use for amplification (also called *negative absorption*) was predicted by Valentin A. Fabrikant in 1939. Optical pumping, i.e., achieving population inversion by the use of another light source, was proposed in 1950 by Alfred Kastler (Nobel Prize for Physics in 1966) as a mechanism for introducing gain. Despite these early discoveries of the basic ingredients of this coherent optical amplifier, it was not until 1960 that a laser was constructed and successfully operated by Maiman. The first maser had been built in 1953 by Charles H. Townes, James P. Gordon, and Herbert J. Zeiger. Townes, Nikolai G. Basov, and Aleksandr M. Prokhorov were awarded the 1964 Nobel Prize for Physics for theoretical work leading to the maser. Masers are now used as low-noise microwave amplifiers for applications such as radio telescopes. In 1958 Arthur L. Schawlow (Nobel Prize for Physics in 1981), together with Townes, suggested a method for extending the stimulated-emission principle of the maser to the optical region of the spectrum.

Albert Einstein	Charles Townes	Arthur Schawlow	Theodore Maiman
(1879--1955)	(1915--2015)	(1921--1999)	(1927--2007)

4.2.4 Laser Materials and Pumping Methods

An enormous variety of materials are used as gain media in lasers and laser amplifiers; these include solids, gases, liquids, and plasmas. The wavelengths of these devices span extended bands of the electromagnetic spectrum, all the way from the microwave to the X-ray region. They make use of a wide variety of resonator configurations and pumping schemes (◘ Fig. 4.7). Pumping may be implemented optically by means of light at a wavelength other than the resonance wavelength, e.g., by a flash lamp or another laser; or it may be implemented by use of an electric current, as is the case in semiconductor laser diodes.

4.3 Optical Resonators and Their Modes

Optical feedback is commonly provided by use of a resonator, which serves as a "container" within which the generated laser light circulates and is built up, and can be stored. Optical resonators take a variety of configurations, as illustrated by the examples in ◘ Fig. 4.7. The most common configuration is two planar or spherical mirrors. Ring lasers use an arrangement of mirrors in a ring configuration or use a closed-loop optical fiber or integrated-optic waveguide. Dielectric waveguides with two cleaved end surfaces are used in semiconductor lasers. In microdisks and microspheres, light circulates inside the rim of the material by total internal reflection at near-grazing incidence, in what are known as whispering-gallery modes. Periodic dielectric structures such as distributed Bragg reflectors

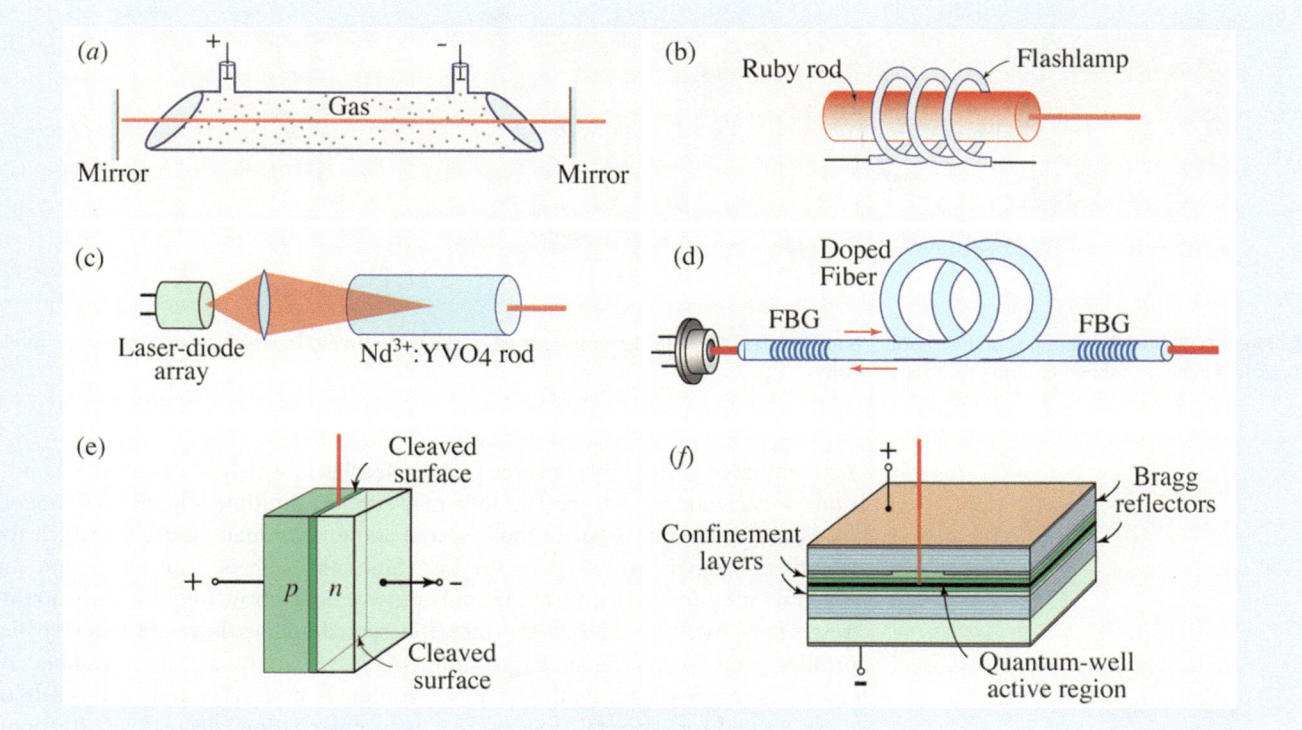

Fig. 4.7 Examples of lasers. (**a**) Gas laser pumped by direct current (DC). (**b**) Solid-state laser, e.g., ruby, optically pumped with a flashlamp. (**c**) Nd:YVO4 solid-state laser optically pumped by a laser diode array. (**d**) Fiber laser (e.g., erbium-doped silica fiber) with fiber Bragg grating (FBG) reflectors, pumped with a laser diode. (**e**) Laser diode (forward-biased *p–n* junction) with cleaved surfaces acting as mirrors, pumped by electric-current injection. (**f**) Quantum-well semiconductor laser, pumped electrically. Charge carriers are restricted to the active region by the confinement layers and Bragg reflectors serve as mirrors

(DBRs) are used as mirrors for trapping the light inside a fiber or in small structures, as in the micropillar resonator. Light can also be trapped in defects within dielectric photonic-bandgap structures, forming photonic-crystal resonators. In microresonators, which are used in microlasers, the size of the resonator can be of the same order of magnitude as its resonance wavelength. Nanoresonators, which can be far smaller than the resonance wavelength have come to the fore in recent years.

The optical resonators are characterized by their quality factor Q, which is the ratio of the resonance frequency to the line width. A low-loss resonator has a large Q, corresponding to sharp resonance and long storage time (in units of optical period). Resonators may have Q as high as 10^8, corresponding to a narrow spectral width of 3 MHz and a storage time of approximately 0.3 μs, for resonance at a wavelength of 1 μm.

4.3.1 Modes

Another foundational principle underlying the laser is that of *modes*. A resonator supports light in specific spatial and longitudinal modes. Modes are fields that self-reproduce as they circulate through the resonator. Spatial modes are spatial distributions that maintain their shape after one round trip. Longitudinal (or spectral) modes are fields with frequencies for which the circulating light arrives in the same phase (or shifted by multiples of 2π) after one round trip.

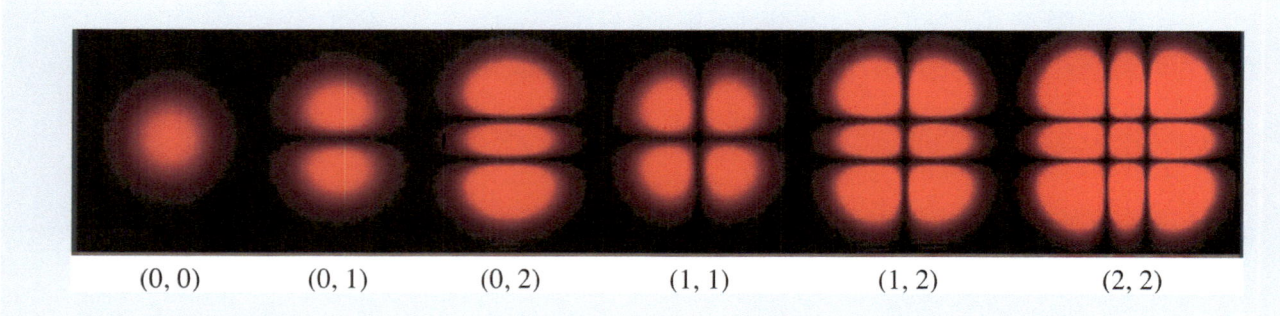

$(0,0)$ $(0,1)$ $(0,2)$ $(1,1)$ $(1,2)$ $(2,2)$

Fig. 4.8 Spatial distributions of the Hermite–Gauss modes of spherical-mirror resonators. The Gaussian mode, which is the lowest-order mode $(0, 0)$, is the most confined around the resonator axis

As mentioned earlier, this ensures positive feedback, which is a necessary condition for oscillation. Each spatial mode may support multiple longitudinal modes.

Laser oscillations occur in those spatial and longitudinal modes for which the round-trip gain is greater than the loss. Since the gain is available within the spectral range defined by the atoms, only those modes whose frequencies lie in this range may oscillate. Moreover, since the spatial modes have different spatial profiles, and therefore undergo different losses, only a finite number of spatial modes oscillate, with the more confined modes favored. And each of these spatial or spectral modes has two polarization degrees of freedom, constituting polarization modes (horizontal/vertical linear polarization or right/left circular polarization).

Ideally, the laser is designed to operate in a single spatial mode with a single longitudinal mode. Such a single-frequency laser is a single optical oscillator with the highest spatial and temporal coherence. However, lasers are also often designed for operation in a single spatial mode with many longitudinal modes. Since these modes oscillate independently, their sum undergoes interference (beating) resulting in amplitude variation and a broader spectrum with reduced temporal coherence. However, such variations are usually on a short time scale, e.g., nanoseconds, so that the average power remains steady when averaged over a longer time, which serves well for certain applications. Operation in multiple spatial modes, each with its own longitudinal modes, can be useful for applications requiring greater power, although such multimode lasers exhibit reduced temporal and spatial coherence.

The spatial modes of the spherical-mirror resonator are the Hermite–Gauss modes illustrated in ◘ Fig. 4.8. The widths of these modes and their divergence angles are determined by the curvatures of the mirrors and their separation.

4.3.2 The Gaussian Beam

The Hermite–Gauss mode with the smallest width is the Gaussian mode. This is responsible for generating the Gaussian beam (◘ Fig. 4.9), which has the smallest angle of divergence for a given width. Its diffraction-limited divergence angle is inversely proportional to the beam radius at its waist W_o, namely $\theta_o = \lambda/\pi W_o$. For example, for $\lambda = 1$ µm and $W_o = 1$ cm, $\theta_o = 3.18 \times 10^{-5}$ radians. At a distance $d = 3.8 \times 10^8$ m (the distance to the moon), this corresponds to a spot diameter $2d\theta_o \approx 24$ km. If the beam radius were increased from 1 cm to 1 m, the spot size at the moon would be only 240 m. The Gaussian laser beam is highly collimated near its waist so that it may be regarded as a planar wave and used for applications requiring plane waves.

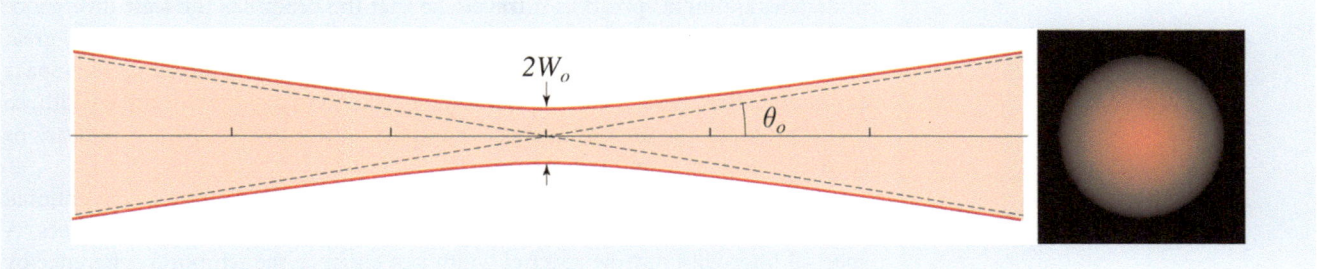

○ Fig. 4.9 The Gaussian beam

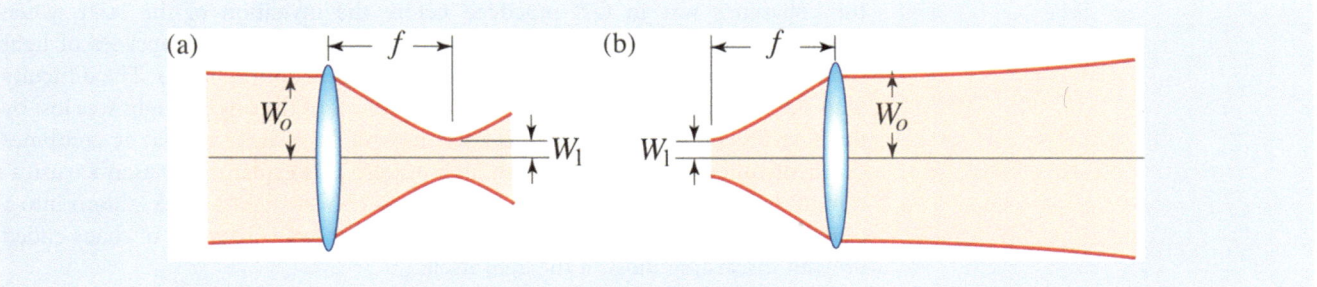

○ Fig. 4.10 Manipulation of a Gaussian beam by use of a lens of focal length f. (**a**) Reduction of the beam waist (focusing). (**b**) Reduction of the divergence angle (collimation)

The angle of divergence, and the associated waist, of the Gaussian beam may be manipulated by the use of lenses. For example, a beam with large waist W_o (and small divergence angle) may be used to generate a beam of smaller waist (and large divergence angle) W_1 by use of a lens of short focal length f. This is important for applications in lithography and laser scanning microscopy or for industrial applications such as cutting and welding. Conversely, a beam of small width and large divergence angle, such as that generated by a laser diode, may be converted into a beam with small divergence angle by use of a lens, as shown in ○ Fig. 4.10. This is used in laser pointers.

4.4   Coherence of Laser Light

The most unique property of the laser is its temporal and spatial coherence. Laser light has a long coherence time (narrow spectrum) and a large coherence area. Conventional light sources have short coherence times (broad spectra) and small coherence area; they are incoherent.

The *coherence time* τ_c is the time duration over which the wave maintains its phase. This quantity is inversely proportional to the spectral width of the light. Laser light has a very narrow spectral distribution, ideally single frequency (a single wavelength), i.e., it is monochromatic or single color. Temporal coherence is also called longitudinal coherence. The *coherence length* $\ell_c = c\tau_c$, where c the speed of light. Long coherence length signifies that the phase of the wave is correlated over a long distance along its direction of propagation. Temporal coherence allows the production of ultrashort pulses of light, as short as a femtosecond or even in the attosecond regime.

Spatial (or transverse) coherence describes the correlation of light fluctuations in the transverse plane. The coherence area is the area within which the wave is correlated. Because of its spatial coherence, laser light in the Gaussian spatial mode undergoes minimal spread as it travels, so that the beam has the least divergence (diffraction-limited), i.e., is highly directional, and the smallest spot size at great distances. This feature is important for various applications including free-space communication. Also, the laser beam can be focused to a spot of minimal width, so that it provides maximal irradiance, an important feature for various applications such as lithography and other industrial applications.

It is essential to note that the coherence properties of light from a conventional incoherent source can be improved by means of spectral and spatial filters. A spectral filter with narrow spectral width can enhance the temporal coherence by filtering out frequencies outside a narrow spectral band. A spatial filter, which may be constructed by sending the light through a pinhole, enhances spatial coherence. Light originating from a point is radiated as a spherical wave, which may be converted into a planar wave with full spatial coherence. Such enhancement of the coherence was in fact practiced before the invention of the laser, when coherent light was generated and used to demonstrate wave properties of light such as interference and diffraction, and to form thin optical beams. The difficulty with such enhancements is that much of the power of the original light was lost by the filtering process. An abiding advantage of laser light is that it combines excellent coherence properties with high power. This explains al-Hasan's frustration in the fictitious story about his attempt to convert light from a lantern into a beam resembling that from the laser. His effort to create order out of chaos ended up with discarding most of the light itself.

Can coherent laser light be distinguished from light generated by a thermal source that is filtered temporally and spatially to have the same coherence time and coherence area as the laser light? The answer is yes! Measurements of other statistical properties of the light intensity and the photons do reveal a difference. One measure is the intensity correlation, introduced by Robert Hanbury Brown and Richard Q. Twiss, fueled major advances in classical and quantum coherence theory. Light from a thermal source exhibits intensity fluctuations whose variance is equal to the squared mean, so that $g^{(2)} = \langle I^2 \rangle / \langle I \rangle^2 = 2$, where I is the intensity, whereas for laser light this variance is zero, which leads to $g^{(2)} = 1$. Photon counts obey Bose–Einstein statistics for thermal light, and Poisson statistics for coherent light from a laser. These counting probability distributions are distinctly different. For a mean number of photons $\langle n \rangle$ detected in a fixed time interval, the count variance is $\langle n \rangle + \langle n \rangle^2$ for thermal light, while it is only $\langle n \rangle$ for coherent laser light.

4.5   Pulsed Lasers

With steady pumping exceeding the threshold required for lasing, the laser output power remains constant over time and the laser is said to be *continuous wave* (CW). Pulsed lasers produce optical power in the form of pulses of certain duration and repetition rate. The creation of short pulses is the temporal equivalent of *focusing* of power in space. Key features of the pulsed laser are the pulse energy, the peak power, and the average power. Higher peak powers may be obtained for shorter pulses with the same pulse energy. Certain applications require high peak power, while others call for high pulse energy. Pulsed operation at a low repetition rate provides adequate time between pulses for the pump to build up a population inversion, thereby allowing the generation of pulses with high energy for the same average power.

Laser pulse durations may be as short as femtoseconds and can be compressed to attoseconds. Pulse-repetition rates extend from hours to more than 10^{11} pulses per second, while peak powers can reach 10 MW. Some gain media are suitable only for use in pulsed lasers since CW operation would require pumping at a steady power so high that it could be impractical or result in excessive heat.

There are two principal schemes for pulsed-laser operation. The first exploits the transient dynamics of the laser system and the energy storage capability of the resonator by on–off switching of the gain, the loss, or the fraction of light extracted from the resonator. These are, respectively, called *gain switching*, *Q-switching*, and *cavity dumping*. The energy is stored during the off-time and released during the on-time. In the second scheme, called *mode locking*, the set of independently oscillating longitudinal modes of the laser are locked together to produce a single periodically pulsed oscillator. Both of these schemes may be used to generate short laser pulses with peak powers far greater than the constant power deliverable by CW lasers.

Gain switching is based on pulsing the pumping source. This is feasible if the pulsing time scale is much slower than time scales governing the lasing process. Examples of gain switching include lasers using electronically charged flashlamps, and semiconductor laser diodes in which the electric current used for pumping is itself pulsed.

Q-switching is loss switching. During the off-time the resonator loss is increased (by spoiling the resonator quality factor Q) using a modulated absorber inside the resonator. Because the pump continues to deliver constant power at all times, energy is stored in the atoms in the form of an accumulated population difference. When the losses are reduced during the on-times, the large accumulated population difference is released, generating an intense short optical pulse.

Cavity dumping is based on storing light in the resonator during the off-times, and releasing it during the on-times. During the off-time, the pump is operated at a constant rate and the generated light is stored in the resonator, which is not allowed to transmit and has negligible losses. The light is subsequently released, or "dumped," as a useful pulse by suddenly removing one of the mirrors altogether (e.g., by rotating it out of alignment), increasing its transmittance to 100 %. As the accumulated light leaves the resonator, the sudden increase in the loss arrests the oscillation, and the process is repeated, resulting in strong pulses of laser light.

Mode-locking. Mode locking is the most important of the various techniques for generating ultrashort laser pulses, from tens of picosecond to less than 10 fs. This is attained by locking the phases of the longitudinal modes together. Since the frequencies of these modes are equally separated, they behave like the Fourier components of a periodic function, and therefore form a periodic pulse train with period equal to the round-trip time between the resonator mirrors (■ Fig. 4.11). The coupling of the modes is achieved by periodically modulating the losses inside the resonator. The pulse width is determined by the number and profile of the

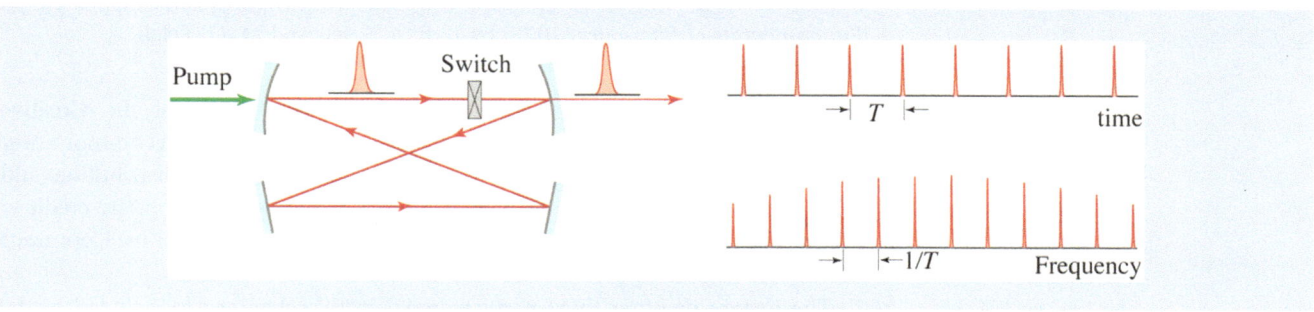

■ **Fig. 4.11** The mode-locked laser. A pulse circulating inside a ring resonator periodically hits the exit mirror and is partially transmitted. The result is a pulse train with period equal to the round-trip time T. The spectrum of the emitted light, which is also periodic with period $1/T$, represents the now-locked longitudinal modes of the resonator. Locking is implemented by modulating the losses inside the resonator using a loss mechanism acting as a switch that lets a pulse out each time period T

spectral components (the modes). In accordance with Fourier theory, the wider the overall spectral width (i.e., the larger the number of modes), the shorter the pulse duration. For example, because of their wide spectral width, Ti:sapphire lasers generate pulses of only a few femtoseconds duration.

4.6 Conclusion

The history of the laser has been both an evolution and a revolution. As described earlier in this article, the development of the laser was an *evolutionary* process that drew together concepts formulated over a period of more than four decades (1917–1960). It also benefitted from the evolution of electromagnetic oscillators with increasing frequency, culminating in the maser. This evolution continues today, with new lasers of ever higher frequencies (wavelengths as short as tens of nanometers for X-ray lasers), pulses of durations shorter than 100 as, and greater optical powers: more than 100 kW for CW operation and peak powers in the GW regime (and as high as PW for laser fusion applications). Focused intensities can be as high as 10^{23} W/m^2.

There is no doubt, however, that within a few years of its invention the laser started a *revolution* in optics and optics-based technologies and applications, which created an abundance of opportunities in the early 1960s for major new scientific discoveries and a proliferation of novel technologies with far reaching applications.

Examples of the new sciences are: nonlinear optics, optoelectronics, laser spectroscopy, femtosecond physics and chemistry, attosecond atomic physics, and quantum optics. Laser applications have expanded to cover all aspects of modern technology:

Applications requiring high power and precision focusing include drilling, cutting, welding, ablation, material deposition, additive manufacturing, directed energy, and fusion.

Applications using the directional precision and focusing capability of the laser include optical disk drives, printers and scanners, barcode scanners, metrology and surveying, lithography, scanning microscopy, and adaptive optics imaging.

Applications utilizing high-speed modulation and switching include fiber-optic and free-space optical communications, optical cables, and interconnects.

Applications to medicine and health care include laser surgery, skin treatments, ophthalmic and cardiovascular diagnostics, laser vision correction (Lasik), and other diagnostic and therapeutic procedures.

The laser is truly a light fantastic!

Acknowledgments Discussions with M. C. Teich are acknowledged. Several of the figures used in this chapter were adapted from figures that appeared in *Fundamentals of Photonics* (2007) by B. E. A. Saleh and M. C. Teich.

Further Reading

1. Townes CH (1999) How the laser happened: adventures of a scientist. Oxford University Press, New York, paperback ed. 2002
2. Maiman T (2000) The laser odyssey. Laser, Blaine, WA
3. Kastler A (1985) Birth of the maser and laser. Nature 316:307–309
4. Bertolotti M (2004) The history of the laser. Taylor & Francis, London
5. Hecht J (2005) Beam: the race to make the laser. Oxford University Press, New York
6. Siegman AE (1986) Lasers. University Science Books, Mill Valley, CA
7. Silfvast WT (2008) Laser fundamentals, 2nd edn. Cambridge University Press, Cambridge
8. Saleh BEA, Teich MC (2007) Fundamentals of photonics, 2nd edn. Wiley, Hoboken, NJ
9. Svelto O (2010) Principles of lasers, 5th edn. Springer, New York
10. Milonni PW, Eberly JH (2010) Laser physics, 2nd edn. Wiley, New York

Solid-State Lighting Based on Light Emitting Diode Technology

Dandan Zhu and Colin J. Humphreys

D. Zhu · C.J. Humphreys (✉)
Department of Materials Science and Metallurgy, University of Cambridge, 27 Charles Babbage Road, Cambridge CB30FS, UK
e-mail: colin.humphreys@msm.cam.ac.uk

© The Author(s) 2016
M.D. Al-Amri et al. (eds.), *Optics in Our Time*, DOI 10.1007/978-3-319-31903-2_5

5

5.1   Historical Development of LEDs

More than 100 years ago in 1907, an Englishman named Henry Joseph Round discovered that inorganic materials could light up when an electric current flowed through. In the next decades, Russian physicist Oleg Lossew and French physicist Georges Destriau studied this phenomenon in great detail and the term 'electroluminescence' was invented to describe this. In 1962, inorganic materials (GaAsP) emitting red light were first demonstrated by Holonyak and Bevacqua [1] at General Electric's Solid-State Device Research Laboratory in Syracuse, New York, although the light emitted was so weak that it could only be seen in a darkened room (by comparison, the efficacy of Thomas Edison's first incandescent light bulb was 10 times greater). Since then, the efficiency of GaP and GaAsP advanced significantly in the 1960s and 1970s. The AlInGaP system was developed later, in the 1980s, and is now the basis of most high-efficiency LEDs emitting in the red-to-yellow visible region. The development of the nitride material system (GaN, InN, AlN and their alloys) in the last two decades has enabled efficient light emission to expand into the blue and green spectral region, and most importantly, allowing the production of white light (blue is the high-energy end of the visible spectrum and therefore enables the production of white light using blue light plus phosphors). Blue LEDs were made possible by a series of key breakthroughs in materials science summarised in ◘ Table 5.1, which will be discussed in greater detail later. In particular, the first bright blue LED was announced at a press conference on November 12, 1993 by Nakamura [2]. The invention of efficient blue LEDs has enabled white light source for illumination. In 1997, white light was demonstrated for the first time by combining a blue gallium nitride (GaN) LED with a yellow-emitting phosphor [3]. Such LEDs are called 'white LEDs'.

Nowadays, solid-state lighting based on LEDs is already commercialised and widely used, for example, as traffic signals, large outdoor displays, interior and exterior lighting in aircraft, cars and buses, as bulbs in flash lights and as backlighting for cell phones and liquid-crystal displays. With the continuous improvement in performance and cost reduction in the last decades, solid-state

◘ **Table 5.1** A summary of the key steps in GaN-based LED development history

1938	Juza and Hahn [84]	The earliest polycrystalline GaN powder was synthesised by reacting ammonia with liquid Ga metal
1969	Maruska and Tietjen [92]	First single crystal GaN film was grown by chemical vapour deposition directly on a sapphire substrate
1972	Pankove et al. [102]	First blue GaN metal-insulator-semicondutor LED was reported
1986	Amano et al. [79]	Crack-fee GaN films with good surface morphology and crystallinity were achieved by growing a thin AlN buffer deposited on sapphire at low temperature before GaN growth
1989	Amano et al. [43]	Amano, Akasaki and co-workers demonstrated that a low-energy electron beam irradiation treatment in a scanning electron microscope could cause a previously highly resistive Mg-doped GaN layer to show distinct p-type conductivity, enabling the first GaN p–n junction LED
1991	Nakamura et al. [38, 94]	Nakamura and co-workers showed that a ~20 nm thick GaN buffer layer deposited at low temperature (~500 °C) before the main GaN growth at ~1000 °C could also be used to grow smooth films on sapphire, including p-type material with good electrical properties
1992	Nakamura et al. [42]	Thermal activation of Mg-doped GaN to achieve p-type conductivity
1993	Nakamura et al. [97]	Blue and violet emitting double-heterostructure (DH) LEDs were successfully fabricated
1993	Nakamura et al. [2]	Nakamura announced the first bright blue LED at a press conference on November 12, 1993
1995	Nakamura et al. [95]	InGaN quantum well LEDs were fabricated
1997	Nakamura et al. [3]	White light was demonstrated for the first time by combining a blue gallium nitride (GaN) LED with a yellow-emitting phosphor

lighting has emerged to be a realistic replacement of incandescent and fluorescent lamps for our homes and offices.

Compared with any other existing lighting technology, solid-state lighting possesses two highly desirable features: (1) it is highly energy efficient with tremendous potential for energy saving and reduction in carbon emissions; (2) it is an extremely versatile light source with many controllable properties including the emission spectrum, direction, colour temperature, modulation and polarisation. The beneficial impact of LEDs on the economy, environment and our quality of life is so evident and well recognised that the 2014 Nobel Prize in Physics was awarded to the inventors of efficient blue LEDs: Isamu Akasaki, Hiroshi Amano and Shuji Nakamura.

5.2 The Importance of Nitride Materials

The main compound semiconductor materials used in LEDs and their bandgap energies are summarised in ◘ Fig. 5.1. For most optoelectronic devices such as light emitting diodes (LEDs), laser diodes, and photodetectors, a direct bandgap is essential for efficient device operation. This is because the optical emission processes in a semiconductor with an indirect bandgap require phonons for momentum conservation. The involvement of the phonon makes this radiative process much less likely to occur in a given timespan, which allows non-radiative processes to effectively compete, generating heat rather than light. Therefore semiconductors with an indirect bandgap are not suitable for efficient LEDs.

Conventional cubic III–V compound semiconductors, such as the arsenides and phosphides, show a direct-to-indirect bandgap transition towards higher energies. Therefore high-efficiency devices can be achieved in the infrared and red-to-yellow visible spectral regions, but the efficiency decreases drastically for

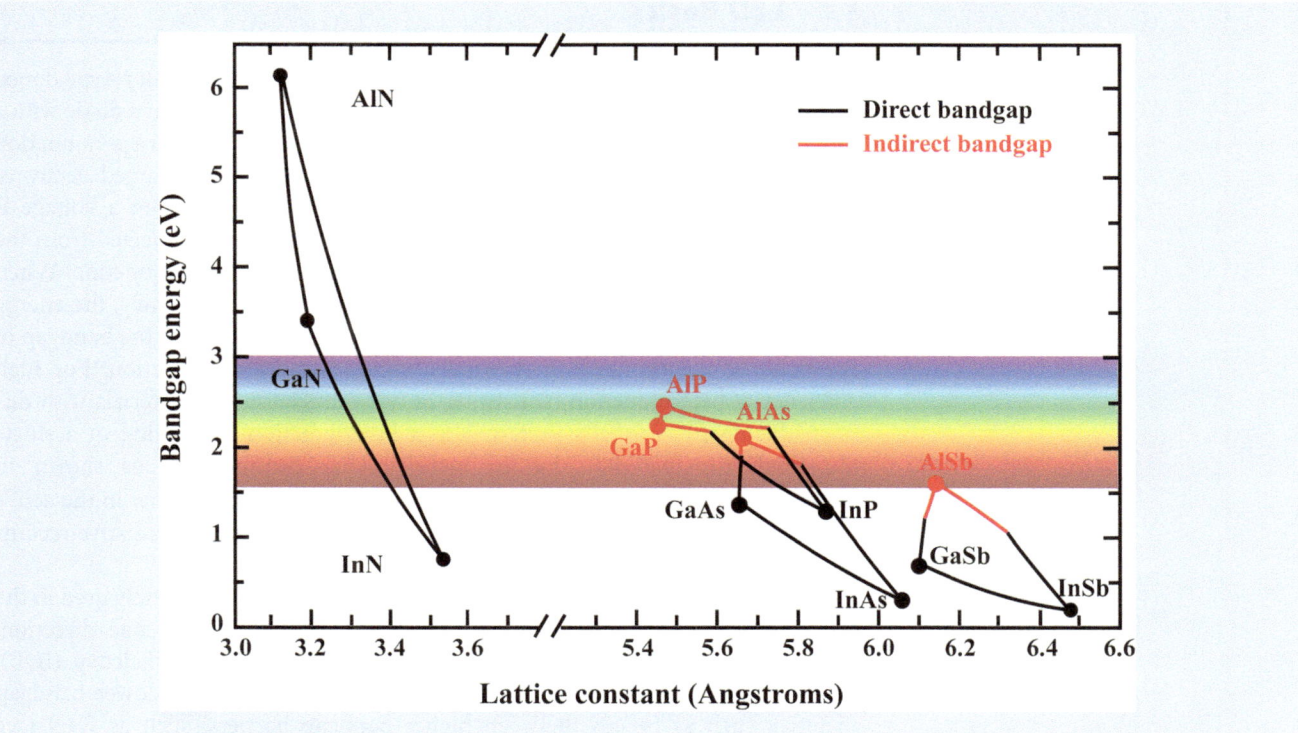

◘ **Fig. 5.1** Bandgap energies at 300 K of III–V compound semiconductors, plotted from data given in Vurgaftman et al. [4] and Vurgaftman and Meyer [5]. For the nitrides, the hexagonal *a* lattice constant has been used. The energy range corresponding to the visible spectrum is also indicated

conventional III–V semiconductors as the bandgap becomes indirect. In contrast, the nitrides have the hexagonal wurtzite structure, and the bandgap remains direct across the entire composition range from AlN to InN, with the bandgap energy covering a wide range from the deep ultraviolet to the infrared region of the electromagnetic spectrum. This makes the group-III nitrides system (consisting of GaN and its alloys with Al and In) particularly suitable for LEDs.

The blue/green and near-UV spectral regions can be accessed using the InGaN alloy, and today, the main application of the nitrides is in blue, green and white emitting LEDs, as well as violet laser diodes used for high-density optical storage in Blu-ray DVDs [6]. Since the InGaN bandgap energy spans the visible spectrum, extending into the infrared to ~0.7 eV for InN, this alloy covers almost the entire solar spectrum, and is thus a potential system for high-efficiency multi-junction solar cells [7].

The wide bandgap of the AlGaN alloy system will enable the fabrication of UV emitters and photodetectors. Possible applications of UV optoelectronics include water purification, pollution monitoring, UV astronomy, chemical/biological reagent detection and flame detection [8, 103].

AlGaN/GaN heterostructures are also suitable for electronic devices such as high electron mobility transistors (HEMTs), which have applications in microwave and radio frequency power amplifiers used for communications technology [9]. Such a wide bandgap materials system also allows device operation at higher voltages and temperatures compared to conventional Si, GaAs or InP-based electronics [10].

Although this chapter will be mainly focused on nitride-based LEDs for lighting applications, it is worth bearing in mind the great potential of nitride materials in other exciting applications mentioned above. And because of their unique materials properties and wide range of applications, group-III nitrides are widely considered to be the most important semiconductor materials since Si.

5.3   LED Basics

The simplest LED structure is a p–n junction, consisting of a layer of p-type doped semiconductor material connected to an n-type doped layer to form a diode with a thin active region at the junction. The principle for light emission in a p–n junction is illustrated in ◻ Fig. 5.2. The n-type region is rich in negatively charged electrons, while the p-type region is rich in positively charged holes. When a voltage is applied to the junction (called forward bias), the electrons are injected from the n-type region and holes injected from the p-type region across the junction. When the electrons and holes subsequently meet and recombine radiatively, the energy released is given out as light with an emission wavelength close to the bandgap of the material incorporated in the active region around the junction. For high efficiency, a heterojunction (consisting of two semiconductor materials with different bandgap) is usually preferred to a homojunction (consisting of a single semiconductor material) due to better carrier confinement, as shown in ◻ Fig. 5.2c, i.e. the electrons and holes are spatially confined together in the active region with lower bandgap energy, which increase the chance of radiative recombination to produce light.

For most high-efficiency LEDs, quantum wells (QWs) are routinely used in the active region, which provide additional carrier confinement in one direction, improving the radiative efficiency, i.e. the internal quantum efficiency (IQE). Quantum wells consist of a very thin (few nm thick) layer of a lower bandgap material, such as InGaN, between higher bandgap barriers, such as GaN (see ◻ Fig. 5.3). The QW active region is sandwiched between two thicker layers of n-type doped and p-type doped GaN for electron and hole injection, respectively.

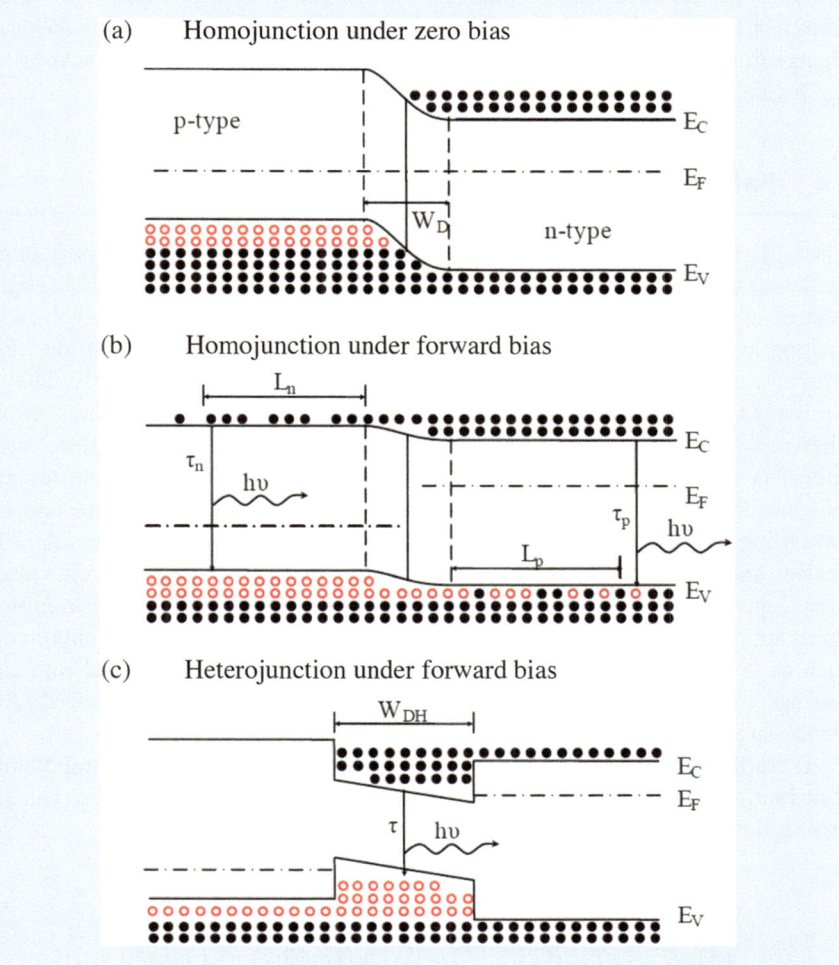

◘ **Fig. 5.2** A *p–n* homojunction under (**a**) zero and (**b**) forward bias. A *p–n* heterojunction under (**c**) forward bias. E_C, E_F and E_v are the conduction band, Fermi and valence band energy. *Filled circle* and *open circle* represent electrons and holes, respectively. In homojunctions, carriers diffuse, on average, over the diffusion lengths L_n and L_p before recombination. In heterojunctions, carriers are confined by the heterojunction barriers (after [11])

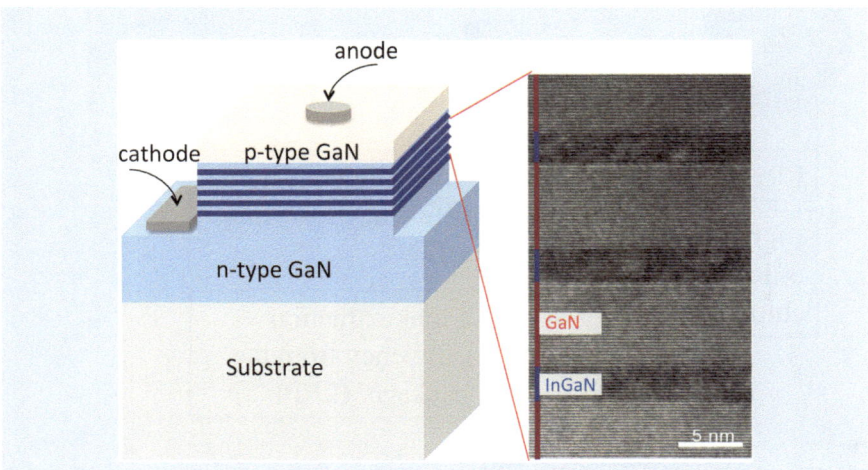

◘ **Fig. 5.3** A schematic InGaN/GaN quantum well LED structure together with a high-resolution transmission electron microscope lattice fringe image of three InGaN quantum wells separated by GaN barriers

The recombination of electron and holes across the InGaN quantum well region results in the emission of light of a single colour, such as green or blue. We can change this colour by varying the composition and/or changing the thickness of the InGaN quantum well.

5.4   Fabrication of an LED Luminaire

The LED structure described above is the essential source of light, but it often makes up only a tiny volume fraction of the final application, such as an LED light bulb or luminaire. ◘ Figure 5.4 illustrates the fabrication procedures involved in making an LED luminaire. The first step is the deposition of the nitride LED structure on a suitable substrate wafer such as sapphire, SiC, Si or GaN. This is performed by crystal growth usually via a process called metal organic vapour phase epitaxy (MOVPE) in a heated chamber or reactor. After deposition, these epiwafers will be processed into LED devices according to the LED chip design, which usually involves several steps including wafer bonding, n and p-type contact patterning, etching, metallisation and surface roughening. The processed LED devices are then separated via cleaving, sawing or laser cutting into individual dies. Depending on the target applications, these individual LED dies are mounted on an appropriate package in a form compatible with other electronic components such as drivers. For white LEDs, phosphors will also be incorporated into the package, together with blue-emitting LED dies in most cases. These packaged LED devices are then ready to be used as the light source in a luminaire.

From the fabrication procedure, we can see that there are many components contributing to the overall efficiency of a packaged LED device. These can be broken down into:

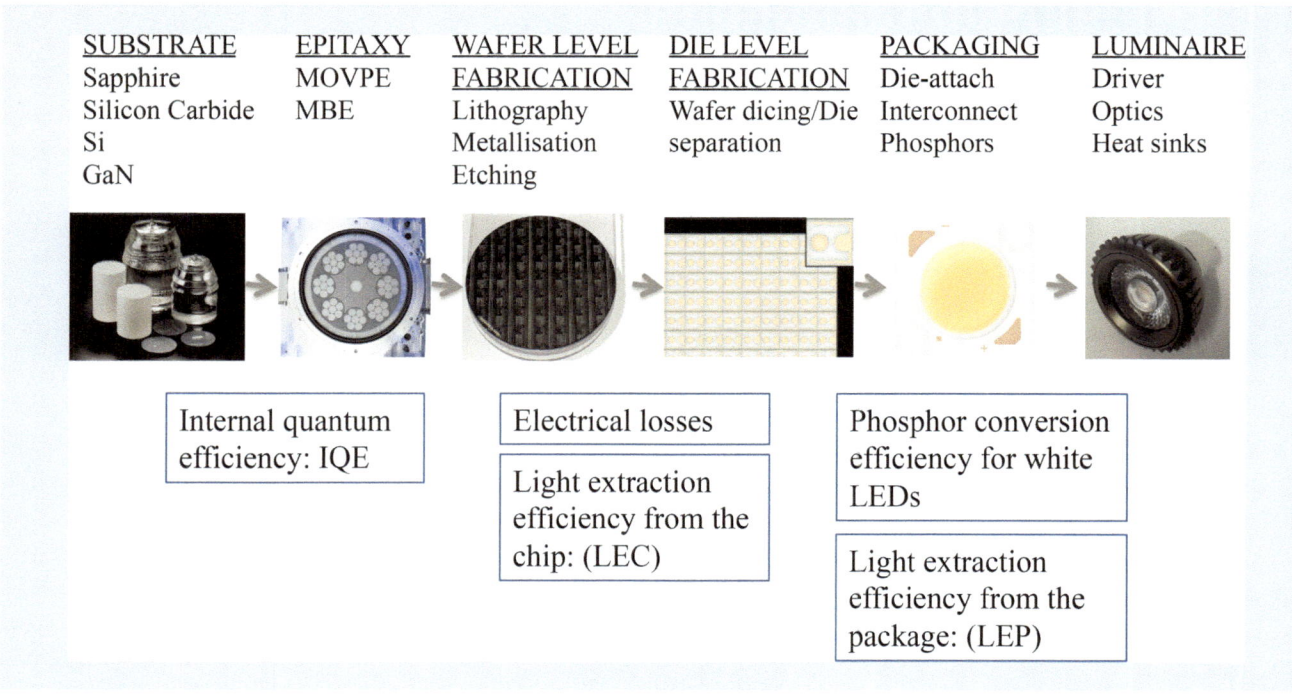

◘ **Fig. 5.4** Illustration of the fabrication procedures involved in making LED luminaries. The corresponding efficiency and losses involved in each procedure are also listed

1. Internal quantum efficiency (η_{IQE})
2. Light extraction efficiency from the chip (η_{LEC})
3. Electrical efficiency (η_{EE})
4. Phosphor conversion efficiency (η_{conv})
5. Light extraction efficiency from the package (η_{LEP})

The IQE is defined as the number of photons emitted from the active region divided by the number of electrons injected into the active region. The IQE is primarily determined by the LED structure design, such as the choice of material compositions, layer thicknesses, doping profile; and for a given structure, the material quality linked to the growth conditions used during the epitaxy procedure. The IQE is also a function of the current density through the LED. At high current density the IQE falls, a phenomenon known as 'efficiency droop'.

The light generated in the quantum well region needs to be extracted from the semiconductor material: most III–V semiconductors have high optical refractive indices (GaN: $n \sim 2.4$; InGaP: $n \sim 3.5$), and only a small portion of the light generated in the quantum well region can escape. This is because much of the light is trapped inside the LED by total internal reflection. Various advanced chip designs have been developed and used during the wafer and die level fabrication procedures to increase the possibility of light extraction from LED chips (LEC) and to minimise the electrical losses caused by the electrical contact and series resistances. Today, an LEC value >85 % is achieved for high performance commercial LED devices with a ThinGaN chip structure, as shown in ◘ Fig. 5.5b [12].

Furthermore LED dies need to be packaged before they can be incorporated with other electronic components in a real application. LED packaging is also critical to achieve high luminous efficiency, dissipate heat generated from the LED chip, improve reliability and lifetime and control the colour for specific requirements, as well as to protect the LED chips from damages due to electrostatic discharge, moisture, high temperature and chemical oxidation. A schematic structure of a high power LED package is shown in ◘ Fig. 5.5a, together with a picture of a commercial white LED package shown in ◘ Fig. 5.5c. The light extraction efficiency from a package (LEP) such as this is as high as 95 %. For white light generation, a yellow-emitting cerium-doped yttrium aluminium garnet (YAG) phosphor plate is added on top of the nGaN layer. To achieve a high phosphor conversion efficiency, the phosphor material is carefully chosen to match the LED emission for optimum excitation.

5.4.1 Efficiency and Efficacy

For a single colour LED such as blue, green and red LEDs, wall-plug efficiency is usually used as a measure of the overall efficiency. The wall-plug efficiency, measured by the light output power (measured in watts) divided by the electrical input (also in watts), is dimensionless and is usually expressed as a percentage. For white LEDs, a different term, efficacy, is usually used instead of efficiency. The unit of efficacy is lumens per watt (lm/W), corresponding to light power output (as perceived by the human eye and measured in lumens) relative to electrical power input (measured in watts). The terms efficiency and efficacy are both widely used in lighting, and care must be taken not to confuse them. The efficacy of a white light source will be explained in more detail later in this chapter. The term efficacy takes into account the sensitivity of the human eye to different colours: it is a maximum for green light at 555 nm.

It should also be noted that the efficiency or efficacy of a luminaire would be lower than the packaged LED devices due to additional losses caused by other

5

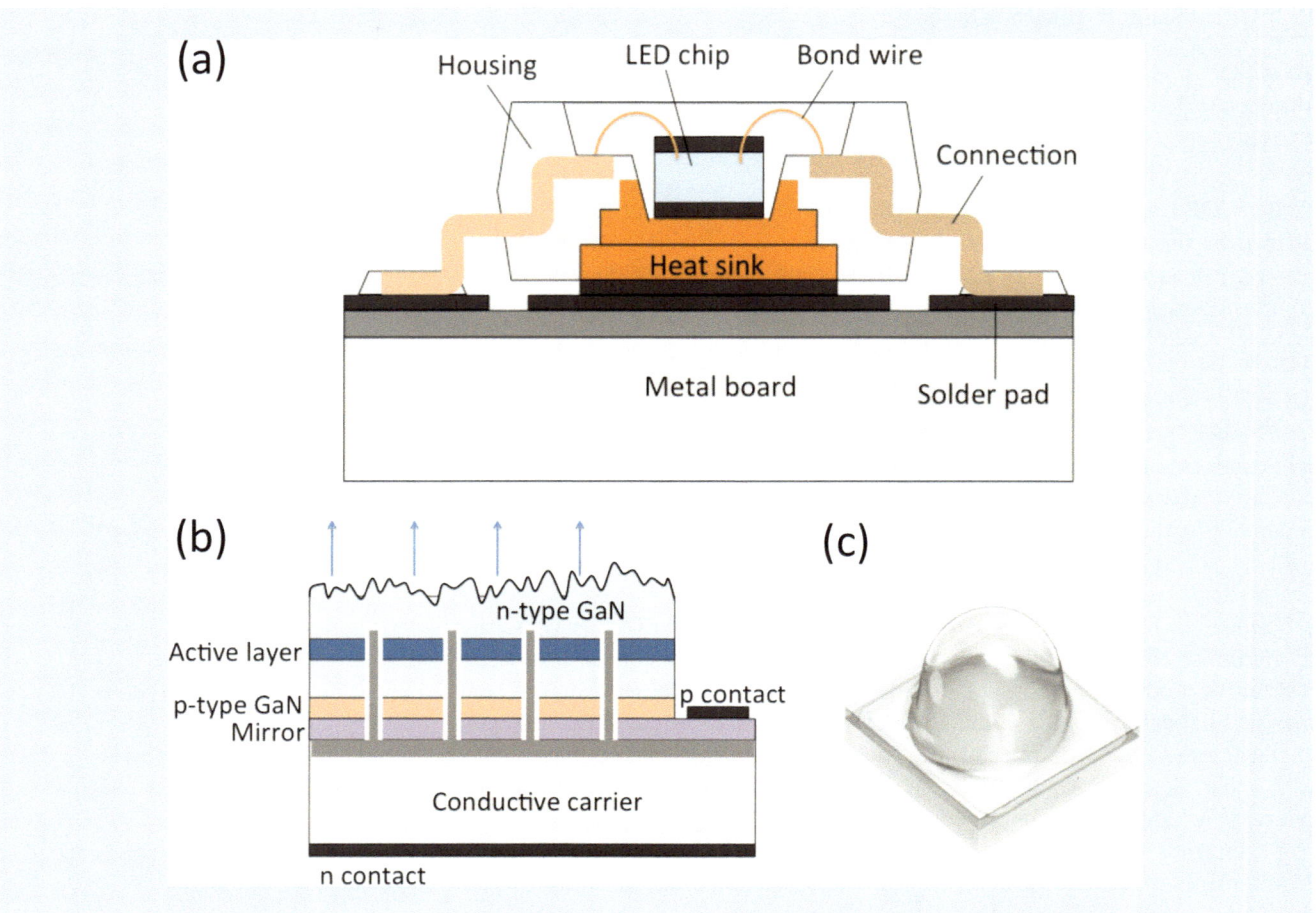

Fig. 5.5 (**a**) The schematic structure of a high-power LED package with good optical efficiency and thermal management, as required for high power LED chips. (**b**) Cross-section of a high power ThinGaN LED chip, illustrating the complex structure of state-of-the-art white LEDs for illumination. (**c**) A picture of a high power white LED package from Osram

components such as optics, heat sinks and electrical drivers. When discussing the efficiency of LED lighting, it is important to be clear about the form of the light source: whether it is a bare die, packaged LED device or luminaire.

The performance of LEDs has improved dramatically over the last decade with sustained improvements in the material quality, LED structure, chip design and packaging. Before moving to the discussions on LED performance and applications, it is worthwhile to first review the historical development of nitride LEDs, in particular the research challenges involved.

5.5   Research Challenges

The research in nitride materials and LED devices is a very broad and interdisciplinary field, spanning crystal growth, physics, materials science and characterisation, device processing, device physics, luminaire design and others. From a materials science point of view, nitride materials are highly defective compared with conventional semiconductor materials such as Si and GaAs, and the remarkable success of nitride-based LEDs is based on a series of wonderful achievements in science and engineering.

5.5.1 Crystal Growth

As with many other semiconductor materials, III-nitrides do not exist naturally, so the crystals need to be grown by some chemical reaction. The predominant growth method for the group-III nitrides is metalorganic vapour phase epitaxy (MOVPE, also called metalorganic chemical vapour deposition, MOCVD), both for research and mass-production of devices such as LEDs and lasers.

It should be noted that one key difference between the nitrides and the other III–V compound semiconductors mentioned earlier in this chapter is the lack of a suitable substrate for heteroepitaxial growth (namely, crystal growth on a different substrate material) of GaN. Bulk substrates of GaAs, GaP and InP can be used for epitaxy of most of the III–Vs and even II–VI compounds. Unfortunately, the nitrides have very high melting temperatures and dissociation pressures at melting, ~2800 K and ~40 kbar, respectively, for GaN, which means that bulk crystals cannot be grown from stoichiometric melts using the usual Czochralski or Bridgman methods [13,14]. Not only have bulk substrates of GaN been unavailable in a sufficient size and at reasonable cost, there is also no other suitable substrate material with a close lattice match to GaN. The properties of the GaN epitaxial layer such as crystal orientation, defect density, strain and surface morphology are to a large extent determined by the substrates used. Most commercial GaN-based LEDs are grown on sapphire or silicon carbide (SiC) substrates. Recently, the use of large area Si substrates has attracted great interest because high quality Si wafers are readily available in large diameters at low cost [106]. In addition, such wafers are compatible with existing sophisticated automated processing lines for 6 inch and larger wafers commonly used in the electronics industry.

Sapphire was the original substrate material, and remains the most commonly used to this day, but it has a lattice mismatch of 16 % with GaN. This is so large that attempts at direct epitaxial growth inevitably result in rough surface morphologies and a very high density of defects called dislocations that thread up through the growing layer: a typical density of such dislocations passing through the active InGaN quantum well region is five billion per square centimetre (5×10^9 cm^{-2}), as shown in ◻ Fig. 5.6.

The development of growth techniques for the reduction of the threading dislocation (TD) density in GaN on sapphire has resulted in considerable improvements. There are numerous methods in the literature, mostly related to the annealing of a low temperature nucleation layer [15], island formation and subsequent coalescence, as detailed in Figge et al. [16] and Kappers et al. [17,18]. An example of TD reduction using an SiN$_x$ interlayer is shown in ◻ Fig. 5.7. The mechanism by which TD density can be reduced is as follows: the thin SiN$_x$ interlayer constitutes a mask containing random holes through which small facetted GaN islands form on regrowth; aided by the inclined facets of the islands, the TDs bend laterally and react with other dislocations to annihilate and form half loops, hence halting their upward propagation, as illustrated in ◻ Fig. 5.7a. It was also found that the growth conditions of the GaN regrowth on top of the SiN$_x$ interlayer have a pronounced effect on the degree of the TD reduction. By using a special 'slow' coalescence method, the TD density of the seed layer (5×10^9 cm^{-2}) was reduced to 5×10^8 cm^{-2} and successively deployed SiN$_x$ interlayers reduce the TD density further to 1×10^8 cm^{-2}, as shown in ◻ Fig. 5.7b.

Dislocations are known to be non-radiative recombination centres [19] that should strongly quench light emission. Indeed, if the dislocation density in other semiconductors, for example, GaAs, exceeds around 1000 per square centimetre (10^3 cm^{-2}), the operation of light emitting devices is effectively killed. However, commercial InGaN blue and white LEDs show high performance despite the fact

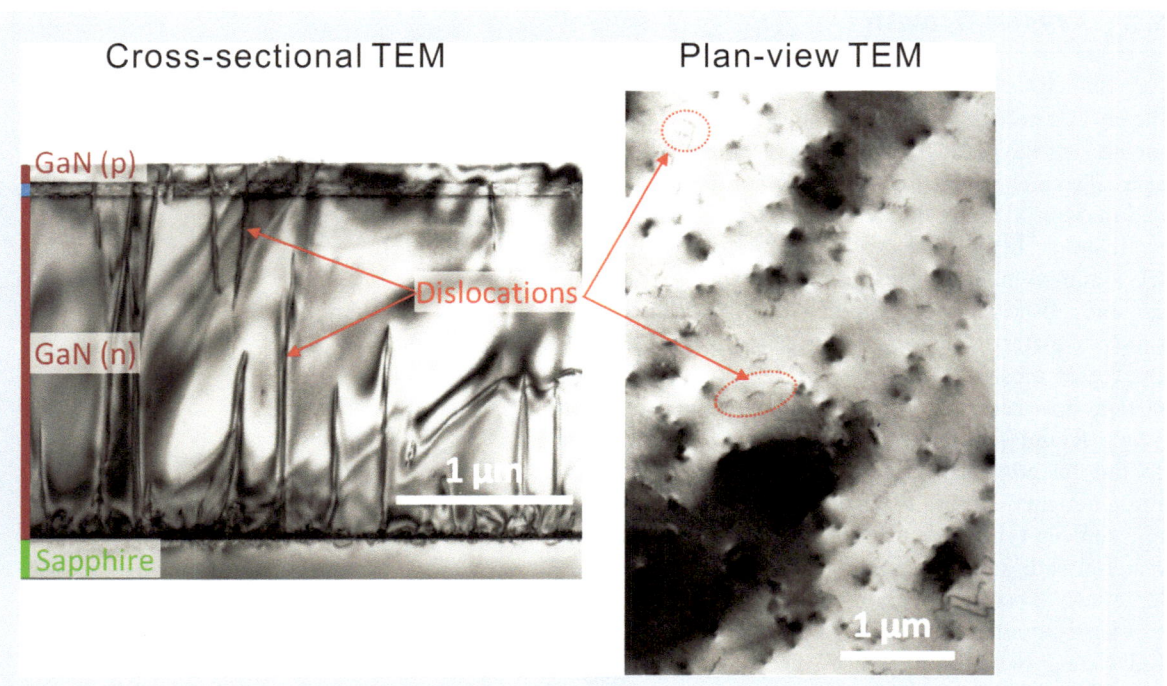

Fig. 5.6 Transmission electron microscopy (TEM) images showing the high density of threading dislocations resulting from the growth of GaN on sapphire substrate. The lattice mismatch between GaN and (0001) sapphire is 16 %, which gives rise to a dislocation density in the GaN of typically 5×10^9 cm^{-2}, unless dislocation reduction methods are used

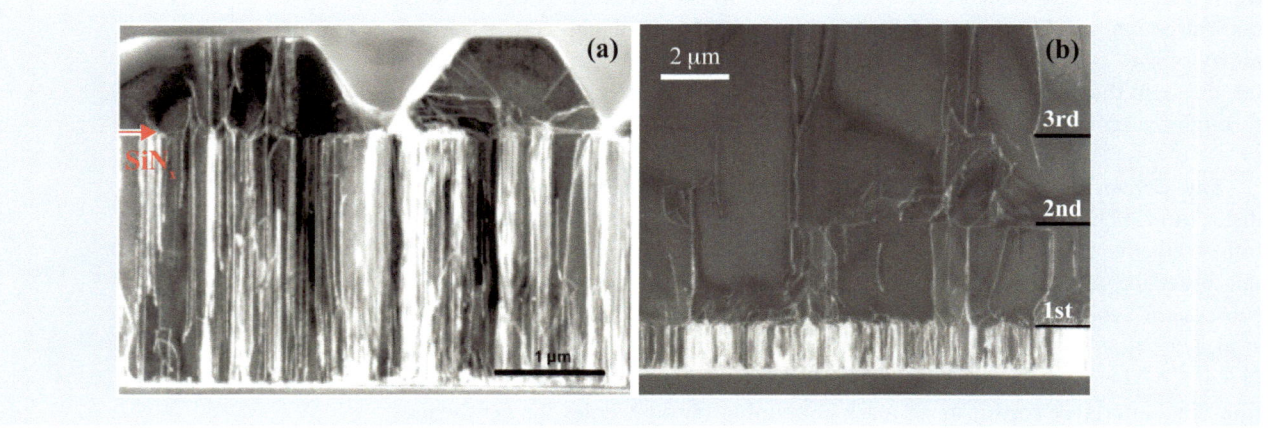

Fig. 5.7 (**a**) Cross-sectional TEM image of an SiN$_x$ interlayer (*arrowed*) deposited on a GaN seed layer followed by the regrowth of GaN islands. Threading dislocations can be observed as bright lines in the image. (**b**) Weak beam dark field TEM image, $g = (11$–$20)$, showing the reduction of edge and mixed TDs with successive SiNx interlayers and a 'slow' coalescence of GaN between the layers

that the TD density of such devices is usually in the range of 10^8 cm^{-2}. The reason that InGaN LEDs are much more tolerant of TDs than other conventional III–V materials is probably due to carrier localisation effects [20–26]. The first contributing factor is the monolayer height interface steps on the InGaN quantum wells. Since the QWs are strained and because of the high piezoelectric effect in GaN, a monolayer interface step produces an additional carrier-confinement energy of about $2k_BT$ at room temperature, where k_B is the Boltzmann constant and T is the temperature. This is sufficient to localise the electrons. Recent three-dimensional atom-probe studies also confirmed that InGaN is a random alloy. Calculations show that random alloy fluctuations on a nanometer scale strongly localise the holes at room temperature. Thus, the above two mechanisms can localise both the electrons and the holes, reducing diffusion to non-radiative defects like TDs. It is interesting to note that the electrons and holes are localised by different mechanisms in InGaN quantum wells.

Although high threading dislocation densities seem to be not very detrimental for InGaN LEDs, laser diodes and AlGaN-based UV-emitters do show a strong dependence of lifetime on dislocation density. Moreover, the growth conditions will also affect many microstructural properties of nitride materials as well as impurity levels and thus the final device properties. Therefore, the research in crystal growth remains highly relevant and important for high performance devices.

5.5.2 Internal Electric Field

The nitrides normally crystallise in the hexagonal wurtzite structure, which is non-centrosymmetric and has a unique or polar axis along a certain direction (the c-axis). Since the bonding is partially ionic due to the difference in electronegativity of the group III and V atoms, a spontaneous polarisation will exist in the crystal because of the lack of symmetry. In addition, most nitride devices involve the use of strained heterojunctions, such as InGaN/GaN. Because the in-plane lattice constant of InGaN is larger than for GaN, the InGaN layer will be under compressive strain perpendicular to the c-axis and under tensile strain along the c-axis when grown epitaxially on GaN. An applied strain along or perpendicular to the c-axis will cause an internal displacement of the metal sublattice with respect to that of the nitrogen, effectively changing the polarisation of the material. This strain effect provides an additional contribution to the polarisation of the material, referred to as the piezoelectric component, and is particularly relevant to strained heterostructures.

Virtually all commercial GaN-based LEDs are grown along the c-axis of the crystal. Since this is a polar direction, there exists an electric field across the InGaN quantum well due to a difference in polarisation for the well and barrier material. The electric field will cause a tilting of the conduction and valence bands in the well, separating the electrons and holes and shifting the quantum well emission wavelength to lower energy, as illustrated in ◘ Fig. 5.8. This is known as the quantum confined stark effect (QCSE).

There are some general observations about the QCSE relevant to nitride QWs: with the presence of an electric field, the transition energy is shifted to a lower value (from $\Delta E_{g,QW}$ to ΔE_{g1}) and this shift is roughly equal to the sum of the shifts of the first electron (ΔE_{e1}) and hole (ΔE_{h1}) levels; it is the hole state that contributes most due to the larger effective mass; electrons and holes are separated from each other spatially by the electric field across the quantum well, resulting in a reduced overlap of electron and hole wave functions and thus a longer radiative lifetime; wider wells (QW2) show more obvious effects of the QCSE and a larger potential drop (ΔE_{E2}) across the well. For a sufficiently wide well, the emission can be lower energy than the bandgap of the quantum well material itself.

The impact of the internal field, especially the piezoelectric field caused by strain, on quantum well recombination behaviour has been confirmed experimentally and reported in various III-nitride-based heterostructures [27–32]. Redshifts of emission energy and lower emission intensity were found in strained quantum wells based on III-nitrides, confirming the strong influence of the strain-induced piezoelectric field. However, with increasing carrier injection, a blue shift of the emission peak was observed by several researchers [33,34] and attributed to the reduction of the QCSE due to the in-well field screening by carriers. Therefore, in

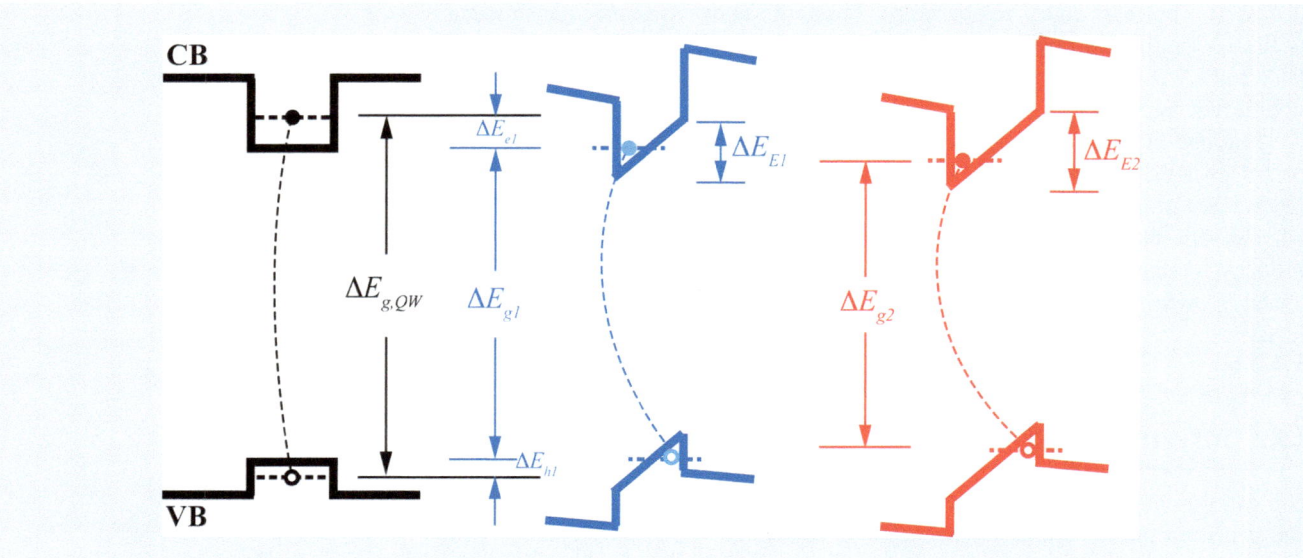

Fig. 5.8 Schematic plot showing the effects of the QCSE on InGaN/GaN quantum wells: *black*: QW1 without electric field; *blue*: QW1 with electric field; *red*: QW2 (thicker quantum well) with electric field

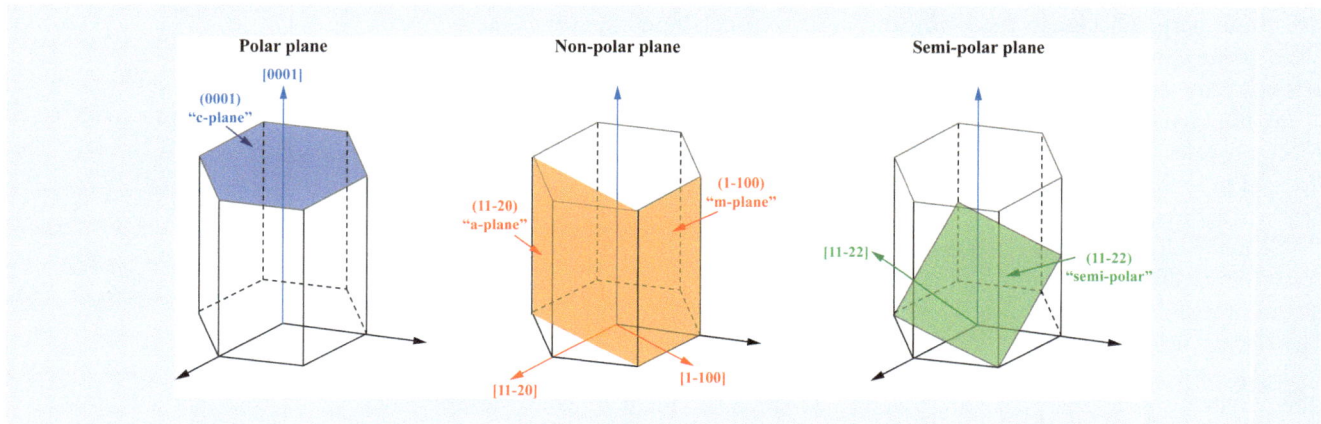

Fig. 5.9 Schematic of the principle polar, non-polar and semi-polar planes of GaN. The QCSE effect should be eliminated by growing along a non-polar direction such as [1–100] and [11–20] or minimised along a semi-polar direction such as [11–22]

an LED structure, the electric field across the quantum wells is not only determined by the polarisation field but also affected by the carrier density and distribution in the quantum well region. The carriers may be from carrier injection (optical or electrical), as well as from doping, either intentional dopants or non-intentional impurities.

From the discussion above, it is obvious that the QCSE is not desirable for LEDs of high efficiency and good colour consistency. ◘ Figure 5.9 shows the main polar, non-polar and semi-polar planes of GaN. In principle, the QCSE should be eliminated by growing along a non-polar direction such as [1–100] and [11–20] or minimised along a semi-polar direction such as [11–22]. The efficiency of non-polar and semi-polar light emitting structures is therefore expected to be enhanced over that of polar.

However, it was found that the defect density is currently much higher in GaN structures grown in such directions [35], unless expensive freestanding non-polar or semi-polar GaN substrates are used [36]. Furthermore, the indium incorporation in the InGaN MQWs grown along non-polar direction is 2–3 times lower than along the *c*-plane for similar growth conditions [37]. The output power of the non-polar LEDs also reduced dramatically when the emission wavelength was longer than 400 nm. Therefore, a non-polar plane is considered not suitable for LEDs with emission wavelengths longer than blue and semi-polar planes are preferred for blue, yellow and red LEDs with reduced internal field, but again high defect densities are a problem. Despite the potential advantages of reduced internal field, non-polar and semi-polar LEDs are currently not commercially viable due to their lower overall performance and the requirement of expensive freestanding GaN substrates.

5.5.3 *p*-Type Doping

For III-nitrides, *p*-type doping is problematic and the realisation of *p*-type conductivity was another major breakthrough in the historical development of nitride-based LEDs. Non-intentionally doped GaN usually shows *n*-type conductivity; however, the improvement in crystal growth methods has managed to reduce this background doping level sufficiently to allow controllable *p*-type doping [38]. Many potential *p*-type dopants have been tried and so far magnesium is the most successful *p*-type dopant for GaN, AlGaN and InGaN with low Al and In mole fractions.

There are two main issues involved in Mg doping: (1) the presence of hydrogen in MOVPE and HVPE growth environments results in the passivation of Mg by forming Mg–H complexes that are electrically inactive; (2) Mg forms relatively deep acceptor states ~160–200 meV above the valence band [39], resulting in only a small fraction activated at room temperature and therefore low conductivity of *p*-type GaN. This means the hole concentration will always be more than an order of magnitude lower than the Mg concentration. Furthermore, heavily Mg-doped GaN is subject to self-compensation due to the formation of donor-like structural defects [40].

The first issue can be solved by thermal annealing under an N_2 ambient at a temperature higher than 700 °C [41,42] or by electron beam irradiation [43] to activate the passivated Mg. The thermal annealing technique has become the standard method for dopant activation because it is straightforward, reliable and can be implemented in-situ, within the MOVPE growth reactor. In contrast, the second issue of a deep acceptor level and self-compensation is intrinsic and is the main reason limiting the hole concentration. ◻ Figure 5.10 shows the

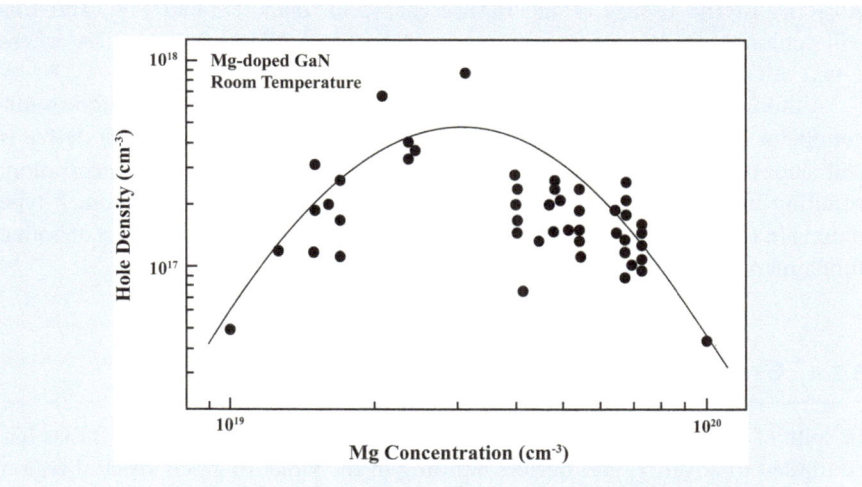

◻ **Fig. 5.10** Hole density in GaN:Mg films as determined by the Hall effect versus the Mg concentration of the films as measured by SIMS. Data from Obloh et al. [44]

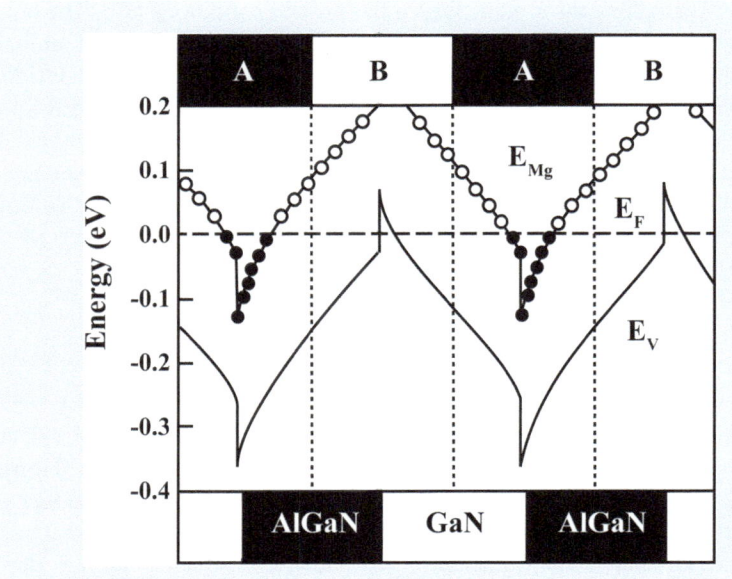

Fig. 5.11 Calculated valence band diagram for the Mg-doped $Al_{0.2}Ga_{0.8}N$/GaN superlattice with spontaneous and piezoelectric polarisation fields taken into account. The *dashed line* indicates the Fermi energy and the *circles* represent the energy of the Mg acceptor with *solid circles* indicating the ionised form. The growth direction for normal Ga-polarity material would be from left to right. Data from Kozodoy et al. [46]

concentration of free holes at room temperature in Mg-doped GaN by MOVPE [44]. The hole concentration reaches its maximum value of about 10^{18} cm^{-3} for a Mg concentration of about 3×10^{19} cm^{-3}, and thereafter decreases with further increase of Mg doping.

A promising method to achieve higher acceptor activation and lower electrical resistivity is to use AlGaN/GaN superlattices. This provides a periodic oscillation in the valence band edge, allowing ionisation of acceptors in the wide bandgap AlGaN layers to provide hole accumulation in the adjacent GaN layers, leading to an overall increase in hole concentration [45]. The principle is illustrated in ▣ Fig. 5.11, where it is apparent that polarisation fields in the nitrides enhance the band edge modulation, leading to parallel sheets of highly concentrated free carriers where the Fermi level intersects the valence band [46]. This can result in spatially averaged hole concentrations in the 10^{18} cm^{-3} range for such superlattices [47,48]. Using the same approach, *p*-type conductivity in $Al_{0.17}Ga_{0.83}N$/$Al_{0.36}Ga_{0.64}N$ superlattices has been demonstrated [49], and this will undoubtedly be a common approach in deep-UV emitting LEDs where *p*-type AlGaN is even more problematic due to wider bandgaps.

Although the development of *p*-type doping has enabled high-efficiency semiconductor devices, the hole carrier concentration in a GaN-based LED device is still about two orders of magnitude lower than the electron concentration, resulting in a large asymmetric carrier distribution in the active region. *P*-type doping in GaN and its alloys with InN and AlN remains a topic of interest at both a fundamental science level and in technological aspects.

5.5.4 Green Gap and Efficiency Droop

In spite of the challenges mentioned above, the performance of nitride LEDs has continued to advance, and devices emitting in the violet to green spectral region have already been commercialised. The highest efficiencies are still achieved for

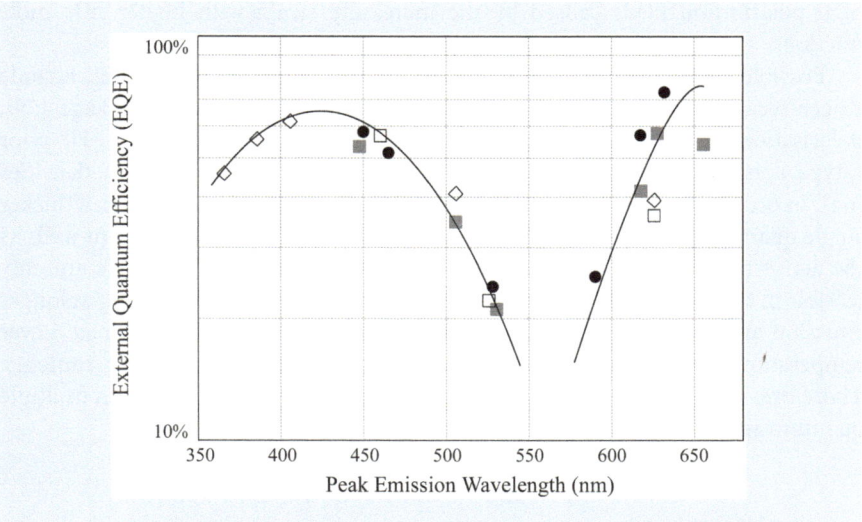

Fig. 5.12 Plot of the external quantum efficiency (EQE) of commercial LED devices measured using EL at 350 mA, showing the issue of the 'green gap'

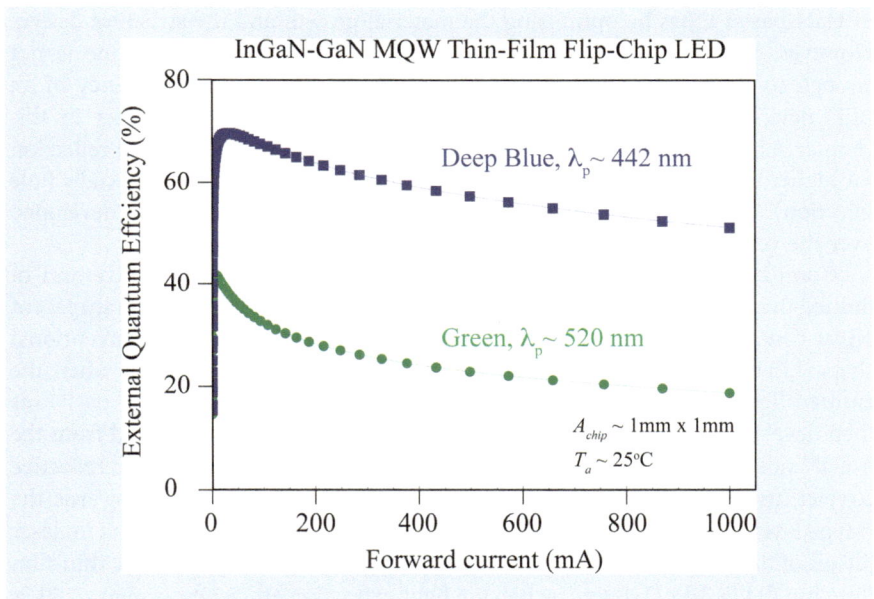

Fig. 5.13 Plot of the external quantum efficiency (EQE) and light output of commercial blue and green emitting LED devices measured using EL at different forward current, showing the issues of 'efficiency droop' at higher current density

blue and violet wavelengths, and despite considerable research efforts (both academic and industrial), a rapid drop in performance towards deep green (the 'green gap') and UV wavelengths remains (Fig. 5.12). Another important problem is that the efficiency of InGaN-based LEDs decreases with increasing current density, an effect known as 'efficiency droop' (Fig. 5.13). Solving the 'green gap' and 'efficiency droop' problems is currently a key focus for research both in academia and industry [12,50–56].

For AlGaInP LEDs, the reason for the lower efficiency at wavelengths shorter than 600 nm is the transition from a direct to an indirect bandgap, as shown in Fig. 5.1. The factors limiting the IQE of nitride LEDs are complex and not well understood. For InGaN, the reason for decreased efficiency in the green spectral region has been attributed to the miscibility gap between GaN and InN [57] and high polarisation fields caused by the increasing strain with higher InN mole fractions.

Possible mechanisms of 'efficiency droop' that have been proposed include Auger recombination [52,56], high defect density [54,58], carrier leakage [59], polarisation-induced built-in electric fields at hetero-interfaces [60,61], poor *p*-type conductivity [62,63] and carrier delocalisation at high current densities [64]. In order to reduce the current density and thus the efficiency droop, a thicker single quantum well has been proposed to replace thin multiple quantum wells as the active region [12]. However, it was found that thicker InGaN QWs are only feasible in the short wavelength range around 400 nm. For LEDs emitting at longer emission wavelengths, the material quality decreases due to growth at lower temperatures and the internal field rapidly increases due to higher In contents. Therefore, most commercial blue- and green-emitting LEDs still use thin multiple quantum wells as the active region.

5.5.5 Chip Design

The discussions above on crystal growth, *p*-type doping, internal fields and efficiency droop are mainly concerned with how to improve the internal efficiency of GaN-based LEDs by optimising the material growth and the structure design. However, improving the generation of light in the active region alone is not enough to achieve an efficient LED device, because the overall efficiency of an LED device is determined by many components as mentioned earlier in this chapter. Chip design is an important area of research to reduce internal reflection for higher light extraction and to enable uniform current injection (especially hole injection). The schematic structures of several different chip designs developed over the years are illustrated in ◘ Fig. 5.14.

Compared with the *n*-type region, the *p*-type layer is very resistive and of limited thickness. To overcome the current spreading problem, a semi-transparent NiAu contact was originally deposited over the *p*-GaN [66] for a conventional shape LED chip. However, this approach results in significant losses when the emitted light passes through the *p*-contact. The 'Flip-chip' (FC) approach was then developed, where the LED chip is inverted and the light is emitted from the *n*-GaN side. In this approach, the NiAu contact is replaced by a thick and reflective contact, usually comprising silver, to reflect back the light emitted towards the *p*-type layer side [67]. In order to overcome the internal reflection problem, laser lift-off of the sapphire substrate and *n*-GaN roughening were used in the thin-film flip-chip (TFFC) LED design, achieving light extraction efficiency as high as 80 % by 2006 [68]. A similar vertical thin-film device (VTF) was also developed, resulting in an estimated light extraction efficiency of 75 % [69]. In recent years, patterned sapphire substrates have become very popular due to the advantages of improved material quality and ease of light extraction. Combining patterned sapphire substrates with an indium-tin-oxide (ITO) current spreading layer, a light extraction efficiency as high as 88 % was estimated [70] for this PSS-ITO approach.

The above approaches all extract the emitted light primarily from the top or bottom side of the LED chip. When a bulk GaN substrate is used, the sidewalls of the LEDs can be used to extract part of the light through geometric die shaping, as shown in ◘ Fig. 5.15. These volumetric LEDs have the potential to achieve even higher light extraction efficiency than thin-film LEDs based on modelling [71]. Today, light extraction efficiencies exceeding 85 % are achieved for high power TFFC InGaN LEDs [12]. When using GaN as a substrate, the light extraction efficiency can also be as high as 90 %.

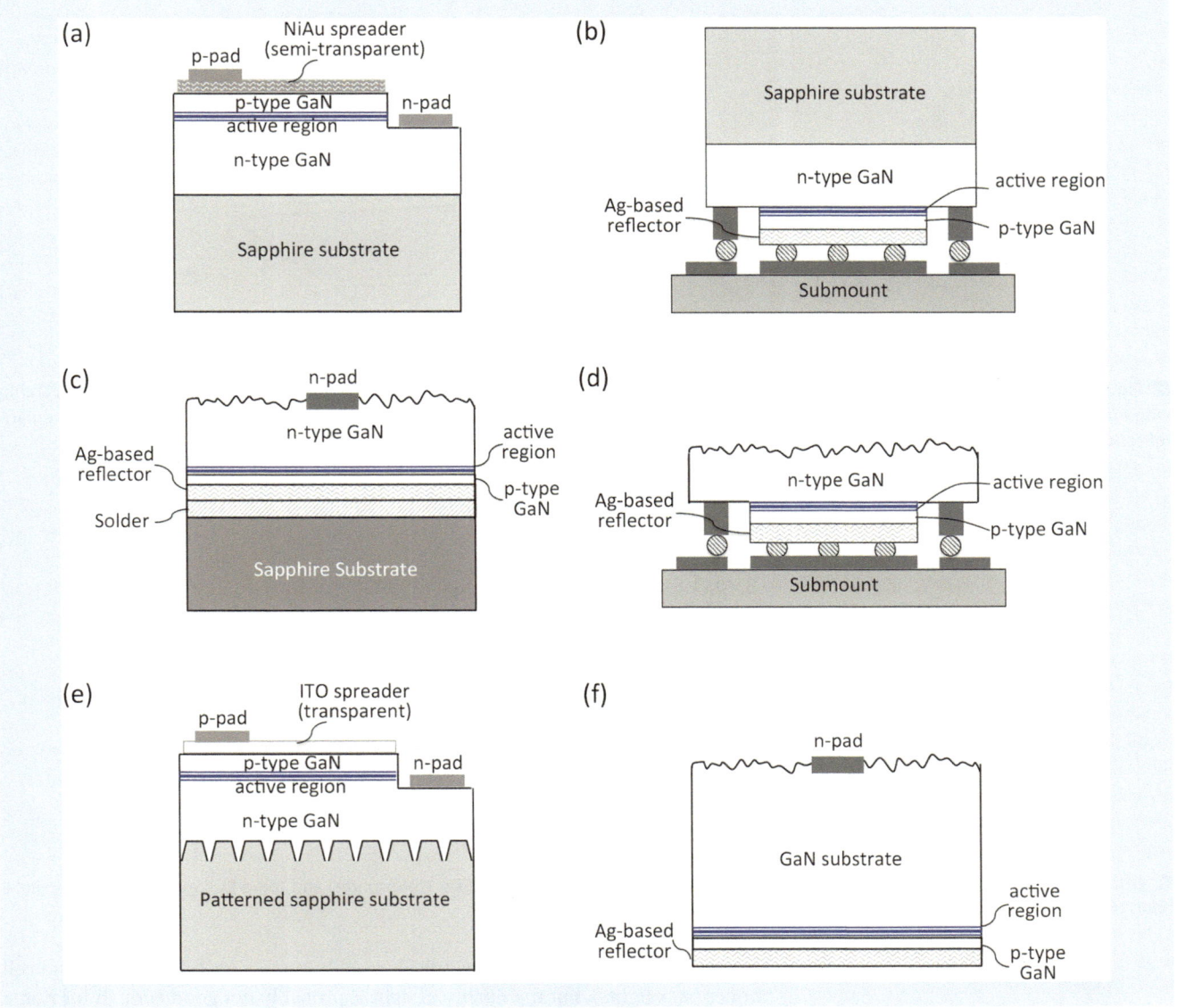

⊙ Fig. 5.14 Schematic cross sections of various GaN-based LED chip designs: (**a**) Conventional chip. (**b**) Flip-chip (FC). (**c**) Vertical thin film (VTF). (**d**) Thin-film flip-chip (TFFC). (**e**) Patterned sapphire substrate combined with ITO contact (PSS-ITO). (**f**) GaN substrate volumetric LED chips. After Nakamura and Krames [65]

5.5.6 Generation of White Light with LEDs

Whereas LEDs emit light of a single colour in a narrow wavelength band, white light is required for a huge range of applications, including LED backlighting for large LCD displays, and general home and office lighting. White light is a mixture of many colours (wavelengths) and there are two main methods to generate white light using LEDs: Phosphor method and RGB method, as illustrated in ⊙ Fig. 5.16.

The first commercially available white LED was based on an InGaN chip emitting blue light at a wavelength of 460 nm that was coated with a cerium-doped yttrium aluminium garnet (YAG) phosphor layer that converted some of the blue light into yellow light [72]. Nearly all white LEDs sold today use this method. The phosphor layer is sufficiently thin that some blue light is transmitted through it, and the combination of blue and yellow produces a 'cool white' light.

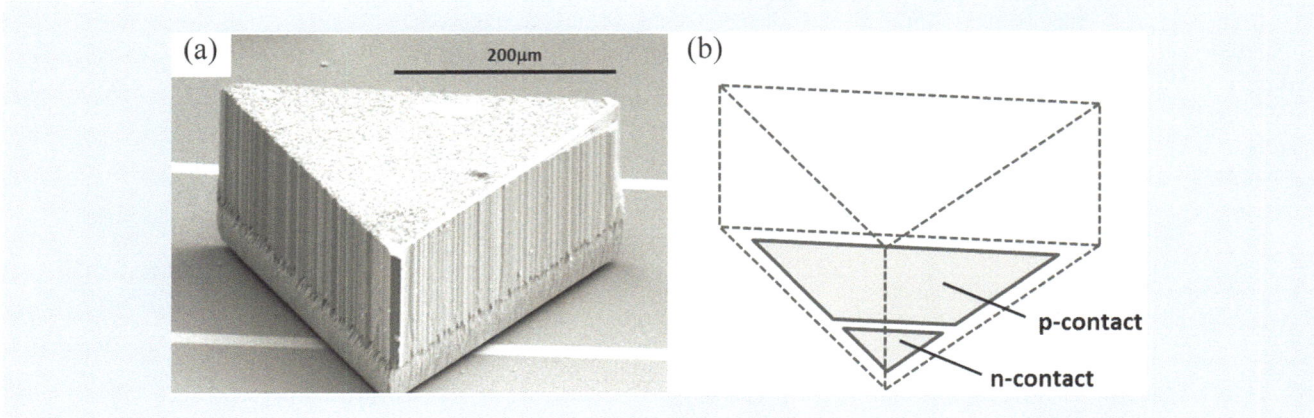

Fig. 5.15 (**a**) Scanning electron microscope image of a fabricated triangular-shaped gallium nitride on gallium nitride (GaN-on-GaN) LED chip with roughened top and side surfaces. (**b**) Corresponding device geometry. Unlike cuboidal shapes, light is not trapped inside by total internal reflection with this geometry. Reprinted with permission from David et al. [71]. Copyright 2014, AIP

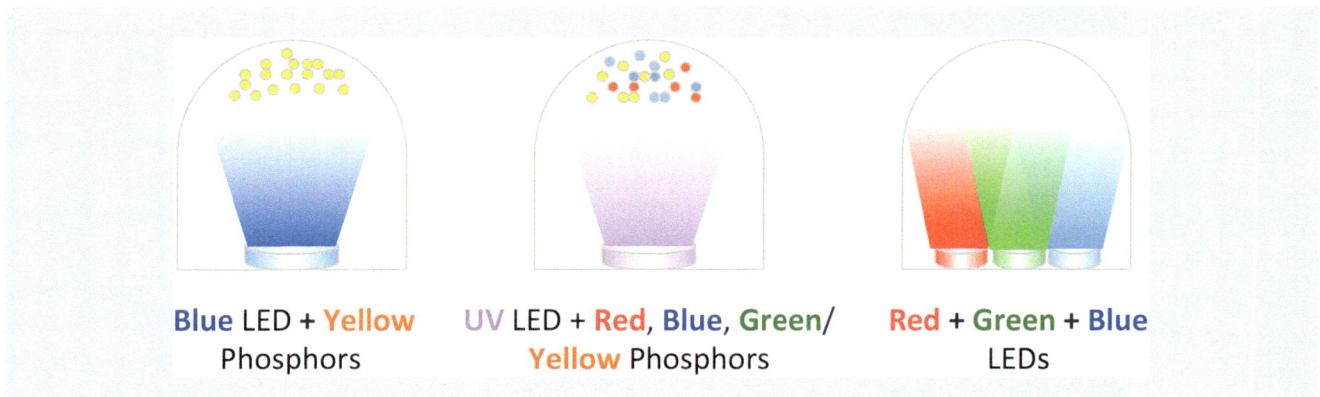

Fig. 5.16 Illustration of white light generation using phosphor method (blue/UV LEDs + phosphors) and using RGB method (Red + Green + Blue LEDs)

This is fine for many applications (displays, lighting in cars, buses, yachts and cell phones back lights), but the quality of light is probably not good enough for home lighting, for which a warmer white light containing some red light is desirable. To generate 'warm white', red phosphors are typically added [73].

Since the efficiency with which existing red phosphors are excited using blue light is much less than that using near-UV light, a better route to generate 'warm white' light might be to use a near-UV LED plus red, green and blue or more coloured phosphors. Thick phosphor layers would be used so that no near-UV light from the LED would be transmitted in much the same way as the phosphor coating on fluorescent tubes and CFLs prevents the transmission of UV light. The drawback of this method is the large intrinsic energy loss from converting a near-UV photon to a lower energy visible photon.

Mixing red and green and blue (RGB) LEDs is an alternative way to produce white light without using phosphors, which is potentially the most efficient. However, there are three basic problems with this method. The first is that the efficiency of green LEDs is much less than that of red and blue, for reasons that are not yet understood (this is known as the 'green gap' problem described earlier). Hence the overall efficiency of this method is limited by the low efficiency of the green. Second, the efficiencies of red, green and blue LEDs change over time at different rates. Hence if a high quality white light is produced initially, over time the quality of the white light could degrade noticeably. However, this process is slow and can be corrected electrically using automatic feedback. Third, because the emission peaks of LEDs are narrower than those of most phosphors, red plus green plus blue LEDs will give a poorer colour rendering than by using phosphors. This problem can be minimised by a careful choice of LED emission wavelengths, and of course, more than three different colour LEDs can be used for better coverage of the visible spectrum. In particular, using four LEDs (red, yellow, green and blue) can give a good colour rendering, although at the expense of increased complexity.

5.5.7 LED Packaging

LED packaging secures and protects the LED chips from damage caused by electrostatic discharge, moisture, high temperature and chemical oxidation. When designing the LED package, the issues involved in optical control, thermal management, reliability and cost need to be addressed simultaneously. The main package components include the LED die/chip, electrodes (anode and cathode), bond wire (connecting the LED die and electrodes), heat sink (removing heat generated by the LED die), phosphor coating (for white light emission) and primary lens (for directing the light beam).

Many solutions have been developed over the years for high power LED packages, as shown in ◻ Fig. 5.17, ranging from single large die packaging with input powers of 1–2 W to chip-on-board and 'Jumbo Die' solutions that can take input powers up to 94 W with lumen flux higher than 10,000 lm from a single package. Depending on the application, different LED package sizes and powers would be required. An interesting trend of LED packaging is to move from chip-based packaging to wafer-level packaging, with advantages of higher packing density, ease of integration on circuit boards, higher current density and higher reliability.

◻ **Fig. 5.17** Wide variety of solutions for high power LED packages. Images from Philips Lumileds, Osram, Cree and Luminus

5.6   LEDs for Lighting

Over the last decade, advances in the material quality, LED structure, chip architecture and package design have improved the performance of LEDs dramatically in terms of light quality, efficiency/efficacy, lifetime and cost. This has enabled LEDs to become a realistic replacement of traditional light sources such as incandescent and fluorescent lamps.

5.6.1   Quality of LED Lighting

People have become used to high quality lighting provided by conventional light sources, especially those installed at home, such as incandescent and halogen lamps. Colour temperature, colour rendering index (CRI) and colour consistency are the main factors when evaluating the quality of a white light source.

The planckian black-body radiation spectrum is used as a standard for white light because its spectrum can be described using only one parameter, namely the colour temperature. The colour temperature (CT) or correlated colour temperature (CCT) of a white light source, given in units of Kelvin, is defined as the temperature of a planckian black-body radiator whose colour is closest to that of the white light source. With increasing temperatures, a planckian black-body radiator glows in the red, orange, yellowish white, white and ultimately bluish white. Therefore, the colour temperate of a white light source can be used to describe its appearance. For conventional lighting technologies, the CCT spans a wide range, from 2700 to 6500 K. 'Warm white' light, such as from incandescent lamps, has a lower colour temperature (2700–3500 K), while 'cool white', which is a more blue–white, has a higher colour temperature (3500–5500 K). 'Warm white' is in the most common lamp colour used in residential lighting in the USA and Europe.

Another important characteristic of a white light source concerns how precisely the different colours of an object show up under illumination from the light source. This is measured in terms of the CRI. Some examples of different light sources and their corresponding spectrum are shown in ◘ Fig. 5.18. An ideal light source, such as sunlight, can reproduce colours perfectly and has a CRI of 100. Natural light LED lamps or full spectrum LED lamps, e.g. white LEDs based on near-UV LEDs plus RGB phosphors technology [107] have a CRI value as high as 95. Therefore the colours under full spectrum LED lamps also appear to be rich and vivid, similar to those under sunlight. 'Warm white' LEDs usually have a CRI higher than 80, which is acceptable to replace conventional light sources for most cases. While for conventional 'cool white' LEDs, the colour reproduction becomes insufficient, similar to fluorescent light.

It is noted that the current CIE colour rendition system of eight test colours to determine the CRI of a light source is designed around conventional light source technology and is not sufficient for LEDs. A new and better method of measuring and rating colour rendition for LED light sources is under development. For lighting professionals, the specific spectrum of a particular light source or the position of the colour points of a light source in relation to the black-body locus is a more accurate way of determining the value of the colour rendition.

Conventional light sources, such as incandescent and halogen lamps, have good colour consistency during their lifespan. For LED lighting, achieving good colour consistency is challenging. The colour distribution of blue LEDs and phosphors may result in greenish, blueish and pinkish white light. Furthermore, the colour of LEDs can shift with temperature and time. LED manufactures have put a lot of efforts into understanding and controlling the colour shift of LEDs.

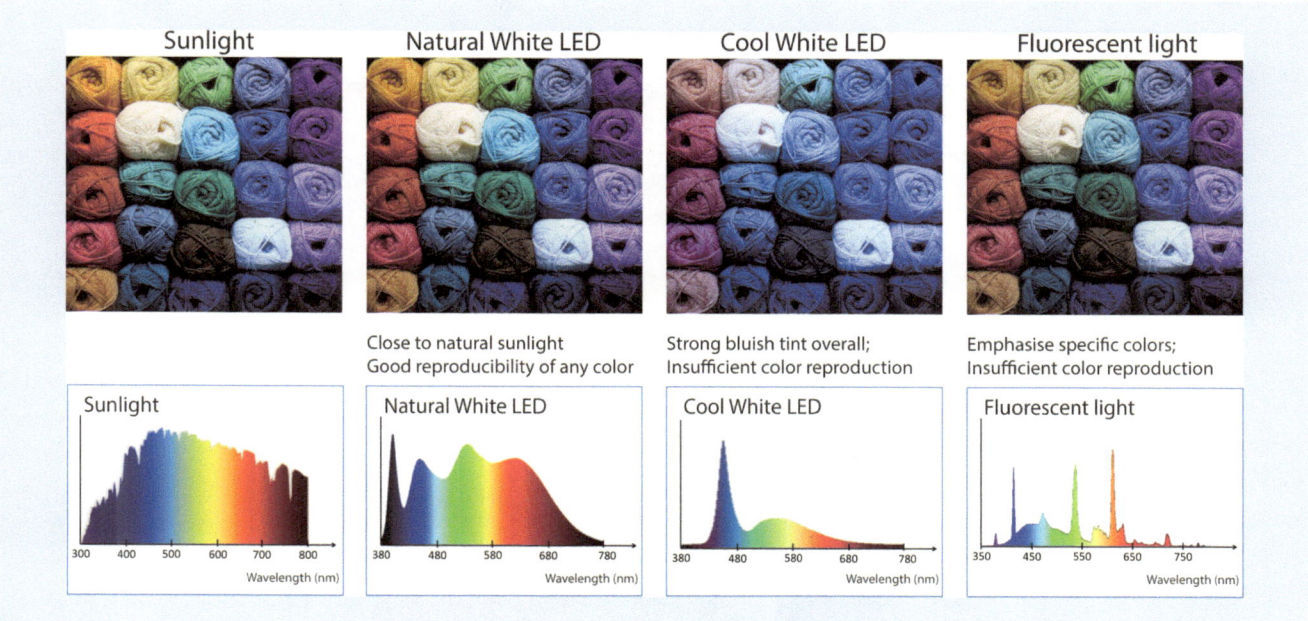

Fig. 5.18 Examples of different light sources and their corresponding spectrum. A broader spectral source more accurately renders colours of illuminated objects Image from online resources

The uniformity of epitaxy, processing and phosphor technologies are improving continuously, enabling a tighter distribution of LEDs in the production process. The LED industry has also adopted a strict binning system to ensure colour consistency between LEDs. Meanwhile, LED industry standards and regulations are being developed. For example, in EU directive (EU-1194/2012), one of the functionality requirements is on colour consistency and a variation of chromaticity coordinates within a six-step MacAdam ellipse or less is required [87]. Some manufactures have implemented LED lighting products that fall within a single three-step MacAdam ellipse to avoid a difference in colour between two sources that may be perceived [74, 90].

Since LEDs have different colours at different temperatures, leading LED manufactures now specify their LEDs at real application temperatures (85 °C), instead of a 25 °C operating temperature, on their datasheet to ensure the customers receive the exact colour intended. Although the colour consistency of LED lighting has improved greatly, the colour shift during its long lifetime remains a large area of concern. The solutions rely on a better understanding of the degradation mechanisms of LED chips and other components with time. Considering the rapid improvement made during the short LED history so far, we have every reason to believe that within a short time LED lighting technology will totally surpass conventional light sources in both quantity and quality.

5.6.2 Efficacy

Radiometric units, such as optical power in watts (W), are used to characterise light in terms of physical quantities. However, the human eye is sensitive only to light in the visible spectrum, ranging from violet (with a wavelength of ~400 nm) through to red (with a wavelength of ~700 nm) and has different sensitivity at different wavelengths, as shown in ◘ Fig. 5.19. The maximum sensitivity of the human eye is to green light with a wavelength of 555 nm. Therefore, to represent the light output of an optical source as perceived by the human eye, photometric

5

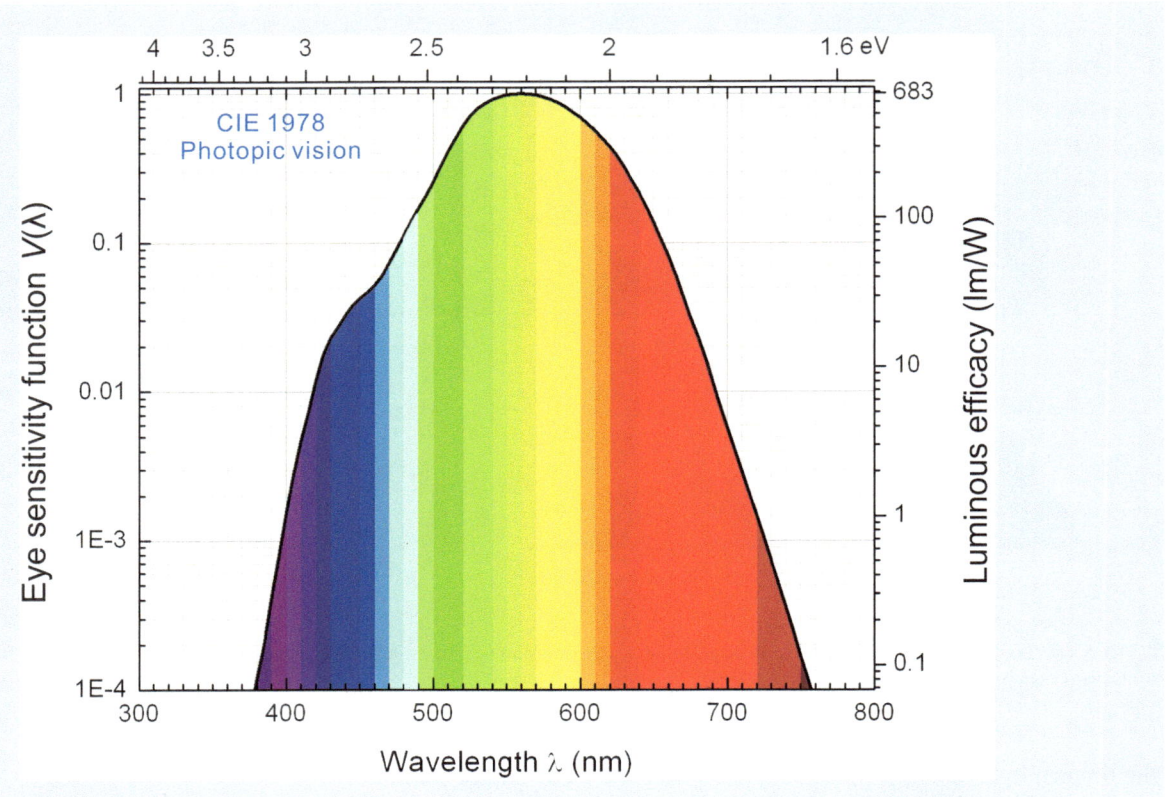

Fig. 5.19 Eye sensitivity function V(λ) and luminous efficacy in lm/W. The maximum sensitivity of the human eye is to green light with a wavelength of 555 nm (Data after 1978 CIE) [81]. It is noted that definition of the luminous efficacy here is the light power output in lumen divided by the optical power in (measured in W), rather than electrical power in

units, such as lumens (lm), are used instead of radiometric units. The efficacy of a light source takes into account the sensitivity of human vision, so that green light contributes more strongly to efficacy than blue or red light, and ultraviolet and infrared wavelengths do not contribute at all. The unit of efficacy is lumens per watt (lm/W), corresponding to light output power (as perceived by the human eye and measured in lumens) relative to electrical input power (measured in Watts).

It should be noted that there is a fundamental trade-off between efficacy and colour rendering [75]. The corresponding colour temperature should also be considered when comparing the efficacy of different white light sources. Generally speaking, a 'warm white' LED source of high CRI usually has lower efficacy compared with a 'cool white' LED source of lower CRI. The highest reported efficacy so far from a packaged LED device is 303 lm/W at a drive current of 350 mA and with a correlated colour temperature of 5150 K [76].

5.6.3 Lifetime

One of the main advantages of LED lighting is its long lifetime that potentially can span to 50,000 h or even 100,000 h. Similar to all electric light sources, LED lighting experiences a decrease in the amount of light emitted over time, a process known as lumen depreciation. For general lighting purpose, the useful life of an LED is defined as the point at which light output has declined to 70 % of initial lumens. The primary cause of LED lumen depreciation is heat generated at the LED junction that will affect the performance of key LED package components as well as materials [77]. Heat management is therefore an important factor in determining the effective useful life of the LED. The lifespan of commercial LED replacement lamps is already longer than 15,000 h (some are longer than 25,000 h). As LEDs become more efficient over time, the problem of heat management will largely disappear and a longer lifetime of LED lighting is expected. The lifetime of LED lamps is also limited by the shorter lifetime of the control electronics used. So more attention is being paid to the development of sophisticated control electronics for LED lighting.

5.6.4 Cost

Cost is probably the major factor limiting the widespread use of white LEDs in our homes and offices. GaN-based LED replacement lamps are significantly more expensive than filament light bulbs or compact fluorescent lamps (CFLs). However, the cost per lumen is continuously decreasing, following the Haltz's law (see ▫ Fig. 5.20).

It should be noted that the total ownership cost of lighting includes energy savings and replacement cost, which makes LEDs more competitive, compared to conventional lighting technologies. Nevertheless, in order to achieve significant market penetration, the initial cost ($/klm) of LEDs needs to be reduced 10 times to be comparable to the cost of CFLs. To achieve the required cost reduction, many aspects of the manufacturing process will need to be addressed in parallel, as illustrated in ▫ Fig. 5.21. This diagram shows that the cost reduction shouldn't be based on sacrificing the three main LED quality factors: efficiency, reliability and customer experience. To make sure LED lighting remains a high quality light source, many aspects including LED materials, chip design, white light generation, component design, power supply circuit, luminaire optical and thermal design need to be taken care of.

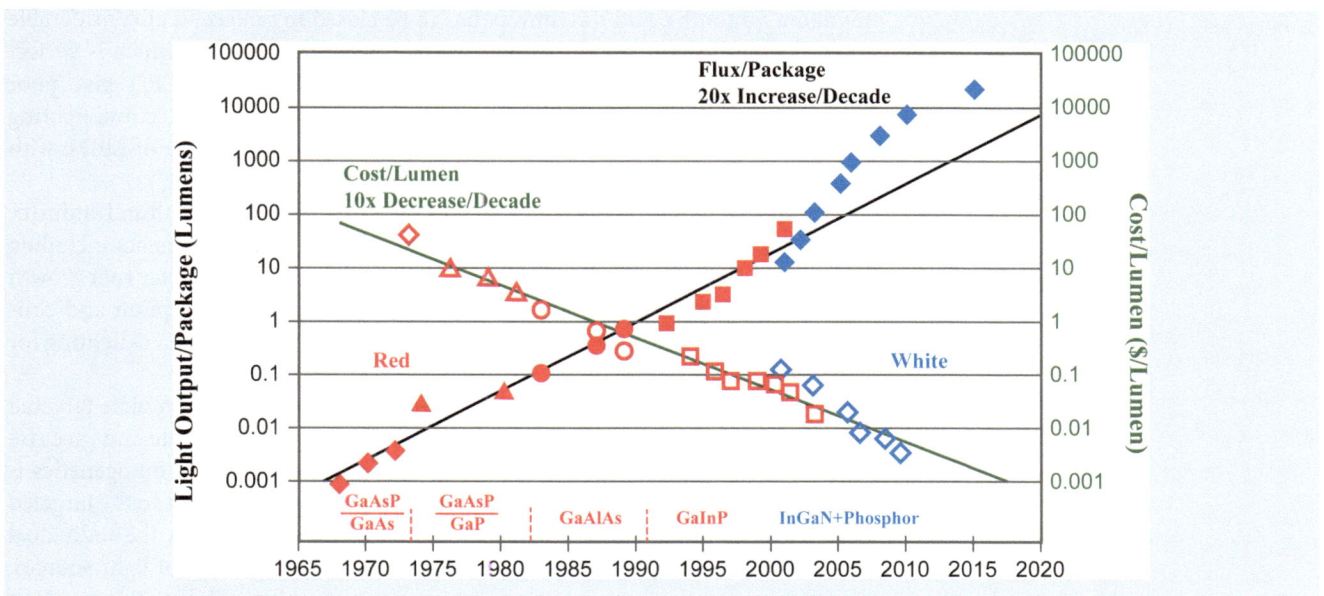

▫ **Fig. 5.20** Haltz's law showing that every decade, the cost per lumen falls by a factor of 10, and the amount of light generated per LED package increases by a factor of 20

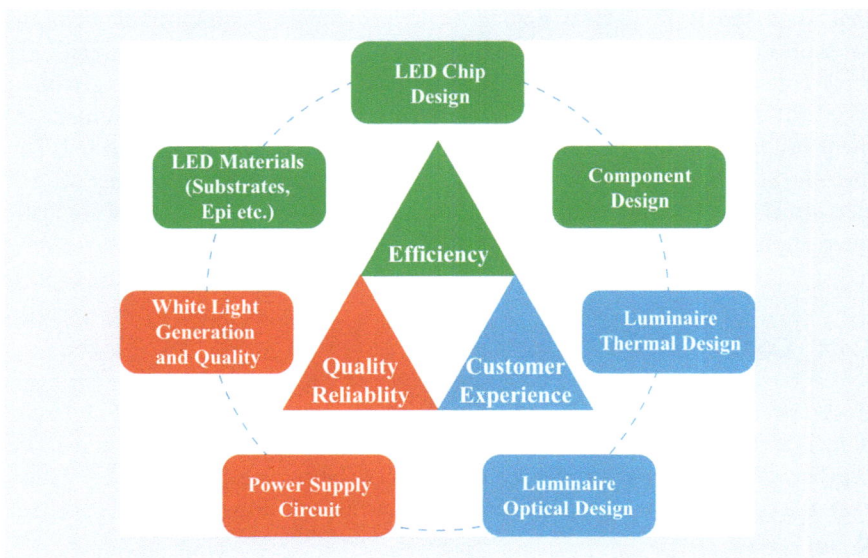

Fig. 5.21 Integrated systems approach to solid-state-lighting manufacturing (after Mark McClear, Cree, Inc., SSL manufacturing workshop, Vancouver, OR, June 2009)

5.7 LED Lighting Applications: The Present and Future

Although significant improvements are still expected, the present performance of nitride LEDs is nevertheless superior in many respects compared with conventional lighting. LEDs are compact, efficient, long-lasting and controllable, and are already widely used, for example (as shown in Fig. 5.22), as traffic signals, in large outdoor displays, as interior and exterior lighting in aircraft, cars and buses, as bulbs in flash lights and as backlighting for LCD TVs cell phones and displays. Due to their long lifetime, LEDs are also being fitted on airport runways, where the operational cost can be significantly lowered: traditional lighting on runways lasts for about 6 months, and the runway has to be closed to replace it, at considerable cost. The performance of LEDs improves at lower temperatures, which is perfect for illuminating refrigerated displays in supermarkets, where CFLs give poor performance because their efficiency is very low when cold. Architectural lighting also favours LEDs, which combine art due to the flexibility in use of LEDs, with energy saving and eco-friendliness.

The research and applications of LED lighting in the horticultural industry (Fig. 5.23) have also attracted a lot of attention [108], with benefits including better control of plant growth, increased yield, earlier flowering, faster root growth and more economical use of space. The lower electricity consumption and controllable light spectrum design are especially attractive features of LED lighting for horticulture applications.

Optogenetics is a new area of neuroscience that uses light to stimulate targeted neural pathways in the brain to uncover how neurons communicate and give rise to more complex brain functions. One key technical challenge in optogenetics is the realisation of a reliable implantable tool to precisely deliver light to the targeted neurons and to simultaneously record the electrical signals from the individual neuron. Such a neural probe requires the successful integration of light sources, detectors, sensors and other components on to an ultrathin cellular-scale injection needle, which can be inserted deep into the brain with minimum damage of tissue. Micro-LEDs are an ideal light source for this application due to the small size and controllable emission wavelength.

Fig. 5.22 Various application examples of LEDs as retail light, as backlighting for LCD TVs, as outdoor street light, as bicycle light, as exterior lighting in cars, as architectural lighting and as airport runway light. Images from Osram, with permission

Fig. 5.23 Philips LED lighting for fast-track growth in horticulture. Image from Philips, with permission

Visible light communication (VLC) technology, more recently referred to as Li-Fi (Light Fidelity), transmits data using light sources that modulate intensity faster than the human eye can perceive. Although still in its infancy, VLC is believed to be a future technology in wireless communication. LEDs are especially suitable for this application due to their fast switch on/off rate and long lifetime. By using an array of micro-LEDs, instead of conventional LEDs, the data transmit rate can be increased to more than 10 Gbps (Gigabits per second). An even bigger picture of this technology is to combine information displays, lighting and

Table 5.2 Comparison of LED light bulbs with conventional classic light bulbs

	Incandescent	Halogen incandescent	Compact florescent	LED light bulbs
Lumen	1100	1200	970	1055
Power (W)	75	70	15	13
Efficacy (lm/W)	15	17	65	81
Colour temperature (K)	2700	2800	2700	2700
Colour rendering index	100	100	81	80
Rated lifetime (h)	750–2000	2000	10,000	15,000
Mercury content (mg)	0	0	≤2	0
Warm-up time to 60 % light	Instant full light	Instant full light	5–40 s	Instant full light
Sales price	Banned [109]	£2.00	£5.00	£10.00

high-bandwidth communications in a single system, which will bring revolutionary solutions for machine-to-machine communications, smart homes and vehicles, mobile communications, imaging systems, personal security, healthcare and so on.

5.7.1 General Illumination and Energy Saving

Among all these exciting applications of LED lighting, those in general illumination, including residential, office, shop, hospitality, industrial, outdoor and architectural lighting, are the most relevant to our daily life and have the greatest energy saving potential. Both LED replacement classic light bulbs and LED fixtures are used for general illumination. A comparison of indoor LED light bulbs with other conventional light bulbs is given in ◘ Table 5.2, showing the advantages of LED lighting in energy saving without sacrificing performance. Due to its high initial cost, the current market penetration of LED lighting products is still very small. However, if the current trends in LED price and performance continue, LED lighting is projected to gain significant market penetration in USA, reaching 48 % of lumen-hour sales of the general illumination market by 2020, and 84 % by 2030.

Global population growth and urbanisation are increasing the overall demand for lighting products and the corresponding energy consumption by lighting. According to a recent US Department of Energy (DOE) report, lighting consumed ~18 % of total US electricity use in 2013, using approximately 609 TWh of electricity, or about 6.9 quads of source energy. LEDs are projected to reduce lighting energy consumption by 15 % in 2020 and by 40 % in 2030, saving 3.0 quads in 2030 alone. Assuming the current mix of generation power stations, these energy savings would reduce green house gas emission by approximately 180 million metric tons of carbon dioxide. Considering the global population growth, resource scarcity and climate change concerns, the development and adoption of LED technology is strategically important for a sustainable society.

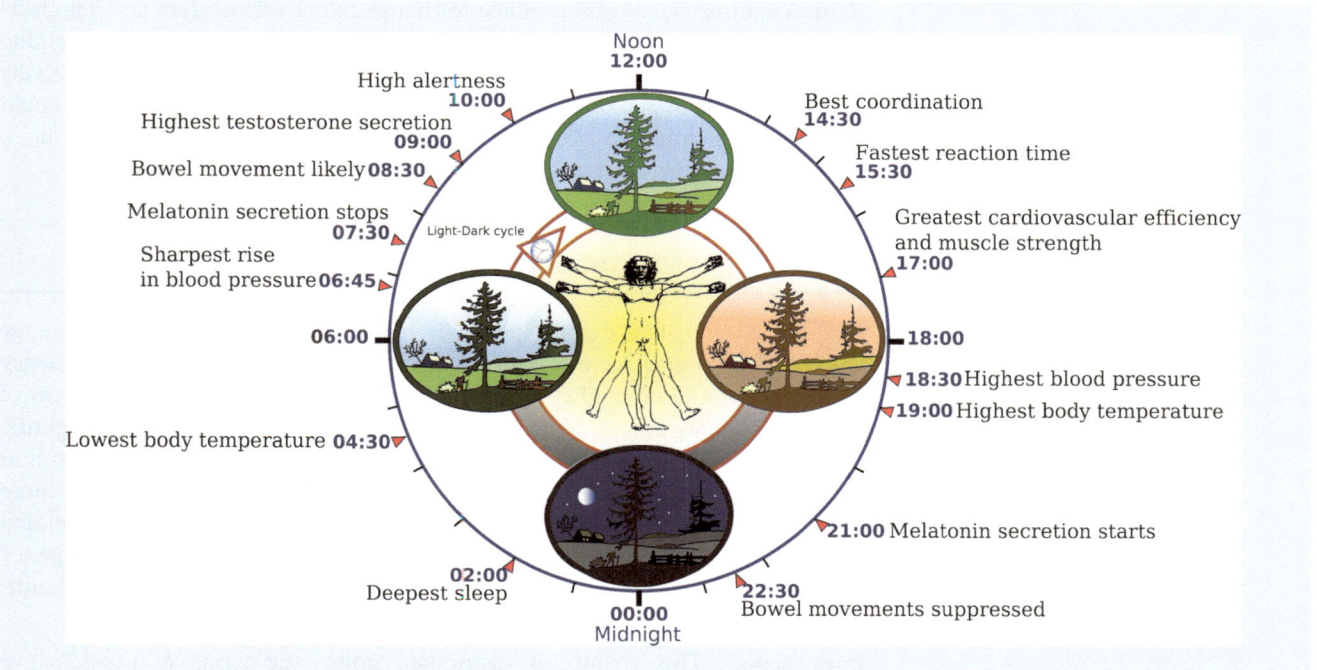

Fig. 5.24 Some features of the human circadian (24-h) biological clock. Image was done by Yassine Mrabet and uploaded by Addicted04 to Wikipedia, under creative CC BY-SA 3.0 free license

5.7.2 Circadian Rhythm Lighting

LED-based solid-state lighting is not just a replacement of traditional illuminations, but rather a multifunctional device we can use to improve our mood, health, productivity and much more. Because it is easily colour-tunable and dimmable, LED lighting is ideal to create circadian rhythm lighting that matches the needs of human biological cycles, or circadian rhythms, in the most effective and appropriate way.

Human beings are governed to some degree by an internal biological clock, called the circadian rhythm, as illustrated in ◘ Fig. 5.24. Light is the most powerful stimulus of the human body clock, and the timing of light exposure during the course of a day is responsible for how circadian rhythms are synchronised with the environment. For example, one of the best cures of 'jet lag' caused by circadian rhythm disruption is exposure to daylight to reset the body clock.

Modern industrialised society heavily relies on artificial lighting. Research tells us that circadian rhythm disruption through inappropriate artificial light causes many physical and mental health issues: fatigue, cancer, obesity, diabetes, depression, mood and sleep disorders, reduced physical and mental performance, reduced productivity and irritability are all related in some shape or form to a circadian system that isn't functioning properly. The most natural light is sunlight, which is dynamic and variable in brightness, colour temperature and spectral distribution during the day. Daylight provides bright blue-rich light in the early morning to deliver an alerting signal as we wake up and a warm, low-level light in the early evening to prepare our body to rest.

The dynamic features and spectral design flexibility of LED lighting enable the creation of personalised lighting to mitigate circadian rhythm disruption, optimise mood and visual experience, and improve our sense of wellbeing, in better ways than ever before. Combined with smart building control systems, LED circadian rhythm lighting can be programmed to change colour temperature and light level automatically, allowing for the indoor reproduction of natural outdoor lighting conditions. Some circadian rhythm lighting products are already commercially available, for example, on aircraft for long-haul flights. In the future, we could expect LED lighting to become more intelligent and closer to natural light, contributing strongly to our health and wellbeing, as well as energy saving.

5.8   Chapter Summary

LED-based solid-state lighting promises to provide a high quality and energy efficient light source for our daily life. With continuous advances in efficiency and reductions in cost, LED lighting is on course to be the dominant form of lighting in homes, offices, cities and transport throughout the world. LED lighting is more than an energy efficient alternative to conventional light sources; it is suitable to create circadian rhythm lighting that can make us healthier and more productive. LED lighting is also intelligent and could interface with building management systems, transmit high-speed wireless data, fine-tune occupancy and functional sensing, and is an important integral part of our future smart home.

References

1. Holonyak N, Bevacqua SF (1962) Coherent (visible) light emission from $Ga(As_{1-x}P_x)$ junctions. Appl Phys Lett 1(4):82–83. doi:▶ 10.1063/1.1753706
2. Nakamura S, Senoh M, Mukai T (1993) High-power InGaN/GaN double-heterostructure violet light emitting diodes. Appl Phys Lett 62:2390–2392. doi:▶ 10.1063/1.109374
3. Nakamura S, Pearton S, Fasol G (1997) The blue laser diode. Springer, Berlin
4. Vurgaftman I, Meyer JR, Ram-Mohan LR (2001) Band parameters for III-V compound semiconductors and their alloys. J Appl Phys 89:5815–5875. doi:▶ 10.1063/1.1368156
5. Vurgaftman I, Meyer JR (2003) Band parameters for nitrogen-containing semiconductors. J Appl Phys 94:3675–3696. doi:▶ 10.1063/1.1600519
6. Saitoh T, Kumagai M, Wang H, Tawara T, Nishida T, Akasaka T, Kobayashi N (2003) Highly reflective distributed Bragg reflectors using a deeply semiconductor/air grating for InGaN/GaN laser diodes. Appl Phys Lett 82:4426–4428. doi:▶ 10.1063/1.1586992
7. Wu J, Walukiewicz W, Yu KM, Shan W, Ager JW III, Haller EE, Lu H, Schaff WJ, Metzger WK, Kurtz S (2003) Superior radiation resistance of In1-xGaxN alloys: full-solar-spectrum photovoltaic materials system. J Appl Phys 94:6477–6482. doi:▶ 10.1063/1.1618353
8. Munoz E, Monroy E, Pau JL, Calle F, Omnes F, Gibart P (2001) III nitrides and UV detection. J Phys: Condens Matter 13:7115–7137. doi:▶ 10.1088/0953-8984/13/32/316
9. Xing H, Keller S, Wu YF, McCarthy L, Smochkova IP, Buttari D, Coffie R, Green DS, Parish G, Heikman S, Shen L, Zhang N, Xu JJ, Keller BP, DenBaars SP, Mishra UK (2001) Gallium

nitride based transistors. J Phys: Condens Matter 13:7139–7157. doi:▶ 10.1088/0953-8984/13/32/317

10. Mishra UK, Parikh P, Wu YF (2002) AlGaN/GaN HEMTs—an overview of device operation and applications. Proc IEEE 90:1022–1031. doi:▶ 10.1109/JPROC.2002.1021567

11. Schubert EF (2006) Light-emitting diodes, 2nd edn. Cambridge University Press, Cambridge

12. Laubsch A, Sabathil M, Baur J, Peter M, Hahn B (2010) High-power and high-efficiency InGaN-based light emitters. IEEE Trans Electron Devices 57:79–87. doi:▶ 10.1109/TED.2009.2035538

13. Popovici G, Morkoç H, Noor Mohammed S (1998) Deposition and properties of group III nitrides by molecular beam epitaxy. In: Gil B (ed) Group III nitride semiconductor compounds: physics and applications. Oxford University Press, Oxford, p 19

14. Porowski S, Grzegory I (1997) Growth of GaN single crystals under high nitrogen pressure. In: Pearton SJ (ed) GaN and related materials. Overseas, Amsterdam, p 295

15. Koleske DD, Coltrin ME, Cross KC, Mitchell CC, Allerman AA (2004) Understanding GaN nucleation layer evolution on sapphire. J Cryst Growth 273:86–99. doi:▶ 10.1016/j.jcrysgro.2004.08.126

16. Figge S, Bottcher T, Einfeldt S, Hommel D (2000) In situ and ex situ evaluation of the film coalescence for GaN growth on GaN nucleation layers. J Cryst Growth 221:262–266. doi:▶ 10.1016/S0022-0248(00)00696-5

17. Kappers MJ, Datta R, Oliver RA, Rayment FDG, Vickers ME, Humphreys CJ (2007) Threading dislocation reduction in (0001) GaN thin films using SiN$_x$ interlayers. J Cryst Growth 300:70–74. doi:▶ 10.1016/j.jcrysgro.2006.10.205

18. Kappers MJ, Moram MA, Zhang Y, Vickers ME, Barber ZH, Humphreys CJ (2007) Interlayer methods for reducing the dislocation density in gallium nitride. Physica B 401–402:296–301. doi:▶ 10.1016/j.physb.2007.08.170

19. Cherns D, Henley SJ, Ponce FA (2001) Edge and screw dislocations as non-radiative centers in InGaN/GaN quantum well luminescence. Appl Phys Lett 78:2691–2693. doi:▶ 10.1063/1.1369610

20. Chichibu SF, Uedono A, Onuma T, Haskell BA, Chakraborty A, Koyama T, Fini PT, Keller S, DenBaars SP, Speck JS, Mishra UK, Nakamura S, Yamaguchi S, Kamiyama S, Amano H, Akasaki I, Han J, Sota T (2006) Origin of defect-insensitive emission probability in In-containing (Al, In, Ga)N alloy semiconductors. Nat Mater 10:810–816. doi:▶ 10.1038/nmat1726

21. Graham DM, Soltani-Vala A, Dawson P, Smeeton TM, Barnard JS, Kappers MJ, Humphreys CJ, Thrush EJ (2005) Optical and microstructural studies of InGaN/GaN single-quantum-well structures. J Appl Phys 97:103508. doi:▶ 10.1063/1.1897070

22. Hammersley S, Badcock TJ, Watson-Parris D, Godfrey MJ, Dawson P, Kappers MJ, Humphreys CJ (2011) Study of efficiency droop and carrier localization in an InGaN/GaN quantum well structure. Phys Status Solidi C 8:2194–2196. doi:▶ 10.1002/pssc.201001001

23. Humphreys CJ (2007) Does in form in-rich clusters in InGaN quantum wells? Philos Mag 87:1971–1982. doi:▶ 10.1080/14786430701342172

24. Oliver RA, Bennett SE, Zhu T, Beesley DJ, Kappers MJ, Saxey DW, Cerezo A, Humphreys CJ (2010) Microstructural origins of localisation in InGaN quantum wells. J Phys D Appl Phys 43:354003. doi:▶ 10.1088/0022-3727/43/35/354003

25. Smeeton TM, Kappers MJ, Barnard JS, Vickers ME, Humphreys CJ (2003) Electron-beam-induced strain within InGaN quantum wells: False indium "cluster" detection in the transmission electron microscope. Appl Phys Lett 83:5419–5421. doi:▶ 10.1063/1.1636534

26. Watson-Parris D, Godfrey MJ, Oliver RA, Dawson P, Galtrey MJ, Kappers MJ, Humphreys CJ (2010) Energy landscape and carrier wave-functions in InGaN/GaN quantum wells. Phys Status Solidi C 7:2255–2258. doi:▶ 10.1002/pssc.200983516

27. Aumer ME, LeBoeuf SF, Bedair SM, Smith M, Lin JY, Jiang HX (2000) Effects of tensile and compressive strain on the luminescence properties of AlInGaN/InGaN quantum well structures. Appl Phys Lett 77:821–823. doi:▶ 10.1063/1.1306648

28. Aumer ME, LeBoeuf SF, Moody BF, Bedair SM, Nam K, Lin JY, Jiang HX (2002) Effects of tensile, compressive, and zero strain on localized states in AlInGaN/InGaN quantum-well structures. Appl Phys Lett 80:3099–3101. doi:▶ 10.1063/1.1469219

29. Aumer ME, LeBoeuf SF, Moody BF, Bedair SM (2001) Strain-induced piezoelectric field effects on light emission energy and intensity from AlInGaN/InGaN quantum wells. Appl Phys Lett 79:3803–3805. doi:▶ 10.1063/1.1418453

30. Leroux M, Grandjean N, Massies J, Gil B, Lefebvre P, Bigenwald P (1999) Barrier-width dependence of group-III nitrides quantum-well transition energies. Phys Rev B 60:1496–1499. doi:▶ 10.1103/PhysRevB.60.1496

31. McAleese C, Costa PMFJ, Graham DM, Xiu H, Barnard JS, Kappers MJ, Dawson P, Godfrey MJ, Humpherys CJ (2006) Electric fields in AlGaN/GaN quantum well structures. Phys Status Solidi B 243:1551–1559. doi:► 10.1002/pssb.200565382

32. Wetzel C, Takeuchi T, Amano H, Akasaki I (1999) Piezoelectric Franz–Keldysh effect in strained GaInN/GaN heterostructures. J Appl Phys 85:3786–3791. doi:► 10.1063/1.369749

33. Kuroda T, Tackeuchi A (2002) Influence of free carrier screening on the luminescence energy shift and carrier lifetime of InGaN quantum wells. J Appl Phys 92:3071–3074. doi:► 10.1063/1.1502186

34. Riblet P, Hirayama H, Kinoshita A, Hirata A, Sugano T, Aoyagi Y (1999) Determination of photoluminescence mechanism in InGaN quantum wells. Appl Phys Lett 75:2241–2243. doi:► 10.1063/1.124977

35. Johnston CF, Kappers MJ, Humphreys CJ (2009) Microstructural evolution of nonpolar (11–20) GaN grown on (1–102) sapphire using a 3D-2D method. J Appl Phys 105:073102. doi:► 10.1063/1.3103305

36. Zhong H, Tyagi A, Fellows N, Wu F, Chung RB, Saito M, Fujito K, Speck JS, DenBaars SP, Nakamura S (2007) High power and high efficiency blue light emitting diode on free-standing semipolar (10-1-1) bulk GaN substrate. Appl Phys Lett 90:233504. doi:► 10.1063/1.2746418

37. Yamada H, Iso K, Saito M, Masui H, Fujito K, DenBaars SP, Nakamura S (2008) Compositional dependence of nonpolar m-plane InxGa1-xN/GaN light emitting diodes. Appl Phys Express 1:041101. doi:► 10.1143/APEX.1.041101

38. Nakamura S, Senoh M, Mukai T (1991) Highly P-typed Mg-doped GaN films grown with GaN buffer layers. Jpn J Appl Phys 30:L1708–L1711. doi:► 10.1143/JJAP.30.L1708

39. Doverspike K, Pankove JI (1998) Doping in the III-nitrides. SEM SEMIMET 50:259–277

40. Kaufmann U, Kunzer M, Maier M, Obloh H, Ramakrishnan A, Santic B, Schlotter P (1998) Nature of the 2.8 eV photoluminescence band in Mg doped GaN. Appl Phys Lett 72:1326–1328. doi:► 10.1063/1.120983

41. Nakamura S, Iwasa N, Senoh M, Mukai T (1992) Hole compensation mechanism of P-type GaN films. Jpn J Appl Phys 31:1258–1266. doi:► 10.1143/JJAP.31.1258

42. Nakamura S, Mukai T, Senoh M, Isawa N (1992) Thermal annealing effects on P-type Mg-doped GaN films. Jpn J Appl Phys 31:L139–L142. doi:► 10.1143/JJAP.31.L139

43. Amano H, Kito M, Hiramatsu K, Akasaki I (1989) P-type conduction in Mg-doped GaN treated with low-energy electron beam irradiation (LEEBI). Jpn J Appl Phys 28:L2112–L2114. doi:► 10.1143/JJAP.28.L2112

44. Obloh H, Bachem KH, Kaufmann U, Kunzer M, Maier M, Ramakrishnan A, Schlotter P (1998) Self-compensation in Mg doped p-type GaN grown by MOCVD. J Cryst Growth 195:270–273. doi:► 10.1016/S0022-0248(98)00578-8

45. Schubert EF, Greishaber W, Goepfert ID (1996) Enhancement of deep acceptor activation in semiconductors by superlattice doping. Appl Phys Lett 69:3737–3739. doi:► 10.1063/1.117206

46. Kozodoy P, Hansen M, DenBaars SP, Mishra UK (1999) Enhanced Mg doping efficiency in Al0.2Ga0.8N/GaN superlattices. Appl Phys Lett 74:3681–3683. doi:► 10.1063/1.123220

47. Kozodoy P, Smorchkova YP, Hansen M, Xing H, DenBaars SP, Mishra UK, Saxler AW, Perrin R, Mitchel WC (1999) Polarization-enhanced Mg doping of AlGaN/GaN superlattices. Appl Phys Lett 75:2444–2446. doi:► 10.1063/1.125042

48. Yasan A, McClintock R, Darvish SR, Lin Z, Mi K, Kung P, Razeghi M (2002) Characteristics of high-quality p-type AlxGa1 − xN/GaN superlattices. Appl Phys Lett 80:2108–2110. doi:► 10.1063/1.1463708

49. Kim JK, Waldron EL, Li YL, Gessmann T, Schubert EF, Jang HW, Lee JL (2004) P-type conductivity in bulk AlxGa1 − xN and AlxGa1 − xN/AlyGa1 − yN superlattices with average Al mole fraction >20%. Appl Phys Lett 84:3310–3312. doi:► 10.1063/1.1728322

50. Cho J, Schubert EF, Kim JK (2013) Efficiency droop in light-emitting diodes: challenges and countermeasures. Laser Photonics Rev 7(3):408–421. doi:► 10.1002/lpor.201200025

51. Galler B, Lugauer HJ, Binder M, Hollweck R, Folwill Y, Nirschl A, Gomez-Iglesias A, Hahn B, Wagner J, Sabathil M (2013) Experimental determination of the dominant type of Auger recombination in InGaN quantum wells. Appl Phys Express 6:112101. doi:► 10.7567/APEX.6.112101

52. Kioupakis E, Rinke P, Delaney KT, Van de Walle CG (2011) Indirect Auger recombination as a cause of efficiency droop in nitride light-emitting diodes. Appl Phys Lett 98:161107. doi:► 10.1063/1.3570656

53. Meyaard DS, Lin GB, Cho J, Schubert EF, Shim H, Han SH, Kim MH, Sone C, Kim YS (2013) Identifying the cause of the efficiency droop in GaInN light-emitting diodes by correlating the onset of high injection with the onset of the efficiency droop. Appl Phys Lett 102:251114. doi:► 10.1063/1.4811558

54. Monemar B, Sernelius BE (2007) Defect related issues in the "current roll-off" in InGaN based light emitting diodes. Appl Phys Lett 91:181103. doi:► 10.1063/1.2801704

55. Rozhansky IV, Zakheim DA (2006) Analysis of the causes of the decrease in the electroluminescence efficiency of AlGaInN light-emitting-diode heterostructures at high pumping density. Semiconductors 40:839–845. doi:► 10.1134/S1063782606070190

56. Shen YC, Mueller GO, Watanabe S, Gardner NF, Munkholm A, Krames MR (2007) Auger recombination in InGaN measured by photoluminescence. Appl Phys Lett 91:141101. doi:► 10.1063/1.2785135

57. El-Masry NA, Piner EL, Liu SX, Bedair SM (1998) Phase separation in InGaN grown by metalorganic chemical vapor deposition. Appl Phys Lett 72:40–42. doi:► 10.1063/1.120639

58. Yang Y, Cao XA, Yan CH (2009) Rapid efficiency roll-off in high-quality green light-emitting diodes on freestanding GaN substrates. Appl Phys Lett 94:041117. doi:► 10.1063/1.3077017

59. Schubert MF, Xu JR, Kim JK, Schubert EF, Kim MH, Yoon SK, Lee SM, Sone CL, Sakong T, Park YJ (2008) Polarization-matched GaInN/AlGaInN multi-quantum-well light-emitting diodes with reduced efficiency droop. Appl Phys Lett 93:041102. doi:► 10.1063/1.2963029

60. Iso K, Yamada H, Hirasawa H, Fellows N, Saito M, Fujito K, DenBaars SP, Speck JS, Nakamura S (2007) High brightness blue InGaN/GaN light emitting diode on nonpolar m-plane bulk GaN substrate. Jpn J Appl Phys 46:L960–L962. doi:► 10.1143/JJAP.46.L960

61. Xu JR, Schubert MF, Noemaun AN, Zhu D, Kim JK, Schubert EF, Kim MH, Chung HJ, Yoon S, Sone C, Park Y (2009) Reduction in efficiency droop, forward voltage, ideality factor, and wavelength shift in polarization-matched GaInN/GaInN multi-quantum-well light-emitting diodes. Appl Phys Lett 94:011113. doi:► 10.1063/1.3058687

62. Kim MH, Schubert MF, Dai Q, Kim JK, Schubert EF, Piprek J, Park Y (2007) Origin of efficiency droop in GaN-based light-emitting diodes. Appl Phys Lett 91:183507. doi:► 10.1063/1.2800290

63. Xie JQ, Ni XF, Fan Q, Shimada R, Özgür Ü, Morkoç H (2008) On the efficiency droop in InGaN multiple quantum well blue light emitting diodes and its reduction with p-doped quantum well barriers. Appl Phys Lett 93:121107. doi:► 10.1063/1.2988324

64. Hammersley S, Watson-Parris D, Dawson P, Godfrey MJ, Badcock TJ, Kappers MJ, McAleese C, Oliver RA, Humphreys CJ (2012) The consequences of high injected carrier densities on carrier localization and efficiency droop in InGaN/GaN quantum well structures. J Appl Phys 111:083512. doi:► 10.1063/1.3703062

65. Nakamura S, Krames MR (2013) History of gallium-nitride-based light-emitting diodes for illumination. Proc IEEE 101(10):2211–2220. doi:► 10.1109/JPROC2013.2274929

66. Nakamura S, Mukai T, Senoh M (1994) Candela-class high-brightness InGaN/AlGaN double-heterostructure blue-light-emitting diodes. Appl Phys Lett 64:1687–1689. doi:► 10.1063/1.111832

67. Steigerwald DA, Bhat JC, Collins D, Fletcher RM, Holcomb MO, Ludowise MJ, Martin PS, Rudaz SL (2002) Illumination with solid state lighting technology. IEEE J Sel Top Quantum Electron 8(2):310–320. doi:► 10.1109/2944.999186

68. Krames MR, Shchekin OB, Mueller-Mach R, Mueller GO, Zhou L, Harbers G, Craford MG (2007) Status and future of high-power light-emitting diodes for solid-state lighting. IEEE J Disp Technol 3(2):160–175. doi:► 10.1109/JDT.2007.895339

69. Fujii T, Gao Y, Sharma R, Hu EL, DenBaars SP, Nakamura S (2004) Increase in the extraction efficiency of GaN-based light-emitting diodes via surface roughening. Appl Phys Lett 84:855–857. doi:► 10.1063/1.1645992

70. Narukawa Y, Ichikawa M, Sanga D, Sano M, Mukai T (2010) White light emitting diodes with super-high luminous efficacy. J Phys D Appl Phys 43:354002. doi:► 10.1088/0022-3727/43/35/354002

71. David A, Hurni CA, Aldaz RI, Cich MJ, Ellis B, Huang K, Steranka FM, Krames MR (2014) High light extraction efficiency in bulk-GaN based volumetric violet light-emitting diodes. Appl Phys Lett 105:231111. doi: ► 10.1063/1.4903297

72. Schlotter P, Schmidt R, Schneider J (1997) Luminescence conversion of blue light emitting diodes. Appl Phys A 64(4):417–418. doi:► 10.1007/s003390050498

73. Mueller-Mach R, Mueller GO, Krames MR, Trottier T (2002) High-power phosphor-converted light-emitting diodes based on III-nitrides. IEEE J Sel Top Quantum Electron 8(2):339–345. doi:► 10.1109/2944.999189

74. MacAdam DL (1943) Specification of small chromaticity differences. J Opt Soc Am 33:18–26

75. Murphy TW (2012) Maximum spectral luminous efficacy of white light. J Appl Phys 111:104909. doi:► 10.1063/1.4721897

76. CREE (2014) ► http://www.cree.com/News-and-Events/Cree-News/Press-Releases/2014/March/300LPW-LED-barrier

5

77. Zhao LX, Thrush EJ, Humphreys CJ, Phillips WA (2008) Degradation of GaN-based quantum well light-emitting diodes. J Appl Phys 103:024501. doi:▶ 10.1063/1.2829781

78. Amano H, Sawaki N, Akasaki I, Toyoda Y (1986) Metalorganic vapor phase epitaxial growth of a high quality GaN film using an AlN buffer layer. Appl Phys Lett 48:353–355. doi:▶ 10.1063/1.96549

79. CIE data of 1931 and 1978 (1978) Available at ▶ http://cvision.ucsd.edu and ▶ http://www.cvrl.org

80. Juza R, Hahn H (1938) Über die Kristallstrukturen von Cu3N, GaN und InN Metallamide und Metallnitride. Z Anorg Allgem Chem 239:282–287. doi:▶ 10.1002/zaac.19382390307

81. Lighting Europe (2013) Guide for the application of the commission regulation (EU) No.1194/2012 setting ecodesign requirements for directional lamps, light emitting diode lamps and related equipment. ▶ http://www.lightingeurope.org/uploads/files/LightingEurope_Guide_-_Regulation_1194_2012_ECODESIGN_Version_1_17_July_2013.pdf

82. MacAam DL (ed) (1993) Colorimetry—fundamentals. SPIE Optical Engineering Press, Bellingham, WA

83. Maruska HP, Tietjen JJ (1969) The preparation and properties of vapor-deposited single-crystal GaN. Appl Phys Lett 15:327–329. doi:▶ 10.1063/1.1652845

84. Nakamura S (1991) GaN growth using GaN buffer layer. Jpn J Appl Phys 30:L1705–L1707. doi:▶ 10.1143/JJAP.30.L1705

85. Nakamura S, Senoh M, Iwasa N, Nagahama S (1995) High-brightness InGaN blue, green and yellow light-emitting diodes with quantum well structures. Jpn J Appl Phys 34:L797–L799. doi:▶ 10.1143/JJAP.34.L797

86. Nakamura S, Senoh M, Mukai T (1993) P-GaN/N-InGaN/N-GaN double-heterostructure blue-light-emitting diodes. Jpn J Appl Phys 32:L8–L11. doi:▶ 10.1143/JJAP.32.L8

87. Pankove JI, Miller EA, Berkeyheiser JE (1972) GaN blue light-emitting diodes. J Lumin 5:84–86

88. Razeghi M (2002) Short-wavelength solar-blind detectors—status, prospects, and markets. Proc IEEE 90(6):1006–1014. doi:▶ 10.1109/JPROC.2002.1021565

89. Zhu D, Wallis DJ, Humphreys CJ (2013) Prospects of III-nitride optoelectronics grown on Si. Rep Prog Phys 76:106501. doi:▶ 10.1088/0034-4885/76/10/106501

90. Soraa (2015) ▶ https://www.soraa.com/news_releases/32

91. Philips (2014) ▶ http://www.philips.com/a-w/about/news/archive/standard/news/press/2014/20140509-Philips-and-Green-Sense-Farms-usher-in-new-era-of-indoor-farming.html

92. ▶ https://en.wikipedia.org/wiki/Phase-out_of_incandescent_light_bulbs

Modern Electron Optics and the Search for More Light: The Legacy of the Muslim Golden Age

Mohamed M. El-Gomati

M.M. El-Gomati (✉)
Department of Electronics, University of York, Heslington, York YO10 5DD, UK
e-mail: Mohamed.elgomati@york.ac.uk

© The Author(s) 2016
M.D. Al-Amri et al. (eds.), *Optics in Our Time*, DOI 10.1007/978-3-319-31903-2_6

6.1   Introduction

This chapter is a brief survey of some of the fundamental premises of electron optics with an emphasis on electron microscopy and its relevance to modern life. The reader is introduced to the basic concept of electron microscopy and the parallels to the more familiar optical microscopes that depend upon the use of light optics. Some recent developments in the technique will be surveyed. The UN General Assembly proclaimed 2015 as the international year of light and light-based technologies, where electron microscopy has played and continues to play a pivotal role in the development of efficient, environmentally friendlier alternative light sources to the incandescent light bulb. These developments follow a scientific method of enquiry that had its roots laid down in the eleventh century by the Arab scholar, Al-Hassan Ibn al-Haytham, known in the west by his Latinised name as 'Alhazen'. Ibn al-Haytham also discovered and correctly explained and described a puzzling effect observed in the field of optics; known as spherical aberration. The correction of this lens-related deficiency in electron microscopy is shown to have been fundamental in developing a new class of efficient Light Emitting Diodes (LED) light bulbs. The occurrence of 'spherical aberration' is also equally important in other imaging devices such as telescopes, and an example showing its pivotal role in the Hubble Space Telescope (HST) is subsequently demonstrated.

6.2   Electron Optics

The field of electron optics is concerned with the formation of a fine beam of electrons to use in a variety of applications as listed below. In one such relevant example; electron microscopy, the electron beam is directed to bombard a solid specimen for the purpose of learning more about the properties of such a sample, but with the particular emphasis on obtaining such information at the smallest possible dimensions. This electron-solid interaction results in an array of signals, or by-products, ranging from electrons to photons. The collection of one of these signals has often led to a specialised technique associated with such a signal. This will be explained in more details later on. There is estimated to be about 100,000 instruments around the world that use electrons to map the features of a given specimen.

Another important technical area concerns the use of a focussed electron beam, which is known as electron beam lithography (EBL). In such applications, the smallest device components like diodes and transistors; being small is essential for high-speed electronics, are drawn using small-diameter electron beams for producing the required masks used in transferring a given pattern composed of these components to a semiconductor wafer. Electron beams have also been used in direct writing on devices, in locally functionalising a surface, as well as in electron-beam-induced deposition of materials, to name but a few relevant examples.

In addition to, the use of a fine beam of electrons to map the surfaces of samples, small-diameter energetic electron beams have also been used in welding metals. The instruments employed for such applications share with the imaging instruments a great deal of commonalities, but the emphasis for welding of metallic materials is on producing an energetic spot of electrons with sufficient current to fuse together the parts of the samples being welded.

Another equally important area that also requires the production of a small-diameter electron beam, albeit not as fine as in electron microscopy, is in conventional television sets and in cathode ray tubes (CRT) used in oscilloscopes for scientific applications. Neither of these methods requires the production of a small

Chapter 6 · Modern Electron Optics and the Search for More Light...

121   **6**

beam diameter, because the human eye can only resolve features of the order of ~200 micro metres (μm) (1 μm is one of a millionth of a metre whilst 1 nm is one billionth of a metre).

As can be appreciated from the foregoing discussion, the subject of electron beam optics is vast in nature and almost impossible to cover all its aspects in a non-specialised book chapter such as this. I will therefore confine the material presented to two main parts: the first is an introductory review of some of the more widely used instruments that are employed to spatially map the constituents (elements) of the solids under investigation. The depth of information gathered in this exercise ranges from the top atomic layer of such a sample down to depths of few microns below its surface. Second, I will briefly turn my attention to some of the latest developments in electron microscopy, and specifically the techniques that have been introduced in the last few years to advance this method. This second part will include some important highlights of the use of electron microscopy in relation to light and light-related techniques as a mark of this year's celebration.

6.3   Parallels with Optical Microscopy

In order to better appreciate electron microscopy, it is helpful to draw a parallel with a much older technique that is also used to image a variety of samples; including solids and liquids; the optical microscope. This instrument is perhaps better known and more widely used, given that many of us can claim to have used one either during science-based education or at work. This fascinating instrument has also over many decades seen extensive research and development to optimise it to the state we are familiar with today. Such advances have in many ways paved the way for the development of the more recent technique, the electron microscope, which has somehow been easier and faster to develop, particularly in its early days. There are, indeed, many distinctive similarities between the propagation of photons (light) and electrons, where examples of ray (photon) optics can be used to illustrate and understand electron optics. Parameters such as focal distance, linear magnification and angular magnification, which define the ray optical systems of optical microscopy, can therefore be understood.

In optical microscopy, we observe an important phenomenon: the way photons behave as they cross a boundary between two media of differing refractive indices causes the incident ray to change direction through a different angle from that of incidence. This phenomenon was formulated, and has come to be known to us, as *Snell's Law* (Willebrord Snellius 1580–1626—with René Descartes having also independently reached the same formulation in his 1637 essay, 'Dioptrics.'). However, recent research findings suggest that this relationship was in fact discovered more than six centuries earlier than the reports of Snellius by an Arab scientist, Abu Sa'd al-Alaa Ibn Sahl (940–1000 CE), commonly known as Ibn Sahl, who resided in Baghdad, capital of today's Iraq. The relatively recent discovery of a manuscript by the French Arab historian of science, Rashed [1] has unveiled the work of Ibn Sahl, who developed this relationship while studying the properties of burning mirrors and lenses. Ibn Sahl showed for the first time the refraction of light by glass lenses, as depicted in ◘ Fig. 6.1. He went on to describe an elaborate instrument design which could be used to manufacture glass lenses of regular and varying shapes [1]. Astonishingly, this discovery went unnoticed and uncredited for almost 1000 years!

Snell's law describes the relationship between the incidence angle α and the refractive angle β and speeds v_1 and v_2 of a photon as it crosses the boundary of two media of differing refractive indices, n_1 and n_2, respectively. This is shown diagrammatically in ◘ Fig. 6.2 and can algebraically be stated as:

6

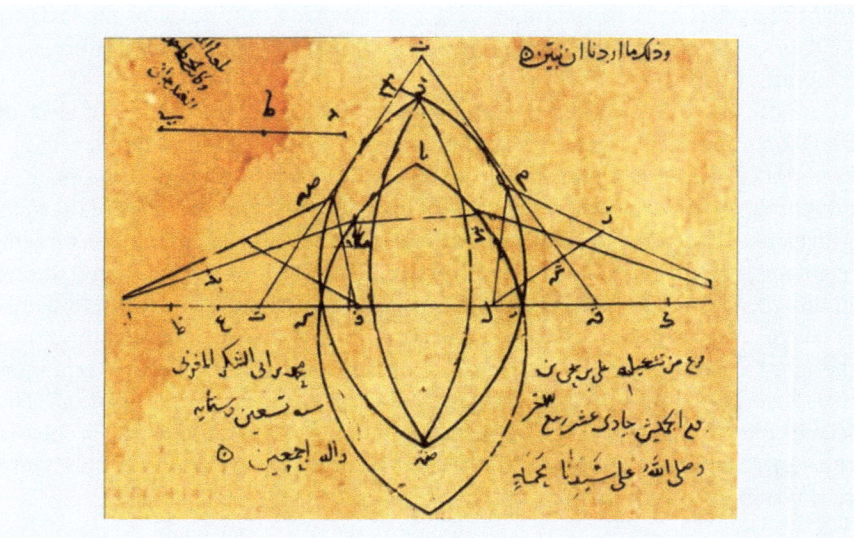

Fig. 6.1 A diagram of the refraction in lenses, by the Arab scientist/mathematician, Abu Sa'd al-Alaa ibn Sahl (d.960CE), who lived in Baghdad, Iraq. The manuscript, dated Thursday 11 Rabi' al-Akhir 690 AH (i.e. 12 April 1291 CE), was discovered and edited by Rashed [1], with permission from the author

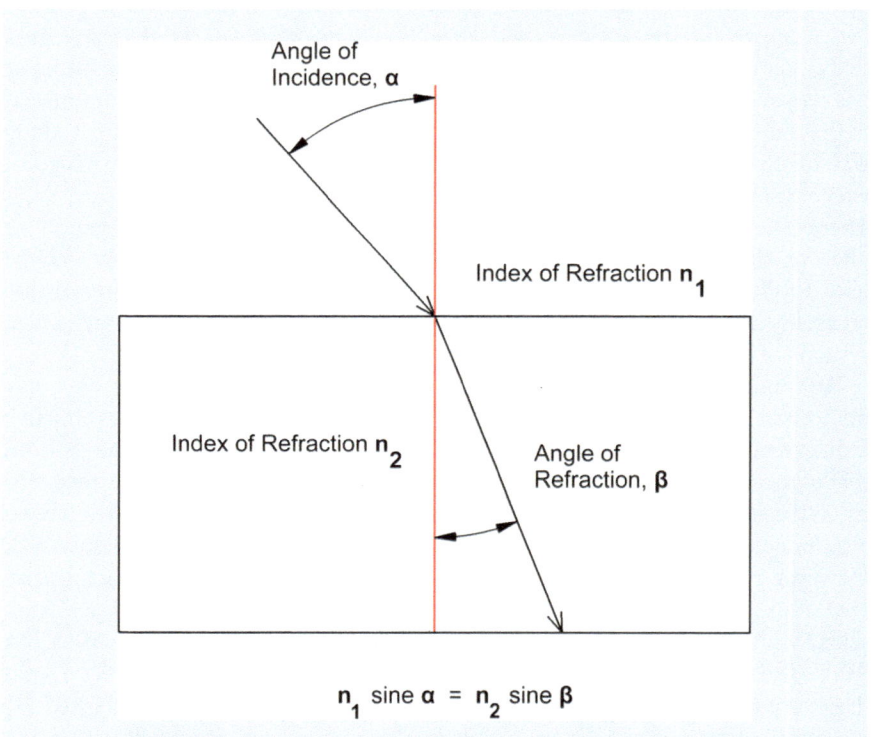

Fig. 6.2 A diagram showing the refraction of light as it crosses a boundary between two media of refractive indices n_1 and n_2

$$\sin \alpha / \sin \beta = n_1/n_2 = v_1/v_2 \tag{6.1}$$

In a simplified way, the behaviour of electrons when moving through a boundary of two areas of differing potentials could also be described by a similar relationship to (6.1) above. Consider the speed $(v_{1,2})$ of an electron of mass (m)

Chapter 6 · Modern Electron Optics and the Search for More Light...

123

6

and charge (*e*) in relation to the potential acting upon it as it moves between two areas of potentials V_1 and V_2:

$$\tfrac{1}{2}\, mv_1^2 + eV_1 = \tfrac{1}{2}\, mv_2^2 + eV_2 \qquad\qquad (6.2)$$

The electric field only applies a perpendicular force on the moving electron. This then causes only the electron perpendicular momentum to be affected; so, as it crosses the (potential) boundary, we can write:

$$mv_1 \sin \alpha_1 = mv_2 \sin \alpha_2 \qquad\qquad (6.3)$$

The above equation can then be rewritten to get:

$$\sin \alpha_1 / \sin \alpha_2 = \sqrt{(V_2/V_1)} \qquad\qquad (6.4)$$

Equations (6.1) and (6.4) have a close resemblance with $\sqrt{(V_{1,2})}$ acting as the refractive index. Such close relationships allow one to reflect on the same optical theory of light optics, and to expect to gain similar results up to a certain extent.

This chapter, however, is mainly concerned with the electron microscope family of instruments. Therefore, whilst there has been a great deal of developments and advancements in optical microscopy in the last few decades, they will not be covered here.

6.4 JJ Thomson and His Discovery, the Electron

It would be unfair, if not incomplete, to discuss electron microscopy without mentioning its pivotal component; *the electron*. Many questions arise: so what is this electron, and how was it discovered and by whom? How do we generate electrons for use in microscopy? Are there different electron sources in use today? What are these and is this important?

The discovery of the electron in 1897 by Joseph John Thomson, widely referred to as JJ Thomson, came amidst intensive research activities in a related instrument to the present electron microscope; namely, the CRT. In Cambridge, where JJ Thomson was just appointed as a Professor of Experimental Physics at the Cavendish Laboratory (the Physics Department at the University of Cambridge, UK), he pursued his research into electrical discharges in CRT. The CRT is a vacuum tube which was made of glass in its early days, with a heated filament acting as the cathode opposite a fluorescent screen acting as the anode. The work carried out in one form of the CRT, called the Crookes tube, has shown these rays to cast a shadow on the glowing walls of the glass tubes and on the fluorescent screens opposite and, as such, move in straight lines. Being charged particles, it meant that they could also be deflected by either electrical or magnetic fields (for more details on the history of the cathode rays, see the Flash of the cathode rays—Dahl [2]).

At that time, there were many opinions on the nature of cathode rays as being waves, atoms or molecules. It was JJ Thomson, however, who carefully designed an experiment to measure the electrical charge to mass of the emitted particles, which established them as unique elementary sub-atomic particles. He called these particles 'corpuscles', which later on were given the name 'electrons' by Fitzgerald as a result of combining the words *electric* and *ion* [3]. In addition, such particles, which can be accelerated, have a wavelength which is shorter than that of light photons by up to 100,000 times (depending on the electron's speed). These properties led researchers later on to accelerate a focussed beam of electrons to illuminate the surface of a specimen—as light is used in optical microscopy—and

6

hence develop electron microscopy as an imaging tool. As a result of this, a resolution imaging limit of about 50 pico-metres (pm) was estimated for electron microscopy, which is yet to be realised, in comparison to an upper limit of some 200 nm for traditional light microscopy. These exciting prospects have encouraged research into developing imaging tools using energetic electron beams. It was in 1931 that Ernst Ruska and Max Knoll [4] finally succeeded in demonstrating the first working electron microscope.

There are two major types of electron microscope; the first of which uses a high energy beam of electrons to penetrate thin samples, and the transmitted electrons are used to form an image of that sample in what has become known as the transmission electron microscope (TEM). The electron beam energy in this case ranges from 50,000 V up to 1,000,000 V, but most widely used electron energies in TEMs today are in the range 100,000–300,000 V. In the second type, known as the 'scanning' type, the incident electron beam is normally in the range 100–30,000 V. These electrons are arranged to impinge the solid surfaces, and the reflected electron signal is used primarily to map the surface topography of the said sample. This type is known as the scanning electron microscope (SEM) and is the most widely used type in industry and academia alike. The major difference between the two instruments is in the collection of the signal from each, in the TEM the signal is of the transmitted electrons through the thin sample, whilst in the SEM the signal is normally of the reflected electrons (i.e. scattered back off the sample's surface). Both types use electrostatic and electromagnetic *electron lenses* to control the electron beam energy, but more importantly for the SEM is to focus it in the smallest possible spot which, when scanned, could be used to form an image of the area the electron impinges.

These electron optical *lenses* are analogous in function to the glass lenses of the optical light microscope. The remainder of this chapter will review the principle of the electron microscope, its major components, and provide a short overview of its applications, particularly in relation to light-related technology, before finally concluding with future trends in electron microscopy.

6.5   The Principle of Electron-Solid Interaction

Imagine an electron beam of an infinitesimally small diameter is incident on a solid surface, as depicted in ◘ Fig. 6.3 below. As these electrons penetrate the solid, they will interact with the sample's constituent atoms. This interaction is referred to as 'electron scattering' and is further divided into two categories: (a) an inelastic scattering, where the incident electron gives up part of its energy as a result of its collision with the solid's atoms; and (b) an elastic scattering, which causes the electron to change its direction of travel with almost no energy loss. It is the first scattering type, however, that gives rise to the signals enumerated in ◘ Fig. 6.3, whilst the second type is normally what determines the shape of the interaction volume of the incident electrons within the solid under study, as depicted in ◘ Fig. 6.4. The shape and size of the interaction volume depends on the incident electron energy, its angle of incidence with respect to the surface and on the average atomic number of the sample under study. If one concentrates on the emitted electron signal alone and plots their number against their energy, one would in principle collect a distribution similar to that shown in ◘ Fig. 6.5.

The collection of one of the signals resulting from the interaction depicted in ◘ Fig. 6.3 has over the years resulted in a specific class of instruments reflecting the type of information gathered from such interaction. For example, if one collects the resulting X-ray photons of a given element making up the solid, the collected image would be a map of the distribution of such an element in the solid, normally

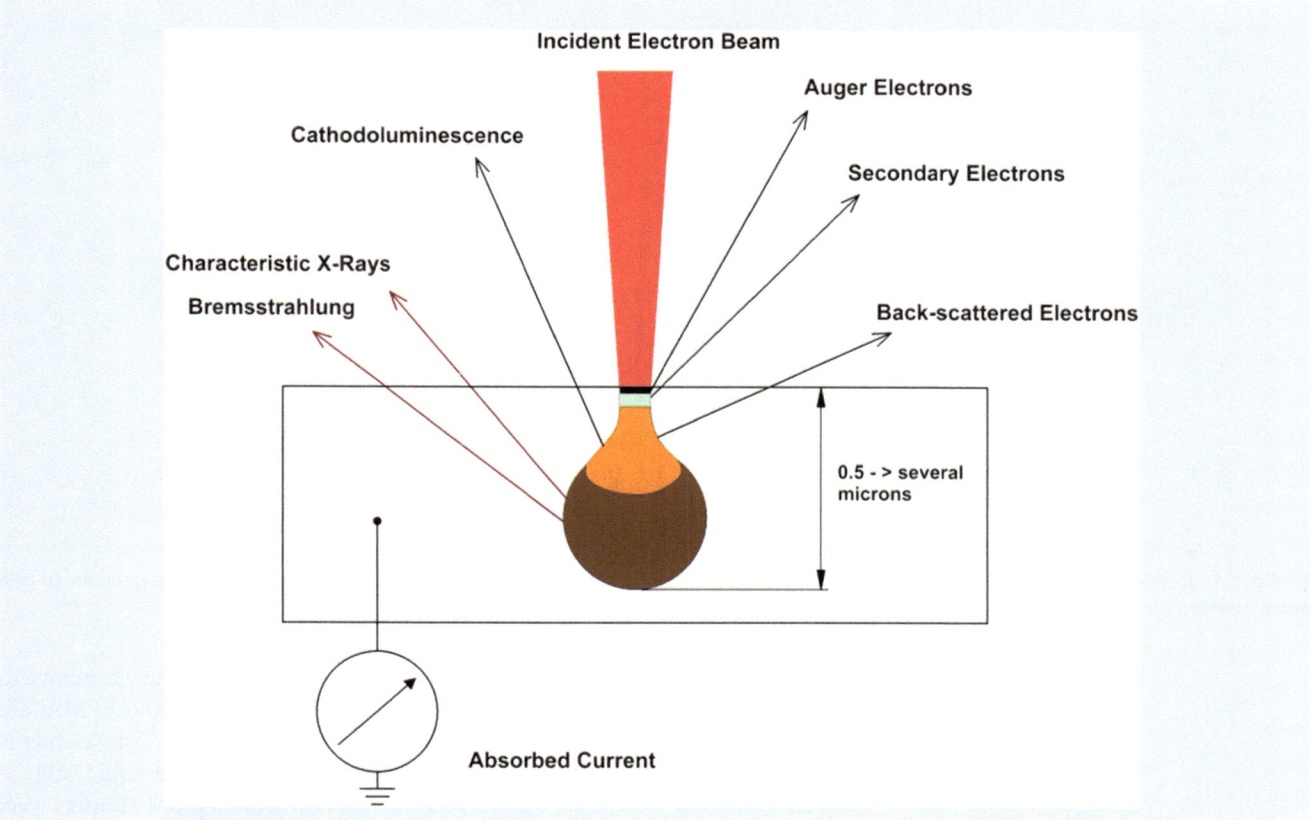

Fig. 6.3 A schematic depicting electron-solid interaction on a solid sample, and the various signals that result from such interaction

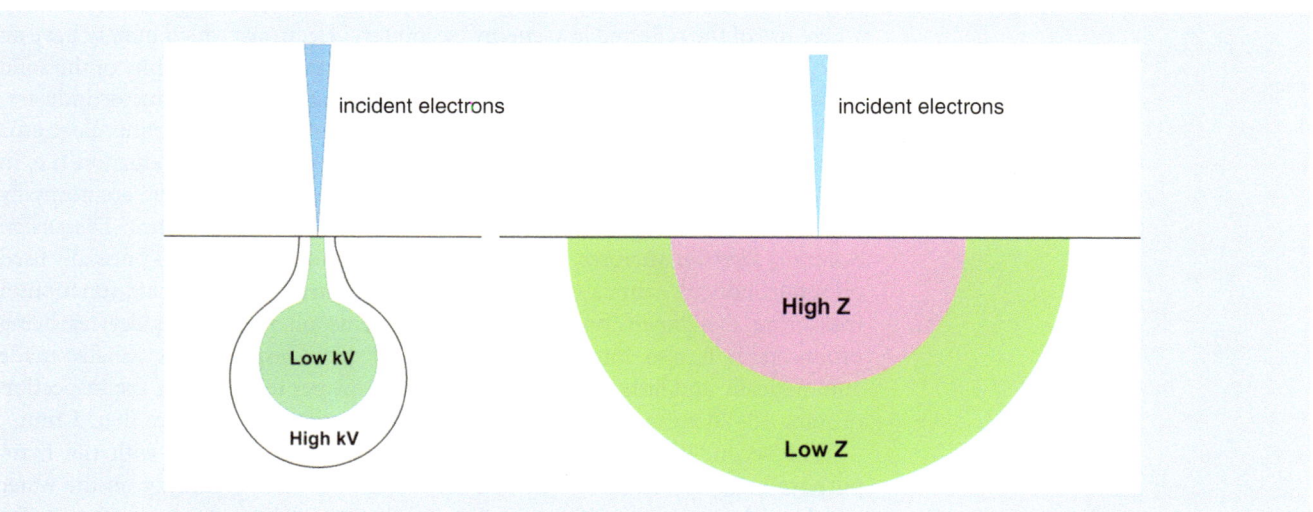

Fig. 6.4 A schematic showing electron-solid interaction where in (**a**) it shows the difference in the overall volume as a function of the incident beam voltage, whilst in (**b**) it shows the difference in volume between high and low atomic number materials

to a depth ranging from ~0.05 μm to few μm depending on the incident electron energy and the sample's atomic number. This technique is referred to as the electron microprobe analysis (EPMA) and is heavily used in material science research and applications. If, however, one uses the Auger electrons of a given element, then these would normally produce a map of the distribution of this element on the very top atomic layers of such a sample—a technique known as

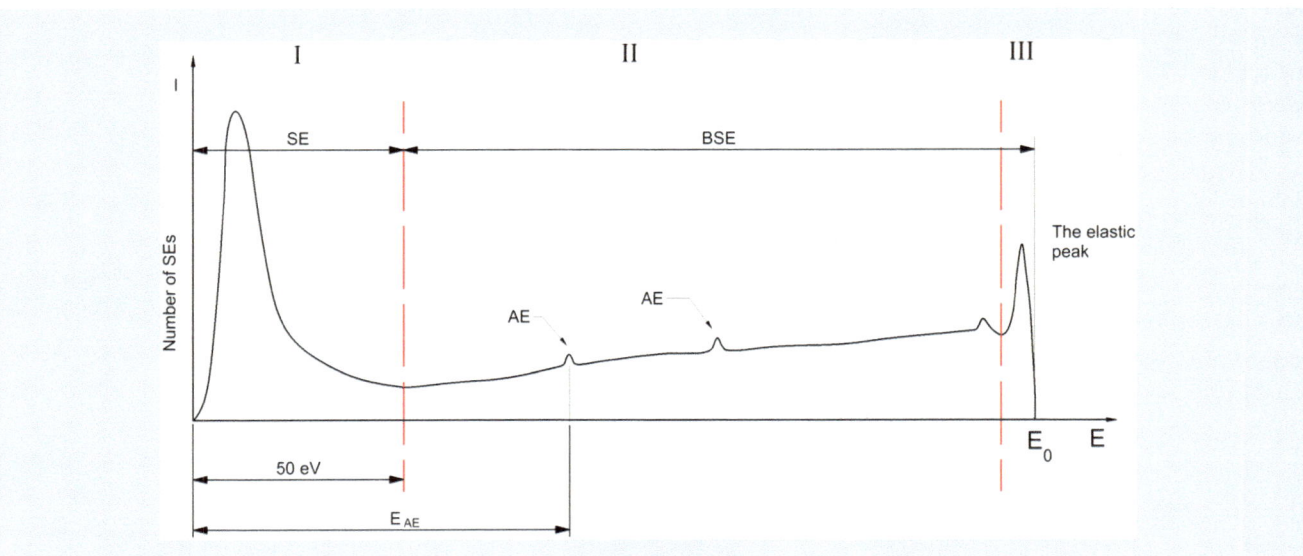

Fig. 6.5 Electron spectrum of a typical distribution in energy and number of electrons that exit the surface of a solid which is bombarded by a beam of energetic electrons. *AE* Auger electrons, *SE* secondary electrons, *BSE* backscattered electrons

Scanning Auger Electron Microscopy (SAM). Auger electrons are an alternative to X-ray emission from atoms and also, like the X-ray photons used in EPMA, are material specific, i.e. almost the equivalent of finger prints of the elements [5]. Whilst X-rays originating a few microns below the surface could still be collected, the collected Auger electrons come from only the top few atomic layers of the solid under study (i.e. the depth of information of the X-rays could be >100 times or more greater than that of the Auger electrons). The collection of other signals would give rise to techniques associated with such other signals.

However, the main and most popular type of electron microscope is the SEM, where use of the reflected low-energy 'secondary' electrons, which mainly have an energy of <50 eV (refer to Fig. 6.5), is made to map the topography of the solid surface. One important application of the SEM is in the semiconductor industry, both during the development of the integrated circuits and the various electronic devices constituting it but, equally as important, during their production (i.e. in what has become known as fabrication lines). In the latter case, there are normally two types of microscope in use. The first is referred to as a *Critical Dimension Scanning Electron Microscope* (CD-SEM). This type of instrument is normally used on semiconductor fabrication lines for quality control, with the main function of measuring the dimensions of some components on the circuits/devices being produced. It is also interesting to note that such inspection is usually made automatically and lasts no more than few seconds per test, making the inspection of some 10–20 areas on a full 300 mm diameter wafer last no more than 1 min.

The second type is normally also used for quality control, with the main purpose being searching for 'defects' or foreign particles appearing on the wafer which could cause the ultimate failure of the circuit built on or next to it, for example, where a particle is found lying between two active parts of the circuit and it unintentionally connects them. These defects can be due to anything ranging from small particles that find their way into the production line as a result of poor practice to failure in procedure. Sometimes these also appear as a result of the use of poor quality or incompatible materials in the fabrication process including, for example, the 'pure or de-ionised' water used for washing the wafers. This class of instruments is referred to as '*defect review SEMs*', which are normally equipped with a number of elemental/chemical detectors used to analyse such defects. The SEM is also equally vital in biological and other physical sciences, such as physics

Chapter 6 · Modern Electron Optics and the Search for More Light. . .

127   6

and engineering. More discussion will be devoted to the SEM later on. In recent years the development of compact and sometimes novel X-ray and various detectors has enabled these methods to be used as add-on techniques to the electron microscope, thus enhancing the instrument's analytical capability and widening its use.

If, on the other hand, the sample under study could be made in the shape of a thin enough section to allow a beam of energetic electrons to penetrate it and be collected from the other side, then a whole new array of signals would result in allowing one to gain more information from such a sample on atomic dimensions. This class of instruments is referred to as TEM. The working of the TEM, however, is quite subtle and different than that of the SEM. To understand the basic principle of TEM, let us investigate the fate of some energetically incident electrons on a thin sample. The transmitted electrons will pass through the thin specimen but, in so doing, will be subjected to one of three possible mechanisms:

1. To pass through with no scattering,
2. To pass through with some angular deflection (elastic scattering) and
3. To lose some energy as it passes through (i.e. inelastic scattering).

The above three possibilities are normally a function of the energy of the incident electrons and the arrangement of the sample's atoms. How we see the varying contrast in the obtained TEM images is quite interesting though.

To better appreciate the underlying principle behind the operation of the TEM, it is useful to consider the following simple example. Imagine a hypothetical specimen made out of four regions of carbon and lead as depicted in ◘ Fig. 6.6 below (where lead is a much higher atomic number material than carbon). If 100 energetic electrons are incident perpendicularly on the surface of this sample and the transmitted electrons are classified according to them being scattered (i.e. change direction of travel) through a small angle of 0.5° or more in comparison with those which pass through without suffering any scattering, or less than 0.5°, then the number of transmitted electrons varies in terms of their scattering

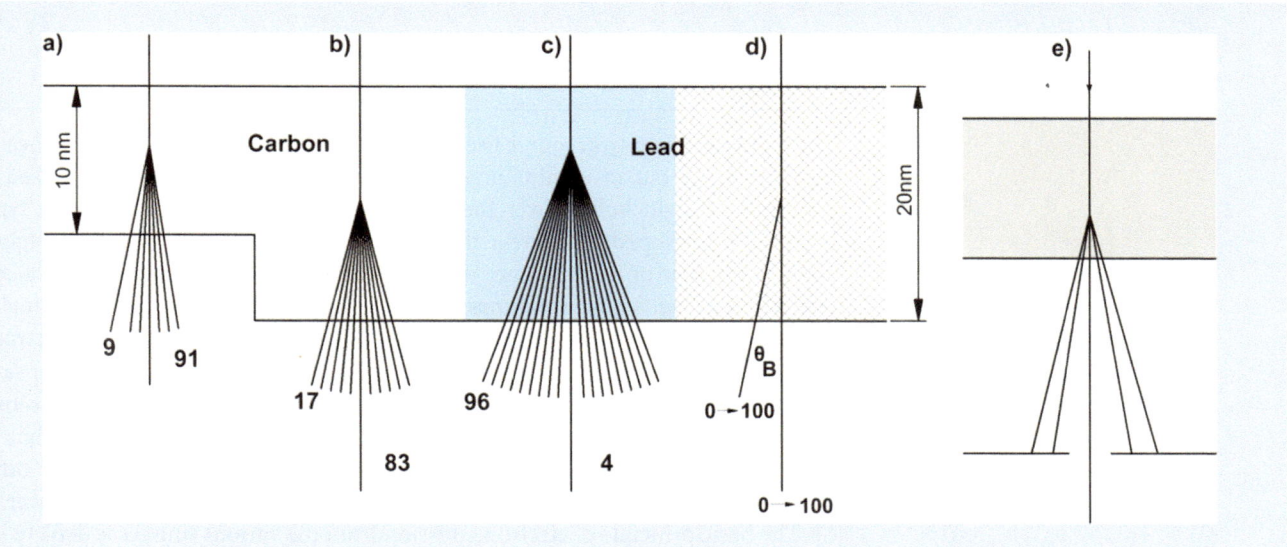

◘ Fig. 6.6 A hypothetical sample used to illustrate the principle of transmission electron microscopy by following the fate of 100 energetically incident electrons, and counting those that scatter through an angle of more than 0.5°. (a) carbon film 10 nm thick of randomly distributed atoms, (b) same carbon film but of 20 nm thickness, (c) 20 nm film of randomly distributed lead atoms, whilst (d) is 20 nm thick lead but of regularly distributed atoms (i.e. crystalline), (e) the objective aperture which sits below the sample in the TEM to stop electrons scattering through more than 0.5° (adopted from [7]) from passing through and hence contributing to the used signal

6

angle in a very interesting way. In part (a) of the sample, nine electrons will scatter through 0.5° or more, in part (b) about 17 will scatter, while part (c) shows a much larger number of about 95 electrons scattering by more than 0.5°. However, in part (d) the number scattered depends on the angle that the incident beam of electrons makes with the sample atoms in what is known as 'Bragg diffraction' [6]. If a small aperture (i.e. a hole cut in a metal plate) is placed below the specimen such that any electrons scattered by 0.5° or more are stopped by this plate whilst those scattered by less than this angle go through, then a number of different 'in-value' electron signals could be collected below the small aperture. The position of this aperture is therefore the key part in the working of the TEM. The detectors used in the TEM also vary from principally detecting the crystallinity of the sample, via the distribution of its atoms; or they determine its atomic number, via the measurement of very small energy losses that the incident electrons suffer in passing through the thin sample.

The TEM is a more complex instrument to manufacture and operate than the SEM. Moreover, its price is normally several times more than that of the SEM. The number of TEM instruments worldwide is perhaps less than 10 % of the installed user base of SEMs. It should, however, be understood that the information gathered from either instrument normally complements the other rather than being an alternative to it.

6.6   The Basic Components of Electron Microscopes

◘ Figure 6.7 depicts a schematic of the two most widely used types of electron microscope: the SEM and the TEM. The major components of either instrument could be divided into the following parts: the electron source, the probe-forming column, the specimen chamber and finally the detectors. A brief coverage of these components will be given below; however, for more detailed discussion of the modelling of these components the reader is referred to [8]. It should also be noted that most, if not all, modern instruments are computer controlled, but this will not be covered here.

6.6.1   The Electron Source

The first type of electron source used in electron microscopes consisted of a heated filament made out of a thin tungsten wire, a similar material to that used in conventional light bulbs. Over the years more and more electron source types have been developed to address the fundamental problem of using the highest possible number of electrons per unit area per unit angle of emission; a concept referred to as the 'source brightness'. The higher the brightness value, the smaller the focused point that can be formed with the same number of incident electrons.

◘ Table 6.1 below summarises the currently available and widely used types of electron sources and their relative properties. The development of field electron emitters as high-brightness electron sources in the last 40 years or so has in particular moved both types of electron microscope discussed here and other probe-forming systems closer towards realising their ultimate resolving power.

The basic principle of electron emission from the various sources is depicted in ◘ Fig. 6.8. The important property that underpins all of these is the material's work function, a property that dictates how much energy is required to liberate an electron from its bound state in the material's atom. For thermionic sources, electrons are liberated by heating the source to 'boiling point', the consequences

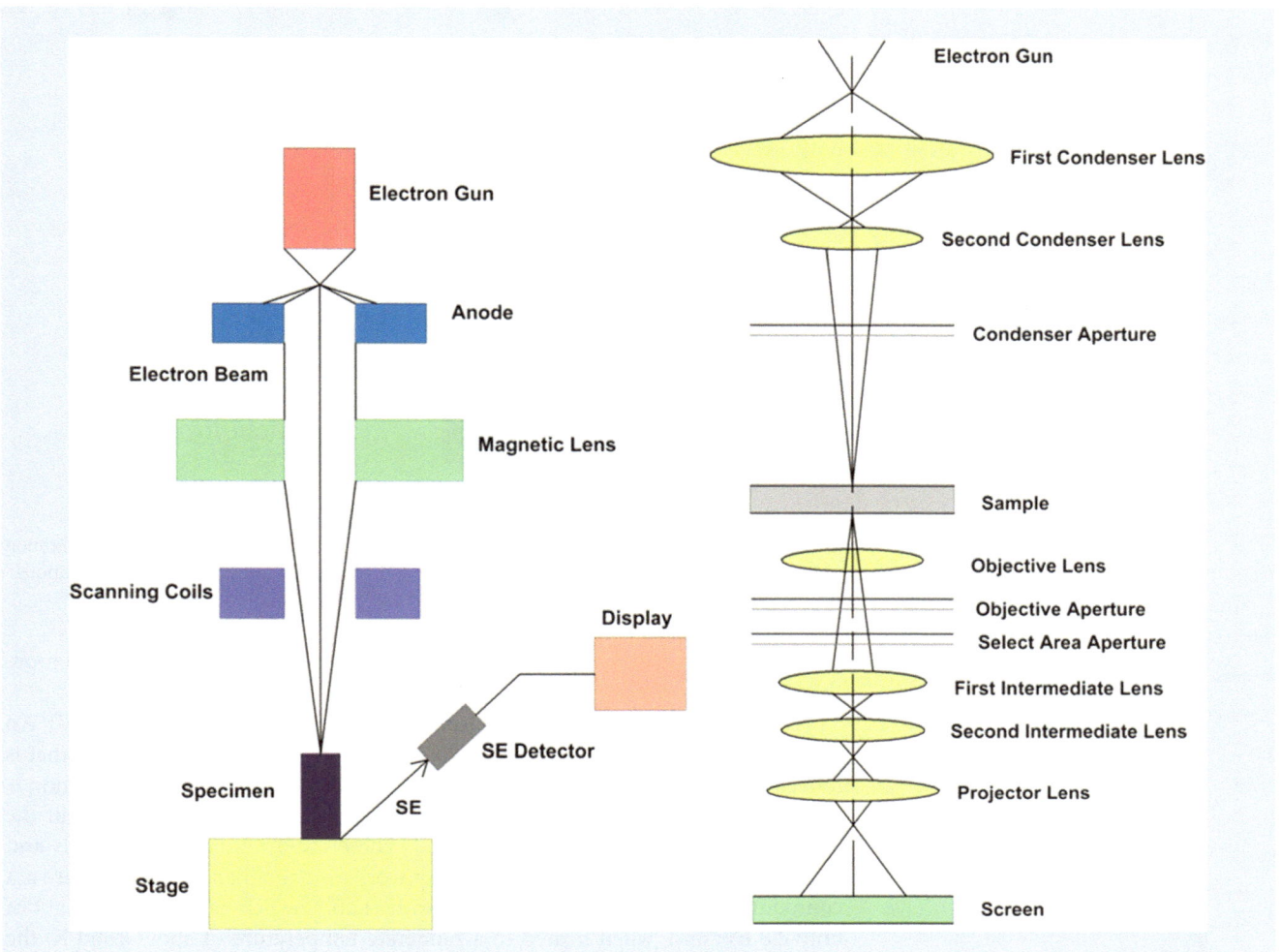

Fig. 6.7 A schematic of the major components of the electron microscopes (**a**) SEM and (**b**) TEM. Note that most microprobes and EBL instruments share the same configuration as the SEM

Table 6.1 A comparison of the characteristics relevant to electron optics for some of the most widely used electron sources (York Probe Sources Ltd, ▶ http://www.yps-ltd.com/)

Emitter type	Thermionic	Thermionic	Schottky FE	Cold FE
Cathode material	W	LaB_6	ZrO/W (100)	W (310)
Operating temperature (K)	2800	1900	1800	300
Effective source radius (nm)	15,000	5000	15 (*)	2.5 (*)
Normalised brightness (A/cm^2 sr kV)	1×10^4	1×10^5	1×10^7	2×10^7
Energy spread @ cathode (eV)	>0.59	>0.50	0.5–0.8	>0.23
Beam noise (%)	1	1	1	5–10
Operating vacuum (mbar)	$<1 \times 10^{-5}$	$<1 \times 10^{-6}$	$<1 \times 10^{-9}$	$<1 \times 10^{-10}$
Typical cathode life (h)	~100	~1000	>10,000	>10,000

6

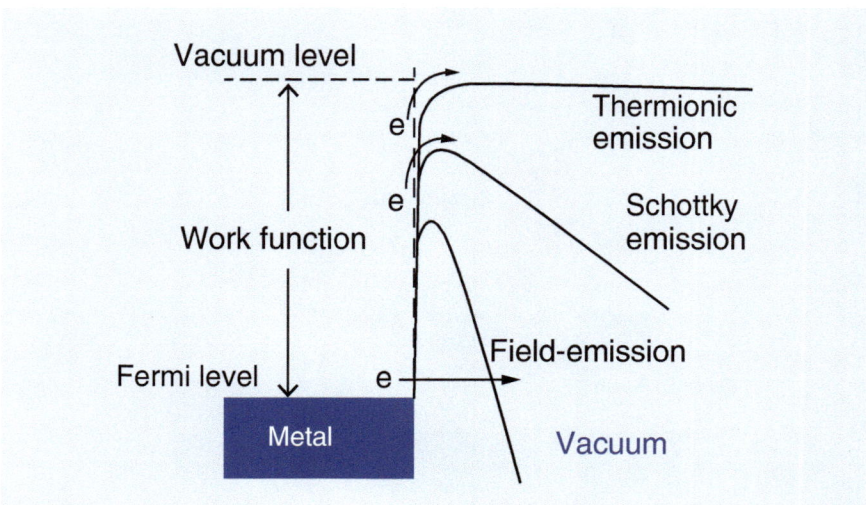

Fig. 6.8 A simplified schematic of the electron potential energy at the solid–vacuum interface showing the modification (*bending*) of the energy barrier as a result of the application of the electric field. For a more detailed description of this phenomenon as applied to various sources, see [10]

of which make the lifetime of this type of source quite limited to mostly no more than few 100 h of use.

The work function of a material is measured in units of electron volts (eV). An eV is the amount of energy gained or lost by the charge of a single electron that is moved across an electric potential of 1 V. For tungsten (W), the work function is about 4.6 eV. The work function acts as a barrier between the electrons in the metal and the vacuum outside the metal. There are a number of materials and combination of materials that have a lower work function than W. One of these is a compound known as lanthanum hexaboride (LaB_6), which has a barrier height of only 2.8 eV; and, when heated to a moderate temperature of about 1800 K, the electrons have enough energy to jump over the barrier thus enabling them to be used as an electron source. The lifetime of this type of source is of a factor of ×3–5 longer than the heated W filament; and their brightness is also more than ×10 higher. The vacuum environment for the operation of this source is, however, more stringent than that of the W counterpart, normally in the region of 10^{-7} mbar.

Field electron emission [9], on the other hand, relies on a totally different mechanism, which exploits the unique wave-particle duality of the electron (see Chap. ▶ 1 for more explanation). This wave nature of the electrons, when utilised, has allowed scientists to develop electron sources with a brightness value of more than 100,000 greater than that obtained from the early tungsten filament counterpart (see ◖ Table 6.1 above). In this method, the electron source is made of a very sharp needle, often less than 1 μm in diameter. When this is subjected to a high electric field, $>10^9$ V/m, the barrier between the electron position in the metal and that of the vacuum level becomes so thin (refer to ◖ Fig. 6.8 [10]) that electrons can tunnel through it and are liberated into the vacuum. Depending on the electron emitter's diameter and its position relative to an electrode positioned in front of it, called the extractor, the high electric field needed for the electrons to tunnel through can be achieved by applying a moderate voltage value between the emitter and the extractor of less than 5000 V.

◖ Figure 6.9 depicts the various materials and configurations of electron sources currently in use in electron optical instruments which give brightness values spanning a wide range, but which for brevity we will not cover here. Suffice it to say, however, that over the last few decades, field electron emitters, and

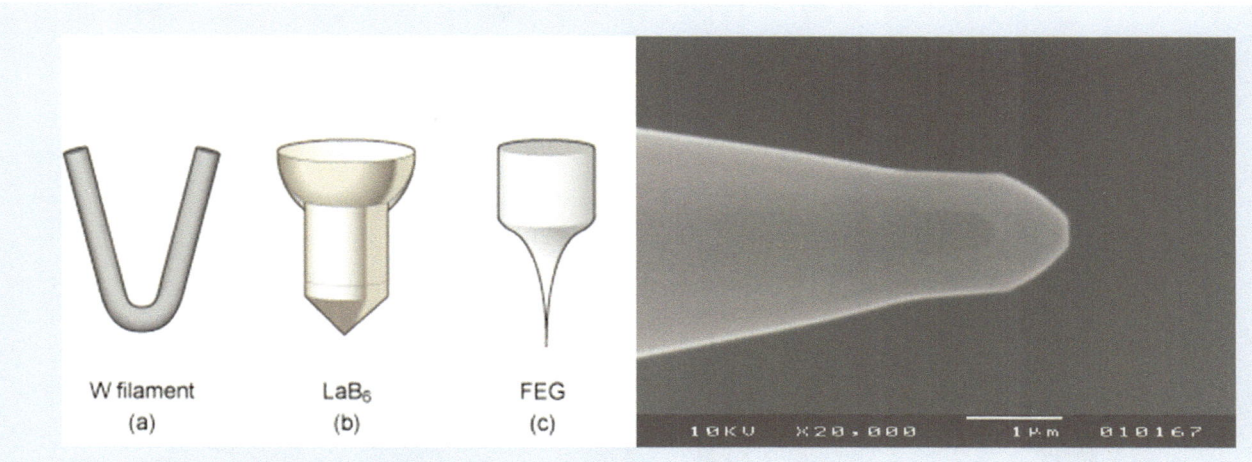

□ **Fig. 6.9** The various electron sources in use: (**a**) the thermionic tungsten (W) filament; (**b**) the low barrier material of lanthanum hexaboride (LaB$_6$); and (**c**) the field electron emitter (see, for example, MyScope at ► http://li155-94.members.linode.com/myscope/sem/practice/principles/gun.php). (**d**) Is a high magnification electron microscope image of a typical Schottky field electron emitter (York Probe Sources Ltd, ► http://www.yps-ltd.com/)

particularly those which are known as thermally assisted or 'Schottky' field electron emitters, have become so popular and relatively easy to use that at least half of the modern SEMs and TEMs are manufactured using this class of electron sources. But the heated tungsten filament, in spite of its clear disadvantages, is still expected to be with us for many years to come because of its relatively low cost and ability to function in poor vacuum conditions.

Recent advances in electron source technology are currently directed at developing small-diameter sources of novel materials in pursuit of even higher brightness. Such materials include carbon nano-tubes (CNT) and nano-rods of conducting materials in general. The CNT is a sheet of carbon atoms rolled to form a cylinder of one or several layers. The diameter of this cylinder ranges from few nm to several tens of nm (see □ Fig. 6.10 for illustration). However, in spite of the potential advantages of such sources in terms of their superior brightness and small source size ([14], it is expected to be sometime before we will see such a source in use due to the great technical challenges in reliably producing and using a stable electron emitter of this type.

6.6.2 The Probe-Forming Column (Electron Lenses)

The probe-forming system consists of a number of 'lenses' that act to form an image of the source of electrons which is used in a focussed spot of electrons. The basic underlying principle of an electron lens is a structure that controls the electron trajectory, ultimately forcing them to converge to a focussed spot. Two types of lens are used in probe-forming systems: an electrostatic one where an electric field is formed by the application of suitable voltage to a specially shaped metal structure which the electrons travel through; or a magnetic lens where an electric current passes through a coil enclosing a carefully shaped iron structure, thus producing a magnetic field that acts on the electrons in a similar fashion to the electrostatic lens. The electrons that pass through such lens structures could then change their path in a diverging or converging manner to a point somewhere away from its starting position (i.e. to form a focused image of the emitted electrons away from its starting point). Such a focused spot could then be used to scan the

6

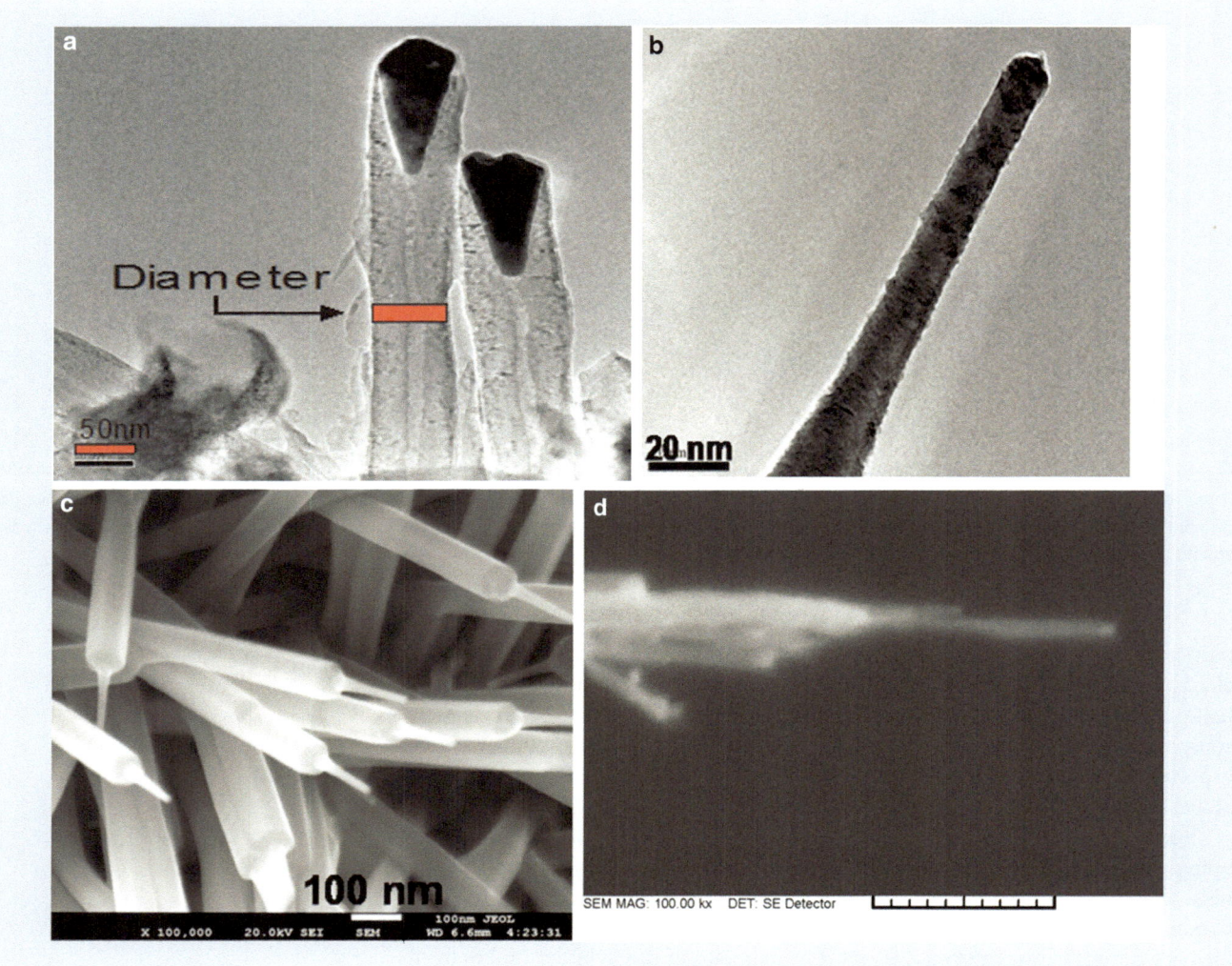

Fig. 6.10 (**a**) TEM image of the end-form of a CNT showing the nickel seed used to grow the CNT. (**b**) TEM image of the ultra-fine tip (nano-needle) of the as-grown In-doped ZnO nano-pencil with a diameter in the range of 13–15 nm. (**c**) SEM image showing the morphology of In-doped ZnO nano-pencils on silicon substrate prepared by thermal evaporation process. The image shows a useful nanostructure made of two parts: ultra-fine tip connected to a base of ZnO nano-rods which could be useful as an electron source. (**d**) Shows a CNT grown on top of a sharp tungsten emitter and the bar measure 1 μm. All of these sources are from the author's research group at the University of York [11–13]

surface of a solid in a raster fashion (i.e. moving the spot across the surface in a straight line and then moving quickly back to the start point and scanning across the next line down) where, at each pixel of the raster, an interaction similar to that depicted in Figs 6.3 and 6.4 takes place. Note that the focussed spot could be smaller or larger than the area where electrons are emitted from depending on the source of electrons used and the applications for which the probe is intended. Note also that the choice to use either an electrostatic or a magnetic lens, or indeed a combination of both, is largely dictated by the intended applications. However, it should be noted that magnetic lenses are more favourable for high resolution applications because of their favourably smaller aberration coefficients (see below for detailed discussion on lens aberrations). The focussed spot diameter (d_T) is therefore composed of a number of contributing sources where their diameters can be simply expressed as the original source of emitted electrons (d_o), the chromatic

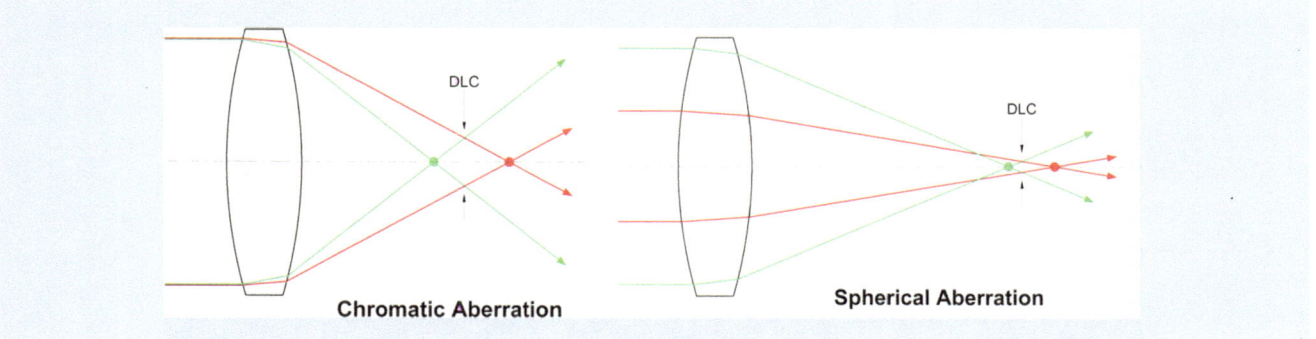

◘ Fig. 6.11 (**a, b**) Schematic diagram illustrating chromatic and spherical aberrations and the disk of least confusion (DLC)

aberration (d_C), spherical aberration (d_S) and diffraction effects (d_D). These contributions could be added in the following manner:

$$d_T = \left(d_o^2 + d_C^2 + d_C^2 + d_D^2\right)^{0.5} \tag{6.5}$$

The lenses within an electron microscope behave in a similar fashion to the glass lenses of optical microscopes and these, too, suffer from a number of fundamental properties (or deficiencies) that adversely affect the quality of the focussed image. These are referred to as lens aberrations which basically blur the focussed image, and would therefore require a correction procedure to resolve. Some of these are inherent to electron microscopy, such as the chromatic aberration, due to the nature of the emitted electrons and the type of source used having a range of different initial energies as they leave the surface of the electron source. The effect of such varying energies is that the focussed electrons would correspondingly land, for each electron energy, at a slightly different distance away from the point of emission, as depicted in ◘ Fig. 6.11a. ◘ Table 6.1 above lists the energy spread typical of the various electron sources in use. Modern electron microscopes, particularly in the case of TEMs, have started to use an energy mono-chromator to correct for the chromatic aberration, but these are not yet commonly used due to the cost they add to the microscope. It is clear from ◘ Table 6.1 that field electron emitters have a smaller energy spread than thermionic sources and these give this class of electron sources a clear advantage for use in electron microscopy and particularly so in the case of the SEM, which only use chromatic aberration correctors in the top of the range instruments.

Another optical effect inherent in all cylindrical or spherical lenses (and mirrors) is the spherical aberration, which in the present case is caused by electrons focussing at different points, depending on how far the electrons are incident from the centre of the lens as it passes through it. The reason for this is that for each of the emitted electrons going through the lens's hole when these are projected back would subtend a range of angles at the point of emission (the source). Again the positions of the focussed electrons would converge differently for each electron trajectory, as the electrons travel through regions of different potentials, and hence will obey the formula given above (Eq. (6.4)). As a result they will focus at a range of distances away from the point of emission. The extent of this focal distance away from the point of emission (or the ideal focal point of such a lens) is a function of the angles subtended, i.e. the size of the lens' hole. This in turn means that the larger the aperture-hole diameter, the worse is the effect. Spherical aberration (C_s) is a feature of all round lenses which causes image distortion and limits the ultimate microscope resolving power. In practice, however, is that the various foci of the electrons past the lens will go through a minimum called the 'disk of least confusion' (DLC), which is normally assumed

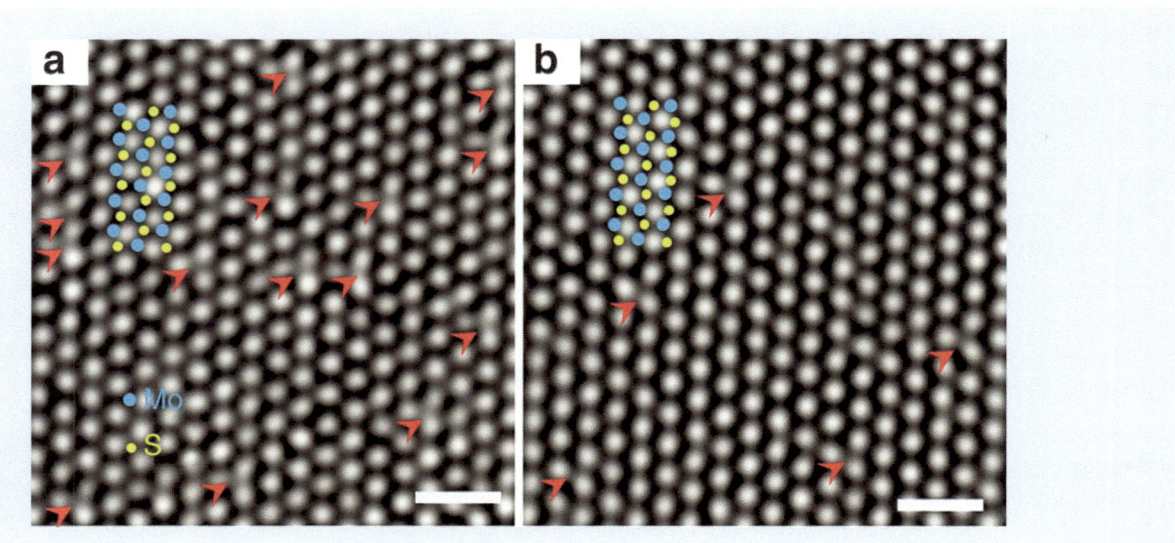

Fig. 6.12 The effect of correcting spherical aberration in a TEM [16]. Note that the *red arrows* refer to points where the separation between the imaged atoms becomes clearer after the spherical aberration correction

as the focal point of the lens. The spherical aberration is schematically illustrated in ◘ Fig. 6.11b.

It is interesting to note that spherical aberration was first noted by the Arab scientist, Ibn al-Haytham (d1040) in his magisterial work on optics, 'Kitab al-Manazer' (see Chap. ▶ 2). Whilst this defect was known early on during the electron microscopy development, an effective resolution was only realised and proposed in 1997 (see [15] for a full account). Again, the effect of spherical aberration tends to be more obvious in the TEM case which, when corrected, allows one to resolve structures a small fraction of a nanometre. Its correction in the SEM is also equally important but only applicable in the top of the range instruments. This has in great part been due to the use of field electron emitters, which has made it easier to reduce its effect in SEMs. ◘ Figure 6.12 shows the effect of correcting spherical aberration in a modern TEM.

Spherical aberration is an effect typically found in imaging instruments, such as cameras and telescopes. Its presence compromises the quality of the resulting image with potential catastrophe, as the scientists in charge of the Hubble telescope found [17]. The HST project cost several billion US dollars and took several years to complete. The first images collected were of very poor quality, as shown in ◘ Fig. 6.13a. The fault causing such a poor image was identified to be due to spherical aberrations of one of the telescope's mirrors and a repair mission was launched to fix it. It consisted of a number of astronauts, one of whom was an experienced experimental scientist who was able to walk in space to repair the fault. ◘ Figure 6.13b shows the effect of correcting the spherical aberration on the quality of the images obtained.

The SEM, on the other hand, has also seen a great deal of advancement to improve the quality of the obtained images as well as increasing its resolving power. This has largely been not only due to the use of Schottky and cold-field electron emitters as electron sources, both having lower energy spread and much higher brightness than its conventional thermionic counterpart, but also due to a range of efficient electron detectors that have recently been developed. As a result most modern SEMs offer high resolution imaging in the range of 1–2 nm. These field emission sources have particularly benefited the use of low voltage imaging where most modern instruments operate down to few 100 eVs and sometimes even less to obtain an image resolution of the order of few nm. This high resolution

Fig. 6.13 Hubble Space Telescope images before and after the STS-61 mission [17]

low-electron energy mode is particularly important in imaging biological and novel and radiation-sensitive materials, as encountered in the field of nanotechnology.

By reducing the incident electron beam energy, the interaction volume between the incident electrons and the specimen also reduces and becomes much closer to the specimen surface, as seen in ☐ Fig. 6.4 above. This extends the use of the SEM to previously unchartered areas. One recently developed field of application is in imaging doped regions of semiconductors, which have traditionally been almost impossible to image using conventional high-electron energy imaging (>5 keV). This is because the current secondary electron detectors used in the SEM are incapable of resolving differences in the atomic number (Z) of the constituting elements of a sample of less than $Z = 1$. However, the amount of doping material used in semiconductor devices is normally less than 0.001 %; and yet by imaging with electron energy of less than, say, 2000 eV one could differentiate between regions of a sample like silicon, which are differently doped, as depicted in ☐ Fig. 6.14. Doping of semiconductors is the crucial step in making electron devices, such as the p-n junction diode; the basic building block of integrated circuits. This low-energy imaging mode is proving valuable in the fabrication lines in the semiconductor industry, where its use is in quality control. The explanation of the mechanism that gives rise to such a contrast is outside the scope of this chapter, and the reader is referred to recent articles discussing this phenomenon [18, 19].

A recent development in SEM technology to cope with the limited space available in modern laboratories is the advent of the table-top electron microscope. A number of microscope manufacturers now specialise in producing such instruments. It is estimated that many thousands of such microscopes have been produced and sold to date. It is expected that further developments will continue to be exerted in this endeavour. One such development from the author's laboratory, depicting a small-size electron column, is shown in ☐ Fig. 6.15. This column also contains a novel electron detector inside the column [20]. One other parallel development has been in using silicon wafer technology to make the various microscope column components, which include the lenses, the deflectors and the stigmator correctors, as depicted in ☐ Fig. 6.16 [21]. The whole column in the latter case measures only a few mm in height making the whole microscope with its control electronics, vacuum system and specimen chamber comparable in size to a laser printer. This SEM is now commercially available.

6

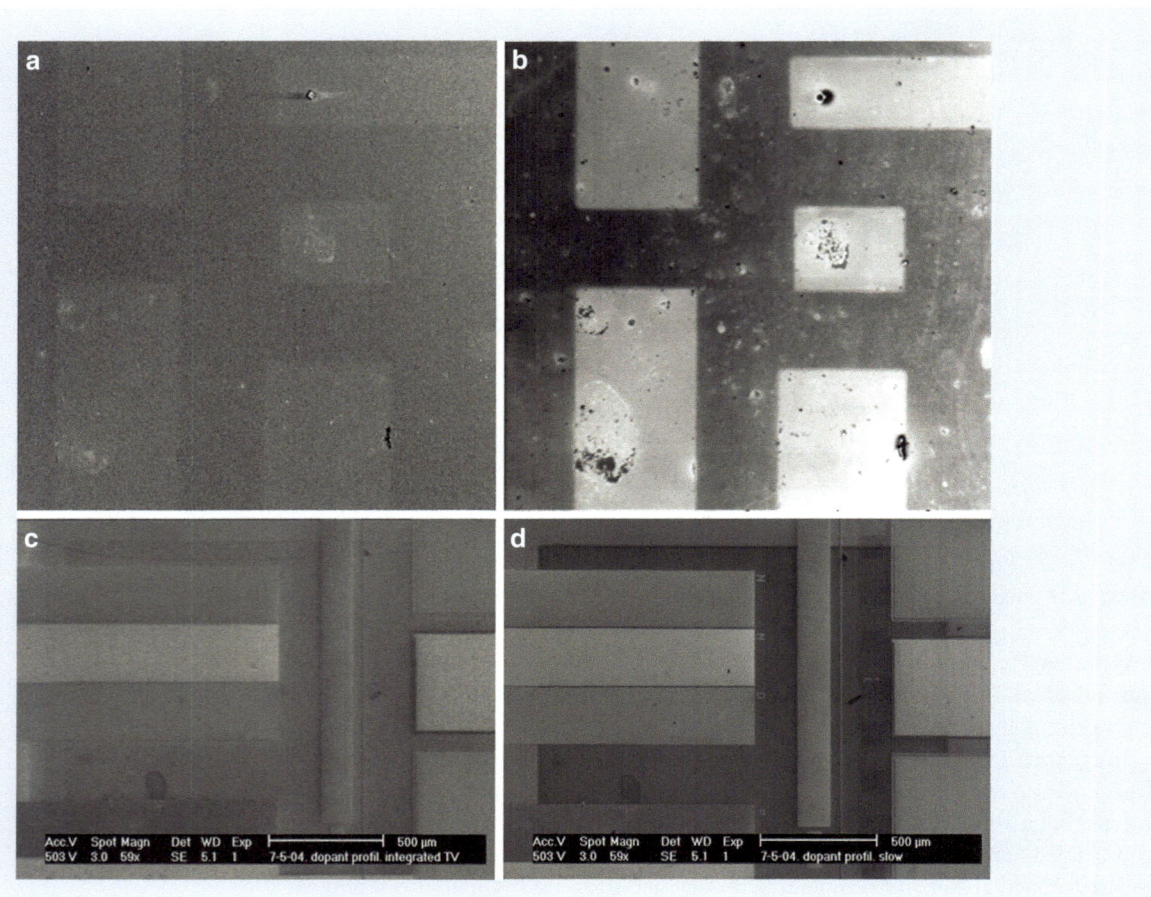

Fig. 6.14 The imaging of doped semiconductors with low-voltage electrons (Image (**a**) collected at 6 keV) and (**b**) collected at only 2 eV). The contrast of the p-regions, which are the bright areas, is increased as the beam energy is decreased. Note in particular the appearance of some small surface particles (contaminants) when imaging at such low-electron energies. In images (**c**, **d**) another effect is shown which is related to the scanning speeds the images are collected (**c**) at high speeds (TV rates) and (**d**) at slow scanning speeds of only 2–5 s per frame [18–20]

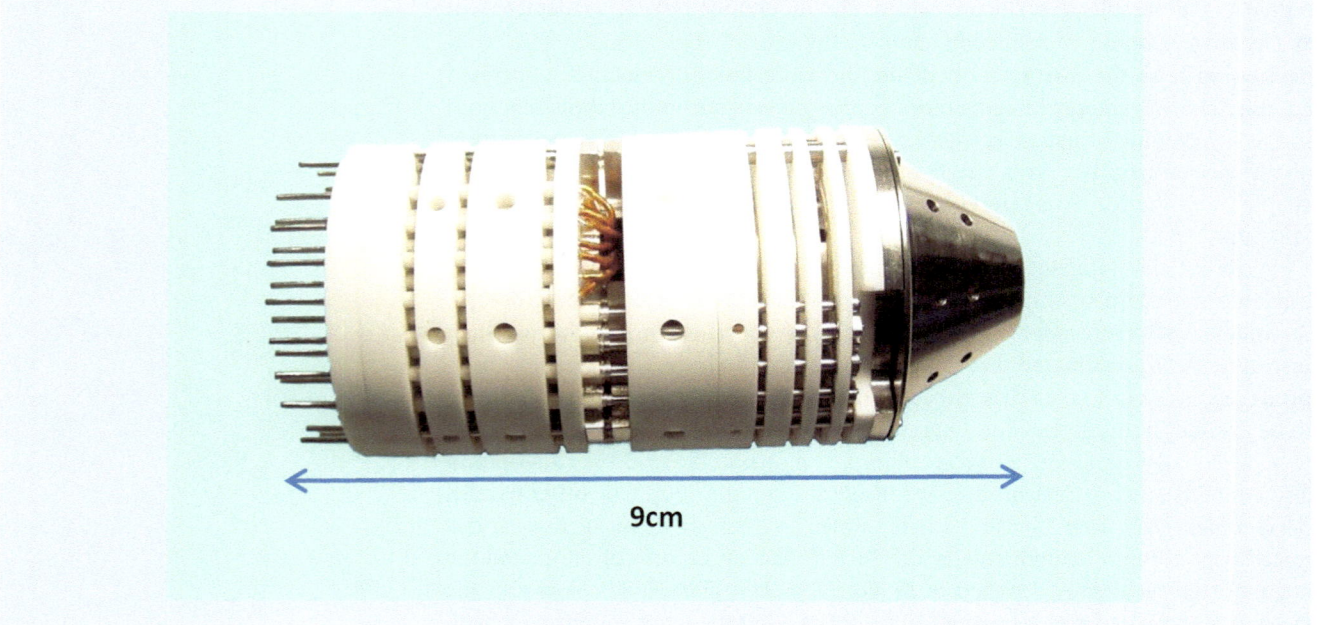

9cm

Fig. 6.15 A whole electron column from the author's laboratory, developed for imaging at ultra-low voltage use, down to 1 eV. It contains all the necessary components to form an image, a Schottky field emission source, the electron lenses and deflectors as well as a novel in-lens electron detector [20]

Fig. 6.16 A fully assembled miniature Si-wafer-based column. The dimensions of the whole column are 30 × 30 × 9 mm [21]

The Specimen Chamber

The specimen chamber of the electron microscope is an airtight metal vessel that is pumped to a pressure of the order of 10^{-5} to 10^{-6} mbar (standard sea level pressure is defined as 100 mbar). Again, the size of the specimen chamber differs greatly between the TEM and the SEM. The former can only accommodate samples measuring a few mms in diameter, where the specimen to be studied occupies only a small fraction of this size. The SEM, on the other hand, could accommodate samples a few cm in diameter or even larger, as in the semiconductor industry where full wafers measuring up to 30 cm in diameter are normally used in critical dimension type SEMs. Often, however, in material science, biology and general engineering, samples seldom measure in excess of several mm in diameter. Therefore most specimen chambers in SEMs measure between 20–40 cm to accommodate the various samples from these disciplines.

6.6.3 The Detectors

It is also customary that the specimen chamber, particularly for the SEM, accommodates a number of detectors to capture the various signals emitted from the sample being studied, as depicted in □ Fig. 6.3. Associated with the choice of a given detector is a requirement to control the total pressure around the sample. For example, if one needs to map the distribution of a given element on the top few atomic layers of a solid sample (as in Auger electron spectroscopy [4]), then the pressure around the sample should be much lower than if the signal being captured is only to reflect the surface topography (i.e. using the low-energy electrons depicted in □ Fig. 6.5). In the former, a pressure in the region of 10^{-9} mbar or lower is needed, whilst for the latter the environment around the sample could be as high as 10^{-5} mbar. There is now a class of instruments, referred to as environmental SEM (ESEM), where the pressure is even higher than conventional SEMs ($>10^{-2}$ mbar). This is to allow for the inspection of insulated or biological specimen, including glass, wood and even meat.

6

The development and use of energy-dispersive X-ray detectors (referred to as either EDS or EDX) has, however, given electron microscopes, and particularly the SEM, an added dimension in analytical power. Traditionally, X-ray analysis of solid samples has been carried out using dedicated instruments that employ wave-dispersive type detectors (WDS) (i.e. where the emitted X-rays are detected according to their wavelength for the WDS rather than according to their energy, as in EDS). Detecting the emitted X-rays was the task of the first electron microprobe for use as an analytical tool, and it is referred to as the electron probe microanalyser (EPMA) [22]. The EPMA is a much more complicated instrument than the SEM due to the special X-ray detectors used, but of course it has more powerful analytical abilities, too. In addition it has a higher X-ray resolving power than the EDS type and thus enables smaller amounts of materials to be detected. However, EDS is cheaper and simpler to operate. It is estimated that up to about half of all electron microscopes in the world are equipped with an EDS detector, which is mainly used for qualitative elemental analysis with detection limits of around one atomic %.

There are many other types of detectors that are also used to gain more information from the sample under investigation, and the semiconductor industry leads in this respect. Some of these are used to check the operation of the integrated circuits being manufactured and are normally employed on fabrication lines for quality control and monitoring during fabrication. The SEM has, indeed, underpinned the progress made in the semiconductor industry over the years and is most likely to continue to play such a role in the future.

In biological applications, the SEM is equally important. However, because the samples are of tissues from living beings and/or plants, these are normally 'specially' treated to withstand the low pressure of the SEM (e.g. in a process called dry-freezing, where the sample is first subjected to a low temperature treatment to freeze-dry it followed by slicing and inspection). This is normally followed by coating its surface with a conducting film to avoid being charged by the incident electrons. These special treatments have by and large been avoided in the ESEM, which is now becoming an important imaging tool in biological applications. But, on the other hand, there are still a lot of instruments which are not of that type in addition to there being some applications which may require inspections in conventional SEMs. In these cases the sample would normally be covered by a conducting metal film prior to inspections. Such films are so thin that they still allow one to observe fine details.

6.7 Fourth-Dimension Electron Microscopy or Time-Resolved Electron Microscopy

In 2009, the Arab laureate, Ahmed Zewail, announced the success of his research group in introducing a new dimension to electron microscopy by utilising femto-second laser pulses to radiate the electron cathode of the microscope, thus releasing single packets of electrons for bombarding the samples instead of a continuous beam. The laser used is carefully chosen to just exceed the work function of the microscope's electron source, which is normally of the field electron emitter type, as discussed in ◘ Sect. 6.6.1 above. The bound electrons at the Fermi level of the electron source are released upon the laser pulse striking the tip of the electron emitter. The released 'packet' of electrons is then accelerated by the electron optical column towards a sample under investigation. Note that these electrons are also focussed by the column in exactly the same way as a continuous stream of electrons would have been; so in effect one is using a focussed packet of electrons to bombard the sample. Zewail called this new microscopy mode ultra-fast electron microscopy, which now has come to be known as 4D UEM (see Chap. ▶ 3).

In these experiments, one can follow the dynamic response of the sample at such incredibly short time scales.

The pioneering work of Zewail is in carefully choosing the wavelength of a laser (which corresponds to a given energy) to slightly exceed the work function of the electron source of the microscope. This then eliminates the need to apply a voltage pulse to the emitter, which is limited to microseconds at best. Laser-pulsing at the femtosecond is now standard in 4D electron microscopy. It should be said that 4D electron microscopy is in its infancy, and there are several research groups around the world who are actively engaged in the development of this exciting 4D microscopy technique (see Chap. ▶ 3 for a full list of references and the various areas of application of the technique). Their efforts are likely to open up new areas of applications and enhance our understanding of materials and biological processes.

6.8 Lensless Electron Microscopy

The 1986 Nobel Prize in Physics was awarded for the development of a new type of electron microscopy called scanning-tunnelling microscopy (STM) and for the development of the first electron microscope, given to one of the survivors Ernst Ruska. The STM was demonstrated by two Swiss scientists, Heinrich Rohrer and Gerd Binnig, who proposed a new method to map surfaces at the atomic scale. In this technique, a voltage-biased sharp-field electron emitter, as discussed above, is brought ever so close to the specimen surface (i.e. in the order of only several atomic layer distances) where electron tunnelling via field emission starts with the application of very small voltages of less than 50 V. The emitter is then scanned in a raster fashion over the surface, and the tunnelling current obtained is used to map the spatial distribution of the surface atoms. The original STM experiment was essentially carried out under ultra-high vacuum conditions (i.e. about 10^{-10} mbar). Today there is an array of different configurations of the same principle in what is referred to as scanning-probe microscopy (SPM). It is ironic that the majority of the SPM methods can be carried out under atmospheric pressure. Scanning-probe microscopy today is used in a variety of disciplines ranging from biological applications to physics and engineering. The reader is referred to recent publications for up to date references [23].

Finally, whilst much work has been invested in developing ever more sophisticated electron lenses to map surfaces at the highest possible resolution, a recent development has demonstrated the possibility of producing high resolution secondary electron images at the sub-10 nm region without the use of an electron lens [24]. There is a great deal of research still to be carried out to optimise the technique and to quantify it, but this has been an interesting development in electron microscopy and could yet lead to further if not different areas of applications of this indispensable instrument.

6.9 Application of Electron Microscopy Towards Light-Producing Devices

In this international year of light and light-based technologies, it is appropriate to cite one area where electron microscopy has been an invaluable research tool in its own advancement. The use of electron microscopy has over the years been fundamental in developing the light-producing devices we have been using, ranging from the incandescent light bulb to the latest development of known as 'light-emitting diodes (LEDs)' (see later on an explanation of LED). There are a number of benefits in developing an efficient 'white' light source like LEDs closely

6

resembling natural light. For example, it is known that sunlight is responsible for replenishing 90 % of the vitamin D in human bodies (via exposing our skins to direct sunlight), compared to only 10 % replenishment via food [25]. This vitamin is of great medical importance to humans. Natural light is also believed to have a protective effect against certain types of cancers (breast and prostate types) by preventing the over-production of cells [26]. In addition, its deficiency could lead to a weak immune system against colds, coughs, etc., and many studies link it to body fatigue (seasonal affective disorder—SAD), broken bones and fracture. The conventional light bulb as well as the fluorescent tubes is very inefficient in resembling sunlight, which makes the search for alternatives all the more important.

In addition to the medical benefits LEDs bring to human life, their development has potential impact on the amount of electricity used worldwide. It is currently estimated (in the year 2015) that lighting amounts to at least 20 % of the total world electricity produced. For a country like the UK, for example, this amounts to the equivalent cost of about $3bn/annum. If white-colour-based LEDs could replace the conventional light bulbs around the house and in offices, then energy saving of more than 80 % could result. The US Department of Energy estimates a projected energy saving amounting to about $30bn/annum, if LED replace incandescent lighting, with the added bonus of avoiding about 1800 million metric tons of carbon emissions [27]. Imagine this being applied worldwide; it would have huge savings on world resources but would particularly allow less developed countries to offer their citizens lighting resources at modest cost. Furthermore, LEDs are more compact in size, contain no mercury—hence safer to use and dispose of—and have more than 30 times longer life (~50,000 h of use).

But what is a light-emitting diode, and what does it consist of? And more crucially what does electron microscopy have to do with its development?

The light-emitting diode is one of many types of the basic diode that are used as the building blocks in semiconductor devices. However, there are some crucial differences between diodes used in general electronic devices, such as the indispensable silicon p-n junction diodes used in most electronic devices, and those employed for lighting. In the former case, a piece of silicon, for example, is designated into two parts. In one part a material like boron (B) replaces the host silicon atoms to a value not exceeding 0.01 % of the total volume which makes the p-region, while on the adjoining part a material like phosphorous (P) is used to a similarly low concentration and this makes the n-region of what becomes a p-n device. The n-region is the one which has more electrons per unit volume than the host silicon whilst the p-region has more 'holes', i.e. less electrons. It is the electron movement within the two regions which causes the flow of electrical signal throughout the electron devices.

The development of white LED technology has eluded scientists for many years. Red and green LEDs have been available for almost 50 years, but it was the blue LED that was required to make up the white light. The case for a successful and functioning blue LED is rather special, though. For brevity, one can summarise the material requirements for producing blue LEDs to be:

1. The semiconductor material has to be of direct type (i.e. when an electron loses energy by falling across the band gap, no phonons—which leads to heat—are also produced.). For more explanation of direct and indirect band gap semiconductors, see Chap. ▶ 10.
2. The energy gap should be intermediate in the range 1.77 and 3.1 eV, which corresponds to a wavelength in the visible spectrum between 400 and 700 nm.
3. The material should be amenable to the formation of p-n junction diodes.

The above requirements eliminate widely used semiconductor materials like silicon and germanium which are both of the indirect type (see [28] for more

Chapter 6 · Modern Electron Optics and the Search for More Light. . .

141

6

discussion of semiconductor fundamentals). Although this is the case, this did not stop researchers from developing LEDs based on silicon compounds, such as silicon carbide (SiC) which has been successfully used to produce commercial LEDs, albeit with very low efficiency.

The search for a material or a compound of materials to satisfy all the above three requirements looks as if it has finally been found in gallium nitride (GaN). This material is a direct III–V semiconductor with an energy gap of 3.36 eV. It is now considered as one of the leading candidates in this technology and commercial devices using this material have already been on sale for the last year or two. The development of this compound material owes heavily to the work and contribution of two groups, *Isamu Akasaki, Hiroshi Amano* of Nagoya University, Japan and *Shuji Nakamura* of the University of Santa Barba Ca, USA, who shared the Nobel Prize for Physics for 2014 'for the invention of efficient blue light-emitting diodes which has enabled bright and energy-saving white light sources'. Whilst they demonstrated the blue-light laser using GaN during the late 1980s and early 1990s, it took many research groups in academia and industry alike all over the world nearly three decades to develop efficient methods for the reliable and large-scale production of GaN. Professor Sir Colin Humphreys of the University of Cambridge, UK (see Chap. ▶ 5) has been one of the world's leading scientists in developing GaN-based LEDs, and his research group's work has built upon the demonstration of the Japanese Nobel Prize winners and developed new methods for the large-scale production of commercial material. The use of electron microscopy has been crucial in this endeavour, and examples from work published by Humphreys' group will be briefly reviewed here, although parallel stories indicative of the crucial role that electron microscopy has played apply in the case of all the other examples and in the various developments by scientists all over the world towards this goal.

The challenge faced by the community researching GaN for use in LED technology, following the Japanese demonstration, was in producing high-quality material free of defects. It is important to appreciate that the role of defects in semiconductors is similar to that of the 'dopant' materials one adds to form the p-n junctions. Further, the small amount of this foreign material or defect that alters the electrical behaviour of the semiconductor is comparable in size to that of a golf ball and a football pitch. However, in terms of the foreign material one adds to make a p- or n-type semiconductor, one knows a priori its characteristics and how much to add to achieve the required results. In the case of material defects, this is uncontrolled with devastating consequences, if their numbers exceeds a certain limit.

One of the challenges faced by Humphreys' group was in using silicon wafers as the base on which to grow the various layers of materials needed to produce the LED device. A schematic of a typical LED device based on GaN is shown in ◘ Fig. 6.17 [26]. The substrates used in this technology and almost all other semiconductor devices, and which do not have an active role in the working of the device, are normally made from silicon wafers to reduce the cost of the more expensive active materials like GaN. However, it is the substrate material that causes some of the dislocation defects seen in the grown GaN materials. Such dislocations reduce the efficiency of the LED and their reduction is therefore of utmost importance in the search for efficient devices.

In addition, depositing GaN on Si causes the former to react with the Si to form a Ga-Si alloy and melt-back etching that alters the concentration of the constituting elements. To avoid this from happening, a layer of aluminium nitride (AlN) is first deposited on the Si as a nucleation layer. The quality of this layer and its interface with the Si substrate are of great importance for the produced LEDs; and the interfacial layer between the Si and the AlN, as depicted in ◘ Fig. 6.18, was only possible to study by using a state-of-the-art TEM with a resolution of 0.1 nm.

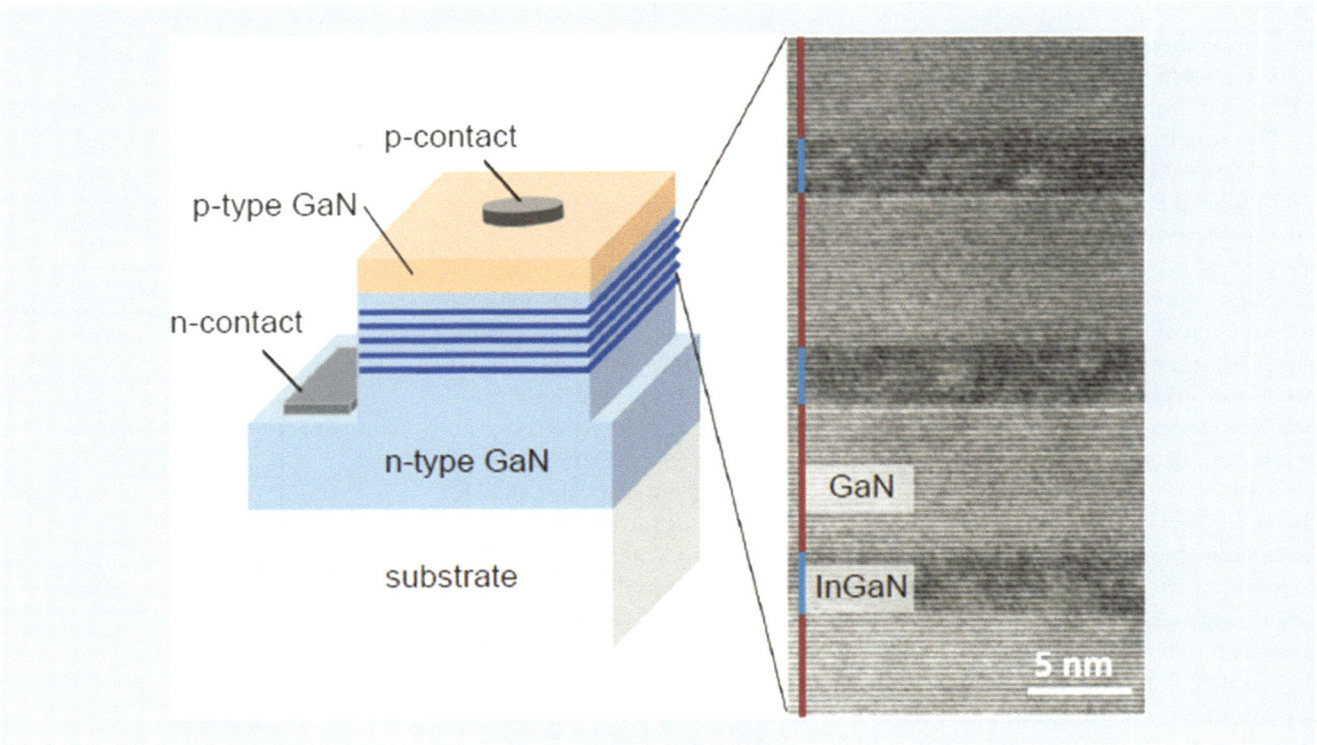

Fig. 6.17 A schematic of an InGaN/GaN quantum well LED (courtesy of Prof. Humphreys) [29]

Fig. 6.18 TEM Image from—Cs corrected instrument showing the lack of crystallinity at the interface between the substrate and the deposited semiconductor (courtesy of Prof. Humphreys) [29]

This TEM resolution could only be achieved after the correction of the microscope's spherical aberration. It is interesting to note that the solution offered by Humphreys' research group has now successfully been taken up by UK industry (Plessey Semiconductors).

It is also interesting to note that whilst most researchers thought that the use of indirect bandgap semiconductors is not suitable for efficient LEDs, nanotechnology is providing a route where this may not necessarily be the case. The work reviewed by Professor Nayfeh in this book demonstrates that indeed there is life for silicon in producing efficient lighting if one uses silicon nano-particles (see Chap. ▶ 10).

Chapter 6 · Modern Electron Optics and the Search for More Light...

143   **6**

6.10 Conclusions

The development of the blue LED, which earnt its developers the Nobel Prize in 2014, and the subsequent tireless work of scientists and engineers in its development as a working device, could not have been realised had they not used state-of-the-art TEMs which are equipped with aberration correctors, but equally if scientists and engineers had not followed the scientific method of enquiry in their research and development. The use of the law of refraction is fundamental in electron and light optics, and the name of Ibn Sahl, the tenth century Arab mathematician, should be on a par with that of Snell. It is also clear that the correction of the spherical aberration defect in electron optical instruments and in other imaging devices is crucial for achieving the ultimate performance of such devices. Whilst Ibn al-Haytham's early discovery of spherical aberration should be correctly acknowledged, as well as his many other pioneering work in optics, his legacy in science goes far beyond these achievements (see Chap. ▶ 2 for more details). More importantly in relation to science and engineering, Ibn al-Haytham again deserves credit and recognition for his role in laying down the foundations of the scientific method of enquiry in his work on optics some 1000 years ago entitled, Kitab al-Manazir. It is fitting therefore for UNESCO to recognise this year and celebrate the contribution of Ibn al-Haytham as a pioneering polymath.

References

1. Rashed R (2005) Geometry and dioptrics in classical Islam. al-Furqan Islamic Heritage Foundation, London
2. Dahl PF (1997) The flash of the cathode rays. CRC, Boca Raton
3. Leicester HM (1971) The historical background of chemistry. Dover Publications Inc, New York
4. Grivet P (1972) Electron optics, 2nd English edn. Oxford, Pergamon.
5. Prutton M, El-Gomati MM (2004) Scaning Auger Electron Microscopy, John Wiley & Sons, Chichester
6. Fultz B, Howe J (2013) Transmission electron microscopy and diffractometry of materials. In: Di Meglio J-M, Rhodes WT, Scott S, Stutzmann M, Wipf A (eds) Graduate texts in physics. Springer, Heidelberg
7. Chescoe D, Goodhew PJ (1984) The operation of the transmission electron microscope. Royal Microscopical Society Microscopy Handbooks 02, Oxford University Press, Oxford
8. Orloff J (ed) (2008) Handbook of charged particle optics, 2nd edn. CRC, Baton Rouge
9. Bronsgeest M (2014) Physics of Schottky electron sources. Pan Stanford Publishing, Pte, Singapore
10. Charbonnier F (1990) Developing and using the field emitter as a high intensity electron source. Appl Surf Sci 94–95:26–43

6

11. Milne WI, Teo KBK, Mann M, Bu IYY, Amaratunga GAJ, De Jonge N, Allioux M, Oostveen JT, Legagneux P, Minoux E, Gangloff L, Hudanski L, Schnell J-P, Dieumegard LD, Peauger F, Wells T, El-Gomati M (2006) Carbon nanotubes as electron sources. Phys Status Solidi 203 (6):1058–1063

12. Wahab H (2012) Metal oxide catalysts for carbon nanotubes growth: The growth mechanism using NiO and doped ZnO. PhD thesis, University of York, UK, York

13. Algarni, Hamed (2013) Synthesis and Characterizations of ZnO Nanostructures for Field Emission Devices. PhD thesis, University of York, UK, York

14. de Jonge N, Lamy Y, Schoots K, Oosterkamp TH (2002) High brightness electron beam from a multi-walled carbon nanotube. Nature 420:393–395

15. Krvanek OJ, Delby N, Murfitt M (2009) Handbook of charged particle optics, 2nd edn. CRC, Boca, Boca Raton

16. Yu Z, Pan Y, Shen Y, Wang Z, Ong Z-Y, Xu T, Xin R, Pan L, Wang B, Sun L, Wang J, Zhang G, Wei Zhang Y, Shi Y, Wang X (2014) Towards intrinsic charge transport in monolayer molybdenum disulfide by defect and interface engineering. Nat Commun 5, 5290

17. The Hubble telescope mission web site. ► https://www.nasa.gov/mission_pages/hubble/main/index.html

18. El-Gomati MM, Zaggout F, Jayakody H, Tear S, Wilson K (2005) Why is it possible to detect doped regions of semiconductors in low voltage SEM: a review and update. Surf Interface Anal 37:901–911

19. Walker CGH, Zaggout F, El-Gomati MM (2008) The role of oxygen in secondary electron contrast in doped semiconductors using low voltage scanning electron microscopy. J Appl Phys 104:123713

20. El-Gomati MM, Mullerova I, Frank L (1998) The electron. IOM communications Ltd, London, pp 326–333

21. Spallas JP, Silver CS, Murray LP, Wells T, El-Gomati MM (2006) A manufacturable miniature electron beam column. Microelectron Eng 83(4–9):984–985

22. Goldstein J, Newbury DE, Joy DC, Lyman CE, Echlin P, Lifshin E, Sawyer L, Michael JR (2003) Scanning electron microscopy and X-ray microanalysis, 3rd edn. Springer, Heidelberg

23. Birdi KS (2003) Scanning probe microscopes, applications in science and technology. CRC, Boca Raton

24. Kirk TL (2010) Near field emission scanning electron microscopy. PhD thesis, Swiss Federal Institute of Technology, Zürich

25. Dowed J (2012) The vitamin D cure. John Wiley & Sons, Hoboken, New Jersey

26. SACN Update on Vitamin D (2007) Public Health England report, TSO, London

27. USA Department of Energy report (2014) Office of communication, Washington

28. Sze SM (1981) Physics of semiconductor devices, 2nd edn. Wiley, New York

29. Humphreys, see chapter 5 this book.

Applications

Contents

The Dawn of Quantum Biophotonics

Dmitri V. Voronine, Narangerel Altangerel, Edward S. Fry, Olga Kocharovskaya, Alexei V. Sokolov, Vladislav V. Yakovlev, Aleksey Zheltikov, and Marlan O. Scully

D.V. Voronine (✉) • A.V. Sokolov • M.O. Scully
Texas A&M University, College Station, TX 77843, USA

Baylor University, Waco, TX 76798, USA
e-mail: scully@tamu.edu

N. Altangerel • E.S. Fry • O. Kocharovskaya • V.V. Yakovlev
Texas A&M University, College Station, TX 77843, USA

A. Zheltikov
Texas A&M University, College Station, TX 77843, USA

Moscow State University, Moscow, Russia

© The Author(s) 2016
M.D. Al-Amri et al. (eds.), Optics in Our Time, DOI 10.1007/978-3-319-31903-2_7

7

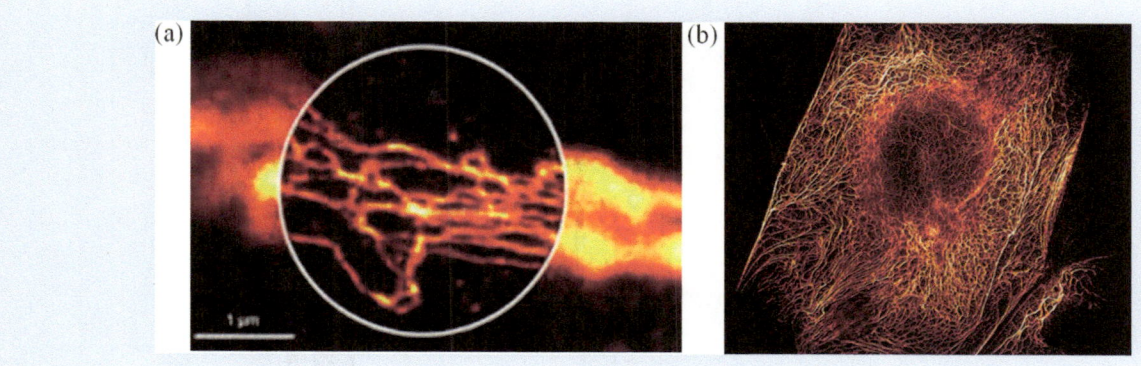

Fig. 7.1 (**a**) STimulated Emission Depletion (STED) microscopy with enhanced resolution (inside the circle) compared with the conventional optical microscopy (outside of the circle). (**b**) REversible Saturable OpticaL Fluorescence Transitions (RESOLFT) image of keratin in cells. Adapted from [1, 2]

7.1 Overview: Toward Quantum Agri-Biophotonics

Quantum mechanics, the crowning achievement of twentieth century physics, is yielding twenty-first century fruit in the life sciences. For example, the broad and multi-faceted field of *Quantum Biophotonics* is exciting and rapidly developing, with many emerging techniques and applications. The broad range of topics includes remote sensing with applications toward plant phenotyping, single cell/virus/biomolecule detection, and superresolution imaging. Progress in the latter was recognized with the 2014 Nobel Prize in Chemistry (Fig. 7.1). New developments in this field are promising and may deliver at least an order of magnitude improvement.

Laser spectroscopy has been widely used for chemical analysis of the living systems. For example, Raman and infrared (IR) spectroscopies can probe the vibrational states of molecules in order to determine, e.g., chemical assay and temperature profile. IR spectroscopy is widely used because it is simple and inexpensive. Raman spectroscopy is more complicated but has many advantages and is a more versatile and powerful tool. Coherent and stimulated Raman techniques can be used to increase the speed and strength of signal acquisition by orders of magnitude. Femtosecond adaptive spectroscopic techniques for coherent anti-Stokes Raman spectroscopy (FAST CARS) was used to detect small amounts of anthrax-type endospores on a nanosecond time scale [3–5] and inside a closed envelope [6]. Fluorescence measurements of fecal matter in water achieved increase of sensitivity by three orders of magnitude [7]. Laser-induced breakdown spectroscopy has been developed to perform studies of plant physiology and phenotyping. Surface-enhanced coherent Raman spectroscopy achieved astonishing results in detection sensitivity [8, 9]. All these techniques are aimed at increasing the speed and reliability of field-based sensing and can be used for improving crop yield. These scientific and technical innovations hold promise for new diagnostic tools bridging the gap between fundamental quantum research and potential agricultural applications.

7.2 Fundamental Light–Matter Interactions and Spectroscopy of Biological Systems

Quantum mechanics provides the most complete description of the light–matter interactions and of the various spectroscopic techniques used for probing the structure–function relations in many fields of science and engineering. These

7

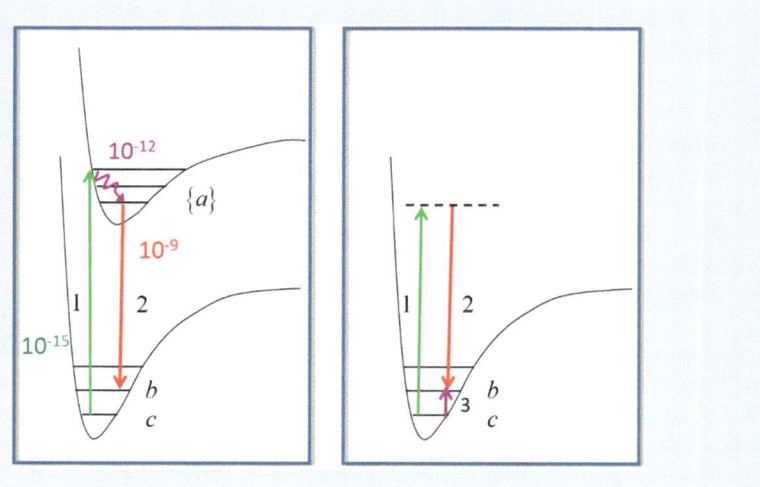

Fig. 7.2 Fundamental light–matter interactions in bio-objects involve absorption, emission, and scattering of light (represented by *arrows*) by molecular systems (represented by *energy level diagrams*). Absorption, relaxation, and fluorescence emission take place on a time scale of 10^{-15} s, 10^{-12} s, and 10^{-9} s, respectively (*left panel*). Non-resonant Raman scattering occurring through a virtual level (*dashed line*) quasi-instantaneously (*right panel*). Infrared absorption (*purple arrow, right panel*) provides complimentary information to Raman spectroscopy

interactions happen on different time scales from femtoseconds [$\sim 10^{-15}$ as in the case of absorption shown by a green arrow 1 in ◘ Fig. 7.2 (left)] to picoseconds [$\sim 10^{-12}$ as in the case of vibrational cooling shown by a purple wavy arrow in ◘ Fig. 7.2 (left)] to nanoseconds [$\sim 10^{-9}$ as in the case of fluorescence emission shown by a red arrow 2 in ◘ Fig. 7.2 (left)] and longer. Some processes are almost instantaneous, for example, Raman scattering [shown by red arrow 2 in ◘ Fig. 7.2 (right)]. Modern pulsed laser sources with pulse durations from nanoseconds to femtoseconds and beyond can be used to investigate and control these processes. Fluorescence and Raman scattering are major laser spectroscopic techniques which can be used for analytical purposes in agricultural, biomedical, environmental, and many other applications. Infrared absorption also provides complimentary information. Below we describe several examples of recent breakthroughs using these and other techniques within the general field of quantum biophotonics.

The Raman spectroscopic technique is a valuable tool which has excellent potential for the analysis of plant and animal agriculture. Raman spectroscopy is a vibrational measurement which can be applied directly to plant and animal tissues yielding characteristic key bands of individual constituent components. The physics of the Raman effect is analogous to scattering off of an oscillating mirror. It relies on inelastic scattering of light from a laser in the visible, near-infrared, or near ultraviolet range off of molecules. The laser light interacts with molecular vibrations resulting in the energy of the laser photons being down shifted (Stokes) or up shifted (anti-Stokes) signals, respectively (◘ Fig. 7.3). The shift in energy gives useful information about the vibrational modes in the system providing molecular "fingerprints" which can be used for identification. Moreover, essentially all molecules have Raman active vibrational transitions.

Unlike optical absorption transitions, it is possible in principle to separate out the Raman emission of a particular target molecule from the many background molecules in the focal volume. This is because the molecular vibrational levels scatter light with distinctive frequency shifts that are often narrowband. Since there are usually many vibrational transitions in a biomolecule, there are many Raman lines to choose from and these can be used to fingerprint the target molecule. Another important advantage of Raman techniques is that the laser

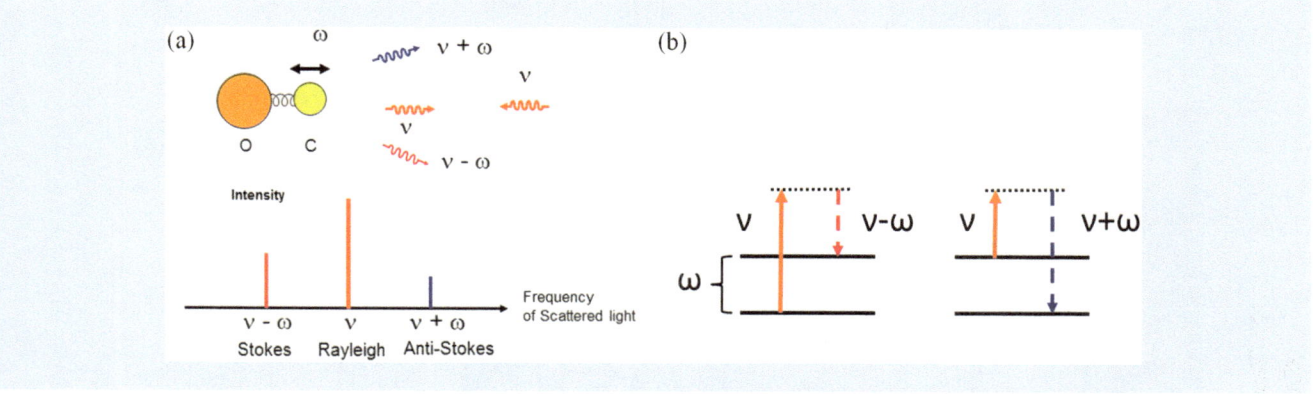

Fig. 7.3 The Raman effect (**a**) and the energy level Stokes and anti-Stokes diagrams (**b**) for a simple molecule. Light at the probe frequency v is inelastically scattered off of the molecule resulting in the characteristic frequency shift which serves as a fingerprint for chemical analysis

does not need to be tuned near the optical transition. Even if the optical transition is in the UV, Raman transitions can be efficiently excited with lasers tuned as far away as the near IR, for example, 1064 nm or longer. For such large detunings, the excited state is essentially unpopulated, and so laser damage effects (such as bleaching) or background from chlorophyll fluorescence is strongly suppressed. These features make the Raman approach particularly useful for plant and animal studies which deal with highly complex samples and their environment. ▫ Figure 7.4 shows an example of application of Raman spectroscopy for the early detection of anthocyanin markers of stress in plants [10].

Brillouin scattering is another light–matter interaction phenomenon which has delivered an emerging biomedical tool that has already been used to study bone, collagen fibers, cornea, and crystalline lens tissue. Unlike Raman spectroscopy, which offers information about the chemical makeup of the sample, Brillouin spectroscopy provides information about the viscoelastic properties of a material, and consequently, can characterize larger bulk changes. Each of these imaging tools offers useful diagnostic information. Therefore a single apparatus that could provide simultaneous measurement of both spectra from the same point would be extremely powerful for sample characterization and analysis [11]. However, if separate instruments are used, the acquisition time for each Brillouin spectrum is very long (~15 min), making such approach impractical. The lack of the same-point detection for both spectra makes the analysis complicated. To overcome these issues, we recently used a single pump laser to generate both Raman and Brillouin spectra [12] and to provide simultaneous imaging from the selected confocal volume. More importantly, we take advantage of the recent advancements in Brillouin spectroscopy to decrease the acquisition time, as any practical implementation of the simultaneous detection requires that the times for both to be comparable. Unlike other approaches that use scanning Fabry Perot cavities, we utilized a virtually imaged phase array (VIPA), which offers a higher throughput efficiency, >80 %, and does not require scanning to extract a complete spectrum. Subsequently, a VIPA-based system drastically cuts down the acquisition time, which was traditionally a limiting factor in Brillouin spectroscopy. The major challenge in Brillouin spectroscopy of biological systems is eliminating the large amount of elastic scattering, which makes it difficult to identify a weak Brillouin peak. We recently reported that this limitation can be overcome by using a molecular/atomic gas cell as a notch filter [13]. Utilizing these advancements, we demonstrate simultaneous Raman–Brillouin microscopy, a potent new tool for bioimaging and analytical characterization.

Both Raman and Brillouin phenomena arise from the inelastic scattering of light, where the scattering causes the frequency of light to shift in accordance with

7

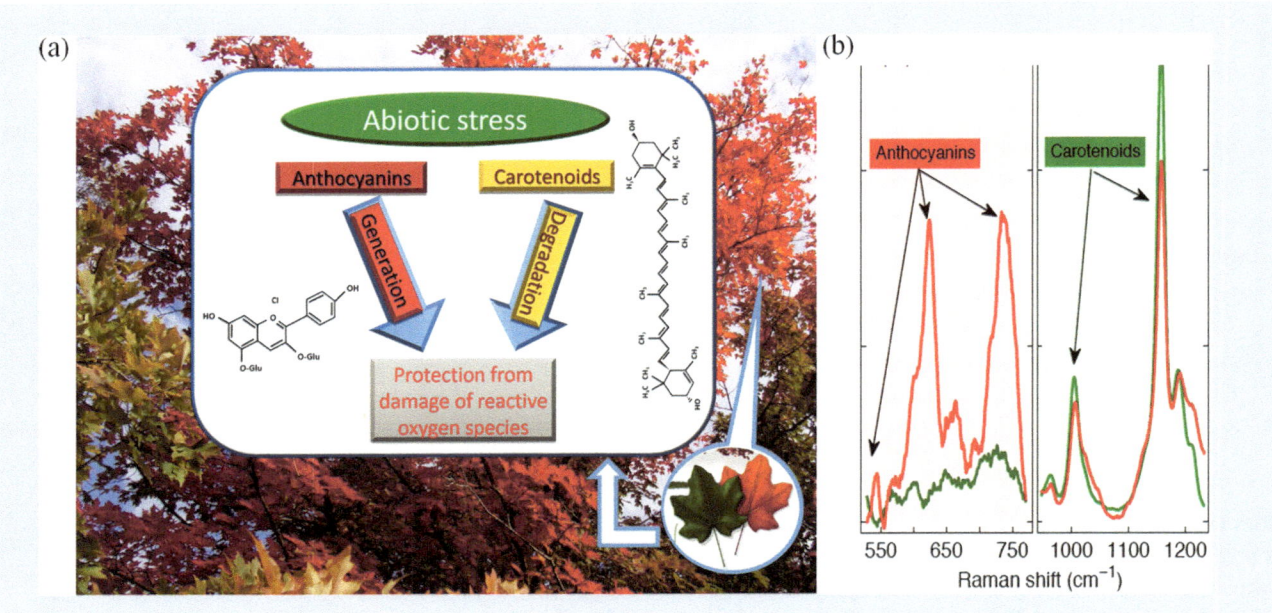

Fig. 7.4 (**a**) The autumn colors of maple trees are due to anthocyanins and are indicative of plant stress. To enhance the survival probability of stressed crops, it is important to identify stressed plants as early as possible to allow for effective intervention. Raman spectroscopy can reveal early spectroscopic signatures of carotenoids and anthocyanins. (**b**) Raman spectra of the unstressed (*green*) and saline stressed (*red*) plants after 48 h. Adapted from [10]

some resonant property. In the case of Raman, the incident light interacts with molecular vibrations causing the light to shift. Similarly, Brillouin scattering is caused by the inelastic interaction of light with periodic fluctuations in a material's index of refraction. From the quantum physics perspective, Brillouin scattering is an interaction between an electromagnetic wave and a density wave (photon–phonon scattering). Thermal motions of atoms in a material create acoustic vibrations, which lead to density variations and scattering of the incident light. These fluctuations carry information about a material's bulk compressibility and viscoelasticity. Whereas Raman scattering can have frequency shifts on the order of 100 THz, Brillouin shifts are only on the order of 10 GHz due to the relatively low energy of the acoustic phonons. The magnitude of the Brillouin shift is dependent upon the collection geometry. Through the measured Brillouin shift we are able to extract mechanical properties of the material, including the speed of sound, adiabatic compressibility, and the longitudinal modulus. While Brillouin scattering is most often used to measure elastic properties of materials, the linewidth of the Brillouin signal also provides information about the viscoelasticity.

The imaging capability of the Raman–Brillouin microscope is shown using a T-shaped sample made of two materials with different mechanical and chemical properties (◘ Fig. 7.5). Cyclohexane and poly(ethylene glycol) diacrylate (PEGDA) hydrogel are model systems, and the hydrogel was cured in a "T"-shaped mold to provide spatial contrast. The body of the T-shaped structure was created using PEGDA. The T-shaped structure was then placed into a solution of cyclohexane, which provided contrast for the image. The corresponding Raman and Brillouin images are shown in ◘ Fig. 7.5. In all cases a high-contrast image was produced, whose spatial accuracy can be confirmed when compared to an optical photograph of the sample.

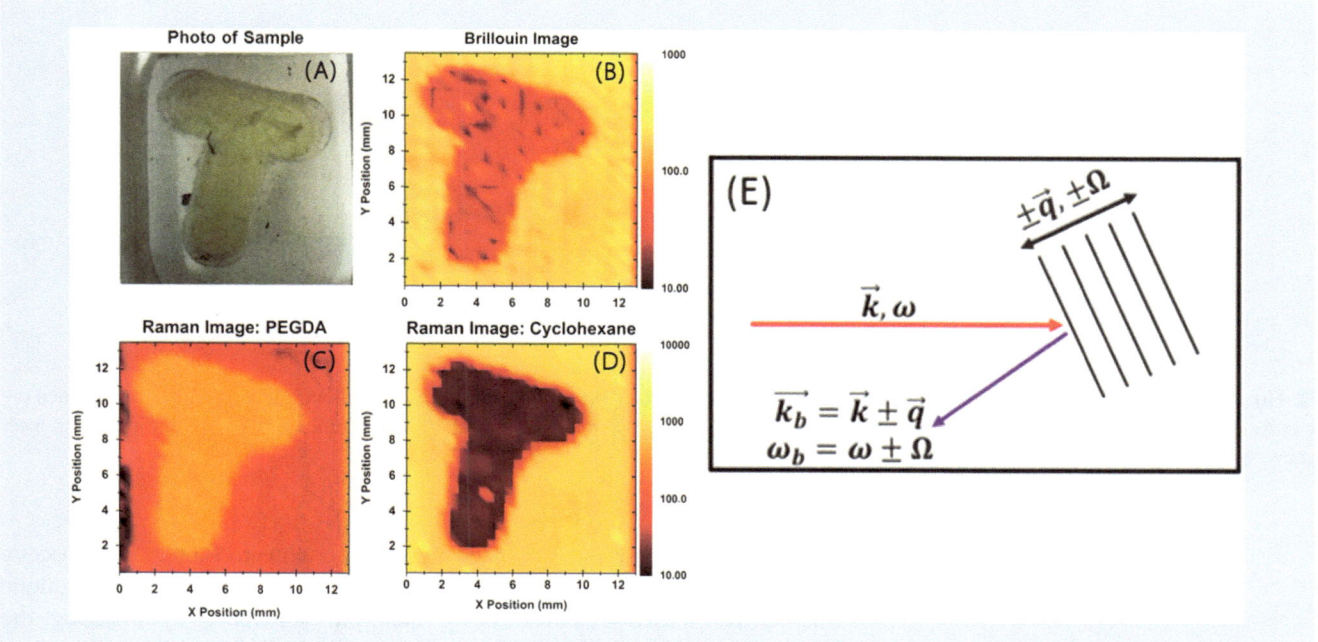

Fig. 7.5 Images of the T-shaped sample: (**a**) Optical photograph; (**b**) Brillouin image using the intensity of the cyclohexane peak at 4.58 GHz as contrast; (**c**) Raman image using the normalized ratio between the PEGDA peaks at ~1400 and ~2800 cm^{-1}; (**d**) Raman image using the cyclohexane peak near 800 cm^{-1} as contrast. (**e**) Schematic diagram of the Brillouin scattering process where incident light interacts with the acoustic field of the material. The magnitude of the frequency shift is dependent upon the direction of light scattered by the acoustic wave. Adapted from [12]

7.3   Quantum-Enhanced Remote Sensing

7.3.1   Anthrax Detection in Real Time

Chemically specific optical imaging and sensing techniques play a pivotal role in developing preventive measures to maintain chemical and biological safety. Frequently, objects to be imaged and identified are not present on a surface, and the light probe has to pass through layers of scattering material. For example, the anthrax-causing bacteria *Bacillus anthracis* can form micron-sized spores which can be present in air, or on the skin of cattle. Brucellosis-causing bacteria *Brucella* spp. can be present in strongly scattering liquids such as milk or blood.

In particular, the detection of anthrax in the farmland environment is an important problem of current interest. Clearly, it would be highly desirable to identify the chemical content of, for instance, air near farm animals, or in wool, and so forth, i.e., identify anthrax spores on-the-fly in air, or image samples of hair in order to chemically distinguish common biological and chemical threats. However, the detection of a chemically specific target in a harsh environment presents several problems. First, targeted molecules are not directly accessible for interrogation, and a remote sensing technique has to be used. On passing through the scattering medium (e.g., wool), both the incident and signal light are substantially diminished, weakening the signal and diffusing the spatial information about the signal's origin. Second, the chemical specificity of anthrax detection has to be maintained in the presence of a substantial background from surrounding chemicals.

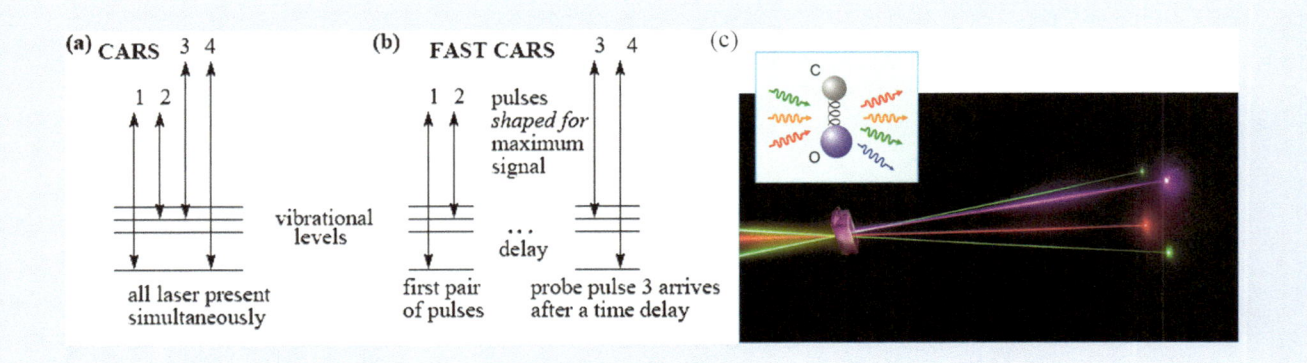

Fig. 7.6 Femtosecond adaptive spectroscopic techniques for coherent anti-Stokes Raman spectroscopy (FAST CARS). Comparison between the energy level schemes and laser configurations for the conventional CARS (**a**) and FAST CARS (**b**). (**c**) Sketch of the laser beam configurations. Inset shows a vibrating molecule emitting signals

The major drawback of the commonly used spontaneous Raman spectroscopy is the weak signal strength. We have developed a new spectroscopic technique (FAST CARS) based on maximizing quantum coherence by breaking the adiabaticity** and laser pulse shaping to optimize it. In FAST CARS spectroscopy the signal is proportional to the number of molecules squared, as opposed to the linear dependence in the spontaneous Raman technique [14]. In addition, FAST CARS was the first proposal for background suppression in precision sensing of minor molecular species within a highly scattering environment [3], and provided an efficient solution to the problem of detecting and identifying anthrax-type bacterial endospores in real time [4, 5].

Figure 7.6 shows the basic idea behind the FAST CARS approach. With standard CARS (Fig. 7.6a) two lasers with frequencies ω_1 and ω_2 (we often refer to these lasers as 1 and 2) are incident on a sample. The difference frequency may be resonant with some molecular vibrational excitation. At the same time, the sample may also absorb a third photon (either a third laser with frequency ω_3, or a second photon from laser 1, $\omega_1 = \omega_3$) and generate light at frequency $\omega_4 = \omega_1 - \omega_2 + \omega_3$. Variations of this scheme are numerous. For example, one of the two lasers 1 or 2 might be resonant with an electronic excitation in the molecule.

The problem with these schemes is that the process is masked by four-wave mixing (FWM) in a non-resonant medium. This produces broadband nonlinear generation that can be much larger than the small, vibrationally resonant CARS process. That is, the FWM generation at the detected wavelength range can be much larger than the signal we want to detect that is resonant with a specific molecular vibration. Various techniques have been proposed for suppressing this non-resonant process, including heterodyne detection of the CARS signal and the use of polarization tricks to suppress the undesired signals. FAST-CARS was developed for suppressing the nonresonant background based on pulse shaping and is shown in Fig. 7.6b. Here, all lasers provide short pulses. Pulses 1 and 2 are applied to the sample first and laser pulse 3 is delayed. When laser number 3 is applied to the sample and lasers 1 and 2 are not, non-resonant processes cannot occur. However, beams 1 and 2 will have excited coherence between the vibrational levels in the molecule for which they are Raman resonant. If this coherence last longer than the delay, laser 3 will scatter from the coherence and still produce a

---

**Two-photon resonant pulses produce $\rho_{bc} \neq 0$ quantum coherence by breaking adiabaticity of the molecular excitation; but off-resonant pulses return the molecule to the ground state as $\rho_{bc} = 0$

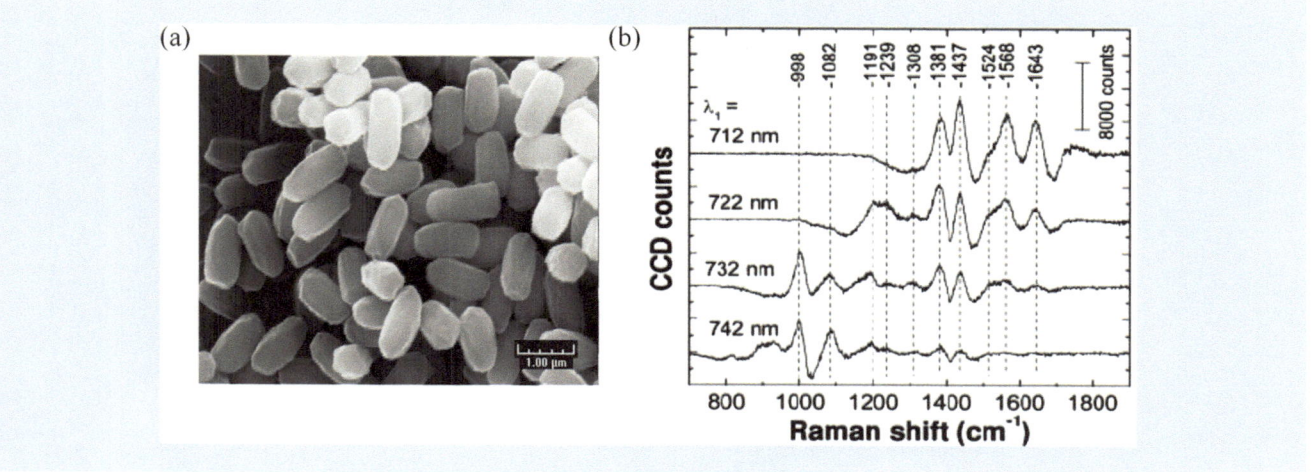

Fig. 7.7 FAST CARS on *Bacillus subtilis* spores (**a**) at various pump wavelengths (λ_1) from 712 to 742 nm (**b**). Adapted from [4]

CARS signal at ω_4. The amount of CARS generation as a function of delay time will show a temporal modulation characteristic of the vibrational energy level structure.

FAST CARS detection of micrometer size *Bacillus subtilis* spores has been performed (■ Fig. 7.7). A combination of ultrafast pump-Stokes Raman excitation and narrow-band probing of the molecular vibrations provides a species-specific signal from spores [4, 5]. In this scheme, the use of spectrally broad preparation pulses leads to the excitation of multiple Raman lines. The narrow-band probing, on the other hand, allows for frequency-resolved acquisition, as in traditional spontaneous Raman measurements. Recording of the whole CARS spectrum at once makes the technique relatively insensitive to fluctuations. The spectral contrast between the broadband preparation and narrow-band probing provides a way to differentiate between the Raman-resonant and non-resonant contributions. It also helps to mitigate the strength of the non-resonant FWM.

Frequently, objects to be imaged and identified are not present on a surface, and the light probe has to pass through layers of scattering material. In particular, the detection of weaponized anthrax in the mail room is an important problem of current interest. It would be highly desirable to identify the chemical content of an unopened envelope (i.e., image through layers of paper and chemically distinguish common biological and chemical threats from nonhazardous materials). However, the detection of a chemically specific target in a scattering medium presents several problems. First, targeted molecules are not directly accessible for interrogation, and a remote sensing technique has to be used. On passing through the scattering medium, both the incident and signal light are substantially diminished, weakening the signal and diffusing the spatial information about the signal's origin. Second, the chemical specificity of anthrax detection has to be maintained in the presence of a substantial background from surrounding chemicals, such as paper. Recently we demonstrated detection of anthrax-like spores inside a closed envelope using coherent Raman microscopy (■ Fig. 7.8) [6].

7.3.2 Stand-Off Spectroscopy

The need for an improved approach and efficient tools for remote optical sensing is high since they would facilitate applications ranging from environmental diagnostics and probing to chemical surveillance and biohazard detection. Present-day techniques rely on collecting incoherently scattered laser light and are often hindered by small signal collection efficiency. Availability of a laser-like

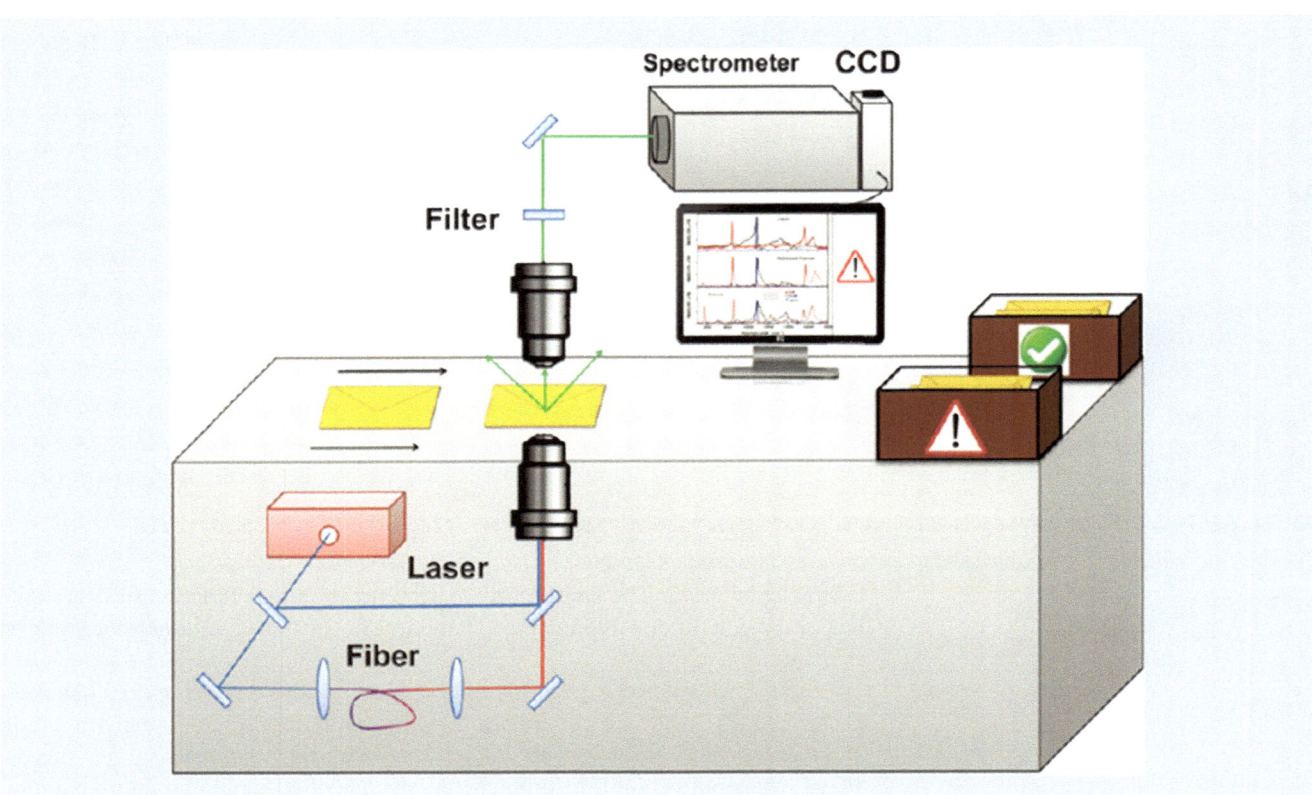

Fig. 7.8 Schematic diagram of the experimental setup for coherent Raman microspectroscopy imaging of anthrax-like spores inside a closed envelope. Adapted from [6]

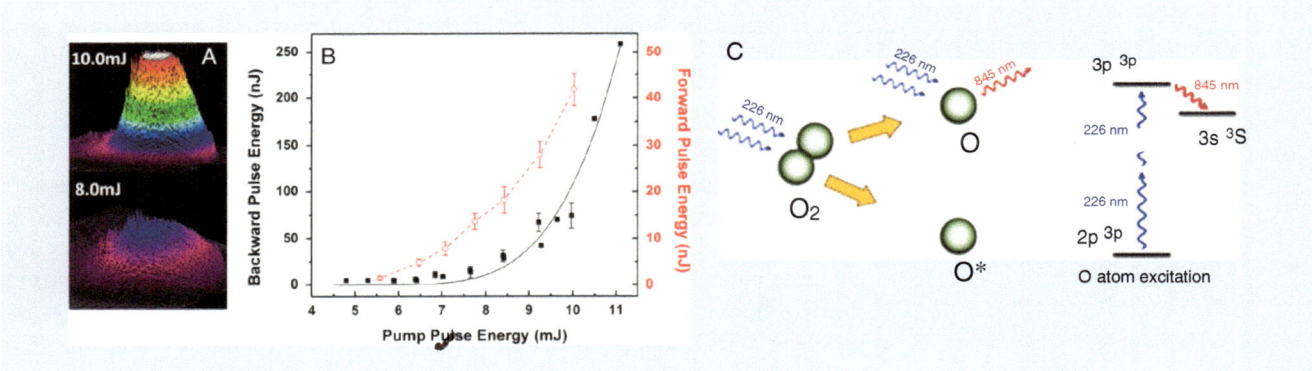

Fig. 7.9 (**a**) Spatial beam profiles of the 845 nm emitted backward pulse at pump energies above (10.0 mJ) and at (8.0 mJ) threshold. (**b**) The energy per pulse of both the forward (*red circles*) and backward (*black squares*) signals versus the pump power. (**c**) Two-photon dissociation of the oxygen molecule and subsequent two-photon resonant excitation of the ground-state oxygen atom fragment result in emission at 845 nm. Adapted from [16, 17]

light source emitting radiation in a controlled directional fashion [15], from a point in the sky back toward a detector, would revolutionize the area of remote sensing. As one example, we have demonstrated the possibility of remote lasing of atmospheric oxygen by using picosecond and nanosecond UV laser pulses that produce atomic oxygen via 226 nm UV pump light, and achieved a bright near-infrared (NIR) laser source at 845 nm wavelength (■ Fig. 7.9) [16, 17]. Nearly two decades after the work on stimulated emission (SE) performed in the context of flame and flow diagnostics, a renewed interest in laser-like emission from open air is motivated by the need for chemically selective stand-off detection of trace gases in the atmosphere [15].

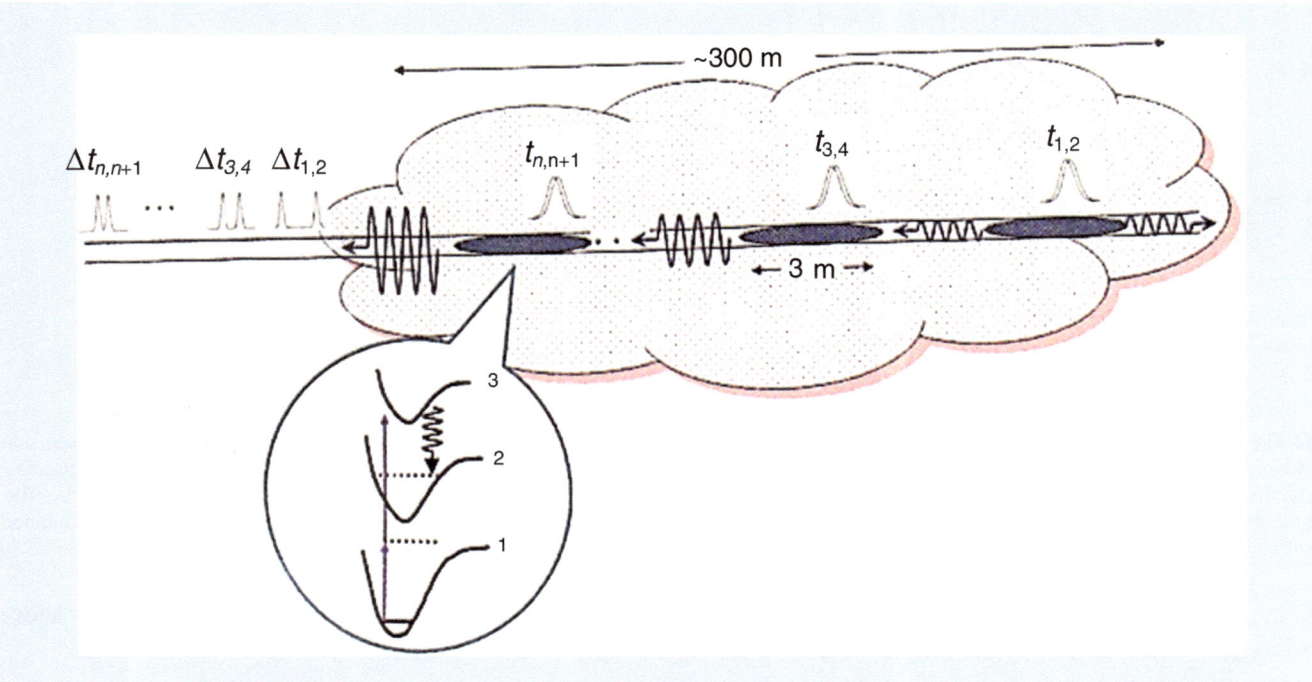

Fig. 7.10 SOS I. Pairs of laser pulses of different colors (e.g., *red* and *blue*) excite a dilute ensemble of molecules in a cloud such that lasing and/or gain-swept superradiance is realized in a direction back toward the observer [18]

Laser-like emission provides a promising tool for a broad class of all-optical stand-off detection methods, as it suggests a physical mechanism whereby a high-brightness, highly directional back-propagating light beam can be generated directly in ambient air. Superradiance can be used to enhance backward-directed lasing in air, using the most dominant constituents such as nitrogen or oxygen. We performed investigations of both the forward and backward-directed emission of oxygen when pumped by nanosecond UV laser pulses. The backward 845 nm beam profile is shown in ◘ Fig. 7.9a using nanosecond pulses approximately 10 mJ/pulse of 226 nm. High-quality, strong coherence-brightened emission was observed (◘ Fig. 7.9b).

Earlier, we proposed to use stand-off spectroscopy (SOS) techniques for detecting harmful impurities in air using gain-swept superradiance [18]. In our first SOS scheme it was demonstrated that by using pairs of laser pulses of different colors (e.g., red and blue) it is possible to excite a dilute ensemble of molecules such that lasing and/or gain-swept superradiance is realized in a direction back toward the out-going laser pulses (◘ Fig. 7.10). This approach is a conceptual step toward spectroscopic probing at a distance, also known as SOS [18].

Another simpler approach was developed on the basis of the backward-directed lasing in optically excited plain air (◘ Fig. 7.11). This technique relies on the remote generation of a weakly ionized plasma channel through filamentation of ultraintense trains of femtosecond laser pulses. Subsequent application of an energetic nanosecond pulse or series of pulses boosts the plasma density in the seed channel via avalanche ionization. Depending on the spectral and temporal content of the driving pulses, a transient population inversion is established in either nitrogen- or oxygen-ionized molecules, thus enabling a transient gain for an optical field propagating back toward the source and observer. This technique results in the generation of a strong, coherent, counter-propagating optical probe pulse. Such a probe, combined with a wavelength-tunable laser signal propagating in the forward direction, provides a tool for various remote sensing

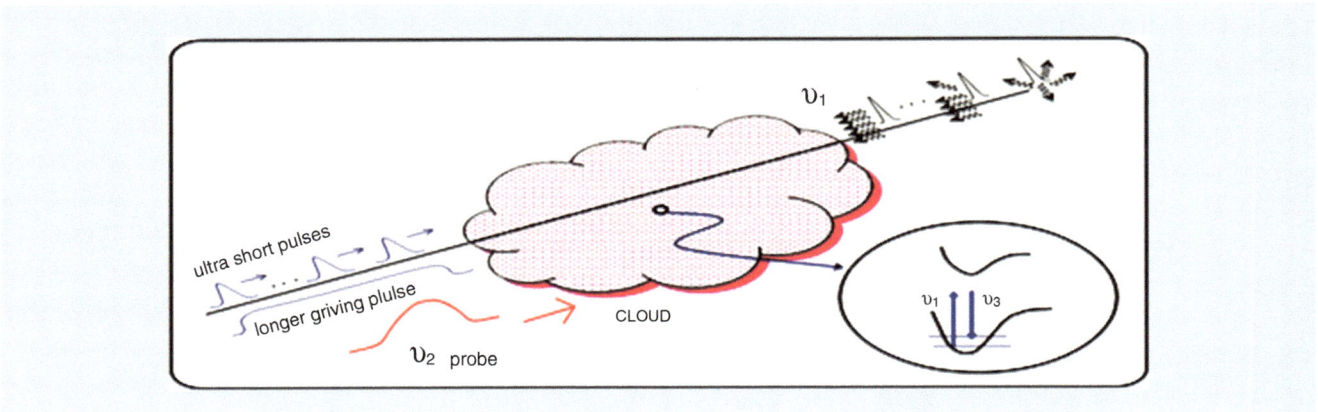

Fig. 7.11 SOS II. An ultrashort laser pulse or sequence of pulses is pre-chirped in such a way that they become compressed by the air dispersion at a pre-arranged distance behind the "cloud." Self-focusing collapse of these pulses results in the generation of weekly ionized "seed" plasma channels. The plasma density in these seed filaments is increased by several orders of magnitude by the application of a longer drive pulse. The properties of the pulses are tailored to produce population inversion in the ionized N_2 and O_2. The air laser radiation at frequency v_1 is combined with an interrogation pulse at v_2 to identify trace amounts of gas in the cloud. Adapted from [15]

applications. The technique could be implemented to probe the air directly above the growing crops and has the ability to pinpoint local areas of infection.

7.3.3 Detection of Plant Stress Using Laser-Induced Breakdown Spectroscopy

Remote sensing applied to detection of stress in plants is a very promising field. Plant stress affects the yield of agriculturally and economically significant plants. Rapid detection of plant stress in the field may allow farmers and crop growers to counter the effects of plant stress and increase their crop return. Plants can be affected by many types of stress including drought, pollution, insects, and microbial infestations which have a negative impact on agriculturally and economically significant florae. Rapid detection with little sample preparation is necessary to scale the sensing technology to the field size. We have recently applied laser-induced breakdown spectroscopy (LIBS) to plant stress detection. LIBS provides the advantages of rapid remote sensing [19]. LIBS measurements are performed on plants by focusing a laser pulse onto the surface of a sample causing ablation and vaporization of the sample material forming a plasma plume. The hot plasma with initial temperatures of up to 100,000 K expands and cools, emitting photons of characteristic frequencies from thermalized atomic constituents. LIBS can also be applied to detect other types of plant stress and other elements for phenotyping. For instance, cotton, a staple crop of Texas, is subject to various pestilences such as Southwestern rust (*Puccinia cacabata*), Alternaria leaf spot (*Alternaria macrospora*), and various types of boll rots. LIBS experiments can be performed using lab-based and portable femtosecond laser systems. We performed LIBS measurements for rapid analysis of the effects of drought stress on gardenia and wheat using a lab-based amplified femtosecond laser system operating at ~800 nm center wavelength, with a pulse duration of ~35 fs. We observed significant differences in the LIBS signals from stressed (not watered) and non-stressed (watered) plants and identified several atomic emission peaks as spectroscopic signatures of plant stress which agreed closely with macro- and micronutrients acquired by plants from the soil and air (Fig. 7.12). The LIBS technology may be able to identify key abiotic stress signals and improve the prediction of abiotic stress response capacity and can be used as a rapid remote sensing platform in the field.

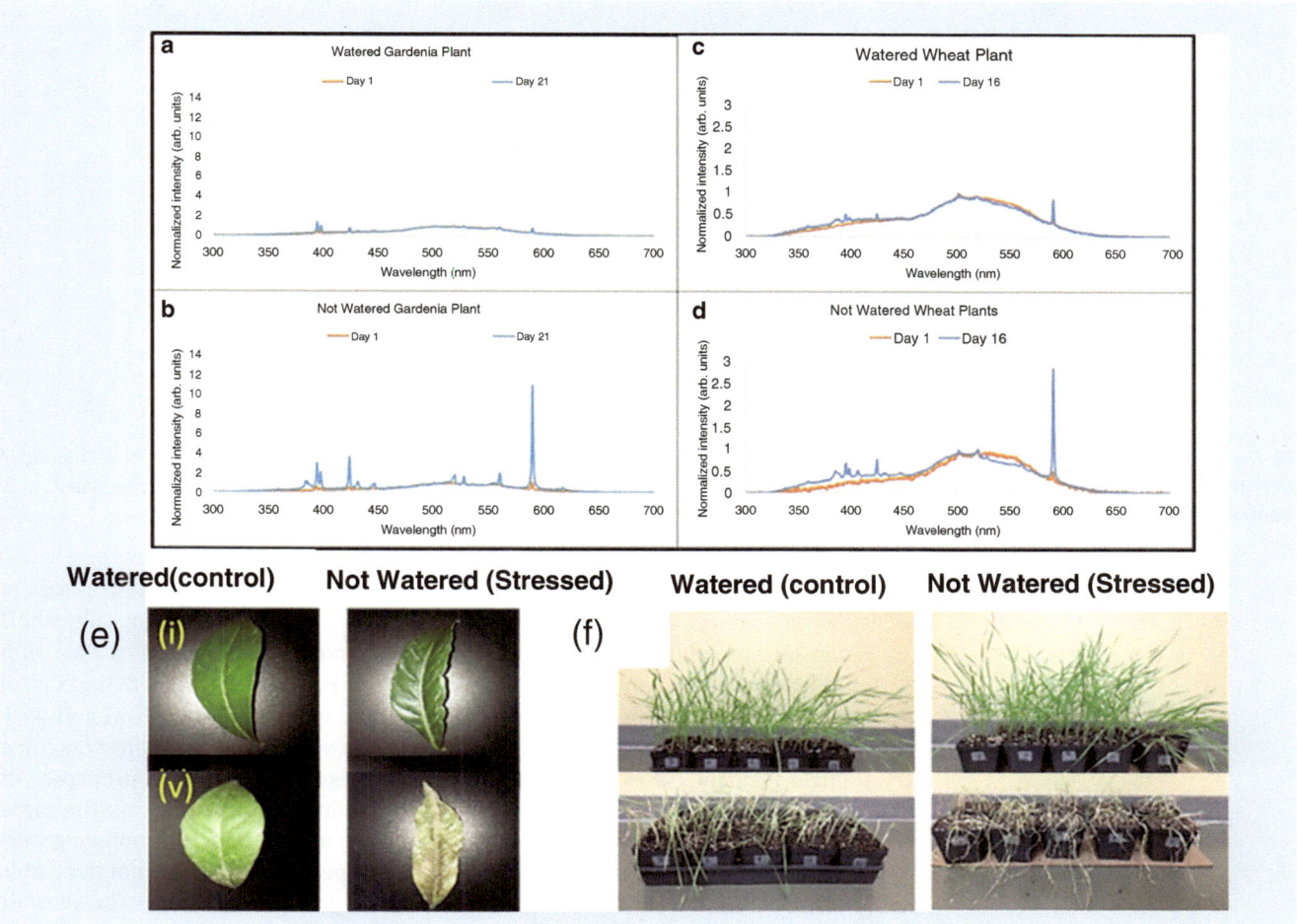

■ **Fig. 7.12** (**a–d**) Averaged LIBS spectra of watered (non-stressed) and not watered (stressed) gardenia and wheat plants. The corresponding photographs of the gardenia leaves (**e**) and wheat plants (**f**) taken on the first and last day of the treatment

7.3.4 Stand-off Detection Using Laser Filaments

Femtosecond filamentation has been observed for various pulse durations (from several tens of femtoseconds to picoseconds) and wavelengths (from UV to IR). Due to the presence of high intensity electromagnetic field $I(r, t)$ the refractive index of the medium has the form $n = n_0 + n_2 I(r, t)$, where the nonlinear Kerr index n_2 leads to an interesting effect of the curvature of wave front acting like a focusing lens ($n_2 > 0$) since the intensity is usually the highest at the center of the beam. The latter leads to self-focusing which can overcome the diffraction and leads to the collapse if the input peak power P_{in} exceeds a critical threshold value P_{cr}. Filaments may be also used to induce rain via water condensation in the atmosphere and to induce lightning via electron condensation.

Filamentation of ultrashort laser pulses in the atmosphere offers unique opportunities for long-range transmission of high-power laser radiation and stand-off detection. With the critical power of self-focusing scaling as the laser wavelength squared, the quest for longer-wavelength drivers, which would radically increase the peak power and, hence, the laser energy in a single filament, has been ongoing over two decades, during which time the available laser sources limited filamentation experiments in the atmosphere to the near-infrared and

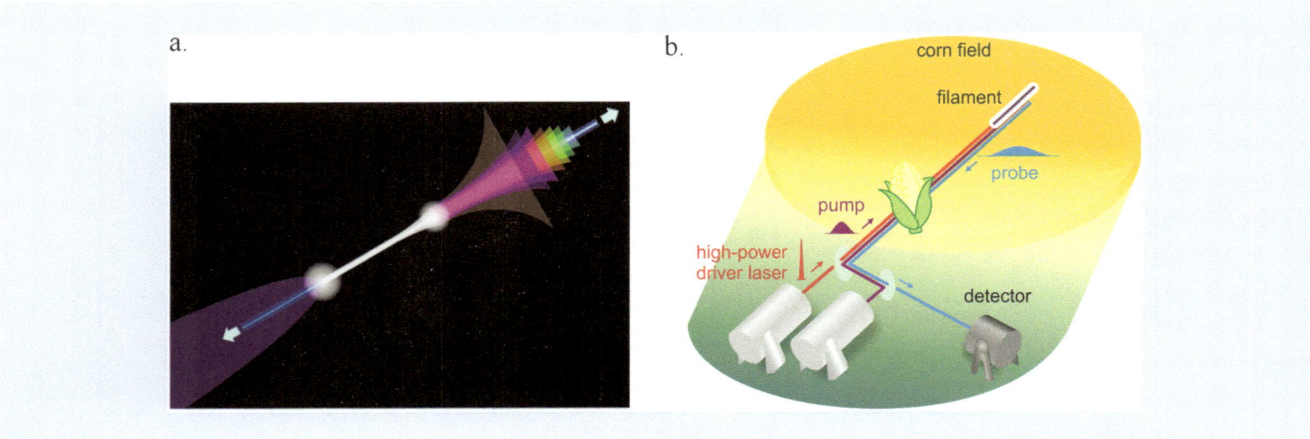

◘ Fig. 7.13 (**a**) Laboratory prototype of a unique mobile high-power laser source of ultrashort pulses in the mid-infrared allows filamentation of ultrashort mid-infrared pulses in the atmosphere. (**b**) This innovative technology offers unprecedented opportunities for long-range signal transmission, delivery of high-power laser beams, and remote sensing in agricultural applications. Adapted from [20]

visible ranges. Recently, a unique high-power laser source of ultrashort pulses in the mid-infrared has been created, allowing filamentation of ultrashort mid-infrared pulses in the atmosphere to be demonstrated for the first time with the spectrum of a femtosecond laser driver centered at 3.9 μm, right at the edge of the atmospheric transmission window, radiation energies above 20 mJ and peak powers in excess of 200 GW that can be transmitted through the atmosphere in a single filament (◘ Fig. 7.13) [20]. These studies revealed unique properties of mid-infrared filaments, where the generation of powerful mid-infrared supercontinuum is accompanied by unusual scenarios of optical harmonic generation, giving rise to remarkably broad radiation spectra, stretching from the visible to the mid-infrared. These thrilling discoveries open new horizons in ultrafast optical physics and offer unique opportunities for long-range signal transmission, delivery of high-power laser beams, and remote sensing of the atmosphere [21, 22]. Filament-induced breakdown spectroscopy may be realized as a remote extension of LIBS which was described in the previous section.

7.4   Quantum Heat Engines

Photonic quantum heat engines typically produce useful work by extracting energy from a high temperature thermal photon source, e.g., the Sun, and rejecting entropy to a low temperature entropy sink, e.g., the ambient room temperature surroundings. Lasing without inversion (LWI) is based on induced quantum coherence in the atoms, molecules, and solid state electrons involved. The early careful analyses of LWI were carried out by Kocharovskaya [23] using dark state, i.e., initial coherence; and Harris [24] using Fano interference, i.e., noise-induced coherence. Another related phenomenon is the correlated emission laser [25] which uses radiatively induced coherence to suppress absorption. It has also been shown that quantum coherence can be used to suppress absorption and obtain LWI [26].

More recently, it has become apparent that quantum coherence can be used to break detailed balance in a photocell and thus suppress recombination. This can increase quantum efficiency [27] and enhance thermodynamic power [28]. This reveals the deep connection between lasers and photovoltaic cells. It was shown that it is possible, in principle, to double the power of a thin medium photocell and/or the power of a laser. Finally it was also shown how quantum noise can be

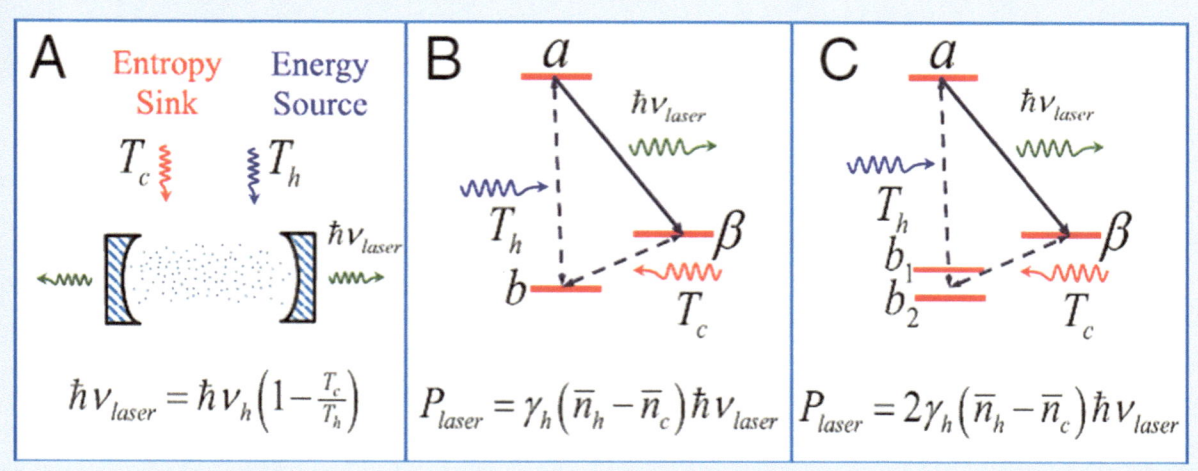

Fig. 7.14 (**a**) Schematic of a laser pumped by hot photons at temperature T_h (energy source, *blue*) and by cold photons at temperature T_c (entropy sink, *red*). The laser emits photons (*green*) such that at threshold the laser photon energy and pump photon energy is related by the Carnot efficiency relation. (**b**) Schematic of atoms inside the cavity. Lower level *b* is coupled to the excited states *a* and *β*. The laser power is governed by the average number of hot and cold thermal photons, $\overline{n}_h$ and $\overline{n}_c$. (**c**) Same as (**b**) but lower level *b* is replaced by two states b_1 and b_2, which can double the power when there is coherence between the levels. Adapted from [28]

used in the spirit of a quantum heat engine to explain the appearance of coherent oscillations and to increase of efficiency in photosynthesis [29].

7.4.1 The Laser and the Photovoltaic Cell as a Quantum Heat Engine

The arch-type example of a quantum heat engine is a laser pumped by hot thermal light as in ▣ Fig. 7.14. Then the frequency of a laser pumped by such narrow band hot light and cooled by narrow band cold light as well as the open circuit voltage of a solar cell or photo-detector obeys the Carnot relations.

However, it is possible to use quantum coherence to go beyond both of these Carnot relations. For example, including quantum coherence in the lower laser state, as in ▣ Fig. 7.14, yields the increase in quantum efficiency for the laser. Likewise it is possible to use quantum coherence to increase voltage quantum efficiency for the photovoltaic cell (▣ Fig. 7.15). However one might well wonder if would that not violate the second law of thermodynamics? The answer is no. Quantum mechanics does allow us to get more energy from a thermal reservoir than a classical Carnot engine can, but at a cost. Overall the Carnot limit applies, but in a subsystem we can do better than the Carnot limit. A clean example of this is the photo-Carnot quantum heat engine which we discuss next.

7.4.2 The Photo-Carnot Quantum Heat Engine

The photo-Carnot engine is simply a piston engine in which photons replace the molecules as the driving fluid, as in ▣ Fig. 7.16. As such, the photons are like the steam molecules of a steam engine. However, the thermal photons are generated in

7

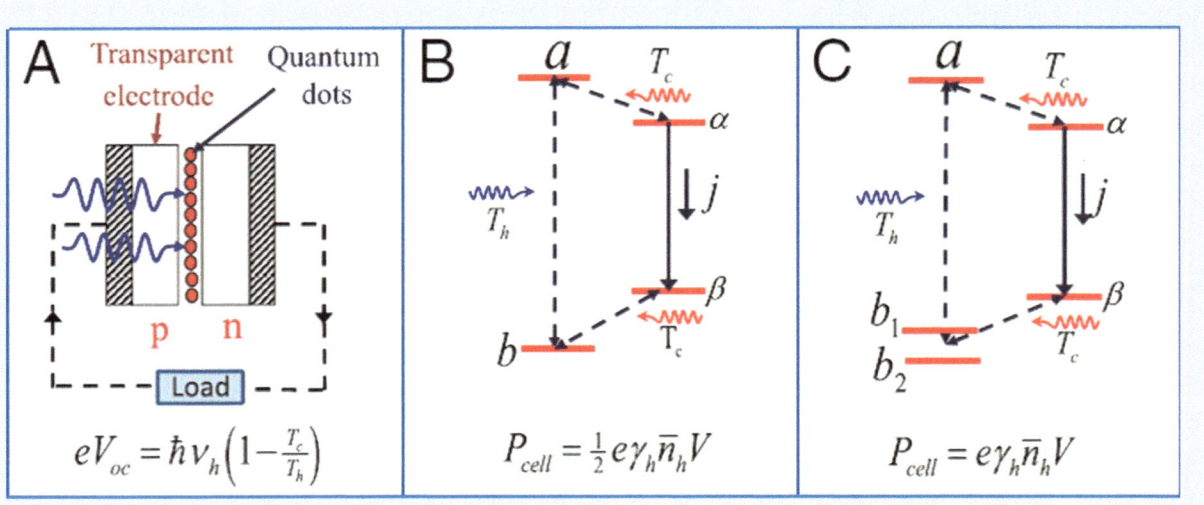

Fig. 7.15 (**a**) Schematic of a photocell consisting of quantum dots sandwiched between p and n doped semiconductors. Open circuit voltage and solar photon energy $\hbar \nu_h$ are related by the Carnot efficiency factor in which T_c is the ambient and T_h is the solar temperature. (**b**) Schematic of a *quantum dot* solar cell in which state b is coupled to a via, e.g., solar radiation and coupled to the conduction band reservoir state α via optical phonons. The electrons in state α pass to valence band reservoir state β via an external circuit, which contains the load. (**c**) Same as (**b**) but lower level b is replaced by two states b_1 and b_2, and when coherently prepared can double the output power. Adapted from [28]

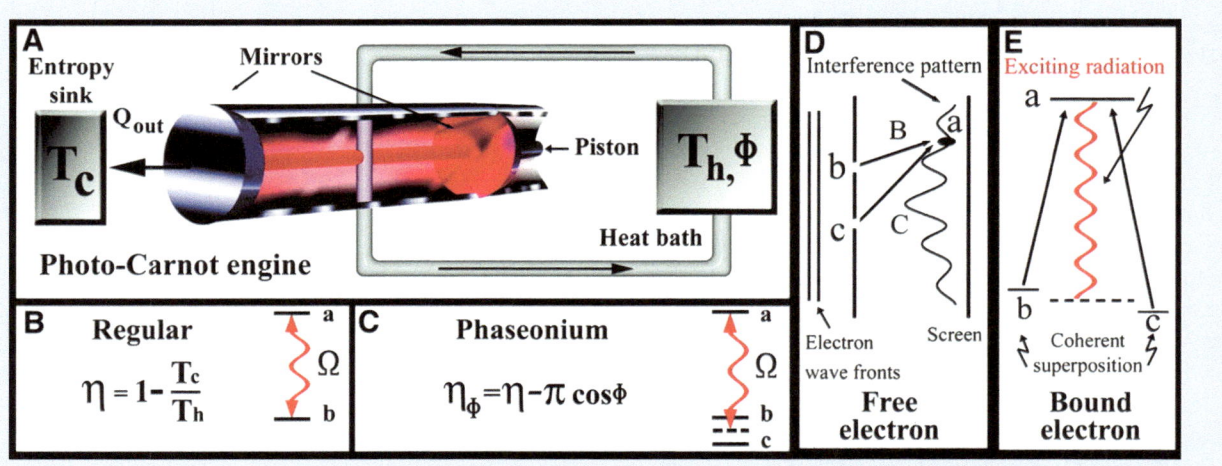

Fig. 7.16 (**a**) Photo-Carnot engine in which radiation pressure from a thermally excited single-mode field drives a piston. Atoms flow through the engine and keep the field at a constant temperature T_{rad} for the isothermal $1 \rightarrow 2$ portion of the Carnot cycle. Upon exiting the engine, the bath atoms are cooler than when they entered and are reheated by interactions with the hohlraum at T_h and "stored" in preparation for the next cycle. The combination of reheating and storing is depicted in (**a**) as the heat reservoir. A cold reservoir at T_c provides the entropy sink. (**b**) Two-level atoms in a regular thermal distribution, determined by temperature T_h, heat the driving radiation to $T_{rad} = T_h$ such that the regular operating efficiency is given by η. (**c**) When the field is heated, however, by a phaseonium in which the ground state doublet has a small amount of coherence and the populations of levels a, b, and c are thermally distributed, the field temperature is $T_{rad} > T_h$, and the operating efficiency is given by $\eta_\phi = \eta - \pi \cos(\phi)$. (**d**) A free electron propagates coherently from holes b and c with amplitudes B and C to point a on screen. The probability of the electron landing at point a shows the characteristic pattern of interference between (partially) coherent waves. (**e**) A bound atomic electron is excited by the radiation field from a coherent superposition of levels b and c with amplitudes B and C to level a. The probability of exciting the electron to level a displays the same kind of interference behavior as in the case of free electrons; i.e., as we change the relative phase between levels b and c, by, for example, changing the phase of the microwave field which prepares the coherence, the probability of exciting the atom varies sinusoidally. Adapted from [30]

the piston cylinder (electromagnetic cavity) of ◘ Fig. 7.16, by the hot atoms; that is, the atoms are the fuel, like the coal in a steam engine. Hence the classical photon driven heat engine must operate (at best) with Carnot thermodynamic efficiency since it is, at bottom, just another heat engine.

However, the plot thickens when we bring quantum coherence into the system. Now the (fuel) atoms can add more heat to the photon flux (working fluid) than was allowed by classical statistical mechanics, i.e., than was allowed by detailed balance. When we use quantum fuel (phaseonium) we can now "beat" the Carnot limit because we can now break detailed balance as per ◘ Fig. 7.16 [30]. This is a useful example to bear in mind when thinking about the quantum PV cell and photosynthesis.

7.4.3 Biological Quantum Heat Engines

Quantum entanglement [31] and other quantum coherence effects, e.g., the photon echo [32–36], have been investigated in a series of interesting photosynthesis experiments. The field of quantum biology of photosynthesis has been rapidly growing since the discovery of quantum coherence effects in the energy transfer process of the photosynthetic green sulfur bacteria [32, 35] and marine algae [33]. However, neither the role of quantum coherence nor the precise mechanism of the highly efficient energy transfer has been identified. Their description requires re-evaluation of the currently used methods and approximations. Also, whether the coherence is generated by coherent laser pulses used in the experiments or whether there is a kind of spontaneous coherence between the quantum levels involved as in the sense of noise-induced Fano-Agarwal interference has been the subject of debate. This latter possibility has been observed in various contexts and is indeed well known in quantum optics as described with applications to laser and solar cell quantum heat engines [27, 28, 37]. However it is not clear whether the quantum optical lore has applicability to photosynthesis for several reasons. High on the list being the question of environmental decoherence.

Recently, we have applied the formalism of the quantum heat engines to photosynthetic complexes such as light-harvesting antennae and reaction centers (RC) (◘ Fig. 7.17). These systems operate as quantum heat engines and their structure is suitable to provide an increase in the efficiency of these processes. Connections between various quantum mechanical effects, namely the coherence/population coupling in photosynthesis [29, 38], and the quantum yield enhancement in laser and solar cell quantum heat engines [27, 28] were investigated. Analogy was drawn with a solar cell operation where the electron transfer efficiency may be increased by a quantum coherence between a doublet of closely lying states. It was proposed that the special pair of molecules in RC has a suitable structure to exhibit similar quantum effects. ◘ Figure 7.17 depicts the proposed schemes in which the efficiency of the electron transfer from the donor molecule (D) to the acceptor (A) may be increased by the quantum coherence between two donor molecules D_1 and D_2 of the special pair. This can provide insight into the structure–function relations of natural molecular architecture and will inspire new nature-mimicking artificial designs.

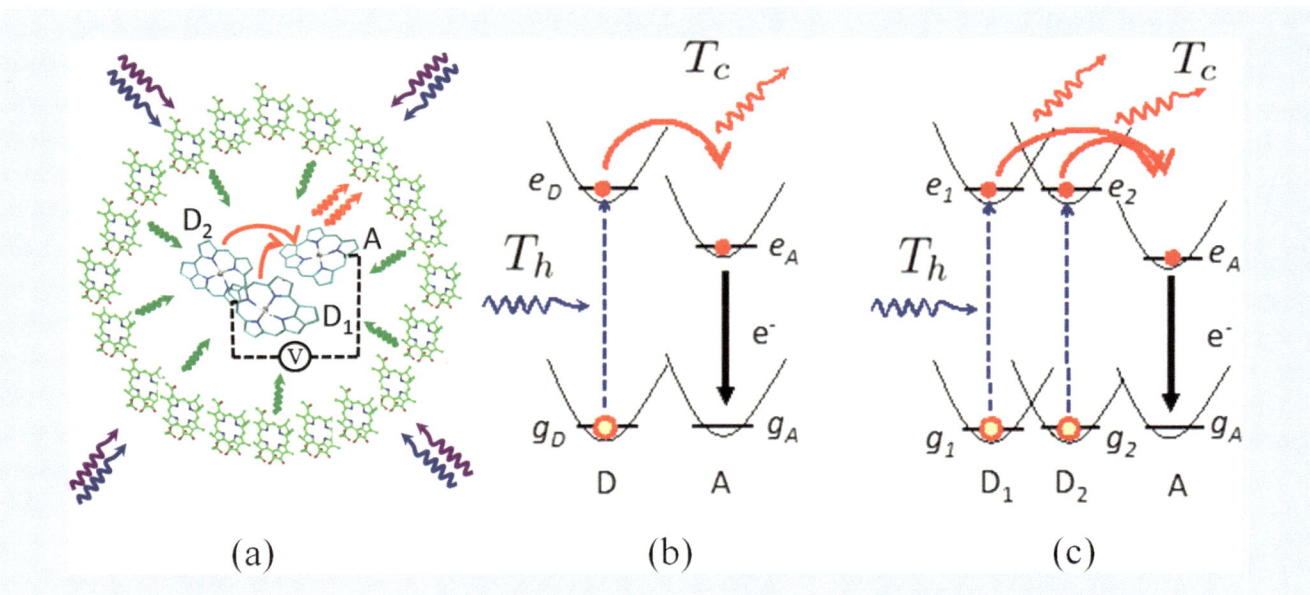

Fig. 7.17 Structure of a reaction center (RC) surrounded by photosynthetic antenna molecules (**a**). Schemes of the charge separation in the toy model of the biological quantum heat engine based on RC (**b**). The narrow band thermal radiation is transferred from the antennae complexes to the RC represented by donor (D) and acceptor (A) molecules. (**c**) Is the same as (**b**) with the upper level a is replaced by two levels a_1 and a_2. Quantum coherence between these levels can increase the power delivered by such a device. Adapted from [29]

7.5   Emerging Techniques with Single Molecule Sensitivity

7.5.1   Coherent Surface-Enhanced Raman Spectroscopy

Biological samples may be analyzed using fluorescence labeling nano-sensing technology such as enzyme-linked immunosorbent assay which is based on antibodies for detection of the presence of specific substances. The principle of this technique is based on the modification of the refractive index of the sensor when the antigen of interest is present. The sensitivity can be further enhanced by placing plasmonic nanoparticles so as to allow surface-enhanced Raman spectroscopy (SERS). These very sensitive techniques can be used to detect extremely small amounts of antigens.

Other techniques such as label-free optofluidic nanoplasmonic sensors have recently shown promise for detection of live viruses in biological media (☐ Fig. 7.18) [39]. Such sensors are based on antibodies immobilized on plasmonic nanostructures such as arrays of small holes in gold or silver chips. These sensors show significantly modified properties of scattered or transmitted light due to the capture of viruses by the attached antibodies. This technology is promising for early diagnosis of pathogens from human blood.

Recently we developed a time-resolved surface-enhanced CARS (tr-SECARS) technique where the gain in signal was attributed to the enhanced electromagnetic fields that are created near the metal particles and in the gaps between the particles or features on the tailored substrates (☐ Fig. 7.19) [8]. Thus the molecules of interest experience these enhanced fields by being attached to or simply near these metal particles or features. This coherent extension of the SERS technique provides an additional signal enhancement due to laser-induced vibrational coherence. It also provides the possibility to study biological systems simultaneously with a high

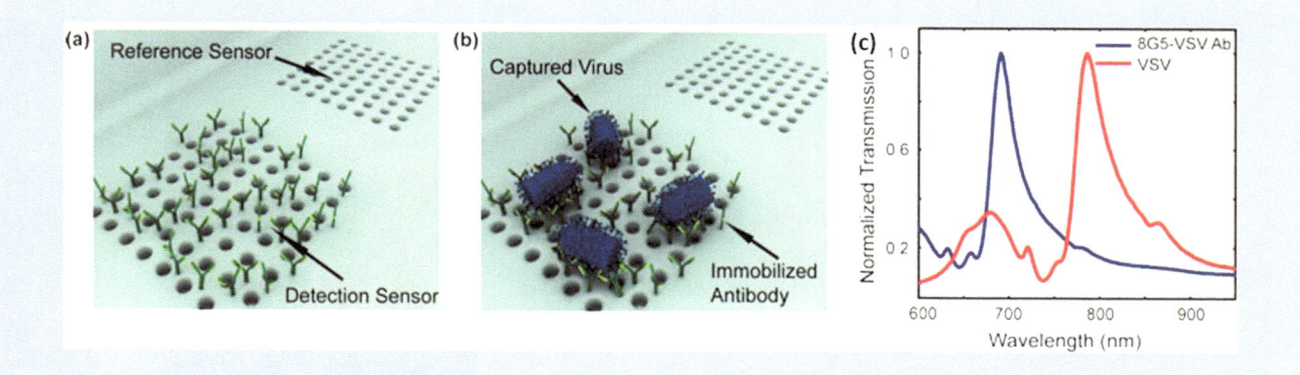

Fig. 7.18 Detection of viruses captured by immobilized antibodies on optofluidic nanoplasmonic sensors (**a**, **b**). Spectra of light transmitted by small holes change due to the capture of viruses (**c**). Adapted from [39]

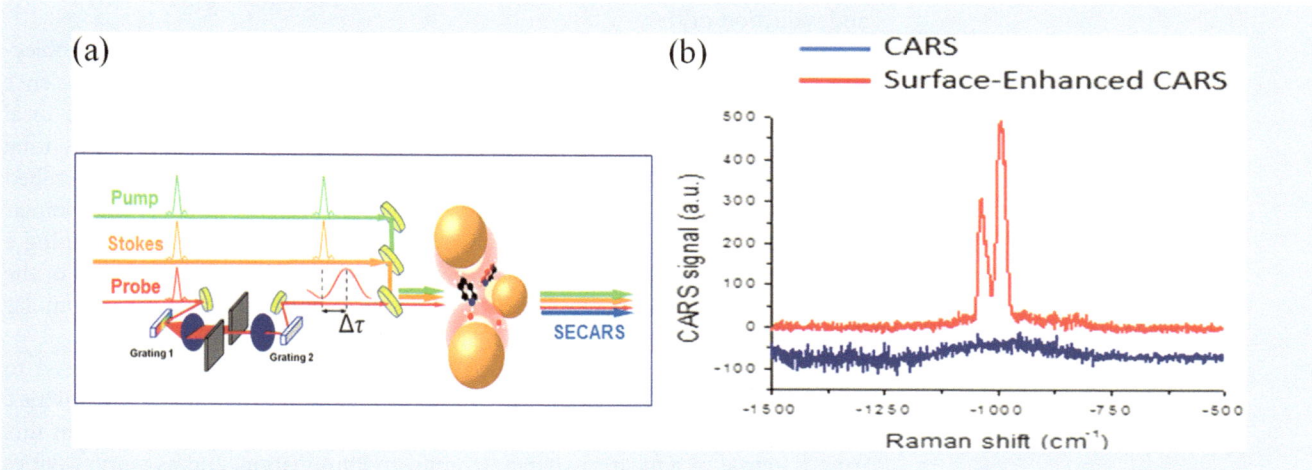

Fig. 7.19 (**a**) Experimental scheme of the time-resolved surface-enhanced anti-Stokes Raman scattering (tr-SECARS) spectroscopy. (**b**) Surface-enhanced CARS (*red*) reveals traces of hydrated pyridine molecules on the surface of gold nanoparticle aggregates with higher sensitivity than the conventional CARS (*blue*). Adapted from [8]

spatial and temporal resolution. Such a SECARS technique has the best-of-both-worlds combination of signal enhancements, i.e., the surface and coherence enhancements. This tr-SECARS technique increased the CARS signal intensity by seven orders of magnitude and was used to detect trace amounts of water on the surface of aggregated gold nanoparticles [8]. Conventional SERS was not able to detect any signal under the same conditions. This was recently reviewed in the literature as having "astonishing" sensitivity [9].

7.5.2 Cavity Ring-Down Spectroscopy

Food and water are essential for life, thus maintaining the safety of these resources is a high priority. The existing problem will only get worse in the years to come, due to limited fresh water supplies and increasing impacts of contaminants on the environment. Importantly, the ability to detect and manage environmental

Fig. 7.20 Integrating cavity setup for ultrasensitive detection of waste products such as urobilin in water. Adapted from [7]

contamination in food and water (in real time) is a critical need for the assessment and reduction of risk.

Traditional epi-illumination fluorescence spectroscopy systems use an objective lens to focus excitation light into the sample and collect the fluorescence emission. In such a configuration, the generated signal is limited to the focal volume of the optics, and is diffusive in nature; only a small fraction of the total emitted light is collected. Because only a small volume of a sample can be probed at any given time with such a configuration, detection of subnanomolar concentrations remains difficult. Thus, a method that could allow for exciting a larger volume of a sample while also providing means for collecting more of the fluorescence emission could greatly enhance the ability to detect sub-picomolar concentrations of, for example, urobilin (**Fig. 7.20**).

To achieve both of these goals, an integrating cavity was recently used to enhance both the excitation and collection efficiency [7]. Integrating cavities, especially spherical cavities, are commonly used to measure the total radiant flux from a source, as a means to generate uniform illumination, and as pump cavities for lasers. The high reflectivity of the cavity walls leads to a very large effective optical path length over which fluorescence excitation occurs; and the result is excitation of the entire volume of any sample placed inside the cavity. The high reflectivity of the cavity walls also means that fluorescence is collected from all directions; the result is the ability to detect a 500 femtomolar concentration of urobilin [7].

An accurate knowledge of the spectral absorption of cells and their constituents is critical to the continuation of progress in the understanding and modeling of biological and biomedical processes. Modeling of highly scattering cellular media, imaging of tissues, cells, and organelles, and laser-based surgical procedures are just a few of the vast number of techniques and procedures that rely on a refined understanding of biological absorption coefficients. Previously absorption measurements have generally been performed using transmission-style experiments in which an attenuation coefficient is measured by observing the decrease in the intensity of a light source as it passes through a sample. While the attenuation coefficient includes the losses due to absorption, it also includes the contribution from any scattering present in the sample. But these problems are avoided by using an integrating cavity for absorption measurements. Specifically, due to the nearly Lambertian behavior of the cavity walls, an isotropic field is created inside the cavity and scattering within the sample cannot change that. The result is absorption measurements that are independent of scattering. The integrating cavity technique was used to obtain what are now widely considered to be the standard reference data for pure water absorption [40, 41]. Cavity ring-down spectroscopy (CRDS) is a technique developed for highly accurate and sensitive measurements of low absorption coefficients. It involves sending a temporally short laser pulse into a high-finesse two mirror cavity and observing the exponential decay, or "ring-down" of the intensity due to absorption (and scattering) inside the cavity. While CRDS is a very powerful technique, it is inherently unable to distinguish the losses due to absorption from losses due to scattering. Much like other transmission-style experiments, it is the attenuation that is being measured. This is clearly a problem for determining weak absorption in highly scattering samples. The new approach is based on CRDS and is called integrating cavity ring-down spectroscopy (ICRDS) where the traditional two mirror cavity used in CRDS is replaced with an integrating cavity. Again, the integrating cavity walls are fabricated from a highly diffuse reflecting material which creates an isotropic light field inside the cavity; the result is that a measurement of the decay time for the optical pulse is inherently insensitive to any scattering in the sample. Thus, ICRDS can provide a direct measurement of the absorption coefficient, even in the presence of strong scattering. However, ICRDS has not previously been implemented because a material with a sufficiently high diffuse reflectivity was not available. Spectralon has been the material with the highest known diffuse reflectivity (99.3 %) [42], but at this level of reflectivity, the decay of the optical pulse due to losses during reflection from the wall is so large that the sensitivity to absorption is low.

A new material has now been developed that has a diffuse reflectivity up to 99.92 % [43]; this is sufficient to make ICRDS a reality. It opens up a plethora of exciting new applications with tremendous impact in, for example, the biomedical area where highly accurate absorption spectra of living cells, tissue samples, liquids, etc., can now be obtained, even when the absorption is very weak compared to scattering in the sample.

As a specific such example, consider human retinal pigmented epithelium (RPE) cells. Among other functions, RPE cells are responsible for absorbing scattered light to improve the optical system and reduce stress on the retina. ◘ Figure 7.21 shows an example in which ICRDS was used to measure the base absorption (i.e., with the pigment removed) of 60 million RPE cells in a 3 mL solution [44]. Also shown is the attenuation (scattering plus absorption) spectrum obtained with the same sample in a spectrophotometer. As an example of the dominance of scattering in the transmission measurement, consider the

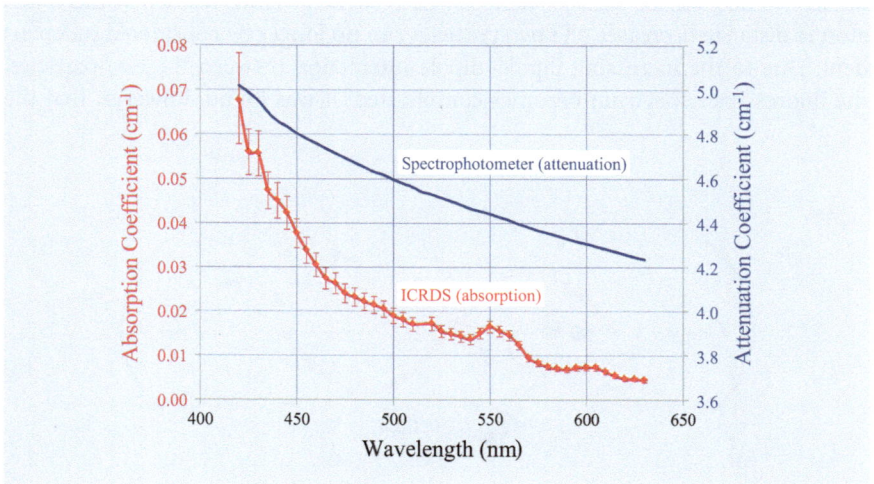

◘ **Fig. 7.21** Absorption coefficient of 6×10^7 human retinal pigmented epithelium (RPE) cells in a 3 mL solution; also shown is the attenuation (scattering + absorption) spectrum. Adapted from [44]

attenuation at 500 nm ($4.6\ \mathrm{cm}^{-1}$) and the absorption ($\sim 0.02\ \mathrm{cm}^{-1}$). This is a factor of 230; the absorption is less than one-half of 1 % of the attenuation. ◘ Figure 7.21 shows biological absorption spectra that were previously inaccessible for study and analysis.

7.6   Superresolution Quantum Microscopy

7.6.1   Subwavelength Quantum Microscopy

The measurement of small distances is a fundamental problem of interest since the early days of science. It has become even more important due to recent interest in nanoscopic and mesoscopic phenomena in biophysics. Starting from the invention of the optical microscope around 400 years ago, today's optical microscopy methodologies can basically be divided into lens-based and lensless imaging. In general, far-field imaging is lens-based and thus limited by criteria such as the Rayleigh diffraction limit which states that the achievable resolution in the focus plane is limited to approximately half of the wavelength of illuminating light. Further limitation arises from out-of-focus light, which affects the resolution in the direction perpendicular to the focal plane.

Many methods have been suggested to break these limits. Lens-based techniques include confocal, nonlinear femtosecond, or stimulated emission depletion microscopy which have been recognized by the Nobel Prize. They have achieved remarkable first results, as shown in ◘ Fig. 7.1. Also non-classical features such as entanglement, quantum interferometry, or multi-photon processes can be used to enhance resolution. However, there is still great interest in achieving nanometer distance measurements by using optical illuminating far-field imaging only.

Recently, new schemes were proposed [45–47] to measure the distance between two adjacent two-level systems by driving them with a standing wave laser field and measuring the far-field resonance fluorescence spectrum, which is motivated by the localization of single atom inside a standing wave field to distances smaller than the Rayleigh limit $\lambda/2$. The basic idea is that in a standing wave, the effective driving field strength depends on the position of the particles (◘ Fig. 7.22). Thus, each particle generates a sharp sideband peak in the spectrum, where the peak position directly relates to the subwavelength position of the particle. As long as the two sideband peaks can be distinguished from each other, the position of each particle can be recovered. However, when the inter-atomic distance decreases, the two particles can no longer be considered independent. Due to the increasing dipole–dipole interaction between the two particles, the fluorescence spectrum becomes complicated. It was found, however, that the

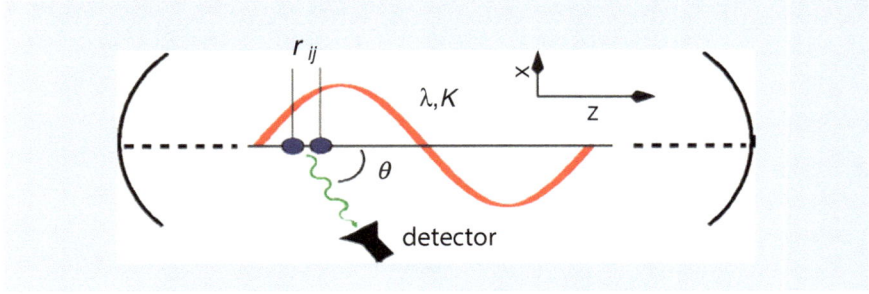

◘ **Fig. 7.22** Two atoms in a standing wave field separated by a distance r_{ij} smaller than half of the wavelength λ of the driving field. The distance of the two atoms is measured via the emitted resonance fluorescence

dipole–dipole interaction energy can directly be extracted from the fluorescence spectrum by adjusting the parameters of the driving field. Since the dipole–dipole interaction energy is distance dependent, it yields the desired distance information. The new schemes showed the applicability to inter-particle distances in a very wide range from $\lambda/2$ to about $\lambda/550$. These schemes can be extended for sensing the inter-molecular distances with applications in molecular and cellular biology, microbiology, and medical research leading to precision resolution in microorganisms such as protozoa, bacteria, molds, and sperms. Other applications would involve mapping of DNA.

7.6.2 Tip-Enhanced Quantum Bioimaging

Understanding the structure–function relationships of molecular constituents of biological pathogens is important for building more precise models and for the design of bacteria inactivating drugs, etc. Pathogenic diseases that are caused by viral pathogens include smallpox, influenza, mumps, measles, chickenpox, ebola, etc. New technology is needed for rapid detection and treatment, and for detailed studies of nanoscale components of pathogens with high spatial and temporal resolution and with simultaneous chemical analysis. Various nano-spectroscopic methods have been used to model pathogens. Direct imaging of the structural dynamics is challenging due to the small size of the biomolecules. Novel quantum optical techniques based on nanoscale sensors with high spatial and molecular-level resolution aim to improve the existing pathogen structure–property models.

Recent exceptional breakthroughs in the spatial resolution (<1 nm) make tip-enhanced Raman scattering (TERS), which is an important variation of SERS, a very powerful tool for in situ chemical analysis on the nanoscale. TERS has recently been applied to imaging biological systems. However, several challenges remain due to weak Raman signals and motion of live cells during long acquisition time imaging. We have performed such nanoantenna-tip-induced bio-sensing of bacteria and nanoscale imaging of biological cells (◘ Fig. 7.23). The goal is to map the molecule–substrate interactions with nanoscale molecular-level spatial resolution in both topography (using atomic force microscopy (AFM)) and chemical identification (using Raman spectroscopy). Both electromagnetic and chemical enhancement effects can be used to identify the molecular biomarkers and nanoscale chemical surface properties of biological systems.

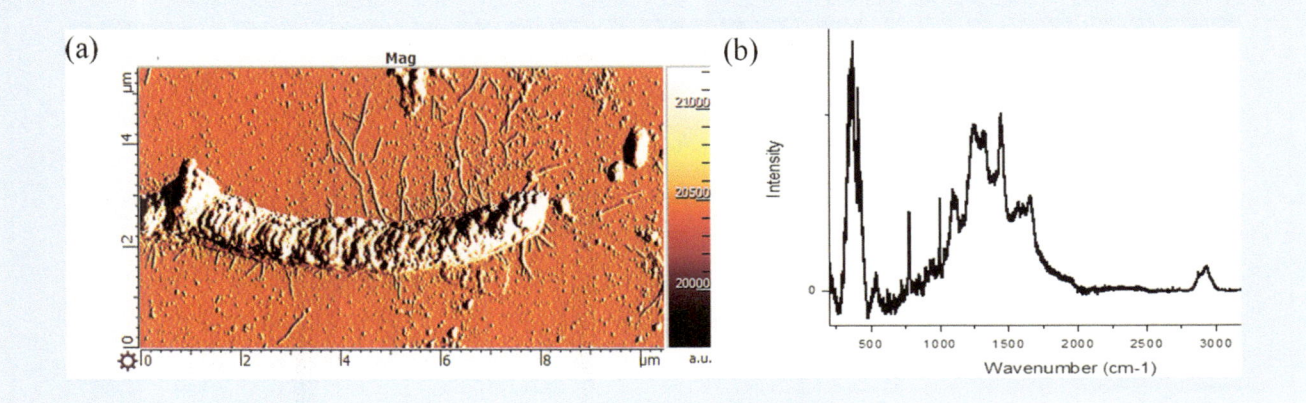

◘ **Fig. 7.23** (**a**) AFM image of a bacterial cell. (**b**) Typical Raman spectrum of bacteria

7.7   Novel Light Sources

7.7.1   Fiber Sensors

Recent research on fiber sensing has been focused on the development and demonstration of advanced fiber components and fiber-based strategies for quantum sensing. Fiber-bundle microprobe sensors have been developed using specifically designed fiber bundles and coupled to a confocal optical microscope to enable multiplex sensing, including multicolor in vivo detection of neuronal activity in a living brain using fluorescent protein biomarkers. Fiber-bundle microprobe sensors have also been used to confront the long-standing issues in practical imaging and sensing, including fiber-based endoscopic Raman imaging (◘ Fig. 7.24).

Recent advances in optical magnetometry pave the ways toward an unprecedented spatial resolution and a remarkably high sensitivity in magnetic field detection, offering unique tools for the measurement of weak magnetic fields in a broad variety of areas from astrophysics, geosciences, and the physics of fundamental symmetries to medicine and life sciences. To unleash the full potential of this emerging technology and make it compatible with the requirements of practical quantum technologies and in vivo studies in life sciences, optical magnetometers have to be integrated with fiber-optic probes. This challenge has been addressed by developing a scanned fiber-optic probe for magnetic field imaging where nitrogen–vacancy (NV) centers are coupled to an optical fiber integrated with a two-wire microwave transmission line. The electron spin of NV centers in a diamond microcrystal attached to the tip of the fiber probe is manipulated by a frequency-modulated microwave field and is initialized by laser radiation transmitted through the optical tract of the fiber probe (◘ Fig. 7.25) [50, 51]. The photoluminescence spin-readout return from NV centers is captured and delivered by the same optical fiber, allowing the two-dimensional profile of the magnetic field to be imaged with high speed and high sensitivity.

7.7.2   Quantum Coherence in X-Ray Laser Generation

The application of the techniques of quantum coherence and LWI to areas such as XUV and X-ray laser generation holds promise. The quantum coherence in atomic

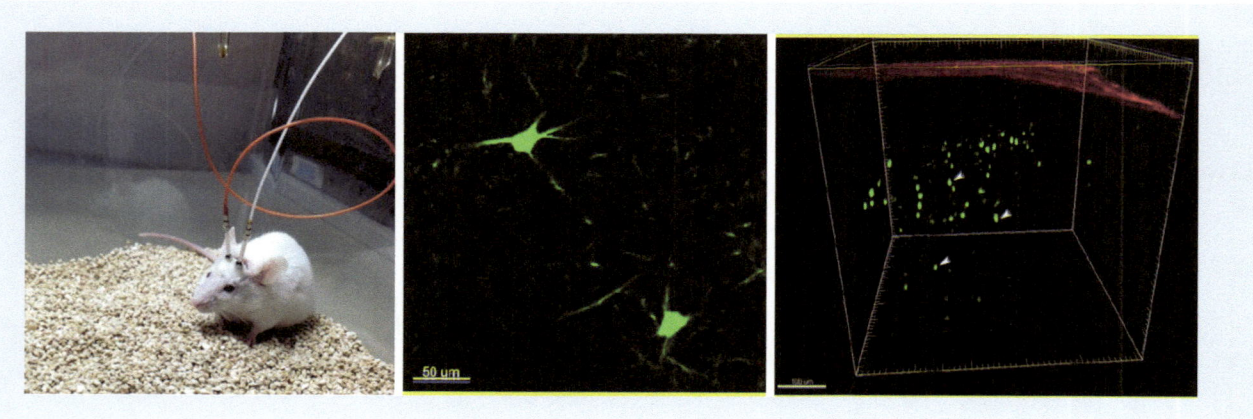

◘ **Fig. 7.24** Fiber-based sensing and imaging for neurophotonics and agricultural applications. Adapted from [48, 49]

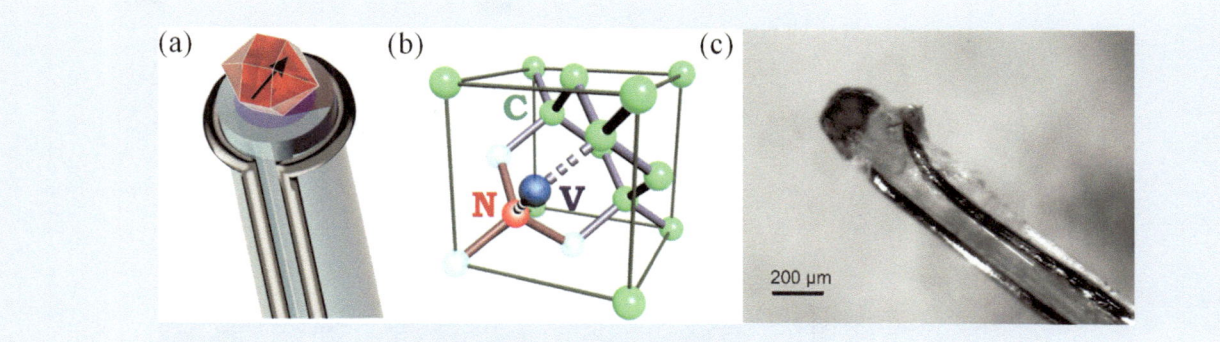

Fig. 7.25 (**a**) A fiber probe integrating an NV-diamond quantum sensor, an optical fiber, and a microwave transmission line. (**b**) A nitrogen atom (N) and a vacancy (V) forming an NV center in a diamond lattice, consisting of carbon (C) atoms. (**c**) Image of a prototype fiber-optic magnetometer and thermometer. Adapted from [50, 51]

and radiation physics has led to many interesting and unexpected consequences. For example, an atomic ensemble prepared in a coherent superposition of states yields self-induced transparency, photon echo, and coherent Raman beats [52, 53]. Another important coherence effect in atoms leads to phenomena of electromagnetically induced transparency, where by preparing an atomic system in a coherent superposition of states, under certain conditions, it is possible for atomic coherence to cancel absorption but not emission (◘ Fig. 7.26). This is the basis for LWI, the essential idea being the absorption cancellation by atomic coherence and interference. Frequently this is accomplished in three- or four-level atomic systems in which there are coherent routes for absorption that can destructively interfere, thus leading to the cancellation of absorption. A small population in the excited state can thus lead to net gain, and this was the subject of substantial theoretical work by us several groups in the 1980s [24–26]. We have carried out the first LWI demonstrations in the mid-1990s [54, 55], and have recently continued in theoretical and experimental fronts to investigate the possibility of coherence driven lasing and lasing in XUV and soft X-ray regions [56, 57].

Existing X-ray laser sources such as X-ray free electron laser (X-FEL) at SLAC provided tremendous excitement for scientists in various disciplines. However, very large cost and size of this great device motivates researchers to search for portable, inexpensive XUV (see for example Suckewer, et al. [56]), and X-ray devices. On the other hand, the table top soft X-ray (SXL) and XUV lasers, due to their compactness, excellent beam quality, and very reliable operation in wavelength range of 10–50 nm hold great promise for tools for high resolution microscopy, micro-holography, very high plasma density measurements, semiconductor surface studies, and nano-lithography. Intensive efforts have been made to develop such compact soft X-ray lasers that are suitable for applications in academic and industrial laboratories as well as in the field biophotonics applications. LWI in the X-ray region would provide appealing opportunities and profound impacts on X-ray laser science as well as studies in the fields such as crystallography, solid materials, health sciences, high resolution microscopy of biological systems, and many more.

7.7.3 Coherent Control of Gamma Rays

Active control of light–matter interactions is an ultimate goal of many quantum biophotonics applications. Coherent control by laser pulse shaping has been a highly active research area for the last two decades addressing a large variety of problems including control of chemical reactions, spectroscopic signals, optical

7

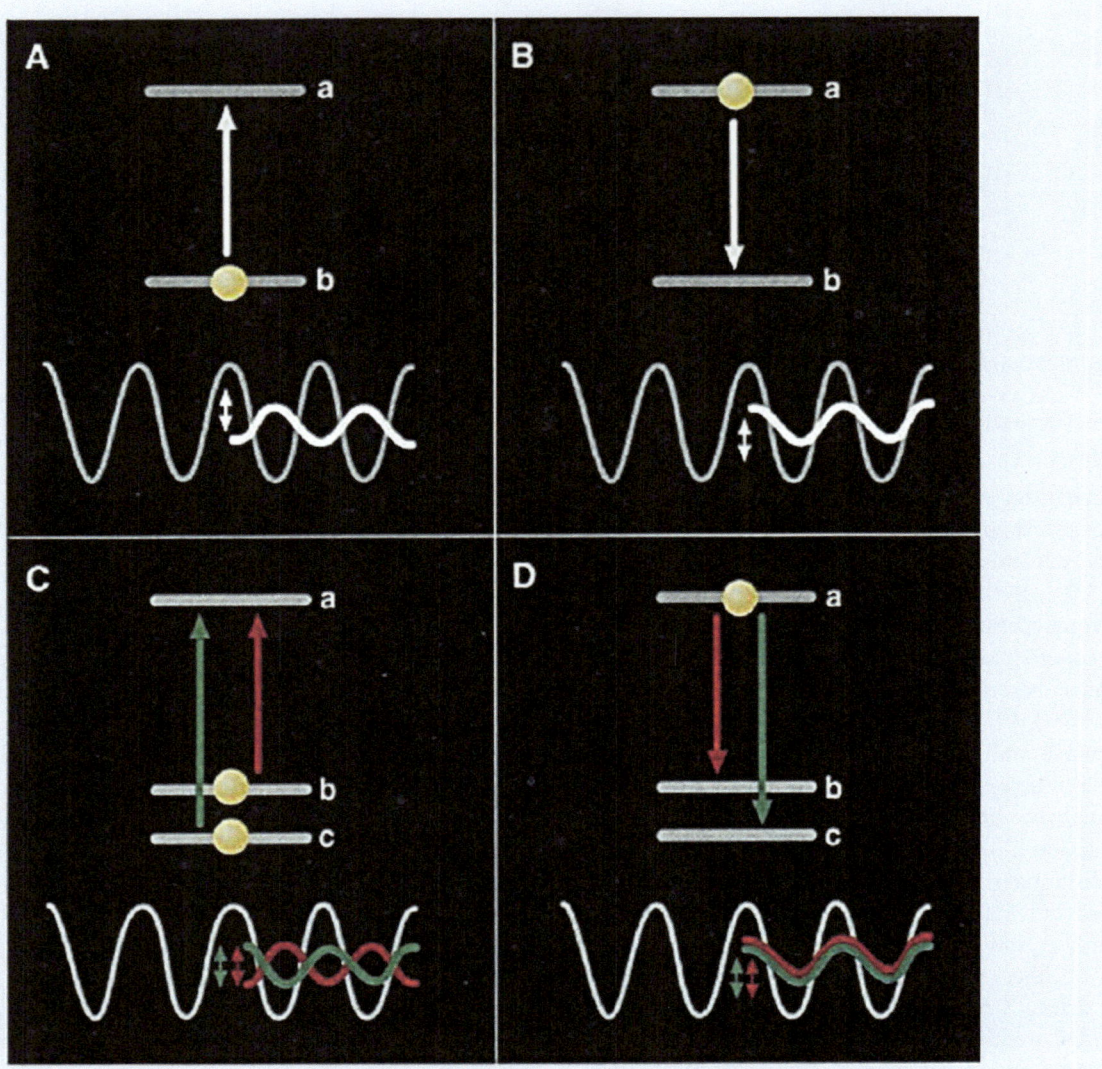

Fig. 7.26 Lasing without inversion (LWI) first experimentally realized at TAMU [54, 55]. Absorption cancellation by atomic coherence and interference is the basis for LWI. Adapted from [58]

sensors, microscopic images, photolithography, optical nanostructures, quantum logic operations, etc. All these applications, however, were aimed at controlling electronic motions, molecular vibrations and rotations using shaped electromagnetic radiation in the frequency range spanning from infrared to optical to ultraviolet range. The X-ray and gamma ray sources with controllable waveforms were up till now unavailable, and therefore there were no related applications.

Now, for the first time, we have an available source of controllable gamma rays [59]. Single ultrashort gamma ray pulses, double pulses, and pulse trains with controllable pulse delays can be produced (◙ Fig. 7.27). Splitting of the single photon into two pulses constitutes the first realization of a time-bin qubit in this range of frequencies. The production of single-photon ultrashort-pulse trains with pulse duration much shorter than the natural lifetime of the emitting nuclear level and with the controllable waveforms provides a unique opportunity for the realization of quantum memories and other nuclear-ensemble—gamma-photons interfaces, opening the prospects for fascinating applications in quantum communication and information.

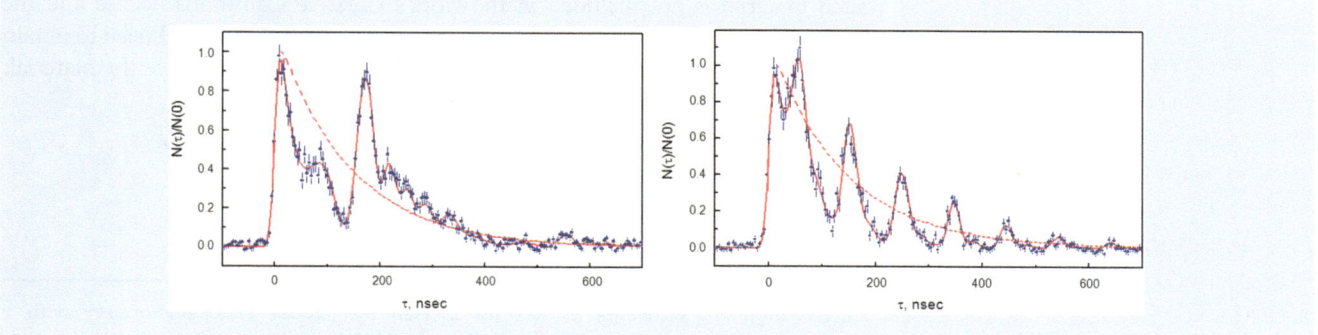

Fig. 7.27 Splitting gamma ray photons into double pulses (*left*) and multiple pulse trains (*right*). Adapted from [59]

The energy of these gamma ray pulses corresponds to nuclear transitions. This may enable, for the first time, coherent control of nuclear reactions. The parameters of the pulses can be widely controlled including the shape and number of pulses, repetition rate (potentially variable in the range from MHz to GHz), and duration (potentially ranged from 100 ns to 100 ps). This allows for a variety of time-resolved experiments (including dynamic X-ray diffraction). Gamma rays can provide an extremely high (potentially sub-Angstrom and currently nanometer) spatial resolution. The simultaneous high spatial and temporal resolution makes this technology promising for nanoscale ultrafast imaging of electronic motions in biomolecules and protein folding dynamics.

7.8   Conclusion

In summary, we have (hopefully) shown that at the interface between quantum optics and biophysics lies the emerging exciting field of quantum biophotonics. With new light sources and quantum techniques, it becomes increasingly possible to apply the techniques of quantum optics to biosciences. This will have major payoff in agriculture, environmental science, and national security, and promises to be the dawn of a new era in biophotonics emphasizing quantum effects.

Acknowledgment   We acknowledge the support of the National Science Foundation Grants No. EEC-0540832 (MIRTHE ERC), No.PHY-1068554, No. PHY-1241032 (INSPIRE CREATIV), No. DBI-1455671, No. DBI-1532188, No. ECCS-1509268 and No. PHY-1307153, the Office of Naval Research grant N00014-16-1-3054, the US DOD awards FA9550-15-1-0517 and N00014-16-1-2578, and CPRIT grant RP160834, and the Robert A. Welch Foundation (Awards A-1261 and A-1547).

such material is not included in the work's Creative Commons license and the respective action is not permitted by statutory regulation, users will need to obtain permission from the license holder to duplicate, adapt or reproduce the material.

References

1. Wildanger D, Rittweger E, Kastrup L, Hell SW (2008) STED microscopy with a supercontinuum laser source. Opt Express 16:9614
2. Chmyrov A, Keller J, Grotjohann T, Ratz M, d'Este E, Jakobs S, Eggeling C, Hell SW (2013) Nanoscopy with more than 100,000 'doughnuts'. Nat Methods 10:737
3. Scully MO, Kattawar GW, Lucht RP, Opatrny T, Pilloff H, Rebane A, Sokolov AV, Zubairy MS (2002) FAST CARS: Engineering a laser spectroscopic technique for rapid identification of bacterial spores. Proc Natl Acad Sci U S A 99:10994
4. Pestov D, Murawski RK, Ariunbold GO, Wang X, Zhi M, Sokolov AV, Sautenkov VA, Rostovtsev YV, Dogariu A, Huang Y, Scully MO (2007) Optimizing the laser-pulse configuration for coherent Raman spectroscopy. Science 316:265
5. Pestov D, Wang X, Ariunbold GO, Murawski RK, Sautenkov VA, Dogariu A, Sokolov AV, Scully MO (2008) Single-shot detection of bacterial endospores via coherent Raman spectroscopy. Proc Natl Acad Sci U S A 105:422
6. Arora R, Petrov GI, Yakovlev VV, Scully MO (2012) Detecting anthrax in the mail by coherent Raman microspectroscopy. Proc Natl Acad Sci U S A 109:1151
7. Bixler LN, Cone MT, Hokr BH, Mason JD, Figueroa E, Fry ES, Yakovlev VV, Scully MO (2014) Ultrasensitive detection of waste products in water using fluorescence emission cavity-enhanced spectroscopy. Proc Natl Acad Sci U S A 111:7208
8. Voronine DV, Sinyukov AM, Xia H, Wang K, Jha PK, Welch G, Sokolov AV, Scully MO (2012) Time-resolved surface-enhanced coherent sensing of nanoscale molecular complexes. Sci Rep 2:891
9. Lis D, Cecchet F (2014) Localized surface plasmon resonances in nanostructures to enhance nonlinear vibrational spectroscopies: towards an astonishing molecular sensitivity. Beilstein J Nanotechnol 28:2275
10. Altangerel N, Ariunbold G, Gorman C, Bohlmeyer D, Yuan J, Hemmer P, Scully MO (2016) Early, in vivo, detection of abiotic plant stress responses via Raman spectroscopy. Conference on Lasers and Electro-Optics OSA Technical Digest (Optical Society of America, 2016), paper SF1H.3. doi:► 10.1364/CLEO_SI.2016.SF1H.3
11. Palombo F, Madami M, Stone N, Fioretto D (2014) Mechanical mapping with chemical specificity by confocal Brillouin and Raman microscopy. Analyst 139:729
12. Traverso AJ, Thompson JV, Steelman ZA, Meng Z, Scully MO, Yakovlev VV (2015) Dual Raman-Brillouin microscope for chemical and mechanical characterization and imaging. Anal Chem 87:7519
13. Meng Z, Traverso AJ, Yakovlev VV (2014) Background clean-up in Brillouin microspectroscopy of scattering medium. Opt Express 22:5410
14. Petrov GI, Arora R, Yakovlev VV, Wang X, Sokolov AV, Scully MO (2007) Comparison of coherent and spontaneous Raman microspectroscopies for noninvasive detection of single bacterial endospores. Proc Natl Acad Sci U S A 104:7776
15. Hemmer PR, Miles RB, Polynkin P, Siebert T, Sokolov AV, Sprangle P, Scully MO (2011) Standoff spectroscopy via remote generation of a backward-propagating laser beam. Proc Natl Acad Sci U S A 108:3130
16. Dogariu A, Michael JB, Scully MO, Miles RB (2011) High-gain backward lasing in air. Science 331:442
17. Traverso AJ, Sanchez-Gonzalez R, Yuan L, Wang K, Voronine DV, Zheltikov AM, Rostovtsev Y, Sautenkov VA, Sokolov AV, North SW, Scully MO (2012) Coherence brightened laser source for atmospheric remote sensing. Proc Natl Acad Sci U S A 109:15185
18. Kocharovsky V, Cameron S, Lehmann K, Lucht R, Miles R, Rostovtsev Y, Warren W, Welch GR, Scully MO (2005) Gain-swept superradiance applied to the stand-off detection of trace impurities in the atmosphere. Proc Natl Acad Sci U S A 102:7806
19. Cremers DA, Radziemski LJ (2006) Handbook of laser-induced breakdown spectroscopy 302. John Wiley, West Sussex
20. Mitrofanov AV, Voronin AA, Sidorov-Biryukov DA, Pugžlys A, Stepanov EA, Andriukaitis G, Flöry T, Ališauskas T, Fedotov AB, Baltuška A, Zheltikov AM (2014) Mid-infrared laser filaments in the atmosphere. Sci Rep 5:8368

21. Malevich PN, Kartashov D, Ališauskas ZPS, Pugžlys A, Baltuška A, Giniūnas L, Danielius R, Lanin AA, Zheltikov AM, Marangoni M, Cerullo G (2012) Ultrafast-laser-induced backward stimulated Raman scattering for tracing atmospheric gases. Opt Express 20:18784

22. Malevich PN, Maurer R, Kartashov D, Ališauskas S, Lanin AA, Zheltikov AM, Marangoni M, Cerullo G, Baltuška A, Pugžlys A (2015) Stimulated Raman gas sensing by backward UV lasing from a femtosecond filament. Opt Lett 40:2469

23. Kocharovskaya O, Khanin YI (1988) Coherent amplification of an ultrashort pulse in a three-level medium without a population inversion. JETP Lett 48:630

24. Harris SE (1989) Lasers without inversion: interference of lifetime-broadened resonances. Phys Rev Lett 62:1033

25. Scully MO (1985) Correlated spontaneous-emission lasers: quenching of quantum fluctuations in the relative phase angle. Phys Rev Lett 55:2802

26. Scully MO, Zhu SY, Gavrielides A (1989) Degenerate quantum-beat laser: lasing without inversion and inversion without lasing. Phys Rev Lett 62:2813

27. Scully MO (2010) Quantum photocell: using quantum coherence to reduce radiative recombination and increase efficiency. Phys Rev Lett 104:207701

28. Scully MO, Chapin KR, Dorfman KE, Kim M, Svidzinsky AA (2011) Quantum heat engine power can be increased by noise-induced coherence. PNAS 108:15097

29. Dorfman KE, Voronine DV, Mukamel S, Scully MO (2013) Photosynthetic reaction center as a quantum heat engine. PNAS 110:2746

30. Scully MO, Zubairy MS, Agarwal GS, Walther H (2003) Extracting work from a single heat bath via vanishing quantum coherence. Science 299:862

31. Sarovar M, Ishizaki A, Fleming GR, Whaley KB (2010) Quantum entanglement in photosynthetic light-harvesting complexes. Nat Phys 6:462

32. Engel GS, Calhoun TR, Read EL, Ahn TK, Mancal T, Cheng YC, Blankenship RE, Fleming GR (2007) Evidence for wavelike energy transfer through quantum coherence in photosynthetic systems. Nature 446:782

33. Collini E, Wong CY, Wilk KE, Curmi PM, Brumer P, Scholes GD (2010) Coherently wired light-harvesting in photosynthetic marine algae at ambient temperature. Nature 463:644

34. Panitchayangkoon G, Hayes D, Fransted KA, Caram JR, Harel E, Wen J, Blankenship RE, Engel GS (2010) Long-lived quantum coherence in photosynthetic complexes at physiological temperature. Proc Natl Acad Sci U S A 107:12766

35. Brixner T, Stenger J, Vaswani HM, Cho M, Blankenship RE, Fleming GR (2005) Two-dimensional spectroscopy of electronic couplings in photosynthesis. Nature 434:625

36. Abramavicius D, Palmieri B, Voronine DV, Sanda F, Mukamel S (2009) Coherent multidimensional optical spectroscopy of excitons in molecular aggregates; quasiparticle versus supermolecule perspectives. Chem Rev 109:2350

37. Kozlov VV, Rostovtsev Y, Scully MO (2006) Inducing quantum coherence via decays and incoherent pumping with application to population trapping, lasing without inversion, and quenching of spontaneous emission. Phys Rev A 74:063829

38. Panitchayangkoon G, Voronine DV, Abramavicius D, Mukamel D, Engel GS (2011) Direct evidence of quantum transport in photosynthetic light harvesting complexes. PNAS 108:20908

39. Yanik AA, Huang M, Kamohara O, Artar A, Geisbert TW, Connor JH, Altug H (2010) An optofluidic nanoplasmonic biosensor for direct detection of live viruses from biological media. Nano Lett 10:4962

40. Fry ES, Kattawar GW, Pope RM (1992) Integrating cavity absorption meter. Appl Opt 31:2055

41. Pope RM, Fry ES (1997) Absorption spectrum (380-700 nm) of pure water: II. Integrating cavity measurements. Appl Opt 36:8710

42. Labsphere (2006) A guide to reflectance coatings and materials. Tech. Rep. ▶ http://www.labsphere.com

43. Cone MT, Musser JA, Figueroa E, Mason JD, Fry ES (2015) Diffuse reflecting material for integrating cavity spectroscopy—including ring-down spectroscopy. Appl Opt 54:334

44. Cone MT, Mason JD, Figueroa E, Hokr BH, Bixler JN, Castellanos CC, Wigle JC, Noojin GD, Rockwell BA, Yakovlev VV, Fry ES (2015) Measuring the absorption coefficient of biological materials using integrating cavity ring-down spectroscopy. Optica 2:162

45. Chang JT, Evers J, Scully MO, Zubairy MS (2006) Measurement of the separation between atoms beyond diffraction limit. Phys Rev A 73:031803

46. Chang JT, Evers J, Zubairy MS (2006) Distilling two-atom distance information from intensity-intensity correlation functions. Phys Rev A 74:043820

47. Liao Z, Alamri M, Zubairy MS (2012) Resonance-fluorescence-localization microscopy with subwavelength resolution. Phys Rev A 85:023810

48. Doronina-Amitonova LV, Fedotov IV, Ivashkina OI, Zots MA, Fedotov AB, Anokhin KV, Zheltikov AM (2013) Implantable fiber-optic interface for parallel multisite long-term optical dynamic brain interrogation in freely moving mice. Sci Rep 3:3265
49. Doronina-Amitonova LV, Fedotov IV, Fedotov AB, Anokhin KV, Zheltikov AM (2015) Neurophotonics: optical methods to study and control the brain. Phys Usp 58:345
50. Fedotov IV, Safronov NA, Shandarov YA, Lanin AA, Fedotov AB, Kilin SY, Sakoda K, Scully MO, Zheltikov AM (2012) Guided-wave-coupled nitrogen vacancies in nanodiamond-doped photonic-crystal fibers. Appl Phys Lett 101:031106
51. Fedotov IV, Doronina-Amitonova LV, Voronin AA, Levchenko AO, Zibrov SA, Sidorov-Biryukov AD, Fedotov AB, Velichansky VL, Zheltikov AM (2014) Electron spin manipulation and readout through an optical fiber. Sci Rep 4:5362
52. Zhu SY, Nikonov DE, Scully MO (1998) A scheme for noninversion lasing for short-wavelength lasers in helium like ions. Found Phys 28:611
53. Rostovtsev Y, Scully MO (2007) Soft X-ray lasing without population inversion in ^{3}He using Pauli principle. J Mod Opt 54:2607
54. Zibrov AS, Lukin MD, Nikonov DE, Hollberg L, Scully MO, Velichansky VL, Robinson HG (1995) Experimental demonstration of laser oscillation without population inversion via quantum interference in Rb. Phys Rev Lett 75:1499
55. Padmabandu GG, Welch GR, Shubin IN, Fry ES, Nikonov DE, Lukin MD, Scully MO (1996) Laser oscillation without population inversion in a sodium atomic beam. Phys Rev Lett 76:2053
56. Sete EA, Svidzinsky AA, Rostovtsev YV, Eleuch H, Jha PK, Suckewer S, Scully MO (2011) Using quantum coherence to generate gain in the XUV and X-ray: Gain-swept superradiance and lasing without inversion. IEEE J Sel Top Quantum Electron 18:541
57. Xia H, Svidzinsky AA, Yuan L, Lu S, Suckewer S, Scully MO (2012) Observing superradiant decay of excited-state helium atoms inside helium plasma. Phys Rev Lett 109:093604
58. Scully MO, Fleischhauer M (1994) Lasers without inversion. Science 263:337
59. Vagizov F, Antonov V, Radeonychev YV, Shakhmuratov RN, Kocharovskaya O (2014) Coherent Control of the waveforms of recoilless gamma-photons. Nature 508:80

7

Optical Communication: Its History and Recent Progress

Govind P. Agrawal

G.P. Agrawal (✉)
The Institute of Optics, University of Rochester, Rochester NY 14627, USA
e-mail: Govind.Agrawal@rochester.edu

© The Author(s) 2016
M.D. Al-Amri et al. (eds.), *Optics in Our Time*, DOI 10.1007/978-3-319-31903-2_8

8.1   Historical Perspective

The use of light for communication purposes dates back to antiquity if we interpret optical communication in a broad sense, implying any communication scheme that makes use of light. Most civilizations have used mirrors, fire beacons, or smoke signals to convey a single piece of information (such as victory in a war). For example, it is claimed that the Greeks constructed in 1084 B.C. a 500-km-long line of fire beacons to convey the news of the fall of Troy [1]. The chief limitation of such a scheme is that the information content is inherently limited and should be agreed upon in advance. Attempts were made throughout history to increase the amount of transmitted information. For example, the North American Indians changed the color of a smoke signal for this purpose. Similarly, shutters were used inside lighthouses to turn the beacon signal on and off at predetermined intervals. This idea is not too far from our modern schemes in which information is coded on the light emitted by a laser by modulating it at a high speed [2].

In spite of such clever schemes, the distance as well as the rate at which information could be transmitted using semaphore devices was quite limited even during the eighteenth century. A major advance occurred in 1792 when Claude Chappe came up with the idea of transmitting mechanically coded messages over long distances through the use of intermediate relay stations (10–15 km apart) that acted as *repeaters* in the modern-day language [3]. ◘ Figure 8.1 shows the inventor and his basic idea schematically. Chappe called his invention *optical telegraph* and developed a coding scheme shown in ◘ Fig. 8.1 to represent the entire alphabet through different positions of two needles. This allowed transmission of whole sentences over long distances. The first such optical telegraph was put in service in July 1794 between Paris and Lille (two French cities about 200 km apart). By 1830, the network had expanded throughout Europe [4]. The role of light in such systems was simply to make the coded signals visible so that they could be intercepted by the relay stations. The opto-mechanical

◘ **Fig. 8.1** Claude Chappe, his coding scheme, and the mechanical device used for making optical telegraphs (licensed under Public Domain via Wikimedia Commons)

communication systems of the nineteenth century were naturally slow. In modern-day terminology, the effective bit rate of such systems was less than 1 bit/s; a bit is the smallest unit of information in a binary system.

The advent of the electrical telegraph in the 1830s replaced the use of light by electricity and began the era of electrical communications [5]. The bit rate B could be increased to a few bits/s by using new coding techniques such as the Morse code. The use of intermediate relay stations allowed communication over long distances. Indeed, the first successful transatlantic telegraph cable went into operation in 1866. Telegraphy employed a digital scheme through two electrical pulses of different durations (dots and dashes of the Morse code). The invention of the telephone in 1876 brought a major change inasmuch as electric signals were transmitted in an analog form through a continuously varying electric current [6]. Analog electrical techniques dominated communication systems until the switch to optical schemes 100 years later.

The development of a worldwide telephone network during the twentieth century led to many advances in electrical communication systems. The use of coaxial cables in place of twisted wires increased system capacity considerably. The first coaxial-cable link, put into service in 1940, was a 3-MHz system capable of transmitting 300 voice channels (or a single television channel). The bandwidth of such systems was limited by cable losses, which increase rapidly for frequencies beyond 10 MHz. This limitation led to the development of microwave communication systems that employed electromagnetic waves at frequencies in the range of 1–10 GHz. The first microwave system operating at the carrier frequency of 4 GHz was put into service in 1948. Both the coaxial and microwave systems can operate at bit rates ∼ 100 Mbit/s. The most advanced coaxial system was put into service in 1975 and operated at a bit rate of 274 Mbit/s. A severe drawback of high-speed coaxial systems was their small repeater spacing (∼ 1 km), requiring excessive regeneration of signals and making such systems expensive to operate. Microwave communication systems generally allowed for a larger repeater spacing but their bit rate was also limited to near 100 Mbit/s.

All of the preceding schemes are now classified under the general heading of telecommunication systems. A telecommunication system transmits information from one place to another, whether separated by a few kilometers or by transoceanic distances. It may but does not need to involve optics. The optical telegraph of Claude Chappe can be called the first optical telecommunication system that spread throughout Europe over a 40-year period from 1800 to 1840. However, it was soon eclipsed by electrical telecommunication systems based on telegraph and telephone lines. By 1950, scientists were again looking toward optics to provide solutions for enhancing the capacity of telecommunication systems. However, neither a coherent optical source nor a suitable transmission medium was available during the 1950s. The invention of the laser and its demonstration in 1960 solved the first problem. Attention was then focused on finding ways for using laser light for optical communication. Many ideas were advanced during the 1960s [7], the most noteworthy being the idea of light confinement using a sequence of gas lenses [8].

Optical fibers were available during the 1960s and were being used for making gastroscope and other devices that required only a short length of the fiber [9]. However, no one was serious about using them for optical communication. The main problem was that optical fibers available during the 1960s had such high losses that only 10 % of light entering at one end emerged from the other end of a fiber that was only a few meters long. Most engineers ignored them for telecommunication applications where light had to be transported over at least a few kilometers. It was suggested in 1966 that losses of optical fibers could be reduced drastically by removing impurities from silica glass used to make them, and that such low-losses fibers might be the best choice for optical

8

communication [10]. Indeed, Charles Kao was awarded one half of the 2009 noble prize for his groundbreaking achievements concerning the transmission of light in fibers for optical communication [11]. The idea of using glass fibers for optical communication was revolutionary since fibers are capable of guiding light in a manner similar to the confinement of electrons inside copper wires. As a result, they can be used in the same fashion as electric wires are used routinely.

However, before optical fibers could be used for optical communication, their losses had to be reduced to an acceptable level. This challenge was taken by Corning, an American company located not far from Rochester, New York where I work. A breakthrough occurred in 1970 when three Corning scientists published a paper indicating that they were able to reduce fiber losses to below 20 dB/km in the wavelength region near 630 nm [12]. Two years later, the same Corning team produced a fiber with a loss of only 4 dB/km by replacing titanium with germanium as a dopant inside the fiber's silica core. Soon after, many industrial laboratories entered the race for reducing fiber losses even further. The race was won in 1979 by a Japanese group that was able to reduce the loss of an optical fiber to near 0.2 dB/km in the infrared wavelength region near 1.55 μm [13]. This value was close to the fundamental limit set by the phenomenon of Rayleigh scattering. Even modern fibers exhibit loss values similar to those first reported in 1979.

In addition to low-loss optical fibers, switching from microwaves to optical waves also required a compact and efficient laser, whose output could be modulated to impose the information that needed to be transmitted over such fibers. The best type of laser for this purpose was the semiconductor laser. Fortunately, at about the same time Corning announced its low-loss fiber in 1970, GaAs semiconductor lasers, operating continuously at room temperature, were demonstrated by two groups working in Russia [14] and at Bell Laboratories [15]. The simultaneous availability of compact optical sources and low-loss optical fibers led to a worldwide effort for developing fiber-optic communication systems [16].

The first-generation systems were designed to operate at a bit rate of 45 Mbit/s in the near-infrared spectral region because GaAs semiconductor lasers used for making them emit light at wavelengths near 850 nm. Since the fiber loss at that wavelength was close to 3 dB/km, optical signal needed to be regenerated every 10 km or so using the so-called repeaters. This may sound like a major limitation, but it was better than the prevailing coaxial-cable technology that required regeneration every kilometer or so. Extensive laboratory development soon led to several successful field trials. AT&T sent its first test signals on April 1, 1977 in Chicago's Loop district. Three weeks later, General Telephone and Electronics sent live telephone traffic at 6 Mbit/s in Long Beach, California. It was followed by the British Post Office that began sending live telephone traffic through fibers near Martlesham Heath, UK. These trials were followed with further development, and commercial systems began to be installed in 1980. The new era of fiber-optic communication systems had finally arrived. Although not realized at that time, it was poised to revolutionize how humans lived and interacted. This became evident only after the advent of the Internet during the decade of the 1990s.

A commonly used figure of merit for communication systems is the bit rate–distance product BL, where B is the bit rate and L is the repeater spacing, the distance after which an optical signal must be regenerated to maintain its fidelity [2]. ◻ Figure 8.2 shows how the BL product has increased by a factor of 10^{18} through technological advances during the last 180 years. The acronym WDM in this figure stands for wavelength-division multiplexing, a technique used after 1992 to transmit multiple channels at different wavelengths through the same fiber. Its use enhanced the capacity of fiber-optic communication systems so dramatically that data transmission at 1 Tbit/s was realized by 1996. The acronym

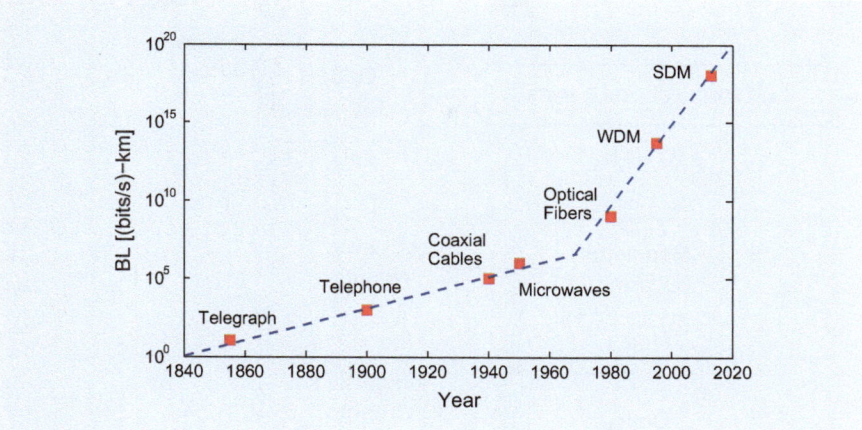

■ **Fig. 8.2** Increase in the *BL* product during the period 1840–2015. The emergence of new technologies is marked by *red squares*. *Dashed line* shows the trend as an aid for the eye. Notice the change in slope around 1977 when optical fibers were first used for optical communications

SDM stands for space-division multiplexing, a technique used after 2010 to further enhance the capacity of fiber-optic systems in response to continuing increase in the Internet data traffic (with the advent of video streaming by companies such as YouTube and Netflix) and fundamental capacity limitations of single-mode fibers (see Sect. 8.5). Two features of ■ Fig. 8.2 are noteworthy. First, a straight line in this figure indicates an exponential growth because of the use of logarithmic scale for the data plotted on the *y* axis. Second, a sudden change in the line's slope around 1977 indicates that the use of optical fibers accelerated the rate of exponential growth and signaled the emergence of a new optical communication era.

8.2 Basic Concepts Behind Optical Communication

Before describing the technologies used to advance the state of the art of fiber-optic communication systems, it is useful to look at the block diagram of a generic communication system in ■ Fig. 8.3a. It consists of an optical transmitter and an optical receiver connected to the two ends of a communication channel that can be a coaxial cable (or simply air) for electric communication systems but takes the form of an optical fiber for all fiber-optic communication systems.

8.2.1 Optical Transmitters and Receivers

The role of optical transmitters is to convert the information available in an electrical form into an optical form, and to launch the resulting optical signal into a communication channel. ■ Figure 8.3b shows the block diagram of an optical transmitter consisting of an optical source, a data modulator, and electronic circuitry used to derive them. Semiconductor lasers are commonly used as optical sources, although light-emitting diodes (LEDs) may also be used for some less-demanding applications. In both cases, the source output is in the form of an electromagnetic wave of constant amplitude. The role of the modulator is to impose the electrical data on this carrier wave by changing its amplitude, or phase, or both of them. In the case of some less-demanding applications, the current injected into a semiconductor laser itself is modulated directly, alleviating the need of an expensive modulator.

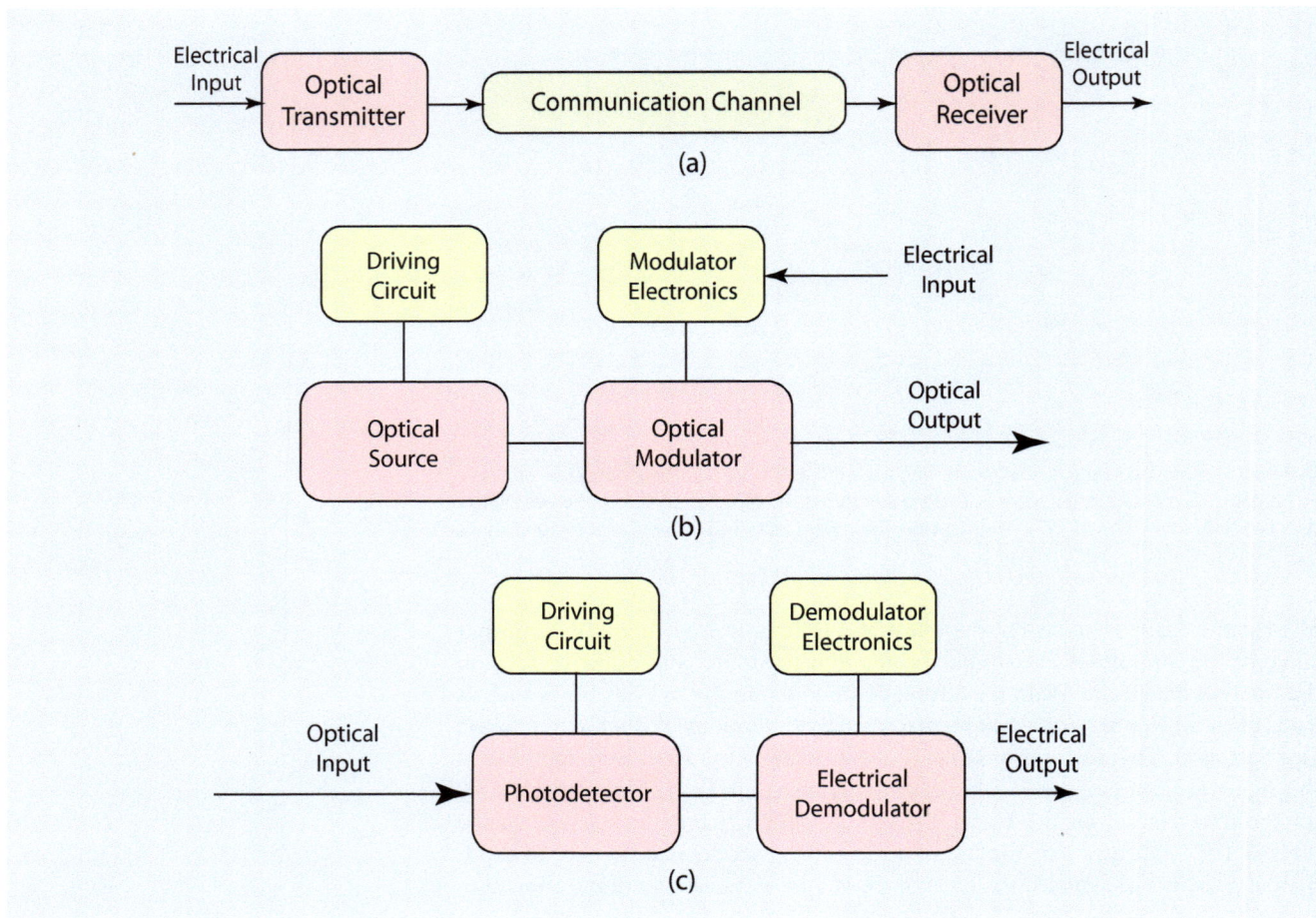

Fig. 8.3 (**a**) A generic optical communication system. (**b**) Components of an optical transmitter. (**c**) Components of an optical receiver

The role of optical receivers is to recover the original electrical data from the optical signal received at the output end of the communication channel. Figure 8.3c shows the block diagram of an optical receiver. It consists of a photodetector and a demodulator, together with the electronic circuitry used to derive them. Semiconductor photodiodes are used as detectors because of their compact size and low cost. The design of the demodulator depends on the modulation scheme used at the transmitter. Many optical communication systems employ a binary scheme referred to as intensity modulation with direct detection. Demodulation in this case is done by a decision circuit that identifies incoming bits as 1 or 0, depending on the amplitude of the electric signal. All optical receivers make some errors because of degradation of any optical signal during its transmission and detection, shot noise being the most fundamental source of noise. The performance of a digital lightwave system is characterized through the bit-error rate. It is customary to define it as the average probability of identifying a bit incorrectly. The error-correction codes are sometimes used to improve the raw bit-error rate of an optical communication system.

8.2.2 Optical Fibers and Cables

Most people are aware from listening to radios or watching televisions that electromagnetic waves can be transmitted through air. However, optical communication systems require electromagnetic waves whose frequencies lie in the visible

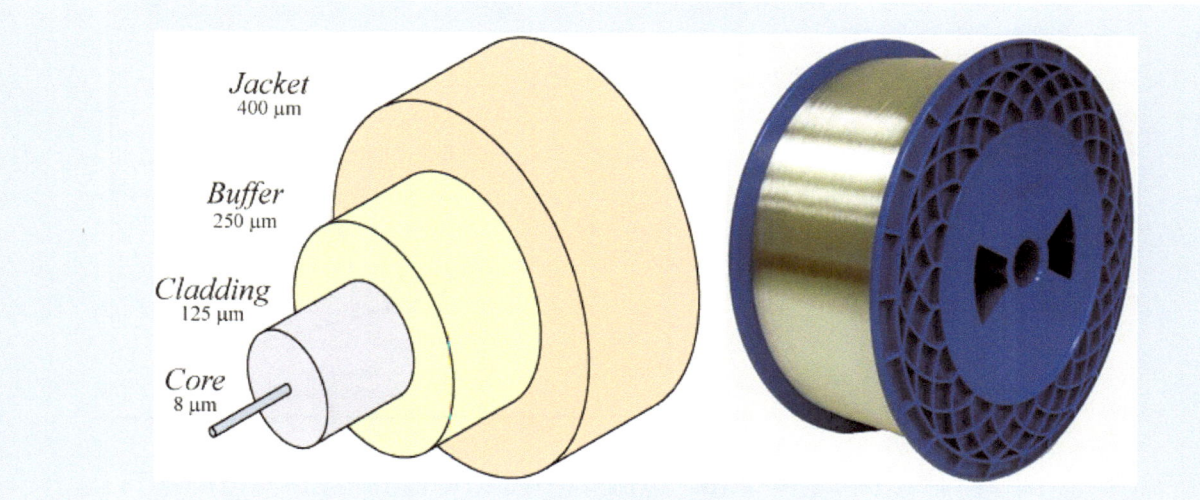

Fig. 8.4 Internal structure of a single-mode fiber; typical size of each part is indicated. A spool of such fiber is also shown (licensed under Public Domain via Wikimedia Commons)

or near-infrared region. Although such waves can propagate through air over short distances in good weather conditions, this approach is not suitable for making optical communication networks spanning the whole world. Optical fibers solve this problem and transmit light over long distances, irrespective of weather conditions, by confining the optical wave to the vicinity of a microscopic cylindrical glass core through a phenomenon known as *total internal reflection*.

Figure 8.4 shows the structure of an optical fiber designed to support a single spatial mode by reducing its core diameter to below 10 μm. In the case of a graded-index multimode fiber the core diameter is typically 50 μm. The core is made of silica glass and is doped with germania to enhance its refractive index slightly (by about 0.5 %) compared to the surrounding cladding that is also made of silica glass. A buffer layer is added on top of the cladding before putting a plastic jacket. The outer diameter of the entire structure, only a fraction of a millimeter, is so small that the fiber is barely visible. Before it can be used to transmit information, one or more optical fibers are enclosed inside a cable whose diameter may vary from 1 to 20 mm, depending on the intended application.

What happens to an optical signal transmitted through an optical fiber? Ideally, it should not be modified by the fiber at all. In practice, it becomes weaker because of unavoidable losses and is distorted through the phenomena such as chromatic dispersion and the Kerr nonlinearity [2]. As discussed earlier, losses were the limiting factor until 1970 when a fiber with manageable losses was first produced [12]. Losses were reduced further during the decade of 1970s, and by 1979 they have been reduced to a level as low as 0.2 dB/km at wavelengths near 1.55 μm. Figure 8.5 shows the wavelength dependence of power losses measured for such a fiber [13]. Multiple peaks in the experimental curve are due to the presence of residual water vapors. The dashed line, marked Rayleigh scattering, indicates that, beside water vapors, most of the loss can be attributed to the fundamental phenomenon of Rayleigh scattering, the same one responsible for the blue color of our sky. Indeed, although water peaks have nearly disappeared in modern fibers, their losses have not changed much as they are still limited by Rayleigh scattering.

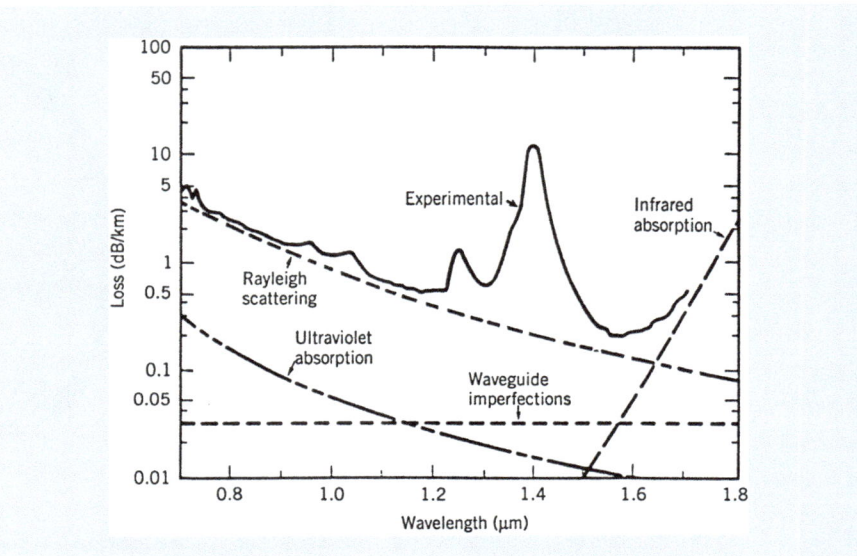

Fig. 8.5 Wavelength dependence of power losses measured in 1979 for a low-loss silica fiber [13]. Various lines show the contribution of different sources responsible for losses (from [2]; ©2010 Wiley)

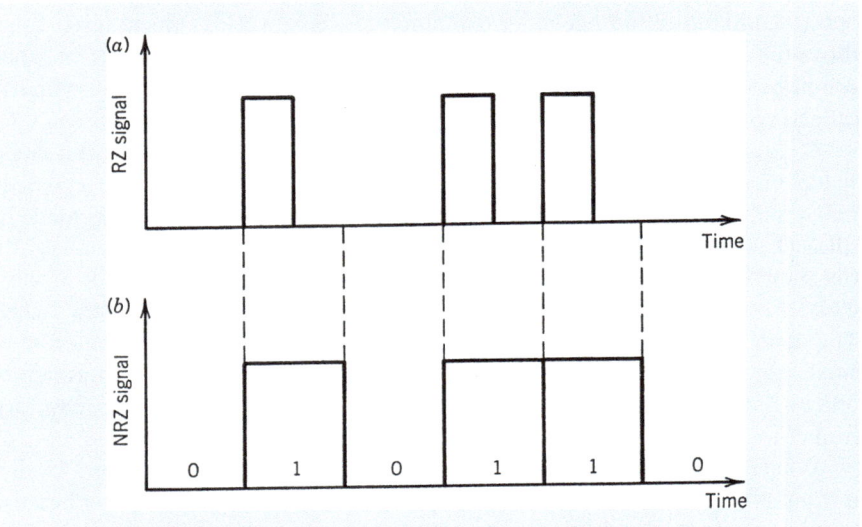

Fig. 8.6 Digital bit stream 010110... coded by using (**a**) return-to-zero (RZ) and (**b**) nonreturn-to-zero (NRZ) formats

8.2.3 Modulations Formats

The first step in the design of any optical communication system is to decide how the electrical binary data would be converted into an optical bit stream. As mentioned earlier, an electro-optic modulator is used for this purpose. The simplest technique employs optical pulses such that the presence of a pulse in the time slot of a bit corresponds to 1, and its absence indicates a 0 bit. This is referred to as on–off keying since the optical signal is either "off" or "on" depending on whether a 0 or 1 bit is being transmitted.

There are still two choices for the format of the resulting optical bit stream. These are shown in Fig. 8.6 and are known as the return-to-zero (RZ) and nonreturn-to-zero (NRZ) formats. In the RZ format, each optical pulse representing bit 1 is shorter than the bit slot, and its amplitude returns to zero before the bit duration is over. In the NRZ format, the optical pulse remains on throughout the bit slot, and its amplitude does not drop to zero between two or more successive 1 bits. As a result, temporal width of pulses varies depending on the bit pattern, whereas it remains the same in the case of RZ format. An advantage of the NRZ format is that the bandwidth associated with the bit stream is smaller by about a factor of 2 simply because on–off transitions occur fewer times. Electrical communication systems employed the NRZ format for this reason in view of their limited bandwidth. The bandwidth of optical communication systems is large enough that the RZ format can be used without much concern. However, the NRZ format was employed initially. The switch to the RZ format was made only after 1999 when it was found that its use helps in designing high-capacity lightwave systems. By now, the RZ format is use almost exclusively for WDM systems whose individual channels are designed to operate at bit rates exceeding 10 Gbit/s.

8.2.4 Channel Multiplexing

Before the advent of the Internet, telephones were used most often for communicating information. When an analog electric signal representing human voice is digitized, the resulting digital signal contains 64,000 bits over each one second duration. The bit rate of such an optical bit stream is clearly 64 kbit/s. Since fiber-optic communication systems are capable of transmitting at bit rates of up to 40 Gbit/s, it would be a huge waste of bandwidth if a single telephone call was sent over an optical fiber. To utilize the system capacity fully, it is necessary to transmit many voice channels simultaneously through multiplexing. This can be accomplished through time-division multiplexing (TDM) or WDM. In the case of TDM, bits associated with different channels are interleaved in the time domain to form a composite bit stream. For example, the bit slot is about 15 μs for a single voice channel operating at 64 kb/s. Five such channels can be multiplexed through TDM if the bit streams of successive channels are delayed by 3 μs. ◘ Figure 8.7a shows the resulting bit stream schematically at a composite bit rate of 320 kb/s. In the case of WDM, the channels are spaced apart in the frequency domain. Each channel is carried by its own carrier wave. The carrier frequencies are spaced more than the channel bandwidth so that the channel spectra do not overlap, as seen in ◘ Fig. 8.7b. WDM is suitable for both analog and digital signals and is used in broadcasting of radio and television channels. TDM is readily implemented for digital signals and is commonly used for telecommunication networks.

The concept of TDM has been used to form digital hierarchies. In North America and Japan, the first level corresponds to multiplexing of 24 voice channels with a composite bit rate of 1.544 Mb/s (hierarchy DS-1), whereas in Europe 30 voice channels are multiplexed, resulting in a composite bit rate of 2.048 Mb/s. The bit rate of the multiplexed signal is slightly larger than the simple product of 64 kb/s with the number of channels because of extra control bits that are added for separating channels at the receiver end. The second-level hierarchy is obtained by multiplexing four DS-1 channels. This results in a bit rate of 6.312 Mb/s (hierarchy DS-2) for North America and 8.448 Mb/s for Europe. This procedure is continued to obtain higher-level hierarchies.

The lack of an international standard in the telecommunication industry during the 1980s led to the advent of a new standard, first called the synchronous

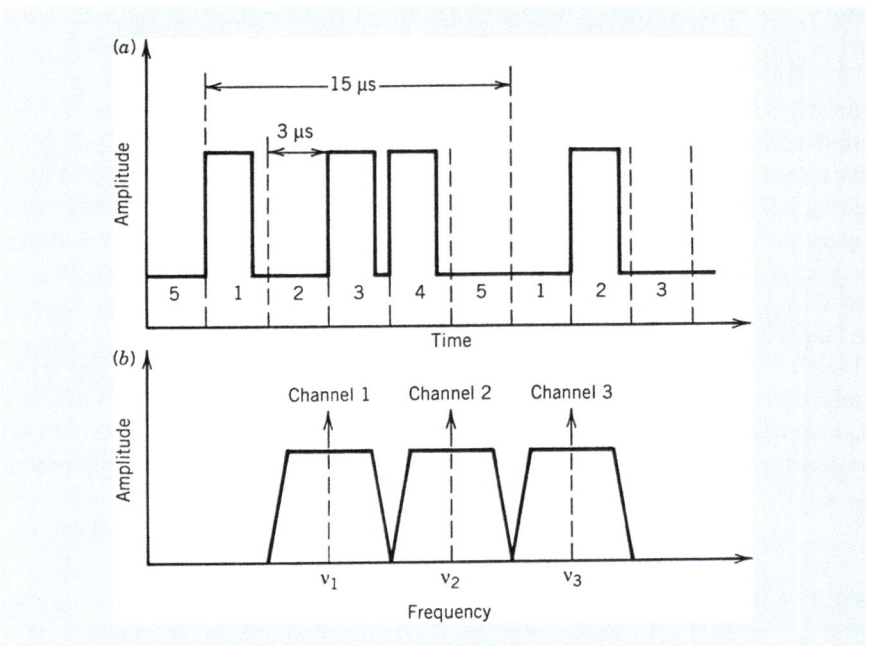

Fig. 8.7 (**a**) Time-division multiplexing of five digital voice channels operating at 64 kb/s. (**b**) Wavelength-division multiplexing of three analog or digital signals

Table 8.1 SONET/SDH bit rates

SONET	SDH	B (Mb/s)	Channels
OC-1		51.84	672
OC-3	STM-1	155.52	2016
OC-12	STM-4	622.08	8064
OC-48	STM-16	2488.32	32,256
OC-192	STM-64	9953.28	129,024
OC-768	STM-256	39,813.12	516,096

optical network (SONET) and later termed the synchronous digital hierarchy (SDH). It defines a synchronous frame structure for transmitting TDM digital signals. The basic building block of the SONET has a bit rate of 51.84 Mbit/s. The corresponding optical signal is referred to as OC-1, where OC stands for optical carrier. The basic building block of the SDH has a bit rate of 155.52 Mbit/s and is referred to as STM-1, where STM stands for a synchronous transport module. A useful feature of the SONET and SDH is that higher levels have a bit rate that is an exact multiple of the basic bit rate. Table 8.1 lists the correspondence between SONET and SDH bit rates for several levels. Commercial STM-256 (OC-768) systems operating near 40 Gbit/s became available by 2002. One such optical channel transmits more than half million telephone conversations over a single optical fiber. If the WDM technique is employed to transmit 100 channels at different wavelengths, one fiber can transport more than 50 million telephone conversations at the same time.

8.3 Evolution of Optical Communication from 1975 to 2000

As mentioned earlier, initial development of fiber-optic communication systems started around 1975. The enormous progress realized over the 40-year period extending from 1975 to 2015 can be grouped into several distinct generations. ◘ Figure 8.8 shows the increase in the *BL* product over the period 1975–2000 as quantified through various laboratory experiments [17]. The straight line corresponds to a doubling of the *BL* product every year. The first four generations of lightwave systems are indicated in ◘ Fig. 8.8. In every generation, the *BL* product increases initially but then begins to saturate as the technology matures. Each new generation brings a fundamental change that helps to improve the system performance further.

8.3.1 The First Three Generations

The first generation of optical communication systems employed inside their optical transmitters GaAs semiconductor lasers operating at a wavelength near 850 nm. The optical bit stream was transmitted through graded-index multimode fibers before reaching an optical receiver, where it was converted back to the electric domain using a silicon photodetector. After several field trials during the period 1977–1979, such systems became available commercially in the year 1980. They operated at a bit rate of 45 Mbit/s and allowed repeater spacings of up to 10 km. The larger repeater spacing compared with 1-km spacing of coaxial systems was an important motivation for system designers because it decreased the installation and maintenance costs associated with each repeater. It is important to stress that even the first-generation systems transmitted nearly 700 telephone calls simultaneously over a single fiber through the use of TDM.

It was evident to system designers that the repeater spacing could be increased considerably by operating the system in the infrared region near 1.3 μm, where fiber losses were below 1 dB/km (see ◘ Fig. 8.5). Furthermore, optical fibers exhibit minimum dispersion in this wavelength region. This realization led to a worldwide effort for the development of new semiconductor lasers and detectors based on the InP material and operating near 1.3 μm. The second generation of

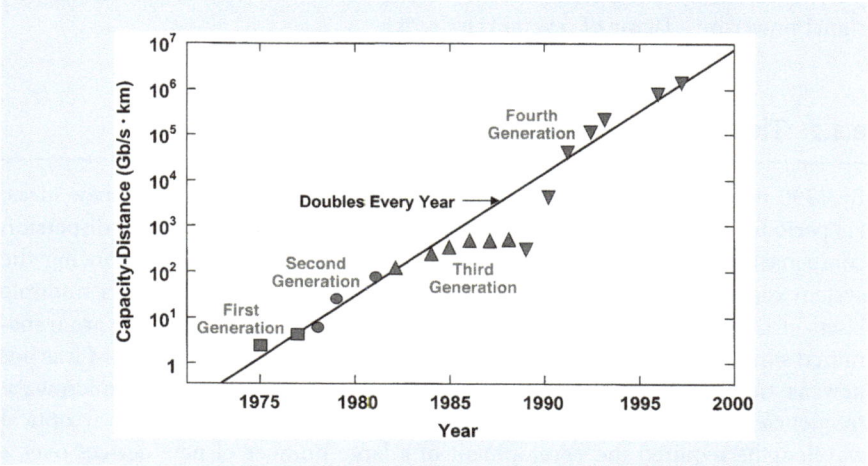

◘ **Fig. 8.8** Increase in the *BL* product over the period 1975–2000 through several generations of optical communication systems. Different symbols are used for successive generations (from [2]; ©2010 Wiley)

8

fiber-optic communication systems became available in the early 1980s, but their bit rate was initially limited to 100 Mbit/s because of dispersion in multimode fibers. This limitation was overcome by the use of single-mode fibers. In such fibers the core diameter is reduced to near 10 μm (see ◻ Fig. 8.4) so that the fiber supports a single spatial mode. A laboratory experiment in 1981 demonstrated transmission at 2 Gbit/s over 44 km of a single-mode fiber. The introduction of commercial systems soon followed. By 1987, such second-generation commercial systems were operating at bit rates of up to 1.7 Gbit/s with a repeater spacing of about 50 km.

The repeater spacing of the second-generation systems was still limited by the fiber loss at their operating wavelength of 1.3 μm. As seen in ◻ Fig. 8.5, losses of silica fibers become the smallest at wavelengths near 1.55 μm. However, the introduction of third-generation optical communication systems operating at 1.55 μm was considerably delayed by a relatively large dispersion of single-mode fibers in this spectral region. Conventional InGaAsP semiconductor lasers could not be used because of pulse spreading occurring as a result of simultaneous oscillation of multiple longitudinal modes. This dispersion problem could be solved either by using dispersion-shifted fibers designed to have minimum dispersion near 1.55 μm, or by designing lasers such that their spectrum contained a dominant single longitudinal mode. Both approaches were followed during the 1980s. By 1985, laboratory experiments indicated the possibility of transmitting information at bit rates of up to 4 Gbit/s over distances in excess of 100 km. Third-generation optical communication systems operating at 2.5 Gbit/s became available commercially in 1990, and their bit rate was soon extended to 10 Gbit/s. The best performance was achieved using dispersion-shifted fibers in combination with lasers oscillating in a single longitudinal mode.

A relatively large repeater spacing of the third-generation 1.55-μm systems reduced the need of signal regeneration considerably. However, economic pressures demanded further increase in its value of close to 100 km. It was not immediately obvious how to proceed since losses of silica fibers at 1.55 μm were limited to near 0.2 dB/km by the fundamental process of Rayleigh scattering. One solution was to develop more sensitive optical receivers that could work reliably at reduced power levels. It was realized by many scientists that repeater spacing could be increased by making use of a heterodyne-detection scheme (similar to that used for radio- and microwaves) because its use would require less power at the optical receiver. Such systems were referred to as coherent lightwave systems and were under development worldwide during the 1980s. However, the deployment of such systems was postponed with the advent of fiber amplifiers in 1989 that were pumped optically using semiconductor lasers and were capable of boosting the signal power by a factor of several hundreds.

8.3.2 The Fourth Generation

By 1990 the attention of system designers shifted toward using three new ideas: (1) periodic optical amplification for managing fiber losses, (2) periodic dispersion compensation for managing fiber dispersion, and (3) WDM for enhancing the system capacity. As seen in ◻ Fig. 8.9, the WDM technique employs multiple lasers at slightly different wavelengths such that multiple data streams are transmitted simultaneously over the same optical fiber. The basic idea of WDM was not new as this technique was already being used at the radio- and microwave frequencies (e.g., by the television industry). However, its adoption at optical wavelengths required the development of a large number of new devices over a time span of a few years. For example, optical multiplexers and demultiplexers that

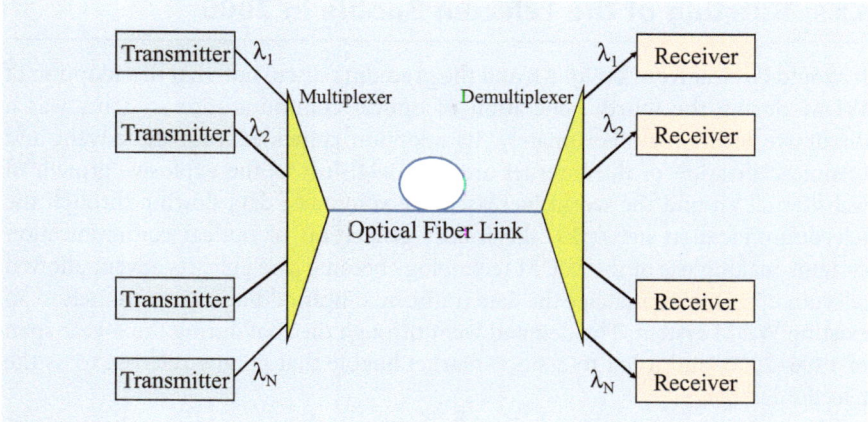

Fig. 8.9 Schematic of a WDM communication system. Multiple transmitters operating at different wavelengths are combined using a multiplexer and all channels are sent simultaneously over the same optical fiber. A demultiplexer at the receiving end separates individual channels and sends them to different receivers

could combine and separate individual channels at the two ends of a fiber link were critical components for the advent of the WDM systems.

The fourth generation of optical communication systems made use of optical amplifiers for increasing the repeater spacing, in combination with the WDM technique for increasing the system capacity. As seen in ◘ Fig. 8.8, the advent of the WDM technique around 1992 started a revolution that resulted in doubling of the system capacity every 8 months or so and led to lightwave systems operating at a bit rate of 1 Tbit/s by 1996. In most WDM systems, fiber losses are compensated periodically using erbium-doped fiber amplifiers spaced 60–80 km apart. Such WDM systems operating at bit rates of up to 80 Gbit/s were available commercially by the end of 1995. The researchers worldwide were pushing the limit of WDM technology. By 1996, three research groups reported during a post-deadline session of the Optical Fiber Communications conference that they were able to operate WDM systems with the total capacity of more than 1 Tbit/s. This represented an increase in the system capacity by a factor of 400 over a period of just 6 years!

The emphasis of most WDM systems is on transmitting as many optical channels as possible over a single fiber by adding more and more lasers operating at different wavelengths. The frequency spacing between two neighboring channels is chosen to be as small as possible but it has to be larger than the bandwidth of each channel. At a bit rate of 40 Gbit/s, a channel spacing of 50 GHz is the smallest that can be used. The standard setting agency, ITU, has assigned a set of fixed frequencies for commercial WDM systems using this 50-GHz channel spacing. All these frequencies lie in the wavelength region near 1550 nm where fiber losses are the smallest. The wavelength range of 1530–1570 nm is called the C band (C standing for conventional), and most commercial WDM systems are designed to work in this band. However, the S and L bands lying on the short- and long-wavelength side of the C band, respectively, are also used if necessary. This approach led in 2001 to a 11-Tbit/s experiment in which 273 channels, each operating at 40 Gbit/s, were transmitted over a distance of 117 km [18]. Given that the first-generation systems had a capacity of 45 Mbit/s in 1980, it is remarkable that the use of WDM increased the system capacity by a factor of more than 200,000 over a period of 21 years.

8.3.3 **Bursting of the Telecom Bubble in 2000**

It should be clear from ◘ Fig. 8.8 and the preceding discussion that the adoption of WDM during the fourth generation of optical communication systems was a disruptive technology. Fortunately, its adoption coincided with the advent and commercialization of the Internet around 1994. Just as the explosive growth of websites all around the world increased the volume of data flowing through the telecommunication networks, the fourth generation of optical communication systems making use of the WDM technology became available. Its advent allowed telecom operators to manage the data traffic by simply adding more channels to an existing WDM system. The demand went through the roof during the 4-year span of 1996–2000, and it led to a stock-market bubble that is now referred to as the telecom bubble.

The formation of the telecom bubble was the result of a rapid growth after 1995 in the telecommunication business. The stocks of companies dealing with the manufacturing and delivery of telecom services soared after 1995. As an example, consider the company JDS–Fitel involved in selling various optical devices needed for telecom systems. Its stock value was around $8 in June 1994, jumped to near $20 in June 1995, and exceeded $70 in June 1996 when the stock was split by 2:1 to bring the price near $35. The stock was split again in November 1997 when its price doubled a second time. In early 1999 the company announced a merger with Uniphase, another fast-growing optics company, resulting in the formation of JDSU. During that year, the stock of JDSU was increasing so rapidly that it was split two more times. ◘ Figure 8.10 shows how the stock price of JDSU varied over the 8-year period ranging from 1996 to 2004 after taking into account multiple splits. A nearly exponential growth during the 1999 indicates the formation of the telecom bubble during that year. A similar growth occurred in the stock price of many other telecommunication and Internet companies.

The telecom bubble burst during the year 2000, and the stock prices of all telecommunication companies collapsed soon after, including that of JDSU as seen in ◘ Fig. 8.10. Several companies went out of business and many surviving were in trouble financially. Commercial WDM systems capable of operating at 1 Tbit/s were still being sold, but their was no buyer for them. The research and development of fiber-optic communications systems slowed down to a crawl as everyone waited for the revival of the telecom industry. It took nearly 5 years before the US economy recovered, only to crash again in August 2008 owing to the formation of another bubble, this time in the real-estate market. One can say that the decade of 2000–2009 has not been a kind one as far as the telecommunication industry is concerned.

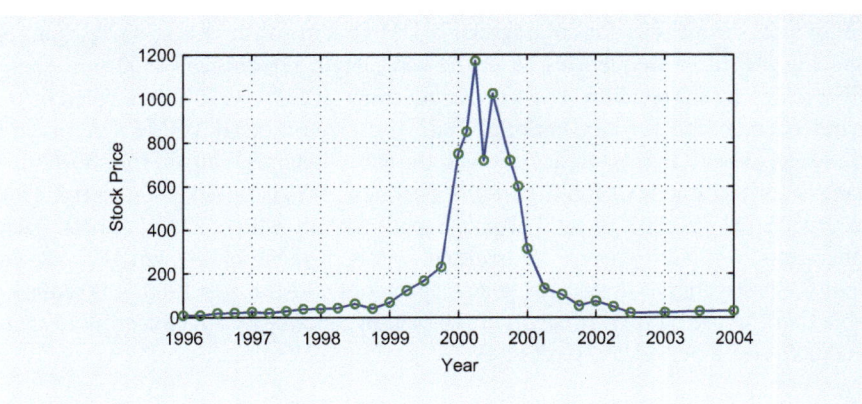

◘ **Fig. 8.10** Price of JDSU stock over a period extending from January 1996 to January 2004. A sharp rise in price during the year 1999 is followed by a sharp decline after July 2001 (*source*: public domain data)

8.4 The Fifth Generation

In spite of the two severe economical downturns, considerable progress has occurred since 2000 in designing advanced optical communication systems, leading to the fifth and sixth generations of such systems. The focus of fifth-generation systems was on making the WDM systems more efficient spectrally. This was accomplished by reviving the coherent detection scheme that was studied in the late 1980s but abandoned soon after fiber-based optical amplifiers became available. Coherent receivers capable of detecting both the amplitude and phase of an optical signal through a heterodyne scheme were developed soon after the year 2000. Their commercial availability near the end of the decade allowed system designers to employ advanced modulation formats in which information is encoded using both the amplitude and phase of an optical carrier.

The basic concept can be understood from ◻ Fig. 8.11, showing four modulation formats using the so-called constellation diagram that displays the real and imaginary parts of the complex electric field along the x and y axes, respectively. The first configuration represents the standard binary format, called amplitude-shift keying (ASK), in which the amplitude or intensity of the electric field takes two values, marked by circles and representing 0 and 1 bits of a digital signal. The second configuration is another binary format, called phase-shift keying (PSK), in which the amplitude remains constant but phase of the electric field takes two values, say 0 and π, that represent the 0 and 1 bits of a digital signal. The third configuration in part (c) of ◻ Fig. 8.11 shows the quaternary PSK (or QPSK) format in which the optical phase takes four possible values. This case allows to reduce the signal bandwidth since two bits can be transmitted during each time slot, and the effective bit rate is halved. Borrowing from microwave communication terminology, the reduced bit rate is called the symbol rate (or baud). The last example in ◻ Fig. 8.11 shows how the symbol concept can be extended to multilevel signaling such that each symbol carries 4 bits or more. An additional factor of two can be gained if one transmits two orthogonally polarized symbols simultaneously during each symbol slot, a technique referred to as polarization division multiplexing.

The concept of spectral efficiency, defined as the number of bits transmitted in 1 s within a 1-Hz bandwidth, is quite useful in understanding the impact of coherent detection in combination with phase-encoded modulation formats. The spectral efficiency of fourth generation WDM systems that employed ASK as the modulation format was limited to below 0.8 bit/s/Hz since at most 40 billion bits/s could be transmitted over a 50-GHz bandwidth of each WDM channel. This value for the fifth-generation systems can easily exceed 3 by using polarization multiplexing in combination with the QPSK format. ◻ Figure 8.12 shows how the spectral efficiency of optical communication systems has evolved since 1990 when its value was near 0.05 bit/s/Hz. Values near 2 bit/s/Hz were realized by 2005 and they approached 10 bit/s/Hz by the year 2010 [19].

The availability of coherent receivers and increased computing speeds led to another advance after it was realized that one can use digital signal processing to

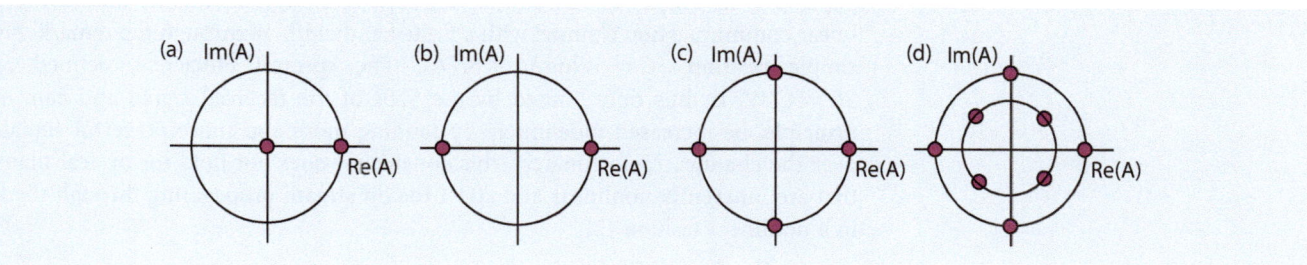

◻ **Fig. 8.11** Constellation diagrams for (**a**) ASK, (**b**) PSK, (**c**) QPSK, and (**d**) multilevel QPSK formats

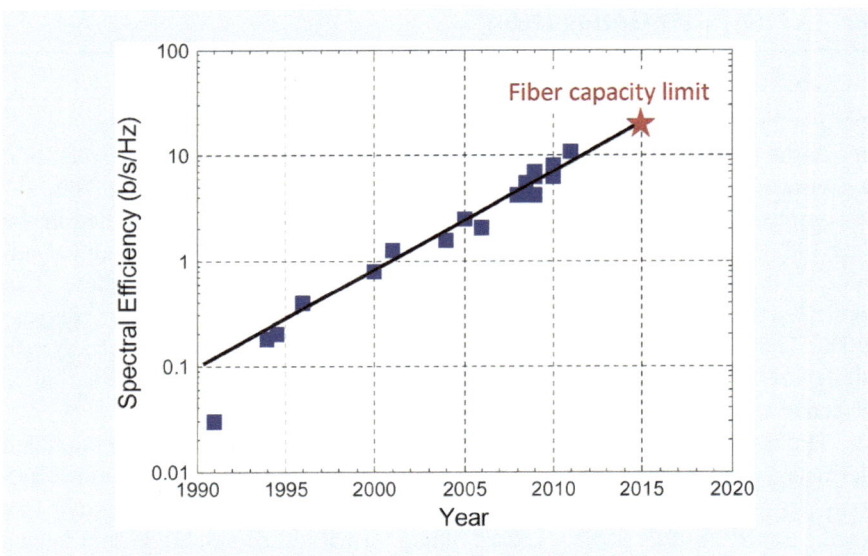

◘ Fig. 8.12 Evolution of spectral efficiency after 1990 through laboratory demonstrations. The *red star* shows the fundamental capacity limit of optical fibers (after [19]; ©2012 IEEE)

improve the signal-to-noise ratio (SNR) of an optical signal arriving at the receiver. Since a coherent receiver detects both the amplitude and the phase of an optical signal, together with its state of polarization, one has in essence a digital representation of the electric field associated with the optical signal. As a result, special electronic chips can be designed to process this digital signal that can compensate for the degradation caused by such unavoidable factors as fiber dispersion. One can also implement error-correcting codes and employ encoder and decoder chips to improve the bit-error rate at the receiver end. A new record was set in 2011 when 64-Tbit/s transmission was realized over 320 km of a single-mode fiber using 640 WDM channels that spanned both the C and L bands with 12.5-GHz channel spacing [20]. Each channel contained two polarization-multiplexed 107-Gbit/s signals coded with a modulation format known as quadrature amplitude modulation. Such techniques are routinely implemented in modern optical communication systems.

8.5   The Sixth Generation

As the capacity of WDM systems approached 10 Tbit/s, indicating that 10 trillion bits could be transmitted each second over a single piece of optical fiber supporting a single optical mode inside its tiny core (diameter about 10 μm), scientists began to think about the ultimate information capacity of a single-mode fiber. The concept of the channel capacity C was first introduced by Shannon in a 1948 paper [21] in which he showed that the SNR sets the fundamental limit for any linear communication channel with a finite bandwidth W through the remarkably simple relation $C = W\log_2(1 + SNR)$. The spectral efficiency, defined as $SE = C/W$, is thus only limited by the SNR of the received signal and can, in principle, be increased indefinitely by sending more and more powerful signals over the channel. Unfortunately, this conclusion does not hold for optical fibers that are inherently nonlinear and affect the bit stream propagating through them in a nonlinear fashion [2].

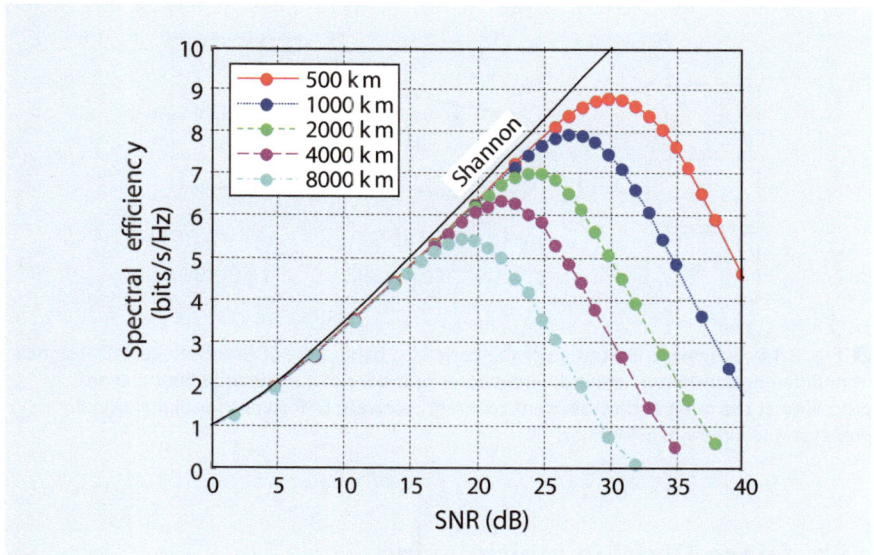

Fig. 8.13 Spectral efficiency as a function of SNR calculated numerically including the nonlinear effects over transmission distances ranging from 500 to 8000 km (after [22]; ©2010 IEEE)

8.5.1 Capacity Limit of Single-Mode Fibers

Considerable attention was paid during the decade of 2000 to estimating the ultimate capacity of single-mode fibers in the presence of various nonlinear effects. In a paper published in 2010 Essiambre et al. were able to develop a general formalism for calculating it [22]. Figure 8.13 shows how the nonlinear effects reduce the spectral efficiency from its value predicted by Shannon's relation, when high signal powers are launched to ensure a high SNR at the receiver. As one may expect, the spectral efficiency depends on the transmission distance, and it becomes worse as this distance increases. However, the most noteworthy feature of Fig. 8.13 is that, for any transmission distance, spectral efficiency is maximum at an optimum value of SNR that changes with distance. For example, the spectral efficiency of a 1000-km-long link is limited to 8 bit/s/Hz for single polarization, irrespective of the modulation format employed. This is in sharp contrast to the prediction of Shannon and reflects a fundamental limitation imposed by the nonlinear effects.

We can use the results shown in Fig. 8.13 to estimate the ultimate capacity of a single-mode fiber. The usable bandwidth of silica fibers in the low-loss window centered around 1550 nm is about 100 nm. This value translates into a channel bandwidth of 12.5 THz. Using this value and a peak spectral efficiency of about 16 bit/s/Hz (assuming polarization-division multiplexing), the maximum capacity of a single-mode fiber is estimated to be 200 Tb/s. This is an enormous number and was thought to be high enough until recently that system designers did not worry about running out of capacity. However, data traffic over fiber-optic networks has experienced a steady growth, doubling every 18 months, since the advent of the Internet in the early 1990s. The growth has even accelerated in recent years owing to the new activities such as video streaming. One way to meet the demand would be to deploy more and more fiber cables. However, this approach will result in a larger and larger fraction of the total electrical power being devoted to supporting optical transport networks. It is estimated that by 2025 the energy demand of modern telecommunication systems will consume a very large fraction of the total US energy budget, unless a way is found to design energy efficient optical networks.

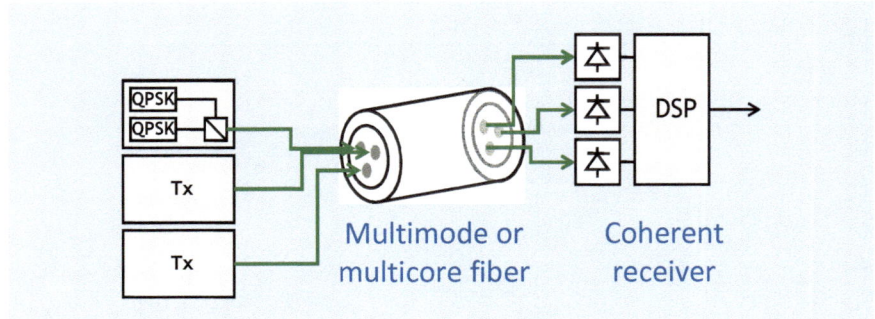

Fig. 8.14 Schematic illustration of the basic idea behind the SDM technique. WDM signals from different transmitters enter different cores or modes of a multimode fiber and are processed at the other end by different coherent receivers; DSP stands for digital signal processing (courtesy of S. Mumtaz)

8.5.2 Space-Division Multiplexing

One proposed solution makes use of space-division multiplexing (SDM) to increase the capacity of fiber-optic communication networks at a reduced energy cost per transmitted bit [23–26]. The basic idea is to employ multimode fibers such that several WDM bit streams can be transmitted over different modes of the same fiber. The energy advantage comes from integrating the functionalities of key optical components into a smaller number of devices. For instance, if a single multimode optical amplifier is used to amplify all spatially multiplexed bit streams, power consumption is likely to be lower compared to using separate amplifiers. For this reason, the SDM technique is attracting increasing attention since 2010, and several record-setting experiments have already been performed. Most of them employ multicore fibers in which several cores share the same cladding. Each core is typically designed to support a single mode but that is not a requirement. ◘ Figure 8.14 shows schematically the basic idea behind SDM using the case of a three-core fiber as an example.

Similar to the case of WDM technology, the implementation of SDM requires not only new types of fibers but also many other active and passive optical components such as mode multiplexers/demultiplexers and fiber amplifiers that can amplify signals in all modes/cores simultaneously. A lot of progress has been made since 2010 in realizing such devices and many laboratory demonstrations have shown the potential of SDM for enhancing the system capacity [23–26]. ◘ Figure 8.15 shows how the capacity of optical communication systems has evolved over a period ranging from 1980 to 2015 and covering all six generations. Single-wavelength systems, employing TDM in the electrical domain, started with a capacity of under 100 Mbit/s in the 1980s and were operating at 10 Gb/s around 1990. The advent of WDM in the early 1990 led to a big jump in the system capacity and subsequent adoption of coherent detection with digital signal processing allowed the capacity to reach 64 Tbit/s by the year 2010 [20]. Further increase in system capacity required the adoption of SDM. In a 2012 experiment, SDM was used to demonstrate data transmission at 1000 Tbit/s (or 1 Pbit/s) by employing a 12-core fiber [24]. Each fiber core carried 222 WDM channels, and each wavelength transmitted a 380-Gbit/s bit stream over a 52-km-long multicore fiber with a spectral efficiency of 7.6 bit/s/Hz.

The simplest SDM case corresponds to a multicore fiber whose cores are far enough apart that they experience little coupling. In this situation, WDM signals in each core travel independently, and the situation is analogous to using separate fibers. Indeed, most high-capacity experiments have employed this configuration

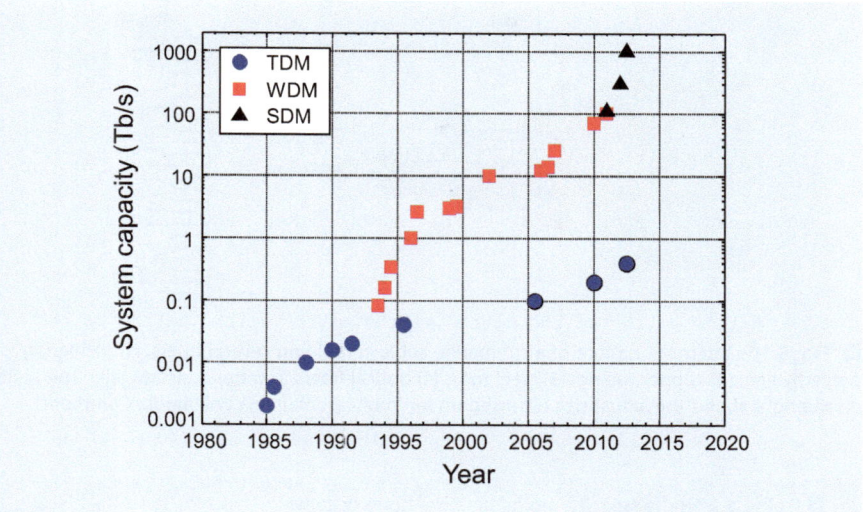

Fig. 8.15 Increase in the capacity of optical communication systems (on a logarithmic scale) realized from 1980 to 2015 using three different multiplexing techniques. Note the change in the slope around 1995 and 2011 when the WDM and SDM techniques were adopted (courtesy of R.J. Essiambre)

through multicore fibers with 7, 12, or 19 cores. In a second category of experiments single-core fibers supporting a few spatial modes are employed [23]. In this case, modes become invariably coupled, both linearly and nonlinearly, since all channels share the same physical path. Degradations induced by linear coupling are then removed at the receiver end through digital signal processing. In a 2015 experiment, a fiber supporting 15 spatial modes was used to transmit 30 - polarization-multiplexed channels over 23 km [27].

8.6   Worldwide Fiber-Optic Communication Network

The advent of the Internet in the early 1990s made it necessary to develop a worldwide network capable of connecting all computers (including cell phones) in a transparent manner. Such a network required deployment of fiber-based submarine cables across all oceans. The first such cable was installed in 1988 across the Atlantic ocean (TAT–8) but it was designed to operate at only 280 Mbit/s using the second-generation technology. The same technology was used for the first transpacific fiber-optic cable (TPC–3), which became operational in 1989. By 1990 the third-generation lightwave systems had been developed. The TAT–9 submarine system used this technology in 1991; it was designed to operate near 1.55 µm at a bit rate of 560 Mb/s with a repeater spacing of about 80 km. The increasing traffic across the Atlantic Ocean led to the deployment of the TAT–10 and TAT–11 cables by 1993 with the same technology. A submarine cable should be strong so that it can withstand biting by large sea animals. ■ Figure 8.16 shows, as an example, the internal structure of a submarine cable containing several fibers for carrying bidirectional traffic. Optical fibers are immersed in a water-resistant jelly that is surrounded with many steel rods to provide strength. Steel rods are kept inside a copper tube that itself is covered with the insulating polyethylene. As the scale on the right side of ■ Fig. 8.16 shows, the outer diameter of the entire cable is still only 1.7 cm.

After 1990, many laboratory experiments investigated whether the amplifier technology could be deployed for submarine cables such that the signal retained its optical form all along the cable length, thus avoiding the use of expensive in-line

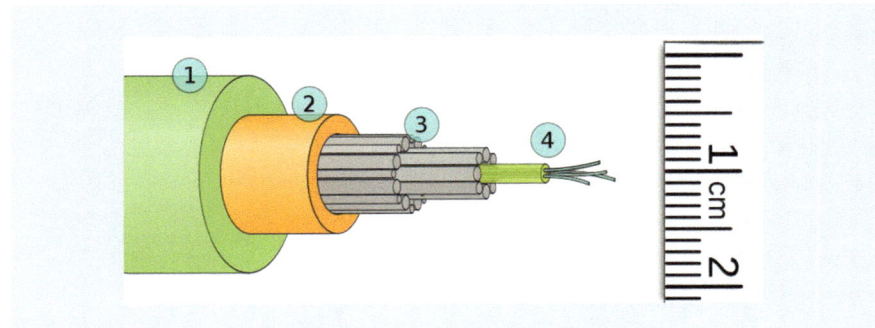

Fig. 8.16 Internal structure of a submarine cable containing several fibers: (1) insulating polyethylene; (2) copper tubing; (3) steel rods; (4) optical fibers in water-resistant jelly. The scale on the right shows the actual size (licensed under Public Domain via Wikimedia Commons)

Table 8.2 High-capacity submarine fiber-optic systems

System name	Year	Capacity (Tb/s)	Length (km)	WDM channels	Fiber pairs
VSNL transatlantic	2001	2.56	13,000	64	4
FLAG	2001	4.8	28,000	60	8
Apollo	2003	3.2	13,000	80	4
SEA-ME-WE 4	2005	1.28	18,800	64	2
Asia–America Gateway	2009	2.88	20,000	96	3
India-ME-WE	2009	3.84	13,000	96	4
African Coast to Europe	2012	5.12	13,000	128	4
West Africa Cable System	2012	5.12	14,500	128	4
Arctic fiber	2015	8.0	18,000	50	4

regenerators. As early as 1991, an experiment employed a recirculating-loop configuration to demonstrate the possibility of data transmission in this manner over 14,300 km at 5 Gbit/s. This experiment indicated that an all-optical, submarine transmission system was feasible for intercontinental communication. The TAT–12 cable, installed in 1995, employed optical amplifiers in place of in-line regenerators and operated at a bit rate of 5.3 Gbit/s with an amplifier spacing of about 50 km. The actual bit rate was slightly larger than the data rate of 5 Gbit/s because of the overhead associated with the forward-error correction that was necessary for the system to work. The design of such lightwave systems becomes quite complex because of the cumulative effects of fiber dispersion and nonlinearity, which must be controlled over long distances.

The use of the WDM technique after 1996 in combination with optical amplifiers, dispersion management, and error correction revolutionized the design of submarine fiber-optic systems. In 1998, a submarine cable known as AC–1 was deployed across the Atlantic Ocean with a capacity of 80 Gb/s using the WDM technology. An identically designed system (PC–1) crossed the Pacific Ocean. The use of dense WDM, in combination with multiple fiber pairs per cable, resulted in systems with large capacities. After 2000, several submarine systems with a capacity of more than 1 Tbit/s became operational (see ▫ Table 8.2). ▫ Figure 8.17 shows the international submarine cable network of fiber-optic communication systems. The VSNL transatlantic submarine system installed in 2001 had a total capacity of 2.56 Tbit/s and spans a total distance of 13,000 km. A submarine

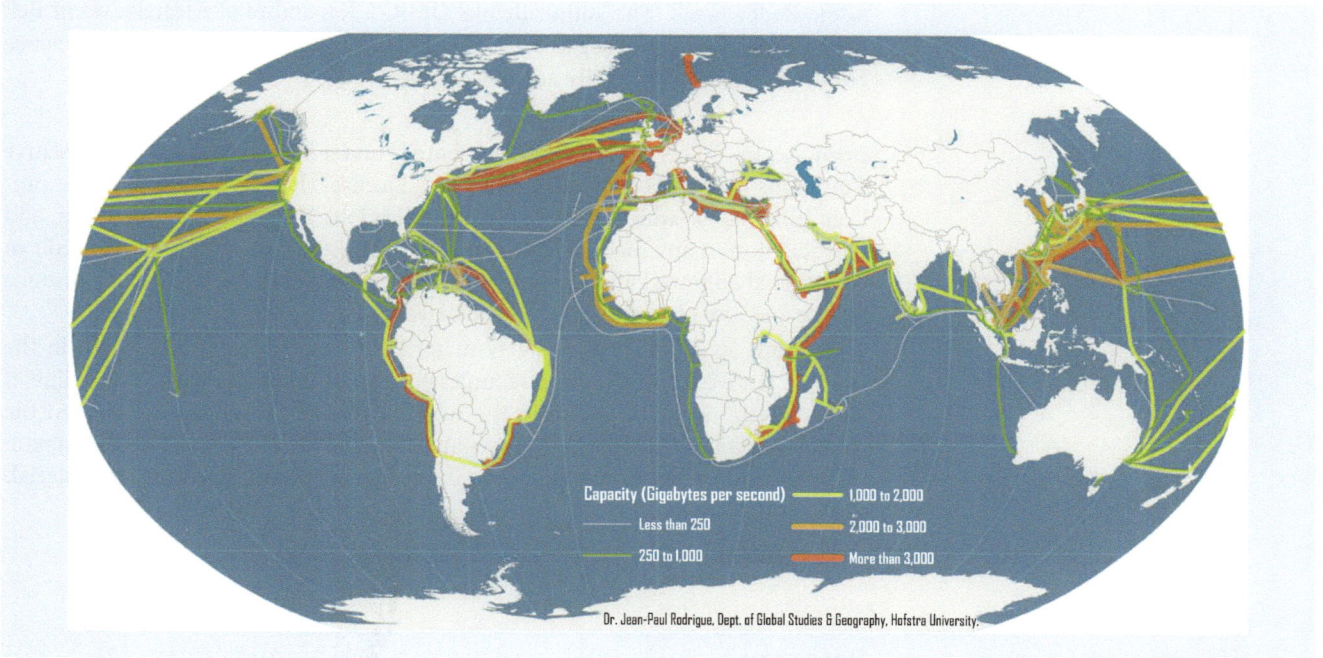

🔲 **Fig. 8.17** International submarine cable network of fiber-optic communication systems around 2015 (*source*: dataset encoded by Greg Mahlknecht, ▶ http://www.cablemap.info)

system, known as India-ME-WE and installed in 2009, is capable of operating bidirectionally at 3.84 Tb/s through four fiber pairs. By 2012, submarine systems with a total capacity of 5 Tbit/s for traffic in each direction became operational. A proposed system, known as Arctic Fiber, will be capable of carrying traffic at speeds of up to 8 Tbit/s by transmitting 50 channels (each operating at 40 Gb/s) over four fiber pairs. It is estimated that more than 400 million kilometers of optical fiber have already been deployed worldwide, a number that is close to three times the distance to sun.

8.7   Conclusions

This chapter began with a brief history of optical communication before describing the main components of a modern optical communication system. Specific attention was paid to the development of low-loss optical fibers as they played an essential role after 1975. I describe in detail the evolution of fiber-optic communication systems through its six generations over a 40-year time period ranging from 1975 to 2015. I also discuss how the adoption of WDM during the 1990s was fueled by the advent of the Internet and how it eventually led in 2000 to bursting of the telecom bubble in the stock markets worldwide. However, the telecommunication industry recovered by 2005, and the researchers have come up with new techniques during the last 10 years. Recent advances brought by digital coherent technology and space-division multiplexing are described briefly in this chapter. 🔲 Figure 8.17 shows the international submarine cable network of fiber-optic communication systems that allows the Internet to operate transparently, interconnecting computers worldwide on demand. Such a global high-speed network would have not been possible without the development of fiber-optic communication technology during the 1980s and the adoption of the WDM technique during the decade of the 1990s. One can only wonder what the future holds, especially if the potential of the SDM technology is realized by the year 2020.

Acknowledgements The author thanks Dr. R.-J. Essiambre of Alcatel–Lucent Bell Laboratories for helpful suggestions. The financial support of US National Science Foundation is also gratefully acknowledged.

References

1. Huurdeman AA (2003) The worldwide history of telecommunications. Wiley, Hoboken, NJ
2. Agrawal GP (2010) Fiber-optic communication systems, 4th edn. Wiley, Hoboken, NJ
3. Chappe I. (1824) Histoire de la télégraphie (in French), University of Michigan Library
4. Holzmann GJ, Pehrson B (2003) The early history of data networks. Wiley, Hoboken, NJ
5. Jones A (1852) Historical sketch of the electrical telegraph. Putnam, New York
6. Bell AG (1876) Improvement in telegraphy, U.S. Patent No. 174,465
7. Pratt WK (1969) Laser communication systems. Wiley, Hoboken, NJ
8. Miller SE (1966) Communication by laser. Sci Am 214:19–25
9. Hecht J (1999) City of light: the story of fiber optics. Oxford University Press, New York
10. Kao KC, Hockham GA (1966) Dielectric-fiber surface waveguides for optical frequencies. Proc IEE 113:1151–1158
11. ▶ http://www.nobelprize.org/nobel_prizes/physics/laureates/2009/
12. Kapron FP, Keck DB, Maurer RD (1970) Radiation losses in glass optical waveguides. Appl Phys Lett 17:423–425
13. Miya T, Terunuma Y, Hosaka T, Miyoshita T (1979) Ultimate low-loss single-mode fiber at 1.55 μm. Electron Lett 15:106–108
14. Alferov Z. (2000) Double heterostructure lasers: early days and future perspectives. IEEE J Sel Top Quant Electron 6:832–840
15. Hayashi I, Panish MB, Foy PW, Sumski S (1970). Junction lasers which operate continuously at room temperature. Appl Phys Lett 17:109–111
16. Willner AE (ed.) (2000) Several historical articles in this millennium issue cover the development of lasers and optical fibers. IEEE J Sel Top Quant Electron 6:827–1513
17. Kogelnik H (2000) High-capacity optical communications: personal recollections. IEEE J Sel Top Quant Electron 6:1279–1286
18. Fukuchi K, Kasamatsu T, Morie M, Ohhira R, Ito T, Sekiya K, Ogasahara D, Ono T (2001) 10.92-Tb/s (273 × 40-Gb/s) triple-band/ultra-dense WDM optical-repeatered transmission experiment. In: Proceedings of the Optical Fiber Communication (OFC) conference, paper PD24
19. Essiambre R-J, Tkach RW (2012) Capacity trends and limits of optical communication networks. Proc IEEE 100:1035–1055
20. Zhou X et al (2011) 64-Tb/s, 8 b/s/Hz, PDM-36QAM transmission over 320 km using both pre- and post-transmission digital signal processing. J Lightwave Technol 29:571–577
21. Shannon CE (1948) A mathematical theory of communication. Bell Syst Tech J 27:379–423
22. Essiambre R-J, Kramer G, Winzer PJ, Foschini GJ, Goebel B (2010) Capacity limits of optical fiber networks. J. Lightwave Technol. 28:662–701
23. Ryf R et al (2012) Mode-division multiplexing over 96 km of few-mode fiber using coherent 6 × 6 MIMO processing. J. Lightwave Technol. 30:521–531

24. Takara H et al (2012) 1.01-Pb/s crosstalk-managed transmission with 91.4-b/s/Hz aggregated spectral efficiency. In: Proceedings of the European conference on optical communications, paper Th3.C.1
25. Richardson DJ, Fini JM, Nelson LE (2012) Space-division multiplexing in optical fibres. Nat Photon 7:354–362
26. Li G, Bai N, Zhao N, Xia C (2014) Space-division multiplexing: the next frontier in optical communication. Adv Opt Commun 6:413–487
27. Fontaine NK et al (2015) 30×30 MIMO Transmission over 15 Spatial Modes. In: Proceedings of the Optical Fiber Communication (OFC) conference, post deadline paper Th5C.1

Optics in Remote Sensing

Thomas Walther and Edward S. Fry

T. Walther
Institute for Applied Physics, TU Darmstadt, Schlossgartenstr. 7, D-64289 Darmstadt, Germany

E.S. Fry (✉)
Department of Physics and Astronomy, Texas A&M University, College Station, TX 77843-4242, USA
e-mail: fry@physics.tamu.edu

© The Author(s) 2016
M.D. Al-Amri et al. (eds.), *Optics in Our Time*, DOI 10.1007/978-3-319-31903-2_9

9.1   Introduction

The observation of light or more specifically the difference between day and night is the very first encounter of physics every human experiences at a very early stage in life. So not surprisingly, optics and the study of the properties of light is one of the oldest in the history of science. The concept of straight propagation of light in homogenous media dates back to the ancient Greeks. This is known as Hero's principle. Although we have learned a lot about light throughout the centuries and have discovered more subtle effects about how light propagates, Hero's principle is the most important feature when applying light to remote sensing. However, we will see that despite its relative early beginnings, modern remote sensing is intimately connected with modern tools of optics such as pulsed lasers, fast detectors, etc.

Optical remote sensing involves the use of light to observe distant objects and to obtain parameters or characteristics of those objects; it can be passive or active. Passive remote sensing would include seeing the light emitted by the object, e.g., a star, the sun, or even the headlights of an approaching car. Remote sensing in the latter case enables one to determine the location of the car and whether it may be on a collision course. Passive remote sensing would also include seeing an object, but via the light from some other source that is scattered/reflected by it; examples include seeing the moon and the planets via the light from the sun that is scattered by them, or just a boy standing in the street and illuminated by the afternoon sun. In either case the light scattered or emitted by the object can also be analyzed to determine some properties of the object; examples include the color of the object and how fast it is moving towards or away from you.

Active remote sensing involves sending out a beam of light and observing the light reflected or scattered by an object. A simple example would be the headlights of a car that illuminate a girl running across the street at midnight. Active remote sensing using a laser to illuminate the object is a technique named LIDAR, a name that arises from a linguistic blend of light and radar. In analogy with the term radar, one might say that it is an acronym for "LIght Detection And Ranging"; but there is no consensus and one can frequently find other definitions such as "Laser Imaging, Detection And Ranging," or "Light Intensity Distance And Ranging." In addition, there is no consensus that it should be an acronym with all capital letters; consequently, it is also variously referenced in publications as LiDAR, Lidar, or just lidar.

This contribution aims at outlining how light is used in remote sensing. First, a short historical overview of remote sensing will be given. This introduction is followed by some of the technical developments leading to modern remote sensing applications such as Lidar. Finally, we will explain some of the applications and advances in the field of Lidar. This article is intended to give a general overview of light in remote sensing while focusing on what is feasible today.

9.2   Historical Overview

The beginning of remote sensing is most likely the invention of triangulation, a technique to measure the distance to some location (possibly remote and inaccessible). Essentially, it involves selecting two positions whose separation is known and measuring the angle subtended at each position by the other position and by the remote location (via light coming from the remote location). Consequently, one has a triangle whose vertices are the two positions and the remote location; the length of one side is known and two angles have been measured, so simple geometry gives the distances to the remote location from each of the two positions. The ancient Greeks already knew that triangles can be used to estimate distances.

Thales had used such a technique in the sixth century BC in order to estimate the height of the Pyramids in Egypt. He simply compared the length of the Pyramid's shadow with his own shadow arriving at the ratio between the height of the Pyramid and his own height. While this knowledge was later somehow lost in Europe, it was noticed in ancient China around 200 AD that triangulation is essential to accurate cartography. The method of using triangles to measure distances was reintroduced to Europe by Arabic scholars around 1000 AD. During the sixteenth century triangulation was widely used in local cartography. The next big step forward was accomplished by Snell in the 1600s; he was the first to establish the required adjustments to the method in order to compensate for the Earth's curvature. Accordingly, local cartography was entirely based on triangulation and many countries established triangular networks over their entire landmass. Today, triangulation on a nationwide scale has been replaced by the global positioning system (GPS). In GPS, satellites continuously send timing information to the GPS receiver. As the exact positions of the satellites are known, the receiver can figure out the runtime differences for the signals to its location and thus determine the position of the receiver relative to that of the satellite. Thus, GPS is basically a modern variant of triangulation.

9.2.1 Speed of Light

In a sense, the measurement of the speed of light historically constituted a form of remote sensing. The ancient Greek philosopher Empedocles had already suspected that the speed of light is finite. However, at the time the theory that the eye actually sends out light was more common and so there was no real need for the notion of the speed of light. The Arabic scholar Alhazen published the first long treatise on optics and postulated that the eye actually only receives light. He then correctly argued that light has a finite speed and that in fact it travels slower in dense media. Since there was no clear idea how to experimentally prove that the speed of light was finite, over the centuries a lot of arguments were exchanged for either view. The first one to experimentally prove that light has a finite speed was Ole Rømer, a Danish astronomer, in 1676. He used an astronomical method in order to arrive at the conclusion that the speed of light was finite by observing the motion of the innermost moon of Jupiter, Io, on its orbit around Jupiter. Based on Rømer's observations, Huygens determined a value of the speed of light which differed from today's value by approximately one third. More refined techniques for measuring the speed of light were developed; these used a long optical path and rotating mirrors that would reflect the light back to the observer if the mirror had rotated to the right angular position by the time the light arrived at it. For a long time this had been the most accurate method until superseded by independent measurements of the wavelength and frequency of the light. The product of these quantities yields the speed of light.

Today, the speed of light has a defined value adopted in 1983. Since the invention of the atomic clock, time is the most accurately measured physical quantity. Consequently, the definition of length arrives at the same level of precision when the speed of light is fixed and distance is defined via the product of time and speed.

9.2.2 Fraunhofer and the Invention of Remote Sensing

Remote sensing in its modern meaning was unknowingly established by Fraunhofer. In 1801, Fraunhofer survived a collapse of the glass-making workshop in which he was working as an apprentice. The Prince Elector of Bavaria was so elated to find at least one survivor that he donated money to the orphan Fraunhofer who was 14 years old at the time. Fraunhofer invested the money in books and learned in the following years to produce optical glass of previously unknown purity and quality. He further invented a spectroscope, essentially a predecessor of today's spectrographs, that can analyze light for its spectral content. In 1814, he observed almost 600 dark lines in the white solar spectrum—far more than Wollaston had observed a couple of years earlier and independently from Fraunhofer. Later Bunsen and Kirchhoff realized that these lines actually were absorption lines of gases in the solar atmosphere. Kirchhoff recognized that a particular set of lines was unique to one special element just like a fingerprint is for a human being. This essentially established spectroscopy as a science and constitutes the first example of remote sensing of the gases in the solar atmosphere. In honor of Fraunhofer, these absorption lines are called Fraunhofer lines. This very specific absorption of any gas (atoms or molecules) will prove very useful when we discuss active remote sensing schemes below.

9.2.3 Passive Remote Sensing

The invention of photography in combination with the ability to fly in unmanned or manned vehicles led to the birth of a broad range of passive remote sensing. The first example is aerial photography. Needless to say, this was very important for military purposes; but it was soon realized that other very valuable information could be extracted from the pictures. This is true in particular for the much higher quality images available from high-flying planes and nowadays satellites. The very sensitive detection devices and optics on board modern satellites allow a variety of remote sensing applications. But, since this article is focussed on optics in remote sensing, the discussion will omit techniques applying acoustic or radar based methods. The most common objectives in the passive satellite based remote sensing programs include meteorological observations for weather forecasting, predictions of hurricane movements, sea surface temperature distributions, military applications and reconnaissance, topographic mapping (stereo photography), or applications in mineralogy, biology, and archaeology.

One of the more sophisticated techniques is hyperspectral imaging in which a variety of images in different spectral regions are obtained; these give a complete picture of the reflective behavior of the Earth's surface. Thus, information such as plant coverage, snow coverage in mountains or the Arctic, and ground-level humidity can be extracted from the data. Such information is especially useful in, for example, analyzing the environmental impact of human activities, or in monitoring other environmental changes.

9.3 The Development of the Laser for Active Remote Sensing

The big leap from passive to active remote sensing required an appropriate light source capable of sending collimated light beams over large distances. In 1960 such a light source, the laser, was operated for the first time; it was invented by T. Maiman after C. Townes and A. Shalow had shown in 1958 that it was theoretically feasible. The acronym laser stands for light amplification by stimulated emission of radiation. In 1918 Albert Einstein had investigated a different phenomenon associated with the interaction of light with matter. Light could be absorbed by an atom, i.e., if a photon, whose energy corresponded to the energy difference between the ground and an excited state of the atom, were to interact with the atom, it could be annihilated and the atom would then be elevated to the excited state of higher energy. The reverse process of an atom in the excited state falling back down to the ground state would then lead to the emission of a photon. Essentially, there are two kinds of emission processes: spontaneous emission occurring by chance without any cause, and stimulated emission occurring when a photon of the correct energy difference interacts with an excited atom and stimulates it to emit. In the latter case, the final result is an atom in the ground state and two photons—the original one and the stimulated one; furthermore, the stimulated photon is an exact copy of the original photon. In essence, this is an amplification process. In the laser this process leads to a large increase in the number of photons; specifically, since the photons are reflected back and forth many times through the medium, many photons are generated. Since the probability an incident photon will be absorbed by a ground state atom is identical to the probability an incident photon will stimulate an excited atom to emit a photon, this laser process works when there are a larger number of atoms in the excited state than in the ground state. This condition is referred to as a population inversion. The many reflections are made possible by the fact that the medium is between two mirrors forming a resonator. One of the mirrors is slightly less reflective than the other so that a small amount of light leaks through it, leaves the laser resonator, and can be used for applications. There are three important properties of this light: First, since the resonator defines an axis of symmetry, the light is highly directed along this very axis; second, since the radiation is closely linked to an atomic transition, the light consists of a single frequency corresponding to that atomic transition frequency; and third, since the photons are generated in a stimulated emission process, the light is coherent. All three properties make up the very unique features of laser light and essentially make it ideal for our purpose of remote sensing. However, just the invention of the laser was not yet enough to fulfill all our needs.

The next necessary step was the invention of the so-called Q-switched laser, which followed in 1962 very soon after the invention of the laser. While a laser in itself can produce very powerful light, the amount of light emanating from it was not enough to perform remote sensing in the atmosphere. The "Q" in Q-switch stands for quality. When the medium of a laser placed inside a cavity is pumped by an energy source achieving population inversion, light is amplified as soon as the gain of the stimulated emission exceeds the losses of the cavity. Consequently, the overall energy in the system is reduced by the stimulated emission (lasing) leaving the medium in the cavity. However, the medium could potentially store much more energy if lasing did not occur.

So, what if at first the cavity was of low quality? Light is not reflected back and forth and no amplification takes place. In this case, almost all of the energy in the pump source is stored in the medium. Once this has happened, the cavity can be suddenly switched into a state of high quality. Now, all of the energy stored in the medium is deposited into a giant pulse with very high energy and a few nanoseconds duration, i.e., some billionths of a second. The sudden switch of the cavity's quality is performed by the Q-switch. In general, this is an electro-optic modulator that manipulates the polarization of the light by applying a short voltage pulse to it.

Inevitably, a pulsed laser has a larger linewidth than a continuous wave laser. The minimum bandwidth is given by the so-called Fourier transform limit. In general, however, a pulsed laser has an even larger bandwidth than this limit. But, since the transitions in typical trace gases are fairly well separated, this linewidth increase does not change the selectivity in detection of trace gases. As stated above, the typical pulse durations of Q-switched lasers are in the nanosecond regime. This duration actually determines the typical spatial resolution one can achieve via time-of-flight measurements.

9

The first laser in 1960 was based on a solid state material known as ruby. The first Q-switched laser was based on the very same material. Ruby as a laser material, however, has a serious drawback. It is only capable of generating laser light around a very well-defined frequency. In fact, most of the early lasers had this limitation, albeit at different wavelengths. Unfortunately, these available laser frequencies were not resonant with any of the transitions of the interesting gases for remote sensing. Thus, another step was necessary in the development of the laser as a valuable tool in remote sensing. The next advancement was the invention of the tunable laser, i.e., a laser whose output frequency could be changed so as to match the transitions of gases of interest. The first tunable laser was the dye laser invented in 1966 independently by Sorokin and by Schäfer. In a dye laser the laser medium consists of an organic dye dissolved in an organic liquid such as methanol and ethanol. In solution, the individual resonances of the dye broaden to a wide band that leads to a broad emission spectrum. By placing additional optical components in the cavity, the laser can be restricted to work at a particular frequency within this band. And, by slightly adjusting the alignment of the components, the output wavelength can be tuned over the entire emission band of the dye. Now, the output of the laser could be tuned to the exact transition needed for a particular gas to be detected. In the following years many dyes were found suitable for such lasers; they cover a broad range of different wavelengths from the IR range, through the visible spectrum, and into the UV regime. And, the Ti:Sapphire laser now provides a solid state alternative that is capable of generating any wavelength in the near IR range.

As we will see later, even shorter pulses can be advantageous for remote sensing. The reason is twofold. First, very short pulses have very high peak intensities that can be directly applied to non-linear detection schemes that are now both available and increasingly important for applications in remote sensing. Second, since pulsed lasers have a larger bandwidth, there is the possibility of detecting several different species at once. Unfortunately, Q-switching only provides pulses in the nanosecond regime, this is essentially due to a combination of the typical size of the laser cavities and the gain of the media involved. However, a laser resonator does not lase on any wavelength, but only on the so-called longitudinal modes. This is comparable to the oscillations of a mechanical string fixed at both ends. The length of the string is an integer multiple of half the wavelength of the possible oscillations; the same is true of the laser oscillations. Thus, in general, the spectrum of a pulsed laser consists of a superposition of these longitudinal modes. There are techniques to synchronize these different modes. When these modes all oscillate with the same phase, something remarkable happens: the radiation in these modes interferes constructively to form giant pulses. The more longitudinal modes are involved the shorter the duration of these pulses. Moreover, the giant pulses oscillate back and forth in the cavity and exit at the corresponding high repetition rate.

9.4 LIDAR

Lidar for active optical remote sensing has a broad range of important applications. Some specific applications to be discussed in the following include: (◼ Sect. 9.4.1) the precision measurement of distances over ranges varying from a few meters to thousands of kilometers, (◼ Sect. 9.4.2) measuring the speed of an object at a distance point, (◼ Sect. 9.4.3) measuring sound speed as a function of depth in the ocean, (◼ Sect. 9.4.4) measuring temperature as a function of depth in the ocean, (◼ Sect. 9.4.5) detecting and identifying underwater objects (fish, mines, etc.), (◼ Sect. 9.4.6) detecting trace impurities in the atmosphere, (◼ Sect. 9.4.7) a quite recent development, a femtosecond-Lidar application for influencing weather phenomena, and (■ Sect. 9.4.8) stand-off super-radiant spectroscopy. A few other important applications that require at least a brief mention include steering a laser beam by inducing refractive index gradients, measuring depth profiles of particulate backscattering in the ocean; obtaining broad areal depictions of the sound speed environment; unraveling sonar signals in regions of varying sound speed; collecting important input for weather forecasting and climate change studies; understanding the behavior of biological populations in the ocean as well as the interactions between oceanic physical and biological structures; and studying/mapping the atmosphere for gases, aerosols, clouds, and temperature.

9.4.1 The Precision Measurement of Distances

The distance d to an object can be measured by sending a short laser pulse to the object and then measuring the time t it takes for the reflected pulse to return. Since the pulse travels a round-trip distance $2d$ in time t, the distance d is given by

$$d = \frac{ct}{2},$$
(9.1)

where c is the speed of light. In practice, one sends a series of pulses, measures the travel time t for each pulse with sub-nanosecond accuracy, and averages the results to then obtain d from Eq. (9.1).

This approach to measuring distances has developed a wide range of applications. It is, for example, very useful in construction and real estate; it essentially replaces the tape measure. Rather than two people with a tape measure to determine the size of a large room, one person with a laser rangefinder can quickly measure it to very high accuracy. Typical of rangefinders for this use is the Bosch Model GLR825 which can measure distances from 50 mm to 250 m with an accuracy of ±1 mm [1]; it presently sells for about $400. Use of a tape measure to determine the width of a 10 m wide room to this kind of accuracy (i.e., one part in 10,000) would be highly problematic.

At the other extreme of LIDAR distance measurements, is the determination of the distance from the earth to the moon (the mean value is 385,000 km). The astronauts placed retroreflectors (corner cubes that reflect incident light by 180°) on the moon during the first lunar landing in 1969; there have been several other retroreflectors placed on the moon since then. The time t for a light pulse to travel from the earth to those retroreflectors and back is then measured and the distance from the surface of the earth to the surface of the moon is determined using Eq. (9.1). Such measurements are quite challenging. As discussed by Dickey et al. [2], the area illuminated on the lunar surface due to the divergence of the transmitted laser beam is ~7 km diameter and the retroreflector only intercepts ~10^{-9} of that area. Due to the diffraction/divergence of the retroreflected beam, a 1 m diameter telescope on the earth will only collect ~2×10^{-9} of it. These two factors lead to a signal collection that is only 2×10^{-18} of the initial laser beam intensity I_0. Other factors such as detector quantum efficiency and mirror reflectivities less than one lead to an overall observed signal of ~10^{-21} I_0. So, if the laser sends out 10^{19} photons in each pulse, the average observed signal would only be one retroreflected photon for every 100 pulses. Nevertheless, the earth–moon distance can be measured with such high accuracy (~1 cm) that the earth–moon system becomes a "laboratory for a broad range of investigations, including astronomy, lunar science, gravitational physics, geodesy, and geodynamics" [2].

Some additional examples in this wide range of applications for laser rangefinder measurements include:

1. Autonomous vehicles—they require knowledge of the distance to nearby objects [3, 4] in order to avoid collisions.
2. Military applications—if the distance to a target is so great that gravitational effects on the projectile must be taken into account when aiming at the target [5, 6], then knowledge of the distance is critical. The MARK VII military rangefinder is an example of a device that meets this need; it operates at ranges of up to 20 km with an accuracy of ± 3 m [7].
3. Sports—in golf, the use of laser ranging can significantly improve hitting accuracy by providing an accurate distance to the target [8].
4. Archaeology—high resolution depth measurements and contour mapping at a site can reveal archaeological features that are otherwise hidden [9].
5. Sub-surface topographical measurements—similar to the archaeology application, the topology of reef substrates (that impact the biology of reef organisms) can be obtained using NASA's Experimental Advanced Airborne Research Lidar (EAARL) [10, 11].

9.4.2 Measuring the Speed of an Object at a Distance Point

If a laser beam is reflected back on itself from an object that is moving in either direction along the same line as the laser beam, then the frequency of the reflected light will be Doppler shifted. By measuring the frequency shift of the reflected light, one can obtain the component of the object's velocity parallel to the laser beam and also determine whether it is approaching or going away. If the object is moving perpendicular to the laser beam, there is no frequency shift in the retroreflected light; but, this component of the velocity could be measured by measuring both the distance and the changing angular direction to the object. The bottom line is that the frequency shift only gives the component of the object's velocity along the direction of the laser beam.

In practice, the system works just as with the rangefinder; a laser pulse is directed to a target and the reflected light is collected. The travel time can still be used to determine the distance to the target; but, in this case the reflected light is also analyzed in frequency to determine the Doppler shift.

To quantify the speed as a function of the frequency shift, recall that the observed frequency due to the motion of a source of electromagnetic waves relative to the observer is given by [12]

$$f = f_0 \sqrt{\frac{c+v}{c-v}}, \tag{9.2}$$

where f is the frequency measured by the observer, f_0 is the frequency in the rest frame of the source, c is the speed of light in the surrounding medium, and v is the relative speed of the source and observer—it is positive (negative) when the source is approaching (moving away from) the observer. Note, v is just the relative speed of source and observer; it does not matter whether the source, observer, or even both are moving.

Now consider the LIDAR application to determine the speed of a target. The laser is the source with frequency f_0, the target is the observer that sees a frequency f_T given by Eq. (9.2) (f on the left-hand side is replaced by f_T). The target is now the source and reflects source light of frequency f_T back towards the laser; the laser position is now the observer and observes a return frequency f_R given by

$$f_R = f_T \sqrt{\frac{c+v}{c-v}} = f_0 \sqrt{\frac{c+v}{c-v}} \sqrt{\frac{c+v}{c-v}} = f_0 \left(\frac{c+v}{c-v} \right) \tag{9.3}$$

If $v \ll c$, this can be approximated as

$$f_R \approx f_0 \left(1 + \frac{2v}{c} \right), \tag{9.4}$$

and the frequency shift due to the Doppler effect is

$$f_R - f_0 \approx \frac{2v}{c} f_0 = \frac{2v}{\lambda_0}. \tag{9.5}$$

Suppose a target is moving with a speed $v = 5$ m/s (~11 miles/hr) and the laser has a wavelength $\lambda_0 = 500$ nm (~blue). Then, the Doppler frequency shift from Eq. (9.5) is

$$f_R - f_0 = 20 \text{ MHz}. \tag{9.6}$$

Measuring such a frequency shift with a good Fabry–Perot is straightforward; a more difficult problem is the laser bandwidth. For a Gaussian laser pulse, the Fourier transform limited time-bandwidth product is

$$\tau \, \Delta f \geq \frac{2 \ln 2}{\pi} \approx 0.44, \tag{9.7}$$

where Δf is the frequency bandwidth and τ is the temporal width (both full width half maximum). Thus, in order to resolve the Doppler shift in Eq. (9.6), the temporal width of the laser pulse must be greater than $\tau = 22 \times 10^{-9}$ s. Of course, for targets moving at higher speeds, this becomes less of a limitation. If the target is moving at 50 m/s, the Doppler frequency shift is 200 MHz and the laser pulse would only be required to have a temporal width greater than 2.2 ns.

9.4.3 Measuring Sound Speed as a Function of Depth in the Ocean

A light pulse propagating in the ocean is scattered by density fluctuations that propagate with the speed of sound. This scattering process is called Brillouin scattering and the remote sensing application, called Brillouin Lidar, has been studied [13–15]. As before, the depth is determined from the travel time of the light pulse; the speed (of sound) is measured by observing the Brillouin frequency shift f_B in the backscattered light. The basic physics of the Brillouin scattering problem is shown in ◘ Figs. 9.1 and 9.2.

◘ Figure 9.1 shows the sound wave fronts for sound waves of wavelength λ_S and wave vector $\vec{k}_S$. A light beam of wave vector $\vec{k}$ is incident at an angle $\theta/2$ with respect to the sound wavefronts. The scattered light has a wave vector $\vec{k}'$; but since $\left| \vec{k}' \right| \approx \left| \vec{k} \right|$ and the angle of incidence on the sound wavefronts is equal to the angle of reflection, the scattered light must be at angle θ with respect to the incident light beam. From the sub-diagrams in ◘ Fig. 9.1, it is clear from momentum conservation that the magnitudes of the wave vectors must satisfy

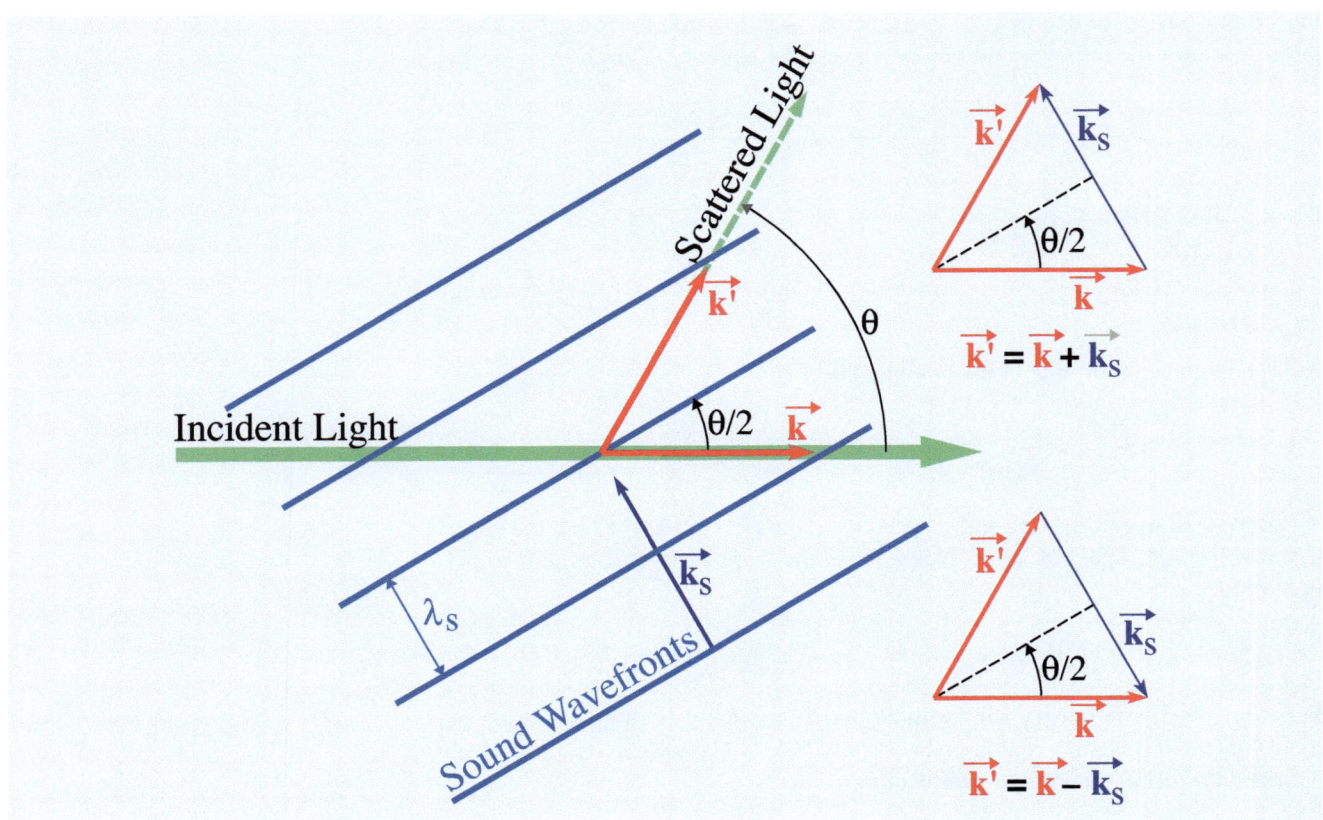

Fig. 9.1 Vector diagram for momentum conservation in Brillouin scattering

$$k_S = 2k\,\sin\frac{\theta}{2}. \tag{9.8}$$

Figure 9.2 shows a sound wavefront that moves a distance $v_S\Delta t$ in a time Δt. Point a on the initial sound wavefront becomes point b after the wavefront has moved the distance $v_S\Delta t$. Since the reflecting surface is a plane wavefront, the object and image distances must be the same. Thus, the distance of the source plane from point a equals the distance of the initial image of the source plane from point a. Similarly, the distance of the source plane from point b equals the distance of the image of the source plane from point b after the time Δt. From the inset in Fig. 9.2 it is clear the distance $V\Delta t$ that the image plane moves in time Δt must be given by

$$V\Delta t = 2\left(v_S\Delta t\,\sin\frac{\theta}{2}\right). \tag{9.9}$$

The relative speed V between the fixed observation plane and the moving image of the light source is therefore

$$V = 2v_S\,\sin\frac{\theta}{2}. \tag{9.10}$$

From Eq. (9.2), the observed frequency due to a source moving with a speed of magnitude V is

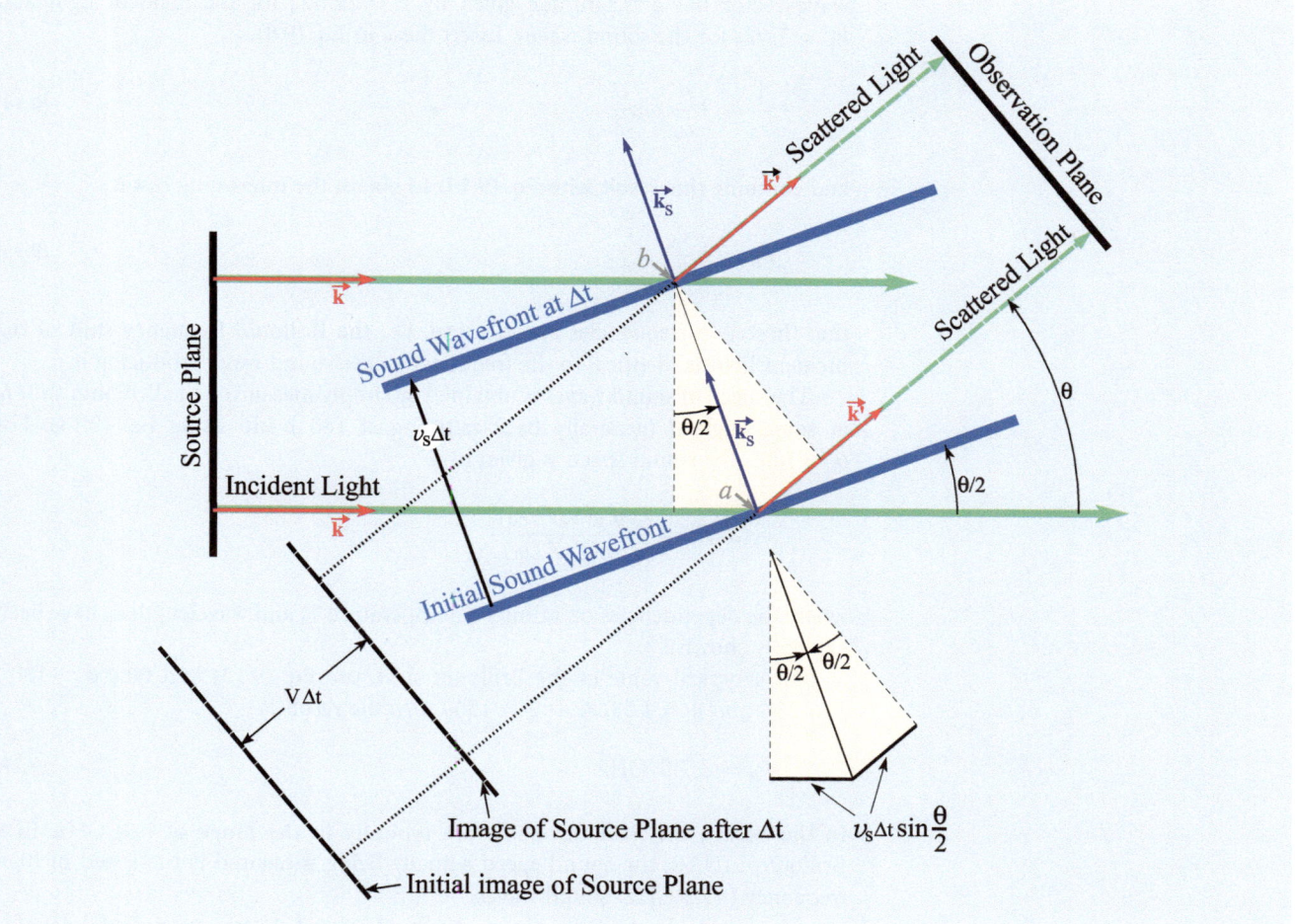

□ **Fig. 9.2** Diagram for determining the Doppler shift

$$f_{obs} = f_0 \sqrt{\frac{c \pm V}{c \mp V}} \approx f_0 \left(1 \pm \frac{V}{c} \right), \tag{9.11}$$

where the approximation in the final step is that $V \ll c$. The upper (lower) sign corresponds to the image of the source plane moving towards (away from) the observation plane. The upper sign would be for $\vec{k}_S$, as shown in □ Fig. 9.2; the lower sign would be for $\vec{k}_S$, as shown in the second of the sub-diagrams of □ Fig. 9.1. The Brillouin frequency shift f_B is then

$$f_B = f_{obs} - f_0 = \pm \frac{V}{c} f_0 = \pm \frac{nV}{\lambda_0}, \tag{9.12}$$

where n is the index of refraction of the medium and λ_0 is the vacuum wavelength of the incident light. Inserting V from Eq. (9.10) gives

$$f_B = \pm v_S 2 \frac{n}{\lambda_0} \sin \frac{\theta}{2}. \tag{9.13}$$

Note that for backscattering ($\theta = 180°$) with $n = 1$, Eq. (9.13) is identical to the Lidar frequency shift of an object moving at speed v_S, Eq. (9.5). Recall that the wave vector has a magnitude given by $k = 2\pi n/\lambda_0$ for the incident light and $k_S = 2\pi/\lambda_S$ for the sound waves. Insert these in Eq. (9.8),

$$\frac{1}{\lambda_S} = 2\frac{n}{\lambda_0} \sin\frac{\theta}{2},$$
(9.14)

and combine this result with Eq. (9.13) to obtain the interesting result,

$$f_B = \pm\frac{v_S}{\lambda_S} \equiv f_S,$$
(9.15)

that these two frequencies are identical, i.e., the Brillouin frequency shift of the incident light is identical to the frequency of the sound wave producing it.

The speed of sound waves is obtained by simply measuring the Brillouin shift f_B at some angle θ (generally backscattering at 180°) and using Eq. (9.13). For $\theta = 180°$, the sound speed is given by

$$v_S(S, T) = \frac{\lambda_0}{2}\frac{|f_B(S, T, \lambda_0)|}{n(S, T, \lambda_0)}$$

where the dependencies on salinity S, temperature T, and wavelength λ_0 have been explicitly shown.

For a typical value of the Brillouin shift, use Eq. (9.13) and take $\theta = 180°$, $\lambda = 530$ nm, $n = 1.33$, and $v_S = 1500$ m/s; the result is

$$f_B = \pm 7.5 \ \text{GHz}.$$
(9.17)

In the oceans the Brillouin shifts are typically in the range of 7–8 GHz. In a Brillouin LIDAR, the sound speed actually being measured is the speed of high frequency (~7.5 GHz) sound waves.

As an example of a Brillouin spectrum, ◘ Fig. 9.3 shows experimental data for Brillouin backscattering by water using the second harmonic of a pulsed Nd:YAG laser [13]. The central peak (at relative frequency 0) is due almost entirely to elastic scattering by suspended particulates. The two peaks offset by ~7.5 GHz are the two Brillouin shifted peaks; they are due only to the water—suspended particulates do not contribute to them. The ratio of the intensity in the central peak to the total intensity in the Brillouin peaks is called the Landau–Placzek ratio [16, 17]. An interesting property of high purity water is that this ratio is so small;

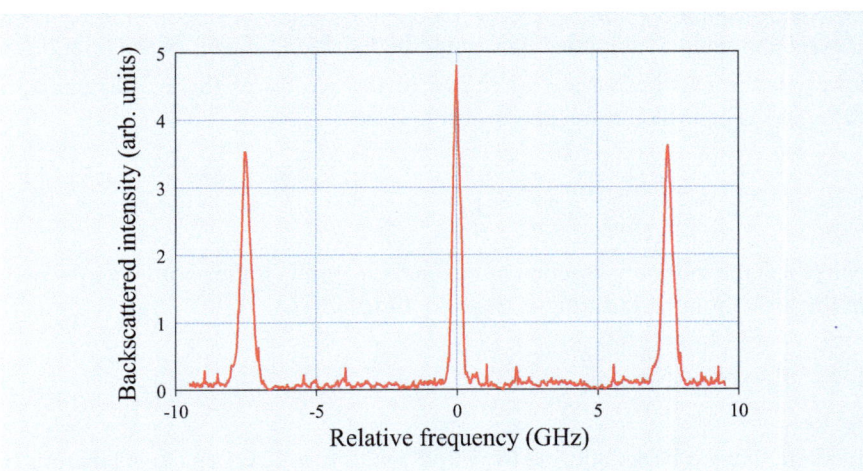

◘ **Fig. 9.3** Experimental data of Brillouin backscattering by water

experimentally and theoretically, it is a minimum and approximately 0–0.04 % at 4 °C, it increases with temperature and is approximately 2 % at 30 °C [17, 18].

A practical implementation of this Brillouin Lidar concept requires a receiver that collects light over an appreciable solid angle, that provides a high frequency resolution of the ~7 GHz frequency shifts, and that provides the measurements with ≈10 ns resolution over a time scale of several hundred nanoseconds (10 ns gives a depth resolution in water of ~1 m).

The first approach to achieve the required resolution might be a Fabry–Perot, but unfortunately the angular acceptance of a suitable Fabry–Perot is too small. For a Fabry–Perot with mirror separation d, the optical path difference between successive rays is $\delta = 2nd \cos \theta$, where n is the index of refraction of the medium between the plates and θ is the angle of incidence on the mirror surface inside the Fabry–Perot. At the transmission peaks the path difference is

$$\delta = 2nd \cos \theta = N\lambda_0, \tag{9.18}$$

where λ_0 is the vacuum wavelength and N is an integer which has its maximum value when $\theta = 0$.

If the free spectral range is $\Delta\lambda$ and the fullwidth at half maximum of the transmission peaks is γ, then the finesse F is defined as $F = \Delta\lambda/\gamma$. As the angle θ increases from the angle of a transmission peak, the path difference decreases and the transmission goes to zero, but rises again at the next transmission peak when the path difference has decreased by λ_0 and Eq. (9.18) is satisfied using $N-1$. If θ is increased from $\theta = 0$, then the maximum angle θ_m at which the transmission moves off the peak occurs when the path difference has decreased by a fraction of λ_0 given by $\gamma/\Delta\lambda$,

$$2nd \cos \theta_m = N\lambda_0 - \frac{\gamma}{\Delta\lambda}\lambda_0 = N\lambda_0 - \frac{\lambda_0}{\mathcal{F}} = 2nd - \frac{\lambda_0}{\mathcal{F}}. \tag{9.19}$$

Solving for θ_m gives

$$1 - \cos \theta_m = \frac{\lambda_0}{2nd\mathcal{F}}, \tag{9.20}$$

and expanding $\cos \theta_m$ for small θ_m gives

$$\theta_m \approx \sqrt{\frac{\lambda_0}{nd\mathcal{F}}}. \tag{9.21}$$

For typical values of $d = 1$ cm, $F = 40$, $n = 1$, and $\lambda_0 = 532$ nm, the angle is $\theta_m \sim 0.07°$; for reference, the angles for the transmission peaks are $0°$, $0.42°$, $0.59°$, $0.72°$, etc. The problem with $\theta_m \sim 0.07°$ can be understood by considering a practical situation in which the Brillouin Lidar receiver mirror has a diameter of 80 cm. The maximum divergence of Lidar signals collected by this mirror from a point source at a distance of 150 m would be $0.31°$. This is already too large compared to $0.07°$, but the problem is even worse. Specifically, a telescope would be required to reduce the diameter of the Lidar return from 80 cm to the few cm diameter of a practical Fabry–Perot; leading to a further significant increase in the beam divergence. The problem has been examined from several aspects by Hickman et al. [14].

The edge technique, a more recent approach, provided the capability to collect the Brillouin scattered light over an appreciable solid angle, to obtain the Brillouin frequency shift as a function of time with ≈10 ns resolution (i.e., 1 m depth resolution), and to do this over a time interval of hundreds of nanoseconds.

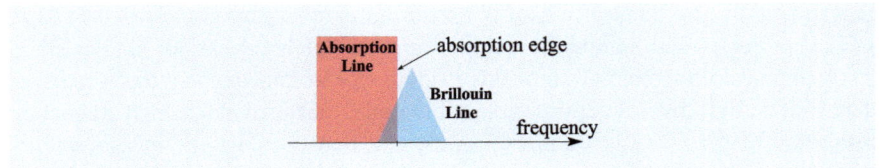

Fig. 9.4 The edge technique concept using an atomic/molecular absorption line

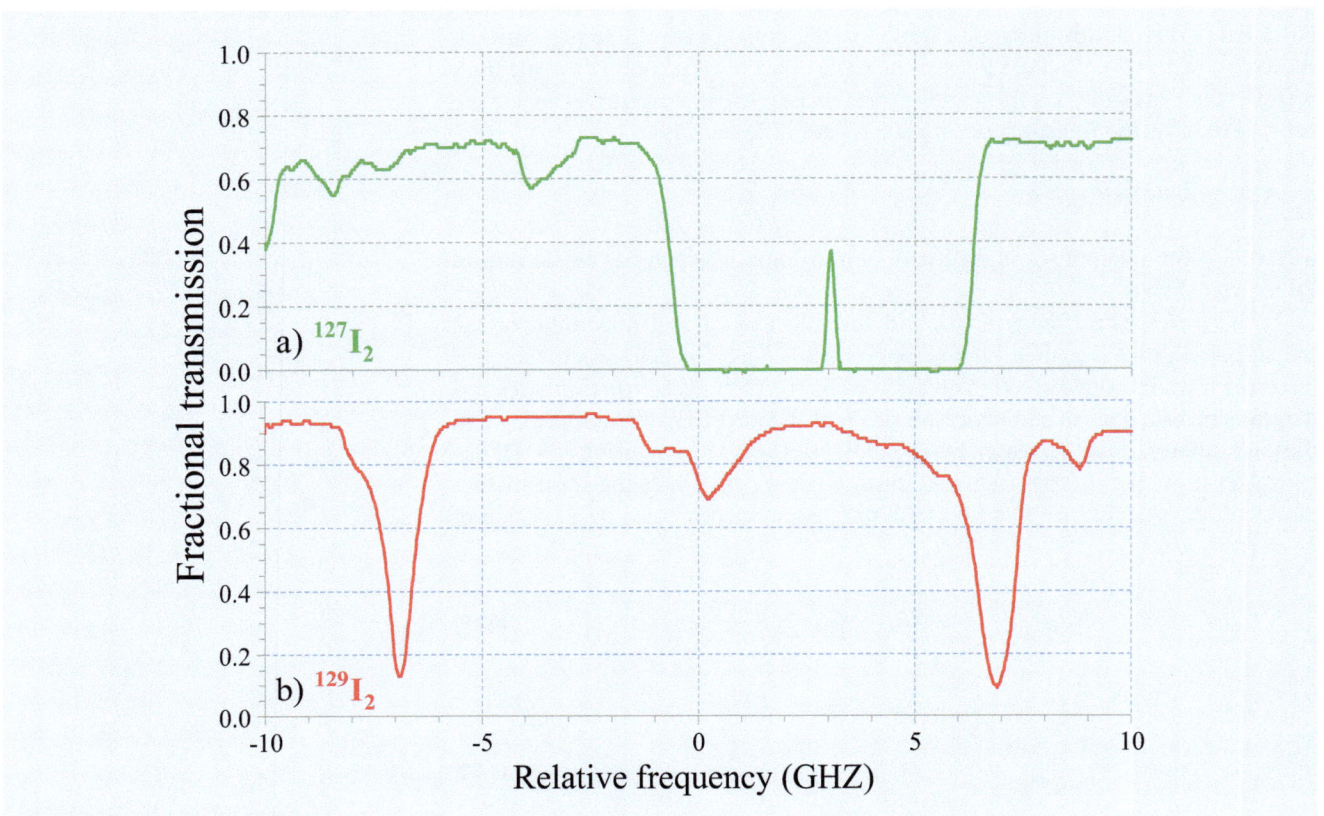

Fig. 9.5 Absorption spectra for the molecules (**a**) $^{127}I_2$ and (**b**) $^{129}I_2$; the zero on the relative frequency axis corresponds to an Nd:YAG laser wavelength of 532.38 nm

The approach is to use an atomic/molecular absorption line and to choose a laser frequency so that a Brillouin shifted component will lie on the edge of the absorption line. ▪ Figure 9.4 illustrates the concept with a fixed frequency rectangular absorption line and a triangular Brillouin line that partially overlaps the absorption line. The Brillouin Lidar return passes through a cell containing the absorbing gas that partially absorbs the Brillouin scattered light. The absorption decreases as the Brillouin shift increases, and vice versa.

Of course, the practical implementation is more complicated because there are two Brillouin lines that must lie on the edges of absorption lines; and there is also the central unshifted line due to scattering by particulates in the water that must be removed. Molecular absorption spectra were examined and a set of absorption lines that meet the requirements were found in $^{127}I_2$ and $^{129}I_2$; their absorption spectra are shown in ▪ Fig. 9.5.

The $^{127}I_2$ has strong absorption at the relative frequency zero (corresponding to the second harmonic of a Nd:YAG laser at 532.38 nm); it is used to remove the central peak. The outside edges of the two $^{129}I_2$ absorption lines are at a relative frequency of approximately ± 7.5 GHz which would be the frequency of the

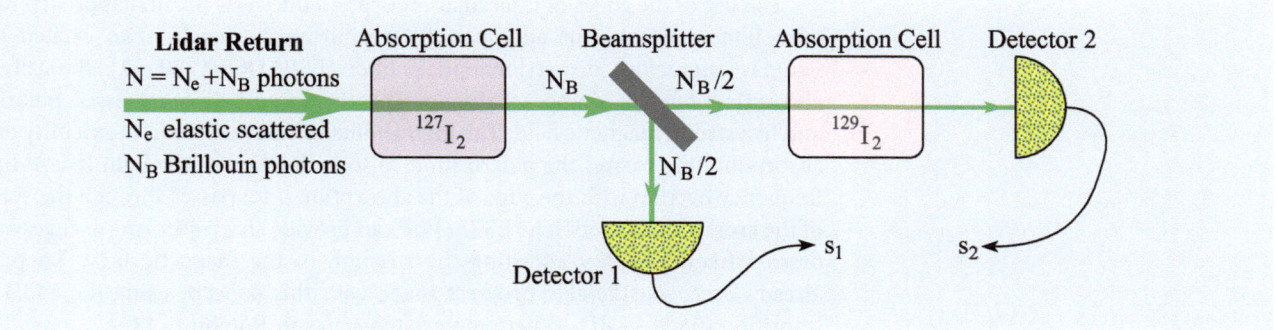

Fig. 9.6 Brillouin Lidar receiver using in $^{127}I_2$ and $^{129}I_2$ absorption cells

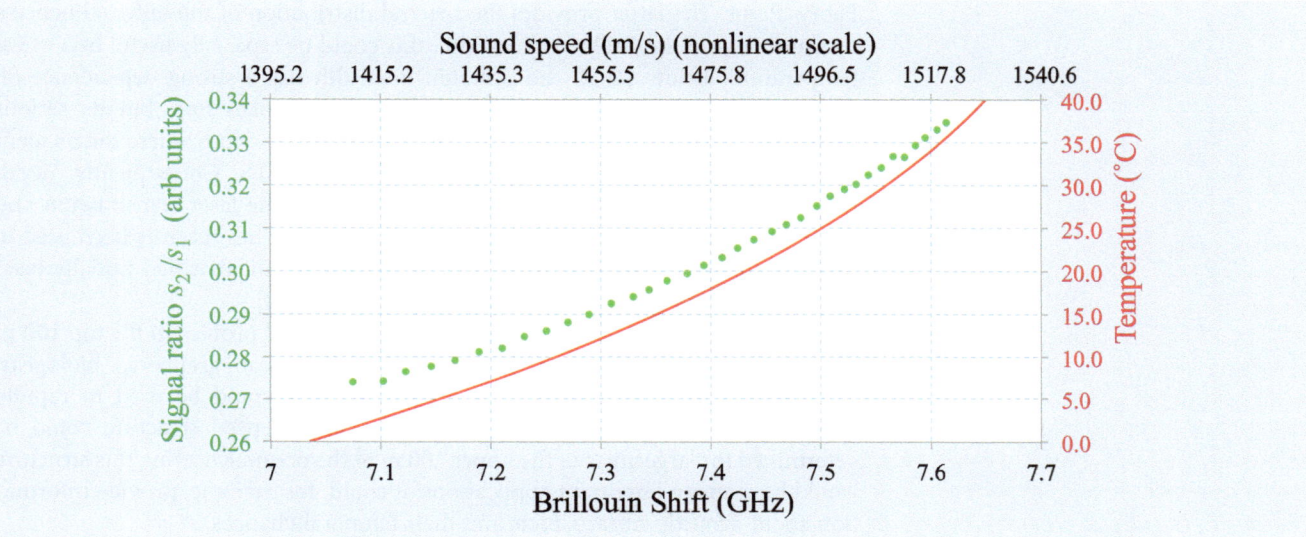

Fig. 9.7 Signal ratio s_2/s_1 (*dots*) and temperature (*smooth curve*) as a function of the Brillouin shift for pure water ($S = 0$); note that the ratio s_2/s_1 is on an arbitrary scale

Brillouin scattered peaks for typical conditions of temperature and salinity in the oceans and other natural waters.

A Brillouin Lidar receiver that implements the edge concept using these molecular absorption lines is shown in **◘** Fig. 9.6. The Lidar return first passes through a $^{127}I_2$ absorption cell that removes the central component and leaves only light that has been Brillouin scattered. Half of this light is then sent to a detector to obtain a reference signal s_1 for the total Brillouin scattered light. The other half passes through a $^{129}I_2$ absorption cell that absorbs a fraction of the Brillouin scattered light based on the extent to which the Brillouin lines overlap the $^{129}I_2$ absorption lines; it provides a signal s_2. After calibration, the ratio s_2/s_1 provides the Brillouin shift.

The efficacy of the system shown in **◘** Fig. 9.6 is demonstrated by its application to obtain the data in **◘** Fig. 9.7 which shows the measured ratio of s_2/s_1 (*dots*) as a function of the Brillouin shift at the laser wavelength, $\lambda_0 = 532.57$ nm, and salinity $S = 0$. The sound speed at each Brillouin shift marker is shown at the top of the plot; this scale is not quite linear because of the temperature dependence of the index of refraction. In practice, the temperature was varied and used to calculate the resulting Brillouin shift and sound speed at each data point [19]. The relation between temperature and Brillouin shift is also shown in **◘** Fig. 9.7 (smooth curve).

The use of the edges of molecular absorption lines was the first approach to the edge filter concept. But an especially promising approach is an excited state Faraday anomalous dispersion optical filter (ESFADOF) [20–22]. Basically, an absorption cell containing an atomic vapor is placed between crossed polarizers and in a strong magnetic field. The high anomalous dispersion in the vicinity of the absorption line rotates the polarization so some fraction of the light (based on its frequency overlap with the edge of the absorption line) passes through the second of the crossed polarizers. The ESFADOF can provide sharp absorption edges at the desired frequencies by adjusting the strength of the magnetic field. There has already been considerable progress made with this concept using the 543.3 nm transition ($5S_{1/2} \rightarrow 8D_{5/2}$) between excited states in Rubidium [23].

Stimulated Brillouin scattering (SBS) has also been used to measure Brillouin shifts [24, 25]. This approach gives a larger signal and since the source of SBS is small, the scattered light can be collimated to within the acceptance angle of a Fabry–Perot. The latter provides the spectral distribution of the SBS and hence a measurement of the Brillouin linewidth; this could be especially useful because at temperatures below 10 °C the Brillouin linewidth has a strong dependence on temperature [26]. SBS does not produce the anti-Stokes line, but its serious drawback is that it makes a measurement at one specific depth where the incident beam is focused to sufficient intensity to produce SBS. Consequently, depth profiling of sound speed requires refocusing of both the laser transmission and the receiver for each depth. In another application, SBS has recently been used to measure the bulk viscosity of water [27]; a measurement that had initially been demonstrated with spontaneous Brillouin scattering [28].

In summary, the very real possibility of sound speed profiles in the top 100 m of the oceans promise vital information to oceanographers, biologists, environmentalists, the military, and others. Aircraft could be used to rapidly obtain sound speed profiles from which the sound speed structure could be determined for large areas of the upper 100 m of the ocean. Knowing this structure would have many important applications; it could, for example, provide information about acoustic surface ducts and their temporal changes.

9.4.4 Measuring Temperature as a Function of Depth in the Ocean

Since over 70 % of the earth's surface is ocean, its temperature distribution on the surface and throughout the top mixed layers plays a major role in our weather and in forecasting our weather. Those temperature distributions in the upper-ocean mixed layers are also of major importance to understanding the physical and biological behaviors of the ocean. Fortunately, experimental depth profile measurements of the Brillouin shift can be converted to depth profiles of the temperature. Actually, this is already clear from ◘ Fig. 9.7—a given ratio s_2/s_1 (*black dot*) corresponds to a unique Brillouin shift (*bottom* axis), a unique sound speed (*top* axis), and a unique temperature (the temperature given by the *smooth curve* for that unique Brillouin shift and sound speed). It is also clear from Eq. (9.13), which in analogy with Eq. (9.16), can be written as

$$f_B(S, T, \lambda_0) = \pm 2 \frac{n(S, T, \lambda_0) v_S(S, T)}{\lambda_0}. \tag{9.22}$$

Since λ_0 is known, if f_B is measured, then Eq. (9.22) is a relationship between salinity S and temperature T. If salinity is known to within 0.1 %, a theoretical analysis shows that an uncertainty of 1 MHz in the Brillouin shift measurement corresponds to a temperature uncertainty of approximately 0.07 °C [19]. If the value of salinity is based on an historical compilation of data [29], it can be expected to be accurate to within a standard deviation of about 1 %; with a 4 MHz uncertainty in the Brillouin frequency shift, the temperature uncertainty would be approximately 0.5 °C [19] (a smaller uncertainty in the Brillouin frequency shift would not have much effect on the temperature uncertainty).

For simplicity, the above discussion neglected pressure effects, but they are fairly easy to include when needed. In principle, the pressure P affects both the index of refraction $n(S,T,\lambda_0,P)$ and the sound speed $v_S(S,T,P)$; but the effects of pressure on the index of refraction are essentially negligible—at 0.0 °C the index increases by less than 0.015 % for a pressure increase of 10 atm [30]. However, pressure can have significant effects on sound speed; it has been well documented and an empirical equation for the dependence is available [31]. This equation must be used in the above calculations for $v_S(S,T,P)$ when the pressure differs from 1.0 atm.

9.4.5 Detecting and Identifying Underwater Objects (Fish, Mines, etc.)

There is an interesting and dramatically simpler version of the Brillouin Lidar that provides the capability of detecting and possibly identifying underwater objects [32, 33]. The point is that the Brillouin Lidar return consists of a central peak due to elastically scattered light and two Brillouin shifted side peaks. Those side peaks only occur if light is being scattered by water at the depth of observation. If there is anything at that depth other than water, the side peaks will not appear.

To detect and identify an underwater object, the laser is tuned to, for example, a strong I_2 absorption line that is chosen based on the condition that I_2 has negligible absorption in the spectral region between 7 and 8 GHz on either side of that absorption line. The Lidar return is then passed through a cell containing I_2 vapor where all elastically scattered light is removed from the Lidar return; only Brillouin scattered light remains. If there is something other than water at the observation depth, there is no light in the Lidar return. It does not matter what that something is, if it is not water, there is no light.

In practice, the Lidar return would pass through range gate optics to define the observation depth in the water, then through the I_2 cell to absorb all elastically scattered light, and then into a camera. The camera would produce a relatively uniform picture if nothing is in the water. But if there is, for example, a fish at the observation depth, the picture would be uniform except for a black region corresponding to the outline of the fish. For a single laser shot, the problem would be for the camera (or array detector) to collect enough light to make a good picture.

9.4.6 Trace Gas Detection

As we have seen earlier, the principle of Lidar can be used to accurately measure distances. Combining this capability with the very specific absorption features of gases provides a method to determine trace gas concentrations in the atmosphere. The idea is simple: tune the pulsed laser to the resonance frequency of the trace gas of interest and fire a short pulse of laser light into the atmosphere. The laser will be scattered back towards the sender. The time of flight of the backwards scattered light will yield the distance at which the pulse was scattered. In this way, a short pulse of nanosecond duration will be broadened to microseconds. But the light returning in a particular nanosecond slice within this broader pulse returns from a well-defined distance. A requirement for this to be true is that the scattering cross-section be small enough such that there are no multiple-scattering events. If there were any, ambiguities in the return time would be introduced. When a certain amount of the trace gas is found in the path of the laser radiation, the light is absorbed more efficiently, indicating that indeed there is a certain amount of the trace gas in the path. In the returning pulse the trace gas would manifest itself as a reduction in the signal. As an example, consider the trace gas to be present only at a height of 10–15 km then obviously light coming back very early, i.e., from heights less than 10 km, no absorption would be present. The same is true for the return signal from heights above 15 km, but of course this light also has to propagate twice through the absorbing layer. Thus, a reduced signal would be observed in the return signal starting at times corresponding to a distance of 10–15 km. This reduction is stronger than the regular signal decay due to a longer path through the atmosphere. However, there is a serious problem. What if clouds are present at a height of 5–10 km? Clearly, the clouds lead to increased levels of scattering. Less light returns to the observer, and the observer not knowing about the clouds would attribute the absence of the signal to absorption, i.e., the presence of the trace gases at a height of 5–10 km as well. Fortunately, there is a solution to this. The technique is called DIAL or DIfferential Absorption Lidar. Essentially, it constitutes a slight modification of the laser system. Instead of just sending out radiation at one wavelength, the laser system now generates radiation at two wavelengths: one right on resonance with the trace gas as before and one slightly off the resonance such that this second beam is not absorbed by the trace gas. The idea is that any cloud in the atmosphere will influence the two wavelengths in the exact same manner. But the trace gas only absorbs radiation at one of the wavelengths. Thus, the difference in the two signal strengths is only caused by the presence of the trace gas. The effect of clouds would show up in both components and could thus be identified as being caused by clouds or any other scatterer.

It is easily conceivable that these types of measurements are quite complex and many challenges had to be met. As an example we want to discuss two issues. First, it would be ideal to make these types of measurements irrespective of the time of day. However, during daytime there is a lot of background radiation due to the sunlight; it easily swamps the small fraction of backscattered light. Thus, it was necessary to develop techniques to suppress the daylight. The solution was essentially narrow bandwidth filters that suppress the very broad spectral content of the sunlight without attenuating the returning laser light. While this constitutes a technical problem, the second issue is more fundamental in nature. It is the so-called inverse problem. Despite the fact that two wavelengths are used for the DIAL system, many effects can contribute to a certain return signal. It might be clouds, water droplets, snowflakes, ice, aerosols, very small volcanic debris from an eruption, etc. Moreover, the concentration distribution of a trace gas might be more complicated than indicated in our example. Thus, the unambiguous extraction of the actual atmospheric layering from the data is actually very complex and intricate. Bohren and Huffman have compared this problem in their book [34] on absorption and scattering by small particles to the problem of a knight hunting a dragon: it is actually quite straightforward to recognize the footprint of a dragon when seeing the dragon; it is very hard to conclude what the dragon might look like from just seeing its footprints.

The DIAL technique was developed in the early 1970s and 1980s as a method of sensitive trace gas detection in the atmosphere. At first DIAL was used to monitor many air pollutants and later it was used to monitor the depletion of the ozone layer. In principle, ozone can be detected using passive remote sensing techniques aboard satellites. The disadvantage is, however, that only the cumulative concentration can be measured. Any height information is lost. While this height information can be extracted via balloon measurements, the laser based DIAL technique provides data much faster, more reliably and up to higher altitudes than the balloon based measurements; it is the superior approach.

9.4.7 Femtosecond-Lidar Application for Influencing Weather Phenomena

As noted earlier, a short pulse has a very large spectral bandwidth; hence, more than one trace gas species could potentially be detected simultaneously. Different spectral parts of the pulse would be absorbed by different species, and one laser with sufficiently short pulses should be able to simultaneously detect an entire range of trace gases. When researchers from France and Germany tried this for the first time at a relatively short distance, the technique indeed worked. However, when they directed the laser into the atmosphere, they discovered a surprise. Instead of a well-defined absorption at particular frequencies, they observed a bright white light extending a few hundred meters into the air. By beam steering and choice of laser parameters the researchers were able to move this channel to a variety of different locations. A thorough analysis revealed that the observation was actually a plasma channel forming due to the high intensity of the pulsed laser. The laser would ionize the air, producing a plasma state of free electrons and ions. The white light then originates from the ions and electrons recombining. In principle such a plasma state should make the laser diverge rapidly due to the contributions of the free electrons to the index of refraction. However, there is a focusing effect that counterbalances it so that propagation of the plasma channel is, in fact, relatively stable over a large distance. This self-focusing of a high intensity laser is known as the Kerr-effect. For high intensities the index of refraction has a non-linear contribution proportional to the intensity of the pulse. Since laser beams usually have a profile with higher intensity in the middle, the index of refraction is higher in the center; this leads to a focusing of the beam. Consequently, the two effects cancel each other and literally form an artificial lightning strike moving up into the atmosphere. Immediately, this sparks the creativity of a physicist. What if such a laser could be used to "direct" a natural lightning strike in a thunderstorm? Instead of more or less random lightning strikes, where it hits could be selected by essentially sending the artificial lightning strike towards an area where the next natural lightning strike is suspected. First experiments already hint at this becoming a reality in the near future. In thinking even further ahead, it is conceivable that using such a plasma channel could produce rainfall from clouds. Today, clouds are seeded by silver iodine in order to improve rain precipitation. In the future, it might become possible to just use such a femtosecond laser system to make rain from a cloud.

9.4.8 Stand-Off Super-Radiant Spectroscopy

Stand-off super-radiant spectroscopy is a promising remote sensing technique that is far more sensitive than Lidar or DIAL [35]. Briefly, the idea for detection of a weak concentration of some species is to use laser pulses to provide optical pumping of the species at a distant location in such a way that it will actually send back a laser beam signal. This is achieved by sending out multiple pairs of pulses along the same path; one pulse in a pair has wavelength λ_1 and the other λ_2. There is a small time delay $\Delta\tau$ between the two pulses in a pair and if $\lambda_1 > \lambda_2$, then the λ_2 pulse must be in the lead. The idea is that due to atmospheric dispersion, the longer wavelength pulse λ_1 will move faster and catch up to the shorter wavelength pulse at a well-defined distance determined by $\Delta\tau$. The wavelengths λ_1 and λ_2 are

9

chosen so that when they both interact with a gas molecule of interest, the molecule will be driven into an appropriate excited state (the upper lasing level) via two photon pumping, or via some Raman process. So, for a given $\Delta\tau$, there will be a well-defined distant point where the molecules of interest will be excited and produce a gain region. By sending out a long sequence of pulse pairs in which $\Delta\tau$ steadily decreases for each pair, there will be a long sequence of these gain regions being created in a direction directly back to the observer. Basically, it is a gain swept amplifier that is lasing straight back to the observer.

This approach will make it possible to detect parts per million concentrations of dangerous gases at distances of several kilometers. An important application of the gain swept amplifier is to actually obtain backward directed lasing from the major constituents of air, N_2 and O_2 [36]. A photon from this backward lasing can combine with a photon from a forward propagating interrogation laser beam to produce a molecular excitation in some trace gas. Then, by measuring the absorption of the backward propagating laser beam as a function of the wavelength of the interrogation laser, the trace gas concentration can be quantified with great sensitivity. Finally, this backward directed lasing could be an important tool in astronomical adaptive optics [37]; it could provide an artificial guide star at any position in the sky.

9.5   Conclusions

Optics and in particular light in remote sensing has evolved in the past decades to be one of the most important applications for learning more about the environment in which we live. The variation of optical methods for remote sensing is very large; it spans the gamut from airplane and satellite based passive remote sensing platforms to active remote sensing using time-of-flight techniques. In active remote sensing, the development of the laser and subsequently the introduction of Lidar have led to particularly striking progress. Lidar can be used for a large variety of tasks that shape knowledge about our environment—ocean temperature, trace gas detection in the atmosphere, air pollution in cities, telemetry, and many other applications. Once more optics and light is demonstrating how important and fundamental it is to our lives.

References

1. Robert Bosch Tool Corporation (2015) Power tools for professionals GLR825. ▸ http://www.boschtools.com/Products/Tools/Pages/BoschProductDetail.aspx?pid=glr825-specs

2. Dickey JO, Bender PL et al (1994) Lunar laser ranging: a continuing legacy of the Apollo program. Science 265(5171):482–490

3. Guizzo E (2011) How Google's self-driving car works. IEEE Spectrum. ▸ http://spectrum.ieee.org/automaton/robotics/artificial-intelligence/how-google-self-driving-car-works

4. Whitwam R (2014) How Google's self-driving cars detect and avoid obstacles. Extremetech. ▸ http://www.extremetech.com/extreme/189486-how-googles-self-driving-cars-detect-and-avoid-obstacles

5. Army Test and Evaluation Command, Aberdeen Proving Ground, MD (1969) Laser rangefinders. Ft. Belvoir Defense Technical Information Center

6. Litz B (2014) Laser rangefinders; Chapter 16 in Modern advancements in long range shooting. Applied Ballistics LLC, Cedar Springs, MI

7. Northrop Grumman Systems Corporation (2013) MARK VII handheld eyesafe laser target locator. ▸ http://www.northropgrumman.com/Capabilities/MarkVII/Documents/markvii.pdf

8. Owen D (2011) Stuff I like: long-distance operators. Golf Dig 62(3):70–71

9. Pappas S (2013) Legend of lost city spurs exploration, debate. Live Science. ▸ http://www.livescience.com/37539-legend-ciudad-blanca-lost-city.html

10. Wright CW, Brock JC (2002) EAARL: a lidar for mapping shallow coral reefs and other coastal environments. In: Proceedings of the 7th international conference on remote sensing for marine and coastal environments, Miami

11. Brock JC, Wright CW et al (2004) LIDAR optical rugosity of coral reefs in Biscayne National Park, Florida. Coral Reefs 23:48–59

12. Young HD, Freedman RA (2014) University physics with modern physics. Pearson

13. Fry ES (2012) Remote sensing of sound speed in the ocean via Brillouin scattering. In: Hou W, Arnone R (eds) Proceedings of SPIE 8372. pp 8372071–8372078

14. Hickman GD, Harding JM et al (1991) Aircraft laser sensing of sound velocity in water: Brillouin scattering. Remote Sens Environ 36:165–178

15. Guagliardo JL, Dufilho HL (1980) Range-resolved Brillouin scattering using a pulsed laser. Rev Sci Instrum 51:79–81

16. Cummins HZ, Gammon RW (1966) Rayleigh and Brillouin scattering in liquids: the Landau—Placzek ratio. J Chem Phys 44:2785–2796

17. Rouch J, Lai CC et al (1976) Brillouin scattering studies of normal and supercooled water. J Chem Phys 65:4016–4021

18. O'Connor CL, Schlupf JP (1967) Brillouin scattering studies of normal and supercooled water. J Chem Phys 47:31–38

19. Fry ES, Emery Y et al (1997) Accuracy limitations on Brillouin lidar measurements of temperature and sound speed in the ocean. Appl Opt 36:6887–6894

20. Schorstein K, Scheich G et al (2007) A fiber amplifier and an ESFADOF: developments for a transceiver in a Brillouin-LIDAR. Laser Phys 17:975–982

21. Popescu A, Walther T (2009) On an ESFADOF edge-filter for a range resolved Brillouin-lidar: the high vapor density and high pump intensity regime. Appl Phys B 98:667–675

22. Rudolf A, Walther T (2012) High-transmission excited-state Faraday anomalous dispersion optical filter edge filter based on a Halbach cylinder magnetic-field configuration, Opt Lett 37:4477–4479

23. Rudolf A, Walther T (2014) Laboratory demonstration of a Brillouin lidar to remotely measure temperature profiles of the ocean. Opt Eng 53:051407, 1–9

24. Shi J, Li G et al (2007) A lidar system based on stimulated Brillouin scattering. Appl Phys B 86:177–179

25. Shi J, Ouyang M et al (2008) A Brillouin lidar system using F–P etalon and ICCD for remote sensing of the ocean. Appl Phys B 90:569–571

26. Fry ES, Katz J et al (2002) Temperature dependence of the Brillouin linewidth in water. J Mod Opt 49:411–418

27. He X, Wei H et al (2012) Experimental measurement of bulk viscosity of water based on stimulated Brillouin scattering. Opt Commun 285:4120–4124

28. Xu J, Ren X et al (2003) Measurement of the bulk viscosity of liquid by Brillouin scattering. Appl Opt 42:6704–6709

29. National Oceangraphic Data Center (NODC), User Services Branch, NOAA/NESDIS E/OC21 (1993) Oceanographic station profile time series

30. The International Association for the Properties of Water and Steam (1997) Release on the refractive index of ordinary water substance as a function of wavelength, temperature and pressure. ▸ http://www.iapws.org/relguide/rindex.pdf

31. Grosso VAD (1974) New equation for the speed of sound in natural waters (with comparisons to other equations). J Acoust Soc Am 56:1084–1091
32. Fry ES, Kattawar GW et al (2000) System and method for detecting underwater objects. Texas A&M University, Patent# 6388246
33. Gong W, Dai R et al (2004) Detecting submerged objects by Brillouin scattering. Appl Phys B 79:635–639
34. Bohren CF, Huffman DR (1983) Absorption and scattering of light by small particles. Wiley, New York
35. Kocharovsky V, Cameron S et al (2005) Gain-swept superradiance applied to the stand-off detection of trace impurities in the atmosphere. Proc Natl Acad Sci U S A 102:7806–7811
36. Hemmer PR, Miles RB et al (2011) Standoff spectroscopy via remote generation of a backward-propagating laser beam. Proc Natl Acad Sci U S A 108:3130–3134
37. Wizinowich PL, Mignant DL et al (2006) The W. M. Keck observatory laser guide star adaptive optics system: overview. Publ Astron Soc Pac 118:297–309

9

Optics in Nanotechnology

Munir H. Nayfeh

M.H. Nayfeh (✉)
Department of Physics, University of Illinois at Urbana-Champaign, 1110 W. Green Street, Urbana, IL 61801, USA
e-mail: m-nayfeh@illinois.edu

© The Author(s) 2016
M.D. Al-Amri et al. (eds.), *Optics in Our Time*, DOI 10.1007/978-3-319-31903-2_10

10.1 Introduction

Optics is one of the most important branches of physics. It involves the study of the behavior and properties of light in vacuum as well as the study of its interactions with matter in the gas, liquid, and solid states [1]. Moreover, the field encompasses the construction of instruments that use or detect light that may serve many other fields, over a wide range of the electromagnetic waves from UV to infrared light.

Nanotechnology or nanoscience and technology, on the other hand, aims at construction, understanding, and putting to use ultrasmall particles [2–4]. Miniaturization of all types of matter including dielectrics, metals, semiconductors, polymers, etc., affords interesting novel properties, especially optical properties. The novel size regime is intermediate between the largest molecules and 100 nm. In this regime, phenomenon may not be as predictable as those observed at larger scales. Using nanoparticles as building blocks to construct advanced devices that exploit their novel properties is at the heart of nanotechnology.

It is to be noted that naturally existing colloids, micelles, polymer molecules, and phase-separated regions in block copolymers, for example, fall in this size regime. More recently, naturally unknown but interesting classes of nanostructures such as carbon nanotubes [5], silicon nanoparticles [6, 7], metal nanoparticles and nanorods [8, 9], and compound semiconductor quantum dots [10] have been designed and fabricated. The application of such extremely small particles can find applications across all fields [4, 11]. In this article we will focus on how miniaturization down to the nanoscale regime impacts the behavior, properties, and interactions of light with matter, especially metals and semiconductors, and how it enables novel advanced devices with application in electronics, photonics, energy and lighting, and biomedicine.

10.2 Optics in Nanometals: Nature of Interaction of Light with Metal

In a dielectric, all electrons are bound in atoms; and each atom interacts with light individually through the interaction of a single bound electron. The total effect in a sample is simply the sum of the individual atomic responses. In a metal solid, some electrons are bound to atoms and others are not bound in specific atoms or ions. Upon interaction with light, two processes can take place in metal. In interband transition, bound electrons, i.e., electrons in the valence band can be promoted due to light absorption to an empty level in the conduction band where they become not bound to a specific ion. Interband absorption forms a significant loss mechanism in metal at optical frequencies. On the other hand, in a metal solid, electrons that are already in the conduction band form a see or cloud of electrons not bound to specific atoms or ions. The cloud can interact with and move under an external electric force collectively at the same time.

10.2.1 Plasma Model

■ Figure 10.1 (left) shows a schematic of the lattice of bulk metal where electrons are subjected to a force created by the electric field of light. Free electrons move opposite to the direction of the electric field while positive ions are stationary, thus causing a shift between the center of distribution of the negative charge and the positive charge. In the absence of the force the centers of the two distributions coincide. ■ Figure 10.1 (right) shows the same situation for a very small sphere of metal, namely a nanoparticle.

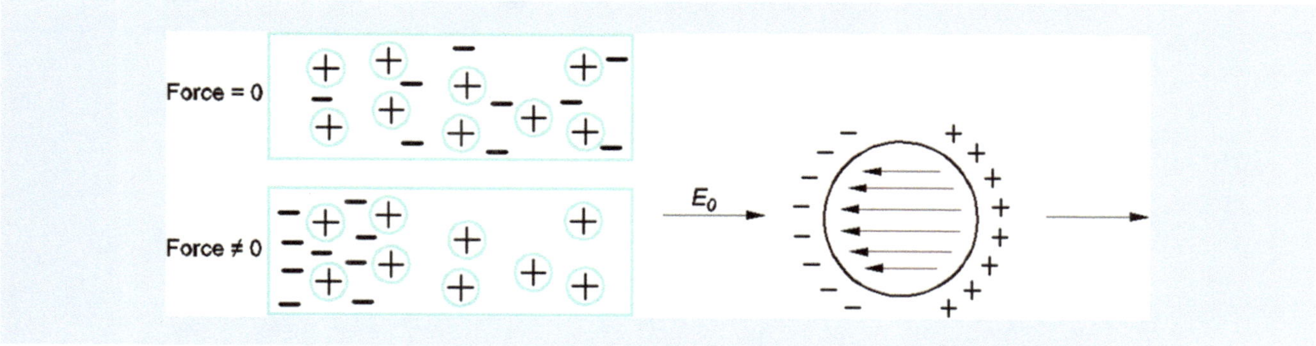

Fig. 10.1 Sketch of metal lattice subjected to a force created by an electric field. (*Left*) bulk and (*Right*) nanoparticle

Electrons in a small sphere are described by a simplified model of a gas of free electrons that moves against a fixed background of positive ion cores [12–14]. This has been called a *plasma model*. In the model, details of the lattice potential and electron–electron interactions are not included explicitly; rather they are simply incorporated into the effective optical mass m of each electron. Under irradiation, the negative electron's see or cloud gets pushed back and forth relative to the fixed background of positive ion at the frequency of oscillation in the electric field of the electromagnetic light wave. In addition, the cloud experiences an effective drag force due to collisions with the positive ion core which tends to slow it down. According to the plasma model, electrons oscillate and their motion is damped with a characteristic collision frequency γ as high as 10^{14} per second (100 THz) (corresponding to the frequency of infrared light). The time elapsed between two consecutive collisions, $\tau = 1/\gamma = 10^{-14}$ s, is known as the relaxation time of the free electron gas. The balance of these two forces gives the response or what is called the real part of the dielectric function of the cloud, $\epsilon(\omega)$, to the light wave:

$$\varepsilon(\omega) = 1 - \frac{ne^2}{\varepsilon_0 m\omega^2}$$

Where n is the density of electrons in the cloud, e and m are the charge and the effective mass of a single of electron, ϵ_0 is the dielectric function of vacuum, and ω is the frequency of the incoming light wave. It is customary to define a useful quantity called plasma frequency or plasmon frequency ω_p in terms of the following group of constants:

$$\omega_p = \sqrt{\frac{ne^2}{\varepsilon_0 m}}$$

In terms of the plasma frequency, the dielectric function takes the simple form $1 - \omega^2/\omega_p^2$. For a metal with a free electron density of $10^{23}/cm^3$, ω_p corresponds to the frequency of ultraviolet light (plasmon energy $\hbar\omega_p \sim 10$ eV) which is 100 or even 1000 times larger than the relaxation frequency γ.

Figure 10.2 plots $\epsilon(\omega)$ as a function of frequency. At resonance with the plasma frequency ($\omega = \omega_p$), $\epsilon(\omega)$ drops to zero. This is a plasma oscillation resonance. For light with a frequency below the plasma frequency, the dielectric function is negative and the light cannot penetrate the sample, rather it is totally reflected. Above the plasma frequency the light waves penetrate the sample as sketched in the figure.

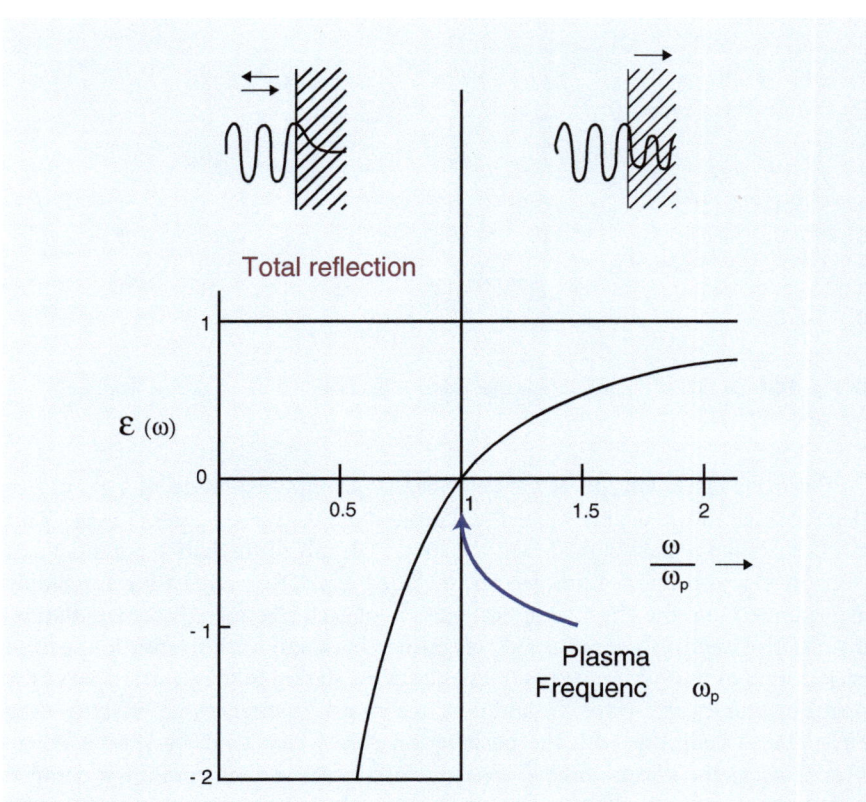

Fig. 10.2 Plot of the dielectric function $\epsilon(\omega)$ as a function of frequency for a metal. $\epsilon(\omega)$ drops to zero at resonance with the plasma frequency ($\omega = \omega_p$) (Image from ▶ https://www.coursehero.com/file/10546609/Plasmonics/)

10.2.2 Miniaturized Metal: Subwavelength Concentration of Light

We now discuss how the interaction of metal with light manifest itself as we reduce the size of the metal sample from large bulk samples to small nanoparticles. We will compare three kinds of structures sketched in ▪ Fig. 10.3: bulk, which is labeled as three-dimensional (3D); a sheet or quantum well, which is labeled as two-dimensional (2D); quantum wire, which is labeled as one-dimensional (1D); and nanoparticle or quantum dot, which is labeled as zero-dimensional (0D).

Bulk Material (3D)

Because free conduction electrons can move over very large distances in bulk metal objects, electrons on average do not oscillate against a certain localized ions with a restoring force. Instead the motion is actually similar to a mass being dragged in a viscous fluid. If the light has a frequency above the plasma frequency (in the ultraviolet (UV) range for metals), electrons will not even oscillate and the light will simply be transmitted or absorbed in interband transitions of the metal. If the light has a frequency below UV, electrons will oscillate out of phase with the incident light, causing a strong reflection. At the plasma frequency, the dielectric function is 0.

Thin Film or Sheet (2D)

When bulk metal is shrunk to a thin film of a few nm thicknesses, electrons cannot move very far in the direction normal to the film. In this case there will be electron oscillations upon light exposure; but they will only exist at the surface of the film.

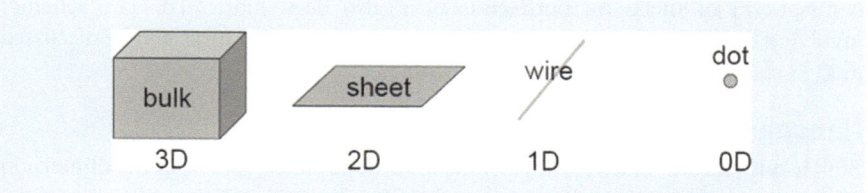

Fig. 10.3 Schematic of four kinds of structures: bulk, a sheet or quantum well, quantum wire, and nanoparticle or quantum dot

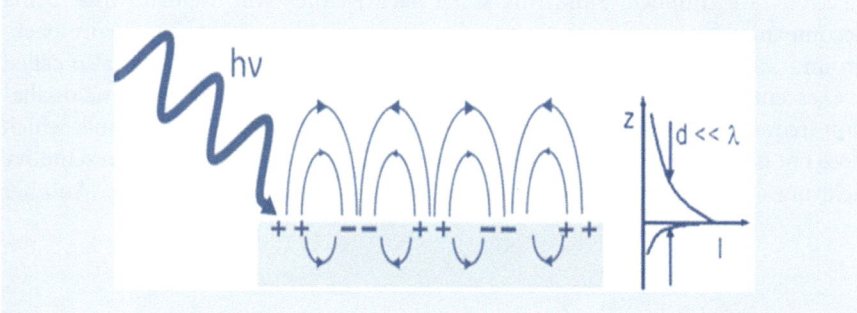

Fig. 10.4 Schematic of light interaction with a few nm thin metal films (*left*) shows electron oscillations at the surface of the film, accompanied with propagation of charge waves along the film. (Adapted from ▶ https://en.wikipedia.org/wiki/Surface_plasmon). (*Right*) Sketch of the field intensity with distance from the surface

This is accompanied with propagation of charge waves along the film. This is known as surface plasmon polaritons (SPPs) as sketched in ◘ Fig. 10.4 (left). In fact, detailed Maxwell's theory shows that the surface waves can propagate along the surface with a broad spectrum of frequencies from 0 up to $\omega_p/\sqrt{2}$, where ω_p is the plasma frequency defined above. At $\omega_p/\sqrt{2}$ the dielectric function is -1 with a charge-wave wavelength shorter than the wavelength of the incident light. The combined effect is a mixed or hybrid light-electron-wave-state. This results in intense light–matter interactions with unprecedented optical response. As shown in ◘ Fig. 10.4 (right), the intense optical field resulting from the hybrid field is a local one, extending outside the thin film into the dielectric only few nanometers, a distance much less than the wavelength of the incident light ($d \ll \lambda$). Thus visible light, which has a wavelength of approximately half a micrometer, can be concentrated by a factor of nearly 100 to travel through metal films just a few nanometers (nm) thick.

Nanowire (1D)

A variety of metal nanowires have been fabricated and studied. They may come as solid nanowire of metal with a surrounding dielectric jacket, i.e., glass or air. For example, chemically prepared silver nanowires ~100 nm diameters were fabricated [15]. The wires were found to support surface plasmon modes propagating along the wires. The wavelength of the propagating charge wave is shortened to about half the wavelength of exciting light. The propagation length of SPP is about 10 μm. The reflectivity at both ends of the wire is about 25 %. Those characteristics are sufficient for the wires to be used as an optical instrument, namely surface plasmon Fabry–Perot resonators. Wires can also be in the form of a cylindrical shell of metal. The shell may have an inner dielectric jacket, such as glass (silicon oxide). The bore of the system may be filled with a semiconductor, such as Si or CdSe. More complicated variations of these architectures have been fabricated to accommodate ports for light entry or extraction. Semiconductor wires with hemispheres of metal on both ends have also been fabricated. The schemes involving hybrid semiconductor-plasmonic (metal) architectures will be discussed in ◘ Sect. 10.4.1 on integrating optics with electronics.

Nanoparticles/Dot (0D)

Bulk is now shrunk in three dimensions to form particles, such that the dimension of the particle is less than the wavelength of incident light. Because the size of the particle is small compared to the wavelength the incident electric field will be constant across the nanoparticle, inducing a uniform displacement of the electron density, making the electron motion in phase (collective cloud motion). Because electrons are confined within the small particle, they will oscillate while being accompanied by a strong restoring force from specific positive ionic core background (◘ Fig. 10.5a). The restoring force leads to non-propagating (also called evanescent) collective oscillation of the surface cloud with a characteristic oscillation frequency similar to a simple harmonic oscillator. This is unlike bulk which does not have a specific oscillation frequency. ◘ Figure 10.5b shows representative field lines, effectively resembling those of an oscillating electric dipole. Another

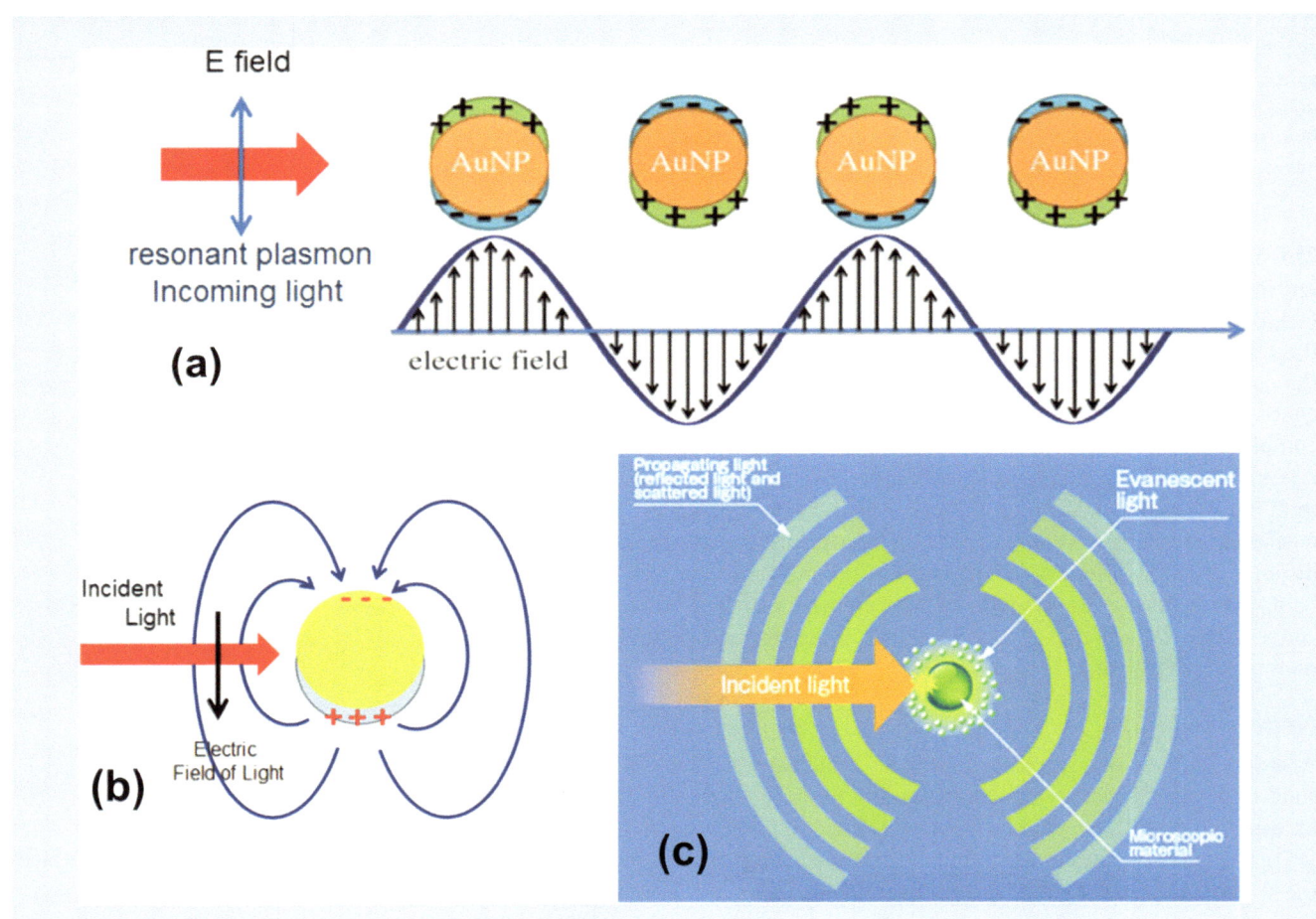

◘ **Fig. 10.5** Schematic of the interaction of a gold nanoparticle with linearly polarized light beam. (**a**) Electrons oscillate while being accompanied by a strong restoring force from specific positive ionic core background (Adapted from E. Yasun, H. Kang, H. Erdal, S. Cansiz, I. Ocsoy, Y-F. Huang, W. Tan, Cancer cell sensing and therapy using affinity tag-conjugated gold nanorods, Interface Focus 3 (2013) 0006: ► http://dx.doi.org/10.1098/rsfs.2013.0006). (**b**) Representative field lines, effectively resembling those of an oscillating electric dipole. (**c**) Region where the non-propagating (evanescent) light is localized, related only to the size of the nanoparticle, not to the wavelength (Image adapted from [16])

striking effect is that the collective oscillations lead to a large absorption and scattering cross section, as well as an amplified local optical electromagnetic field. The electric field has its maximum strength just outside the nanoparticle and drops rapidly with distance extending only to ~30 nm. For small particles less than ~15 nm, light absorption dominates; whereas for nanoparticles greater than ~15 nm scattering dominates. In fact, detailed Maxwell's theory shows that surface waves can be resonantly sustained with frequency at $\omega_p/\sqrt{3}$, which corresponds to dielectric function of -2. We should stress because the evanescent light is non-propagating it is localized within the approximate radius of the particle, even if the size of that particle is much smaller than the incident light wavelength (◘ Fig. 10.5c). The region where the non-propagating evanescent light is localized is related only to the object's size and not to the wavelength, which can be considered as light with no diffraction limit. The nanoparticle practically confines light into super intense "hot-spot."

10.2.3 Miniaturization-Induced Coloration of Metals

◘ Figure 10.6 (top left) shows a large block of gold under illumination by white light. When light strikes the large block, the red and green components are reflected; while the blue component is absorbed and scattered. Interband transitions of bulk gold give it a yellow color. But a metallic luster is also added

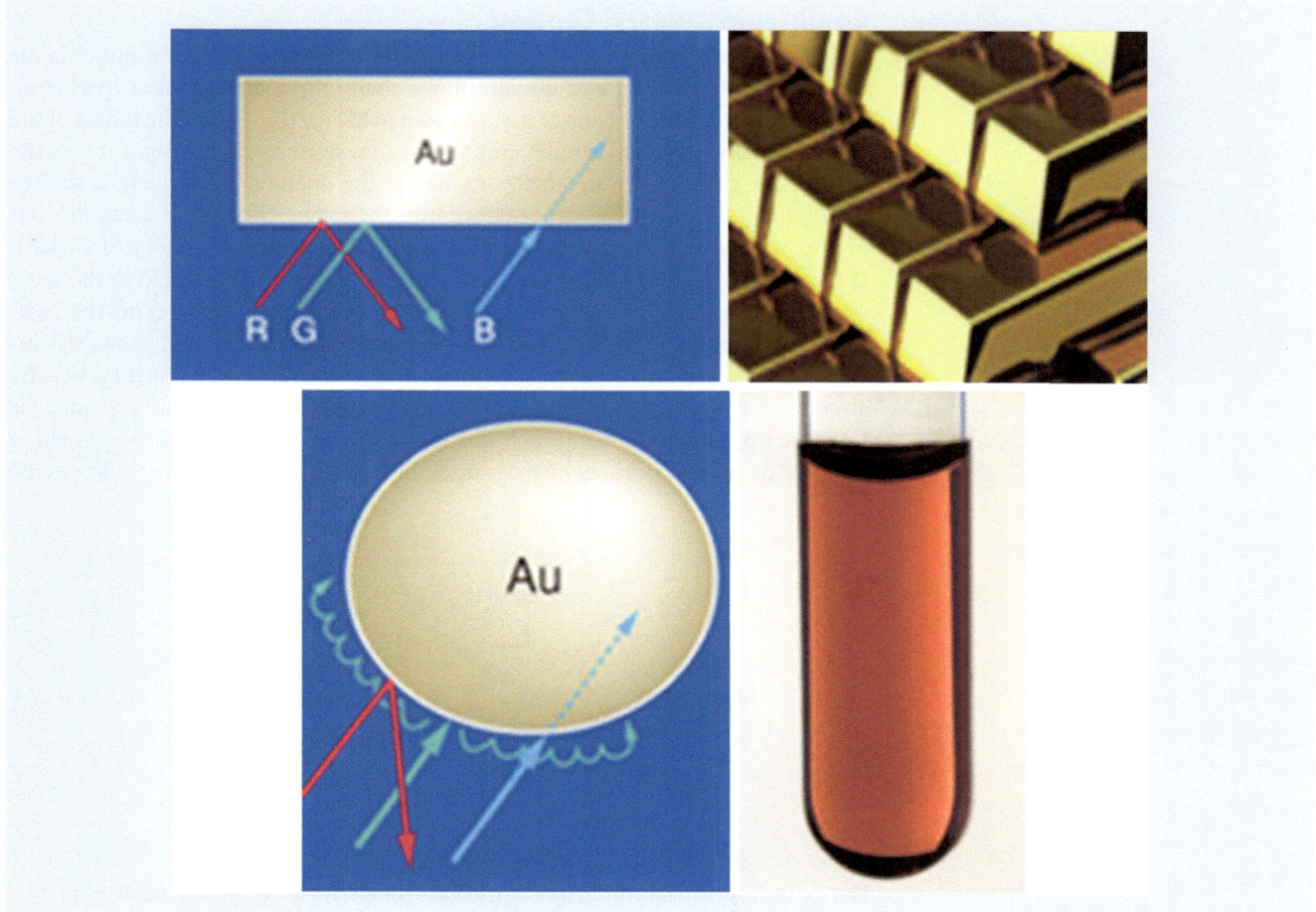

◘ **Fig. 10.6** Schematic of the interaction of gold with white light, showing the individual responses to red, green, and blue (RGB) components, as well as the resulting color as seen by the naked eye (*top row*) large block of gold and (*bottom row*) colloid of gold nanoparticles. (Adapted from [16])

to the yellow color, which appears as a golden color (■ Fig. 10.6 top right). Charge oscillations in bulk do not have a specific frequency to cause specific coloration.

When white light strikes, on the other hand, the nanoparticle shown in ■ Fig. 10.6 (bottom left), the color will be different. This is due to a specific plasmon resonance due to a confined collective oscillation of electrons. A plasmon resonance in gold nanoparticles occurs over a thin slice of light at a frequency in the green. Thus not all of the components of incoming white light resonate and are absorbed. Among the RGB components, only the G (green) component resonates with the electrons and is absorbed in the gold nanoparticle. The B (blue) light is absorbed and scattered, and only the remaining R (red) component gets reflected or passes through. This is the reason that stained glass mixed with gold nanoparticles appears red to the naked eye. ■ Figure 10.6 (bottom right) demonstrates that, when light strikes a gold nanoparticle colloid, only the red color is reflected or passes through [16].

10.2.4 Plasmonic Lenses

Having a negative refractive index is the basic principle behind plasmonic lenses. This is a very common property for noble metal at specific frequency. There are two types of such lenses. One type is based on confinement enhancement. Another is based on transmission enhancement of evanescent waves. In this section we present examples of metal-based lenses.

Confinement-Based Lensing

We discuss two configurations for confinement-based lensing: a continuous conic waveguide concentrator and discontinuous chain of nanoindentation (particles). (1) Consider the hollow cone of metal shown in ■ Fig. 10.7 (left). The radius of the cone gradually decreases from 50 nm to 2 nm, for example. When light strikes the cone and a plasmonic resonance is excited at the opening of the cone, a surface plasmonic polariton (SPP) propagates towards the tip. This causes accumulation of energy resulting in a giant local field at the tip, as shown in ■ Fig. 10.7 (left). ■ Figure 10.7 (right) displays the amplitudes of the local optical field in the cross section of the cone for the normal and longitudinal (with respect to the axis) components of the optical electric field. The magnitude of the field grows significantly as the oscillations approach the tip. The transverse x component grows by an order of magnitude, while the longitudinal z component, which is very small far from the tip, grows relatively much stronger at the tip. This causes the local field to increase by nearly 3 orders of magnitude in intensity and four orders in energy

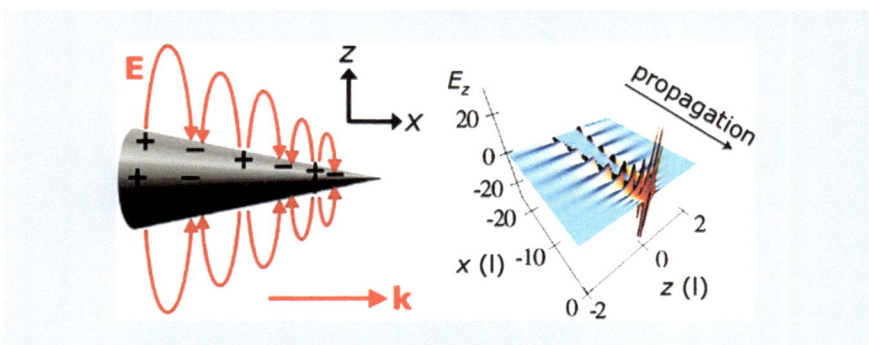

■ **Fig. 10.7** (*Left*) Geometry of a conic nanoplasmonic waveguide showing propagation of a charge oscillation wave. (*Right*) Snapshot of instantaneous E_z fields in the longitudinal cross section (*xz*) plane, normalized to far-zone (excitation) field. (Adapted from "Stockman MI (2004) Phy Rev Lett 93: 137404")

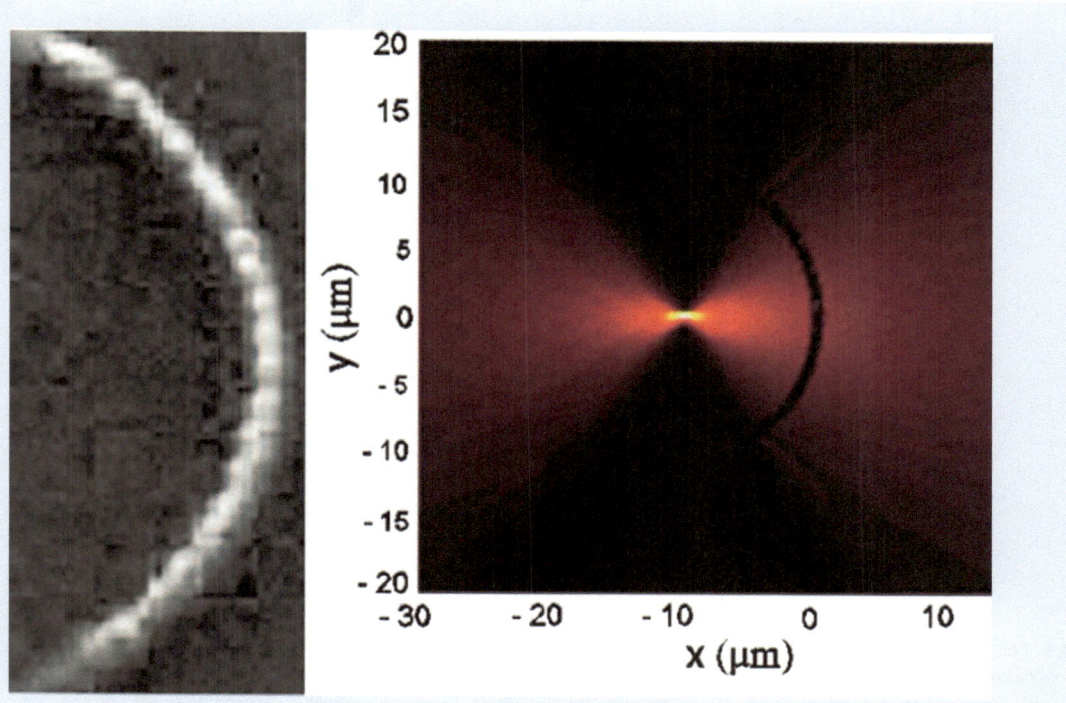

Fig. 10.8 Focusing of light by a curved chain of spheroid gold nanoparticles indented in a thin gold film (*left*) top view image of the chain using a scanning electron microscope (SEM), (*right*) Magnitude of scattered electric field calculated above the gold surface. The illuminating beam has a wavelength of 800 nm and is incident perpendicular to the gold surface and polarized along *x*-direction (Adapted from "Evlyukhin AB, et al. (2007) Opt Exp 15: 16667–16680")

density. It should be noted that the propagation is slowed down and asymptotically stopped when it approaches the tip. It never actually reaches the tip (the travel time to the tip is logarithmically divergent) [17]. The cone as such may represent a tapered plasmonic waveguide. (2) Consider a chain of nanoparticles shown in ◘ Fig. 10.8 (left). Fabrication starts with a thin gold film. Nanoindentations (nanoparticles) are made on the film in a parabolic chain configuration. The radius of curvature of the chain is 10 μm. The particle diameter and inter-particle distance are ~350 and 850 nm, respectively, while the particle height is 300 nm. When light, in the wavelength range of 700–860 nm, for example, is normally incident on the thin film on the right of the chain, charge oscillations are induced on the surface of the film which travel or propagate to the left towards the chain (surface plasmon polariton or SPP). When the waves hit the gold nanoparticles they get focused to a submicron spot, as shown in ◘ Fig. 10.8 (right) [18].

Transmission-Based Lensing

Another concept of lensing involves transmission through holes or slits. An array of holes is drilled in an opaque metallic film in an arc formation, as shown in ◘ Fig. 10.9a. The diameter of the holes is of nanoscale. For a plane wave incident upon such a structure, the phase shift experienced by light as it passes through each individual hole is sensitive to either the length, width, or even the materials inside the hole. With the adjustment of the properties of individual holes, it becomes possible to achieve a focusing action [19].

A slit-based lens is built by first depositing a 400 nm thick flat gold film on silicon oxide substrate. Then an ion beam is used to mill 13 rectangular slits in the thin film, as shown in the left panel of ◘ Fig. 10.9b. The slits increase in widths from 80 nm at the center slit to 150 nm on the left or right end. ◘ Figure 10.9c

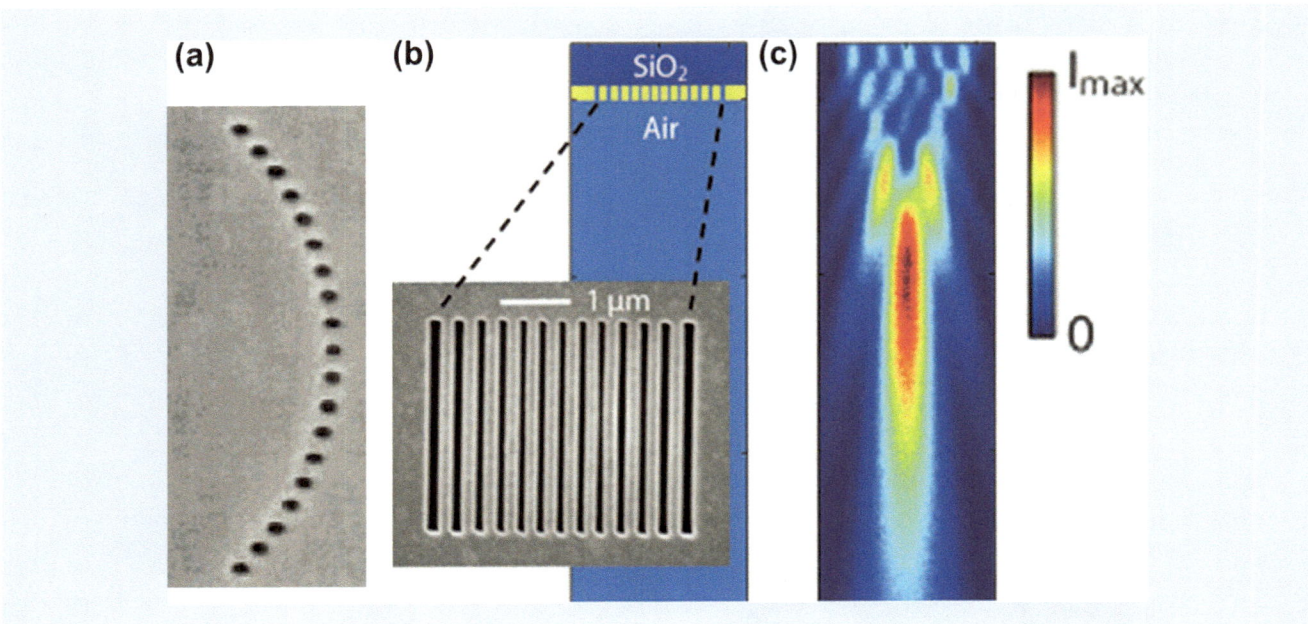

Fig. 10.9 Focusing of light by transmission through holes or slits. (**a**) Chain of holes in a metal plate. (**b**) Nanoscale slit array drilled in gold film on a fused silica substrate (*dark blue*). The film is 400 nm thick (*yellow*). The air slits are different in widths (80–150 nm) (*light blue*). The *inset* shows a scanning electron micrograph of the structure as viewed from the air-side. (**c**) Focusing pattern measured by confocal scanning optical microscopy (CSOM) (Adapted from "Lieven Verslegers et al., Nano Lett. 9, 235–238 (2009)")

gives the measured field intensity in a cross section through the center of the slits (along the *x*-direction). The measurement demonstrates focusing of the wave. Thus this configuration acts as a far-field cylindrical lens for light at optical frequencies [20].

10.2.5 Metamaterials: Negative Refractive Index

The above discussion shows that metal dielectric interface, with features smaller than the wavelength of light separated by distances smaller than the wavelength of light, is very special. At plasmon resonances they have a negative index. In fact those may be considered as a class of more general material called metamaterials that exhibit properties beyond those found in nature. In 1967 [21] Victor Veselago theorized that material with negative refractive index would exhibit optical properties opposite to those of dielectrics, such as glass or air. Contrary to dielectrics, when light propagates in metamaterial (1) light refracts on the other side of the normal to the interface, that is, energy is transported in a direction opposite to the dielectric case, as shown in Fig. 10.10; (2) light produces negative pressure, which pulls metamaterial towards it instead of positive pressure which pushes away as in conventional material. The basic principle behind all of the opposite effects is closely related to the above plasmonic effect. They are due to the collective interaction of the light with the electron clouds at the surface of the conductor. This photon–plasmon interaction generates intense, localized optical fields. The waves are confined to the interface between metal and insulator. This narrow channel serves as a transformative guide that, in effect, traps and compresses the wavelength of incoming light to a fraction of its original value.

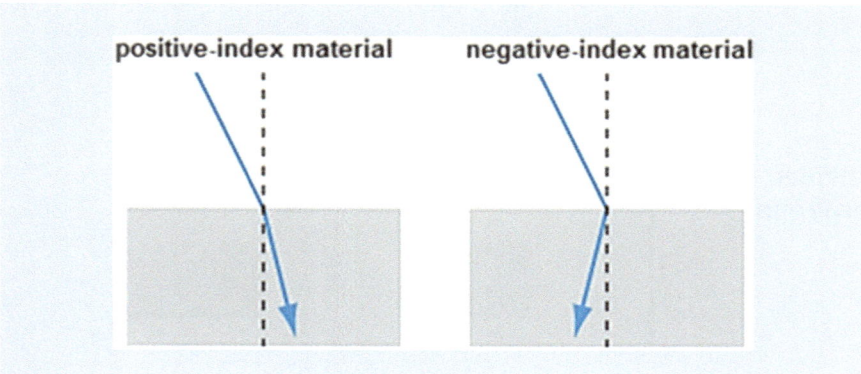

positive-index material   negative-index material

☐ **Fig. 10.10** Schematic of light refraction at a plane interface of dielectric—dielectric interface and at a dielectric—metamaterial (negative index) interface

10.2.6 Heat Loss: Are Plasmonic-Based Devices Practical?

One pivotal problem inherit in plasmonic technology is heat loss. Because plasmonic devices involve light interaction with metal, it necessarily involves energy loss due to heat dissipation. Metals are plagued by large losses due to strong electronic interband absorption, especially in the visible and UV spectral ranges. Among all metals, noble metals of gold and silver show the least energy loss. That is why gold and silver are used in most common plasmonics. Even in gold and silver, losses in the optical range including near- and mid-infrared (IR) regions are effectively still too high to make practical plasmon-based devices. Therefore, the search is continuing to find different approaches, such as creation of alloys and composites to make plasmonic materials that exhibit lower losses [22].

10.3 Optics in Nanosemiconductors

The interaction of semiconductor nanocrystals with light is fundamentally different from their bulk counterpart. Quantum dots, which were first discovered in 1980, are tiny particles or nanocrystals of a semiconducting material with diameters in the range of 2–10 nm. In this size regime, additional quantum effects start to play an important role that can significantly alter some properties of the original material, such as optical activity. The most apparent result of this is that nanocrystals can fluoresce in distinctive colors determined by the size of the particles.

10.3.1 Bandgap and Excitons

Unlike metals, semiconductors do not electrically conduct, i.e., have no free electrons in the conduction band. All of the electrons are bound to atoms, i.e., they are in the valence electronic bands. This is because in metals the conduction and valence bands overlap but in semiconductors the edges of bands are separated by an energy gap called bandgap, as shown in ☐ Fig. 10.11a [23].

In semiconductors an external agent such as external light is needed to impart enough energy to a bound electron to overcome the bandgap energy, as shown in ☐ Fig. 10.11b. When this takes place the electron is placed in the conduction band accompanied with its separation from the positive charge. Separation of the electron from the positive charge leaves behind in the valence band what is called

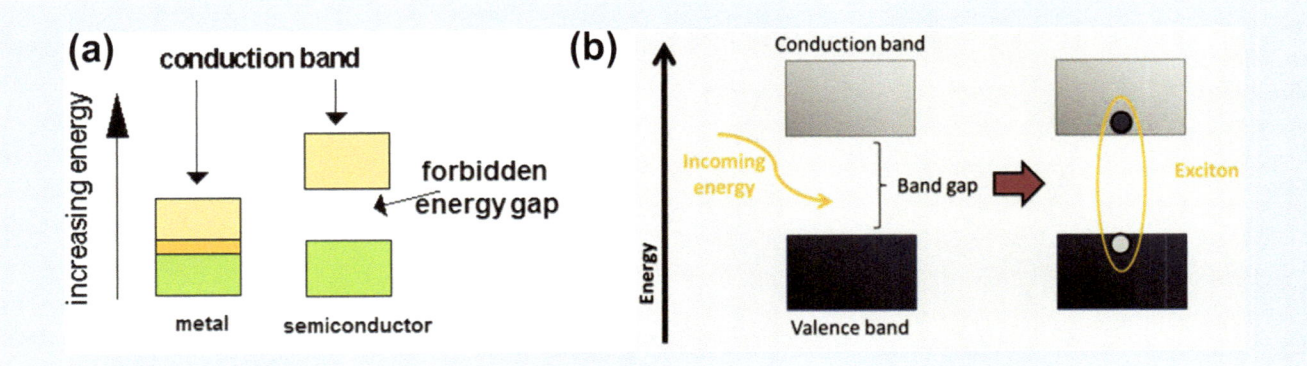

Fig. 10.11 (**a**) Conduction and valence bands for metal and semiconductor (Adapted from ▶ http://webs.mn.catholic.edu.au/physics/emery/hsc_ideas_implementation.htm). (**b**) Light excitation of semiconductor elevating an electron from the valence band to the conduction band creating an exciton (Adapted from [23])

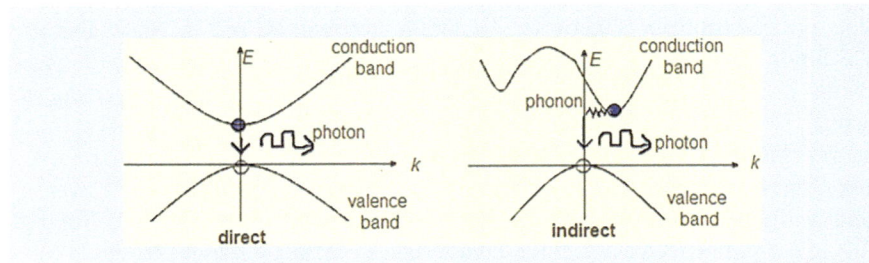

Fig. 10.12 Schematic of the band structure showing the conduction and valence bands for direct bandgap material and indirect bandgap material

a positive hole. The electron as well as the positive hole are free to move away from their original common site; but they move together as an electron–hole pair or exciton, with the electron orbiting around the hole at an average distance or exciton Bohr radius. In silicon, for example, the Bohr radius is 4.2 nm. The exciton structure is actually an "atom" which in some respect resembles the structure of a hydrogen atom, but 100–200-fold larger. Because of the large size, excitons are less bound and more fragile.

10.3.2 Direct and Indirect Bandgap Materials

The interaction (absorption or emission) of light with matter must satisfy the conservation of energy and momentum. The ability of meeting these conservations laws depends on the type of material. Two classes of semiconductors are interesting in this regard: indirect bandgap materials such as silicon and direct bandgap semiconductor such as CdSe (see ▪ Fig. 10.12). The bandgap is called "direct" if the momentum of electrons and holes is the same in both the conduction band and the valence band. In this case an electron can directly emit a photon, which conserves energy while the momentum is automatically conserved, as shown in ▪ Fig. 10.12 (left). In an "indirect" gap material, the momentum of electron and hole are not the same; hence a photon cannot be readily emitted because, to conserve momentum, the electron must pass through an intermediate state and transfer momentum in the form of phonon quanta to the crystal lattice, as shown in ▪ Fig. 10.12 (right). The probability of coincidence of three particles: electron, hole, and phonon with specific properties is significantly lower than the coincidence of only two such particles (electron and hole). Thus, the probability of emission of a photon is much lower in indirect bandgap semiconductors than in direct bandgap ones. Bulk silicon is therefore a very poor light emitter, while in most cases, the direct bandgap semiconductors are good light emitters. Direct bandgap material is therefore used in light emitting devices, while indirect bandgap material is used in electronic devices.

10.3.3 Enhancing and Blue Shifting of Luminescence by Quantum Confinement

As we shrink the size of direct semiconductor material, novel optical properties begin to emerge as we approach a certain characteristic size scale. The characteristic scale is the electron–hole distance (the Bohr radius) of the material. For example, for CdSe the Bohr radius is 5.6 nm. In this regime, the interaction of light with the material gets modified because quantum quantization of the energy levels of the electron as well as that of the hole according to Pauli's exclusion principle becomes important. A simplified treatment considers the energy of the electron and the hole as the energy of the charge in an infinite well (see ▫ Fig. 10.13). The top well is for an electron and the bottom inverted well is for a hole.

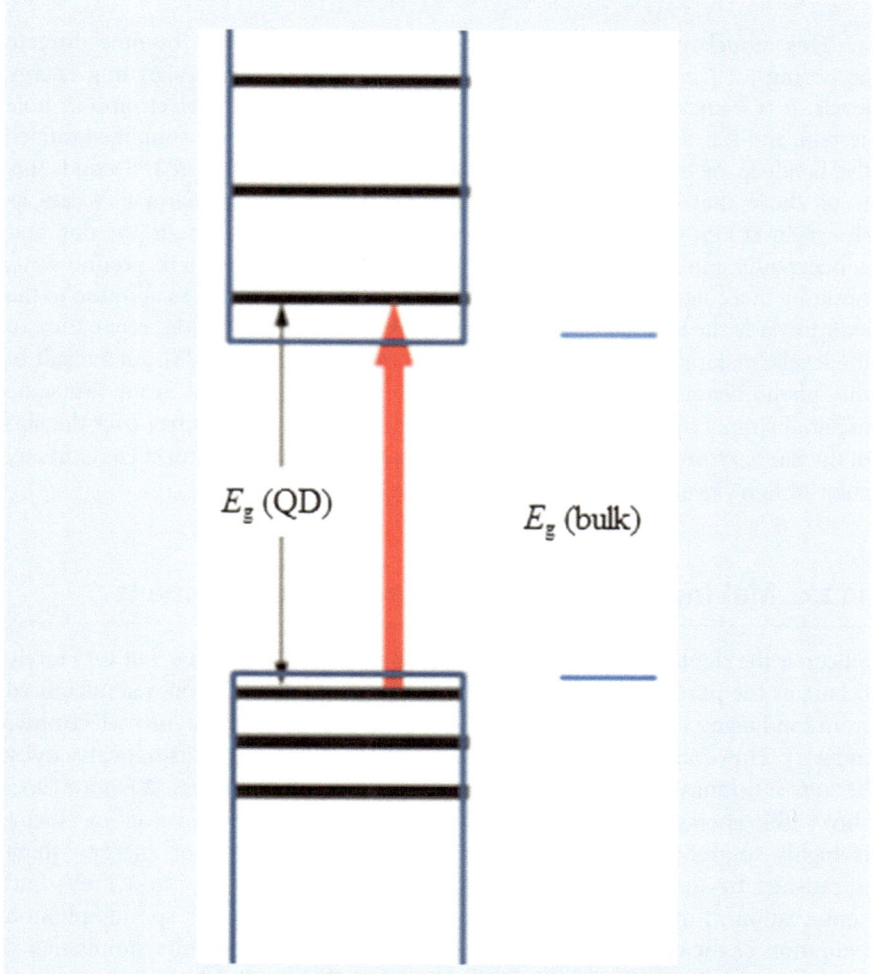

▫ Fig. 10.13 Simplified infinite well model of the energy of the electron and the hole in a quantum dot. The top well is for an electron and the bottom inverted well is for a hole

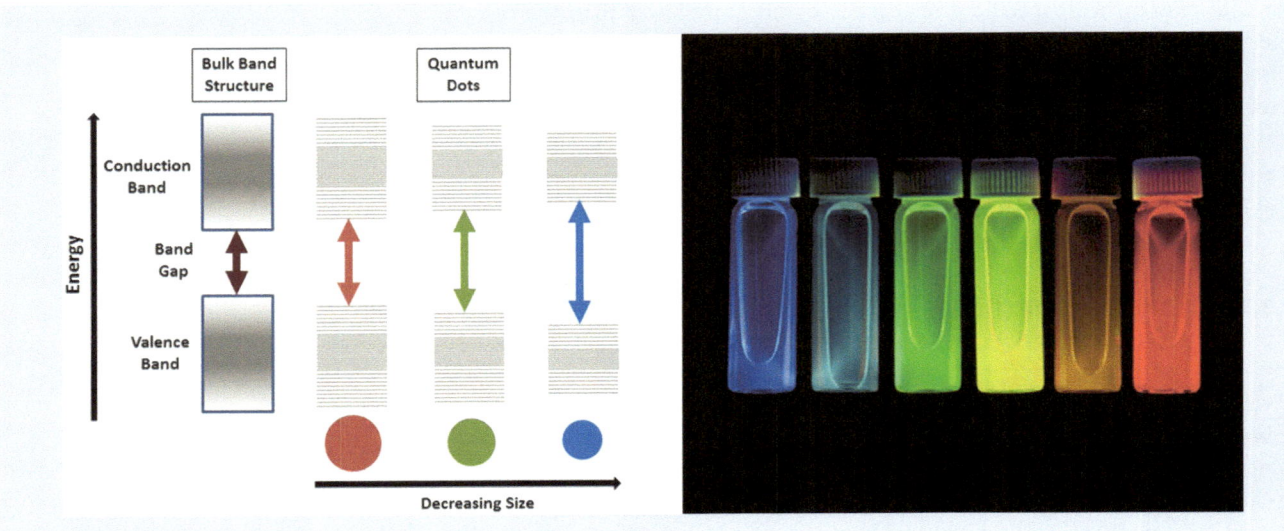

Fig. 10.14 (*Left*) The conduction and valence bands with decreasing size of a semiconductor nanoparticle (quantum dot), showing splitting of energy levels due to the quantum confinement effect. The bandgap increases with decrease in size of the nanocrystal (Images from (▶ http://www. sigmaaldrich.com/materials-science/nanomaterials/quantum-dots.html)). (*Right*) Vials of quantum dot colloids of increasing average size from left to right emitting light with color from blue to red respectively (Image from ▶ http://nanocluster.mit.edu/research.php [24])

10

This model results in three effects. First the energy levels become discrete according to: $E_n = (h^2 n^2)/(8 m_c R^2)$ where n is an integer designating energy levels, h is Planck's constant, m_c is the effective mass of the electron and hole system, and R is the radius of the quantum dot (particle). In this simplified model, the bandgap of the material widens to $E_g = E_{g0} + (h^2)/(8 m_c R^2)$. Second, this result shows that as the particle decreases in size, the bandgap energy increases, as shown in ◘ Fig. 10.14 (left). More energy is then needed to excite the dot, and concurrently, more energy is released when the crystal returns to its ground state, resulting in a color shift towards blue in the emitted light. Third, in addition to the shift towards the blue the emission becomes stronger due to the discrete nature of the levels, making the nanoparticles much brighter than bulk [25]. As a result of this phenomenon, quantum dots can emit any color of light from the same material simply by changing the dot size. Additionally, with control over the size of the nanocrystals, quantum dots can be tuned during manufacturing to emit any color of light, as shown in ◘ Fig. 10.14 (right).

10.3.4 Making Silicon Glow: Quantum Confinement

Silicon is the eighth most common element in the universe by mass, but very rarely occurs as the pure free element in nature. Monocrystalline silicon, manufactured from sand using sophisticated technology, is the backbone of the microelectronics industry. However silicon is the dullest material with regard to its optical activity because it belongs to the class of indirect bandgap semiconductors. ◘ Figure 10.15 shows the energy-momentum diagram for silicon. Light emission in silicon is highly improbable because it requires both conservation of energy, which is satisfied by the emission of the appropriate photon energy at 1.1 eV, and conservation of momentum, which is satisfied by emission of a specific phonon (vibration of the crystal) of 55 meV energy. Processes that require simultaneous emission of light and vibrations are highly unlikely and take a long time to happen, providing emission lifetimes on the order of milliseconds. In direct

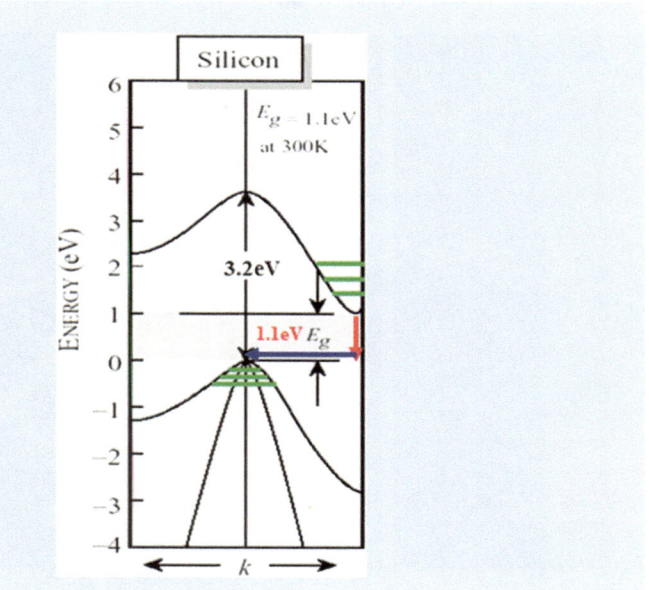

Fig. 10.15 The energy–momentum diagram of the band structure of silicon. Emission of light requires both conservation of energy, which is satisfied by the emission of the appropriate photon energy at 1.1 eV (*red vertical arrow*), and conservation of momentum, which is satisfied by emission of a specific phonon (vibration of the crystal) of 55 meV energy (*horizontal blue arrow*). The *upper set of green lines* are discrete levels due to confinement of an electron and the *lower set of green lines* are due to the confinement of a hole

semiconductors, light emission proceeds readily with a lifetime of nanoseconds as the momentum is automatically satisfied without vibration and emission of phonons.

Recent developments enabled by nanotechnology are starting to change the picture. Some silicon nanostructures provide interactions with light now approach or even exceed the performance of equivalent direct bandgap materials, which promise to take silicon into the realm of optics. A significant deviation from bulk properties was found in 1990 when L. T. Canham noticed visible photoluminescence in porous silicon he produced via chemical etching of bulk silicon in HF acid [26].

But porous silicon is just an interconnected nanoscale network of silicon skeletal (sponge-like structure). The first report of micro- and nanoparticles prepared from porous Si came from the Sailor group at University of California in San Diego [27]. Strong ultrasound was used to shatter porous silicon into micro- and nanoparticles. But the particles do not have specific configuration or uniformity because silicon is a hard material and the particles are basically a result of mechanical stress and shattering of the interconnected skeletal backbone. In 1997, a new self-limiting etching procedure was developed by the Nayfeh's group at the University of Illinois which produced on the silicon wafer disconnected individual spherical nanoparticles of preferred or magic configuration or sizes, which can be softly retrieved and stored in a liquid of choice [28–32]. The particles are protected by mono-hydride coating and can be produced in commercial amounts and stored for later use for many years. The smallest of these particles is 1 nm and fluoresces in the blue. Other sizes include 1.7 nm, fluorescing in the green; 2.15 nm fluorescing in the yellow–orange; and 2.9 nm fluorescing in the orange–red.
Figure 10.16 (left) shows the luminous of colloids of 1 and 2.9 nm particles in alcohol using excitation at a wavelength of 365 nm. The 1-nm Si nanoclusters are amenable to testing and accurate first principle simulations because they consist of a manageable number of Si and H atoms and are produced in macroscopic

◘ Fig. 10.16 (*Left*) Blue and red luminescence of colloids of 1 and 2.9 nm particles in alcohol, respectively, with excitation at a wavelength of 365 nm. (*Right*) A prototype structure of 1-nm particle has a configuration of $Si_{29}H_{24}$. Silicon atoms in *orange*. Hydrogen atoms in *white*

10

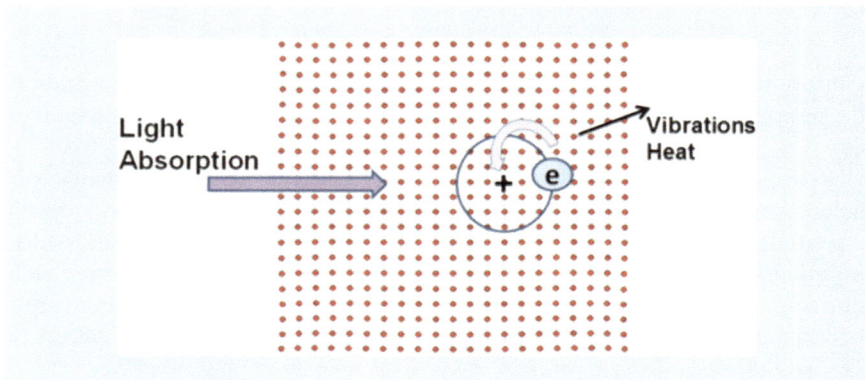

◘ Fig. 10.17 Response of bulk silicon to light. Exciton forms followed by recombination to produce vibrations and heat before it can emit a photon

amounts. Those studies showed that the 1-nm particle simulated by Lubos Mitas has a configuration of a super molecule $Si_{29}H_{24}$, as shown in ◘ Fig. 10.16 (right) [33–37].

One important characteristic of these particles is that the energy and momentum conservation rules governing light interaction get modified. In bulk silicon crystals or large particles, photo produced excitons move freely in all directions. Because it is not easy to conserve the momentum in indirect gap material when light is emitted, they recombine again before light is emitted with energy turning into vibrations and heat imparted to the crystal, as sketched in ◘ Fig. 10.17.

When the particle is reduced to a size smaller than the exciton radius, especially in the sub 3-nm regime, the particles become less rigid allowing the silicon atoms to move, relax, or adjust especially those on the surface. Moreover, excitons get more strongly confined spatially, causing the momentum of the electron and hole to spread out appreciably according to the Heisenberg uncertainty principle, such that their momentum distributions can overlap. In the overlapping region, the momentum conservation may be readily satisfied; thus recombination of the

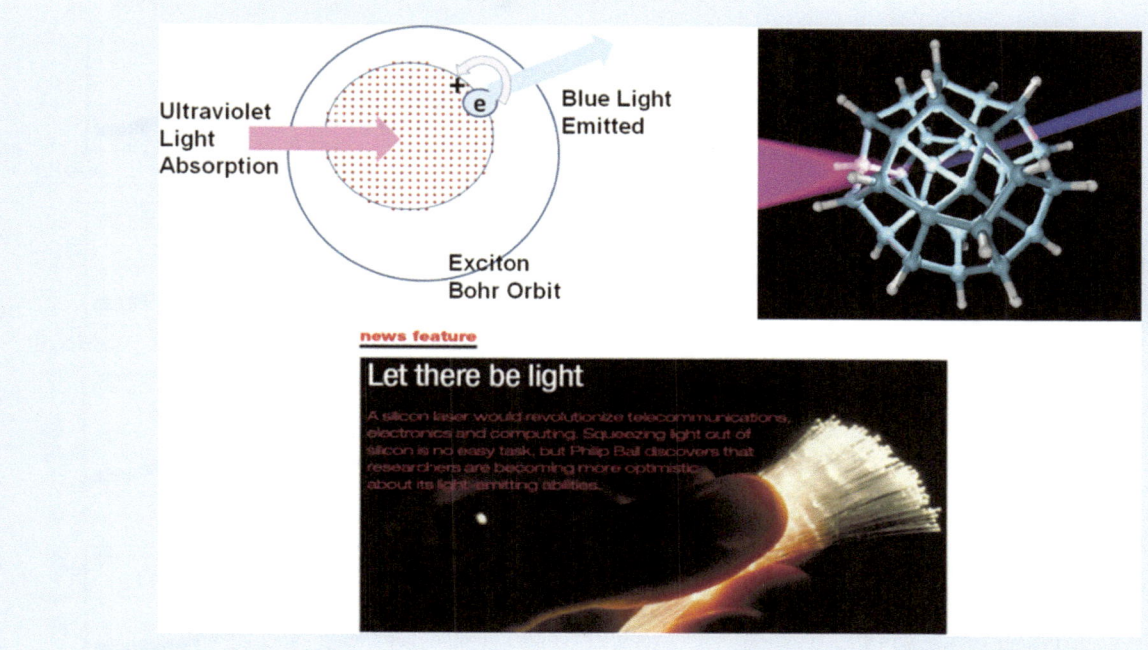

Fig. 10.18 (*Top left*) Schematic of response of 1-nm silicon nanoparticle to light. In 1-nm nanoparticle, much smaller than the Bohr radius, recombination produces a blue photon very fast much before vibrations and heat is produced. (*Top right*) Computer simulation image of 1-nm nanoparticle from [34]. (*Bottom*) Nature magazine report on the brightness of the silicon nanoparticles in terms of a biblical theme "Let there be light" (Image from [38])

electron and hole to turn the excess energy into light becomes stronger and faster than producing vibration and heat. In this case a photon (light) is produced, with the energy of the emitted photon (color of the light emitted) depending on the size of the particle.

For a nanoparticle size of 1 nm, for example, emission of blue photons dominates the production of vibration and heat making it very bright with a performance that matches particles of direct bandgap material as in ◘ Fig. 10.18 (top left) [39]. ◘ Figure 10.18 (top right) is a computer simulation image of 1-nm nanoparticle carried out by researchers at Livermore National Laboratory [35]. Making silicon the optically dullest material in the universe glow is highly dramatic. Nature magazine reported this effect using the theme "Let there be light," as depicted in ◘ Fig. 10.18 (bottom); it is a biblical theme of creation, signaling the transformation of the universe from the dark state to the bright state [38].

10.3.5 Optical Nonlinearity in Nanosilicon

The structure of a silicon crystal has cubic symmetry which exhibits inversion symmetry (centrosymmetry). But when the crystal becomes very small, it loses its rigidity and atoms especially the ones on the surface re-adjust, which exerts strain across the entire particle that alters the cubic structure and breaks the symmetry. Breaking the symmetry produces fundamental effects on its interaction with light. ◘ Figure 10.19 (top left) displays one theoretical process that takes place in 1-nm particle in which some of the surface atoms lose their hydrogen atoms, move as much as 1.5 Å each towards each other followed by reconstruction/connection to form Si–Si dimer-like systems.

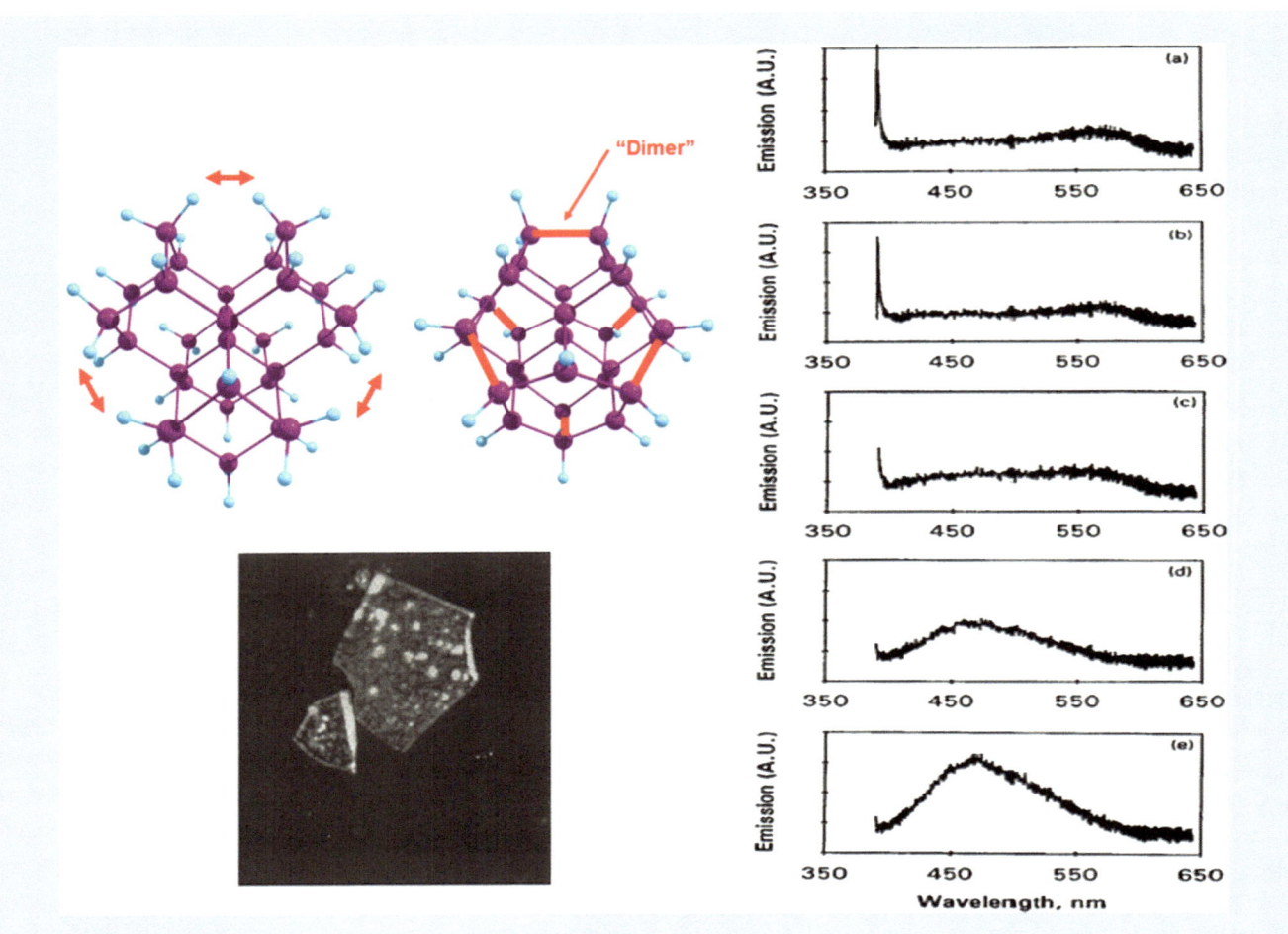

Fig. 10.19 Response of microcrystallites of 1-nm silicon particles to 780-nm near-infrared femtosecond excitation. (*Top left*) Model of surface reconstruction that breaks the crystal symmetry of silicon. (*Left bottom*) Image of reconstituted microcrystals. (*Right*) Shows strong blue luminescent band indicating two-photon excitation as well as competing sharp radiation at half the wavelength of the incident beam. The intensity of the second harmonic anti correlates with the wide band luminescence

Blue luminescent 1-nm nanoparticles were used to test this mechanism. Particles prepared in water solvent were dried on device-quality silicon, which allowed them to form microcrystallites films, as shown in ▣ Fig. 10.19 (left bottom). The microcrystallites were then excited by high intensity near-infrared femtosecond laser at 800-nm wavelength. ▣ Figure 10.19 (right) displays several frames of the emission from different crystallites. It shows strong wide blue luminescent band as well as competing sharp radiation at half the wavelength of the incident beam (second harmonic). The intensity of the second harmonic anticorrelates with the wide band luminescence as seen in the figure. Both the sharp and wide band response indicates very strong coupling with light and the emergence of second-order nonlinear phenomena. The excitation of the blue luminescence band indicates a two photon excitation process, while the second harmonic indicates the breaking of the centrosymmetry. Other theoretical simulations showed that the mechanical strain exerted is extremely high, corresponding to a pressure of several Gpa [40]. Because atoms on the surface are under high strain, they can undergo large-amplitude molecular-like vibrations to relieve the strain [41]; and hence can also couple efficiently to thermal and mechanical stimuli. Second-order nonlinearity in silicon opens up other prospective applications in optics, including modulation, amplification, gain and laser action, and signal processing [42].

10.3.6 Optical Gain in Nanosilicon-Based Material

Not only miniaturization afforded silicon strong luminescence and optical non-linearity, but there are also some preliminary reports indicating optical gain and laser action in films of silicon nanoparticles [43]. In one report [44], silicon nanoparticles were created inside a glass slab by silicon ion implantation followed by high temperature annealing that allows the silicon atoms to condense into silicon nanoparticles of ~3-nm diameter. The slab acted as an optical amplifier of weak light beams. In other reports 1-nm silicon nanoparticles were reconstituted into microcrystallites [45]. The crystallites were irradiated by infrared femtosecond pulses of a laser beam with high peak power. Microscopic directed blue emission was observed. Microscopic red emission was also observed from clusters of 2.9-nm nanoparticles under excitation of strong incoherent green light [46].

10.4 Applications of Optics in Nanotechnology

Miniaturization has triggered strong light–matter interactions. Nanotechnology allowed researchers to study light–matter interactions at the nanoscale and to launch the subfield of nano-optics. A considerable amount of basic knowledge as well as novel functions of nanomatter has accumulated over the past 20 years. Some applications have already reached the practical level, while others are futuristic. In this section we briefly present some practical applications of optics in nanotechnology in service of fields as diverse as electronics, opto- and photo-electronics, elementary particles, biomedicine, energy harvest and lighting, and art as in stained glass and lusterware pottery.

10.4.1 Integration of Optics and Electronics

Bulk semiconductors, especially silicon, form the backbone of modern electronics and computing. Bulk silicon, however, is a very dull material, being an especially poor emitter of light, turning added energy into heat. This makes integrating electronic and photonic circuits a challenge. There have been several proposals to alleviate this problem. One approach involves doping silicon with other materials. However, the emitted light is not in the visible rather in deep infrared. Moreover, the emission is not very efficient and can degrade the electronic properties of silicon. Another approach is to use nanotechnology by making silicon devices that are very small, such as using luminescent nanoparticles, five nanometers in diameter or less as introduced above. As we have seen above, at that size quantum confinement effects allow the device to emit light. But making electrical connections at that scale is not currently feasible and may compromise the optical activity of the nanomaterial [47], as well as afford very low electrical conductivity.

Another scheme involves subjecting bulk silicon to concentrated fields of plasmonics. The high fields afforded by plasmonic nanometal material can cause distortion and modification of the crystal structure, hence the interaction with light [48–50]. We describe briefly an architecture that demonstrates some of these effects. First a pure silicon nanowire is wrapped with a coating of glass, as shown in ◻ Fig. 10.20 (left). Then it is mounted on a glass substrate. Then it is coated with silver. Because of the glass substrate, silver does not wrap completely around the wire, effectively making the silver coating take the shape Ω. This leaves a narrow transparent glass window through which a laser beam can be sent in and emitted light can be extracted [48]. Practically, this narrow window did not compromise

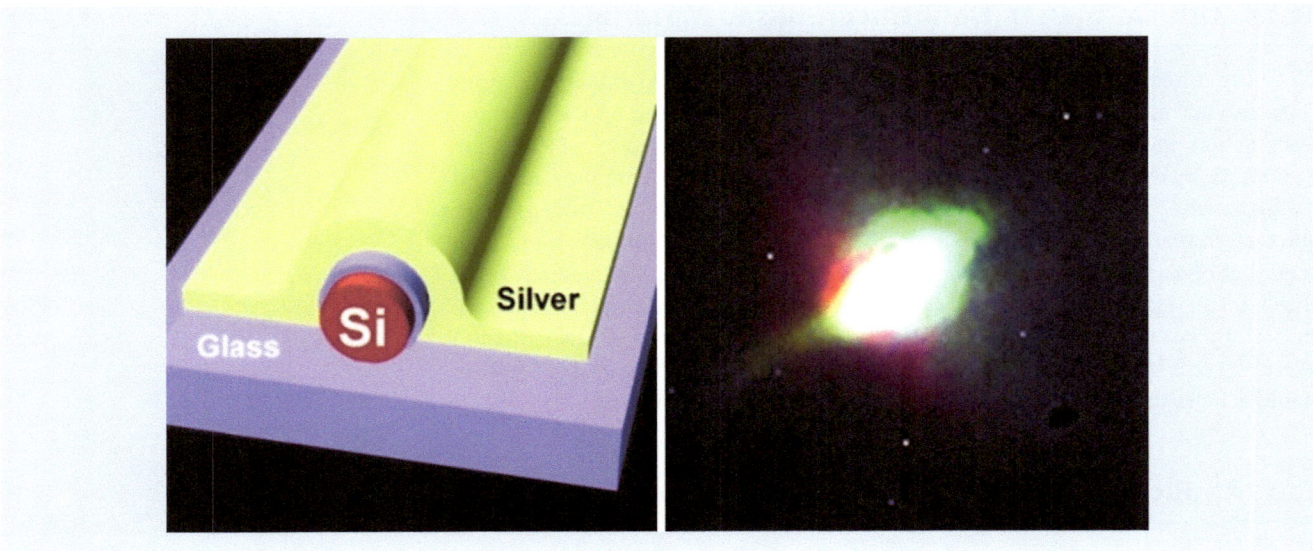

Fig. 10.20 Schematic of a hybrid silicon–silver nanowire system. (*Left*) Silicon nanowire wrapped with a glass coating followed by silver coating in the shape of Ω, on a glass substrate. (*Right*) white light emission when a laser beam irradiates the system through the Ω window. (From [48])

the silver coating from acting as a plasmonic cavity. When a narrow band blue laser beam is sent through the window, the silicon wire produces white light that spans the visible spectrum, as shown in ◘ Fig. 10.20 (right). It is to be noted that bare silicon wires do not produce such white light when excited by the blue laser, indicating plasmonic effects. The broad bandwidth light provides good potential for operation in photonic or optoelectronic devices.

10.4.2 Confined Light in Service of Substance Detection

Understanding interactions between strong light and matter is central to many fields. Deeper understanding of strong light coupling with matter is expected to afford the creation of advanced tailored applications [50]. The designation "strong field" applies to an external electromagnetic field that is sufficiently strong to cause significant alterations in atomic or molecular structure and dynamics of the material. Present-day laser technology as well as light enhancement due to plasmonic effects has made it possible to study the behavior of atoms and molecules in fields that have peak electric field strength of the order of atomic fields inside atoms or molecules. Under these conditions, even tightly bound ground states must be greatly altered by the presence of the field [51, 52].

Understanding interactions between light and nanomatter is central to many fields, providing invaluable insights into the nature of matter as well as the nature of light. Indeed, greater understanding of light–matter coupling has enabled creation of tailored applications, resulting in a variety of devices such as microscopic lasers, switches, sensors, modulators, fuel and solar cells, and detectors. As discussed above, hybrid-plasmonic monolithic nanowire optical cavities highlight recent progress made in tailoring light–matter coupling strengths.

Detailed Maxwell's theory was used to calculate the enhancement in intensity of light and the corresponding electric fields of light in the proximity of gold nanoparticles [53]. The results for 10-nm nanoparticles are given in ◘ Fig. 10.21a, using a gray/white color code for incident light of 1 V/m electric field. The enhancement based on these theoretical estimations can be as large as a factor of 10, which provides enhancements of two orders of magnitude in intensity.

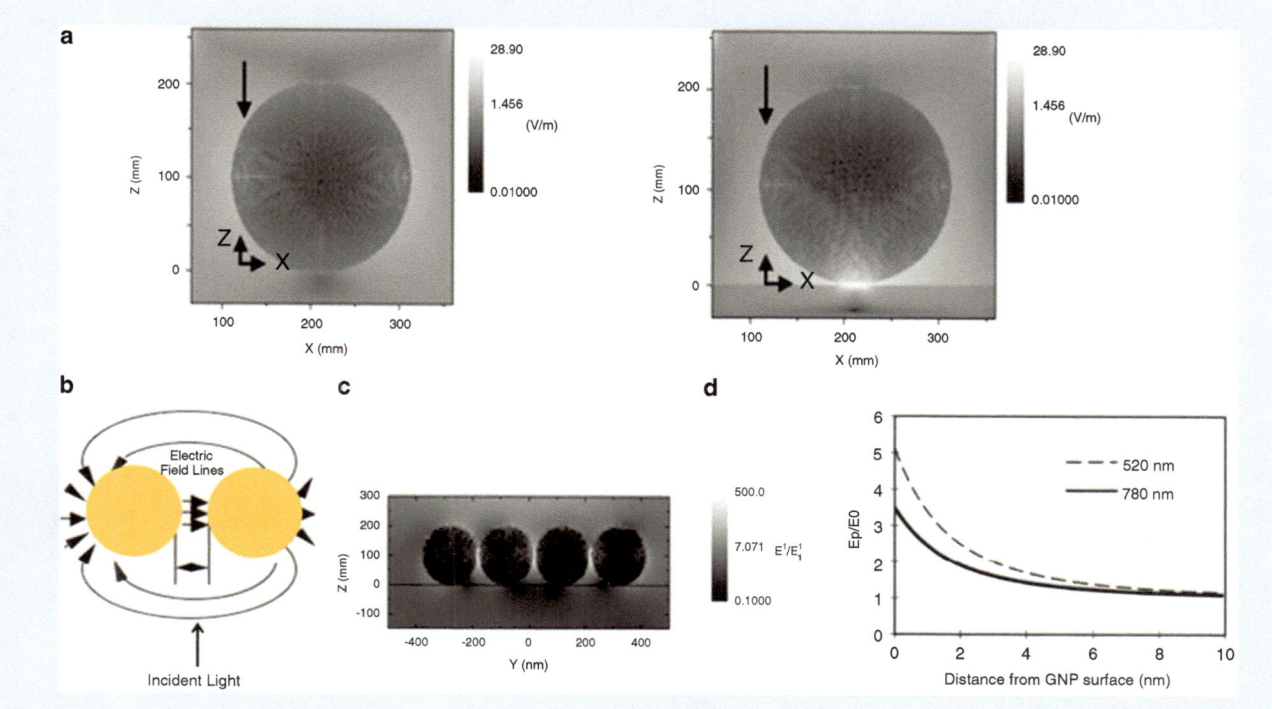

Fig. 10.21 Detailed Maxwell's theory was used to calculate the intensity of light and the corresponding electric fields of light in the proximity of 10-nm gold nanoparticles. (**a**) Near a single nanoparticles (**b**) schematic of the field lines between two particles as well as (**c**) actual results. *Gray/white color code* is used for incident light of 1 V/m electric field (**d**) detailed dependence, as an example, of the electric field near a gold nanoparticle illuminated with light of two different wavelengths. (From [53])

Enhancements can be even more dramatic when there is more than one metal particle. ■ Figure 10.21b, c displays a schematic of the field lines between two particles as well as actual results using the gray/white color code, respectively. It shows, in space between two closely placed plasmonic nanoparticles the electric field may be enhanced by nearly a factor of 30, providing nearly three orders of magnitude enhancement in intensity. ■ Figure 10.21d gives detailed dependence on the wavelength of the light. As an example, the figure gives the electric field near a gold nanoparticle illuminated with light of two different wavelengths [53]

Thus the confined light to nanometal spaces is short range (dies out exponentially) and does not propagate to large distances. Because of the short range of the field, it interacts only with nanostructures next to it, within ~10–30 nm. Thus it provides spatial nanoresolution. This opens up many applications, especially those optical processes or phenomena that depend very sensitively on the magnitude of the electric field, such as fluorescence, Raman scattering, and infrared absorption, resulting in plasmon-enhanced fluorescence, surface-enhanced Raman scattering, and surface-enhanced infrared absorption spectroscopy.

The Raman scattering effect, for example, depends on the fourth power of the field; thus they are enhanced if scattering is performed with the confined light near a metal nanoparticle rather than with ordinary propagating light. The Raman scattering effect is used for sensing and identification of substances. When a substance is irradiated with light, it scatters light called Raman light at a wavelength or frequency slightly shifted from the wavelength of the original irradiated light by an amount equal to the natural frequency of molecules that make up the substance. So, detecting the Raman light and analyzing its spectrum allows the

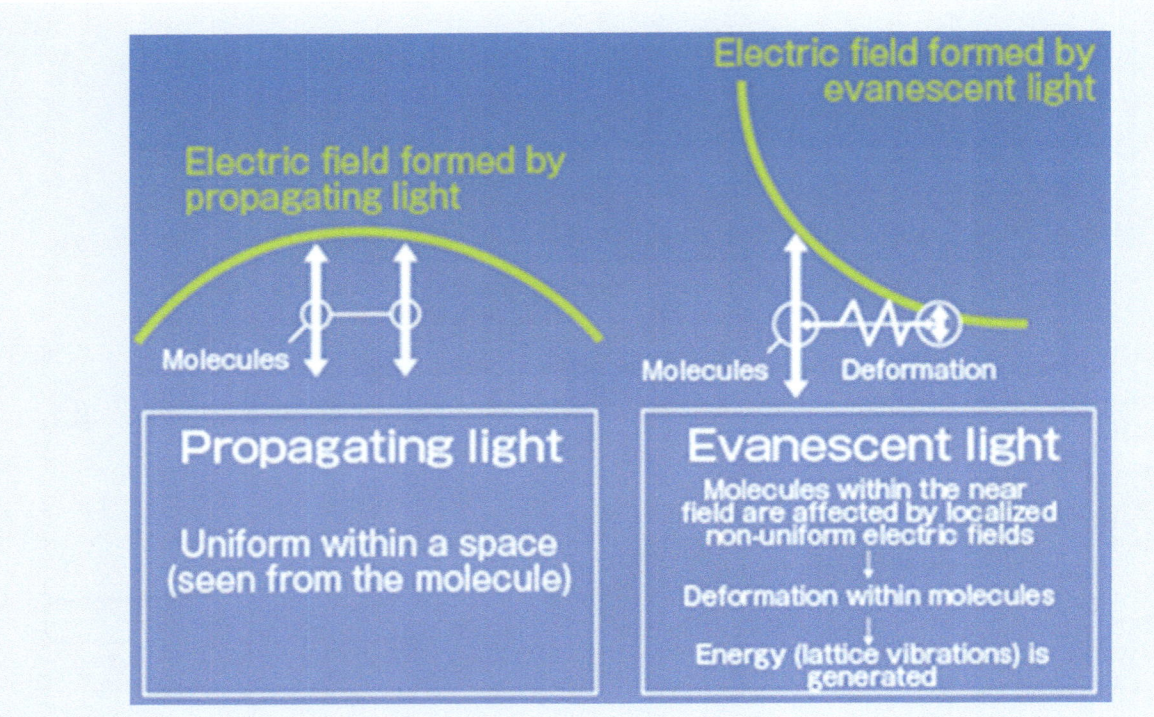

Fig. 10.22 Schematic of a large molecule in the proximity of metal nanoparticle. It shows the profile of the plasmonic electric field. Different parts of the molecule experience different electric field strengths, introducing strong distortion within the molecules (From [16])

identification of the substance. Since Raman light is usually very weak, detecting its intensity directly is quite difficult. If the electrical field strength is increased 10 times by surface plasmon resonance, then the Raman light is intensified 10^4 times so the intensity becomes 10,000 times higher.

Another interesting phenomenon of strong fields is the ability to optically manipulate the structure of molecules. Since the electric field in the vicinity of the plasmonic nanoparticles drops very rapidly over 10 nm, different parts of large molecules will experience different electric field strengths hence different electric stress. This causes distortion within the molecules and consequently causes new ways to deliver energy to the molecule including vibrations, nonlinearities, and electronic structure, as shown in ◘ Fig. 10.22. The additional energy supplements the energy of the incident photon [16].

10.4.3 Nanofabrication and Nanolithography

Atoms or molecules are placed near the atomically sharp metal tip (tungsten or gold tip) of a scanning tunneling microscope (STM), as shown in ◘ Fig. 10.23 (left top) [54, 55]. Below the tip is a conducting substrate at a distance of 1 nm. When the tip is biased by 1–3 V, an electric field is setup in the gap which can reach more than 100 MV/cm. An intense pulsed laser light bathes the metal tip and the gap and the substrate. The electric field in the vicinity of the gap is very strong due to the dc field of the tip as well as to plasmonic fields. The combined effect of the light and electric field is confined to a nanospace and will be very intense on atoms or molecules falling in this space near the tip. The intense pulsed laser light can be tuned to resonate with a certain state in an atom or molecule, promoting them to high-lying excited states while the combined electric field can assist in stripping the excited electron to produce a single ionization event of a free electron and a free ion.

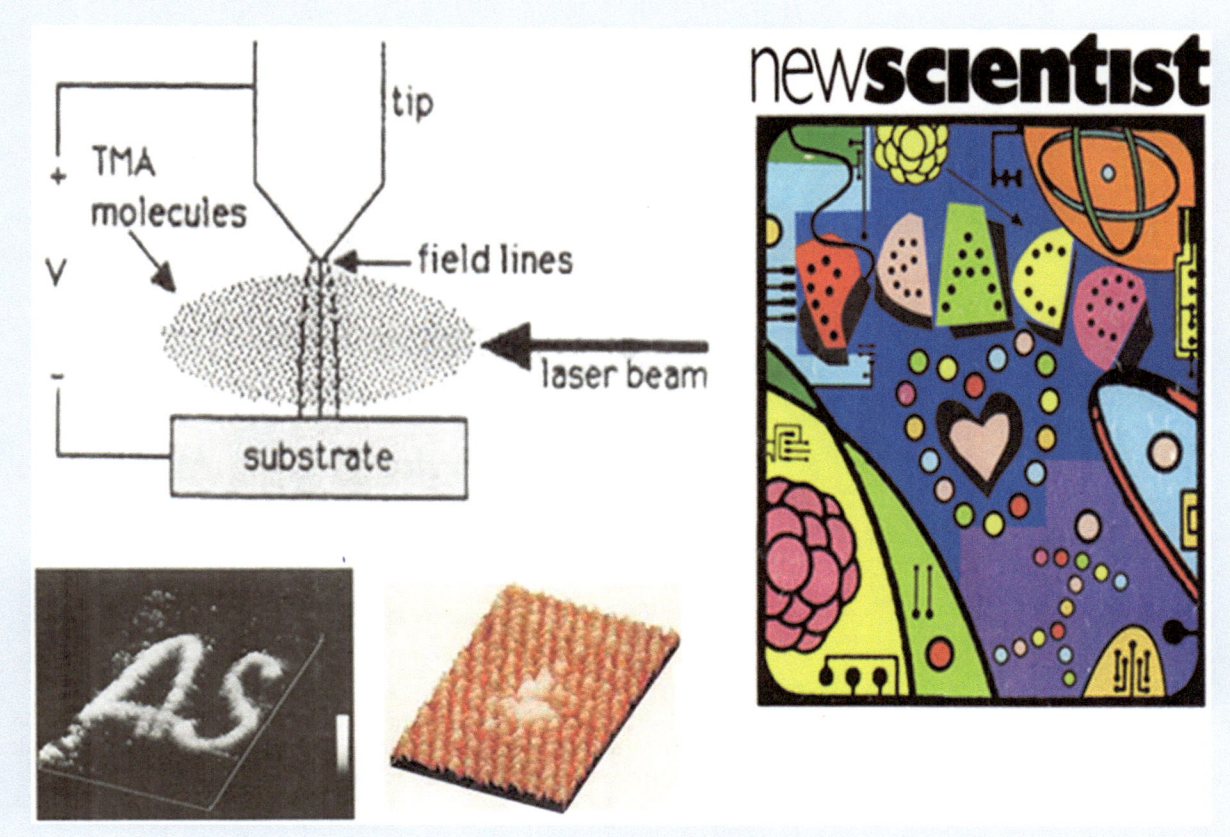

🔹 **Fig. 10.23** (*Left top*) Schematic of a tungsten tip of a scanning electron microscope bathed with a laser beam in a chamber with a certain pressure of a molecular gas. The metal tip is biased at 1–3 V. Molecules, in the gap or near the gap, are subjected to the combined effect of the laser field and the localized field of the metal tip excites, dissociates, ionizes, and pins down ions on the counter surface. (*Left bottom*) Magnified image of the initials of Arthur Schawlow "*AS*" written with trimethyl aluminum molecules on a silicon wafer. A single molecule of trimethyl aluminum is picked on a graphite surface. (*Right*) Cover of The New Scientist showing displaying a nanofabrication pattern in the form of a "heart" performed by the process depicted in the top left schematic (From [54]) along with patterns in the form of the "molecular man" by IBM and the word "PEACE" by Hitachi, both written using STM tips only

Since the electric field is focused or confined to nanoscale then the atomic ion is accelerated, pushed, and guided to the surface with nanoscale resolution. Programming the position of the tip allowed the Nayfeh's group to make nanopatterns on the surface with atomic resolution. Since light can resonate with specific atoms or molecules then the process is highly selective and can pick a certain atom from a mixture of atoms or molecules, as shown in 🔹 Fig. 10.23 (left bottom). In one of the earliest news media reports on nanotechnology under the title the "Smallest Graffiti in the World," and subtitle "Nanotechnology Rules," the British magazine, "new Scientist" in 1992 covered the process and displayed several patterns ("heart") on its cover along with the work from IBM ("molecular man"), and Hitachi of Japan ("peace") (🔹 Fig. 10.23 (right)) as symbols of nanotechnology [54]. In 1992 on 70th birthday of Arthur Schawlow, the inventor of the laser the author presented him with the nanoscale of his initials "*AS*" written with trimethyl aluminum molecules on a silicon wafer, shown in 🔹 Fig. 10.23 (left bottom). STM-based processing has been proposed by the community for nanolithography in the electronics industry, but since it relies on the movement of mass for patterning it was concluded that it is very slow and as such unpractical for mass production.

The electric field can also be confined in the vicinity of a very thin tungsten wire, such as that of a Geiger counter used in detecting and counting charged

continued on page 20

Fig. 10.24 (*Left*) Report in search and discovery of Physics Today about the use of intense light combined with an electric field to identify and selectively detect single atoms (From [57]). (*Right*) cover of The Sciences displaying Alhazen conducting some laboratory experiments. Inside, the issue reports on the detection of single atoms using the process described in the ▣ Fig. 10.24 top left sketch (From [58])

particles for nuclear radiation research [56]. This configuration has been used in 1975–1977 for identifying and counting single atoms by Hurst, Nayfeh, and Young at Oak Ridge National Laboratory [57]. It is to be noted in 1975 William Fairbank Jr., Theodor Hansch, and Arthur Schawlow at Stanford University reported detecting as low as 100 atoms/cm^3 using cw resonance fluorescence in sodium vapor [57]. The Oak Ridge results were covered in the search and discovery news of Physics Today, as shown in ▣ Fig. 10.24 (left) [57], and in The Sciences magazine published by the New York Academy of Sciences shown in ▣ Fig. 10.24 (right) [58]. The cover of same issue of "The Sciences" shows a sixteenth century painting of Islamic astronomers at work in their observatory. One of the scientists shown is Alhasan Ibn Alhaytham (Alhazen). Because the process constitutes the ultimate sensitivity in analytic detection of matter, namely a single atom, the work was also covered in Britannica Yearbook of Science and the Future (1979), McGraw Hill Yearbook of Science and Technology (1979), The World Book Science Annual (1978), and more.

10.4.4 Photovoltaics and Photocurrent

One hot area involving the interaction of light with nanosemiconductor in the presence of an external electric field is the generation of voltage that can be stored. A thin film of silicon nanoparticles or capsules of silicon nanoparticles, for example, are placed on top of a silicon-based p-n junction (amorphous, polycrystalline, or monocrystalline silicon solar cell). When light strikes the nanoparticles some light is absorbed. As a result, electron hole (e–h) pairs (excitons) are produced in the nanoparticles. If the electron and hole are separated from each other completely before they could recombine to produce light (luminescence) or

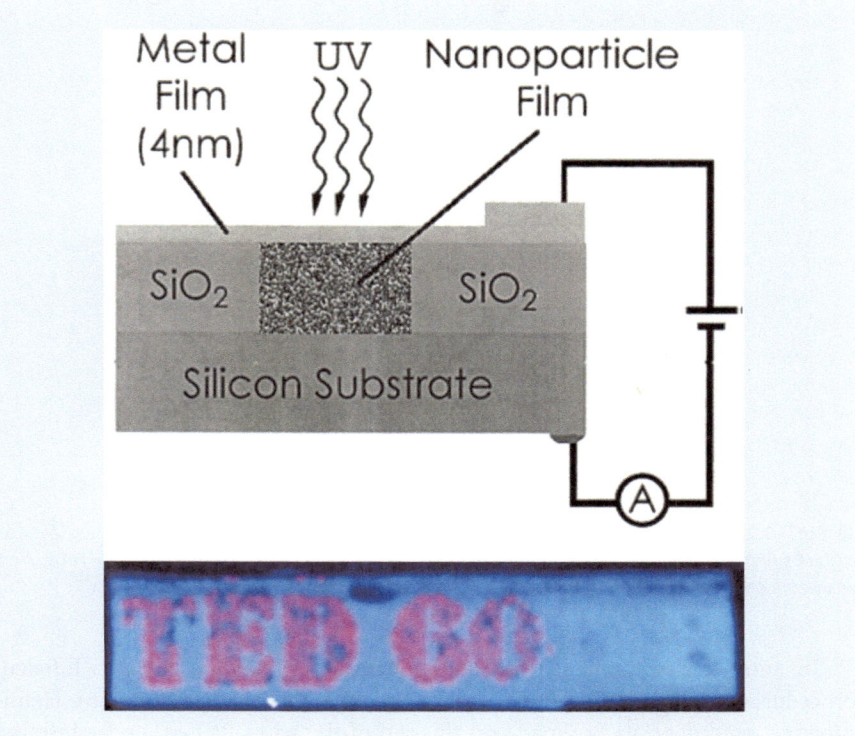

◻ Fig. 10.25 (*Top*) Highly sensitive photodetectors for various applications consisting of holes filled with silicon nanoparticles and covered by an aluminum thin layer (*bottom*) highly magnified image of a pattern of holes in the shape of the words "TED 60" filled with red luminescent silicon nanoparticles under UV excitation was presented to Theodor "Ted" Hansch on his 60th birthday celebration in Munich [62]

recombine non-radiatively to produce vibrations and heat then the electrons and holes may be transported, collected, and stored on two external electrodes appropriately constructed and positioned. The voltage difference between the electrodes can be harnessed at a later time as a battery. The electric field in a p-n junction plays a pivotal role in charge separation and collection. This architecture using 2.9-nm and 1-nm silicon nanoparticles showed relative power enhancements of the efficiency of the under lying solar cell in the UV and in the visible [59, 60, 61].

If the nanoparticles are placed simply on a simple conducting substrate instead of a p-n junction, then a continuous flow of electrons may proceed which may act as an instantaneous current source. A thin oxide layer is grown on a silicon wafer, followed by etching out a hole or multiple holes using a mask. The holes are filled with silicon nanoparticles and an aluminum thin layer is grown as a cap/electrode, as shown in ◻ Fig. 10.25 (top). Architectures of this type have been used as highly sensitive photodetectors for various applications [62]. In one application, it provided a sensitive detector for UV radiation while being blind to visible radiation, property which is useful for elementary particle collisions in neutrino, dark matter, and rare decay experiments. In those weak UV Cherenkov radiation is produced that gives vital information about momentum and geometry of collision [63]. Following the same procedure using a mask in the shape of the words "TED 60", a red luminescent pattern was created, as shown in ◻ Fig. 10.25 (bottom). In November 2001, the silicon wafer was presented to Theodor "Ted" Hansch on his 60th birthday celebration in Munich. Among the guests are four physics Nobel Laureates: Norman Ramsey, Steven Chu, Claude Cohen-Tannoudgi, and Carl Weiman. The image was published in 2005 in the IEEE Transactions on Nanotechnology just after Hansch received the Nobel Prize [62].

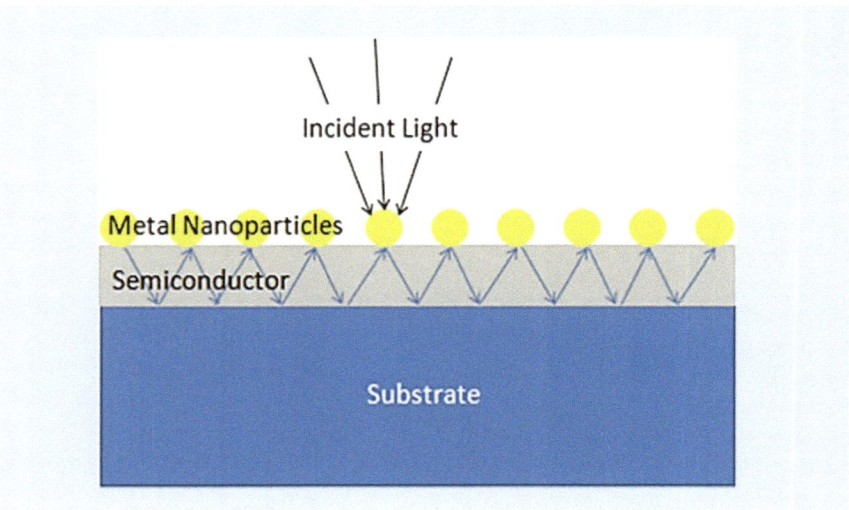

Fig. 10.26 Schematic of a plasmon solar cell. It consists of a thin silicon-based active layer on a glass substrate. Gold nanoparticles are placed on the active layer. (Adapted from ▶ https://en. wikipedia.org/wiki/Plasmonic_solar_cell)

In another development, silicon nanowire arrays were used [64]. Efficient procedures for fabrication of nanowires were recently developed [65, 66]. Hemispherical gold deposits are made on the end of the wires. When a near-infrared optical field falls onto the gold deposits, it excites plasmon resonances, which remarkably amplify the intensity, effectively making them like antennas. The same gold deposit can form a p-n like Schottky junction with the nanowire, which enhances charge collection. Thus the system acts as an effective near-infrared photodetector [64]. Silicon-based sensitive UV photodetectors have military as well as commercial applications. Military applications include missile warning systems, biological attack warning systems, and jet engine sensors.

Light interaction with nanometal is also emerging as a useful intermediate agent for improving light coupling to thin film solar cells; hence improving their efficiency [67]. Thin film solar cells utilize 1–2 μm thick semiconductor material placed over substrates of cheap material compared to silicon, such as glass, plastic, or steel. Metal nanoparticles are deposited on the top of the semiconductor, as shown in ◼ Fig. 10.26. When light hits these metal nanoparticles at their surface plasmon resonance, light is scattered in many different directions. This allows light to travel along the active semiconductor and bounce between the substrate and the nanoparticles enabling the semiconductor to absorb more light even though it is a thin layer [67].

10.4.5 Solid State LED White Lighting

The objective of using electronic chips for lighting is to manufacture light bulbs that are superior to conventional bulbs yet with characteristics that match the sun's white light spectrum, given in ◼ Fig. 10.27 (top). Light emitting diodes (LED) are chips powered by electricity. They produce light of specific color in the UV, blue, green, or red light range. White light bulbs for domestic use can be manufactured by combing light from three LEDs: blue, green, and red. Because the emission of each LED is not wide enough, the mix produces "finger shaped" spectrum, not smooth as normal sun light (◼ Fig. 10.27 middle). The problem has been somewhat alleviated and simplified in recent years by using a single blue LED source and a wide band green/orange phosphor converter, which combines to

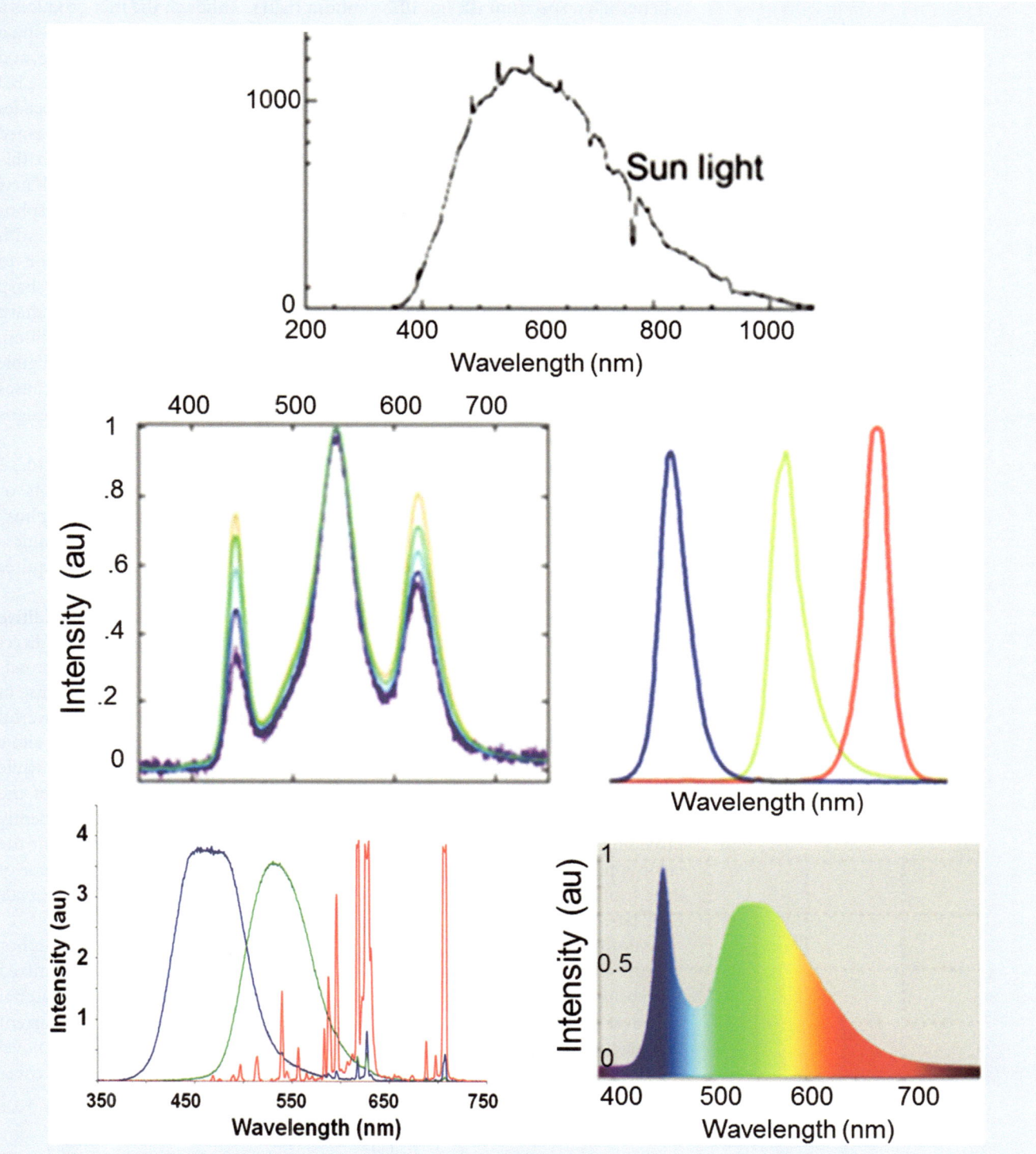

Fig. 10.27 (*Top*) Spectrum of sun light (*middle*) mixing light from three LEDs: *blue*, *green*, and *red*, producing "finger shaped" spectrum not smooth as normal white sun light (*bottom right*) single blue LED source and a green/orange phosphor converter, which combines to a "hand shaped" spectrum not filled enough (*bottom*) single UV LED and three phosphors (blue, green, and red phosphor); red phosphor converters, based on Europium (Eu) provides spectra dominated by *sharp red lines*

10

a better filled spectrum (■ Fig. 10.27 bottom right). Although the mix produces a "hand-shaped" spectrum and short of full match with sun light as it is missing a red component, the developers of this process Isamu Akasaki, Hiroshi Amano, and Shuji Nakamura were awarded the 2014 Nobel Prize in Physics. The first LED (red) was invented by Nick Holonyak at the University of Illinois several decades earlier (▶ http://www.led50years.illinois.edu/nicks-story.html), but it was deemed that the more recent blue LED (invented by the awardees) and its use in this configuration afforded a very important service, namely filling the world with new "white" light. A third solution is a UV LED and three wide band phosphor converters (blue, green, and red phosphor). However, there is a problem with red phosphor converters as they are based on europium (Eu). In addition to problems of availability and stability, those provide spectra dominated by sharp red lines (■ Fig. 10.27 bottom left), which produces a mixed spectrum with sharp lines not smooth enough as normal white sun light (■ Fig. 10.27 top). In addition, phosphor films are found to have appreciable reflectivity causing a non-negligible fraction of the pumping LED light to be reflected backwards. This in turn causes heating and requires more power input. More energy efficient and rigorous designs must be used to minimize heating.

The novel optical properties of nanomaterial discussed above in ■ Sect. 10.3.3 may alleviate many of these problems. In fact, CdSe-ZnSe quantum dots or nanoparticles have been demonstrated as potential substitute for "red phosphor" for use in near UV pumping [68]. However, due to their direct nature, luminescence of a given nanoparticle size is sharply dependent on its size, which requires the use of an appropriate size distribution to produce broadened emission.

Silicon nanoparticles have also been demonstrated as substitute for or additive to Eu-based "red phosphor" [69], as shown in ■ Fig. 10.28. Because of the indirect nature of silicon, individual silicon nanoparticles produce inhomogeneous broadened luminescence (■ Fig. 10.28 top), avoiding sharp-line mixing; and because of strong UV absorption they are versatile for pumping with the emerging powerful UV LEDs. It is hoped that UV-pumped "nano-phosphor" converters would allow manufacturing of high quality white bulbs that cover much larger areas while affording high color rendering index which is the quantitative measure of the ability of the bulb to reveal the colors of objects faithfully, while independently providing a correlated color temperature (CCT), which is a specification of the color appearance of the light emitted by the bulb relating its color to the color of light from a black body radiator heated to the same temperature. Commercials CCT labels include warm, daylight, and cool labels of temperature.

Another advantage of using nanomaterial as a component in the phosphor mixture stems from the fact that nanomaterial reduces the reflectivity of the mixed composite material so as not to harm the pumping LED chip. Also nanomaterial improves heat dissipation, thus prolonging the lifetime of the bulb. Thus current manufacturing process of white light can be developed further using the novel interaction of light with nanomatter to produce better efficiency bulbs that cover smoothly the solar white light range, as well as handle larger areas.

10.4.6 Plasmonic Hyperthermic-Based Treatment and Monitoring of Acute Disease

One serious problem inherit in plasmonics is energy loss due to heat generation resulting from strong absorption especially in the infrared—visible and UV spectral ranges [70, 71]. So far, the losses are effectively too high to make practical plasmon-based electronic or photonic devices. However, these losses turned out to be a blessing for other applications, especially for cancer therapy. In this treatment,

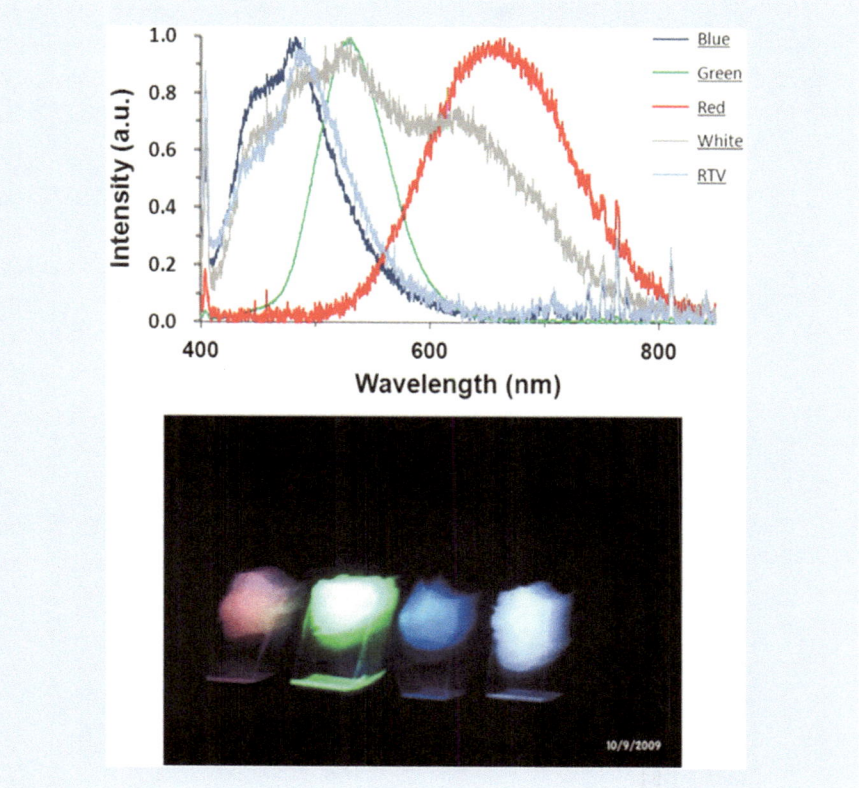

Fig. 10.28 Demonstration of 2.9-nm Si nanoparticles as equivalent to a "red phosphor" for LED technology. (*Top*) Normalized emission spectra of blue phosphor ZnS:Ag (*in blue*), green phosphor ZnS:Cu,Au,Al (*in green*) and red Si nanoparticles (*in red*) individually dispersed in RTV under the excitation of 365 nm radiation. The *gray spectrum* is due to a mixture of the three components dispersed in RTV under excitation of 365 nm. Normalized spectrum of pure RTV (*in light blue*). (*Bottom*) from *left* to *right* the corresponding luminescent images of the individual components and of the mixture under excitation of 365 nm radiation

a near-infrared laser is used, which penetrates deep into the tissue, heating implanted nanoparticle to about 49 °C, as shown in ▪ Fig. 10.29 [72]. This is the temperature level which is needed to kill many targeted cancer cells. This results in a threefold increase in killing cancer cells and a substantial tumor reduction within 30 days. Current hypothermic techniques are not selective, i.e., they involve applying heat to the whole body, which heats up cancer cells and healthy tissue, alike. Thus, healthy tissue tends to get damaged. By using gold nanoparticles, which amplify the low energy heat source efficiently, cancer cells can be targeted better and heat damage to healthy tissues can be minimized.

Improvement of the sensitivity of Raman analysis due to the enhanced electric field in the proximity of gold nanoparticles (as discussed above in ▪ Sect. 10.4.2.1) has recently been used to monitor and confirm death of cancer cells. The extra sensitivity allows following changes in the molecular content inside cells, including destruction or formation of molecules in cancerous cells during their death [70].

10.5 Plasmon Effect in Ancient Technology and Art

Red glasses have been found in Italy that dated back to the late Bronze Age. The phenomena stayed essentially unsolved till the last decade. It is now believed that the colors are attributed to white light excitation of plasmon surface modes of electron oscillations in different metal nanoparticles incorporated during the

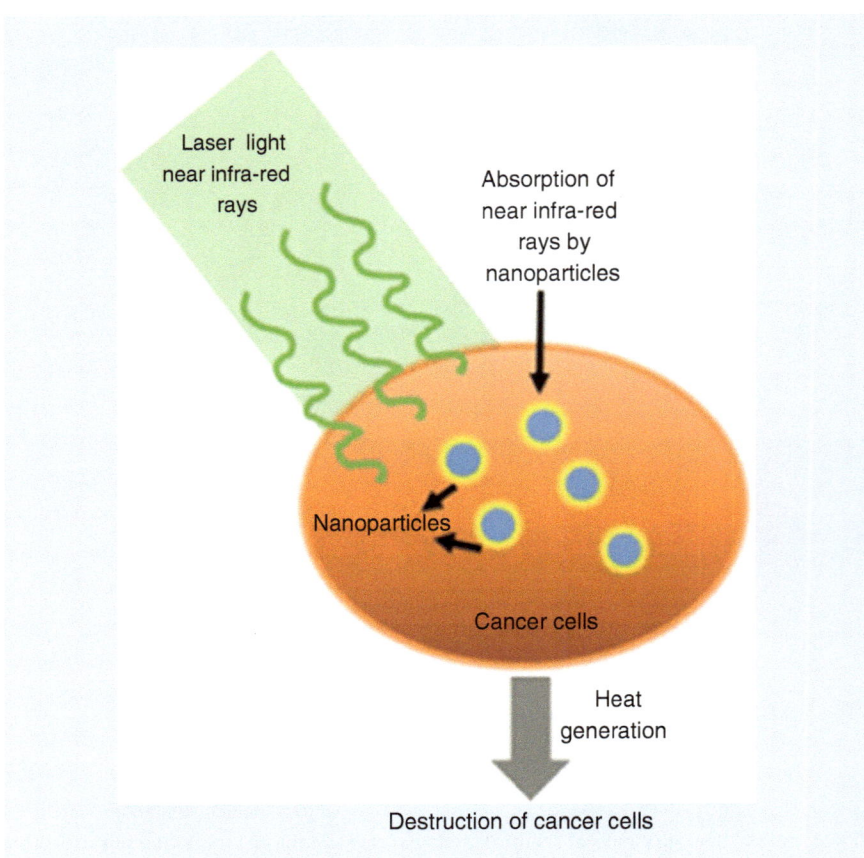

Fig. 10.29 Schematic of plasmonic nanoparticle-based photothermal therapy for cancer treatment. A non-consequential weak infrared light gets concentrated in gold nanoparticles, heating them enough to kill cells (Adapted from [69])

manufacturing process [73, 74]. In addition, gold nanoparticles were identified in some red tesserae of Roman times [75]. A vivid example is the well-known Roman Lycurgus Cup in glass dated from the fourth century CE and currently exhibited in the British Museum, as shown in ▣ Fig. 10.30 (left) [76], and the elaborate and complex multicolor stained glass window shown in ▣ Fig. 10.30 (right) [77].

As was discussed above in ▣ Sect. 10.2.3, the phenomenon can be easily understood in terms of color mixing effects. When white light shines only the G (green) component resonates with the electrons in the gold nanoparticles and is absorbed, hence removed. The B (blue) light is weakened by suffering scattering in all direction. The remaining R (red) component passes through. This is the reason that stained glass mixed with gold nanoparticles appears red to the naked eye.

In the Middle Ages exploitation of the outstanding optical properties of metallic nanoparticles became more sophisticated. Abbasid and Fatimid Islamic potters were able as early as 836–883 CE to develop new chemical technology that allowed them to create complex nanometal-based multilayer structures. First they were incorporated in ceramics, which resulted in the emergence of lusterware, a special type of glazed ceramics, with striking optical effects. Secondly, the distribution of nanoparticles was programmed to include multilayers such that luster decoration color can change depending on the angle from which it is observed. An example of these types of ceramic decorations is given in ▣ Fig. 10.31 [78]. The color change is spectacular and brilliant. A variety of very intense colored metallic shine including golden-yellow, blue, green, pink, etc., can be observed. Recent studies in 2005 concluded that the multilayer features give rise to light interference

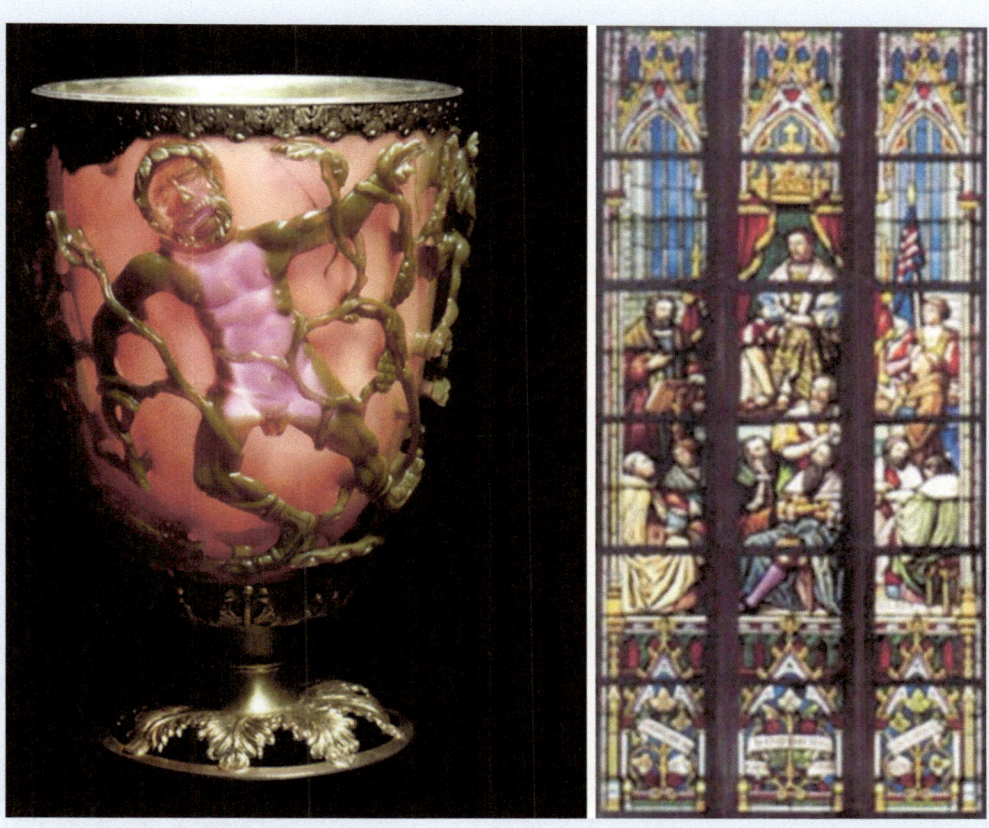

Fig. 10.30 (*Left*) The well-known Roman Lycurgus Cup in glass dated from the fourth century CE. It is currently exhibited in the British Museum (*Right*) Medieval stained glass windows (Courtesy: NanoBioNet and ▶ www.nano.gov] [76])

phenomena and scattering through rough interfaces, which adds to the surface plasmon effect and strongly contributes to the observed color [79].

Several material analysis studies [80–83] confirmed that the color of the luster decorations come from metallic nanoparticles. For instance, the presence of metal particles was directly confirmed by conducting high resolution transmission electron microscopy and imaging (TEM) along with material analysis using electron energy loss spectroscopy. Generally, copper and silver clusters were incorporated by applying a mixture of a paint, which contained copper and silver salt powders onto a glazed ceramic. This was followed by annealing in a reducing atmosphere. It is now believed that the basic mechanism involves the two processes of ion exchange and crystallization (nucleation and crystal growth) of copper and silver metallic nanoparticles inside the glassy matrix [84].

The ingenious technology, however, was way ahead of its time hence it was based on empirical chemical means, which made the technology vulnerable to extinction. In fact the technology has nowadays partially been lost. Vincent Reillon and collaborators [85] performed elaborate optical modeling, which confirmed the role of multilayer interference and interface scattering in the color of metallic shine reflection (specular direction). On the experimental side, modern artists [81, 83] conducted experiments to re-create luster decorations. They used modern kilns, which allowed alternative oxidization (oxygen flux) and reduction (CO flux) phases during the firing [73]. The best recreations achieved exhibited partial multilayer structure with a weaker organization.

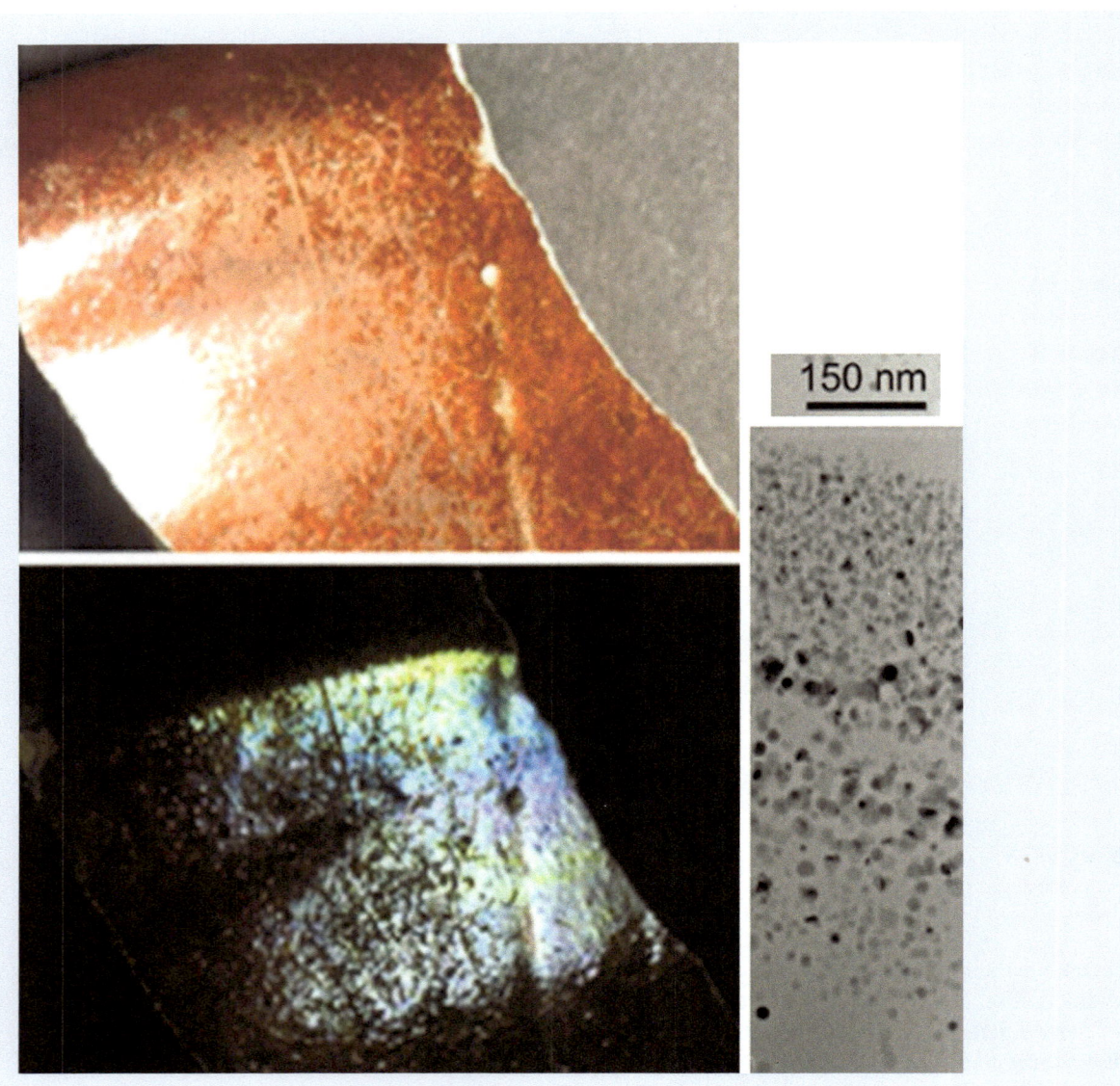

◘ Fig. 10.31 A nineth century CE lusterware from Mesopotamia (Susa). (*Top left*) Lusterware appears *red*. (*Bottom left*) It appears *blue/green* when viewed from different angle. (Micrographs, courtesy D. Chabanne), (*right*) TEM image showing the presence of metal nanoparticle (Adapted from [77, 78])

10.6 Alhasan Ibn Alhaytham (Alhazen) and the Nature of Light and Lusterware

The Abbasid and Fatimid era are considered as two of most creative golden ages of Muslim civilization. The era saw many fascinating advances in science and technology. The period witnessed an aggressive and stimulating advanced chemical and manufacturing technology of glass and pottery, exploiting advanced lighting, color and optical effects of metal nanostructures produced from mixed silver and gold powders. Not only the chemical industry was advanced, material and metallurgical industry including thermo-mechanical forging and annealing was sophisticated. Steel cakes were subjected to such protocols to refine the steel to its exceptional quality. Domestic and military products were produced. One of the products of the industry, the famous Damascus blades, was pivotal in the fight against the Crusaders. Those were forged directly from small cakes of steel (named '*wootz*') imported from ancient India [86, 87]. The blades were remarkably hard, sharp, and light weight. In fact recent high resolution transmission electron microscopy of samples of the blades revealed the presence of carbon nanotube, which afforded the products exceptional material properties.

Alhasan Ibn Alhaytham (Alhazen) was born in Basra, Iraq in this flourishing era (965). He was not an exception; he contributed to several fields including astronomy, mathematics, medicine, and especially optics. He ventured into pioneering experimental scientific research, with methodology approaching what we use today in the finest research institutes. His studies examined the nature of light and color which resulted in several revolutionary breakthroughs. With limited instrumentation he embarked on thorough examination of the passage of light through various media and its dispersion into its constituent colors, which led him to challenge and refute several adopted notions in optics including:

- Constructing the first generation of pinhole cameras for experimental research
- Establishing, using his pinhole camera, that the human eye is not a source but a sensor or detector of light and color; and accurately described the mechanism of sight and the anatomy of the eye;
- Classifying light sources as luminescent and scattering sources [88, 89];
- Giving the first complete (*unique*) presentation of the law of reflection, by putting the incident ray and the reflected ray in the plane of incidence [90];
- Making the laws of refraction *unique* by introducing the pivotal principle of putting the incident and the refracted rays in the same plane (incidence plane) [90];
- Introducing the reciprocal law of refraction, which Kepler used to discover the law of total reflection [90];
- Proving light travels in straight line and that the speed of light is not infinite rather finite with different speeds in matter, moving more slowly in dense media, which was pivotal to the development of Fermat principle [91];
- Concluding that light is a mix of different colors by conducting experiments on the dispersion of light into its constituent colors; and the discovery that rainbows are caused by refraction of light; those were the first experiments to be conducted before the more detailed experiments conducted by Newton using advances in optical equipment, such as prisms to break up white light into a range of colors [92];
- Studying atmospheric refraction, and using it to discover that twilight only ceases or begins when the sun is 19° below the horizon and to deduce the height of the atmosphere to be 55 miles, which is basically correct (▶ http://www.jackklaff.com/hos.htm);
- Laying the foundation for the scientific method [93];
- Writing the first comprehensive book on optics, which got translated into Latin, powering research in optics in Europe for many centuries;

These developments show that the quest of Alhazen revolutionized the study of optics at the time and laid the foundation for the scientific method [93]. Alhazen, however, did not try to solve the optical phenomenon of stained glass or luster-ware. But those were problems for future generations. For centuries to follow, artists kept mixing silver and gold powder with glass or ceramics to fabricate colorful glass or pottery while no body was able to produce a scientific reason as to how these ingredients worked. The first step towards explaining the phenomenon had to wait until 1908 when Gustav Mie developed a theory of the optical properties of metallic colloids [94]. He showed that the color of a metal nanoparticle depends on its size as well as on the optical properties of the precursor metal and the adjacent dielectric materials. But full understanding of stained glass and lusterware required even more time and nineteenth to twenty-first century developments including electromagnetic theory, electromagnetic wave nature, and interference/diffraction and scattering of light, solid state theory, plasma theory, nanoscale phenomenon, advanced material characterization technologies, such as electron-based imaging and material analysis. In fact, it was only in the last two decades the scientific community has succeeded in solving the problem [95]. Those medieval artists or chemical industrialists were actually "nanotechnologists" synthesizing metal nanoparticles and harnessing what we today call plasmonics: a new field based on electron oscillations in metals called plasmon.

The recognition of the work of Alhazen may have been slow in the Middle Ages, but it has been coming strong in recent years. In medieval Europe, he was honored as *Ptolemaeus Secundus* ("Ptolemy the Second") or simply called "The Physicist." In the twentieth to twenty-first centuries he has been given several titles including "the First Scientist"; "Hero of Science"; "The Father of Modern Optics"; Pioneering Scientist; Pioneering Scientific Thinker, Rare Genius in Physical Research, The Optical Scientist, etc. In fact, year 2015 marks the 1000th anniversary since the appearance of the Alhazen's remarkable seven volume treatise on optics "Kitab al-Manazir." The year 2015 has also been adopted as the year of light and the United Nations through its arm UNESCO launched "2015 international year of light (IYL2015)" as a global initiative intended to raise awareness of how optical technologies promote sustainable development and provide solutions to worldwide challenges in energy, education, agriculture, communications, and health. One of the major scientific anniversaries that will be celebrated during the 2015 International Year of Light is the works on optics by Alhazen (Ibn Al-Haytham) (1015). UNESCO and the IYL2015 in partnership with the UK-based organization 1001 Inventions launched a high-profile international educational campaign celebrating Ibn al-Haytham. Moreover, 1001 Inventions and the King Abdulaziz Center for World culture in partnership with UNESCO and the IYL2015 are planning to produce a short film on his work. King Abdulaziz City for Science and Technology is also celebrating this occasion by publishing an edited book on optics for a very broad audience, to which this article is contributed.

10.7 From Alhazen to Newton to the Trio: Dispersion of Light

In 1015 Alhazen was concerned with understanding the different color components of light through natural phenomenon, such as propagation of light in material, reflection and refraction, and the rainbow effect. Newton 600 years later used manufactured glass prisms which became available then to deliberately disperse light into its components and to introduce the seven color names red, orange, yellow, green, blue, indigo, and violet for segments of the spectrum. Nearly 300 years after Newton Theodor "Ted" Hansch and Arthur Schawlow of Stanford University and John "Jan" Hall of the University of Colorado set out to achieve big strides in the quest to disperse light and isolate extremely narrow components using much more advanced electronics and optics capabilities, such as prisms, gratings, interference filters, spectrographs, and telescopic light expanders. But such a highly pure color components would have extremely weak intensity, which would necessarily require amplification if it is to be a practical light source. In 1972, the Stanford group dispersed the red fluorescence from a chemical dye into much finer components while simultaneously being amplified using the very advanced light–matter interaction concepts of stimulated emission (introduced by Einstein) and laser gain (Schawlow inventor) to produce a narrowband directed red laser (light amplification by stimulated emission of radiation). Pulses or flashes of red light of 8 ns durations with a record color definition (bandwidth) of 0.0004 Å (or 0.001 cm^{-1} or 30 MHz) and energy of ~1 nJ per pulse were achieved (Hansch laser). In addition to being narrow band, the laser allows for change (tuning) of wavelength with good control and precision, a pivotal property for matching (resonating with) and studying electronic structure of atoms and molecules [96]. John Hall, on the other hand, used a different approach by starting out with the red light of a helium–neon gas laser [97, 98]. This red laser light at a given time is narrow band (very pure in color) but it is practically not as narrow over longer times because its frequency or wavelength drifts as a result of drift in ambient conditions (temperature, vibrations, etc.). Using advanced optical and electronic equipment he succeeded in making the frequency or wavelength practically stand still for long times. This high finesse light source is extremely useful but it is not tunable in wavelength.

The two groups used these advanced devices in which the color of light has been defined much more precisely to perform high resolution laser spectroscopy measurements of unprecedented accuracy and intrinsic physical interest. Jan Hall used his source to measure accurately the speed of light, allowing a re-definition of the SI meter. It should be mentioned here that the first version of the helium–neon laser was invented in 1960 by Ali Javan (Azerbajani born in Tehran, Iran) at Bell Labs [99]. However the laser produced invisible light at 1.15 μm. Only 18 months later in 1962, Javan's colleagues White and Rigden of Bell Lab constructed the visible He–Ne laser which is more exciting, captivating, and convenient [100].

The Stanford group utilized the tunability of the Hansch laser in a Doppler free technique that was introduced theoretically earlier by Russian scientists [101]. The technique allows the laser to be blind to all moving atoms in a gas sample and see only the ones that happen to be stationary along the direction of the red light beam, which unmasks the true structure of the atoms. ◘ Figure 10.32 (left top) is a photo from a celebration at Stanford recognizing Schawlow and Hansch as California Scientist of the year for these developments and the use of the narrowband tunable laser in precise spectroscopic measurements. Issa Shahin, a doctoral student in the Stanford group, was involved in using the narrow bandband dye laser to study the structure of sodium and hydrogen [96]. Munir Nayfeh, a doctoral student under the superposition of Hansch and Schawlow reproduced a variation to the Jan Hall's stable He–Ne light source (◘ Fig. 10.32 left bottom) [103] and utilized it as a wavelength standard along with the narrowband red laser to unmask and measure the hyperfine structure and binding energy of the simplest of all atoms, the hydrogen atom, whose binding energy is the corner stone of all fundamental constants in nature, namely the Rydberg constant. The study achieved then a record accuracy of 9 parts per billion, which allowed the improvements of all other fundamental constants [104]. The importance of improving the accuracy of the Rydberg constant was evident in 1995 when, on the occasion of the 100 year anniversary of the top American Journal Physical Review, the journal listed this achievement as one of the seminal papers in the 100 year life of the Journal, as shown in ◘ Fig. 10.32 (right) [102].

Alhazen and Newton and the Trio were concerned with breaking light into its frequency components; other scientists, on the other hand, focused on the time domain. Short pulses of light of duration of micro (10^{-6}), nano (10^{-9}), pico (10^{-12}), femto (10^{-15}), and now atto (10^{-18}) seconds have been isolated. In this regard, we mention that the Egyptian-born Ahmad Zewail used femtosecond pulses of light to capture the very real-time dynamics inside molecules, studies that earned him in 1999 the Nobel Prize in chemistry [105]. In the last decade the interest ventured into the atto second regime, providing capabilities to make real-time observations of valence electron motion in solids [106].

Schawlow, Hansch, and Hall were recognized by Nobel Prizes. Arthur Schawlow received the 1981 Nobel Prize in Physics for his work on development of laser light. Theodor W. Hänsch and John Hall shared with Roy Glauber (for his

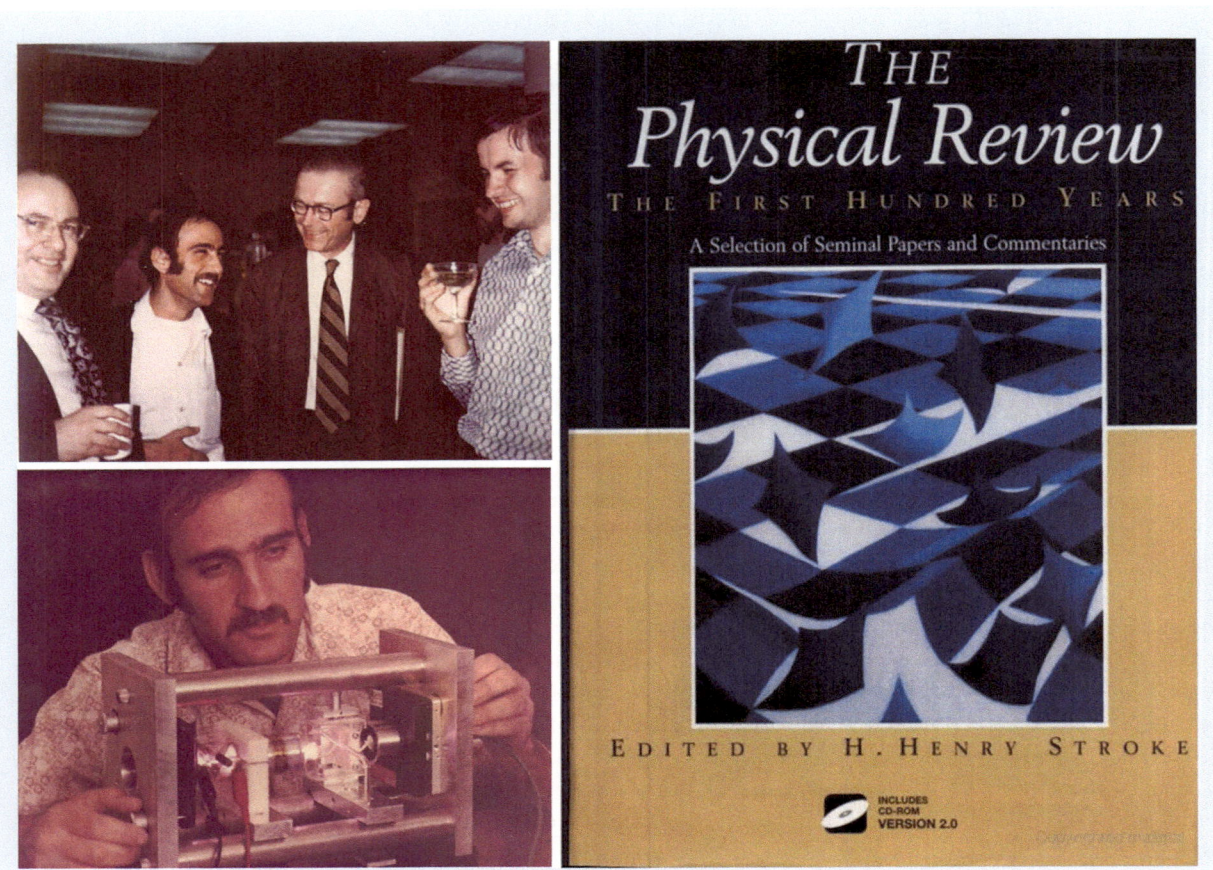

Fig. 10.32 (*Left top*) Celebration at Stanford from left to right: Arthur Schawlow, Munir Nayfeh, Stanford President, and Theodore Hansch for the choice of Schawlow and Hansch as California Scientist of the Year for the high resolution spectroscopy achievements. (*Left bottom*) Home built He–Ne laser by Hansch and Nayfeh after [96] stabilized to 10 parts per billion. It has iodine vapor cell placed in the cavity. It is stabilized by locking it to the nth hyperfine component of isotropically pure I^{129} at 633 nm. The photo decorated the office of Arthur Schawlow (*right*) cover of the book celebrating the 100 year anniversary of the American Journal Physical Review [102]

work on coherence of light) the 2005 Nobel Prize in Physics for their work in precision spectroscopy using laser light sources.

It is to be noted that Newton's fame actually comes from gravity (as the father of) rather than from light. Alhazen fame is beginning to come in light, emerging as "the father of modern optics." There were many sources describing Ibn al-Haytham (Alhazen) as such. One of those is "Impact of Science on Society— Volumes 26–27—(1976) Page 140, a prestigious UNESCO publication whose first edition came out in 1950. The study stated that "one name stands out as that of a rare genius in physical research: Abu Ali Al-Hasan Ibn Al-Haytham (965–1039) of Basra (Iraq), without question the father of modern optics" [107].

Figure 10.33 presents selected clips and images from the mass media giving tribute to the contributions of Alhazen. There is one more thing that can be said about Alhazen. If the Nobel Prize was in place in the nineth century or its regulations did not exclude deceased recipients, this author believes Alhazen would have certainly been a recipient for his revolutionary breakthroughs in advancing our understanding of the nature of light.

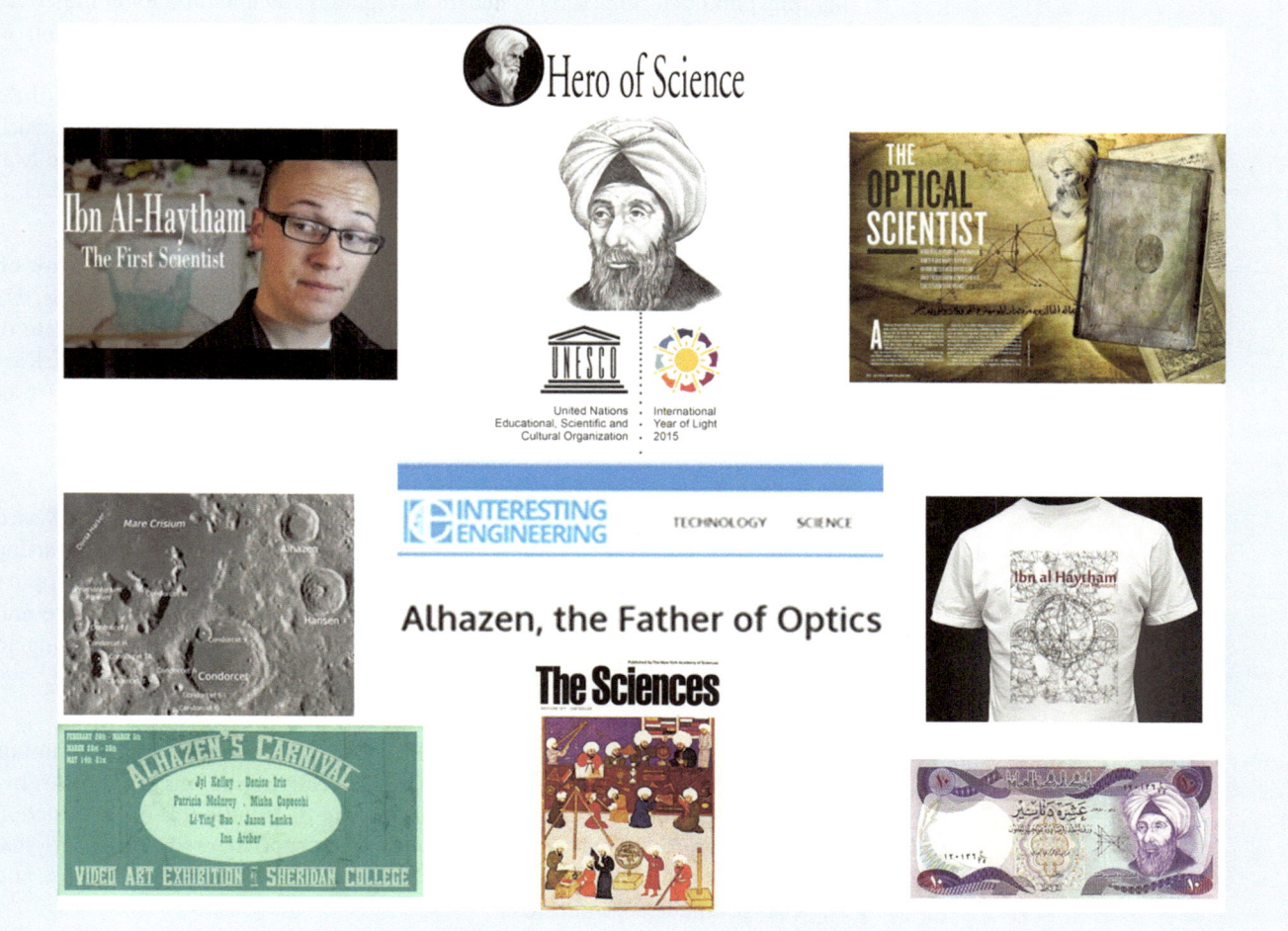

Fig. 10.33 Selected clips and images from the mass media giving tribute to Alhazen contributions

10.8 Conclusion

We focused in this article on how miniaturization impacts light interaction and propagation in metal and semiconductors. Unlike semiconductors, coupling of light to oscillating free electrons in metals allows metal nanostructures to concentrate or focus light to spots limited only by their size, much beyond the wavelength of incoming light, which enables novel concepts of metal-based lenses. Moreover, strong concentration, absorption, and scattering induce drastic color change as well as temperature rise and interactivity with the surrounding environment. On the other hand, miniaturization of semiconductors affords the material strong size-dependent luminescence and color, to the degree that it can take indirect material, such as silicon from being the dullest material to a glowing material. Integration of plasmonic and semiconductor effects in hybrid architectures is promising synergetic applications.

These novel and unprecedented nanoscale optical functionalities enable a variety of exciting applications including a variety of futuristic applications [108]. But there are novel exciting applications which have been demonstrated and they may be around the corner for the consumer use, such as hyperthermia treatment of cancer, monitoring cell death through molecular changes, substance identification and detection, photovoltaic thin film solar cells, solid state white lighting, photodetectors, subwavelength waveguides and antennas for commercial, military and elementary particles applications, nanofabrication, integration of optics and electronics, and art of stained glass and lusterware.

Metal plasmonic devices face, however, significant challenges because of heat losses at visible and other high frequencies important for telecommunication. Such heat losses have seriously limited their practical use in electronics; but the heat effect was turned into an asset in the fight on acute disease at the cellular level.

Acknowledgment We acknowledge financial grants that supported our work on Nanoengineered light emitting silicon, including the University of Illinois, US National Science Foundation, US Army, US Office of Naval Research, State of Illinois, Grainger Foundation, Beckman Foundation, King Fahd University, King Saud University and King Abdulaziz City for Science and Technology, Sharp of America, and SunGen of Canada, and Saudi ARAMCO.

▪ ▪ Occasion

2015 has been declared as the International Year of Light and Light-based Technologies. This year celebrates many milestones in the history of optics starting from the 1000 year anniversary of Ibn Al-Haytham's achievements in optics among which is his great book on Light. King Abdulaziz City for Science and Technology in Riyadh, Saudi Arabia is celebrating this occasion by publishing an edited book on topics in optics for a very broad audience.

References

1. Editor (1993) McGraw-Hill encyclopedia of science and technology, 5th edn. McGraw-Hill, New York, NY
2. Drexler KE (1986) Engines of creation: the coming era of nanotechnology. Doubleday, New York
3. Feynman R (2009) There's plenty of room at the bottom. Nat Nanotechnol 4:781. doi:▶ 10.1038/nnano.2009.356. ▶ http://www.nature.com/nnano/journal/v4/n12/full/nnano.2009.356.html
4. National Nanotechnology Initiative (2016) What is nanotechnology? ▶ http://www.nano.gov/nanotech-101/what/definition; Benefits and applications, ▶ http://www.nano.gov/you/nanotechnology-benefits
5. Iijima S, Ichihashi T (1993) Single-shell carbon nanotubes of 1-nm diameter. Nature 363:603–605
6. Nayfeh MH, Rogozhina E, Mitas L (2002) Silicon nanoparticles: next generation of ultrasensitive fluorescent markers. In: Baratron M-I (ed) Synthesis, in functionalization, and surface treatment of nanoparticles. American Scientific Publishers, Stevenson Ranch

7. Nayfeh MH, Mitas L (2007) Silicon nanoparticles: new photonic and electronic material at the transition between solid and molecule. In: Kumar V (ed) Nanosilicon. Elsevier, Amsterdam

8. Lohse SE, Murphy CJ (2013) The quest for shape control: a history of gold nanorod synthesis. Chem Mater 25:1250–1261

9. Babak Nikoobakht B, El-Sayed MA (2003) Preparation and growth mechanism of gold nanorods (NRs) using seed-mediated growth method. Chem Mater 15:1957–1962

10. Murray CB, Kagan CR, Bawendi MG (2000) Synthesis and characterization of monodisperse nanocrystals and close-packed nanocrystal assemblies. Annu Rev Mater Res 30:545–610

11. Foresight Institute (2016) Applications of nanotechnology. ▶ https://www.foresight.org/nano/applications.html

12. Pines D, Bohm D (1952) A collective description of electron interactions: II. Collective vs individual particle aspects of the interactions. Phys Rev 85:338

13. Messiah A (1999) Quantum mechanics. Dover, Minneola

14. Maier SA (2007) Plasmonics: fundamentals and applications. Springer, New York

15. Ditlbacher H, Hohenau A, Wagner D, Kreibig U, Rogers M, Hofer F, Aussenegg FR, Krenn JR (2005) Silver nano wires as surface plasmon resonators. Phys Rev Lett 95:257403

16. Hamamatsu Photonics KK (2016) Nanophotonics. ▶ http://www.hamamatsu.com/eu/en/technology/innovation/nanophotonics/index.html

17. Stockman MI (2004) Nanofocusing of optical energy in tapered plasmonic waveguides. Phys Rev Lett 93:137404

18. Evlyukhin AB, Bozhevolnyi SI, Stepanov AL, Kiyan R, Reinhardt C, Passinger S, Chichkov BN (2007) Focusing and directing of surface plasmon polaritons by curved chains of nanoparticles. Opt Express 15:16667–16680

19. Yin L, Vlasko-Vlasov VK, Pearson J, Hiller JM, Hua J, Welp U, Brown DE, Kimball CW (2005) Sub wavelength focusing and guiding of surface plasmons. Nano Lett 5:1399–1402

20. Verslegers L, Catrysse PB, Yu Z, White JS, Barnard ES, Brongersma ML, Fan S (2009) Planar lenses based on nanoscale slit arrays in a metallic film. Nano Lett 9:235–238

21. Veselago VG (1968) The electrodynamics of substances with simultaneously negative values of ε and μ. Sov Phys Usp 10:509–514 (Russian text 1967)

22. West P, Ishii S, Naik G, Emani N, Boltasseva A (2010) Identifying low-loss plasmonic materials, SPIE Newsroom. ▶ http://www.academia.edu/1037839/Identifying_low-loss_plasmonic_materials

23. White Noise (2014) ▶ http://whitenoise.kinja.com/dimensions-in-semiconductors-can-something-be-zero-dim-1543310309

24. Bawendi M (2016) ▶ http://nanocluster.mit.edu/research.php

25. Trwoga PF, Kenyon AJ, Pitt CW (1998) Modeling the contribution of quantum confinement to luminescence from silicon nanoclusters. J Appl Phys 83:3791

26. Canham LT (1990) Silicon quantum wire array fabrication by electro-chemical and chemical dissolution of wafers. Appl Phys Lett 57:1046

27. Heinrich JL, Curtis CL, Credo GM, Kavanagh KL, Sailor MJ (1992) Luminescent colloidal Si suspensions from porous Si. Science 255:66

28. Yamani Z, Thompson H, AbuHassan L, Nayfeh MH (1997) Ideal anodization of silicon. Appl Phys Lett 70:3404

29. Yamani Z, Ashhab S, Nayfeh A, Nayfeh MH (1998) Red to green rainbow photoluminescence from unoxidized silicon nanocrystallites. J Appl Phys 83:3929

30. Ackakir O, Therrien J, Belomoin G, Barry N, Muller J, Gratton E, Nayfeh M (2000) Detection of luminescent single ultrasmall silicon nanoparticle using fluctuation spectroscopy. Appl Phys Lett 76:1857–1859

31. Belomoin G, Therrien J, Smith A, Rao S, Chaieb S, Nayfeh MH (2002) Observation of a magic discrete family of ultrabright Si nanoparticles. Appl Phys Lett 80:841

32. Nielsen D, Abuhassan L, Alchihabi M, Al-Muhanna A, Host J, Nayfeh MH (2007) Currentless anodization of intrinsic silicon powder grains: formation of fluorescent Si nanoparticles. J Appl Phys 101:114302

33. Mitas L, Therrien J, Twesten R, Belomoin G, Nayfeh MH (2001) Effect of surface reconstruction on the structural prototypes of ultrasmall ultrabright Si_{29} nanoparticles. Appl Phys Lett 78:1918

34. Allan G, Delerue C, Lannoo M (1996) Nature of luminescent surface states of semiconductor nanocrystallites. Phys Rev Lett 76:2961

35. Draeger EW, Grossman JC, Williamson AJ, Galli G (2003) Influence of synthesis conditions on the structural and optical properties of passivated silicon nanoclusters. Phys Rev Lett 90:167402

36. Sundholm D (2003) First principles calculations of the absorption spectrum of $Si_{29}H_{36}$. Nano Lett 3:847

37. Lehtonen O, Sundholm D (2005) Density-functional studies of excited states of silicon nanoclusters. Phys Rev B 72:085424
38. Ball P (2001) Let there be light. Nature 409:974
39. Smith A, Yamani Z, Turner J, Habbal S, Granick S, Nayfeh MH (2005) Observation of strong direct-like oscillator strength in the photoluminescence of 1 nm silicon nanoparticles. Phys Rev B 72:205307
40. Mantey K, Zhu A, Boparai J, Nayfeh M, Marsh C, Alchaar G (2012) Observation of linear solid-solid phase transformation in silicon nanoparticles. Phys Rev B 85:085417
41. Rao S, Mantey K, Therrien J, Smith A, Nayfeh M (2007) Molecular behavior in the vibronic and excitonic properties of hydrogenated silicon nanoparticles. Phys Rev B 76:155316
42. Nayfeh M, Akcakir O, Belomoin G, Barry N, Therrien J, Gratton E (2000) Second harmonic generation in microcrystallite films of ultrasmall Si nanoparticles. Appl Phys Lett 77:4086
43. Kanemitsu Y (2003) Luminescence from Si/SiO2 nanostructures. In: Pavesi L et al (eds) Towards the first silicon laser. Kluwer Academic, Dordrecht, pp 109–122
44. Pavesi L, Dal Negro L, Mazzoleni C, Franzò G, Priolo F (2000) Optical gain in silicon nanocrystals. Nature 408:440–444
45. Nayfeh MH, Barry N, Therrien J, Akcakir O, Gratton E, Belomoin G (2001) Stimulated blue emission in reconstituted films of ultrasmall silicon nanoparticles. Appl Phys Lett 78:1131
46. Nayfeh MH, Chaieb S, Rao S, Barry N, Therrien J, Belomoin G, Smith A (2002) Observation of laser oscillation in aggregates of ultrasmall silicon nanoparticles. Appl Phys Lett 80:121
47. Hilliard JE, Nayfeh HM, Nayfeh MH (1995) Re-establishment of photoluminescence in Cu quenched porous silicon by acid treatment. J Appl Phys 77:4130
48. Cho C-H, Aspetti CO, Park J, Agarwal R (2013) Silicon coupled with plasmon nanocavities generates bright visible hot luminescence. Nat Photonics 7:285–289
49. Gu Z, Liu S, Sun S, Wang K, Lyu Q, Xiao S, Song Q (2015) Photon hopping and nanowire based hybrid plasmonic waveguide and ring-resonator. Nat Sci Rep 5:917
50. Piccione B, Aspetti CO, Cho C-H, Agrawal R (2014) Tailoring light–matter coupling in semiconductor and hybrid-plasmonic nanowires. Rep Prog Phys 77:086401
51. Nicholaides C, Nayfeh MH, Clark CW (eds) (1989) Atoms in strong fields. Plenum, New York
52. Glab W, Nayfeh MH (1985) Stark induced resonances in the photoionization of hydrogen. Phys Rev A 31:530–532
53. Nedyalkov N, Imamova S, Atanasov P, Obara M (2010) Gold nanoparticles as nanoheaters and nanolenses in the processing of different substrate surfaces. J Phys Conf Ser 223:012035
54. Clery D (1992) Nanotechnology rules, OK! New Sci 1811:42
55. Yau S-T, Saltz D, Nayfeh MH (1990) Laser-assisted deposition of nanometer structures using scanning tunneling microscopy. Appl Phys Lett 57:2913
56. Hurst GS, Nayfeh MH, Young JP (1977) One-atom detection using resonance ionization spectroscopy. Phys Rev A15:2283
57. Lubkin G (1977) Resonance electron spectroscopy detects single atoms. Phys Today 30:17
58. Quanta (1977) One-atom chemistry. The Sciences. The New York Academy of Sciences, Quanta, p 5, May/June1977
59. Stupca M, Alsalhi M, Al Saud T, Almuhanna A, Nayfeh MH (2007) Enhancement of polycrystalline silicon solar cells using ultrathin films of silicon nanoparticle. Appl Phys Lett 91:063107
60. Maximenko Y, Elhalawany N, Yamani Z, Yau S-T, Nayfeh MH (2013) Polyaniline – Si nanoparticles nanocapsules as a dual photovoltaic sensitizer. Mater Res Soc Symp Proc 1500
61. Chowdhury Fl, Nayfeh MH, Nayfeh AM (2016) Enhancened performance of thin film silicon solar cells with a top film of silicon nanoparticles due to down-conversion and near resonance charge transport. J Sol Energy 125:332–338
62. Nayfeh M, Rao S, Nayfeh O, Smith A, Therrien J (2005) UV photodetectors with thin film Si nanoparticle active mediaum. IEEE Trans Nanotechnol 4:660
63. Magill S, Xie J, Nayfeh M, Yu H, Fizari M, Malloy J, Maximenko Y (2015) Enhanced UV light detection using wavelength-shifting properties of Silicon nanoparticles. J Instrum 10: P05008
64. Jee S-W, Zhou K, Kim D-W, Lee J-H (2014) A silicon nanowire photodetector using Au plasmonic nanoantennas. Nano Convergence 1:29
65. Qiu T, Wu XL, Mei YF, Wan GJ, Chu PK, Siu GS (2005) From Si nanotubes to nanowires: synthesis, characterization, and self-assembly. J Cryst Growth 277:143

66. Mantey K, Shams S, Nayfeh MH, Nayfeh O, Alhoshan M, Alrokayan S (2010) Synthesis of wire-like silicon nanostructures by dispersion of SOI using electroless etching. J Appl Phys 108:124321

67. Catchpole KR, Polman A (2008) Plasmonic solar cells. Opt Express 16: 21793–21800. ▶ http://www.opticsinfobase.org/oe/abstract.cfm?URI=oe-16-26-21793

68. Song H, Lee S (2007) Red light emitting solid state hybrid quantum dot–near-UV GaN LED devices. Nanotechnology 18:255202

69. Stupca M, Nayfeh OM, Hoang T, Nayfeh MH, Alhreish B, Boparai J, Aldwayyan A, AlSalhi M (2012) Silicon nanoparticle-ZnS nanophosphors for UV- based white LED. J Appl Phys 112:074313

70. Kirui DK et al (2013) Targeted near-IR hybrid magnetic nanoparticles for in vivo cancer therapy and imaging. Nanomed Nanotechnol Biol Med 9:702–711

71. Kang B, Mackey MA, El-Sayed MA (2010) Nuclear targeting of gold nanoparticles in cancer cells induces DNA damage, causing cytokinesis arrest and apoptosis. J Am Chem Soc 132:1517–1519

72. Chen J, Keltner L, Christophersen J, Zheng F, Krouse M et al (2002) New technology for deep light distribution in tissue for phototherapy. Cancer J 8:154–163

73. Angelini I, Artioli G, Bellintani P, Diella V, Gemmi M, Polla A, Rossi A (2004) Chemical analyses of bronze age glasses from Frattesina di Rovigo, northern Italy. J Archaeol Sci 31:1175–1184

74. Artioli G, Angelini I, Polla A (2008) Crystals and phase transitions in protohistoric glass materials. Phase Transit 81:233–252

75. Colomban P, March G, Mazerolles L, Karmous T, Ayed N, Ennabli A, Slim H (2003) Raman identification of materials used for jewelry and mosaics in Ifriqiya. J Raman Spectrosc 34:205–213

76. Freestone I, Meeks N, Sax M, Higgitt C (2007) The Lycurgus cup - a Roman nanotechnology. Gold Bull 40:270–277

77. Mallmann M (2008) Medieval stained glass window. Science in School. Courtesy: NanoBioNet and, ▶ www.nano.gov. ▶ http://www.scienceinschool.org/2008/issue10/nanotechnology

78. Mirguet C, Roucau C, Sciau P (2009) Transmission electron microscopy a powerful means to investigate the glazed coating of ancient ceramics. J Nano Res 8:141–146

79. Chabanne D (2005) Le décor de lustre métallique des céramiques glaçurées (IXème-XVIIème siècles), Matériaux, couleurs et techniques. PhD thesis, University Bordeaux 3

80. Bobin O, Schvoerer M, Miane JL, Fabre JF (2003) Colored metallic shine associated to luster decoration of glazed ceramics: a theoretical analysis of the optical properties. J Non Cryst Solids 332:28–34

81. Colomban P (2009) The use of metal nanoparticles to produce yellow, red and iridescent color, from bronze age to present times in luster pottery and glass: solid state chemistry, spectroscopy and nanostructure. J Nano Res 8:109–132

82. Lafait J, Berthier S, Andraud C, Reillon V, Boulenguez J (2009) Physical colors in cultural heritage: surface plasmons in glass. C R Phys 10:649–659

83. Cizer S (2010) History, technique and art of Luster. Dokuz Eylul University, Narhdere, Izmir

84. Roqué J, Molera J, Cepria G, Vendrell-Saz M, Perez-Arantegui J (2008) Analytical study of the behaviour of some ingredients used in luster ceramic decorations following different recipes. Phase Transit 81:267–282

85. Reillon V, Berthier S (2006) Modelization of the optical and colorimetric properties of lustered ceramics. Appl Phys Mater Sci Process 83:257–265

86. Sanderson S (2006) Materials: carbon nanotubes in an ancient Damascus sabre, sharpest cut from nanotube sword. Nature 444:286

87. Reibold M, Paufler P, Levin AA, Kochmann W, Pätzke N, Meyer DC (2006) Materials: carbon nanotubes in an ancient Damascus sabre. Nature 444:286

88. Simon G (1996) Vision according to Alhazen: Sciences et savoirs aux XVIe et XVIIe siècles. Presses universitaires du Septentrion: 15

89. Le Guet Tully F (2012) Brief history of astronomical optics. ▶ https://lise.oca.eu/spip.php?rubrique37

90. Mach E (2003) The principles of physical optics: an historical and philosophical treatment. ▶ https://books.google.com/books?isbn=0486495590

91. Høg E (2008) 650 Years of optics: from Alhazen to Fermat and Rømer, Astrometry and optics during the past 2000 years - arXiv.org. ▶ https://arxiv.org/pdf/1104.4554. www.astro.ku.dk/~erik/HoegAlhazen.pdf

92. Color Spaces - color phenomena (2010) ▶ www.color-theory-phenomena.nl/08.00.html

93. Toler P (2012) Alhazen: the first true scientist? Wonders and Marvels. ▶ http://www.wondersandmarvels.com/2012/08/alhazan-the-first-true-scientist.html

94. Mie G (1908) Beiträge zur optik trüber medien, speziell kolloidaler metallösungen. Ann Phys 25:377–445
95. Wagner H-P, Kaveh-Baghbadorani M (2015) Plasmonics: revolutionizing light-based technologies via electron oscillations in metals. ► http://phys.org/news/2015-06-plasmonics-revolutionizing-light-based-technologies-electron.html
96. Hänsch TW, Shahin IS, Schawlow AL (1971) High-resolution saturation spectroscopy of the sodium D lines with a pulsed tunable dye laser. Phys Rev Lett 27:707
97. Baer T, Kowalski FV, Hall JL (1980) Frequency stabilization of a 0.633-μm He–Ne longitudinal Zeeman laser. Appl Opt 19:3173–3177
98. Schweitzer WG Jr, Kessler EG Jr, Deslattes RD, Layer HB, Whetstone JR (1973) Description, performance, and wavelengths of iodine stabilized lasers. Appl Opt 12:2927
99. Javan A, Herriott D, Bennett W (1961) Population inversion and continuous optical maser oscillation in a gas discharge containing a He-Ne mixture. Phys Rev Lett 6:106
100. White AD, Rigden JD (1962) Continuous gas maser operation in the visible. Proc IRE 50:1697
101. Vasilenko LS, Chebotayev VP, Shishaev AV (1970) JETP Lett 12:113
102. Stroke HH (ed) (1995) The physical review: the first 100 years: a selection of seminal papers and commentaries. Springer/AIP press, New York
103. Nayfeh MH (1974) Precision measurement of the Rydberg by saturated spectroscopy. PhD thesis, Stanford University
104. Hänsch TW, Nayfeh MH, Lee SA, Curry SM, Shahin IS (1976) Precision measurement of the Rydberg constant by laser saturation spectroscopy of the Balmer line in hydrogen and deuterium. Phys Rev Lett 32:1336
105. Zewail AH (1990) The birth of molecules. Sci Am 263:76
106. Reiter F, Graf U, Serebryannikov EE, Schweinberger W, Fiess M, Schultze M, Azzeer AM, Kienberger R, Krausz F, Zheltikov AM, Goulielmakis E (2010) Route to attosecond nonlinear spectroscopy. Phys Rev Lett 105:243902–243904
107. UNESCO (1976) Impact Sci Soc 26–27:140
108. Focus/Feature (2015) Nano-optics gets practical. Nat Nanotechnol 10:11–15

10

Optics and Renaissance Art

Charles M. Falco

C.M. Falco (✉)
College of Optical Sciences, University of Arizona, Tucson, AZ 85721, USA
e-mail: falco@email.arizona.edu

© The Author(s) 2016
M.D. Al-Amri et al. (eds.), *Optics in Our Time*, DOI 10.1007/978-3-319-31903-2_11

11.1 Introduction

An extensive visual investigation by the artist David Hockney [1] lead to the discovery of a variety of optical evidence in paintings as described in a number of technical papers [2–8]. This work demonstrated European artists began using optical devices as aids for creating their work early in the Renaissance well before the time of Galileo. These discoveries show that the incorporation of optical projections for producing certain features coincided with the dramatic increase in the realism of depictions at that time. Further, it showed that optics remained an important tool for artistic purposes continuing until today.

Our earliest evidence of the use of optical projections is in paintings of Jan van Eyck and Robert Campin in Flanders c1425, followed by artist including Bartholome Bermejo in Spain c1474, Hans Holbein in England c1530, and Caravaggio in Italy c1600, to name a few. Significantly, the optical principles of the camera obscura were described the eleventh century Arab scientist, philosopher, and mathematician, Abu Ali al-Hasan ibn al-Haytham, known in the West as Alhazen or Alhacen (b.965 Basra d.1039, Cairo). This is important for the present discussion because by the early thirteenth century al-Haytham's writings on optics had been translated into Latin and incorporated in the manuscripts on optics of Roger Bacon (c1265), Erasmus Witelo (c1275), and John Peckham (c1280).

Concurrent with the growing theoretical understanding optics were practical developments, such as the invention of spectacles in Italy around 1276. Pilgrims carried small convex mirrors into cathedrals to use as wide-angle optics to enable a much larger area of the scene to be visualized, showing how common the uses of optics had become by this time. As described below, evidence within paintings shows that at some point during this period someone realized replacing the small opening in a *camera obscura* with a lens resulted in a projected image that was both brighter and sharper. One lens from a pair of reading spectacles allows projection of images of the size, brightness, and sharpness necessary to be useful to artists, although with the optical "artifact" of having a finite depth of field (DOF). It is important to note that concave mirrors also project images, but with the advantage for an artist that they maintain the parity of a scene. For this reason it seems likely that, at least in the initial period, artists used them rather than refractive lenses.

The earliest visual depiction of lenses and concave mirrors of which I am aware are in Tomaso da Modena's 1352 paintings of "Hugh of Provence" and "Cardinal Nicholas of Rouen."[1] Either the spectacles or the magnifying glass in these paintings would have projected an image useful for an artist. His "St. Jerome" and "Isnardo of Vicenza" both show concave mirrors as well. This shows that the necessary optics to project images of the size and quality needed by artists were available 75 years before the time of Jan van Eyck.

The examples in what follows are selected from several well-known European artists. As will be shown, in each case features are shown in portions of their works that are based on optical projections.

---

1   These paintings are located in the Chapter House of the Seminario building of the Basilica San Nicolo in Treviso, Italy.

11.2 Analysis of Paintings

11.2.1 Jan van Eyck, *The Arnolfini Marriage*, 1434

One of the earliest examples we have found of a painting that exhibits a variety of evidence that the artist-based portions of it on optical projections is shown in ◘ Fig. 11.1. Several different types of optical analysis demonstrate the chandelier, enlarged in ◘ Fig. 11.2, is based on an optical projection.

The advantage of an optical projection of a real chandelier for an artist even of the skill of van Eyck is it would have allowed him to mark key points of the image. Even without tracing most of the image this would have enabled him to obtain the level of accuracy seen for this complex object that never had been previously achieved in any painting. The use of a lens results in an optical base for certain of the features, even though a skilled artist would not have needed to trace every detail in order to produce a work of art even as convincing as this one.

Since an optical projection only would be useful for certain features of any painting, and not for others, it is important to analyze appropriate aspects of the chandelier to determine whether or not they are based on optical projections. After establishing an optical base it would have been easier for van Eyck to "eyeball" many of the features [1]. As a result, paintings like the *Arnolfini Marriage* are collages consisting of both optical and non-optical elements, with even the optical elements containing eyeballed features as well [1]. Another important point is that all paintings of three-dimensional objects reduce those objects to two dimensions and, in doing so, lose some of the spatial information.

Elsewhere, based on the size of the candle flame, we estimated the magnification of the chandelier is 0.16 [6]. This means the outer diameter of the original chandelier was approximately 1 m which is consistent with the sizes of surviving chandeliers of that period. This magnification is small enough that the DOF for a lens falling within any reasonable range of focal lengths and diameters would be over 1 m. Because of this, van Eyck would have seen the entire depth of the real chandelier in the projected image without needing to refocus. Hence, if based on an optical projection the positions of the tops of each of the six candle holders should exhibit something close to perfect hexagonal symmetry after correcting for perspective. However, even if he had carefully traced a projected image there should be deviations from ideal symmetry due to the imperfections of any such large, hand-made object. If, instead, he had painted this complex object without the aid of a projection, and without the knowledge of analytical perspective that was only developed many decades later [9], larger deviations in the positions of these candle holders would be expected.

Marked with dots in ◘ Fig. 11.3 are the positions of the tops of each of the candle holders. The six-sided shape connecting them is an ideal hexagon that has been corrected for perspective. As can be seen, the agreement of the positions of the candle holders with the points of a perfect hexagon is remarkable. The maximum deviation of any of the candle holders from a perfect hexagon is only 7°, corresponding to the end of that half-meter-long arm being bent only 6.6 cm away from its "ideal" hexagonal position. Importantly, this analysis shows the arms are bent away from their "ideal" positions, but that none of them is either longer or shorter than the others. This is just what would be expected for a real chandelier. The deviations from perfect hexagonal symmetry are all on a circle, with the root-mean-square deviation only 4.1 cm. Although we shouldn't expect a hand-made fifteenth century chandelier to exhibit accuracy greater than this, some or all of the deviations could have resulted from slight bends during fabrication, transportation, hanging, or subsequent handling.

11

Fig. 11.1 Jan van Eyck, *The Arnolfini Marriage*, 1434

◻ **Fig. 11.2** Jan van Eyck, The Arnolfini Marriage (detail), 1434

Although the overall chandelier is three dimensional, the individual arms are two dimensional. We devised an analysis scheme based on this, as shown in ◻ Fig. 11.4 [3, 6]. In this figure we individually corrected each of the six arms of the chandelier for perspective and overlaid them to reveal similarities and differences. Where a complete arm is not shown in the figure it is because it is partially obscured by arms in front of it. While the loss of spatial information when projecting a three-dimensional object into two dimensions introduces ambiguities, the scheme we used to analyze this chandelier avoids this limitation.

After transformation of the arms to a plan view of each the main arcs are identical to within 5 % in width and 1.5 % in length. That they are the same length is consistent with our independent analysis of the radial positions of the candle holders described above [6]. However, since it would have been easier for van Eyck to eyeball many aspects of this chandelier, rather than to trace the entire projected image, it is not surprising that there are variations in the positions of the decorative features attached to those arcs.

From this evidence and other that we published [1–3, 6] we can conclude with a high degree of confidence that van Eyck's chandelier is based on an optical projection of a real chandelier. Further, the small differences provide insight into the artistic choices van Eyck made to deviate from simply tracing the projection. However, the most important point is that the unprecedented realistic perspective of this complex object is a result of an optical projection that was made over a century earlier than previously thought possible [9].

11

🔹 **Fig. 11.3** *The Arnolfini Marriage* (detail). As can be seen, a perspective-corrected hexagon fits the positions of the tops of the candle holders to a remarkable accuracy, with small deviations from ideal symmetry consistent with a large, hand-made fifteenth century object

11.2.2 **Lorenzo Lotto, *Husband and Wife*, 1523–1524**

"Family Portrait" by Lorenzo Lotto (1523–1524) shown in 🔹 Fig. 11.5 provides considerable quantitative information about the lens that optical evidence indicates Lotto used in creating this painting. 🔹 Figure 11.6 is a detail from *Husband and Wife* showing an octagonal pattern on an oriental carpet that appears to go out of focus at some depth into the painting. Overlaid on this painting are three segments of a perspective-corrected octagon whose overall fit to the pattern is seen to be excellent, and whose quantitative details we calculate below.

As we have shown elsewhere [3, 6], based on the scale of the woman in the painting the magnification is approximately $M = 0.56$. Any optical projection at such a high magnification intrinsically has a relatively shallow DOF, the value of which depends on the focal length and diameter of the lens as well as the magnification. To change the distance of sharp focus requires physically moving the lens with respect to the subject and the image plane. To refocus an image on a region further into a scene from its original plane of focus requires moving the lens further away from the scene. This movement of the lens to refocus results in a small decrease in the magnification of the projected scene, as well as in a slight change in the vanishing points. Although such effects are fundamental characteristics of images projected by lenses, they are extremely unlikely to occur in a painting if an artist had instead laid out patterns using sighting devices or

■ **Fig. 11.4** This figure contains the outlines of all six arms on the chandelier after correcting for perspective with the arms to the viewer's right flopped horizontally to overlay on the arms to the left. The main arc of all six arms is the same to within 1.5 % in length and 5 % in width. Variations are consistent with the decorative features on the main arc having been hand attached to the original chandelier as well as having been eyeballed when creating the painting

following geometrical rules first articulated in the fifteenth century [9]. Since we already have discussed several aspects of this painting elsewhere, here we summarize our previous analysis [3, 6–8].

The distance across the wife's shoulders in the painting, compared with measurements of real women, provides an internal length scale that lets us determine the magnification to be $M \approx 0.56$. This in turn allows us to determine the repeat distance of the triangular pattern on the actual carpet to be 3.63 cm. Since the first place where the image of the carpet changes character is approximately 4–5 triangular-repeats into the scene, we calculate the depth of field to be DOF = 16 ± 1.5 cm. We now can use geometrical optics to extract quantitative information from this painting.

The focal length (FL) and magnification (M) are given by the following equations from geometrical optics: [3]

$$1/FL = 1/\left(d_{\mathrm{lens\,B\,subject}}\right) \,+\, 1/\left(d_{\mathrm{lens\,B\,image}}\right) \tag{11.1}$$

and

$$M \;=\; \left(d_{\mathrm{lens\,B\,image}}\right)/\left(d_{\mathrm{lens\,B\,subject}}\right) \tag{11.2}$$

11

◘ Fig. 11.5 Lorenzo Lotto, *Husband and Wife*, c1523–c1524

As indicated by the overlays in ◘ Fig. 11.6, there are three regions of this octagonal pattern. These regions are the result of Lotto having refocused twice as he exceeded the DOF of his lens. We label these Regions 1, 2, and 3, with Region 1 the closest to the front of the painting. Thus, for the first two Regions,

$$1/\mathrm{FL} = 1/\left(d_{\mathrm{lens\,B\,subject1}}\right) + 1/\left(d_{\mathrm{lens\,B\,image1}}\right) \tag{11.3}$$

and

$$1/\mathrm{FL} = 1/\left(d_{\mathrm{lens\,B\,subject2}}\right) + 1/\left(d_{\mathrm{lens\,B\,image2}}\right) \tag{11.4}$$

However, the measured DOF is 16 ± 1.5 cm, so for Region 2

$$d_{\mathrm{lens\,B\,subject2}} \approx d_{\mathrm{lens\,B\,subject1}} + 16\ \mathrm{cm} \tag{11.5}$$

and thus

$$1/\mathrm{FL} = 1/\left(d_{\mathrm{lens\,B\,subject1}} + 16\ \mathrm{cm}\right) + 1/\left(d_{\mathrm{lens\,B\,image2}}\right) \tag{11.6}$$

Because Region 2 is further into the scene it is at a slightly lower magnification than is Region 1 so its DOF will be somewhat larger than 16 cm. We can calculate DOF_2 from

▣ Fig. 11.6 *Husband and Wife* (detail). The overlays are perspective-corrected sections of an octagonal pattern that we fit to the painting. As described in the text, the details of this portion of the painting are in excellent qualitative and quantitative agreement with the three-segment, perspective-corrected octagon that is predicted by the laws of geometrical optics for such a projected image

$$\text{DOF}_2 = 2\,C \times f\# \times (1 + M_2)/\,M_2{}^2 \tag{11.7}$$

where C is the circle of confusion, $f\#$ is the lens diameter/focal length, and M_2 is the magnification of Region 2. Hence,

$$\text{DOF}_2 = \text{DOF}_1 \times (1 + M_2)\,/\,(1 + M_1) \times (M_1/M_2)^2 \tag{11.8}$$

Region 3 of the pattern thus starts at a depth of 16 cm + DOF$_2$ into the scene, so

$$d_{\text{lens B subject3}} = d_{\text{lens B subject1}} + 16\ \text{cm} + \text{DOF}_2 \tag{11.9}$$

and

$$1/\text{FL} = 1/\!\left(d_{\text{lens B subject1}} + 16\ \text{cm} + \text{DOF}_2\right) + 1/\!\left(d_{\text{lens B image3}}\right) \tag{11.10}$$

The magnifications M of the three regions are given by:

$$0.56 = d_{\text{lens B image1}}\,/\,d_{\text{lens B subject1}} \tag{11.11}$$

$$M_2 = d_{\text{lens B image2}}\,/\,\left(d_{\text{lens B subject1}} + 16\ \text{cm}\right) \tag{11.12}$$

$$M_3 = d_{\text{lens B image3}}\,/\,\left(d_{\text{lens B subject1}} + 16\ \text{cm} + \text{DOF}_2\right) \tag{11.13}$$

11

This analysis gives us seven Eqs. (11.3), (11.6), (11.8), (11.10), (11.11), (11.12), and (11.13) and eight unknowns: FL, $d_{lensBsubject1}$, $d_{lensBimage1,2,3}$, DOF_2, $M_{1,2}$. If we make a single assumption about any one of these unknowns we can then solve these equations uniquely for the other seven unknowns using simple algebra. Assuming that the distance from the lens to the carpet was at least 1.5 m, but not greater than 2.0 m (i.e., $d_{lensBsubject1} = 175 \pm 25$ cm) we find

focal length $= 62.8 \pm 9.0$ cm
$M_2 = 0.489 \pm 0.9$
$M_3 = 0.423 \pm 1.5$

The magnification when moving from Region 1 to Region 2, as measured from our fit of a perspective-corrected octagon, decreases by 13.1 % from the original 0.56 of the painting, in excellent agreement with the -12.6 ± 1.5 % calculated from the above equations. Similarly, the measured magnification decreases by a further 13.3 % when going to Region 3, again in excellent agreement with the calculated value of -13.5 ± 1.6 %.

From Eq. (11.7),

$$f\# = \left[DOF_1 \times M_1{}^2\right] / \left[2\, C\, (1 + M_1)\right]$$

If we assume the simple lens available to Lotto resulted in a circle of confusion on the painting of 2 mm, we find $f\# \approx 22$, and hence a diameter of 2.9 ± 0.4 cm. As we have confirmed with our own experiments, a lens or concave mirror with these properties projects a quite useful image of a subject that is illuminated by daylight.

To summarize, using only the measured magnification of this painting (0.56, i.e., roughly half life size, as determined from the size of the wife), and making a reasonable assumption about the distance Lotto would have positioned his lens from the carpet (175 ± 25 cm), equations from geometrical optics uniquely determine both changes in magnification, −13.1 and −13.3 %, of the central octagonal pattern, as well as the focal length and diameter of the lens, 62.8 ± 9.0 cm and ~3 cm, respectively, used to project this image. The three sets of vanishing points exhibited by the octagonal pattern, as well as the depths into the painting where they occur, are a direct consequence of the use of a lens to project this portion of the painting. Other quantitative information extracted from this painting is discussed elsewhere [3, 6–8].

Recently we developed a portable high-resolution digital camera that allows us to acquire important information about paintings without needing to remove them from museums for detailed study [7]. Since infrared light penetrates many pigments further than does visible light it often can be used to reveal "underdrawings" or other features not apparent in the visible [10, 11]. ◘ Figure 11.7 is an infrared (IR) "reflectogram" of the Lotto painting captured in situ where it was located on the wall of the Hermitage State Museum in St. Petersburg. Although many features are revealed in this image, one immediate observation is we can see that Lotto used a different pigment for the woman's dark dress than he used for the man's jacket. This provides us with previously unknown information about the artist's working technique.

◘ Figure 11.8 shows the octagonal pattern of the table covering in greater detail. As can be seen by comparison with ◘ Fig. 11.5, the red and yellow pigments Lotto used are largely transparent in the IR so this image provides an uncluttered view of the black lines he used to create this feature on the painting.

Three distinct types of markings can be clearly seen for the lines making up the triangular pattern of the border of this feature. Well-defined lines are in the region nearest the front of the image, consistent with tracing a projected image. These "traced" lines abruptly change to tentative lines in the middle region, at just the

■ **Fig. 11.7** Lorenzo Lotto, Husband and Wife, c1523–c1524. Infrared (IR) reflectogram

depth into the scene where our previous analysis showed the magnification was reduced by 12.6 ± 1.5 % due to having to refocus because of exceeding the depth-of-field. Because of this, Lotto faced significant difficulty to create a plausible match for this geometrical pattern after refocusing. His abrupt change to tentative lines reflects this difficulty. After re-establishing a plausible freehand sketch form of the geometrical pattern by the rear of this central region, the quality of the lines again abruptly changes to only short dashes in the region farthest into the scene, where our previous analysis shows the magnification was reduced by an additional 13.5 ± 1.6 % due to having to refocus a second time after again reaching the limit of the depth-of-field. These results from IR reflectography provide important insights into the actual working practices of an artist, revealing quite specific details about how he made use of projected images 75 years prior to the time of Galileo.

Our analysis of this painting found a change in the vanishing point that takes place part way back in the pattern in the border of the carpet to the right, quantitatively consistent with the change that is caused by the shift in position of a lens as it is refocused. ■ Figure 11.9 shows the IR reflectogram of this portion of the painting. Overlaid to the left are seven units of a perfectly repeating structure that replicates the geometrical pattern of the border. As can be seen, after

11

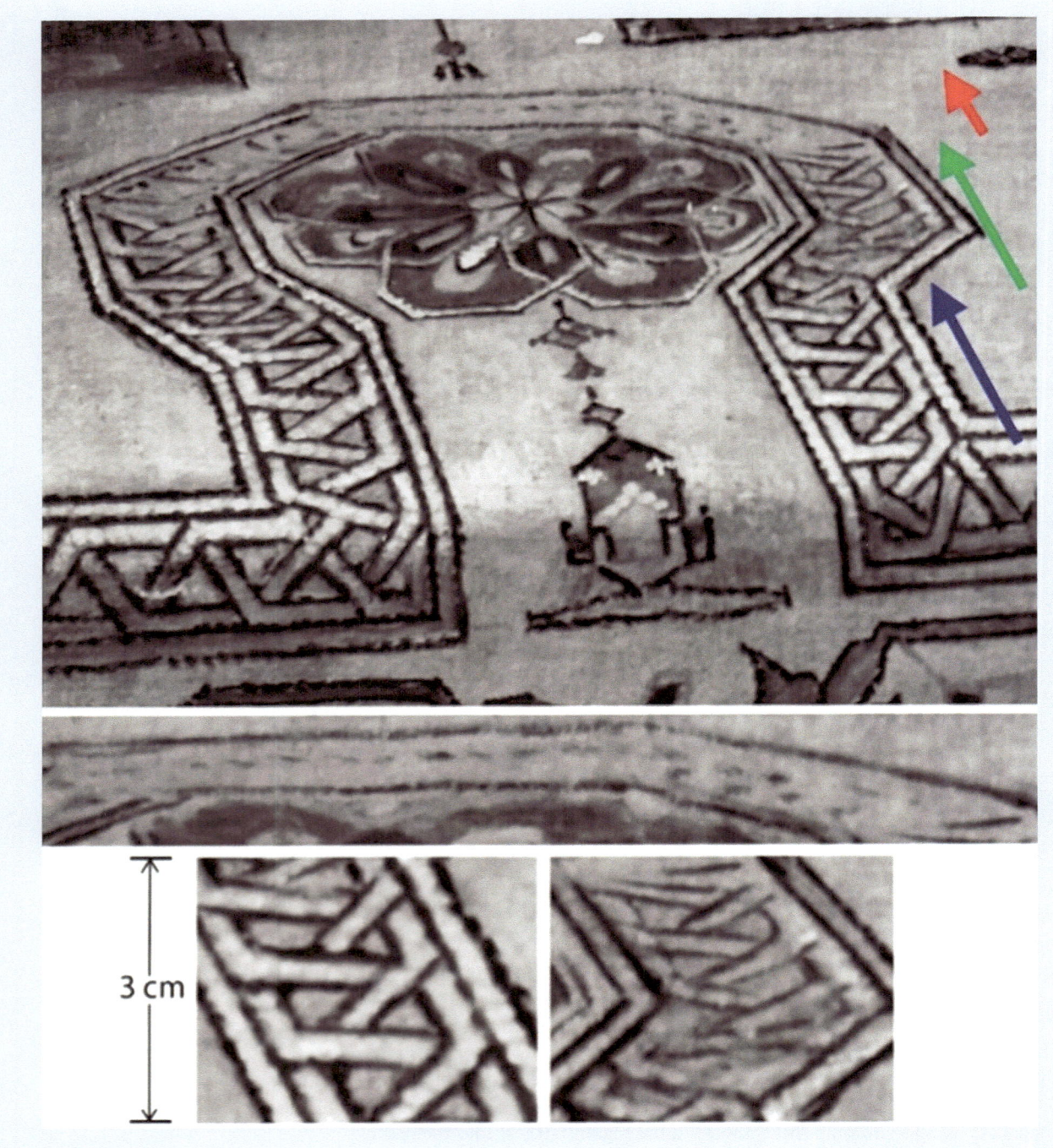

Fig. 11.8 Shows the octagonal pattern of the table covering in greater detail. As can be seen by comparison with ▪ Fig. 11.5, the *red* and *yellow* pigments Lotto used are largely transparent in the IR, providing us with a clear view of the black lines he used to create this feature on the painting

correcting for perspective, this structure is an excellent fit to the repeating pattern near the front of the carpet. The maximum deviation from a "perfect" fit is consistent with the degree of perfection found in the hand-made carpets of this type. Although an eighth unit of the structure does not fit at all, a small change in optical perspective makes the same repeating structure fit at the rear, again to within better than 2 mm. This change in perspective occurs at the same depth into the painting where our previous analysis found a shift in vanishing point, as

Fig. 11.9 IR reflectogram of the border pattern of ☐ Fig. 11.5. Overlay at left is seven segments of a repeating structure. When corrected for perspective, this is seen to be an excellent fit to the pattern at the front of the table covering. Changing the perspective, as happens when a lens is moved to refocus, gives an excellent fit to the pattern at the back. The maximum deviation of the perfect repeating structure from the pattern on the painting is 2 mm

happens when a lens is repositioned to focus further into a scene. Further, not only does the perspective change where a lens would have had to have been moved to refocus, the painting is missing a half-segment of the repeating pattern at this location. This is consistent with Lotto attempting to create a plausible match between two segments of a repeating structure after refocusing had caused the magnification and perspective to change. All of these detailed findings from IR reflectography are consistent with our other work showing this portion of the painting is based on the optical projection of an actual hand-made carpet [2, 3, 6, 7].

☐ Figure 11.5 is the full image of this painting in the visible captured in situ using a standard digital camera with a 35 mm f/2 lens. This image reveals some of the difficulties with in situ image capture in a museum environment. The painting was illuminated by a combination of indirect sunlight from windows to the left, and overhead tungsten lights, each having its own color temperature. The shadows visible along the left and top borders were cast by the ornate frame in which the painting is mounted. The roughly equal darkness of these shadows indicates that the level of illumination from both types of sources was approximately equal. However, closer inspection shows that the illumination across the surface of the painting is not uniform. This can be most easily seen in the region of the man's chest, which is too bright due to a partial specular reflection of one of the light sources that could not be eliminated by repositioning the camera within the constraints of the room.

Figure 11.7 is an IR reflectogram of the full 96 × 116 cm painting, captured under the less than ideal lighting conditions described in the previous paragraph. Although many features are revealed by this IR reflectogram, one immediate observation is that Lotto used a different pigment for the woman's dress than he used for the man's jacket, providing us with previously unknown information about the artist's working technique.

Again, all of these detailed findings from IR reflectography are consistent with our earlier work that showed this portion of the painting is based on the optical projection of an actual hand-made carpet. I note that we have used fourteenth century optical technology (i.e., one lens of a pair of reading spectacles, as well as a metal concave mirror we fabricated following descriptions in texts of the time) to accurately reproduce all of the effects we have found in this carpet, as well as in all of the other paintings we have shown to contain elements based on optical projections, including projecting such patterns directly on a screen of the same shade of red used in this painting. Even on such a colored screen, the projected images are quite distinct and easy to trace.

11.2.3 Hans Holbein the Younger, *The French Ambassadors to the English Court*, 1532

A prominent feature of *The French Ambassadors to the English Court* by Hans Holbein is the anamorphic skull at the bottom of the 1532 painting. This feature is shown in Fig. 11.10. The way this appears to someone viewing it at a grazing angle is shown by linearly compressing it by 6× in Fig. 11.11 (Right), with a real skull for comparison in Fig. 11.11 (Left).[2] Very obvious differences include that the jaw of Holbein's skull is much longer than the real skull, the slope of the top of the skull is steeper, and the eye sockets and nose are much more pronounced as well as aimed more in the direction of the viewer.

To see if optical projections may account for the appearance of this skull in the painting, we used a concave mirror of focal length 41 cm to project the image of a real skull onto a screen at a grazing angle in order to produce an anamorphic image. Figure 11.11 (Left) is a photograph of the real skull taken from precisely the location of that concave mirror after the mirror had been removed from its holder. However, because of the limited depth of focus of the projected image on

---

2   The anamorphic skull is 106 cm long and 14.4 cm high. To visually compress its length to be the same as its height so that it appears approximately like Fig. 11.11 (Right) requires viewing the painting at a grazing angle of $\sin^{-1}(14.4/106) \approx 8°$. At this angle the far end of the anamorphic feature is over 100 cm further away from the viewer than is the near end, so that for reasonable viewing distances the magnification of the far end is significantly less than that of the near end. Also, since for any reasonable viewing distance the depth of the feature is greater than the depth of field of the eye, it requires the viewer to scan back and forth through the feature, with their eyes constantly refocusing when doing so, in order to "construct" a composite image in their mind that does indeed strongly resemble Fig. 11.11 (Right). Although our analysis shows that this anamorphic feature was constructed with the aid of optical projections, the multiple positions of the lens needed to generate it, coupled with the multiple movements and refocusing of the eye needed to view it, along with the mental compositing need to construct the final image of it in the brain, results in an underlying complexity to Fig. 11.11 (Right). For these reasons, because Fig. 11.11 (Right) was generated by a linear transformation, it only approximately reproduces what the feature looks like to the viewer when examining the painting from a grazing angle.

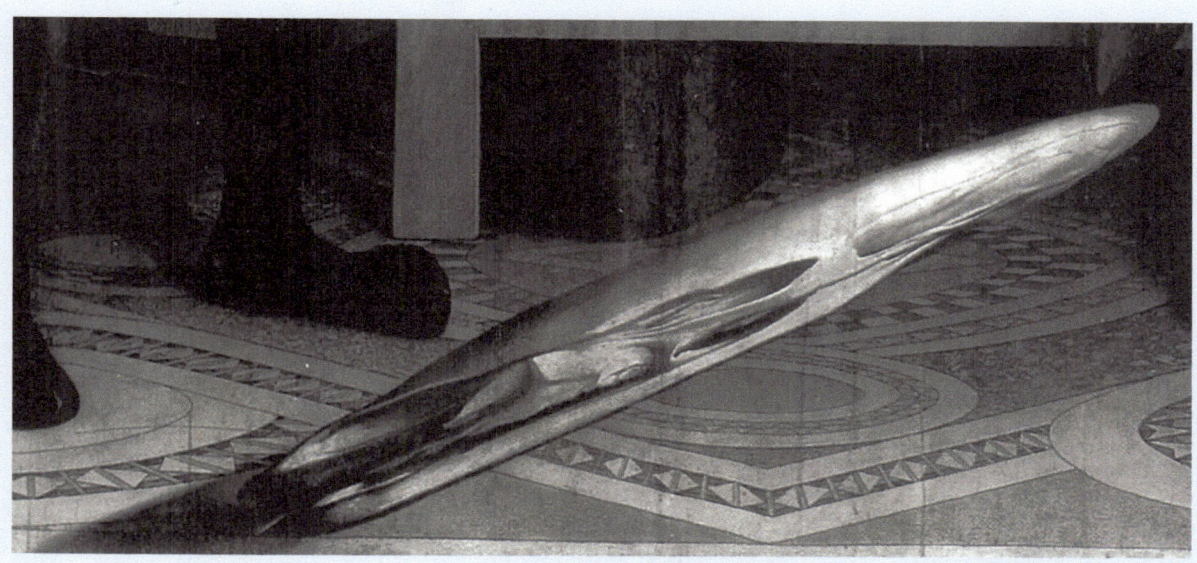

■ **Fig. 11.10** *The French Ambassadors to the English Court* (detail). This detail shows the unusual feature at the bottom of Holbein's painting. Viewed from a grazing angle to visually compress it, this feature appears as shown in ■ Fig. 11.11 (*Right*). Possibly not apparent in this small B&W reproduction is that this anamorphic skull does not occupy the same visual space as the rest of the painting

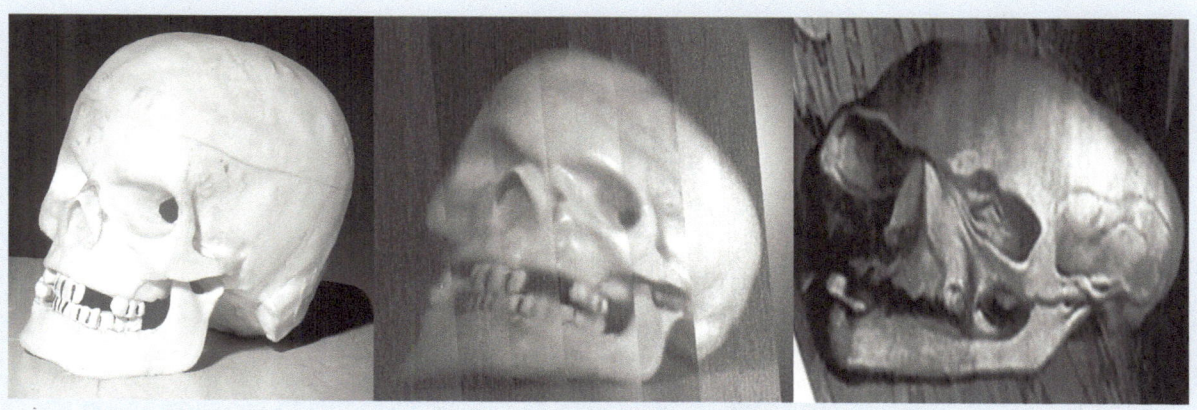

■ **Fig. 11.11** (*Left*) Photograph of a skull taken from the position of the concave mirror used to project its image onto a tilted screen to form an anamorphic image. (*Center*) Composite of the individual in-focus segments of the projected anamorphic image of the skull after linearly compressing it horizontally. (*Right*) Anamorphic skull in *The French Ambassadors* after linearly compressing it horizontally

the tilted screen, it was necessary to refocus the concave mirror a number of times in order to generate the composite anamorphic image that we have compressed linearly to produce ■ Fig. 11.7 (Center).

The segments of each of the in-focus images are visible in this composite. What is striking about ■ Fig. 11.11 (Center) is how well it reproduces the very unusual visual appearance of the linearly compressed skull from Holbein's painting. Although mathematical and graphical methods can be used to construct anamorphic images, the optics-produced composite of ■ Fig. 11.11 (Center) is far more complex than is obtained from any such construction. The magnification of each segment in the anamorphic photographic composite is linear in the vertical direction, but is proportional to 1/sin of the grazing angle in the horizontal. The overall composite of ■ Fig. 11.11 (Center) is thus the result of a nonlinear,

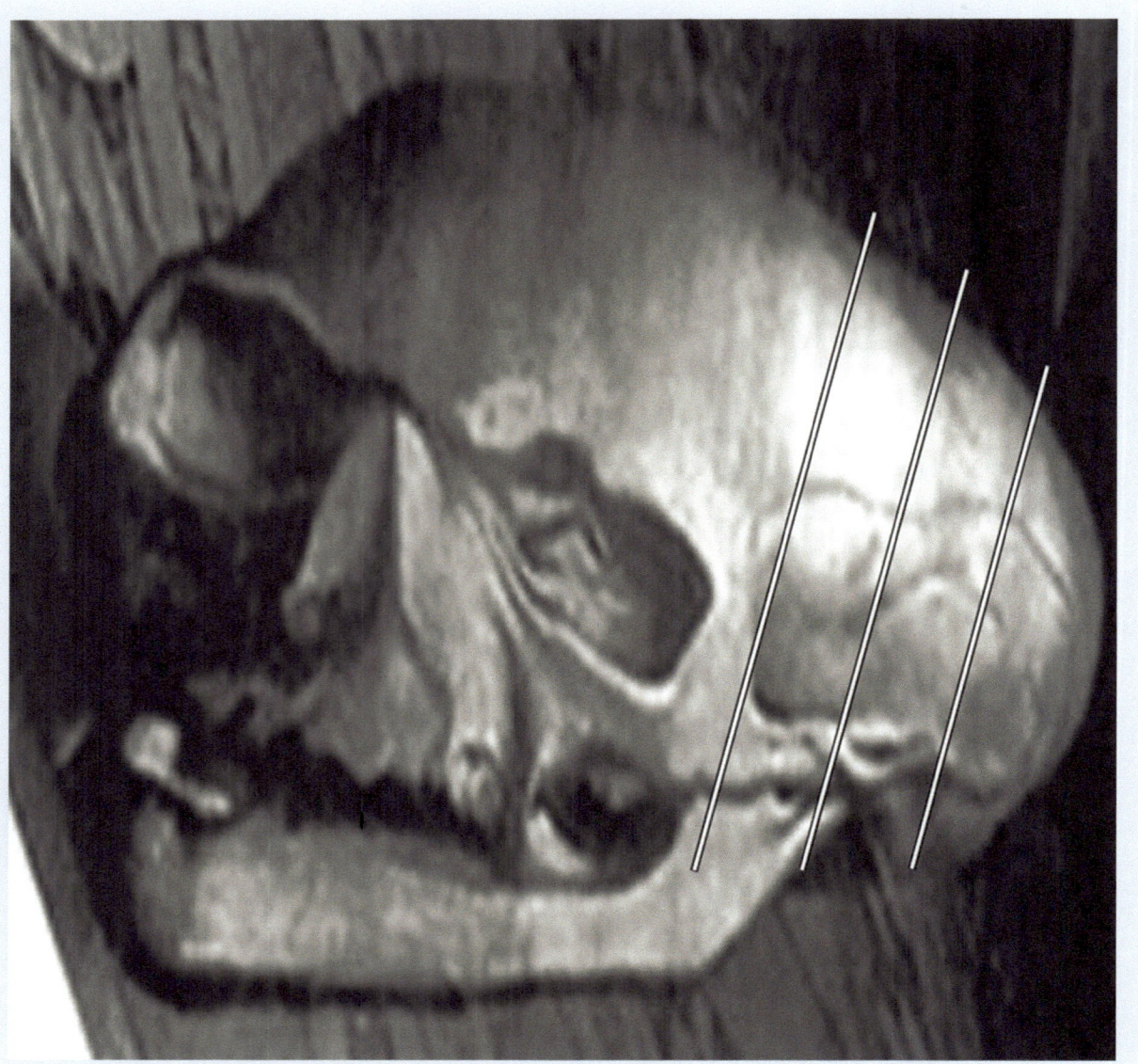

Fig. 11.12 Anamorphic skull in *The French Ambassadors to the English Court*. For this figure we have rotated the feature in ■ Fig. 11.10 clockwise by 25° and then linearly compressed it by 6×. The height of the skull in this image compared to a real one gives a magnification $M = 0.71 \pm 0.5$. The lines indicate two regions where it can be seen Holbein duplicated features (notably, the two dark depressions just above the jaw, and the double-humped line midway up the skull). A discontinuity in the slope of the top of the skull is also visible at the left edge of the leftmost marked region

piecewise-segmented transformation. Although this complex transformation was naturally produced by the optical projection, it would be quite implausible to have resulted from any sort of a graphical or mathematical construction [9]. We conclude that the probability is extremely small that Holbein could have accidentally reproduced these complex features without having projected them with a lens.

■ Figure 11.12 shows ■ Fig. 11.11 (Right) at a larger scale. Marked on this figure are two regions where we observed that Holbein has duplicated features of the skull. Because the lens and canvas (or, less likely, the skull) has to be moved a number of times when piecing together an anamorphic image from segments projected at such a high magnification, it is very easy to accidentally duplicate a region, so its presence provides additional evidence that Holbein had to refocus a lens. The duplicated segment corresponds to a region 3.0 ± 0.5 cm wide on a real skull. That same region corresponds to a width of 8.2 cm on the actual painting which gives us an approximate lower limit measure for the depth of focus. From the results of our experiments shown in ▢ Fig. 11.11 (Left) and (Center), that region of the skull is at an angle of $25° \pm 5°$ with respect to the perpendicular to the axis of the lens, so its depth into the scene is 1.3 ± 0.5 cm. Although a more accurate value for the depth of focus can be obtained by convoluting this measured DOF into the calculation, for our purposes here the approximate value 8.2 cm will suffice. Using this value, along with a circle of confusion of 2 mm and the measured $M = 0.71$, we calculate as a lower limit

$$f\# \geq \text{Depth of Focus} /[2C \times (M+1)]$$
$$= 12.0$$

Because we have neglected the DOF in the calculation shown here, this value for the $f\#$ of Holbein's lens is somewhat smaller than the actual value, as well as represents a lower limit. However, this calculation is sufficient to show that the $f\#$ of Holbein's lens is consistent with the values we obtained for Lotto's and Campin's lenses (22 and 25.2, respectively).

11.2.4 Robert Campin, *The Annunciation Triptych (Merode Altarpiece)*, c1425–c1430

Robert Campin was a contemporary of Jan van Eyck and they are documented to have known each other. The center and right panels of Robert Campin's *Merode Triptych* of c1425$_B$28 contain the earliest evidence we have found to date of the use of direct optical projections. A detail of the right panel is shown at the lower left of ▢ Fig. 11.13. As we previously showed, this portion of the painting exhibits the same complex changes in perspective seen in Lorenzo Lotto's *Husband and Wife*, resulting from Campin also having refocused his lens twice [4, 6].

The upper right in ▢ Fig. 11.13 shows one of the two sets of slats (the set that is numbered on the lower inset), with each slat individually rotated to be vertical and expanded horizontally by a factor of $3.5\times$ to accentuate any deviations from being straight. Marked on the slats are the locations of "kinks" exhibited by each of them, with those kinks connected by lines. The positions of the lines connecting the kinks are shown on the inset at the lower left. Comparing with ▢ Fig. 11.2 of Reference 5 it can be seen that the slats are kinked at the same two depths into the painting where we previously showed, with a different type of analysis using different data, that Campin had to refocus due to the DOF of his lens. Geometrical constructions can be devised which exhibit kinks, but not in the overall configuration of this painting. The complex perspective exhibited by the latticework in this portion of the painting is a direct and inevitable outcome from the DOF of a lens, but would be extremely unlikely to have resulted from any geometrical construction, or from the use of a straightedge.

Using the height of the head in the full painting as a scale, the magnification of this portion of the painting is $M \approx 0.27$. If we assume a circle of confusion of 1 mm Eq. (11.7) yields $f\# = 25.2$. We can obtain an estimate for the focal length with the assumption the lens or concave mirror had a diameter of 3 cm, in which case the focal length $\text{FL} = f\# \times 3 \text{ cm} = 76 \text{ cm}$, which is quite reasonable.

11

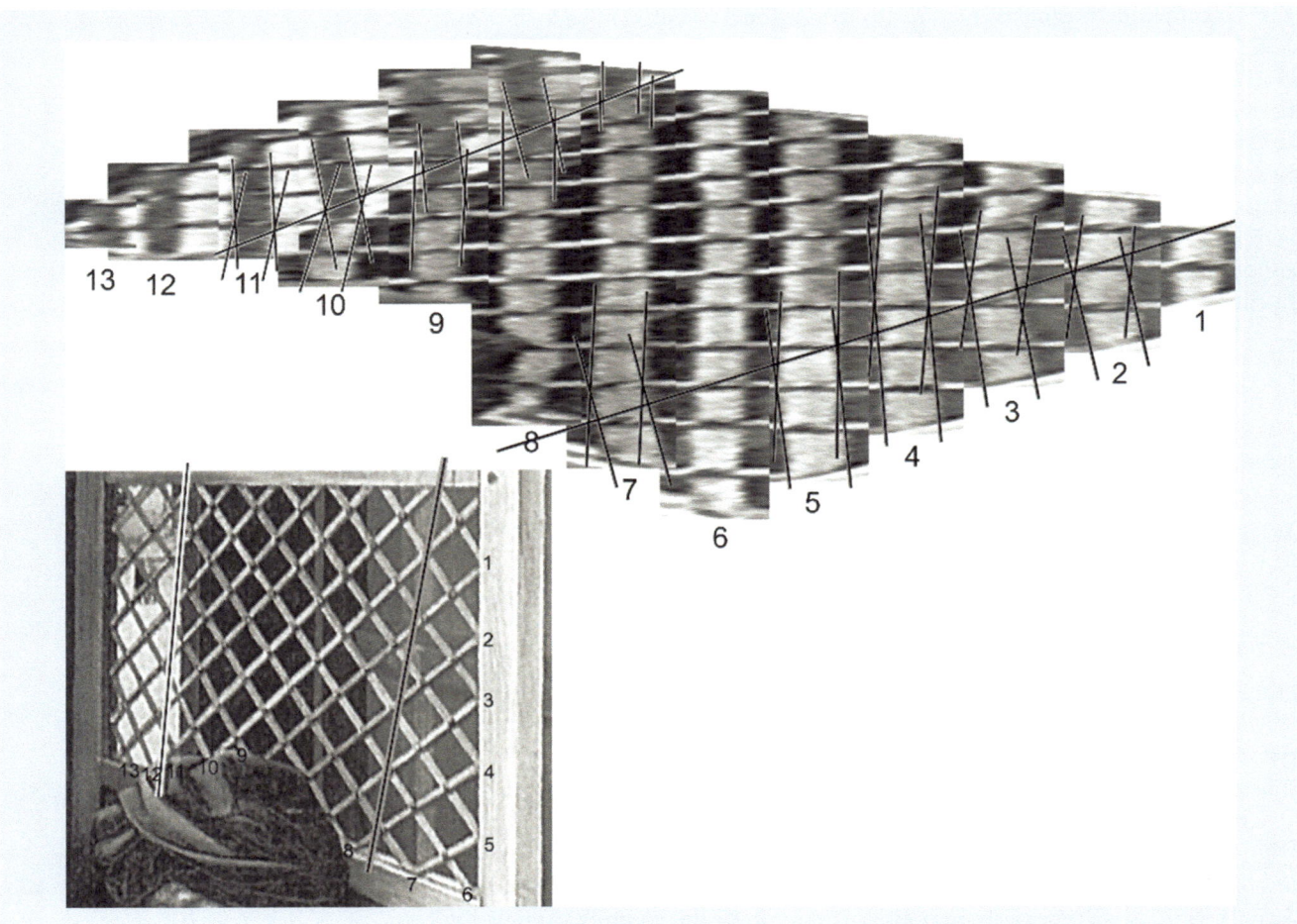

Fig. 11.13 (*Lower Left*) Detail of the *Merode Altarpiece* with one set of slats numbered. (*Upper Right*) Slats rotated to be vertical and expanded horizontally by 3.5×. We have connected the "kinks" that are apparent in the slats by lines, the locations of which are shown in the detail at the *lower left*

11.3 Conclusions

These discoveries demonstrate that highly influential European artists used optical projections as aids for producing some of their paintings early in the fifteenth century, at the dawn of the Renaissance, at least 150 years earlier than previously thought possible. In addition to van Eyck and Lotto we have also found optical evidence within works by well-known later artists including Bermejo (c1475), Holbein (c1500), Caravaggio (c1600), de la Tour (c1650), Chardin (c1750), and Ingres (c1825), demonstrating a continuum in the use of optics by artists, along with an evolution in the sophistication of that use. However, even for paintings where we have been able to extract unambiguous, quantitative evidence of the direct use of optical projections for producing certain of the features, it does not mean that these paintings are effectively photographs. Because the hand and mind of the artist are intimately involved in the creation process, understanding these images requires more than can be obtained from only applying the equations of geometrical optics. As to how information on optical projections came to these artists, evidence points to it having come via the Cairo-based scholar Ibn al Haytham [12].

11.4 Acknowledgments

I am very pleased to acknowledge my collaboration with David Hockney on all aspects of this research. Also, we have been benefited from contributions by Aimée L. Weintz Allen, David Graves, Ultan Guilfoyle, Martin Kemp, Nora Roberts (neé Pawlaczyk), José Sasián, Richard Schmidt, and Lawrence Weschler.

References

1. Hockney D (2001) Secret knowledge: rediscovering the lost techniques of the old masters. Viking Studio, New York
2. Hockney D, Falco CM (2000) Optical insights into renaissance art. Opt Photonics News 11 (7):52–59
3. Hockney D, Falco CM (2003) Optics at the dawn of the renaissance. In: Technical digest of the optical society of America, 87th annual meeting, Optical Society of America, Washington DC
4. Hockney D, Falco CM (2004) The art of the science of renaissance painting. In: Proceedings of the symposium on 'effective presentation and interpretation in museums', National Gallery of Ireland, Dublin, Ireland, p 7
5. Hockney D, Falco CM (2005) Optical instruments and imaging: the use of optics by 15th century master painters. In: Proceedings of the SPIE photonics Asia, Bellingham, Washington, vol 5638, p 1
6. Hockney D, Falco CM (2005) Quantitative analysis of qualitative images. In: Proceeding of the IS&T-SPIE electronic imaging, SPIE, Bellingham, Washington, vol 5666, p 326
7. Falco CM (2009) High resolution digital camera for infrared reflectography. Rev Sci Instrum 80:071301–071309
8. Hockney D, Falco CM (2012) The science of optics: recent revelations about the history of art. In: Proceedings of the SPIE, Bellingham, Washington, vol 8480, p 84800A
9. Kemp M (1992) The science of art. Yale University Press, New Haven, CT
10. van Asperen de Boer JR (1968) Infrared reflectography: a method for the examination of paintings. Appl Opt 7(9):1711–1714
11. Faries M (2002) Techniques and applications, analytical capabilities of infrared reflectography: an art historian s perspective. In: Barbara B et al (eds) Scientific examination of art: modern techniques in conservation and analysis. National Academy of Sciences, Washington, DC
12. Falco CM, Weitz Allen A (2009) Ibn al-Haytham's contributions to optics, art, and visual literacy. In: Beckinsale M (ed) Painted optics symposium. Fondazione Giorgio Ronchi, Florence, Italy, p 115
13. Fendrich L (2002) Traces of artistry. Chron High Educ 53(36):B20

The Eye as an Optical Instrument

Pablo Artal

P. Artal (✉)
Laboratorio de Óptica, Universidad de Murcia, Campus de Espinardo (Edificio 34), 30100 Murcia, Spain
e-mail: pablo@um.es

© The Author(s) 2016
M.D. Al-Amri et al. (eds.), *Optics in Our Time*, DOI 10.1007/978-3-319-31903-2_12

12

12.1 Introduction

For us, humans, vision is probably the most precious of senses. Although our visual system is a remarkably sophisticated part of our brain, the process is initiated by a modest optical element: the eye. ◘ Figure 12.1 shows a schematic and simplified example of the visual process. The eye forms images of the visual world onto the retina. There, light is absorbed in the photoreceptors and the signal transmitted to the visual cortex for further processing. The eye, the first element in the system, is a simple optical instrument. It is composed of only two positive lenses, the cornea and the crystalline lens, that project images into the retina to initiate the visual process. In terms of optical design complexity and compared with artificial optical systems, often formed by many lenses, the eye is much simpler. However, despite this simplicity and the relative poor imaging capabilities, the eye is adapted to the requirements of the visual system. ◘ Figure 12.2 shows a schematic illustration of the eye as compared to a photographic objective that is composed of many single lenses.

Optical systems use transparent materials as glass or plastics with refractive index selected to bend the light rays to form images. In the case of the human visual system, our eyes have to form images of a large field of view for objects placed at different distances with high resolution at least at a central area of the retina. And these tasks have to be accomplished using living tissues.

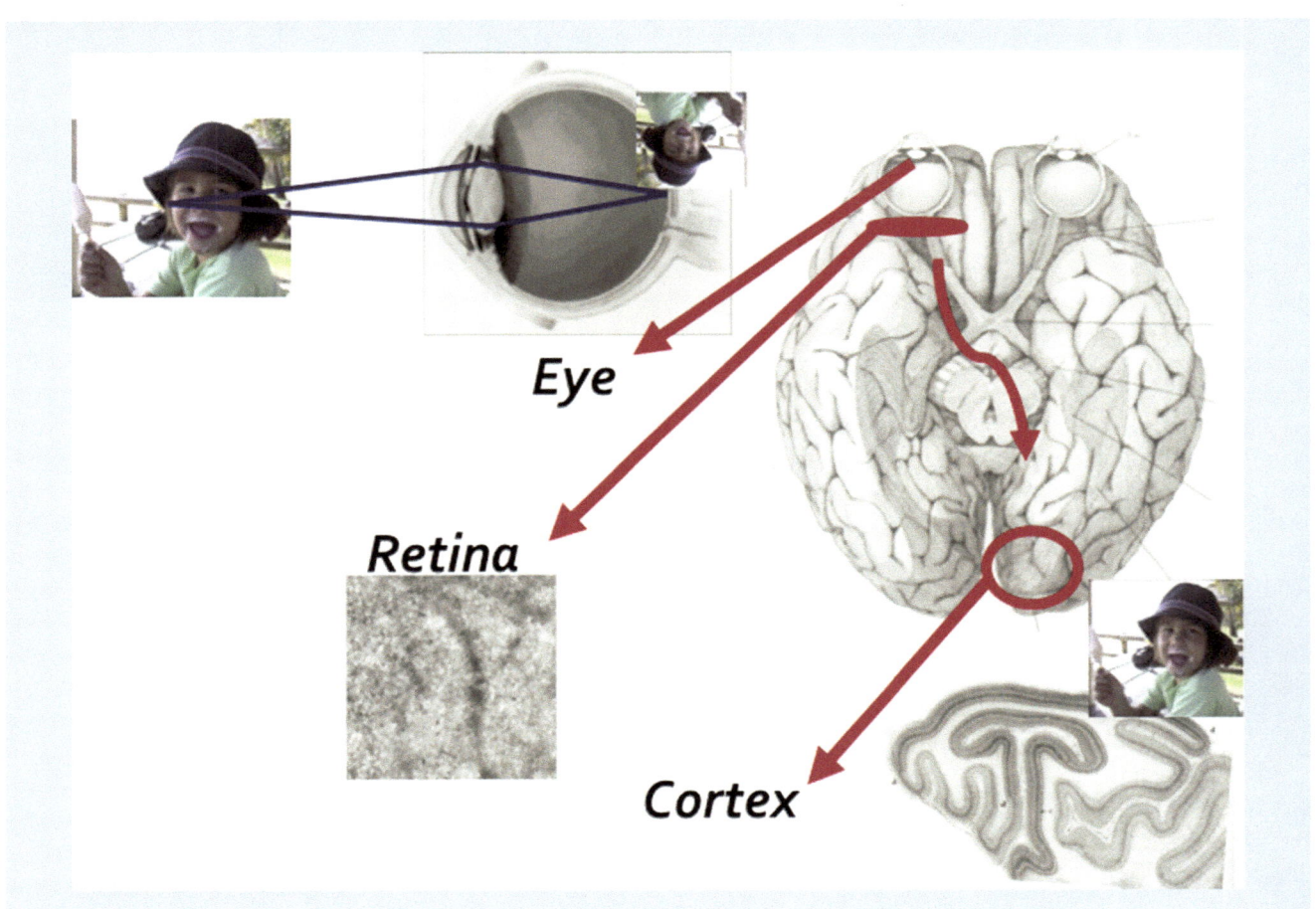

◘ **Fig. 12.1** Schematics of the visual system. The eye forms images of the world on the retina. There, optical images are sampled by the photoreceptors, converted into electrical signals and transferred to the visual cortex for further processing. The eye, although is the simplest part, since it is placed first in the visual cascade may impose fundamental limits

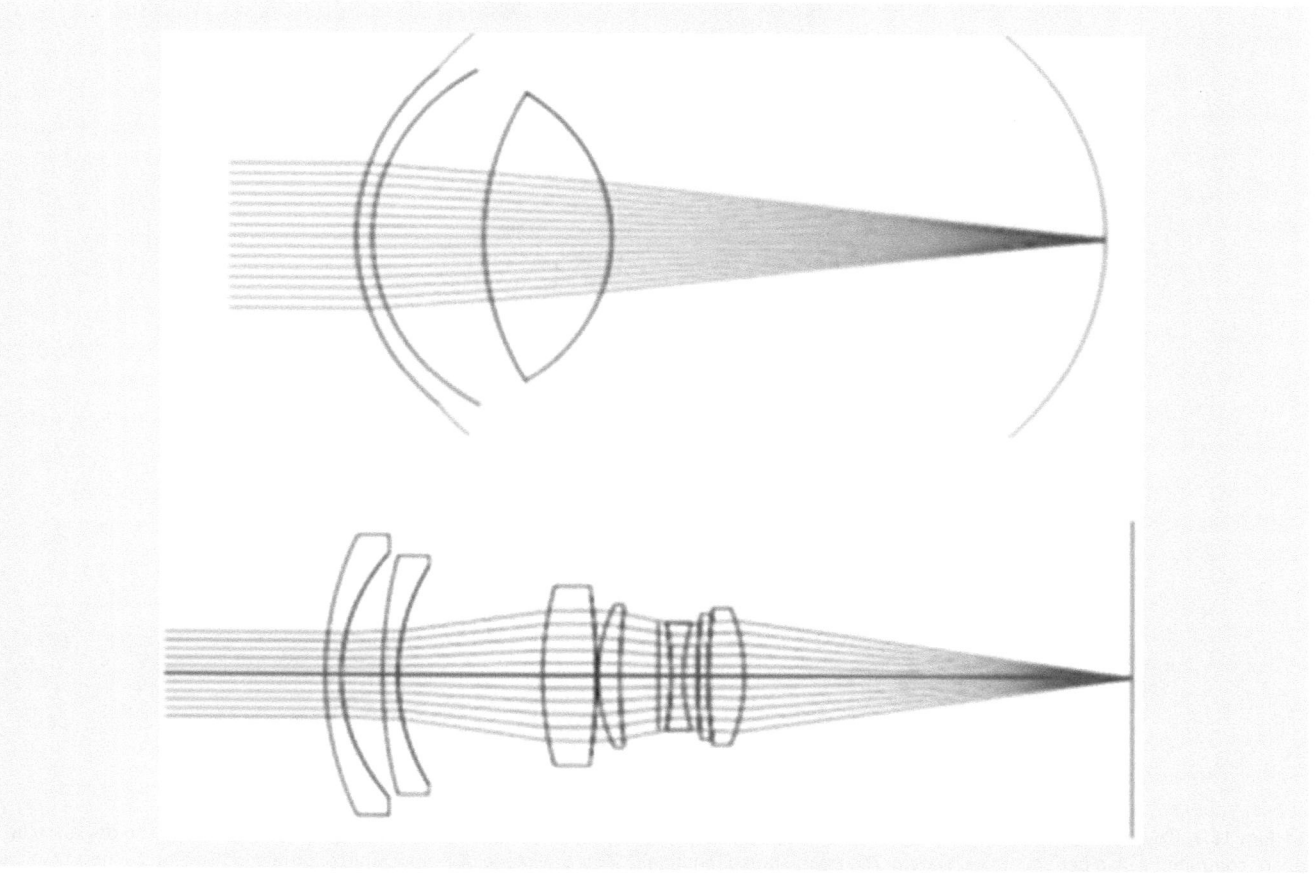

Fig. 12.2 The eye is only composed of two lenses, cornea and crystalline lens, as compared with artificial systems, such as camera objective that may have many single lenses

The eye as an optical instrument is extremely important because our vision is only good when the images formed on the retina are of sufficient quality. If the retinal images are too blurred, the visual system will not work properly. The opposite situation is not true since there are retinal and neural diseases that may impair vision even when the eye forms good quality retinal images.

The intrinsic nature of the light is somehow responsible for some of the characteristics of the eye. Or equivalently, the eye is adapted to transmit visible light and form images on the retina. The sensitivity of the retina is also optimized in the central part of visible spectrum and is similar to the solar emission spectrum. Light may be considered as a transverse electromagnetic wave. Monochromatic light waves have electric fields with sinusoidal oscillation perpendicular to the traveling path. Visible light has wavelengths ranging from approximately blue (400 nm) to red (700 nm), what is a small fraction of the electromagnetic spectrum. A simpler geometrical description of the light as rays pointing along the direction of wave propagation is often used to describe some of the image properties of the eye.

It is interesting to note that also the particle nature of light may have a role in vision under particular conditions. Absorption of light by matter only can be interpreted if the light is considered as a particle, called a photon. Photon absorptions occur in the photoreceptors following the rules of a random process, discontinuously in discrete quanta. Specifically, the light intensity reaching each photoreceptor only determines the probability of a photon being absorbed. This imposes another fundamental limit to vision related to the photon statistics. However, this is restricted to very low luminance conditions after dark adaptation.

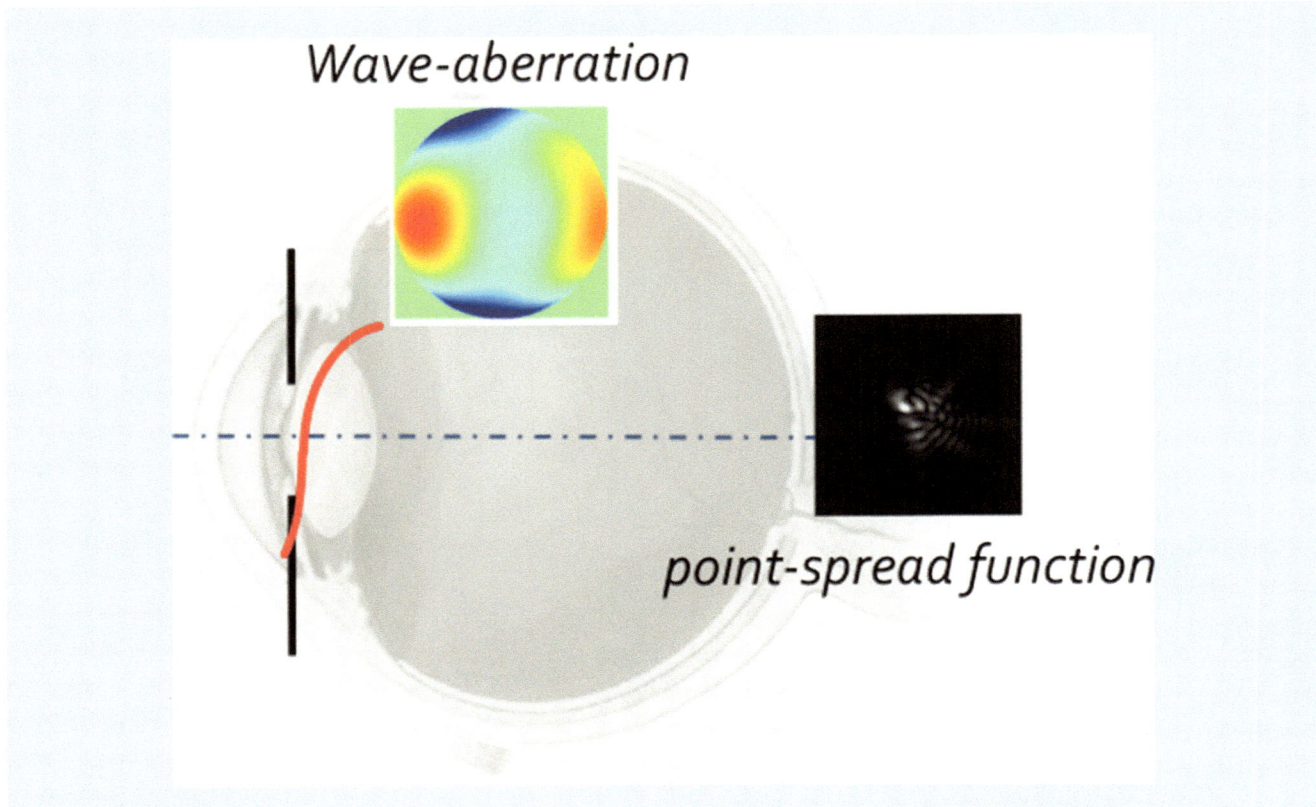

Fig. 12.3 The main functions describing the optical quality of the eye. The wave-aberration on the pupil plane represents the deviation in phase compared with a perfect optical system. The point-spread function (PSF) is the image of a point source formed in the retina. A good eye will form compact and symmetrical images, while in an aberrated eye the PSF will be more extended and asymmetric

Under most normal viewing conditions, the quality of the retinal images is governed by the wave-like nature of the light. The functions used to describe the quality of any optical instrument are showed in ◘ Fig. 12.3. The wave-aberration function is defined as the difference between the perfect (spherical) and the real wave-front for every point over the pupil. It is commonly represented as a two-dimensional map, where color level represents the amount of wave-aberration, expressed either in microns. The image of a point source is called the point-spread function, PSF. An eye without aberrations has a constant, or null, wave-aberration and forms a perfect retinal image of a point source that depends only on the pupil diameter. By performing a convolution operation, it is possible to predict the retinal images of any object. This can be easily understood as placing a weighted PSF onto each point of the geometrical image. Readers interested in more information on the nature of light and/or the functions describing image quality could read some general optics references [1–3].

12.2 The Anatomy of the Eye

The human eye can be described as a fluid-filled quasi-spherical structure. Anatomically consists essentially of three tissue layers: an outer fibrous layer (the sclera and cornea), an inner layer consisting largely of the retina, but including also parts of the ciliary body and iris, and an intermediate vascular layer made up of the choroid and portions of the ciliary body and iris. The eye in adult humans is approximately a sphere of around 24 mm in diameter. It is made up of a variety of cellular and non-cellular components derived from ectodermal and mesodermal germinal sources. Externally it is covered by a resistant and flexible tissue called the sclera, except in the anterior part where the transparent cornea allows the light to pass into the eye. Internal to the sclera are two other layers: the choroid to provide nutrients and the retina, where the light is absorbed by the photoreceptors after image formation. The eye moves due to the action of six external muscles permitting fixation and the scanning of the visual environment. The light reaching the eye is first refracted by the cornea, a thin transparent layer free of blood vessels of about 12 mm in diameter and around 0.55 mm thickness in the central part. An aqueous tear film on the cornea assures that the first optical surface is smooth to provide the best image quality. After the cornea, the anterior chamber is filled with the aqueous humor, a water-like substance. The iris, two sets of muscles with a central hole whose size depends on its contraction, acts as a diaphragm with characteristic color depending on the amount and distribution of pigments. The aperture is the opening in the center of the iris, and limits the amount of light passing into the eye. The entrance pupil is the image of the iris through the cornea and the exit pupil the image of the aperture through the lens. The aperture size changes with the ambient light, from less than 2 mm in diameter in bright light to more than 8 mm in the dark. The pupil controls retinal illumination and limits the rays entering the eye affecting the retinal image quality. After the iris, the crystalline lens, in combination with the cornea, form the images on the retina. The crystalline lens is an active optical element. It changes its shape modifying its optical power. The lens is surrounded by an elastic capsule and attached by ligaments called zonules to the ciliary body. The action of the muscles in the ciliary body permits the lens to increases or decreases power.

The retina has a central area, the fovea, where photoreceptors are densely packed to provide the highest resolution. The eyes move continuously to fixate the desired details into the fovea. The peripheral parts of the retina render lower resolution but specialize in movement and object detection in the visual field. The typical field covered by the eye is quite large as compared with most artificial optical system, at least $160 \times 130°$.

The cornea is approximately a spherical section with an anterior radius of curvature of 7.8 mm, posterior radius of curvature of 6.5 mm, and refractive index of 1.3771. Since the largest difference in refractive index occurs from the air to the cornea (actually the tear film), this accounts for most of the refractive power of the eye, on average over 70 %. The lens is a biconvex lens with radii of curvature of 10.2 and −6.0 mm for the anterior and posterior surfaces. The internal structure of the lens is layered, which produces a non-homogeneous refractive index, higher in the center than in the periphery and with an equivalent value of 1.42. The refractive indexes of the aqueous and vitreous humors are 1.3374 and 1.336, respectively. More detailed information on different aspects of the eye's geometry and its optical properties can be found in references [4–6].

An average eye with these distances: 3.05, 4, and 16.6 mm for the anterior chamber, lens, and posterior chamber, respectively, will have a total axial length of 24.2 mm and will image objects placed far from the eye precisely in focus into the retina. This situation is called emmetropia. However, most eyes are affected by refractive errors since they do not have the adequate optical properties or the dimensions required for perfect focus. Refractive errors are classified as myopia, when the images of distant object are focused in front of the retina, and hypermetropia, when distant objects are focused behind the retina. In addition, the eye is not rotationally symmetric, being a common manifestation the presence of astigmatism: the retinal image of a point source consists of two perpendicular lines at different focal distances. ◘ Figure 12.4 shows an example of a myopic eye and the degradation founds in its retinal image.

The ocular media filter the wavelengths reaching the retina. There are a good matching between transmission and photoreceptor sensitivity. The cornea and the

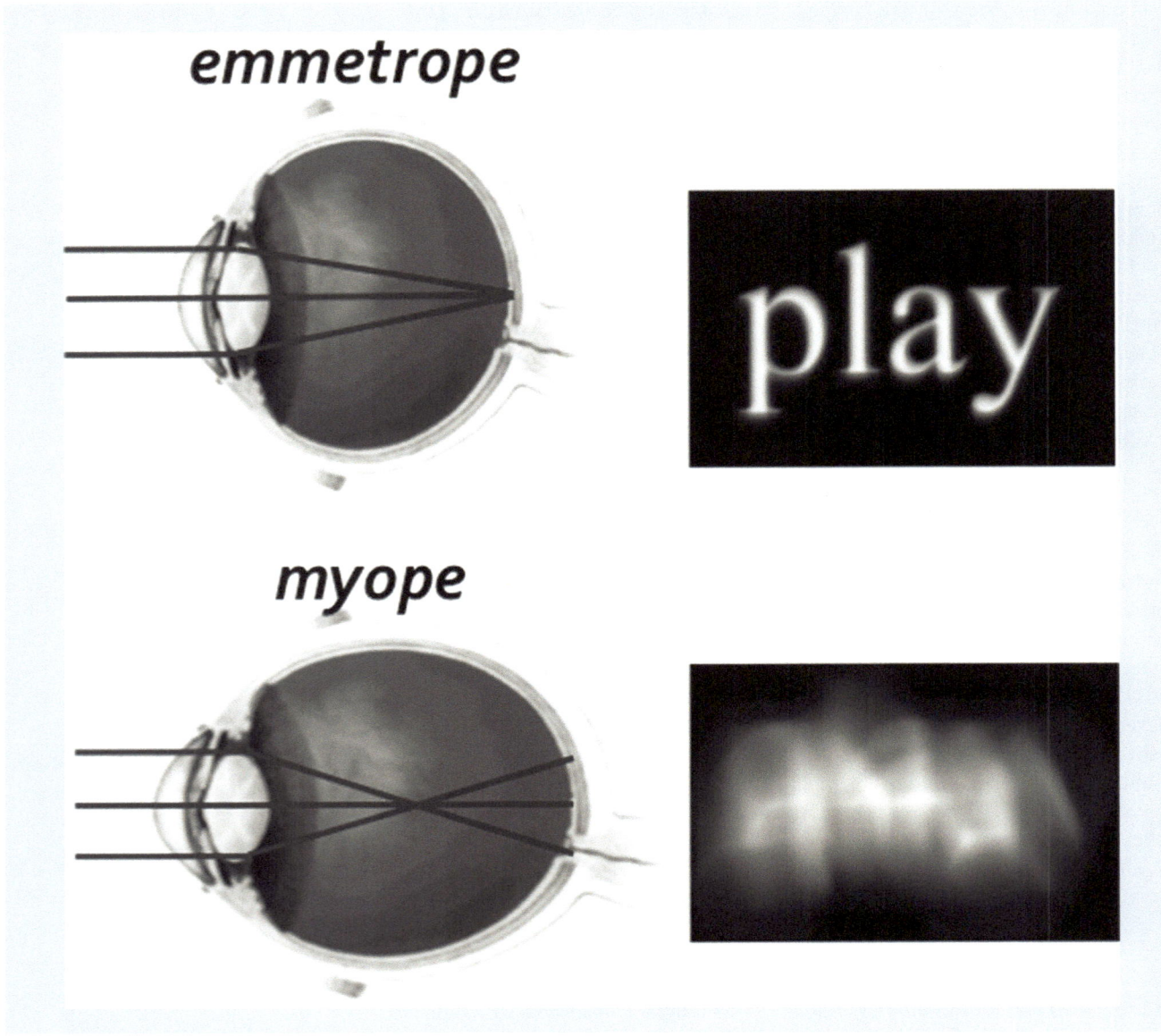

■ **Fig. 12.4** Examples of an emmetrope and a myopic eye and the image formed in the retina of the word "play." In myopes, the image is formed behind the retina and the images are blurred

vitreous have bandwidths that exceed the visible spectrum, but the lens absorbs light in the short wavelength (blue) part of the spectrum. The retina has also pigments that filter the light reaching the photoreceptors. The main filter in the retina is the yellow macular pigment located within the macular region near the fovea. It has been suggested that the macular pigment may protect the retina from degenerative diseases and also improve vision by removing blue light.

12.3   The Quality of the Retinal Image

Even when the eyes are at perfect focus, as in the case of an emmetrope, they do not produce completely perfect images. This means that the retinal image of a point source is not another point, but an extended distribution of light. Several factors are responsible for the degradation of the retinal images: diffraction of the light in the eye's pupil, optical aberrations, and intraocular scattering. Diffraction blurs the images formed through instruments with a limited aperture due to the wave nature of the light. The effect of diffraction is usually small and only can be noticed with small pupils. The impact of the ocular aberrations in the eye's image quality is more significant for larger pupil diameters. The pupil of the eye varies diameter from around 2–8 mm in diameter. This corresponds approximately to an aperture range from f/8 to f/2, values which can be compared with the typical values in a camera objective. ◱ Figure 12.5 shows an example of realistic retinal images of letters for the same eye for small (3 mm) and a larger (7 mm) pupil. Note how aberration degrades the image for larger pupils.

The amount of aberrations for a normal eye with about 5 mm pupil diameter (f/4 aperture) is approximately equivalent to less than 0.25 D of defocus, a small error typically not corrected when dealing in the clinic with refractive errors.

The particular shapes of the eye's lenses, refractive index distribution, and particular geometry are responsible for the limited optical quality of the eye compared with artificial optical systems. A normal eye has at least six times lower quality than a good (diffraction-limited) artificial optical system. Each eye produces a peculiar retinal image depending on the optical aberrations present. This can be demonstrated by how a point source is projected in the retina. For example, the shape of stars would depend on our image quality. ◱ Figure 12.6

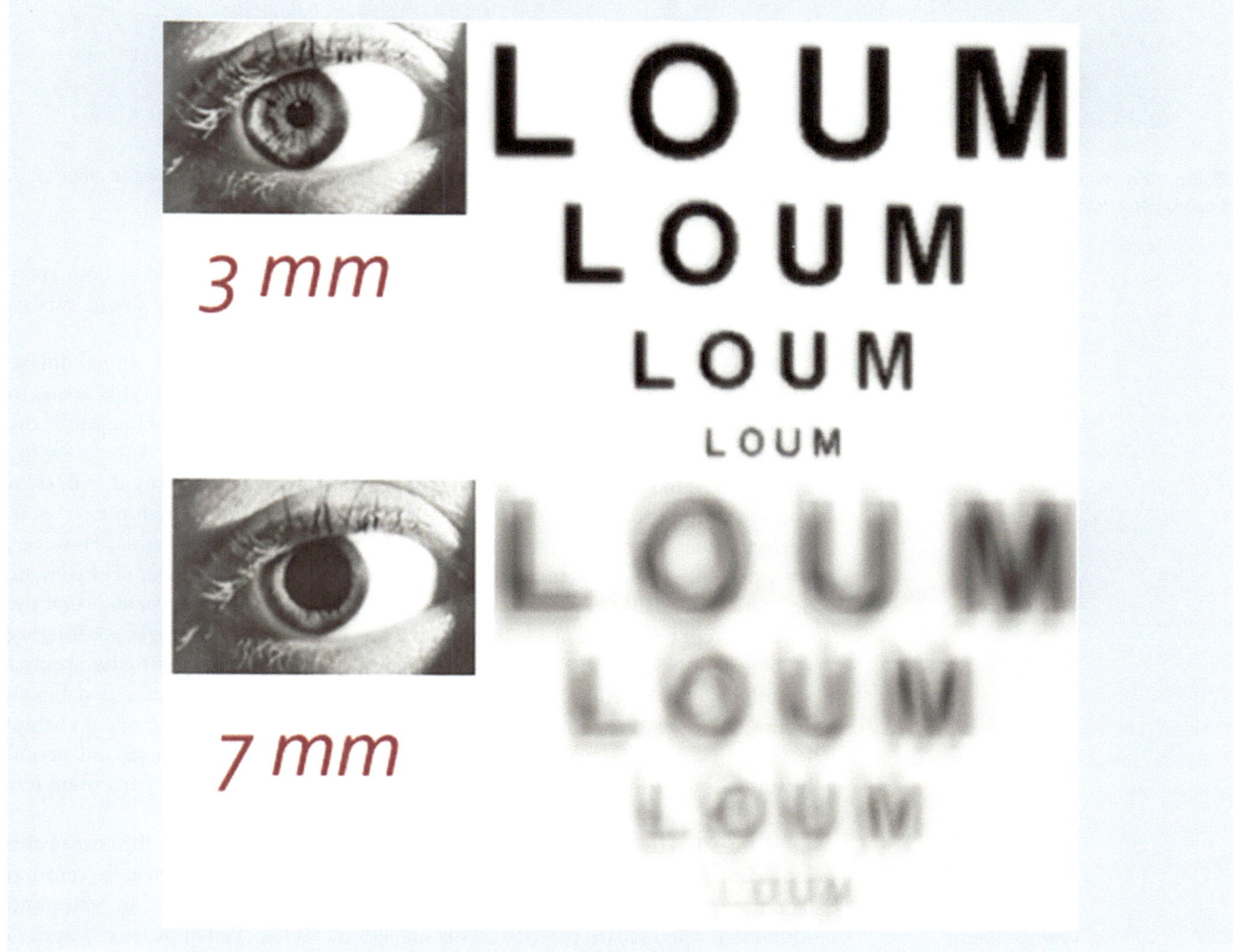

◱ **Fig. 12.5** Example of the effect of the pupil diameter in the retinal image quality. Aberrations affect more with larger pupils

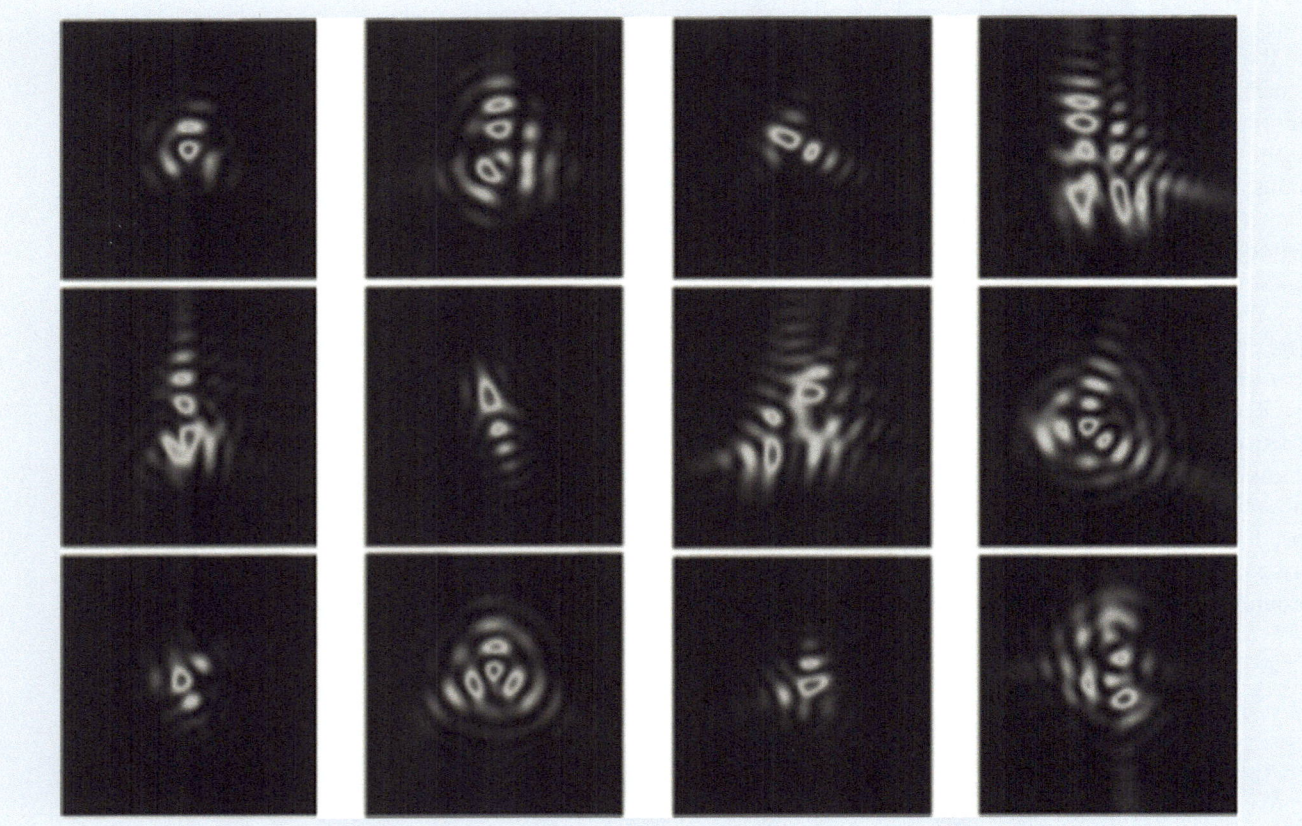

Fig. 12.6 Retinal PSFs measured in a group of normal young subjects. Each eye has particular optical characteristics that produce a unique PSF. It can be understood as if each one of us sees point objects (for instance, stars or distant lights) differently

shows PSFs for a group of normal eyes. This could be understood as how every subject sees an individual start. All are different in shape and size, so our experience of point objects is quite personal.

In addition, chromatic effects also contribute to reduce the retinal image quality since real scenes are usually polychromatic (in white light). This is due to the dependence of refractive index on wavelength that produces changes of the power of the eye with wavelength [7]. The chromatic difference in defocus for the eye from red to blue is large: around 2 diopters. This can be understood as if when you see simultaneously two letters, one red and one blue, and when the red is in perfect focus, the other would be defocus by nearly 2D in your retina. However, your perception of color images is not like that since the real impact of chromatic aberration is smaller than the equivalent of 2D defocus blur. The reason is that the visual system has mechanisms to minimize the impact. The relative larger filtering of blue light in the lens and the macular pigment, together with the spectral sensitivity of the retina, reduce the contribution of the most defocused bluish colors. Figure 12.7 shows an example of the appearance on the retina of a white letter a in a normal eye. It is important to note that due to retinal and neural factors, the actual impact of this chromatic blur is reduced and our perception less affected.

A question that attracted the interest of many scholars was how the cornea and lens contributed to the eye's optical quality. Early in the nineteenth century, Thomas Young neutralized the cornea by immersing his own eye in water and found that astigmatism persisted. This suggested that the crystalline lens itself have some degree of astigmatism. Recent experiments have also shown that the lens

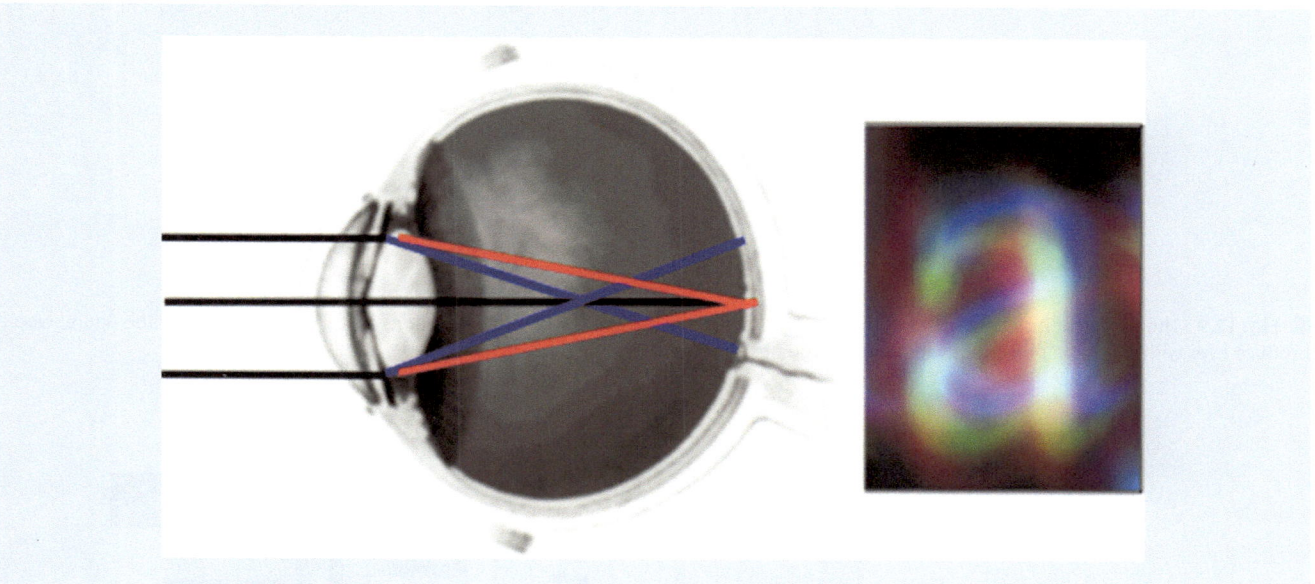

Fig. 12.7 Chromatic aberration in the eye. Due to the chromatic dispersion of the ocular media, if an eye is perfectly focused for red light, it will be myopic (up to 2 diopters) for blur light. This affects the quality of white light images, as the realistic simulation of an image of letter "a" shows (on the *right*)

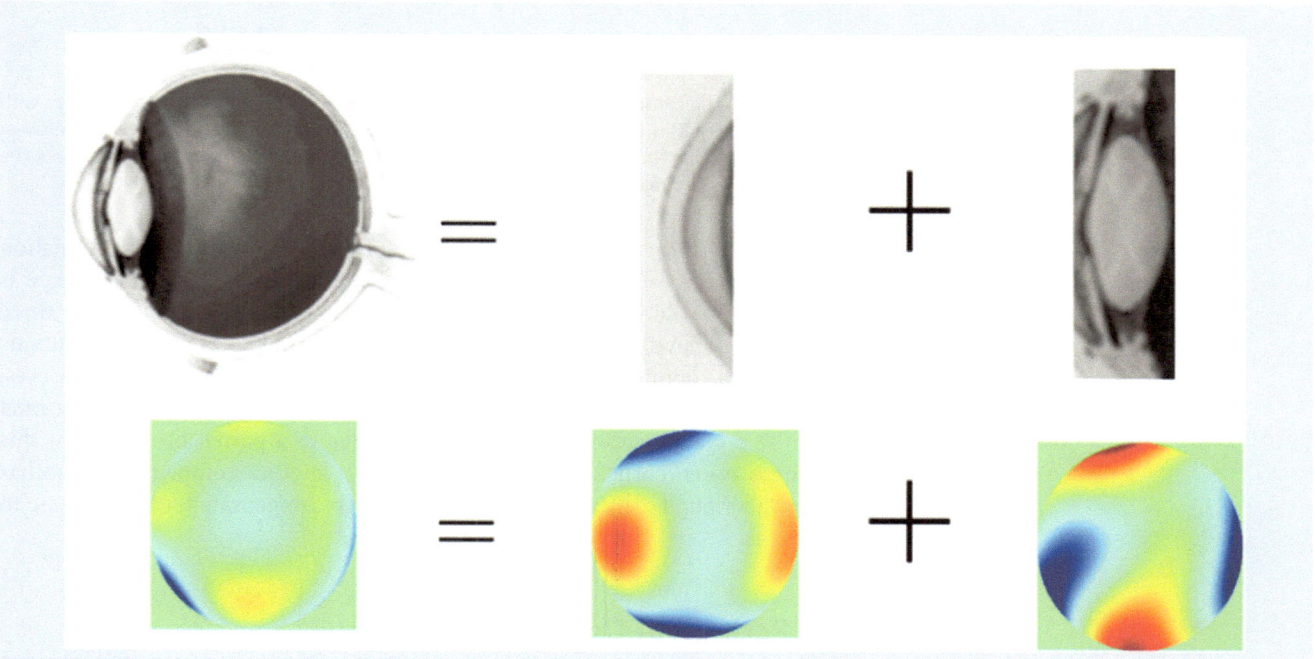

Fig. 12.8 The cornea and lens in young eyes have similar aberrations but with opposite signs rendering the whole eye with better quality than each component isolated. The eye behaves like an aplanatic optical system with partial correction of spherical aberration and coma

compensates not only for some moderate amounts of corneal astigmatism, but also spherical aberration and coma. ▪ Figure 12.8 shows as an example the aberrations for one author's eye for the anterior cornea, internal optics (mostly the lens), and the complete eye. The aberrations of the cornea and the lens are somehow opposite rendering an eye with improved optics. The eye as an optical system presents an aplanatic design of the eye, with partial correction of the spherical aberration and coma [8, 9]. This may help to maintain a rather stable optical quality independent of some alignment ocular variables. The reason for this compensation is found in

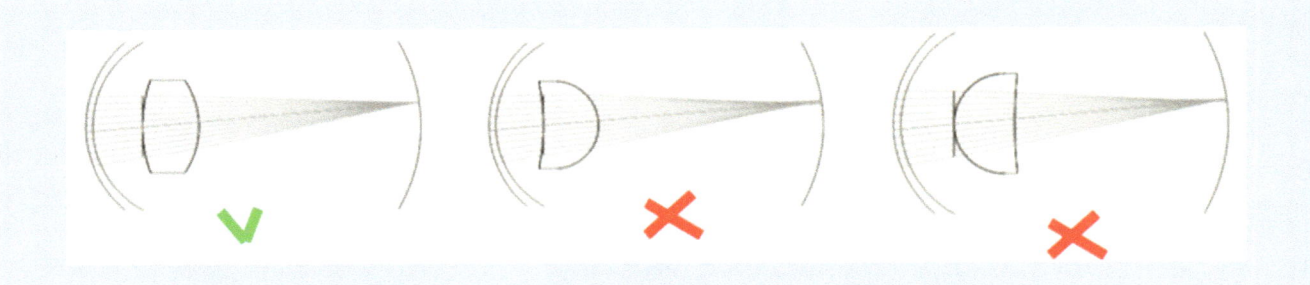

Fig. 12.9 The particular biconvex shape of the lens is an optimized design. Other possible options (as those on the right of the figure) would produce eyes with the same power but with worst image quality

Fig. 12.10 The center of the visual field (fovea) has the highest resolution and declines with retinal eccentricity. This figure shows a practical example. When fixating at the smallest letter at the left when placing this page at 30 cm distance, the larger letters should appear visible at the corresponding eccentricities. This shows the degradation in acuity with eccentricity

the particular shape of the cornea and lens. They have form factors (a relation between their curvature radii) of opposite sign. This means that their shape is optimized by evolution. ■ Figure 12.9 shows an example. Although the three schematic eyes have the same power, so could be considered as plausible solutions in a design, the optimum one is the biconvex lens that actually present in our eyes.

However, this optimized design is only present in younger eyes. During normal aging, the eye's aberrations tend to increases due to a partial disruption in this coupling between cornea and lens [10]. There is another compensatory mechanism: smaller pupil diameters in older eyes tend to compensate for this increase in aberrations.

12.4   Peripheral Optics

The central visual field (the fovea) has the highest spatial resolution; however, the periphery of the retina also plays a crucial role in our visual system. We use the peripheral parts of the visual field to detect objects of interest that we may bring to our fovea more detailed information. Ocular movements change fixation accordingly.

The optics of the eye has a different behavior when the images are formed eccentrically. The oblique incidence of light on the eye produces off-axis aberrations. The ability to discriminate small objects decreases severely with eccentricity. For example, while the normal resolution in the fovea is 1 min of arc, it will increase to 2.5, 5, and 10 at 10, 20, and 30° of eccentricity, respectively. ■ Figure 12.10 shows an example. When fixating to the smallest letter with the

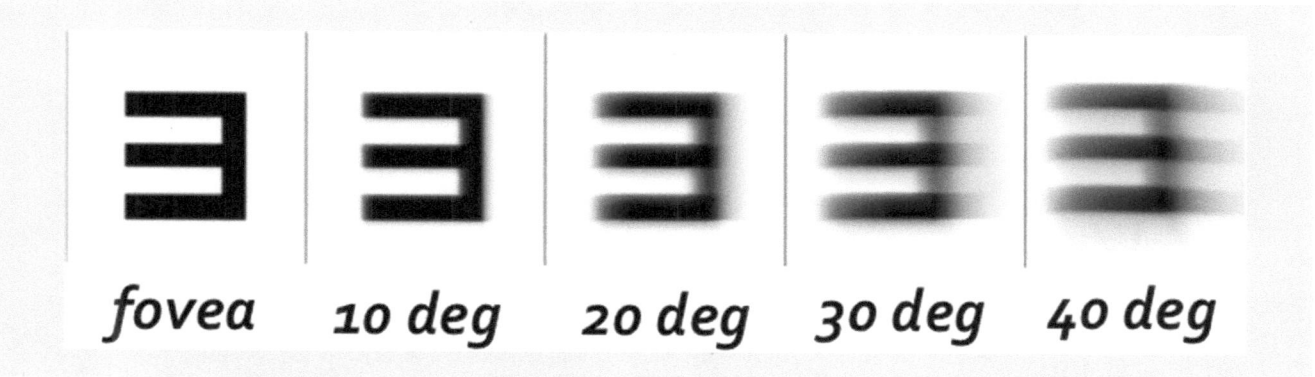

◨ Fig. 12.11 Retinal images of a letter "E" in a normal eye for different angles at the retina, fovea, 10, 20, 30, and 40°. At larger eccentricities, the off-axis aberrations further degrade the quality of the retinal images

fovea placing the book, or the screen, at around 30 cm, the larger letters at the different eccentricity have the correct size to be still legible.

This resolution reduction is due to both optical and neural factors: the eccentric angular incidence induces optical aberrations, which lower the contrast of the retinal images, and the density of cones and ganglion cells also decline with eccentricity, resulting in sparse sampling of the image. In the fovea (central vision) the optics is in many cases the main limiting factor for vision, in the periphery vision is limited by neural factors. The optics is degraded for eccentric angles by distortion, field curvature, astigmatism, and coma [11]. Field curvature is a defocus for off-axis objects and implies that the best image is not formed on a plane but on a parabolic surface. In the eye, the screen is the retina, which has a spherical shape constitutes a curved image plane that in most cases compensates for field curvature. Astigmatism off-axis induces a significant optical degradation in the periphery.

◨ Figure 12.11 shows examples of retinal images of a letter for different eccentricities. Despite the poor optics, visual acuity in the periphery cannot be improved with optical corrections. However, it is interesting to know that our peripheral optics is also optimized by the gradient index structure of the crystalline lens. This was demonstrated by comparing the peripheral image quality in the eyes of a group of patients with one eye implanted with an artificial intraocular lens and the fellow eye still with the natural pre-cataract lens. The eyes implanted had more astigmatism in the periphery than the normal eyes. This result suggests that the crystalline lens provides a beneficial effect also to partially compensate peripheral optics.

12.5 Conclusions

The eye is a simple and robust optical instrument that is fully adapted to serve our visual system. Although the optical quality is not as good in the eye as in the best artificial optical systems, it matches what is required by most of the visual capabilities. There are also a number of compensating mechanisms in the visual system that renders some of the potential optical limitations as invisible. For instance, the large potential deleterious effect of chromatic defocus is limited by proper color filters and the band-pass spectral sensitivity.

An interesting discussion in the last decades has been the possibility to correct for the aberrations of the eye using adaptive optics [12]. This is now technically possible in the laboratory and also, although partially, with correcting devices such

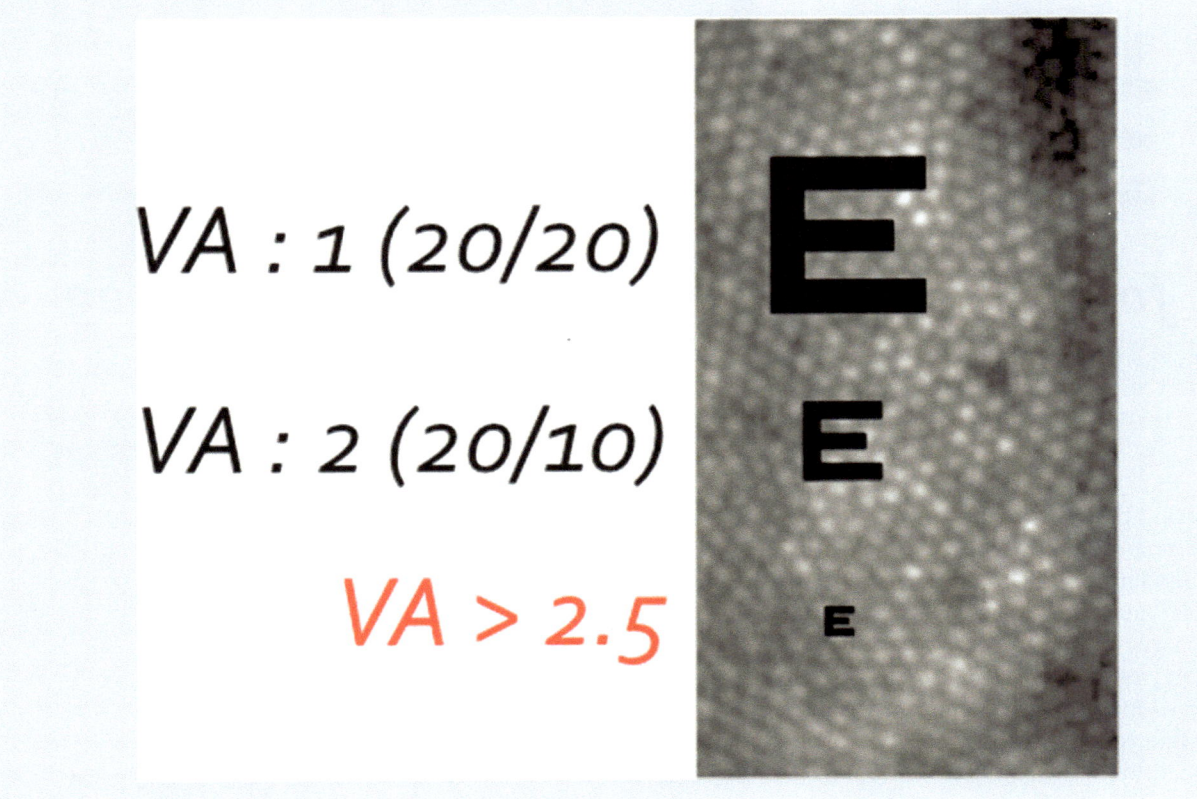

Fig. 12.12 The sampling of the images the retina by the photoreceptors imposes the second limit to vision after the optics. Images of letters corresponding to decimal visual acuity 1, 2, and larger than 2.5 are represented. At the limit, a very small letter even if well projected into the retina would cover one single receptor. Then, the details of the letter will not be detectable. This suggests that even if the optics were perfect, visual acuity would not be improved

as intraocular lenses. The correction of the eye's aberrations may improve vision in some subjects but there are fundamental limitations that cannot be surpassed. The first is the sampling of the retinal images by the photoreceptors. Even if sharp images are projected into the retina, the smallest letter to be perceived will require several photoreceptors across to be properly interpreted. ◘ Figure 12.12 shows this schematically at a correct scale. Images of letters smaller than those corresponding visual acuity (decimal) two will not be discriminated even if the letter is resolved by the eye's optics.

However, as was pointed out, the main cases for optical degradation are not higher order aberrations, but defocus and astigmatism. In that context, the manipulation of the eye's optics by different devices has been a successful technological development since the correction of defocus in the thirteenth century to the use of cylindrical lenses to correct astigmatism in the nineteenth century. Today, it is also possible to correct and induce also higher order aberrations in contact lenses, intraocular lenses, or laser refractive surgery procedures.

The future of correcting the eye's optics is both exciting and promising. And photonics and light technology will surely play a key role. The use of advanced optoelectronics would allow new prostheses to restore accommodation in the presbyopic eye. Two-photon interaction in the cornea by using femtosecond lasers may offer the possibility of changing the optical properties without the need to remove tissue, as is the case in the current ablation-based procedures. Optical technology is also fundamental in new diagnosis instrument. New swept-source optical coherence tomography allows full three-dimensional imaging in real time of the eye in an unprecedented manner. And ophthalmoscopes equipped with adaptive optics obtain high-resolution images of the retinal structures in vivo. Optics and photonics are now, more than ever, at service to help our eyes to see well.

References

1. Pedrotti FL, Pedrotti LS (1993) Introduction to optics, 2nd edn. Prentice Hall, New Jersey
2. Goodman JW (2005) Introduction to fourier optics, 4th edn. McGraw-Hill, New York
3. Smith WJ (1990) Modern optical engineering, 2nd edn. McGraw-Hill, New York
4. Le Grand Y, El Hage SG (1980) Physiological optics. Springer, Berlin
5. Smith G, Atchison D (1997) The eye and visual optical instruments. Cambridge University Press, New York
6. Atchison DA, Smith G (2000) Optics of the human eye. Butterworth-Heinemann, Oxford
7. Bedford RE, Wyszecki G (1957) Axial chromatic aberration of the human eye. J Opt Soc Am 47:564–565
8. Artal P, Benito A, Tabernero J (2006) The human eye is an example of robust optical design. J Vis 6:1–7
9. Artal P, Tabernero J (2008) The eye's aplanatic answer. Nat Photonics 2:586–589
10. Artal P, Berrio E, Guirao A, Piers P (2002) Contribution of the cornea and internal surfaces to the change of ocular aberrations with age. J Opt Soc Am A 19:137–143
11. Jaeken B, Artal P (2012) Optical quality of emmetropic and myopic eyes in the periphery measured with high-angular resolution. Invest Ophthalmol Vis Sci 53:3405–3413
12. Fernández EJ, Iglesias I, Artal P (2001) Closed-loop adaptive optics in the human eye. Opt Lett 26:746–748

Optics in Medicine

Alexis Méndez

A. Méndez (✉)
MCH Engineering LLC, Alameda, CA 94501, USA
e-mail: alexis.mendez@mchengineering.com

© The Author(s) 2016
M.D. Al-Amri et al. (eds.), *Optics in Our Time*, DOI 10.1007/978-3-319-31903-2_13

13

13.1 Introduction

13.1.1 Why Optics in Medicine?

Unlike the present time, medical practitioners of the ancient world did not have the benefit of sophisticated instrumentation and diagnostic systems, such as X-rays, ultrasound machines, or CT scanners. Visual and manual auscultations were the tools of the day. Hence, since the early days of medicine, optics has been a useful and powerful technology to assist doctors and all forms of healthcare practitioners carry out examination and diagnosis of their patients. This is so because one of the fundamental aspects of medicine is observation and physical examination of the patience's general appearance. Hence, anything that can help "see" better the condition of a patient will be of aid. As such, optics, as the science that studies the behavior and manipulation of light and images, is an ideal tool to assist doctors gain better visual examination capabilities by providing improved illumination, magnification, access to small or internal body cavities, among others. But it is in reality light and its interaction with living tissues that is at the center of what makes optics in medicine possible. Light possesses energy and is capable of interacting with biological cells, tissues, and organs. Such interaction can be used to probe the state of such living matter for diagnostics and analytical purposes or, it could be used to induce changes on the same living systems and be exploited for therapeutic purposes. The science of light generation, manipulation, transmission, and measurement is known as *photonics*. The application of photonics technologies and principles to medicine and life sciences is known as *biophotonics*.

Nowadays, it is not only optics but also photonics that are used extensively in a myriad of medical applications, from diagnostics, to therapeutics, to surgical procedures. Hence, when we use the term medical optics, we are referring to biomedical optics and biophotonics as well. The interrelation between optics and light in medicine is ever present and it could be said that more significant advances in biophotonics are now due to the availability of more powerful, concentrated, and multi-spectral light sources which have been available only in the last 50 years. Historically, ambient light was the illumination source, which precluded performing exams late in the day or during certain hours in the winter time. Oil candles in the ancient world gave way to wax ones and alcohol burning lamps in the fifteenth through the nineteenth centuries until the development of electricity and the introduction of the electric lamp by Edison. Then, in the 1960s, with the development of semiconductor lasers, light emitting diodes (LEDs) and lasers, modern medical optics began to take shape and, coupled with the availability of optical fibers, a new generation of medical instruments and techniques began to be developed.

Fiber optics has been used in the medical industry even before their adoption and subsequent explosion as the technology of choice for long haul data communications [1]. The advantages of optical fibers have been recognized by the medical community long ago. Optical fibers are thin, flexible, dielectric (non-conductive), immune to electromagnetic interference, chemically inert, non-toxic, and of course, small in size. They can also be sterilized using standard medical sterilization techniques. Their major advantage lies in the fact that they are thin and flexible so they can be introduced into the body for both remotely sense, image and treat. Their initial and still most successful biological/biomedical application has been in the field of endoscopic imaging. Prior to the development of such devices, the only method of inspecting the interior of the body was through invasive surgery. Many patients owe their lives today to the existence of fiberoptic endoscopes. Optical fibers are not only useful for endoscopes, but can also be used to transmit light to tissue regions of interest either to illuminate the tissue so that it

Table 13.1 Medical industry trends that promote the use of optical fibers

- Drives towards minimally invasive surgery (MIS)→Need for disposable probes and catheters
- Miniaturization, Automation and Robotics→Need for instrumented catheters

- Sensors compatible with MRI, CT, PET equipment as well as thermal ablative treatments involving RF or microwave radiation→Need for fiber sensors

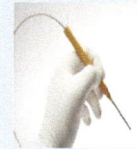

- Increased user of lasers→Need for fiber delivery devices
- Increased use of optical imaging and scanning techniques→Need for fiber OCT probes

can be inspected, or if much higher power laser light is used, to directly cut or ablate it. Hence, they are used extensively as laser-delivery probes, as well as imaging conduits in optical coherence tomography (OCT).

Optical fibers have revolutionized medicine in many ways and continue to do so thanks to the advent of new surgical trends, as summarized in Table 13.1. One such trend is the advent of minimally invasive surgery (MIS) where the trend now is to avoid cutting open patients and instead, perform small cuts and incisions through which a variety of different surgical instruments, such as catheters and probes, are inserted through these small opening, thus minimizing the postoperative pain and discomfort. Furthermore, there is today growing use of surgical robots where a surgeon operates them remotely using control arms to do a surgical procedure from the comfort of his office while the patient is at a remote hospital location. However, one of the issues with these types of systems is the fact that the surgeon loses the actual manual feedback and does not have sensitivity of the force needed to apply to a scalpel or other surgical tools. This is called *haptic* feedback. These "robotic surgeons" operate using very small tools and catheters, and in order to make sensing elements compatible with such slender instruments, fiber optics represent an ideal solution to provide shape, position, as well as force-sensing information to the remote surgeon's controls.

Fiber optic and photonic devices are also being exploited as sensing devices for patient monitoring during medical imaging and treatment using radiation devices such as MRI, CT, and PET type scan systems that involve the use of high-intensity electromagnetic fields, radiofrequencies, or microwave signals. Because the patient's risk of an electric shock conventional electronic monitoring devices and instrumentation cannot be used in these applications. Instead, patient monitoring is performed using optical fiber sensors.

Based on the above arguments it becomes evident the need for and benefits of optics (and photonics) in medicine. Table 13.2 summarizes the key general applications for optics in medicine. In general, it could be said that optics has been and will continue to be an enabling technology to further the development and advancement of medicine and the healthcare industry as a whole.

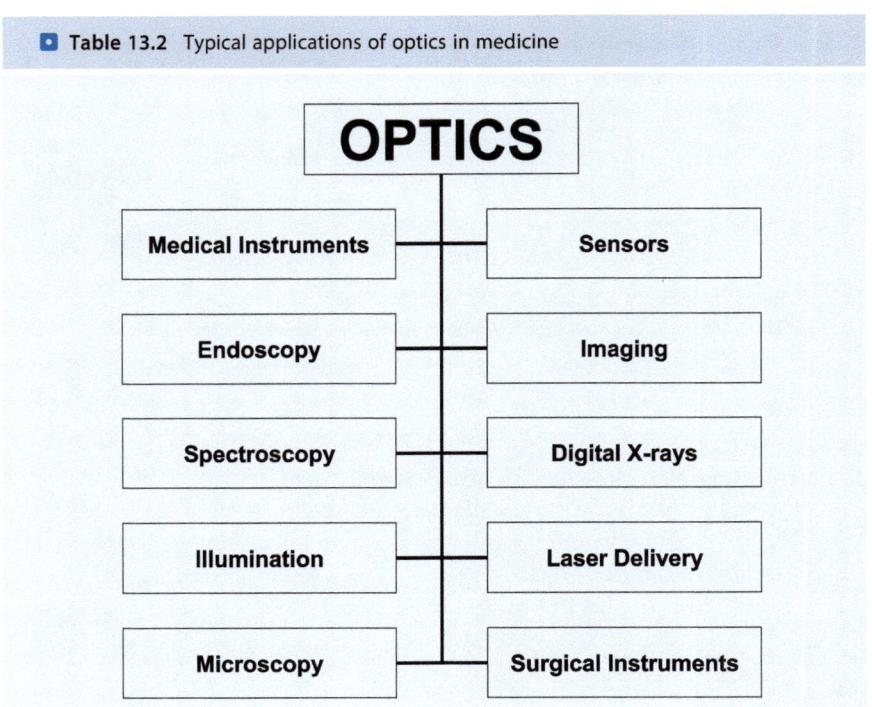

■ **Table 13.2** Typical applications of optics in medicine

■ **Table 13.2** Typical applications of optics in medicine

13.1.2 Global Healthcare Needs and Drivers

We all need medical care, from the day we are born, until the day we die. However, this need for medical care has now been affected and accentuated by a convergence of social, demographic, economic, environmental, and political global trends that have been developing over the last few decades. It is a world full of challenges that impact how to effective deliver healthcare in an effective, affordable, and sustainable fashion. On one hand, average lifestyles have changed drastically in the past century resulting in a more sedentary lifestyle with lack of exercise, poor diet, smoking, and excessive alcohol consumption that have resulted in a growing number of chronic diseases such as obesity, arteriosclerosis, diabetes, and cancer that have become leading causes of death and disability. On the other hand, the entire global population keeps growing. According to a recent United Nations Department of Economic and Social Affairs (DESA) report the world's population is estimated to be in excess of 7.3 billion and growing at ~1.1 % annual rate, and expected to reach 8.5 billion by 2030, 9.7 billion in 2050, and 11.2 billion in 2100 [2]. As illustrated in ■ Fig. 13.1, the world population has experienced continuous growth since the end of the Great Famine and the Black Death back in 1350, when the total population stood at merely 370 million. Nowadays, total annual births are approximately 135 million/year, while deaths are around 56 million/year, but expected to increase to 80 million/year by 2040.

Add to this the fact that in certain parts of the world the population is aging, including the USA, Japan, and parts of Europe. Globally, the number of persons aged 65 or older is expected to reach to nearly 1.5 billion by 2050. An aging population puts additional demands on healthcare since older people are more vulnerable to illness and chronic diseases. Furthermore, life expectancy at birth has increased significantly. The UN DESA estimates a 6-year average gain in life expectancy among the poorest countries, from 56 years in 2000–2005 to 62 years in 2010–2015, which is roughly double the increase recorded for the rest of the world. Another key trend and global challenge is the expected shortage of medical doctors and physicians available to meet the healthcare needs of a growing world population.

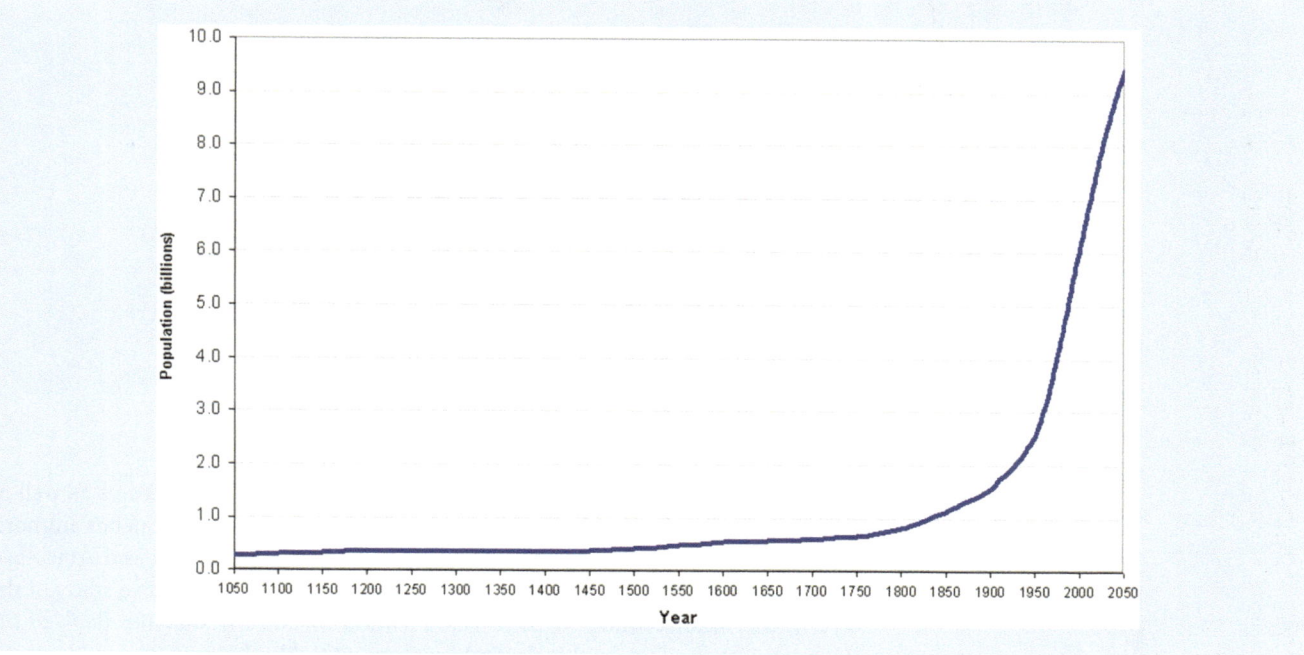

■ **Fig. 13.1** Historical global population growth

A large global population requires more doctors, medical devices, medical supplies, clinics, hospitals, and overall healthcare infrastructure to address the needs of people needing immunizations, or getting sick or injured. Hence, there is and will continue to be an overall growth and expansion of the health care industry on a global basis, that continuous to demand more medical instruments and technical innovations that can facilitate and expedite medical examinations, while reducing costs. Historically, optics has been an enabling technology for the design and development of such medical devices and instruments.

Another relevant and converging present trend is how biomedical devices and instruments are so extremely pervasive across the healthcare industry today. We may not realize it, but whenever we get our blood pressure tested, monitor our blood sugar, or when a expectant mother is being monitored by her doctor, an instrument or sensing device is needed which, often times, is based on the use of an optical technique or based on the use of optical components. Couple this with the fact that in many parts of the underdeveloped world there is not enough doctors, hospitals, clinics, and instrumentation available to support local populations. Hence, it becomes critically important to develop simple, practical, effective, and inexpensive medical devices that can be used in rural and remote areas by non-professionals to examine and treat patients.

13.1.3 Historical Uses of Optics in Medicine

Mankind has always been fascinated with light and the miracle of vision, dating back to the first century when the Romans were investigating the use of glass and how viewing objects through it, made the objects appear larger. However, most of the significant developments of optics for medical diagnosis and therapy started occurring in the nineteenth century. Before that, the vast majority of the known published works on optics and medicine dealt mostly with the anatomy and physiology of the human eye. For instance, the Greek anatomist, Claudius Galen (130–201) provided early anatomical descriptions of the structure of the human

13

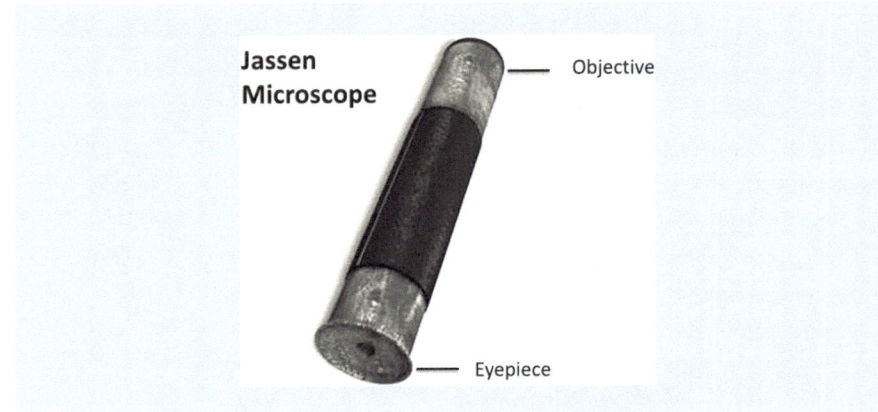

Jassen Microscope — Objective — Eyepiece

Fig. 13.2 Photograph of the Jansen compound microscope (c. 1595)

eye, describing the retina, iris, cornea, tear ducts, and other structures as well as defining for the first time the two eye fluids: the vitreous and aqueous humors. Subsequently, Arab scholars Yaqub ibn Ishaq al-Kindi (801–873) and Abu Zayd Hunayn ibn Ishaq alIbadi (808–873) provided a more comprehensive study of the eye in the ninth century in their *Ten Treatises on the Eye and the Book of the Questions of the Eye*. In the eleventh century Abu Ali al-Hasan ibn al-Haytham (965–1040)—known as Alhazen—also provided descriptions of the eye's anatomy in his Book of Optics (Kitab al-Manazir).

It is around this time that the so-called reading stones are being used as magnifying lenses to help read manuscripts. The English philosopher Robert Bacon (1214–1294) described in 1268 in his *Opus Majus* the mechanics of a glass instrument placed in front of his eyes. Then, in the thirteenth century, Salvino D'Armate from Italy made the first eye glass, providing the wearer with an element of magnification to one eye.

With the advent of the optical telescope optics took a significant step forward towards the development of one of the first early medical instruments—the microscope [3]. The compound microscope was developed around the late 1590s by Hans and Zacharias Janssen, a father and son team of Dutch spectacle makers, who experimented with lenses by placing them in series inside a tube and discovered that the object near the end of the tube appeared greatly enlarged (see ◘ Fig. 13.2).

A seminal optical medical instrument development came in 1804 when the German born physician Philipp Bozzini (1773–1809) developed and first publicized his so-called light conductor (*Lichtleiter*), which enabled the direct view into the living body [4]. The *lichtleiter* was an early form of endoscope which consisted of an open tube with a 45° mirror mounted at the proximal end with a hole in it. Illumination was provided by a burning alcohol and turpentine lamp was shone to a speculum mounted on the distal end and made to fit to the specific anatomy of the desired body opening to be inspected (see ◘ Fig. 13.3). In December 1806 Bozzini's light conductor was presented to the professors of the Josephinum, the "Medical-Surgical Joseph's Academy" in Vienna.

A period of significant activity and innovation in medical optics occurred from the mid-1800s through the early 1900s, when a variety of early medical instruments such as otoscopes, ophthalmoscopes, retinoscopes, and others, as well as improved illumination systems were developed. In 1851 German scientist and physician Hermann L. F. von Helmholtz (1821–1894) used a mirror with a tiny aperture (opening) to shine a beam of light into the inside of the eyeball [5]. Helmholtz found that looking through the lens into the back of the eye only produced a red reflection. To improve on the image quality, he used a condenser

■ **Fig. 13.3** Bozzini's original light conductor with specula (c. 1806)

■ **Fig. 13.4** Helmholtz ophthalmoscope (c. 1851)

lens that produced a 5× magnification (■ Fig. 13.4). He called this combination of a mirror and condenser lens an *Augenspiegel* (eye mirror).

The term ophthalmoscope (eye-observer) did not come into common use until later. Helmholtz also invented the ophthalmometer, which was used to measure the curvature of the eye. In addition, Helmholtz studied color blindness and the speed of nervous impulses. He also wrote the classic Handbook of Physiological Optics.

In 1888 Prof. Reuss and Dr. Roth of Vienna used bent solid glass rods to illuminate body cavities for dentistry and surgery. This would be the earliest idea to

13

use a precursor of an optical fiber for medical applications. Decades later, in 1926, J. L. Baird of England and Clarence W. Hansell of the RCA Rocy Point Labs, propose independently of each other fiber optic bundles as imaging devices. A few years later, German medical student Heinrich Lamm assembles the first bundles of transparent optical fibers to carry the image from a filament lamp, but is denied a patent. Then in 1949, Danish researchers Holger M. Hansen and Abraham C. S. van Heel begin investigating image transmission using bundles of parallel glass fibers. Prof. Harold H. Hopkins from Imperial College in London begins work in 1952 to develop an endoscope based on bundles of glass fibers. University of Michigan Medical professor Basil Hirschowitz visits Imperial College in 1954 to discuss with Prof. Hopkins and graduate student, Narinder Kapany, about their ideas for imaging fiber bundles. Hirschowitz hires undergraduate student Larry Curtis to develop a fiber optic endoscope at the University of Michigan. Curtis fabricates the first clad optical fiber from a rod-in-tube glass drawing process. Prof. Hirschowitz tests first prototype fiber optic endoscope using clad fibers in February of 1957, and then introduces it to the American Gastroscopic Society in May of the same year.

The first solid-state laser was built in 1960 by Dr. T. H. Maiman at Hughes Aircraft Company. Within the year, Dr. Leon Goldman, chairman of the Department of Dermatology at the University of Cincinnati, began his research on the use of lasers for medical applications and later established a laser technology laboratory at the school's Medical Center. Dr. Goldman is known as the "father of laser medicine." He is also the founder of the American Society for Lasers in Medicine and Surgery [6]. However, the first medical treatment using a laser on a human patient was performed in December 1961 by Dr. Charles J. Campbell of the Institute of Ophthalmology at Columbia–Presbyterian Medical Center, who used a ruby laser that is used to destroy a retinal tumor. Since then, lasers have become an integral part of modern medicine [7].

During the 1980s and 1990s, extensive research was conducted to develop fiber-optic-based chemical and biological sensors for diverse medical applications [8].

OCT is a newer optical medical imaging technique, first introduced in the early 1990s, that uses light to capture micrometer resolution, three-dimensional images from within biological tissue based on low-coherence, and optical interferometry [9]. OCT is a technique that makes possible to take sub-surface images of tissues with micrometer resolution. It can be thought of as the optical equivalent of an ultrasound scanning system. This is an active area of medical research at the moment.

13.1.4 Future Trends

Optics and photonics, as mentioned earlier, are powerful, versatile, and enabling technologies for the development of present and future generations of medical devices, instruments, and techniques for diagnostic, therapy, and surgical applications.

Given the present R&D activity worldwide based on optical and photonic techniques it should be no surprise to expect a broader utilization of optically based solutions across the healthcare industry and medical profession. In the future, advances in the development of ever smaller and thinner medical probes and catheters should be expected, as well as broad utilization of OCT devices to become as common as ultrasound scanning devices are in today's society. There will also be a proliferation of laser-based treatments and therapies. Endoscopy, for its part, will continue to evolve and more sophisticated and smaller devices will be

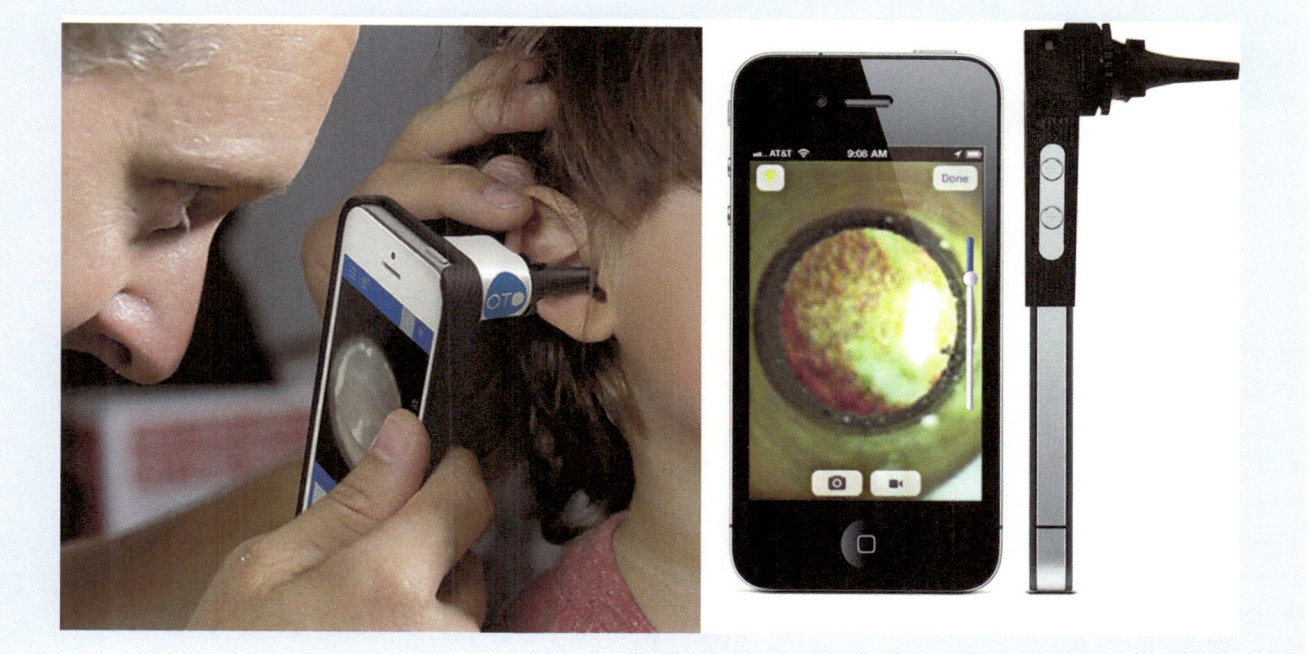

Fig. 13.5 A smartphone otoscope

developed that will combine more functions (from the standard illumination and visualization) with direct tissue analysis and laser treatment. Optical imaging techniques will continue to advance along with digital X-rays to make non-invasive examination and diagnosis safe, fast and with greater resolution and pinpoint accuracy.

Other future capabilities brought on by optics will be in the form of the so-called lab-on-a-fiber or LOF for short [10], where optical fibers are combined with micro- and nano-sized functionalized materials that react to specific physical, chemical, or biological external effects and can thus serve as elements to build multi-function, multi-parameter sensing devices. Light would remotely excite the functionalized materials which are embedded in the fiber's coating material. These materials in turn will react to specific biological or chemical substances (*analytes*) and induce an optical signal change proportional to the given analyte concentration.

Some future innovations can already be witnessed today in the form of optical devices used in combination with smart portable cell phones [11, 12]. For example, several new companies have now developed accessories for attachment to smartphones, which turn them into electronic video equivalents of conventional medical examination instruments such as otoscopes (to view inside ears), ophthalmoscopes (to view the inside of eyes), or even simple microscopes. Such devices are passive, optical elements that couple images from the patient to the video lens onto the smartphone's digital camera transform it into a fully functioning, network-connected medical instrument, capable of sending images and video remotely to a consulting doctor. ◘ Figure 13.5 depicts a cell phone otoscope in use, while ◘ Fig. 13.6 depicts a smartphone version of an ophthalmoscope and a dermal loupe.

Another such smartphone innovation is the so-called CellScope developed by researchers at the University of California at Berkeley [13]. The CellScope is a microscope that attaches to a camera-equipped cell phone and produces two kinds

13

Fig. 13.6 Examples of an ophthalmoscope and a dermal loupe attached to a smartphone

of microscopy imaging: brightfield and fluorescence. The idea is that such device can then be used in the field (on remote locations or those where little medical infrastructure is available) and take snap magnified pictures of disease samples and transmit them to medical labs via mobile communication networks, and screen for hematologic and infectious diseases in areas that lack access to advanced analytical equipment.

13.2 Early and Traditional Medical Optical Instruments

As discussed earlier, optics has been used throughout the centuries as a technology to assist medical doctors perform examinations of patients. Many of the medical instruments in use today rely on optics and optical components to perform their intended function. In particular, there a set of very basic but very popular and common medical instruments that were developed in the nineteenth century and continue to be used in the medical profession of today. Among these optical instruments we have the *otoscope*, the *ophthalmoscope*, *retinoscope*, *laryngoscope*, and even basic devices such as the head mirror.

In general, many of the basic optical medical instruments have in common the goal to provide both a more direct illumination and optical magnification of the area under examination. Conceptually, these optical instruments are similar to a telescope or microscope, but their optical design is different. Typically, a medical instrument consists of a tubular structure fitted with an objective lens on the distal (patient) end, and an objective lens on the viewing (doctor) end, represented as (1) and (2) in Fig. 13.7.

This lens arrangement produces a magnification of the object under inspection on the objective side (distal end), which has a size Y, and is positioned a distance P from the entrance pupil of the objective lens. The visual magnification factor M_v is calculated as Eq. (13.1):

$$M_v = \theta' D / Y \tag{13.1}$$

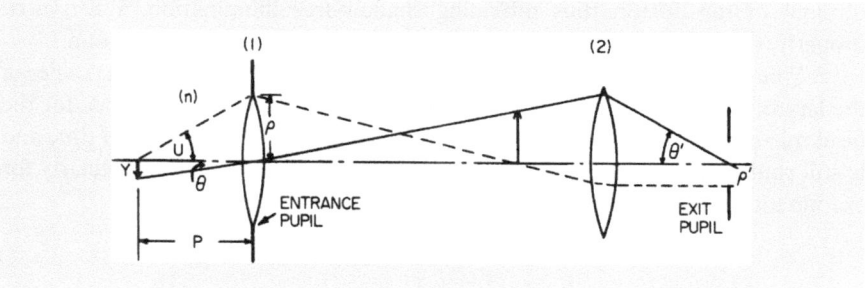

Fig. 13.7 Schematic of a basic medical optical instrument

Fig. 13.8 A medical head mirror and common placement on a doctor's head

where θ' is the angle of the light ray from the eyepiece, D is the viewing distance from the observer to the eyepiece. Hence, the magnification factor M is inversely proportional to the working distance P.

In the sections to follow, we shall describe the basic optical operating principles and uses of such devices. Our discussion of these devices is by no means exhaustive, but is intended to provide the reader with an overall idea on the utilization of optics in medicine and brief introduction on the subject of medical optical instruments [14].

13.2.1 Head Mirror

The most basic optical medical instrument is the so-called head mirror (see **Fig. 13.8**). A head mirror has historically been used by doctors since the eighteenth century for examination of the ear, nose, and throat. It consists of simple circular concave mirror—made of glass, plastic, or metal—with a small opening in the middle, and mounted on an articulating joint to a head strap made of leather or fabric. The mirror is positioned over the physician's eye of choice, with the concave mirror surface facing outwards and the hole directly over the physician's eye.

In use, the patient sits and faces the physician. A bright lamp is positioned adjacent to the patient's head, pointing towards the physician's face and hence towards the head mirror. The lamp's light gets concentrated by the curvature of the mirror and reflected off it towards the area of examination, and along the line of sight of the doctor, thus providing shadow-free illumination. When used properly, the head mirror thus provides excellent shadow-free illumination.

A French obstetrician named Levert, who was fascinated with the intricacies of the larynx and dabbled with mirrors, is credited with conceiving the idea for the head mirror back in 1743. Today's head mirror has withstood the test of time and is still routinely used by ophthalmologists and otolaryngologists, particularly for examination and procedures involving the oral cavity.

13.2.2 Otoscope

An *otoscope* is a hand-held optical instrument with a small light and a funnel-shaped attachment called an ear speculum, which is used to examine the ear canal and eardrum (tympanic membrane). It is also called *auriscope*. The otoscope is one of the medical instruments most frequently used by primary care physicians [15]. Health care providers use otoscopes to screen for illness during regular check-ups and also to investigate ear symptoms. Ear specialists—such as otolaryngologists and otologists—use otoscopes to diagnose infections of the middle and outer ear (otitis media and otitis externa).

The design of a modern otoscope is very simple [16]. It consists of a handle and a head (▣ Fig. 13.9). The handle is long and texture for easy gripping and contains batteries to power an integrated light. The head houses a magnifying lens on the eyepiece with a typical magnification of 8 diopters; a cone-shaped disposable plastic speculum at the distal end; and an integrated light source (either lamp bulb, LED, or fiber optic). The doctor inserts a disposable speculum into the otoscope, straightens the patient's ear canal by pulling on the ear, and inserts the otoscope to peer inside the ear canal. Some otoscope models (called pneumatic otoscopes) are provided with a manual bladder for pumping air through the speculum to test the mobility of the tympanic membrane.

The most commonly used otoscopes in emergency rooms and doctors' offices are monocular devices. They provide only a two-dimensional view of the ear canal. Another method of performing *otoscopy* (visualization of the ear) is use of a binocular microscope, in conjunction with a larger metal ear speculum, with the patient supine and the head tilted, which provides a much larger field of view and depth perception, thus affording a three-dimensional perception of the ear canal.

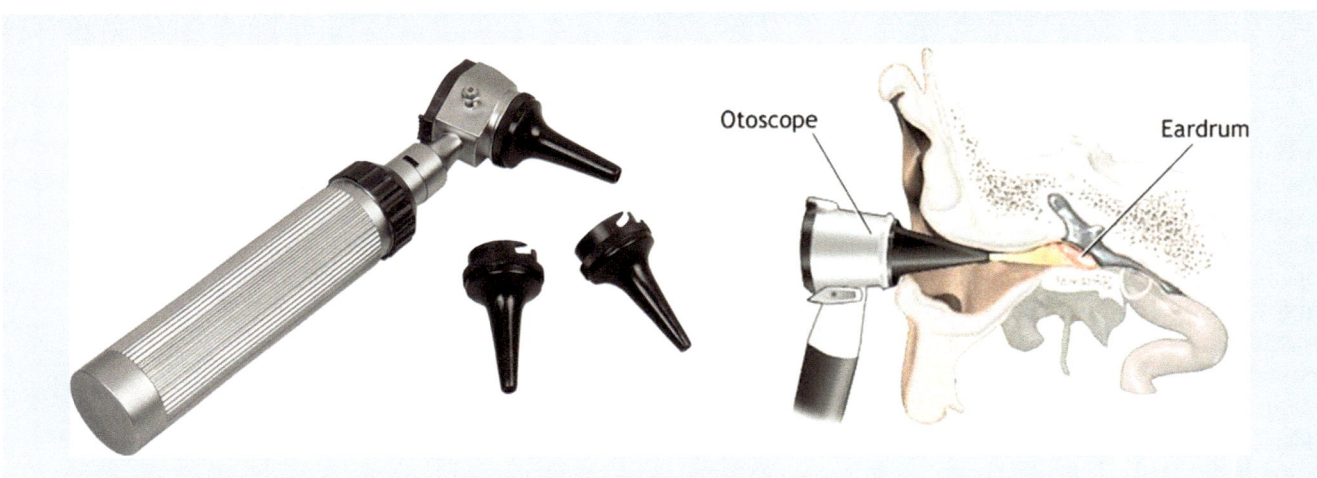

▣ **Fig. 13.9** Otoscope for visual inspection inside the ear canal

Fig. 13.10 Ear examination in the nineteenth century

The microscope has up to 40× power magnification, which allows for more detailed viewing of the entire ear canal and eardrum.

The otoscope is a valuable tool beyond its primary role as an examination tool for detecting ear problems. It can also be used for transillumination, dermatologic inspection, examination of the eye, nose, and throat and as an overall handy light source.

13.2.2.1 History of the Otoscope

Early ear examinations were performed by direct observation of the ear canal during daylight. As a consequence, examinations were limited to times of the day and year when there was adequate bright daylight. Furthermore, a device was needed to gain more direct access to the ear canal and to keep it open and provide direct illumination inside. Hence, over the years, the use of a *speculum* (a conical shape device that can be safely inserted into the ear) was adopted. In 1363 Guy de Montpellier in France described the first aural and nasal specula [17]. However, some means or direct illumination was needed in order to perform more effective ear examinations. The next major requirement was for an adequate method of directing concentrated natural daylight into the depths of the ear canal, which was accomplished by using a perforated mirror mounted either on a handle or on the head, which shone light directly into the ear canal. This allowed the doctor to look down the center of the beam of light, thus eliminating shadow effects and parallax (difference in the apparent position of an object viewed along two different lines of sight) (◘ Figs. 13.10 and 13.11).

Von Troltsch is generally credited with popularizing the use of a mirror in *otoscopy* after he showed it in 1855 at a meeting of the Union of German Physicians in Paris. He ultimately fastened the mirror to his forehead as is still currently practiced by some doctors. The size and focal length of the mirror was not standardized for some time. In an attempt to catch more light, used huge mirrors and only gradually was a diameter of 6–7 cm eventually adopted. A further

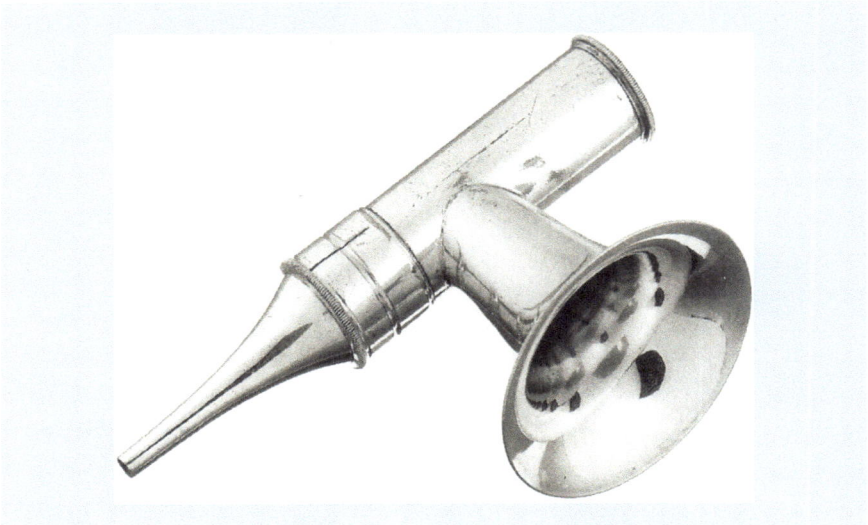

Fig. 13.11 Bunton's Auriscope (c. 1880). It can be observed the metal tip speculum, the rear objective lens for viewing, as well as the middle horn used to direct light from a candle or lamp

improvement to Von Troltsch's early auriscope is Brunton's device which was first described in an 1865 Lancet article. This auriscope combined mirror and speculum into a single instrument and worked on the principle of a periscope: light from a candle or lamp was concentrated by a funnel and then reflected by a plane mirror set at an angle of 45° into the ear canal. The mirror had a central perforation through which the doctor could view the ear. Brunton's auriscope was fitted with a magnifying lens for the observer and could also be sealed with plain glass at the illuminating end. These were the first otoscopes to be electrically illuminated.

13.2.3 Ophthalmoscope

An *ophthalmoscope* is an optical instrument for examining the interior of the eyeball and its back structures (called the *fundus*) through the pupil by injecting a light beam into the eye and looking at its back-reflection. An ophthalmoscope is also referred to as a *funduscope*. The fundus consists of blood vessels, the optic nerve, and a lining of nerve cells (the retina) which detects images transmitted through the cornea, a clear lens-like layer covering of the eye. Ophthalmoscopes are used by doctors to exam the interior of eyes and help diagnose any possible conditions or detect any problems or diseases of the retina and vitreous humor. For instance, a doctor would look for changes in the color the fundus, the size, and shape of retinal blood vessels, or any abnormalities in the *macula lutea* (the portion of the retina that receives and analyzes light only from the very center of the visual field). Typically, special eyedrops are used to dilate the pupils and allow a wider field of view inside the eyeball.

A modern ophthalmoscope (■ Fig. 13.12) consists essentially of two systems: one for illumination and another for viewing. The illuminating system is comprised of light source (a halogen or tungsten bulb), a condenser lens system, a reflector (a prism, mirror, or metallic plate) to illuminate the interior of the eye with a central hole through which the eye is examined. The viewing system is made of a sight hole and a focusing system, usually a rotating wheel with lenses of different powers. The lenses are selected to allow clear visualization of the structures of the eye at any depth and compensate for the combined errors of refraction between patient and examiner.

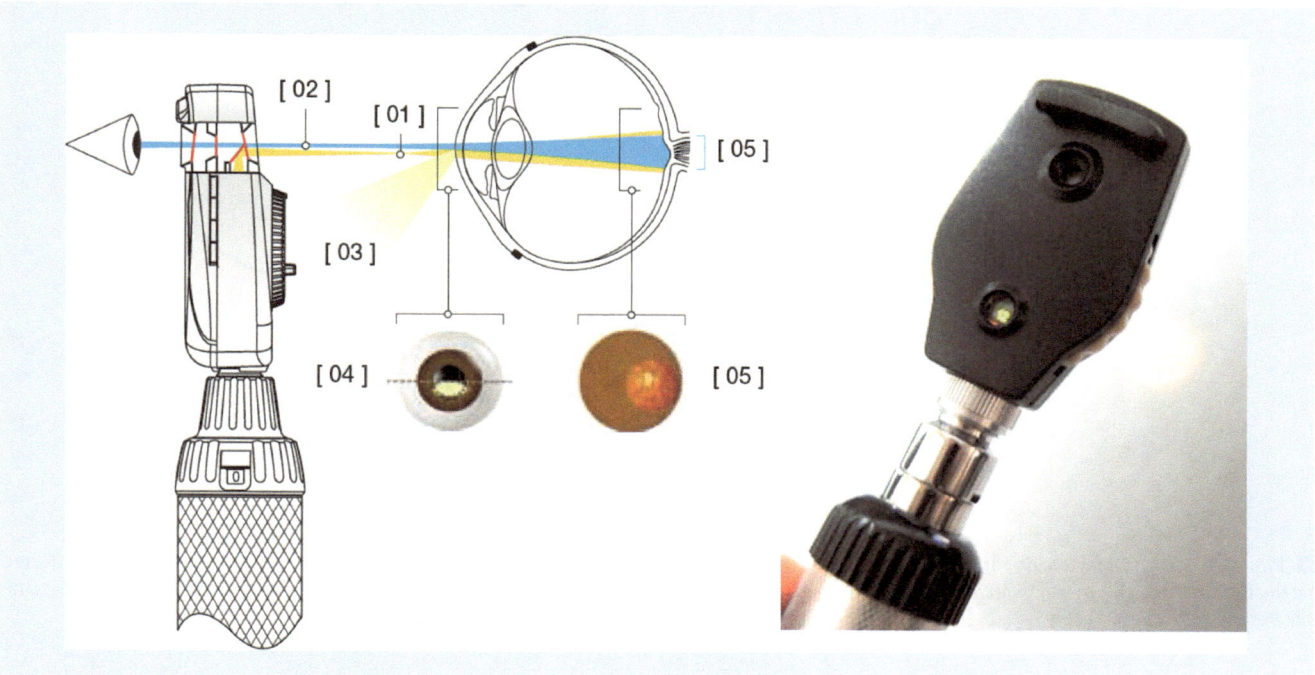

Fig. 13.12 Aspect of a modern ophthalmoscope. A light beam is projected into the eye (*1*). The medical examiner has a direct line of sight into the back of the eye (fundus) (*2*). Path of light reflected of the cornea and iris (*3*). Image observed at the pupil (*4*). Image observed at the back of the eye (*5*) (Images courtesy of Heine)

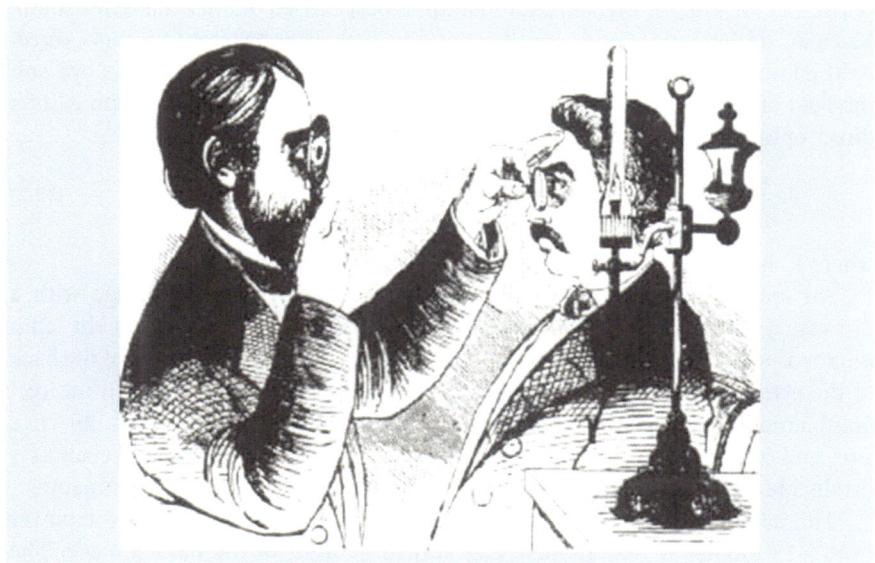

Fig. 13.13 Nineteenth century illustration of Helmholtz original ophthalmoscope

German physician Hermann von Helmholtz is credited with the invention of the ophthalmoscope back in 1851, which he based on an earlier version developed by Charles Babbage in 1847. Helmholtz original ophthalmoscope (see **Fig. 13.13**) was very basic (made or cardboard, glue, and microscope glass plates) but it allowed him to place the eye of the observer in the path of the rays of light entering and leaving the patient's eye, thus allowing the patient's retina to be seen. In 1915, Francis A. Welch and William Noah Allyn invented the world's

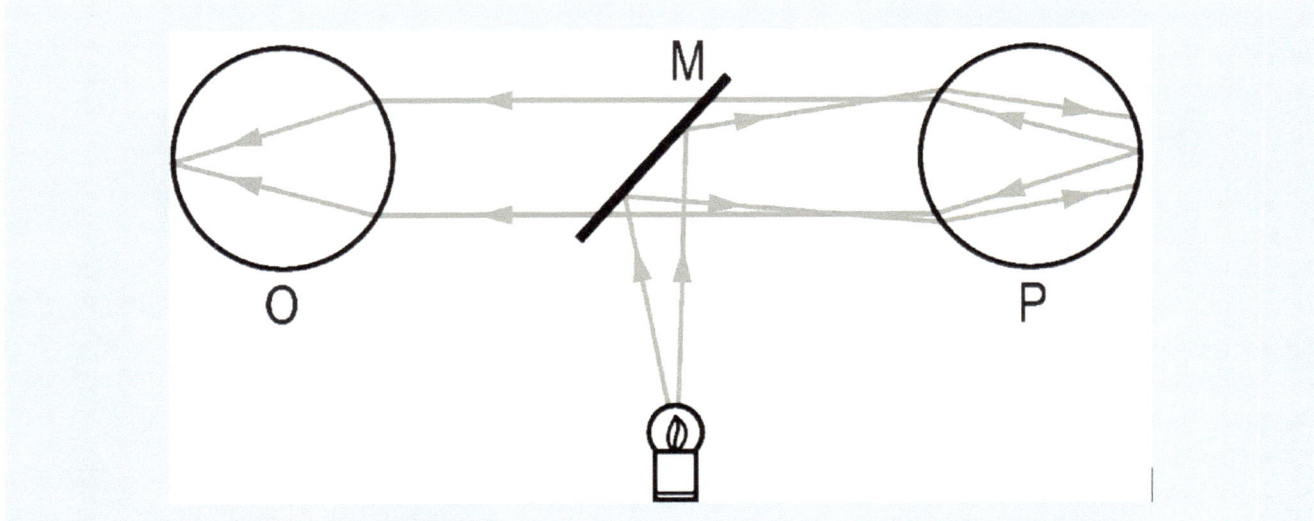

Fig. 13.14 Optical raytracing for a direct ophthalmoscope. Light from the illuminating source is reflected into the eye and then back-reflected by the fundus through a mirror (with either a hole through it or with partial reflectivity). O is the observer's eye, while P is the patient's eye; M, semi-silvered mirror. After [18]

first hand-held direct illuminating ophthalmoscope, and resulted in the formation of the Welch Allyn medical company—still in business today.

There are two types of ophthalmoscope: *direct* and *indirect*. A *direct ophthalmoscope* produces an upright (unreversed) image with 15× magnification. The direct ophthalmoscope is used to inspect the fundus of the eye, which is the back portion of the interior eyeball. Examination is best carried out in a darkened room. Macular degeneration and opacities of the lens can be seen through direct ophthalmoscopy. The instrument is held at close range to the patient's eye and the field of view is small (less than 10°) (■ Fig. 13.14). The magnification M of a direct ophthalmoscope is equal to:

$$M = F_e/4 \tag{13.2}$$

where F_e is the power of the eye.

An *indirect ophthalmoscope* produces an inverted (reversed) image with a 2–5× magnification and formed. A small hand-held lens and either a slit lamp microscope or a light attached to a headband are used to form an image of the back of the eye in space, at approximately arm's length from the doctor. An indirect ophthalmoscope provides a stronger light source, a specially designed objective lens, and opportunity for stereoscopic inspection of the interior of the eyeball. It is invaluable for diagnosis and treatment of retinal tears, holes, and detachments.

This aerial image is usually produced by a strong positive lens ranging in power from +13 diopter to +30 diopter that is held in front of the patient's eye. The practitioner views this aerial image through a sight hole with a focusing lens to compensate for *ametropia* and accommodation. This instrument provides a large field of view (25–40°) and allows easier examination of the periphery of the retina. This instrument has been supplanted by the binocular indirect ophthalmoscope (■ Fig. 13.15). The magnification of an indirect ophthalmoscope M is equal to:

$$M = F_e/F_c \tag{13.3}$$

where F_e and F_c are the powers of the eye and of the condensing lens, respectively.

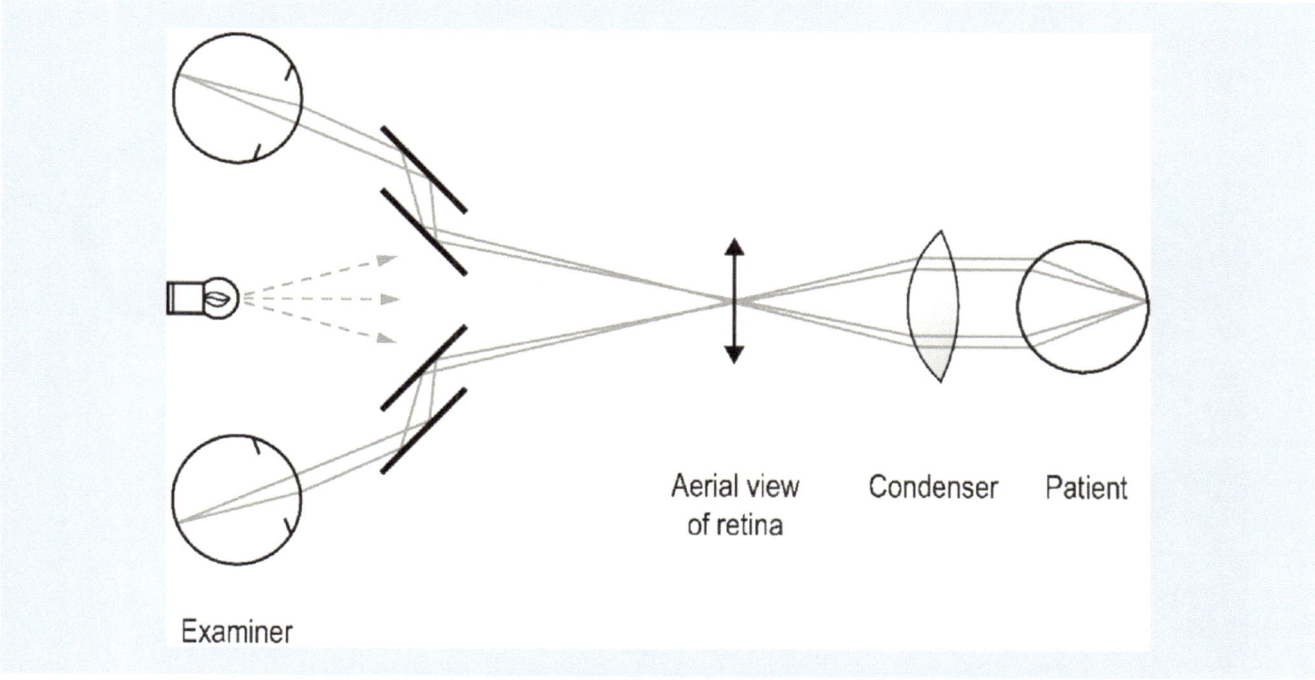

◘ **Fig. 13.15** Optical raytracing for a binocular indirect ophthalmoscope. The light source mounted on the doctor's head illuminates a hand-held condenser lens which forms an inverted stereoscopic image of the retina in free space (aerial image). After [18]

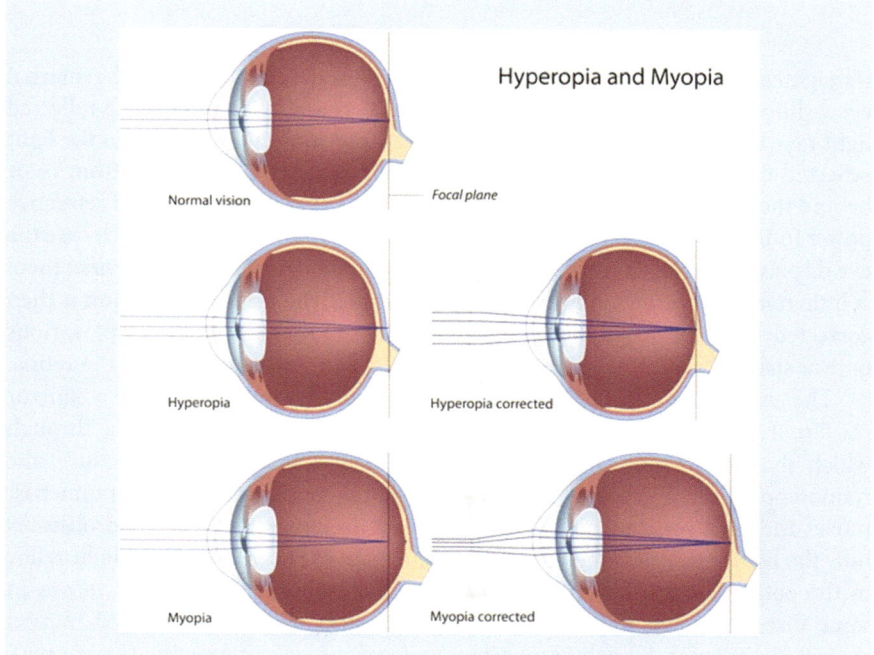

◘ **Fig. 13.16** Types of human vision and associated corrective optical lenses

13.2.4 **Retinoscope**

A *retinoscope* is an optical hand-held device used by optometrists to measure the optical refractive power of the eyes and whether corrective glasses might be needed and the associated prescription value. As shown in ◘ Fig. 13.16, a person can have normal vision (*emmetropia*), *myopia* (nearsightedness), *hyperopia*

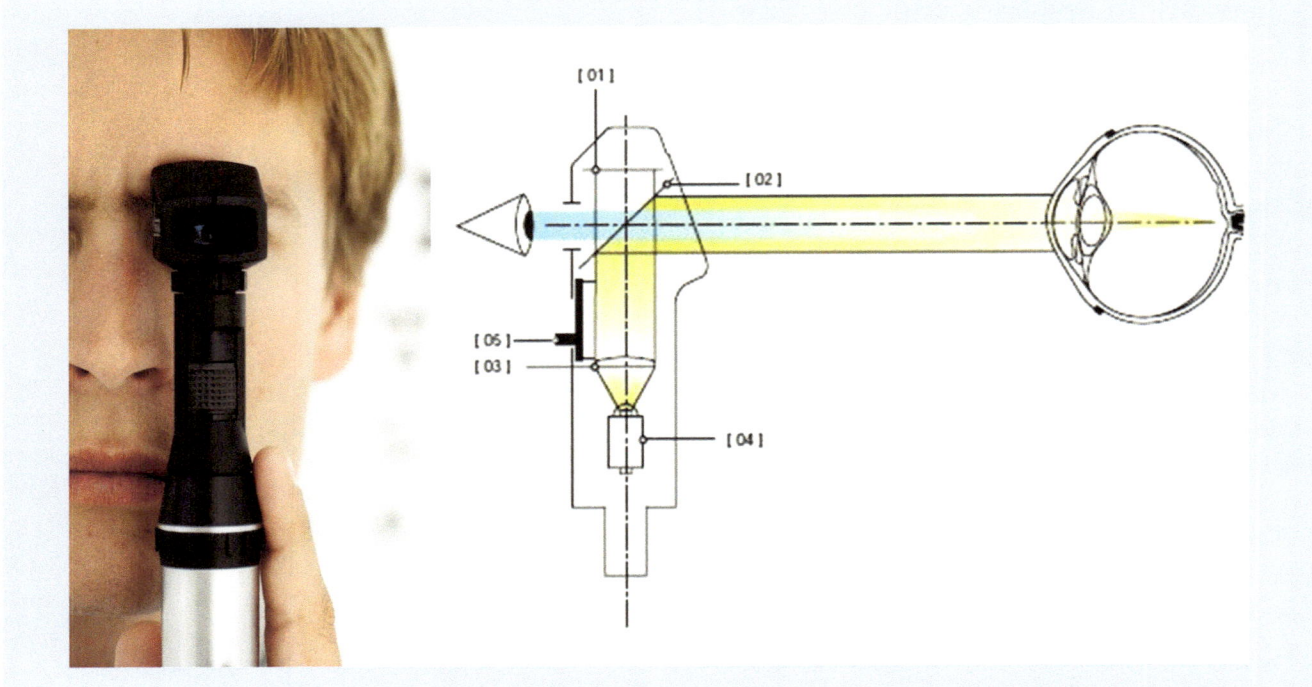

Fig. 13.17 A modern retinoscope. An integrated lamp or LED light source (*4*) shines light through a collimating lens (*3*) onto a partially reflective mirror (*2*), which directs the light to the eye. The back-reflected light from the fundus and the cornea is examined by the doctor through the eyepiece (*1*) and focus adjusted using the lens dial (*5*) (Image courtesy of Heine)

(farsightedness), or astigmatism. The retinoscope is used to illuminate the internal eye (while the patient is looking a far fixed object) and observe how the reflected light rays by the retina (called the *reflex*) align and move with respect to the light reflected directly off the pupil [19]. If the input light beam focuses in front of or behind the retina, there is a "refractive error" of the eye. A high degree of refractive power indicates that the light focus remains in front of the retina, in which case the eye displays myopia. Conversely, if the focal spot happens behind the retina, there is little refractive power and the eye has hyperopia. The error of refraction is then corrected by using a *phoropter*, which introduces a series of lenses of various optical strengths until the retinal reflex focuses at the right position on the retina.

The retinoscope consists of a light, a condensing lens, and a mirror (■ Fig. 13.17). The mirror is either semi-transparent or has a hole through which the practitioner can view the patient's eye. During the procedure, the retinoscope shines a beam of light through the pupil. Then, the optometrist moves the light vertically and horizontally across the patient's eye and observes how the light reflects off the retina (see pictures in ■ Fig. 13.18). If the light reflex in the patient's pupil moves "with" or "against" motion. If the reflex moves in same direction, then the correction requires plus power (myopia) and motion against direction of the retinoscope, means negative power correction (hyperopia).

To determine the corrective refractive lens power needed, lenses of increasing refractive power are placed in front of the eye and the change in the direction and pattern of the reflex is observed. The optometrist keeps changing the lenses until reaching a lens power that provides adequate focusing on the retina, which manifests as alignment of the reflex with the streak light image outside of the pupil.

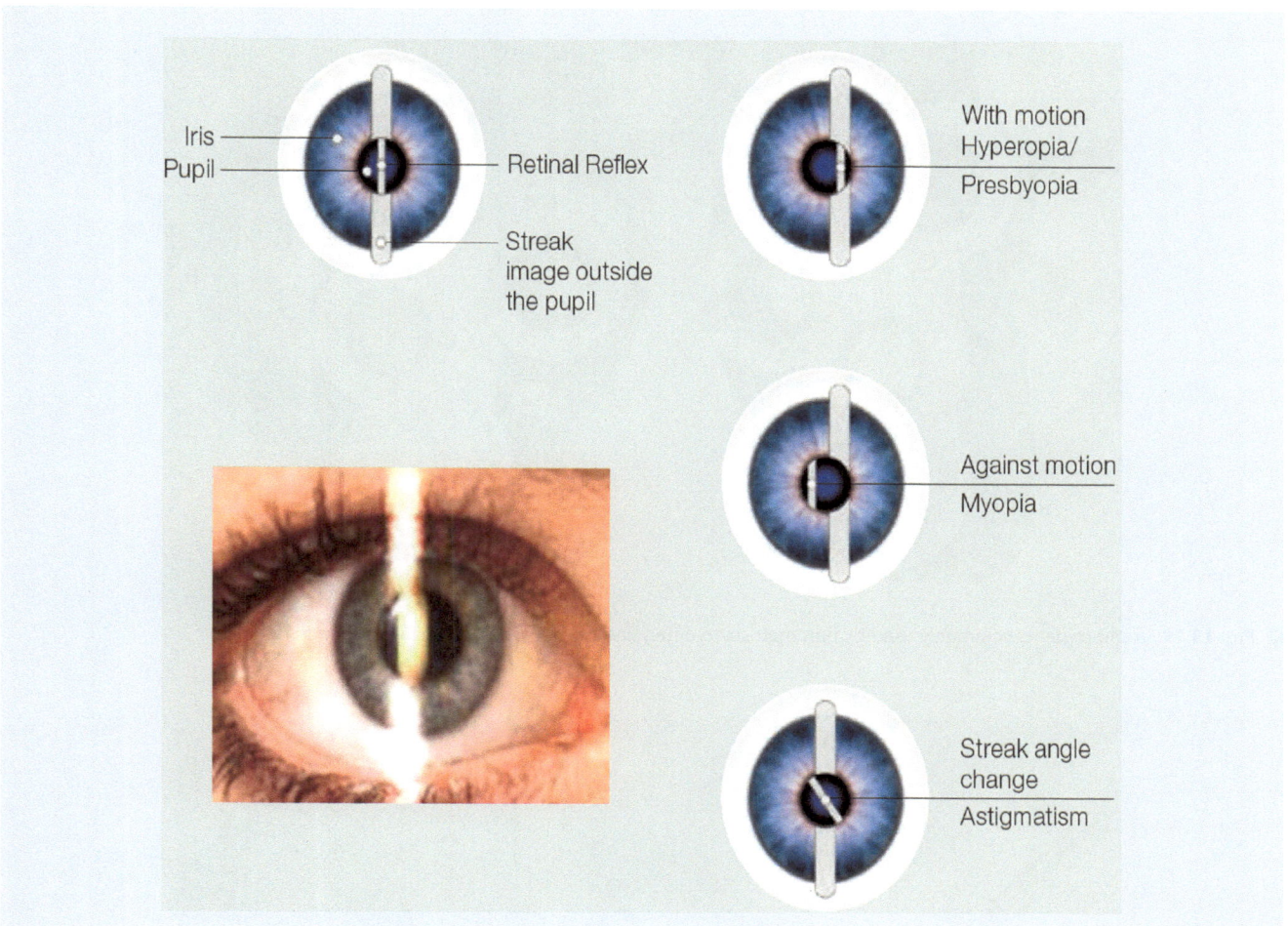

🔹 **Fig. 13.18** Aspect of reflex images from the human eye seen by a doctor using a retinoscope. If the reflex moves in same direction, then myopia is detected; if reflex is noted on motion against direction of the retinoscope, hyperopia is present. If the reflex line is oblique instead of vertical, then astigmatism is present. An aligned reflex means correct vision (*Source*: Heine)

13.2.5 Phoropter

A *phoropter* is an ophthalmic binocular refracting testing device, also called a *refractor*. It is commonly used by ophthalmologists, optometrists, and eye care professionals during an eye examination to determine the corrective power needed for prescription glasses. It is commonly used in combination with a retinoscope.

🔹 Figure 13.19 shows a photograph of phoropter which consists in double sets (one for each eye) of rotating discs containing convex and concave spherical and cylindrical lenses, occluders, pinholes, colored filters, polarizers, prisms, and other optical elements. The patient sits in front of the device and the lenses within a phoropter refract light in order to focus images on the patient's retina at the right spot to compensate for each individual eye refractive errors. The optical power of these lenses is measured in 0.25 diopter increments. By changing these lenses, the examiner is able to determine the spherical and cylindrical power, and cylindrical axis necessary to correct a person's refractive error. These instruments were first devised in the early to mid-1910s.

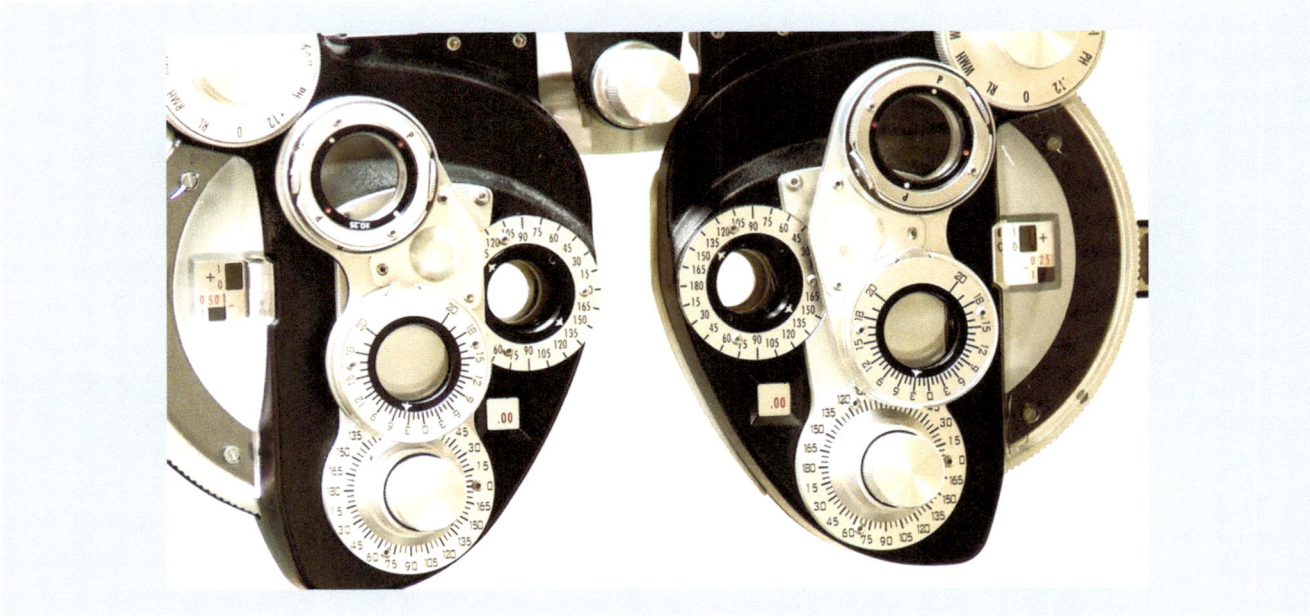

Fig. 13.19 A phoropter is commonly used by optometrists to determine the necessary corrective lens prescription

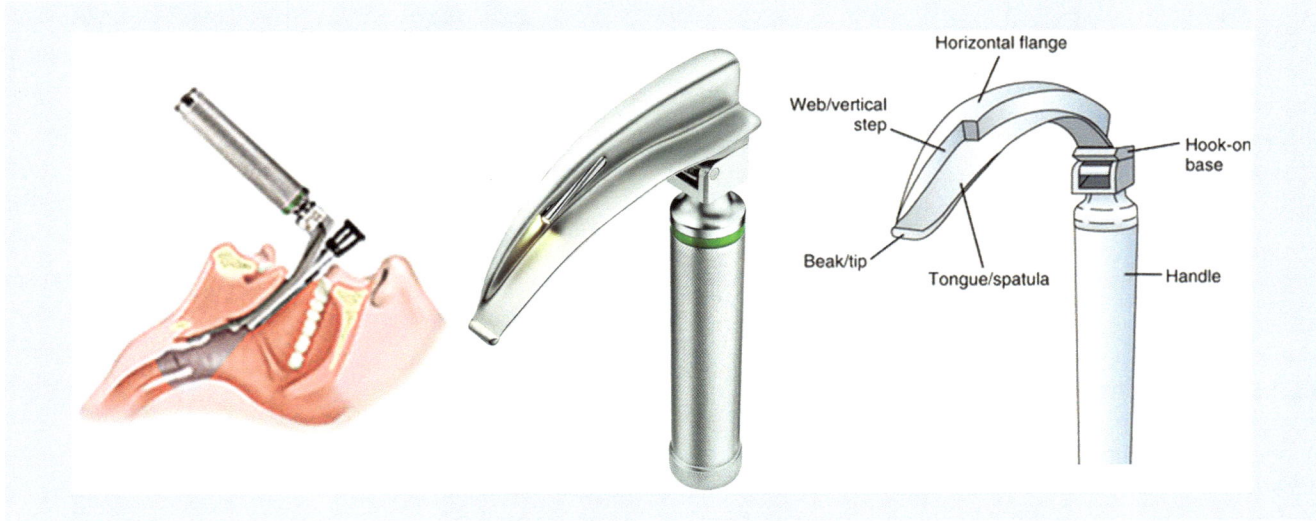

Fig. 13.20 A direct laryngoscope and its insertion into a human throat for examination

13.2.6 **Laryngoscope**

A *laryngoscope* is an optical instrument used for examining the interior of the larynx and structures around the throat. There are two types of laryngoscopes: *direct* and *indirect* laryngoscopes. A direct laryngoscope (■ Fig. 13.20) consists of a handle containing batteries, an integrated light source, and a set of interchangeable blades for easy reach and placement into a patient's throat. Besides being used for visualization of the glottis and vocal cords, a direct laryngoscope may also be used during surgical procedures to remove foreign objects in the throat, collect tissue samples (biopsy), remove polyps from the vocal cords, perform laser treatments and, very commonly, as a tool aid to facilitate tracheal intubation during general anesthesia or in cardiopulmonary resuscitation.

The blades in a laryngoscope help provide leverage to open wide the mouth and throat, as well as to keep the tongue in place and avoid a gag reflex. There are two basic styles of laryngoscope blades most commonly used: curved and straight. The Macintosh blade is the most widely used of the curved laryngoscope blades, while the Miller blade is the most popular style of straight blade. Blades come in different sizes, to accommodate different patients.

An *indirect* laryngoscope consists of a combination of a small mirror mounted at an angle on a long stem and a light source. The mirror is usually circular in form and made in various sizes, but is small enough to be placed in the throat behind the back of the tongue. The source of light is either a small bright lamp worn on the forehead of the observer, or a concave mirror, also worn on the forehead, for the purpose of concentrating light from some other source. Light is reflected to the back of the throat by the mirror and directed to illuminate up the interior of the larynx. The mirror also serves to reflect back to the doctor an image of the throat, to appreciate the structure of the glottis and vocal cords.

Some historians credit Benjamin Guy Babington (1794–1866), with the invention of the laryngoscope back in 1829 [20], who called his device the *glottiscope*. However, Manuel Garcia (1805–1906)—a Spanish tenor and singing maestro—experimented back in 1854 with a combination of throat mirror and light to observe the action of his own vocal cords and larynx when producing tones and sounds. His observations were published in the Royal Philosophical Magazine and Journal of Science in 1855 [21], and they constitute the first physiological records of the human voice as based upon observations in the living subject. For this, he is also recognized as the original inventor of the laryngoscope. ◘ Figure 13.21 shows a photograph and illustration of his original laryngoscope device.

Mirror-based laryngoscopy for the investigation of laryngeal pathology was pioneered back in 1858 by Johann Czermak, a professor of physiology at the University of Budapest. Czermak applied an external light source and a head-mounted mirror to improve visualization. During this period of time, a laryngoscopic examination was made as depicted in ◘ Fig. 13.22. The patient opens his mouth as widely as possible, protruding his tongue. The doctor, with a small napkin takes the protruded tongue between his thumb and forefinger and holds it in place, so as to enlarge opening of the mouth as much as possible. The laryngeal mirror is next inserted and dexterously positioned to the back of the mouth to direct the light from the external light source (mirror or lamp) into the back of the throat. An image of the lower throat is reflected back by the mirror for the doctor to view and assess the condition of the larynx.

All previous observations of the glottis and larynx had been performed under indirect vision (using mirrors) until 1895, when Alfred Kirstein (1863–1922) of Germany performed the first direct laryngoscopy in Berlin, using an *esophagoscope* he had modified for this purpose, calling device an *autoscope*, and the modern, direct laryngoscope was born [22].

13.3 Fiber Optic Medical Devices and Applications

The field of fiber optics has undergone a tremendous growth and advancement over the last 50 years. Initially conceived as a medium to carry light and images for medical endoscopic applications, optical fibers were later proposed in the mid-1960s as an adequate information-carrying medium for telecommunication applications. Ever since, optical fiber technology has been the subject of considerable research and development to the point that today light wave communication systems have become the preferred method to transmit vast amounts of data and information from one point to another.

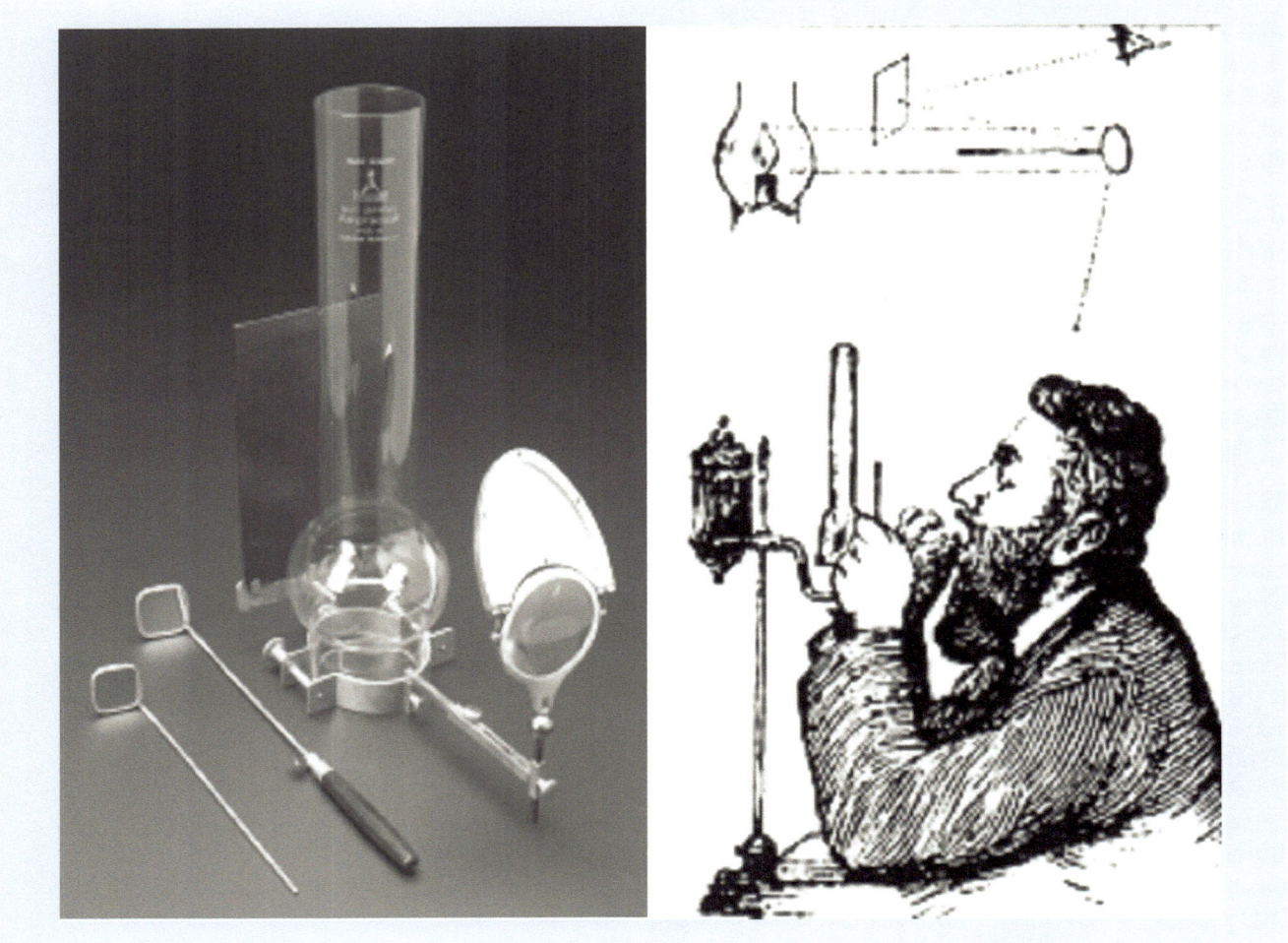

Fig. 13.21 Original indirect laryngoscope developed by Manuel Garcia to view the movement of his won vocal chords (c. 1870)

Given their EM immunity, intrinsic safety, small size and weight, autoclave compatibility and capability to perform multi-point and multi-parameter sensing remotely, optical fibers and fiberoptic-based devices are seeing increased acceptance and new uses for a variety of biomedical applications—from diverse endoscopes, to laser-delivery systems, to disposable blood gas sensors, and to intra-aortic probes. This section illustrates—through several application and product examples—some of the benefits and uses of biomedical fiber sensors, and what makes them such an attractive, flexible, reliable, and unique technology.

13.3.1 Optical Fiber Fundamentals

At the heart of this technology is the optical fiber itself. A hair-thin cylindrical filament made of glass (although sometimes are also made of polymers) that is able to guide light through itself by confining it within regions having different optical indices of refraction. A typical fiber structure is depicted in ■ Fig. 13.23. The central portion—where most of the light travels—is called the core. Surrounding the core there is a region having a lower index of refraction, called the cladding. From a simple point of view, light trapped inside the core travels along the fiber by bouncing off the interfaces with the cladding, due to the effect of the total internal

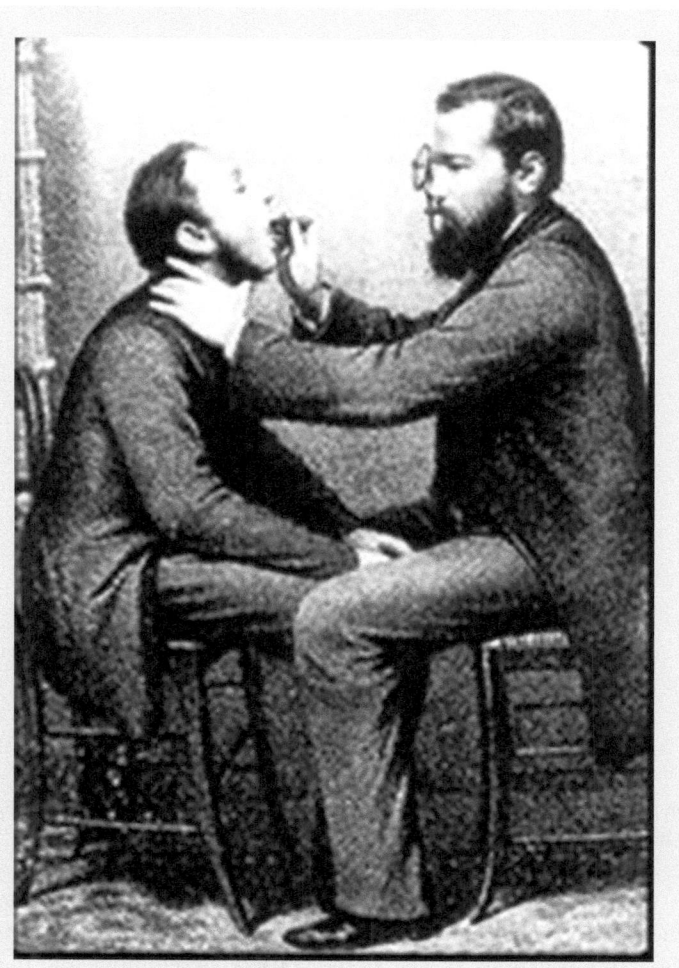

Fig. 13.22 Nineteenth century illustration of a mirror-based laryngoscope examination of a patient's throat

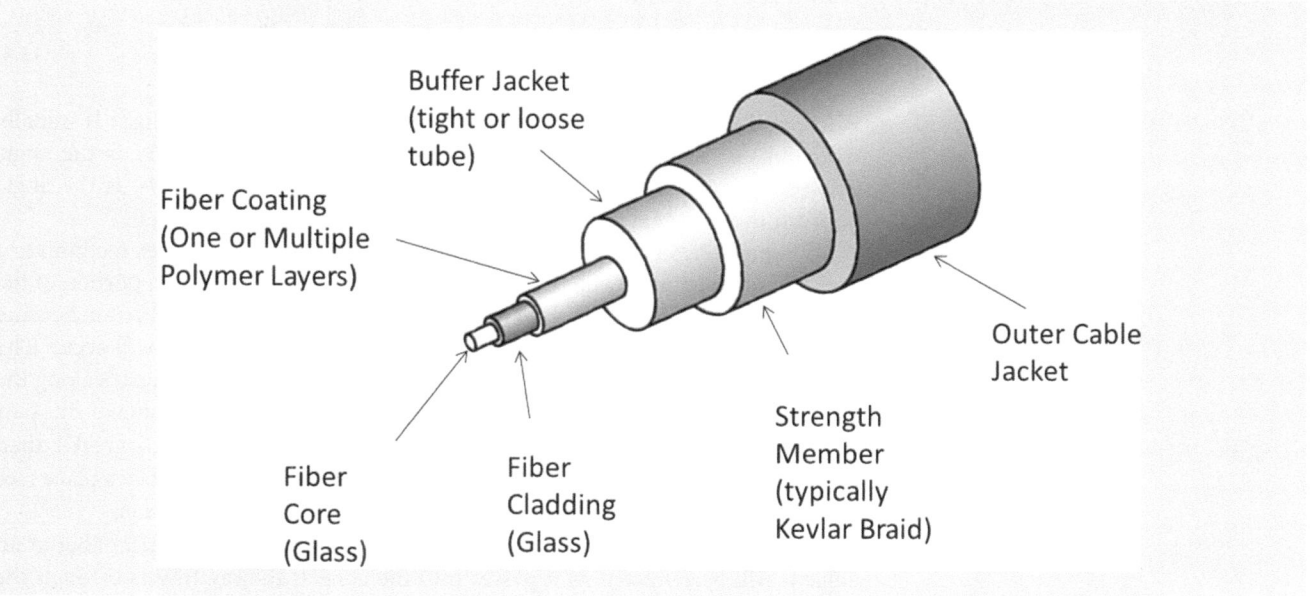

Fig. 13.23 Schematic of an optical fiber

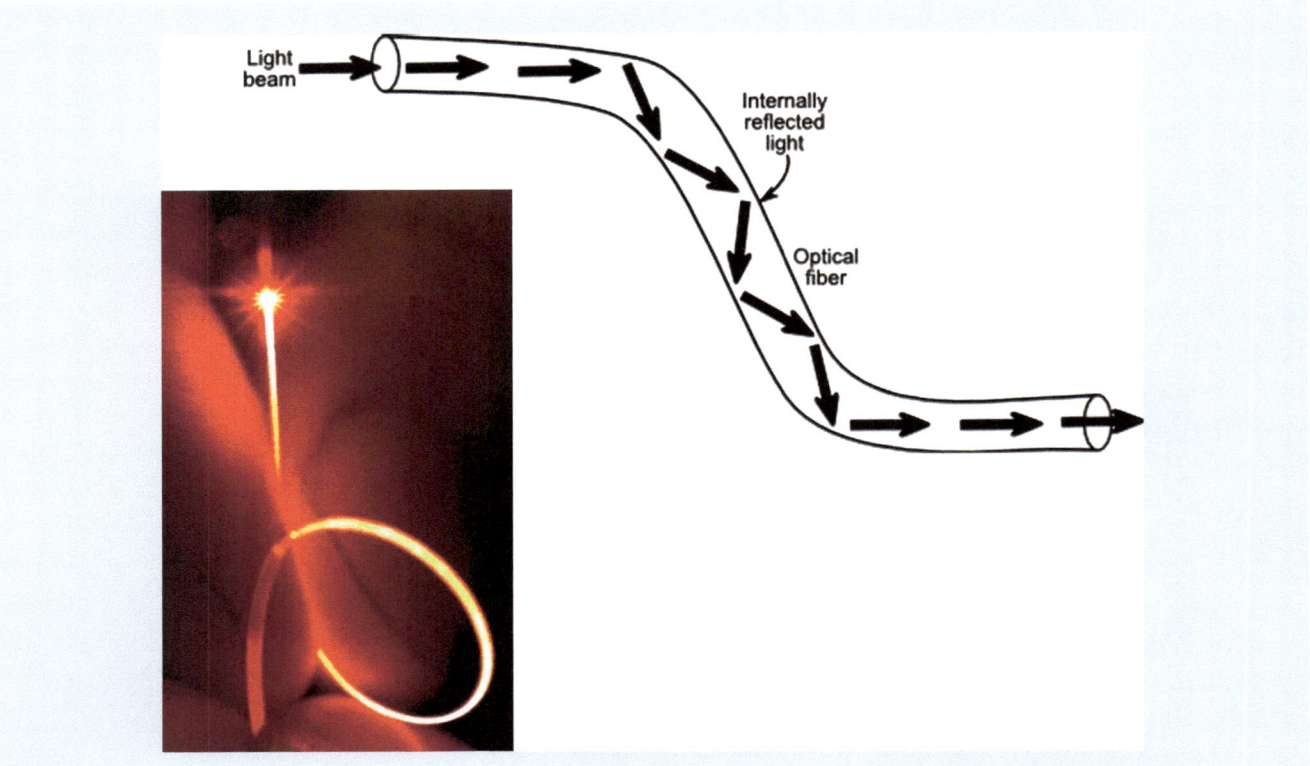

Fig. 13.24 An optical fiber is able to guide light through the principle of total internal reflection. This allows the transmission of light energy (and signals) through any patch or shape taken by the optical fiber

reflection occurring at these boundaries (■ Fig. 13.24). In reality though, the optical energy propagates along the fiber in the form of waveguide modes that satisfy Maxwell's equations as well as the boundary conditions and the external perturbations present at the fiber.

Refraction occurs when light passes from one homogeneous isotropic medium to another; the light ray will be bent at the interface between the two media. The mathematical expression (Eq. (13.4)) that describes the refraction phenomena is known as Snell's law,

$$n_0 \sin \phi_0 = n_1 \sin \phi_1 \tag{13.4}$$

where n_o is the index of refraction of the medium in which the light is initially travelling, n_1 is the index of refraction of the second medium, Φ_o is the angle between the incident ray and the normal to the interface, and Φ_1 is the angle between the refracted ray and the normal to the interface.

■ Figure 13.25a shows the case of light passing from a high-index medium to a lower-index medium. Even though refraction is occurring, a certain portion of the incident ray is reflected. If the incident ray hits the boundary at ever-increasing angles, a value of $\phi_0 = \phi_c$ will be reached, at which no refraction will occur. The angle ϕ_c is called the critical angle. The refracted ray of light propagates along the interface, not penetrating into the lower-index medium, as shown in part ■ Fig. 13.25b. At that point, $\sin \phi_c$ equals to unity. For angles ϕ_0 greater than ϕ_c, the ray is entirely reflected at the interface, and no refraction takes place (see ■ Fig. 13.25c). This phenomenon is known as *total internal reflection.*

In ■ Fig. 13.26, a ray of light incident upon the end of the optical fiber at an angle θ will be refracted as it passes into the core. If the ray travels through the high-index medium at an angle greater than Φ_c it will reflect off of the cylinder

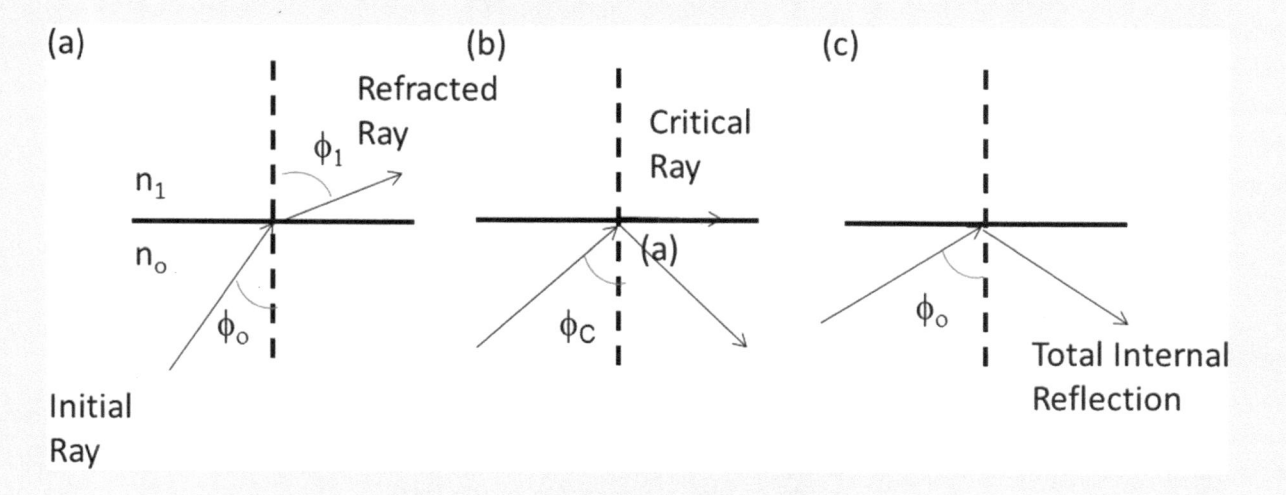

Fig. 13.25 Light reflection and refraction between two media with different indices of refraction

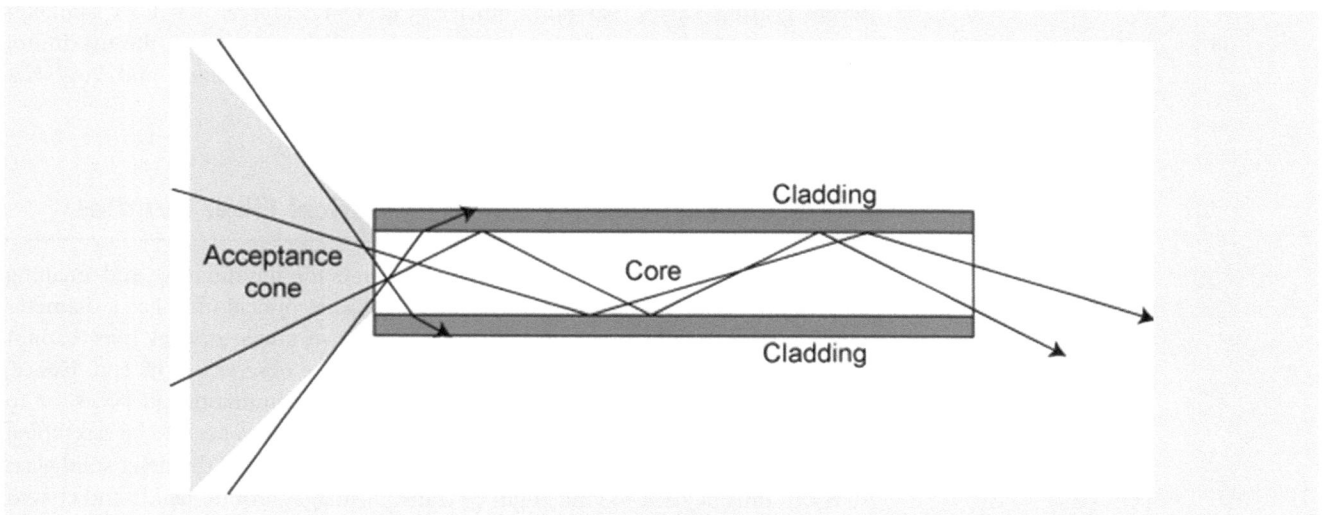

Fig. 13.26 Light ray propagation along the core of an optical fiber. At the entrance to the fiber, a conical region is defined by the so-called acceptance angle, which is the region in space where light can be effectively collected and coupled into the fiber for guiding

wall, will have multiple reflections, and will emerge at the other end of the optical fiber. For a circular fiber, considering only meridional rays, the entrance and exit angles are equal. Considering Snell's law for the optical fiber, core index n_0, cladding index n_1, and the surrounding media index n,

$$
\begin{aligned}
n \sin \theta &= n_0 \sin \theta_0 \\
&= n_0 \sin \left(\frac{\pi}{2} - \phi_c \right) \\
&= n_0 \left[1 - (n_1 - n_0)^2 \right]^{1/2} \\
&= \left(n_0^2 - n_1^2 \right)^{1/2} = \text{Numerical Aperture.}
\end{aligned}
\tag{13.5}
$$

The term $n\sin\theta$ is defined as the *numerical aperture* or NA for short. The NA is determined by the difference between the refractive index of the core and that of

13

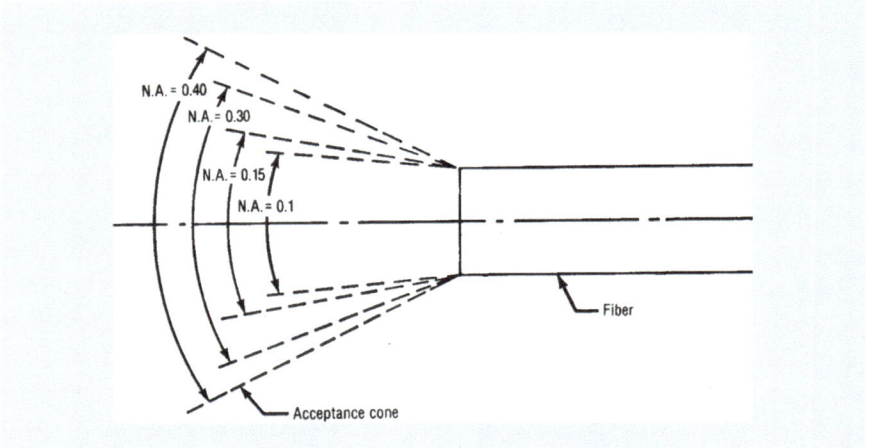

Fig. 13.27 The greater the NA of an optical fiber, the bigger the acceptance cone, and broader the angle of capture of light by the fiber

the cladding. It is a measure of the light-acceptance capability of the optical fiber. As the NA increases, so does the ability of the fiber to couple light into the fiber, as shown in Fig. 13.27. The larger NA allows the fiber to couple in light from more severe grazing angles. Coupling efficiency also increases as the fiber diameter increases, since the large fiber can capture more light. Therefore, the maximum light-collection efficiency occurs for large-diameter-core fibers and large-NA fibers.

13.3.2 Coherent and Incoherent Optical Fiber Bundles

In medicine, optical fibers have been considered for illuminating and imaging applications since the 1920s. Typically, a single glass optical fiber has a diameter ranging from 1 mm down to ~8 μm. However, a single optical fiber cannot transmit an image—only a bright light spot would be observed at its end. Hence, in order to carry a reasonable amount of light for illumination purposes, or to transmit and image, hundreds to thousands of optical fibers need to be assembled into bundles. Bundles of multiple single optical fibers of small diameter solid glass rods can thus be used to guide light or transmit images around bends and curved trajectories.

Glass optical fiber bundles are of two types: *incoherent* and *coherent*. An *incoherent* bundle consists of a collection of fibers randomly distributed in the bundle and is typically intended for illumination purposes only. In contrast, a *coherent* optical fiber bundle has an ordered array of fibers in which the relative position of each individual fiber at its input and output with respect to the bundle is maintained. That is to say, the position of individual fibers is at same locations over the cross section of both bundle ends as depicted in Fig. 13.28. In between the ends, the fibers need not have a fixed orientation and can move flexibly. Coherent bundles are used for conveying an image from one end to the other by the effect created by the grouping of the individual light conducted by each fiber which is perceived in the eye of the observer as a full image. To achieve better image quality and resolution, a large number of small diameter fibers are need for a given bundle diameter. Typically, fibers used in bundles have diameters on the order of 8–12 μm and their count can range from about 2000 up to 40,000 [23]. In the case of imaging bundles, larger diameter fibers are used of 30–50 μm in diameter.

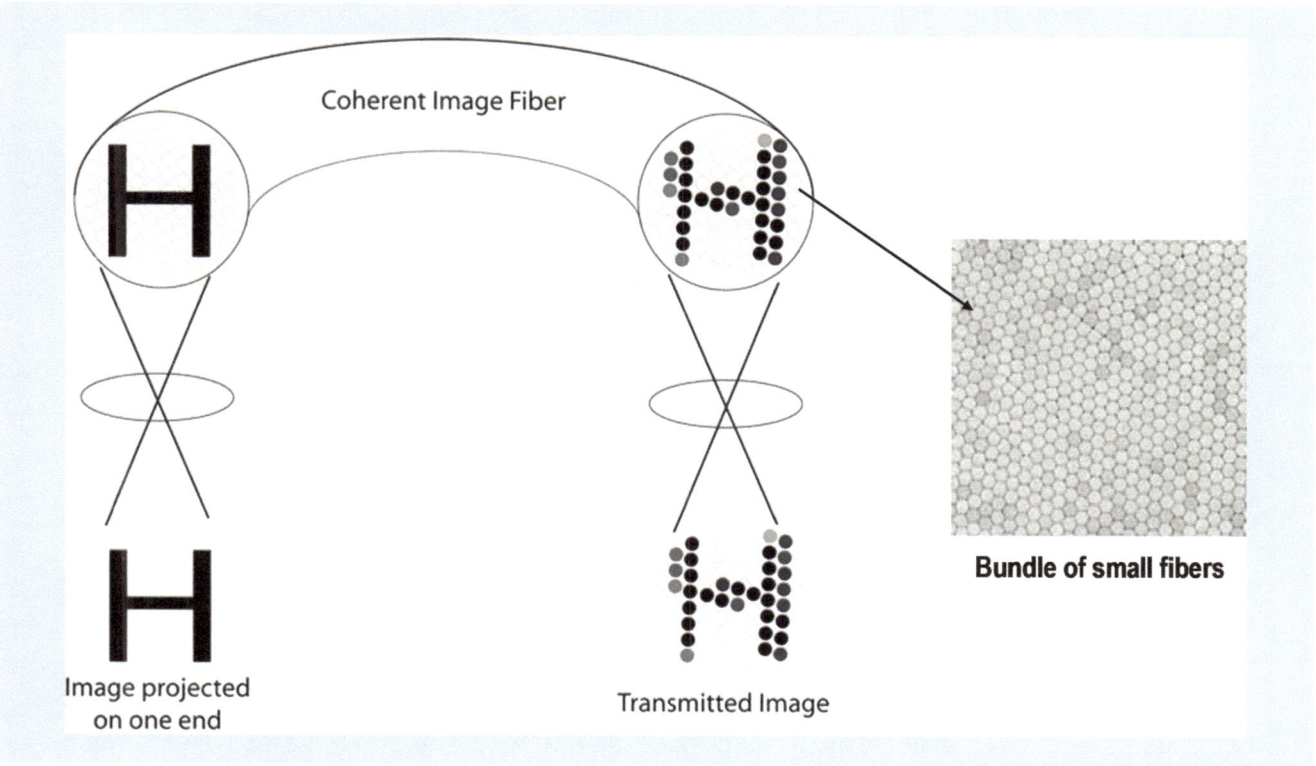

Fig. 13.28 A coherent optical fiber bundle. Images are accurately transmitted by preserving the relative position of the fibers at each end of the bundle. The bundle consists of a multitude of individual glass fibers of a small diameter (~12 μm) that create a lattice effect

Fabrication of illumination (non-coherent) and imaging (coherent) bundles is based on the same processes of drawing optical fibers or glass rods through heating furnaces and doing repeated draws of multi-stack sets, to achieve arrays with the desired quantity of fibers of the appropriate diameter. There are three common to fabrication methods for coherent bundles: fused image bundles, wound image bundles, and leached image bundles. ☐ Figure 13.29 illustrates the three steps needed to fabricate fused as well as leached image fiber bundles.

An individual fiber (or rod) is made by starting with a so-called perform made by the tube-in-rod technique where a single glass rod (which will become the fiber's core) is inserted into a tube made of glass with a lower refractive index (cladding). In the case of a leached bundle, an additional glass jacket made of a leachable glass is used. This glass perform is placed in an electric heating furnace that runs at a temperature close to the softening point of the glass. The heat causes the solid glass road to soften. Once soft, the glass is pulled down into a thin filament by a pulling mechanism. The final diameter of the filament is controlled by the ratio of the speeds between the advancing preform and the drawn fiber. Typically, the initial drawn fiber is more of a solid rod with a 2 mm diameter. In the next drawing stage, a multitude of mono fibers are stacked together and drawn in the furnace to produce a multi-fiber rod. The drawn filament from a multi-fiber preform consists of several 100 monofilament fibers. In the third stage, several multi-fiber rods are stacked together to perform the so-called multi-multi drawing process. The multi-multi stack assembly is fed through the furnace and drawn into a filament of rod of the desired diameter. Such filament will be composed of thousands of individual glass fibers. As shown in ☐ Fig. 13.30, imaging multi-fiber arrays of square, circular, or hexagonal shape and in different sizes can be fabricated with this process.

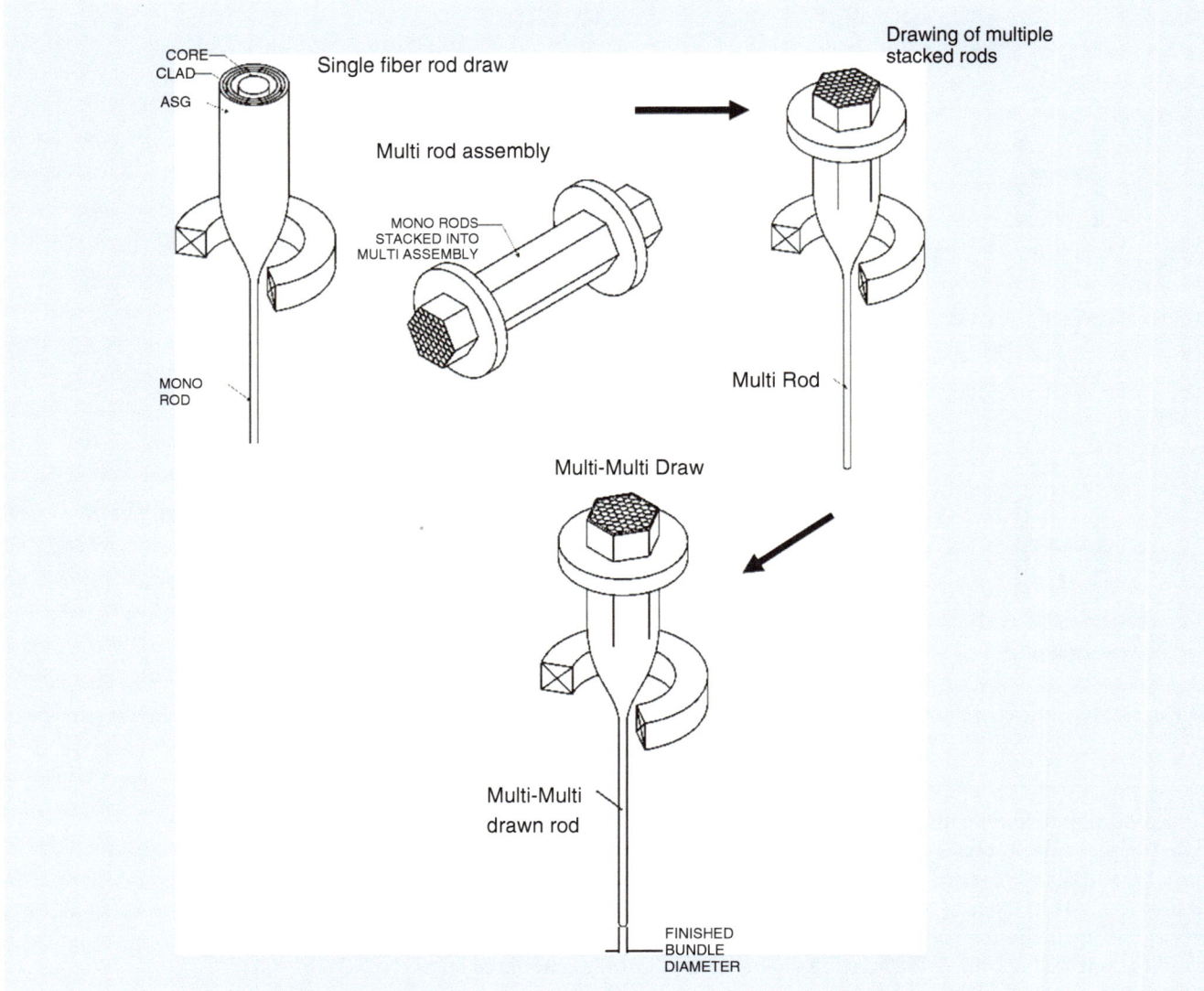

Fig. 13.29 Fabrication process to make fiber bundles

In the particular case of leached fiber bundles, each bundle end is properly secured and the entire bundle is soaked in an acid solution which will dissolve the leachable glass, allowing the fibers to move freely between the bundle ends.

Wound imaging bundles are made by winding a multi-fiber array as a single layer on a drum, and then stacking the desired number of layers manually in a laminating operation.

13.3.3 **Illuminating Guides**

Fiber optic illuminating guides are non-coherent and are used primarily to guide light to a desired point to provide illumination and enhance visual clarity. Imaging bundles are typically made of 30–50 μm diameter fibers, with NA values around 0.6. Most commonly, illuminating bundles are used as part of fiberscopes, endoscopes, and personal lights for surgeons. As seen in ◘ Fig. 13.31, when surgeons are operating on a patient, they need cool, bright light to help then see better tissues and organs—the closer the direct illumination to the operating field,

◙ **Fig. 13.30** Photograph of different styles and shapes of drawn coherent optical fiber bundles

the better. Rigid, light-guiding rods are also made (from single solid glass rods or from multi-core rods) for applications in dentistry and light therapy (see ◙ Fig. 13.32).

13.3.4 Fiberscopes and Endoscopes

An endoscope is an optical instrument used for direct visual inspection of hollow organs or body cavities. Typically, an endoscope is generally introduced through a natural opening in the body (◙ Fig. 13.33), but it may also be inserted through an incision. Instruments for viewing specific areas of the body include the bronchoscope, colonoscope, cystoscope, gastroscope, laparoscope, proctoscope, and several others. Although the design may vary according to the specific use, all endoscopes have similar construction and elements: an objective lens (distal end), illuminating fiber bundle, imaging coherent fiber bundle, fixed or articulating handle, and an eyepiece (proximal end). Accessories that might be used for diagnostic or therapeutic purposes include irrigation channels, suction tips, tubes, and suction pump; forceps for removal of biopsy tissue or a foreign body; biopsy brushes; an electrode tip for cauterization; as well as a video camera, video monitors, and image recorder. Many modern endoscopes have also articulating ends, that are remotely controlled by the doctor using knobs on the handle that adjust pull wires inside the body of the endoscope. ◙ Figure 13.34 shows a modern, flexible, and fiber-optic endoscope.

Endoscopes can be rigid or flexible as depicted in ◙ Fig. 13.35. Modern endoscopes (both flexible and rigid) make use of fiber optic imaging bundles to achieve image transmission. However, earlier models relied on miniature flat or rod lenses to guide images from the objective end to the eyepiece as shown in ◙ Fig. 13.36. Hippocrates II (460–377 BC) reported using catheters and primitive

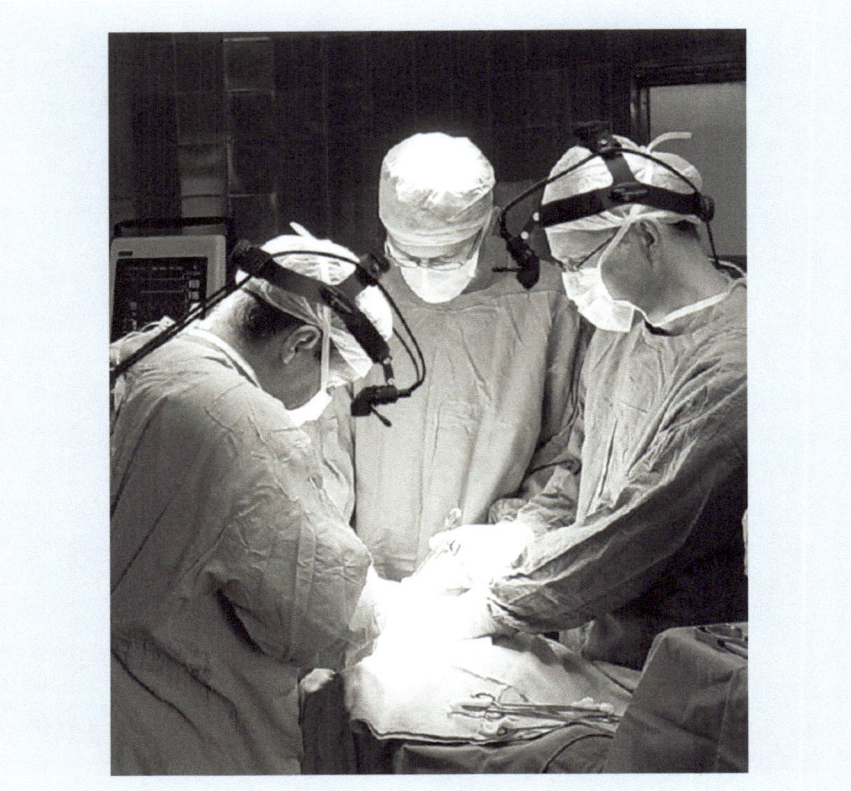

Fig. 13.31 Fiber optic illuminators used in the operating room by surgeons

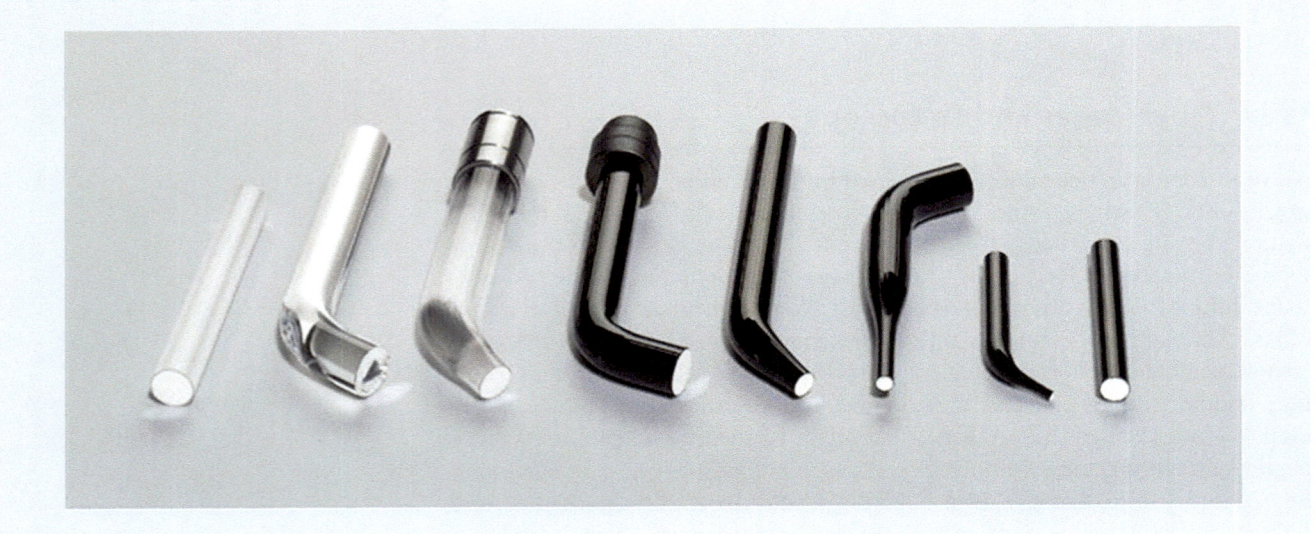

Fig. 13.32 Solid fiber optic illuminating rods

forms of visualization tubes over two millennia ago. In the nineteenth century, endoscopy was very rudimentary and relied on the insertion of long, rigid metal tubes into body cavities. In 1910 Victor Elner used a gastroscope to view the stomach, while in 1912 the first semi-flexible gastroscope was developed. Then, Heinrich Lamm was the first person to transmit images through a bundle of optical

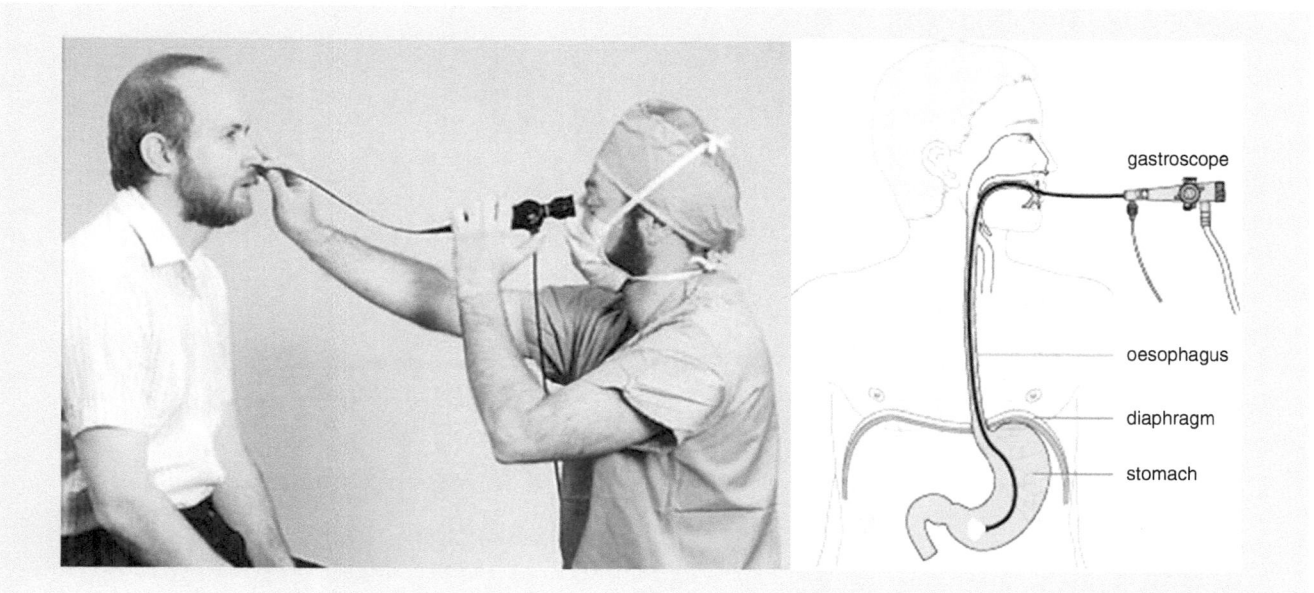

Fig. 13.33 Endoscopes are commonly used by doctors to inspect patients' internal organs through body cavities, such as the nose or throat

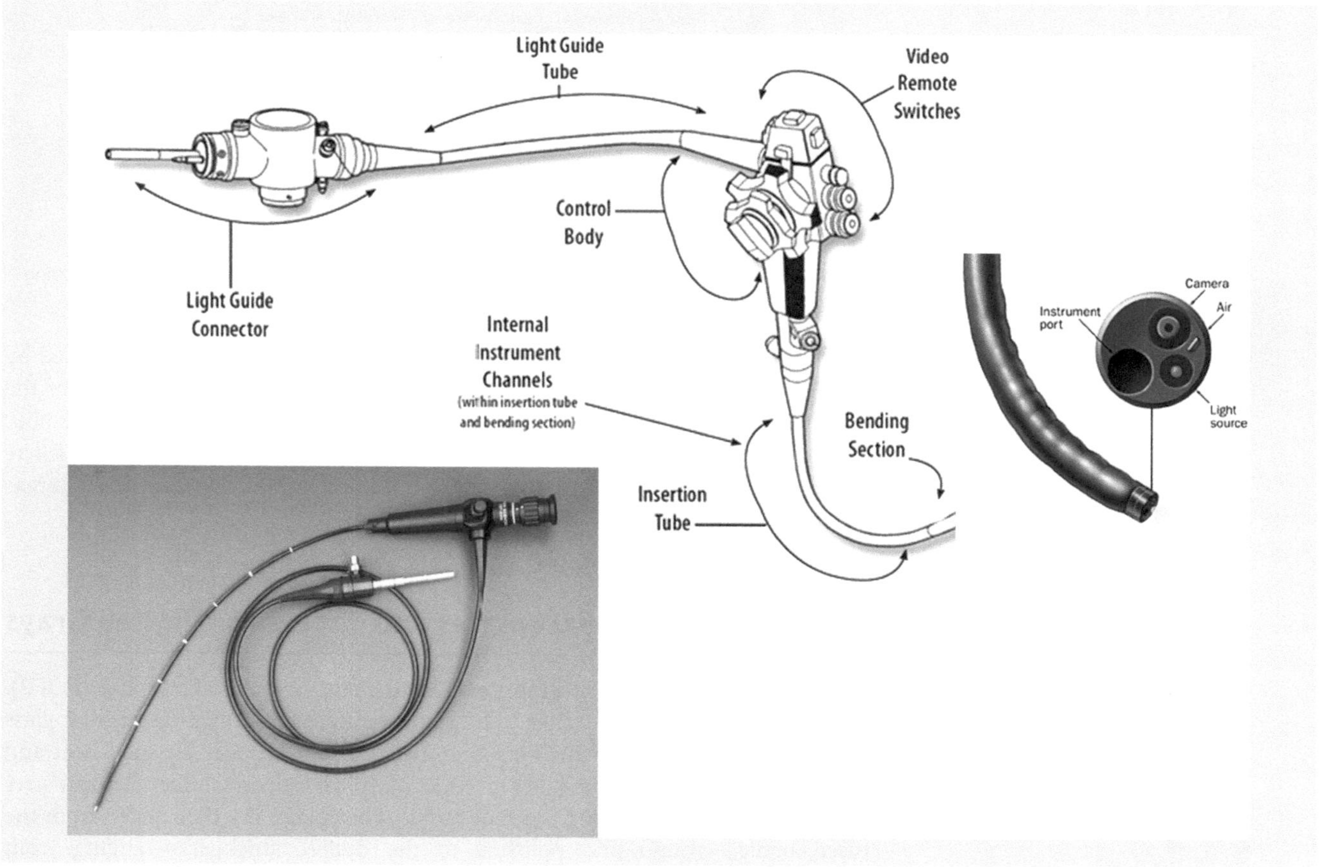

Fig. 13.34 Aspect of a modern, flexible fiberscope fitted with articulating knobs, camera lens, and instrument port on the distal end

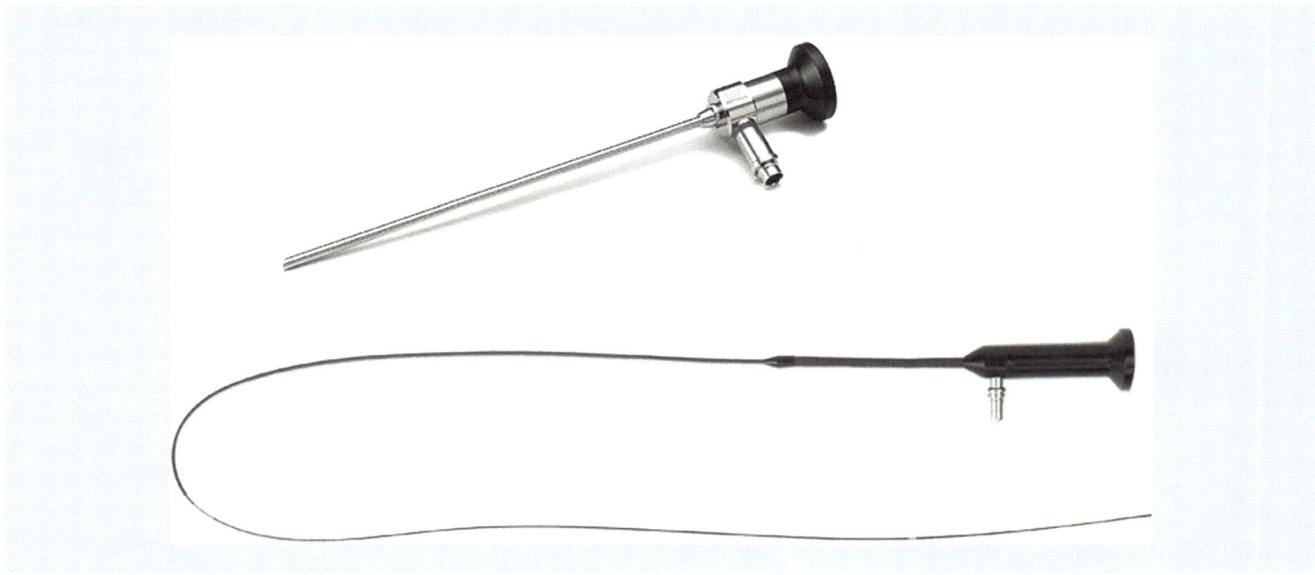

Fig. 13.35 Examples of a rigid (*upper*) and flexible (*lower*) endoscopes

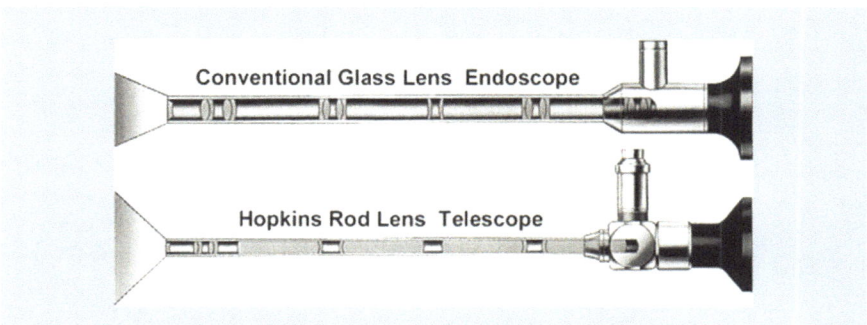

Fig. 13.36 Early designs of rigid endoscopes using rod or glass lenses for image relaying

fibers in 1930. In 1957, clad optical fibers were first proposed and developed by Lawrence Curtis as a graduate student at the University of Michigan, under the supervision of Dr. Basil Hirschowitz, who in 1957 demonstrated the first fiber optic endoscope [24]. From then on, the devices became known as *fiberscopes*. The fiberoptic endoscope has great flexibility, reaching previously inaccessible areas and has become the norm in medicine.

13.3.5 Fused Fiber Faceplates and Tapers for Digital X-rays

Another type of coherent imaging conduit is the fiber optic fused faceplate (FOFP). FOFPs are made as pre-arranged blocks of multiple pre-drawn multi-fiber glass rods (known as boules), which are then fused together under elevated heat and pressure to form a solid piece (Fig. 13.37). Typical individual fiber element sizes range from as small as 4 to 25 μm or larger. Thin plates are then sliced from the fused boule, ground and polished to the desired thickness—ranging from ~100 mm down to a practical limit of 50 μm. Typical shapes are round or rectangular. Depending on the intended application, the FOFP end faces can be coated with a specific spectral filtering, phosphorescent, or anti-reflective coating.

◘ Fig. 13.37 A fused fiberoptic faceplate (*back*) and taper (*front*). Photo courtesy of Schott glass

Optically, an FOFP behaves as zero-thickness optical window transferring an image, fiber by fiber, from one face of the plate to the other. Image magnification or reduction can be achieved by tapering the cross section of the bulk plate during the manufacturing process. In this case, the boule is drawn down and a neck region is formed with an hour-glass shape piece. The piece is cut into two pieces, machined and the ends polished resulting in a fused fiber optic taper.

Faceplates and tapers also function as dielectric barrier and mechanical interface and are optically used as a two-dimensional image conduit for energy conversion, field-flattening, distortion correction, and contrast enhancement. They are typically used for imaging applications bonded to cathode ray tubes (CRT) and LCD displays, image intensifiers, charged coupled device (CCD) or complementary metal-oxide semiconductor (CMOS) detectors, image plane transfer devices, X-ray digital detectors, among others.

In the medical area, fiber optic tapers and faceplates have found widespread use for both dental and medical digital radiography (such as mammography, fluoroscopy, intra-oral, panoramic, or cephalometric) where instead of using conventional film to obtain the X-ray images, an electronic photosensitive device such as a CCD or CMOS detector chip is used to convert the X-ray energy into electronic pixel signals via the use of an intermediate faceplate. Digital radiography offers high-resolution images while greatly reducing patient and sensor exposure to harmful X-rays by using low-dose X-ray sources. In addition, digital X-ray imaging speeds the availability of images for diagnostic, while also making the viewing, sharing, transmitting, and storing of X-ray patient data so much easy and compatible with modern electronic record systems. Furthermore, faceplates also provide a critical X-ray absorbing barrier between the X-ray emitter and the semiconductor detector device, prolonging their service life and reducing background noise.

As shown in ◘ Fig. 13.38, when an X-ray source emits radiation energy (that would pass through the patient) the transmitted energy impinges on a scintillator plate which converts the radiation rays into visible photons. The scintillating coating—e.g., cesium iodide (CSI) or gadolinium oxysulfide (Gadox) doped with Tl or Eu—is deposited directly on the large end of a fused fiber-optic taper. The light is then transferred and reduced through the taper and coupled to a digital CCD chip where a black and white image is formed which can then be viewed on a computer screen or monitor and readily archived as an electronic image file.

13

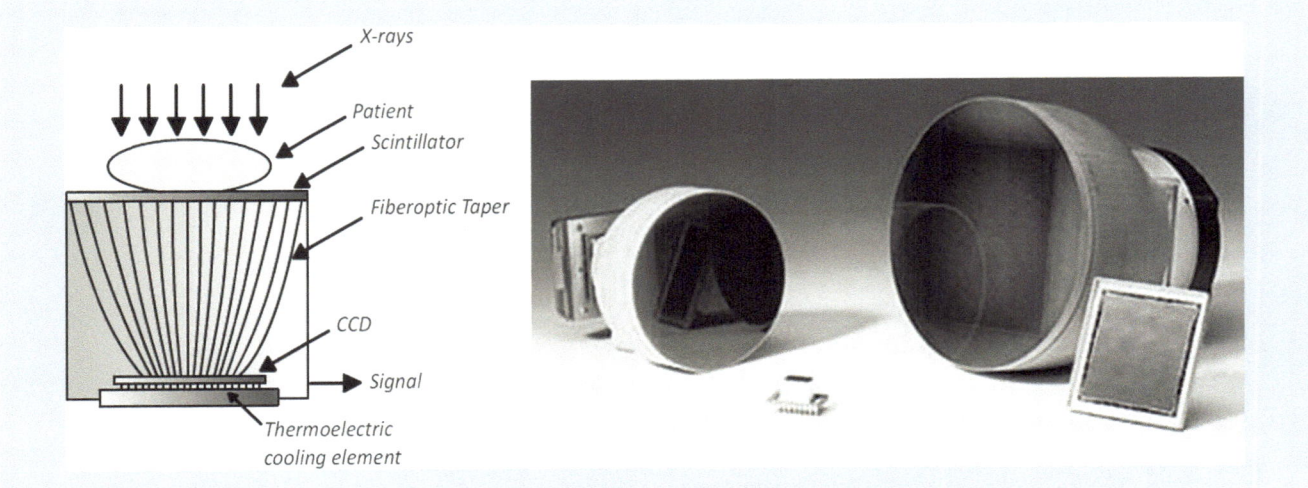

Fig. 13.38 In digital X-rays, a fused fiberoptic taper is used to guide the light image from a scintillator an electronic CDD detector array for processing and visualization

13.4 Conclusions

As discussed in this section, optics is a useful, practical, versatile, and powerful technology that, throughout history, has helped human kind perform visual examination, diagnostics, and therapeutics on both the sick and healthy. Optics technology and optical components are at the core in a variety of modern-day optical devices and instruments such as endoscopes, patient monitoring probes, and sensors, as well as in advanced robotic assisted surgery systems.

The harnessing power of light, and its interaction with living matter, is extremely useful and beneficial for a variety of medical purposes and treatments ranging from laser procedures for tattoo removal, to eye surgery to vessel and tissue ablation and coagulation, up to modern photodynamic therapy treatments. We have seen how the field of optics is in itself a subset of a more complex and interdisciplinary area of research known as Biophotonics.

New advancements in optics and photonics are driving the development of a new generation of imaging tools—such as optical coherence and photo-acoustic tomography—that can readily provide two and three-dimensional images of diverse human body tissues and organs.

Optics has, and will continue to be, an *enabling technology* for the advancement of medicine promoting unimaginable new devices, techniques, and applications to happen in the not too distant future.

References

1. Katzir A (1990) Selected papers on optical fibers in medicine, SPIE milestone series, vol MS 11. SPIE, Bellingham, WA
2. United Nations DESA (2015) World population prospects: the 2015 revision. United Nations, New York
3. Bradbury S (1967) The evolution of the microscope. Pergamon Press, Oxford
4. Bozzini P (1806) Lichtleiter, eine Erfindung zur Anschauung innerer Teile und Krankheiten. J Prakt Heilkunde 24:107–124
5. Von Helmholtz HLF (1856) Handbuch der physiologischen Optik. L. Voss, Leipzig
6. History of ASLMS. American Society for Laser Medicine and Surgery. ► http://www.aslms. org/aslms/history.shtml
7. Choy DS (1988) History of lasers in medicine. Thorac Cardiovasc Surg 36(Suppl 2):114–117
8. Mignani AG, Baldini F (1979) Fibre-optic sensors in health care. Phys Med Biol 42(5):967–979
9. Fujimoto JG et al (2002) Optical coherence tomography: an emerging technology for biomedical imaging and optical biopsy. Neoplasia 2(1–2):9–25
10. Cusano A, Consales M, Crescitelli A, Ricciardi A (eds) (2015) Lab-on-fiber technology. Springer, Berlin
11. Lakshminarayanan V et al (2015) Smartphone science in eye care and medicine. Opt Photonics News 26:45–51, Optical Society of America
12. Maamari RN et al (2014) Novel telemedicine device for diagnosis of corneal abrasions and ulcers in resource-poor settings. JAMA Ophthalmol 132:894–895
13. Breslauer DN et al (2009) Mobile phone based clinical microscopy for global health applications. PLoS One 4, e6320
14. Hett JH (1969) Medical optical instruments. In: Kingslake R (ed) Applied optics and optical engineering, Vol. 5, optical instrument. Academic, New York
15. Devroey D et al (2000) Do general practitioners use what's in their doctor's bag? Scand J Prim Health Care 20:242–243
16. Kravetz RE (2002) The otoscope. Am J Gastroenterol 97:470
17. Feldmann H (1996) History of the ear speculum. Images from the history of otorhinolaryngology, highlighted by instruments from the collection of the German Medical History Museum in Ingolstadt. Laryngorhinootologie 75:311–318
18. Millodot M (2009) Dictionary of optometry and visual science, 7th edn. Butterworth-Heinemann, Edinburgh
19. Millodot M (1973) A centenary of retinoscopy. J Am Optom Assoc 44:1057–1059
20. Koltai PJ, Nixon RE (1989) The story of the laryngoscope. Ear Nose Throat J 68(7):494–502
21. García M (1855) Observations on the Human Voice. Proc R Soc Lond 7:399–410
22. Hirsch NP, Smith GB, Hirsch PO (1986) Alfred Kirstein Pioneer of direct laryngoscopy. Anaesthesia 41(1):42–45
23. Kawahara I, Ichikawa H (1987) Fiberoptic Instrument technology. In: Sivak MV (ed) Gastrointestinal endoscopy. W. B. Saunders, Philadelphia, pp 201–241
24. Hirschowitz B (1979) A personal history of the fiberscope. Gastroenterol 76;(4):864–869

Quantum Optics

Contents

Atom Optics in a Nutshell

Pierre Meystre

P. Meystre (✉)
Department of Physics and College of Optical Sciences, University of Arizona, Tucson, AZ 85721, USA
e-mail: pierre.meystre@optics.arizona.edu

© The Author(s) 2016
M.D. Al-Amri et al. (eds.), *Optics in Our Time*, DOI 10.1007/978-3-319-31903-2_14

14.1 Introduction

One of the most counter-intuitive aspects of quantum mechanics, the fundamental theory of nature that was developed starting in the early twentieth century, is the concept of wave-particle duality.

We are all familiar with the notions of waves and particles. We have observed water waves when throwing pebbles in ponds as children, and we have learned in high-school that sound and light consist of waves as well. What we may not have learned, though, is that light sometimes behaves as particles instead. Even more unsettling is the fact that atoms, and all massive particles for that matter, sometimes behave as waves.

Wave-particle duality is the central tenet of atom optics: since atoms, very much like light, behave sometimes as waves, sometimes as particles, it is possible in principle to do with them pretty much everything that can be done with light. One can build a broad variety of atom optical instruments such as mirrors, beam splitters, and lenses. One can develop techniques for imaging, microscopy, diffraction, interferometry, and more. One can even realize atom analogs of lasers.

This chapter sketches selected aspects of atom optics, a few of its recent developments, and some of its promise. We start with a brief historical overview of some of the milestones that have lead to our current understanding of atoms, from the Greek philosophers of antiquity to the development of quantum mechanics and the key experiments of the early twentieth century that confirmed the wave-particle duality of atoms and other massive particles. We then discuss how the wave nature of atoms becomes increasingly more evident as their temperature is decreased. For that reason it is oftentimes advantageous to work at extremely low temperatures, close to absolute zero, in applications such as atom microscopes and atom interferometers, or to build atom optics "analogs" of the laser, Bose–Einstein condensates. After outlining the basic ideas behind these devices we conclude with a brief overview of some current and future applications, with an emphasis on the role of atom optics in helping answer fundamental physics questions.

The bibliography is limited to a few milestone papers and is certainly not meant to be comprehensive. It also does not attempt to give proper credit to all research groups who have contributed significant advances to atom optics, sometimes within weeks of the work by research groups mentioned here. Due to their advanced technical content these papers will likely be of limited use to the casual reader beyond their historical interest. The excellent review [6] gives a comprehensive list of references through 2009. The elegant set of lecture notes of [18] also discusses some of the extraordinary promise of atom optics for tests of fundamental physics at a level appropriate for advanced graduate students and experts in the field.

14.2 Particles or Waves?

This short chapter is not the place to give a comprehensive review of our historical understanding of the nature of light. For our purpose it is sufficient to review a few of the key steps that resulted in that understanding. We then draw a similar sketch of the historical development of our understanding of atoms. This will set the stage for a discussion of the close parallels that have guided the development of atom optics.

The central idea that we will need to become somewhat comfortable with is the concept of "wave-particle duality," the co-existence of particle and wave properties in objects that we are used to think of as one or the other, but not both. From everyday experience we are quite familiar with particles and waves, for instance, from watching the surf rolling on a beach, or tiny grains of sand on that same beach. What we need to grasp, though, is the rather counter-intuitive concept of both light and atoms behaving sometimes as particles, and sometimes as waves. *Why* this is the case is not a question that physics answers—it is a question perhaps best left to philosophers—but *how* this is the case is something that we now understand well. This is described beautifully and with extraordinary predictive power by modern quantum physics.

14.2.1 Light

Perhaps a good place to start is with the great Greek philosophers and mathematicians Pythagoras (c. 570–c. 495 BC), Plato (c. 428–c. 348 BC), and Euclid (c. 325–c. 265 BC). They thought that light consists of rays that travel in straight lines from the eye to the object, and that the sensation of sight is obtained when these rays touch the object, much like the sense of touch. Plato's student Aristotle (384–322 BC), though, had a different theory, considering instead that light travels in something like waves rather than rays. The understanding that light travels from the eye to the object remained largely unchallenged until it was finally disproved more than a thousand years later by Alhazen (965–1039), one of the earliest to write and describe optical theory. He studied in particular light and the nature of vision with the combined use of controlled experiments and mathematics.

Meanwhile the debate between the corpuscular and the wave nature of light already apparent in the conflicting views of the Pythagorean School and Aristotle continued unabated for centuries. Isaac Newton (1642–1726), who performed numerous experiments on light toward the end of the seventeenth century and whose extraordinary contributions include the understanding of the color spectrum and of the laws of refraction and reflection, argued that those effects could only be understood if light consisted of particles, because waves do not travel in a straight line. However, the corpuscular theory failed to explain the double-slit interference experiments carried out by Thomas Young (1773–1829)—we will return to these experiments at some length later on. It was replaced in the nineteenth century by Christiaan Huygens' (1629–1695) wave theory of light. Finally, James Clerk Maxwell (1831–1879) developed the equations that unify electricity and magnetism in a theory that describes light as waves of oscillating electric and magnetic fields. This is the culmination of the classical theory of light, and one of the greatest, if not the greatest achievement of nineteenth century physics. At that point, it appeared that light was indeed formed of waves, and the corpuscular theory seemed ruled out once and for all.

However things changed again at the beginning of the twentieth century in a way that revolutionized physics and profoundly transformed our understanding of nature. In 1900, Lord Kelvin gave a celebrated talk entitled "Nineteenth Century Clouds over the Dynamical Theory of Heat and Light" in which he stated with remarkable insight that [16]

» The beauty and clearness of the dynamical theory, which asserts heat and
   light to be modes of motion, is at present obscured by two clouds.

He went on to explain that the first of these two clouds was the inability to experimentally detect the "luminous ether"—the medium that was thought to be vibrating to create light waves; and the second was the so-called ultraviolet catastrophe of blackbody radiation—the fact that Maxwell's theory utterly failed to predict the amount of ultraviolet radiation emitted by objects as a function of

14

their temperature. As it turns out, these two clouds led to two earthshaking revolutions in physics: relativity theory and quantum mechanics.

Quantum mechanics reopened the centuries-old wave-particle debate, but it resolved it with a very unexpected and dramatic new answer: light behaves sometimes as waves, and sometimes as particles. In trying to understand how the radiation emitted by an object depends on its temperature, Max Planck (1858–1947) advanced the revolutionary idea that energy comes up in tiny discrete lumps, or quanta. With this ad hoc assumption, he was able to explain the experimental data that Maxwell's theory failed to explain. Following on that work, Albert Einstein (1879–1955) proposed that light also comes in small lumps of energy, now called photons. This allowed him to correctly characterize how electrons are emitted from surfaces of metal irradiated by light. (The other cloud mentioned by Lord Kelvin, the absence of a luminous ether, leads to Einstein's theory of relativity.)

What modern quantum theory teaches us is that light sometimes behaves as waves, as in the Young double-slit interference experiment, and sometimes as particles, as in the photoelectric effect. Although photons are massless they carry both energy and momentum that can be used to alter the motion of massive objects. We will talk about all this quite a bit more in this chapter, but first let's turn for a moment to what quantum theory has to say about atoms.

14.2.2 Atoms

As with light, a good place to start is again in ancient Greece. This is where Democritus (c. 460–370 BC) and Leucippus (fifth century BC) developed the theory of atomism, the idea of an ultimate particle and that everything was made out of indivisible "atoms." The first experiments that showed that matter does indeed consist of atoms are due to John Dalton (1776–1844). He recognized the existence of atoms of elements and that compounds were formed from the union of these atoms, and put forward a system of symbols to represent atoms of different elements. [The symbols currently used were developed by Jöns Jacob Berzelius (1779–1848).] In a major breakthrough, in 1897 J.J. Thompson (1856–1940) discovered the electron and advanced the so-called plum pudding model of the atom. In that description the volume of the atom was composed primarily of the more massive (thus larger) positive portion (the plum pudding), with the smaller electrons dispersed throughout the positive mass like raisins in a plum pudding to maintain charge neutrality.

Early in the twentieth century Ernest Rutherford (1871–1937) carried out a number of experiments that suggested that atoms consist instead of a tiny, positively charged nucleus, with electrons orbiting around it at relatively large distance, and discovered the existence of positively charged protons. In 1920 he further proposed the existence of the third atomic particle, the neutron, whose existence was experimentally confirmed quite a bit later, in 1932, by James Chadwick (1891–1974). Rutherford's experiments led to the development of the so-called Bohr–Sommerfeld model of the atom.

In a groundbreaking development and with extraordinary insight, Prince Louis-Victor de Broglie (1892–1987) then postulated that if light exists both as particle and wave then atoms, and all massive particles, should be the same (◘ Fig. 14.1). This was the key missing piece of the puzzle [9]. This property, now known as the *wave-particle duality*, guided Erwin Schrödinger (1887–1961), Werner Heisenberg (1901–1976), and others in developing quantum mechanics, the theory that has led to a wealth of extraordinary inventions from the internet to cell phones, from GPS to medical imaging, and to a myriad other developments that impact just about every aspect of modern life.

Fig. 14.1 Prince Louis-Victor de Broglie, who came up with the idea that massive particles have a wave character

The first experiments confirming the wave nature of massive particles were carried out by Clinton J. Davisson (1881–1958) and Lester H. Germer (1896–1971), who in 1928 observed the diffraction of electrons by a crystal of Nickel [8]. The first experiments demonstrating the wave nature of atoms and molecules followed soon thereafter, in 1930, in Helium experiments performed by Immanuel Estermann (1900–1973) and Otto Stern (1888–1969) [10], thereby fully confirming the de Broglie hypothesis.

To briefly complete the story as we currently understand it, we now know that protons and neutrons are actually not elementary particles. They belong to a family of particles called baryons, made up of three elementary constituents called quarks (two "up" quarks and one "down" quark for the proton, and one "up" quark and two "down" quarks for the neutron) bound together by the nuclear force. Together with another family of particles called mesons, made up of two quarks, they form the hadrons family.[1] The electrons, by contrast, are believed to be true elementary particles and belong to a family called leptons. They interact with the atomic nuclei via the electromagnetic force, whose "force particle" is the photon. The Standard Model of elementary particle physics comprises two additional types of interactions: the weak interaction, responsible for radioactive decay and nuclear fission, and gravitation, which allows massive particles to attract one another in accordance with Einstein's theory of general relativity (■ Fig. 14.2).

However, to break atoms into their subatomic constituents requires very large energies, much larger than normally considered in atom optics experiments. For the purposes of this chapter it is therefore sufficient to consider atoms as essentially "elementary particles" that interact with each other via relatively weak electric and magnetic fields, most importantly for us with light fields.

14.2.3 Particles and Waves

By optical waves one usually means those electromagnetic waves that are visible to the human eye. They have very short wavelengths, of the order of a millionth of a meter or less, and very high frequencies, of the order of 100,000 billion of oscillations

---

1 In an exciting new development announced in summer 2015, experiments carried out at the CERN Large Hadron Collider near Geneva, Switzerland provided evidence for the existence on pentaquarks, a new type of hadrons consisting of five quarks.

Fig. 14.2 Clinton J. Davisson and Lester H. Germer (Lucent Technologies Inc./Bell Labs, courtesy AIP Emilio Segrè Visual Archives)

per second.[2] Blue light consists of waves of higher frequency and shorter wavelength than green light, and green light consists of waves of higher frequency and shorter wavelength than red light. Past blue light and toward shorter wavelengths blue is followed by ultraviolet light, X-rays, and gamma rays. These waves are invisible to the human eye. On the other side of the spectrum and moving toward longer wavelengths, red is followed by infrared, microwaves, and radio waves, all also invisible to us. The wavelength of light is usually denoted by the Greek letter lambda, with symbol λ, and its frequency by the Greek letter "nu," written ν.

The particles of light are called photons. They are massless, and their energy E is proportional to their frequency. The proportionality constant is called Planck's constant, This is a fundamental constant that appears in the description of all quantum phenomena.[3] It traditionally denoted by the letter h, so that

$$E = h\nu. \tag{14.1}$$

Photons also carry a momentum, denoted by the letter p, which is inversely proportional to their wavelength,

---

2   We recall that waves are characterized by an amplitude, a wavelength, and a frequency: The wave amplitude is defined as half the vertical distance from a trough to a crest of the wave, the wavelength is the distance between two crests of the wave, and the frequency is the number of crests that an observer at rest sees passing in front of her eyes every second.

3   The value of the Planck constant is extremely small, $h = 6.62606957 \times 10^{-34} \; m^2 \; kg/s$, hinting at the fact that quantum mechanics is especially important in the atomic and subatomic worlds.

$$p = h/\lambda. \tag{14.2}$$

In vacuum the product of the wavelength of light and its frequency is equal to the speed of light $c = 299, 792, 458$ m/s, $\lambda\nu = c$.

It might come as a surprise that a massless particle such as the photon carries momentum. We recall that the momentum of an object of mass M is the product of its mass times its velocity v, $p = Mv$.[4] Momentum is a very important quantity in physics: Newton's second law of motion $F = Ma$, where a is the acceleration— the change in velocity v—tells us that the force F required to change the velocity of the object is proportional to its mass (more precisely, that the change in momentum of the object is equal to the force acting on it). This is why it is harder to stop a freight train than a bicycle!

How, then, can a massless object carry momentum? To properly understand why this is the case requires invoking Einstein's special relativity theory. The basic idea is twofold: First, one needs to know that nothing can move faster than the speed of light c, and that the only particles that can move at that speed must be massless. This is because it would take an infinite amount of energy to bring any massive particle to that velocity. Second, the description of classical mechanics embodied in Newton's laws does not apply to particles moving at extremely high velocities, near the speed of light. Their motion must be described instead in the framework of the theory of relativity.[5] Unfortunately in that extreme regime of velocities our intuition tends to fail us, so we will simply take Einstein at his word and accept that photons do carry momentum, a property that has been confirmed in numerous experiments. Remarkably, the fact that light can modify the trajectory of a massive particle was already conjectured during the Renaissance, but without a sound theoretical basis, by none other than the great mathematician and astronomer Johannes Kepler (1571–1630) who observed that the tail of comets always points away from the sun and concluded [17] that

>> The direct rays of the Sun strike upon it [the comet], penetrate its substance, draw away with them a portion of this matter, and issue thence to form the track of light we call the tail.

14.2.4 Atoms as Waves

We will soon come back to the photon momentum and its importance in atom optics. But before doing so we turn to the other actor in our story, the atoms, and sketch how they are described when they behave as matter waves, or de Broglie waves.

Very much like any other type of waves, they are characterized by a frequency and a wavelength, as first postulated by Louis de Broglie and then formalized in the framework of quantum theory. The de Broglie wavelength λ_{dB} of a non-relativistic massive particle of mass M is related to its momentum p by the equation

$$p = h/\lambda_{dB}, \tag{14.3}$$

---

4 Unfortunately the Greek letter ν used for frequencies and the roman letter v used for velocities look quite similar.

5 In the theory of special relativity the energy E of a particle is related to its momentum p by the equation $E^2 = p^2c^2 + M^2c^4$. For a massless particle, $M = 0$, this reduces simply to $E = pc$. For photons the energy is $E = h\nu$, and we have seen that in vacuum $\lambda\nu = c$, from which it follows that $p = h/\lambda$. As it turns out, the familiar definition of the momentum $p = Mv$ is only approximate. It holds for non-relativistic massive particles, that is, for particles moving much more slowly than the speed of light.

in complete analogy to the situation with photons. However, its kinetic energy takes the familiar form $E = Mv^2/2$ or, remembering that for atoms $p = Mv$ and with the relationship between p and λ_{dB},

$$E = h^2/(2M\lambda_{dB}^2).$$ (14.4)

So, while there are important similarities between light waves and matter waves, as evidenced by the relationships between momentum and wavelength of Eqs. (14.2) and (14.3), there are also important differences due to the fact that photons are massless objects while atoms have a mass. For optical waves, the energy is proportional to the momentum, $E = pc$, while for (non-relativistic) matter waves the energy is proportional to the square of the momentum, $E = p^2/2M$. This has important implications for atom optics.

Under everyday circumstances we experience atoms just as particles, not as waves. To understand why this is so let us estimate the size of the de Broglie wavelength. To do so, we need to figure out the momentum of an atom. Since the masses of the various atoms are known and can easily be found in a number of reference books or the internet all we need to do is determine their typical velocity. Let's imagine for a moment a box filled with some atomic gas, maybe Lithium or Sodium, at a temperature T. If it were possible to observe the individual atoms under a microscope, we would see that they move in random directions, going left or right or up or down or forward or backward, some faster, some more slowly, like little kids on a playground. Denoting the average velocity of all these atoms by the symbol $\langle v \rangle$, we would find that it is equal to 0, $\langle v \rangle = 0$: there are lots of atoms in the container, billions of billions of them, and just about as many of them are moving at a given velocity in one direction as in the opposite direction.

But if we took the square of all the individual velocities and averaged the result, call it $\langle v^2 \rangle$, we would find that it is different from zero. This is because the square of any number, be it positive or negative, is a positive number, and the average of a bunch of positive numbers is also positive. Importantly we would also discover that the higher the temperature T of the sample, the larger $\langle v^2 \rangle$, and that T is proportional to $\langle v^2 \rangle$. This is in fact precisely how temperature is defined: it is (in some units) the kinetic energy, or average energy of motion $M\langle v^2 \rangle/2$ of the atoms. The temperature at which all atoms cease to move is absolute zero, $T = 0$. It is impossible to cool anything below that temperature since the atoms cannot move more slowly than not moving at all![6]

Equation (14.3) teaches us that the de Broglie wavelength is inversely proportional to p—the smaller p, the larger λ_{dB}. Near absolute zero the atoms move extremely slowly. They have a very small momentum p, and hence a large de Broglie wavelength. At higher temperatures the atoms move faster, their momentum is larger, and their de Broglie wavelength is therefore smaller. This decrease is proportional to the square root of the temperature, or, in mathematical terms, λ_{dB} is proportional to $1/\sqrt{T}$. At room temperature, one finds that it is of the order of a tenth of a billionth of a meter, or a tenth of a nanometer, 10^{-10} m (this is 0.0000000001 m), a size comparable to the radius of an atom. This is why it is so difficult to observe the atoms as waves: their de Broglie wavelength is simply too small to be observable under normal circumstances.

A good strategy to investigate and exploit the wave nature of atoms is therefore to work at very low temperatures, where their de Broglie wavelength is more easily observable. For a typical atom cooled to a millionth of a degree above absolute zero

---

6   This is the classical physics view of things. The situation is somewhat more subtle in quantum mechanics, which teaches us that atoms still move a tiny bit at $T = 0$, but there is no need for us to worry about this here.

the de Broglie wavelength is of the order of a micron, a millionth of a meter. A millionth of a meter is still very small, so even at extremely low temperatures the wave nature of atoms is quite elusive—except that one micron also happens to be close to the wavelength of visible light. This is an important coincidence because as we know, to measure the size of any object we need to have an appropriate "measuring stick," not too big and not too small, just right. Because the wavelength of visible light turns out to be perfectly matched to the de Broglie wavelength of ultracold atoms, it can serve as that perfectly matched measuring stick. For this and several other reasons to which we will return the combination of visible light and ultracold atoms is a marriage made in heaven.

14.2.5 Cold Atoms and Molecules

Over the years optical and atomic scientists have developed exceedingly sophisticated methods to control the way light interacts with atoms. It is possible to exploit this know-how to prepare and manipulate atoms with extraordinary sophistication, and in particular to cool atoms to temperatures only a minute fraction of a degree above absolute zero.

At first sight, using light to cool atoms doesn't seem to make much sense: our intuition tells us that when we shine light on an object it becomes warmer, not colder. Therefore to use laser light to cool atoms requires one to be rather clever and to understand in detail the way they interact. For a simple qualitative discussion of the basic idea, though, it is sufficient to recall that atoms can both absorb photons (as happens, for instance, in your eyes) or emit them (for instance, in a light bulb.) We also need to keep in mind that all physical processes must satisfy two fundamental laws of nature: conservation of energy—one cannot create energy out of nothing; and conservation of momentum—a moving object will keep moving in a straight line at constant velocity unless one applies a force to it. This is Newton's first law. So, when a photon is absorbed by an atom both its energy and its momentum are transferred to the atom. In the reverse process, when an atom emits light, it loses the momentum h/λ that the photon carries away by changing its velocity.[7]

Remembering that the temperature of a gas is a measure of the energy of random motion of the atoms, this suggests a way by which light can be used to cool atoms: If we can somehow arrange for the moving atoms to predominantly absorb photons propagating *toward* them, then the momentum transferred to them by the photons will be opposite to their direction of motion. They will be pushed back and slowed down. The trick is of course that this needs to be done to all atoms, whether they move up or down, left to right or right to left, and backwards or forward. This is important: one needs to avoid as much as possible having atoms absorbing photons propagating in the same direction as they move, since this would accelerate them rather than slow them down. It turns out that this can actually be achieved by using six different light beams with just the right wavelengths. This mechanism, called Doppler cooling [14, 24], allows to cool atomic gases very significantly, down to roughly a thousandth of a degree above absolute zero. One can do even better by using more complex arrangements of light beams and by cleverly exploiting the internal structure of the atoms. Combining a variety of techniques it is now possible to cool atomic samples to within a

---

7   It turns out that the bulk of energy conservation is achieved via transitions of an atomic
     electron between different orbits around the nucleus, while the bulk of momentum con-
     servation is normally achieved by changing the velocity of the atom.

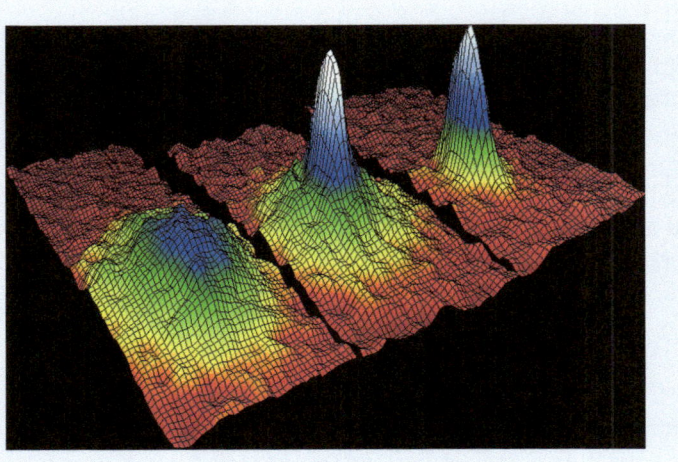

Fig. 14.3 The first atomic BEC ever created by the JILA group of E. Cornell and C. Wiemann in 1995 (Image credit Michael Matthews, JILA)

billionth of a degree above absolute zero or even colder, to a point where their de Broglie wavelength is of the order of a fraction of a millimeter !

A major experimental milestone resulting from the use of such cooling techniques was reached in 1995 (Fig. 14.3) by the groups of Carl Wieman and Eric Cornell at JILA [2], and soon thereafter by Wolfgang Ketterle and coworkers at MIT [7]. They succeeded in realizing atomic Bose–Einstein condensates, a tour-de-force for which they were awarded the 2001 Physics Nobel Prize. A Bose–Einstein condensate is a state of matter where all atoms "condense" into a single quantum object where they are all in the same state, some sort of a "super atom." Ideally the atoms form then a single macroscopic quantum wave, much like photons in a laser behave collectively as a single entity. This exotic object was predicted as early as 1924 by Albert Einstein, expanding on work by the Indian physicist Satyendra Nath Bose, but it is not until atomic samples could be cooled to the extraordinarily low temperatures now possible that it could be produced and observed in its almost pure form.

14.3 Atomic Microscope

We mentioned earlier that if we had at our disposal a microscope that could track individual atoms, we would be able to observe their random thermal motion. While this is not possible at room temperature at this time, the availability of ultracold atomic systems has now made such devices a reality at temperatures approaching absolute zero.

The key idea is that because ultracold atoms carry very little energy of motion it is possible to trap them in extremely shallow potentials, in particular in the periodic potentials that can be produced by standing optical waves—these are the waves produced by two light beams of the same wavelength propagating in opposite directions. Atoms can be trapped in the troughs of these waves in such a way that if they try to escape, then radiation pressure pushes them back down, somewhat like a ball always rolls down to the bottom of a slope. Using standing waves along two or three directions the landscape in which the atoms are trapped resembles an egg crate. It is called an optical lattice potential, or simply optical lattice, and atoms can be trapped at its local minima, as sketched in Fig. 14.4. In 2009 Marcus Greiner and his collaborators at Harvard University devised a microscope that successfully imaged individual atoms localized in such a tightly

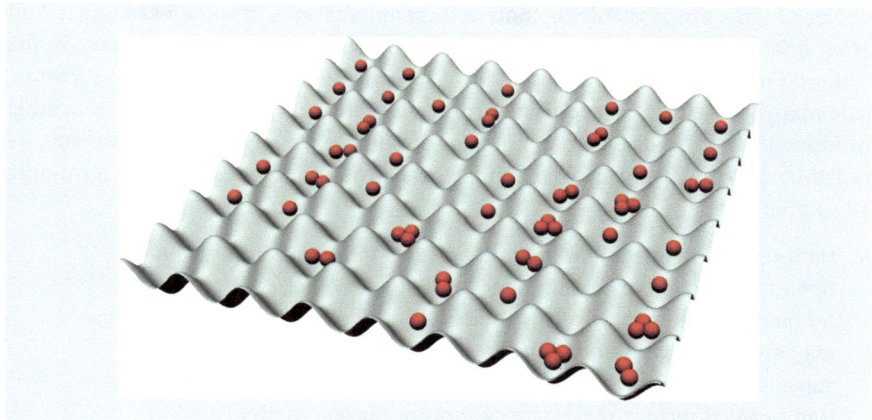

Fig. 14.4 Artist rendition of the way ultracold atoms can be trapped in an optical lattice (Image credit Andrew Daley, University of Strathclyde)

Fig. 14.5 Illustration of the way atoms can be manipulated and probed individually in an atom microscope. In this case, the system is modified to observe the transition from a Bose–Einstein condensate (*left*) to the so-called Mott insulator (© Immanuel Bloch, Max-Planck Institute for Quantum Optics)

spaced optical lattice [3]. This was soon followed by a second microscope (◘ Fig. 14.5) developed by Immanuel Bloch's group at the Max Planck Institute for Quantum Optics [23]. In these groundbreaking experiments the atoms trapped and individually imaged were bosons[8] but more recently that same technique has also been extended to fermions [5, 19].

Ultracold atoms trapped in optical lattices provide a powerful proving ground to study a number of effects in manybody physics, the situations dominated by the collective behavior of large ensembles of constituents. This is a broad and challenging area of research that is central to the understanding of many phenomena in fields ranging from condensed matter physics to nuclear physics. For example, the collective behavior of electrons in crystal structures is key in understanding their electrical and optical properties.

However experiments in solids can be challenging, in part because it is difficult to control the strength of inter-particle interactions. In contrast, a number of

---

8   Atoms come in two classes, bosons and fermions. Bosons are characterized by the fact that identical bosons can, and like to, occupy the same quantum state in unlimited number. In stark contrast, two identical fermions cannot be in the same quantum state.

powerful tools are available to control these interactions in ultracold atoms. And atom microscopes even permit to address and manipulate individual atoms in the system. For these reasons they provide a remarkable tool to simulate and investigate manybody effects in exquisitely controlled situations. They offer considerable promise to help understand a number of complex manybody phenomena. As noted by Martin Zwierlein [25], whose MIT group developed the first fermionic atom microscope [5],

> » High-resolution imaging of more than 1,000 fermionic atoms simultaneously would enhance our understanding of the behavior of other fermions in nature, particularly the behavior of electrons. This knowledge may one day advance our understanding of high-temperature superconductors, which enable lossless energy transport, as well as quantum systems such as solid-state systems or nuclear matter.

14.4 Interferences

We can easily observe the interference of waves when we drop a pair of pebbles in a quiet pond. Each pebble is the source of a small wavelet that propagates away in regular circles, and when the two meet they interfere to produce a complex pattern of crests and troughs.

Similar interferences are also familiar in optics, most famously perhaps in the Young double-slit experiment mentioned earlier. In that case, an optical wave propagates from one side to the other of an absorbing screen through either one or two parallel slit openings (or even more simply one or two pinholes). In the case of a single slit, after it passes through the hole the light wave begins to spread much like the wave generated by a single pebble—the narrower the slit, the larger the angle of spread. With two slits the situation is then akin to what happens with the two pebbles: As they spread spatially the light beams originating at the two slits begin to overlap and interfere, much like the wavelets in the pond. This results in a pattern of alternating dark and bright regions, the analog of the crests and troughs. The more pure the color of the light, the higher the contrast between the bright and dark fringes. This interference phenomenon, perhaps the most direct demonstration of the wave nature of light, is what led Huygens to develop his wave theory of light.

Remarkably, interferences still occur if the light beam is so feeble that only one photon at a time flies past the screen, perhaps one every second, or one every minute, or even one per month! If one waits long enough for the successive photons to slowly build an image, say, on a photographic plate or a CCD camera, that image will still exhibit the same precise interference pattern as if the beam were intense and produced an image in the blink of an eye. The interference pattern builds up one photon at a time!

This should seriously bother you, because one would expect that each individual photon goes through either one slit or the other, but not both, and the situation should then be completely analogous of the one pebble case. *Obviously*, we should not observe interferences in that case. But this is not so: "Obvious" is obviously not a good characterization of what can happen in the quantum world.

14.4.1 Atom Interferences

The situation may seem even more bizarre with atoms, which we are used to think of as particles. But since they obey the same wave-particle duality as photons it is

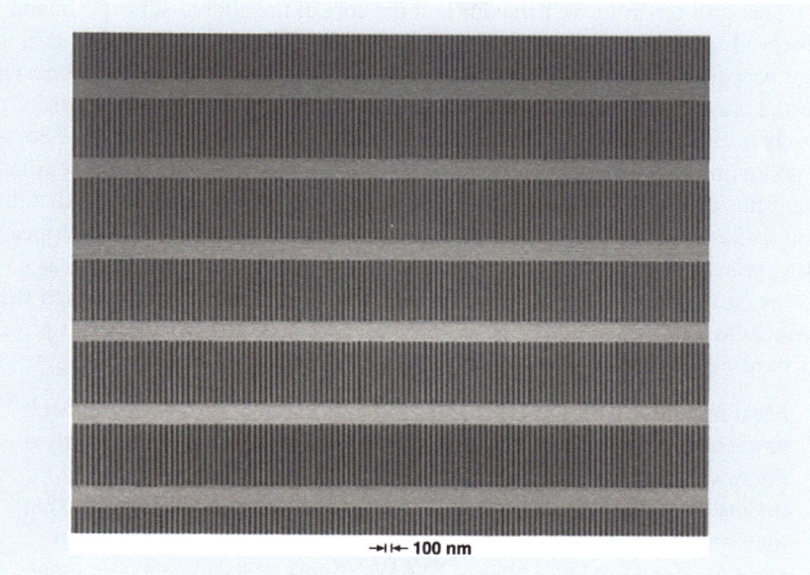

possible to produce and observe the interference of matter waves as well. A simple way to do so is to mimic Young's optical double-slit experiment. Practical challenges are that for interferences to be easily observable the width of the slits must be much narrower than de Broglie wavelength, and also that the slit separation should typically be of the order of that wavelength. Modern nanotechnology has solved this problem and makes it possible to fabricate a variety of combinations of holes and slits through which matter waves can propagate. For example, one can pass a beam of atoms through a large array of parallel slits. This is an atom optics analog of the diffraction gratings widely used in optics. An important and useful property of such gratings is that the interferences of the individual wavelets result in different wavelengths (colors) exiting the grating at different angles. Likewise, a nanofabricated mechanical grating can redirect an atomic beam, or even a single atom, in a direction that depends on its energy and momentum. As a result, properly designed gratings can act as mirrors or as beam splitters for atoms (■ Fig. 14.6).

One can also use light instead of nanofabricated elements to achieve that goal. In the discussion of laser cooling we mentioned that if a photon is absorbed by an atom, then its momentum must be transferred to that atom because of momentum conservation. So, if a photon propagating from left to right is absorbed, then that atom must experience a small velocity kick in that same left to right direction. If, however, the photon propagates from right to left, the velocity kick to the atom will be from right to left as well. And if the atom interacts simultaneously with *two* light fields, one propagating to the left and the other to the right, then it suffers both a velocity kick to the left *and* a velocity kick to the right. As a result the atom "goes in both directions," or more precisely the atomic matter wave is split into two partial waves, one propagating to the left and the other to the right. Acting together, the two light beams act as an atomic beam splitter.

Much like the observation that optical interferences build up "one photon at a time," this is his very strange. Loosely speaking, the atom can move in two directions at the same time, and be in two places at the same time. In the classical world such a behavior would be impossible: The atom would go either to the left *or* to the right, but not to the left *and* to the right.

This counter-intuitive behavior is at the core of the double-slit experiment: The observed interferences can only be understood if the atom is described as a wave that propagates simultaneously through both slits, so as to produce the interfering partial waves, just like with the two pebbles. Yet, if the atom is a particle, then surely it must go through either one or the other slit, but not both, right? So, what is going on? Can we not place small detectors near the slits, and measure which of them the atom went through? The answer is that one can certainly do that, but if one makes this "which way" determination, then the interferences disappear! In other words, if we don't ask "which way" the atom went then it behaves as a wave and produces interferences, but if we measure which slit it went through then it behaves as a particle, with no interferences.[9] How can that be? The great physicist Richard Feynman put it beautifully when he wrote [11]

» Because atomic behavior is so unlike ordinary experience, it is very difficult to get used to, and it appears peculiar and mysterious to everyone – both to the novice and to the experienced physicist. Even the experts do not understand it the way they would like to, and it is perfectly reasonable that they should not, because all of direct, human experience and of human intuition applies to large objects. [. . .] We choose to examine a phenomenon which is impossible, absolutely impossible, to explain in any classical way, and which has in it the heart of quantum mechanics. In reality, it contains the only mystery.'

14.4.2 Atom Interferometry

Optical interferometry is a remarkably powerful technique that uses the wave nature of light to measure small distances or displacements with extraordinary accuracy and sensitivity. As an example, the LIGO interferometers[10] built in Louisiana and the state of Washington to detect gravitational waves are able to measure length changes of one part in 10^{21} (this is one followed by 21 zeros!). Obviously, not all optical interferometers are that sensitive (or that expensive) but because of their remarkable properties, they are ubiquitous in R&D laboratories and industrial settings.

Optical interferometers come in many variations, but the basic idea is always pretty much the same. They rely on some combination of beam splitters and mirrors to divide a light beam into two or more partial beams that propagate in different environments where they are subjected to different forces and fields before being recombined to produce an interference pattern. For instance, one of the beams could go through an atomic vapor while the other propagates through vacuum, or one beam could have travelled a longer distance than the other; or perhaps one beam bounces off a moving mirror while the other is reflected by a mirror at rest. The key point is that the spatial and temporal features of the resulting interference pattern contain a great deal of information about the different environments that the partial beams propagated through (◘ Fig. 14.7).

Especially since the invention of the laser, optical interferometers have found countless uses in fields as diverse as physics, astronomy, engineering, applied science, remote sensing, seismology, telecommunications, biology, medicine, and manufacturing, to list just a few examples. Applications range from the measurement of extraordinarily small distances to the precise determination of specific

---

9   The same is also true for the photons of the previous section, and for any quantum particle.
10   The LIGO acronym stands for Large Interferometer Gravitational Wave Observatory.

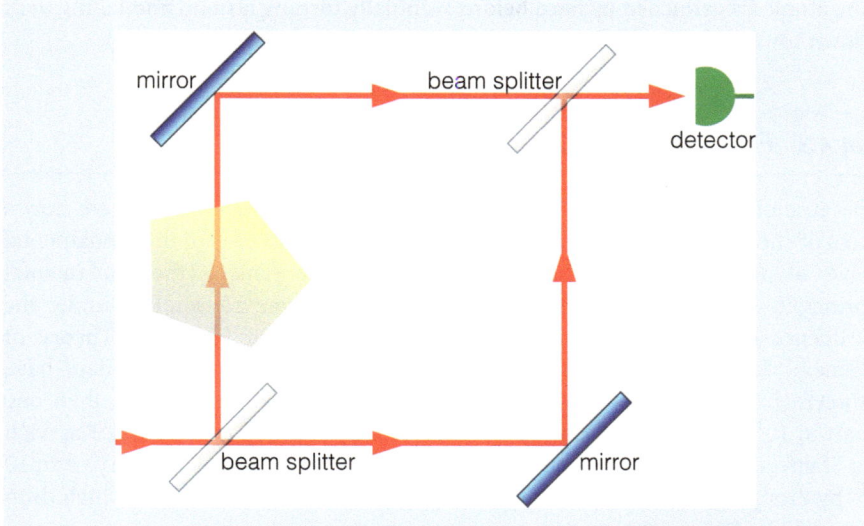

◘ **Fig. 14.7** Schematics of an interferometer, with two beam splitters to first separate and then recombine the partial beams. One of the beams propagates through some kind of an environment that modifies that interference fringes observed at the detector

atomic or molecular properties, from navigation and guidance to tests of the fundamental laws of physics, from medical imaging to electronic chip fabrication, and much more.

Despite all these successes there are situations where there is considerable benefit in using matter-wave interferometry instead. This is because not surprisingly, massive particles are orders of magnitude more sensitive than photons when it comes to measuring accelerations. For example, one finds that everything else being equal, interferometric gyroscopes (called Sagnac interferometers) using atoms rather than photons have a sensitivity that is larger by the ratio of their rest energy Mc^2 to the energy $h\nu$ of a photon,[11] that is, by $Mc^2/h\nu$. For visible red light and a typical atom such a Cesium this is a factor of about 10^{10}, or 10 billions! This is why atom interferometers are so well adapted to the precise measurement of rotations and accelerations, and can also serve as sensors for other forces and fields [6].

Gravimeters are one example of a practical device that can benefit significantly from atom interferometry. They are important in oil and mineral exploration, where they rely on the fact that different types of rocks or liquids have different densities. They determine the local value of gravity by measuring the acceleration of a free falling mass. Atom interferometers permit in principle to significantly increase the precision of these measurements over other methods. Using small atomic samples as free masses, they operate by splitting their atomic matter waves into two partial waves of different velocities. After some time during which the atoms are free falling the velocities of the two partial waves are then interchanged by a matter wave "mirror." Finally they are recombined to produce an interference pattern from which one can infer the acceleration with high precision and accuracy. To take advantage of the fact that the precision increases with the free fall duration one sometimes uses "atomic fountains" to increase that time. In that case

---

11　The rest energy of a massive particle is its energy when it is not moving. For a particle at rest, $p = 0$, the relativistic energy equation $E^2 = p^2c^2 + M^2c^4$ of footnote 5 reduces to $E = Mc^2$, the equation for the rest energy of a massive particle famously associated with Einstein.

14

the atoms are launched upward before eventually turning around and falling back down toward the earth (for a brief history of atomic fountains see [15]).

14.4.3 Fundamental Studies

Because of their remarkable potential sensitivity atom interferometers are now a tool of choice not just in practical applications, but also in tests of the fundamental laws of physics, such as the Equivalence Principle. This is the fundamental principle which states that all objects fall with the same acceleration under the influence of gravity. It forms the foundational basis of Einstein's Theory of General Relativity. The best tests of the Equivalence Principle to date have shown that the accelerations of two falling objects differ by no more than one part in 10^{13}—this is one followed by 13 zeros [21]. A group led by Mark Kasevich at Stanford University aims for an improved test of this principle to one part in 10^{15} by dropping atoms of two different isotopes of rubidium[12] in a 10 m high drop tower [22] (◘ Fig. 14.8).

◘ **Fig. 14.8** Stanford 10 m tower used for tests of the equivalence principle (courtesy Mark Kasevich, Stanford University)

---

12   Isotopes are variations of a chemical element that all have the same number of protons and electrons, but differ by the number of neutrons in their nucleus, and hence have different masses.

In another example, atom interferometry offers great promise for the development of a new mass standard. Surprisingly perhaps, the kilogram is the last physical unit that is defined by an artifact, the International Prototype Kilogram. It is the mass of a block of platinum–iridium alloy stored in an environmentally controlled vault in the basement of the International Bureau of Weights and Measures in Sèvres, near Paris. In addition to being subject to damage, this standard presents the fundamental issue of not being based on a physical law.

One proposal is to define the unit of mass in terms of a frequency. This would be possible provided that one assigns to the Planck constant h a fundamental value, much like the speed of light c is assigned a fundamental value that allows to connect lengths to times.[13] The basic idea is that the momentum imparted on atoms by light is proportional to the frequency of that light, and inversely proportional to the mass of the atom, the proportionality factor being given by Planck's constant. If it were assigned a fundamental value, then one could connect masses to frequencies extraordinarily accurately. Future space-borne atom interferometers might then allow the measurement of atomic masses anywhere on Earth better than 1000 times more accurately than is presently the case.

A third example of a basic science application of atom interferometry is in gravitational wave detectors. Four centuries after Galileo (1564–1642) used telescopes to study and revolutionize our understanding of the Universe they remain our most powerful tool to learn about it, whether they detect radio waves, sub-millimeter waves, infrared radiation, visible light, ultraviolet radiation or X-rays. However, it is believed that extremely significant additional information would be provided by the detection and characterization of the gravitational waves produced by the motion of massive objects, in particular closely orbiting compact massive objects such as neutron stars or black holes binaries, merging supermassive black holes, collapsing supernovae, or pulsars. Gravitational waves might also provide information on the processes that took place in the early Universe, shortly after the Big Bang. However they interact only extremely weakly with matter, and so far they have remained elusive.[14]

It is expected that in the near future Advanced LIGO, the upgraded version of the LIGO gravitational wave antennas, will be sensitive enough to detect gravitational waves at the rate of maybe a few events per year. To further increase sensitivity and the frequency of observations, future systems will likely need to be space-based, one example being the proposed Laser Interferometer Space Antenna (LISA). Mark Kasevich and his coworkers at Stanford have proposed an alternative space-based hybrid approach (◘ Fig. 14.9) combining optical methods and atom interferometry [13]. Their proposal draws on the use of an optical method to measure the differential acceleration of two spatially separated, free falling atom interferometers whose mirrors and beam splitters are produced by light pulses sent back and forth between them through space. Comparing the matter-wave interference fringes in the two interferometers would provide a record of the effect of gravitational waves on the travel time a laser pulse linking the two atom interferometers. It is argued that using atoms instead of mirrors as test masses would reduce a number of systematic errors.

---

13 State-of-the-art atomic clocks can measure times to accuracies in excess of one part in 10^{17}, so that reducing the determination of a physical quantity to a measurement of time or frequency is particularly favorable.

14 Note added: This is no longer the case! A few months after this article was completed the LIGO Scientific Collaboration and the Virgo Collaboration reported the first direct observation of a gravitational wave signal, resulting from the collision of two massive black holes [1]. This historic breakthrough, 100 years after Einstein's 1916 prediction, opens the way to a new era in observational astronomy.

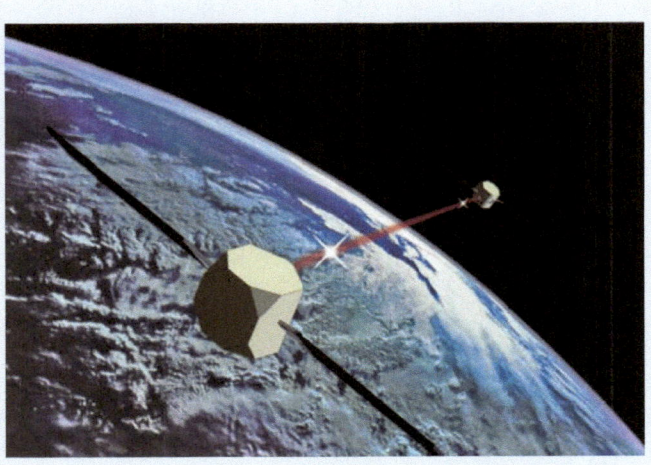

Fig. 14.9 A proposed space-borne atom interferometer gravitational wave detector. Two widely separated atom interferometers are controlled by light pulses sent back and forth between them. Passing gravitational waves will modify the travel time of the light pulses between the interferometers, resulting is a shift on their interference fringes that provide a record of the gravitational wave (courtesy Mark Kasevich, Stanford University)

14.4.4 BEC Atom Interferometers

In conventional optics it is often favorable to use lasers rather than conventional light sources, if only because they can have very high photon fluxes within a very narrow range of wavelengths. So one might ask whether the same is true in atom optics, and whether it would be advantageous to use an "atom laser," or Bose–Einstein condensate, rather than a regular beam of non-condensed ultracold atoms.

The simple answer is that this is typically not the case, because of an important difference between photons and atoms: Two light beams propagating in different directions in free space can cross without perturbing each other, because photons don't directly interact. In contrast, atoms do collide. Collisions are random events that result in uncontrolled changes in the interferences between the atom matter waves. This leads to additional detection noise that can significantly limit the sensitivity and accuracy of measurements. As collisions become more frequent when the atomic flux is increased this limits the applicability of Bose–Einstein condensates in atom interferometry.

This problem can, however, be circumvented to some extent by reducing the collisions between atoms. This can sometimes be achieved in ultracold atomic samples by using magnetic fields to control the collision rate. In the best cases it is possible to almost completely suppress collisions, leading to the potential for high precision interferometry with Bose–Einstein condensates. Alternatively, under appropriate conditions other quantum effects can be exploited to increase the sensitivity of the system and the precision of measurements, using, for example, the so-called squeezed states or number states of the matter waves.[15] An atom interferometer based on this principle was recently realized in the group of Jörg

---

15   Quantum objects are subject to fundamental random fluctuations called quantum noise—this is why even at $T = 0$ atoms are not completely still, see footnote 6. But it is sometimes possible to prepare atoms or photons in such a way that this noise is "squeezed away," or more precisely transferred to a place where it does not add imprecision to a specific type of measurement.

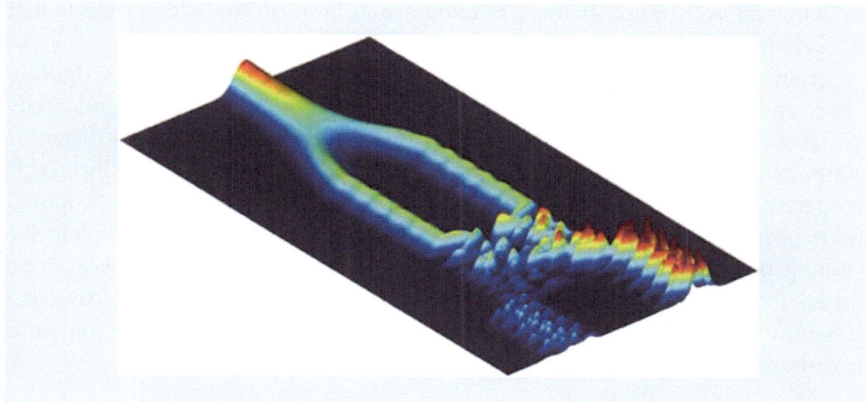

Fig. 14.10 BEC atom interferometer using squeezed states of the matter waves to reduce the noise below the so-called shot noise limit of conventional systems (courtesy Jörg Schmiedmayer, Technical University Vienna)

Schmiedmeyer at the University of Vienna [4], see ◼ Fig. 14.10. Another possible approach involves the use of fermionic atoms instead of bosons, although the interference contrast tends to be reduced in that case.

14.5 **Outlook**

It is widely accepted that quantum mechanics is the fundamental theory of nature. It has been and continues to be put to numerous, increasingly elaborate tests that it has so far passed with flying colors. Yet, in everyday life we don't observe the remarkable quantum effects that we can achieve with small ensembles of atoms or with photons under exquisitely controlled conditions. We cannot make a car be "in two places at the same time," or, in the famous example of Schrödinger, we cannot have a cat that is both alive and dead at the same time. Our everyday world seems to be most definitely governed by the laws of classical physics, not by quantum mechanics. This is extremely puzzling, because if the quantum mechanical description of nature is more fundamental than its classical description, then quantum mechanics should govern not just the microscopic world, but the macroscopic world as well.

Why and how macroscopic systems lose their quantum features and become essentially classical are challenging questions that are being addressed by a number of researchers, both theoretically and experimentally. On the theoretical side, proposed explanations range from relatively mundane mechanisms, such as increasingly fast decoherence resulting from the contact of objects of increasing size to their environment, to speculations about the role of gravity in washing out quantum features in massive objects.

On the experimental side, there are exciting efforts to observe quantum interferences in increasingly macroscopic objects, with the goal of improving our understanding of the physical mechanisms that wash out quantum features in objects of increasing complexity. For example, a group around Markus Arndt at the University of Vienna has succeeded in demonstrating the wave nature of large organic molecules, from the "buckeyball" C_{60} to the very large molecule TPPF152 ($C_{168}H_{94}F_{152}O_8N_4S_4$,) which contains 430 atoms and has a thermal de Broglie wavelength of about one picometer, a millionth of a millionth of a meter [12]. It is hoped that eventually such experiments will help determine whether the quantum to classical transition is a practical and relatively mundane issue or a truly fundamental occurrence. Is there a fundamental limit on the size of objects that can behave as de Broglie waves, or are the challenges only practical?

In an ambitious proposal, Oriol Romero-Isart, Markus Aspelmeyer, Ignacio Cirac, and coworkers have recently proposed a method to prepare and verify spatial quantum superpositions of a nanometer-sized object separated by distances comparable to its size [20]. It is hoped that such experiments will eventually be able to operate in a parameter regime where it will be possible to test various proposed mechanisms beyond quantum mechanics that have been advanced to explain the washing out of quantum properties in macroscopic objects. It will be exciting indeed to see these proposed experiments being realized and start answering questions that have surrounded quantum mechanics and its interpretation since its early days, nearly 100 years ago.

Acknowledgements I wish to thank the colleagues and collaborators, too numerous to list individually, with whom I have exchanged ideas on various aspects of atom optics, quantum optics, and the strangeness of quantum mechanics over the years. This work was supported by the DARPA QuASAR and ORCHID programs through grants from AFOSR and ARO, the US Army Research Office, and NSF.

References

1. Abbott BP et al (2016) Observation of gravitational waves from a binary black hole merger. Phys Rev Lett 116:061102
2. Anderson MH, Ensher JR, Matthews, MR, Wieman, CE, Cornell, EA (1995) Observation of Bose-Einstein condensation in a dilute atomic vapor. Science 269:5221
3. Bakr, WS, Gillen, JI, Peng, A, Foelling, S, Greiner, M (2009) Quantum Gas microscope detecting single atoms in a Hubbard regime optical lattice. Nature 462:74
4. Berrada, T, van Frank, S, Bücker, R, Schumm, T, Schaff, J-F, Schmiedmayer, J (2013) Integrated Mach-Zehnder interferometer for Bose-Einstein condensates. Nat Commun 4:2077
5. Cheuk, LW, Nichols, MA, Okan, M, Gersdorf, T, Ramasesh, VV, Bakr, WS, Lompe, T, Zwierlein, MW (2015) Quantum-gas microscope for fermionic atoms. Phys Rev Lett 114:193001
6. Cronin, AD, Schmiedmayer, J, Pritchard, DE (2009) Optics and interferometry with atoms and molecules. Rev Mod Phys 81:1051 gives a comprehensive discussion of atom optics and interferometry, including a review of scientific advances and a broad range of application of atom interferometers
7. Davis, KB, Mewes, M-O, Andrews, MR, van Druten, NJ, Durfee, DS, Kurn, DM, Ketterle, W (1995) Bose-Einstein condensation in a gas of sodium atoms. Phys Rev Lett 75:3969
8. Davisson C, Germer, LH (1927) Diffraction of electrons by a crystal of nickel. Phys. Rev. 30:705
9. de Broglie L (1925) Recherche sur la Théorie des Quanta. PhD thesis, University of Paris

10. Estermann E, Stern O (1930) Beugung von Molekularstrahlen. Z Phys 61:95
11. Feynman RP (2011) In: Leighton RB, Sands, M (eds) Feynman lectures on physics, Chap. 1, vol. III. Free to read online edition available at ▶ http://www.feynmanlectures.caltech.edu/
12. Gerlich S, Eibenberger S, Tomandl M, Nimmrichter S, Hornberger K, Fagan PJ, Tüxen J, Mayor M, Arndt M (2011) Quantum interference of large organic molecules Nat Commun 2:263
13. Graham PW, Hogan JM, Kasevich MA, Rajendran S (2013) New method for gravitational wave detection with atomic sensors. Phys Rev Lett 110:171102
14. Hänsch TW, Shawlow AL (1975) Cooling of gases by laser radiation. Optics Commun 13:68
15. Kasevich MA, Riis E, Chu S (1989) Atomic fountains and clocks. Optics News 15:31
16. Kelvin L (1901) Royal Institution lecture. Nineteenth-century clouds over the dynamical theory of heat and light. Philosophical magazine, series 6, vol. 2, p 1
17. Kepler J, as quoted in *A Comet Called Halley*, by I. Ridpath, Cambridge University Press (1985), see ▶ http://www.ianridpath.com/halley/halley2.htm
18. Müller H (2012) Quantum mechanics, matter waves, and moving clocks. In: Proceedings of the international school of physics "Enrico Fermi". Vol. 188. Atom interferometry, pp 1339–418. IOS Press, available to read on ▶ http://arxiv.org/pdf/1312.6449.pdf
19. Parsons MF, Huber F, Mazurenko A, Chiu CS, Setiawan W, Wooley-Brown K, Blatt S, Greiner M (2015) Site-resolved imaging of fermionic ^{6}Li in an optical lattice. Phys Rev Lett 114:213002
20. Romero-Isart O, Pflanzer AC, Blaser F, Kaltenbaek R, Kiesel N, Aspelmeyer M, Cirac JI (2011) Large quantum superpositions and interference of massive nanometer-sized objects. Phys Rev Lett 107:020405
21. Schlamminger S, Choi KY, Wagner TA, Gundlach JH, Adelberger EG (2008) Test of the equivalence principle using a rotating torsion balance. Phys Rev Lett 100:041101
22. See the M. Kasevich group website at Stanford University. ▶ http://web.stanford.edu/group/kasevich/cgi-bin/wordpress/?pageid=11.
23. Sherson JF, Weitenberg C, Endres M, Cheneau M, Bloch I, Kuhr S (2010) Single-atom-resolved fluorescence imaging of an atomic Mott insulator. Nature 467:68
24. Wineland DJ, Dehmelt H (1975) Proposed $10^{14} \Delta v < v$ laser fluorescence spectroscopy on Tl$^+$ mono-ion oscillator III. Bull. Am. Phys. Soc. 20:637
25. Zwierlein M, as quoted in ▶ http://www.sci-news.com/physics/science-microscope-fermionic-atoms-02799.html.

Slow, Stored and Stationary Light

Michael Fleischhauer and Gediminas Juzeliūnas

M. Fleischhauer (✉)
Department of Physics and research center OPTIMAS, University of Kaiserslautern, 67663 Kaiserslautern, Germany
e-mail: mfleisch@physik.uni-kl.de

G. Juzeliūnas
Institute of Theoretical Physics and Astronomy, Vilnius University, Saulėtekio 3, LT-10222 Vilnius, Lithuania
e-mail: gediminas.juzeliunas@tfai.vu.lt

© The Author(s) 2016
M.D. Al-Amri et al. (eds.), *Optics in Our Time*, DOI 10.1007/978-3-319-31903-2_15

15.1 Introduction

Since the experiments of Michelson and Morely and their brilliant explanation by Albert Einstein more than 100 years ago which have laid the foundation for the theory of relativity, we know that light propagates in empty space with the largest possible velocity. This speed of about 300,000 km/s is so fast that we can have a phone conversation around the globe without noticing that an electromagnetic signal has to be transmitted for every bit of information. When we look through a window or a prism of quartz we see that light gets refracted. Refraction is due to the fact that light propagates in a transparent medium at a slightly lower speed than allowed by the universal traffic laws of nature. This speed, called phase velocity depends on the color of light and the variation of the phase velocity in media, is what causes the beauty of a rainbow or the bright fan of colors produced by a prism. Yet the change of the velocity of light in water, in glass or even in diamond is small, it is typically less than a factor of 2. But what if this factor is 10^7, a ten with 6 extra zeros, i.e. 10,000,000? Such light can truly be called ultra slow. As opposed to propagation faster than the vacuum speed of light, this is not forbidden by Einstein's theory of relativity, but for a long time did not seem feasible. It did so until the late 1980s and early 1990s, when Steve Harris from Stanford University pointed out that an effect he termed electromagnetically induced transparency (EIT) [1, 2] can lead to a massive reduction of the effective speed of pulsed light [3]. When we talk about 'slow' light we talk about the speed of *pulses* of light, called group velocity, which needs to be distinguished from the phase velocity mentioned above.

Although a number of experiments have seen evidence of velocity reduction in EIT media, it took until 1999 [4–6] that slow light received a great deal of attention. In 1998 the group of Lene Hau at the Rowland Institute for Science together with Steve Harris managed to decelerate the propagation of light in an atomic gas to 17 m/s, i.e. almost 20 million times slower than in vacuum. The cover page of the journal Nature (■ Fig. 15.1), where this experiment was

■ **Fig. 15.1** *Slow light:* Cover page of the 18th February 1999 issue of the journal Nature illustrating an experiment on slow light by the group of Lene Hau at the Rowland Institute for Science. Using an ultracold gas of atoms the physicists managed to slow down a pulse of light to a velocity of 17 m/s (Reproduced with permission of the journal Nature)

published in 1999, illustrates the achievement showing that a trained cyclist could even outrace such a light pulse. This is of course only a figurative way to demonstrate how slow the light was compared to the usual. Actually there was no cyclist involved in the experiment. The light was propagating in a tiny cloud of ultracold atoms contained in a vacuum chamber over a very small distance, as one can see by closer inspection of the figure. This spectacular result then triggered a rapidly growing activity in the field leading to many fascinating applications.

So, what is slow light and what is it good for? How can we understand the physics of it and how can we practically make light go so slow? These are the questions we want to answer in the following using simple pictures, on the one hand, and supplementing them with a little bit of details, on the other hand, for those who want to go slightly deeper. Yet we will avoid math as much as possible and refer those who seek more detailed information to the specialized literature [7–11].

15.2 Slow Light, Stopped Light and Stationary Light: A Simple Picture

How can one slow down light to such extremely low velocities? Imagine a fast racing car (◘ Fig. 15.2). If a heavy trailer is attached to the car, its engine has now also to pull the trailer. This slows down the car considerably. Something similar happens with light in a specially arranged atomic medium used in EIT experiments. Light is composed of photons—tiny particles which are very fast, so one can visualize them as fast racing cars. When entering the atomic medium, most of the photons are converted into a special kind of atomic excitations (which we here call spin excitations) which cannot move on their own, and thus behave like heavy trailers. The atomic excitations generated in this way are coupled to the small number of remaining photons which have to pull a vast number of immobile spin excitations while travelling in the medium. In this way, the propagation of the whole pulse of light is slowed down dramatically. The possibility to convert 'fast cars' into 'immobile trailers' is a small, but important difference to usual cars and trailers we encounter in real life. When the crawling light pulse reaches the end of the medium, the atomic excitations (trailers) are converted back to photons (fast cars), so the light exiting the medium becomes fast again.

Now imagine that the number of photons converted into atomic excitations (i.e. fast cars converted into trailers) can somehow be increased at will. This means there is an even lesser number of remaining cars to pull the whole bunch of trailers.

◘ **Fig. 15.2** *A simple picture of slow light:* Imagine a bunch of racing cars that enter a parking lot where heavy trailers get attached to them. Since the racing cars have to pull the trailers, they get slowed down considerably. When they reach the end of the parking lot, the trailers get detached, and the cars can move on with their full speed. Slow light is almost like this, except that cars get partially converted into trailers at the entrance to the parking lot and converted back at the end

15

And now imagine further that the conversion between cars and trailers can be changed while the fast cars are going through the trailer park. What if all of them are converted and no racing car is left to pull? The pulse would stop! This is the essence of stopped, or more precisely stored light, theoretically predicted in [12] and soon after experimentally verified in [13, 14]. The important difference of this kind of light storing and using a black piece of paper, which just absorbs the light, is that here the information carried by the photons is still present in the medium, in our analogy in the form of heavy trailers. Thus in principle all information about the original photons stored in the atomic excitations (trailers) can be converted back into photons (fast cars) either completely or in part. When the slow-light pulse reaches the end of the medium, the atomic excitations can no longer be dragged along and are fully converted back into light. In this way the stored light pulse can be fully retrieved.

Light storage is of particular interest in information technology especially in quantum information science. Light is an ideal carrier of information be it classical information which we use in every-day life or be it quantum information which may encounter at some day in a quantum network. Yet in the second case it is rather difficult to store information without loosing the quantum character, referred to as quantum coherence. Here light storage is an extremely useful method to build what is called a quantum memory for light. In fact first proof-of-principle demonstrations of quantum memories for photons based on light storage have already been made in a number of labs [15, 16].

It is noteworthy that by storing a light pulse all its photons (i.e. all the racing cars) are converted into immobile atomic excitations (trailers). Yet there is another way to make photons immobile where the photons are still present in the medium. This is called stationary light. It is formed when two counter-propagating pulses of light are driving the *same* spin excitations of a properly prepared atomic medium [17–21]. This corresponds to having two types of racing cars, one going from the left to the right, and another one from the right to the left. Both types of cars are trying to pull the same immobile trailers in opposite directions, as illustrated in ▪ Fig. 15.3. The forces compensate, so the cars and the trailers remain at rest. More precisely stationary light behaves like massive quantum particles with zero average velocity. Note that in the quantum world physical quantities such as the particle velocity fluctuate and thus we need to talk about averages here.

One can also produce a situation where two counter-propagating pulses of light drive *different* spin excitations of the atomic medium. If the two types of spin excitations are coupled to each other in the right way, two-component slow light is formed which has a more complex structure resembling what is known in quantum physics as a particle with a spin degree of freedom [22–25]. This is like having two types of racing cars going in opposite directions, each pulling different types of

▪ **Fig. 15.3** *The principle of stationary light:* Imagine racing cars entering a parking lot with heavy trailers from opposite sides. When attaching trailers to the cars they are pulled in opposite directions with equal forces and thus don't move at all. In this way the racing cars can be brought to halt even without converting them completely into trailers as is the case for light storage

Fig. 15.4 *Multi-component slow light:* When racing cares moving in opposite directions pull different types of trailers, both types of cars would slow down independently of each other. However, when coupling the trailers together in a proper way a situation is created that corresponds in physics to quantum particles with an internal degree of freedom

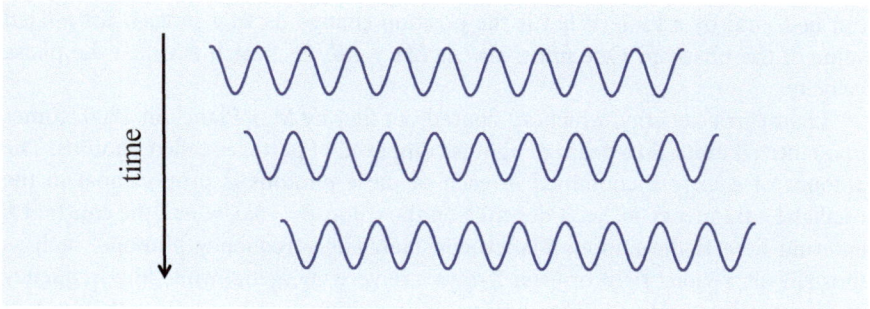

Fig. 15.5 *Light waves:* Light are waves of the electric field oscillating in space with a certain period, the wavelength λ. The 'hills' and 'valleys' of the wave, i.e. the points of maximum and minimum wave amplitude propagate in space with phase velocity c, such that at a fixed point in space the electric field oscillates in time with frequency $\omega = 2\pi c/\lambda$

trailers, as shown in ■ Fig. 15.4. If the trailers were not coupled to each other, the two types of cars would slowly move in opposite directions pulling their respective trailers independently from each other. Yet if there is a coupling between the two sorts of trailers, the oppositely moving cars and trailers influence each other, making a more complex dynamics, resembling that of a relativistic quantum particle.

15.3 A Microscopic Picture of Light Propagation in a Medium

In order to understand the mechanism behind slow light we first have to talk about the microscopic physics of light propagation in a medium. In particular we will discuss what the physical origin of absorption and refraction is, two phenomena which we are familiar with in every-day life.

15.3.1 Absorption, Emission and Refraction

Light is nothing else than an electromagnetic wave build up from oscillating electric and magnetic fields. The color of light is determined by the oscillation frequency $\omega = 2\pi/T$, given by the inverse of the temporal period T of oscillations. The electric field of a plane wave propagating along say the x axis of some coordinate system has a sinusoidal form depicted in ■ Fig. 15.5. It is characterized by the frequency ω, and a corresponding wavelength λ, which is the spatial period of the wave. This can be written in the following form:

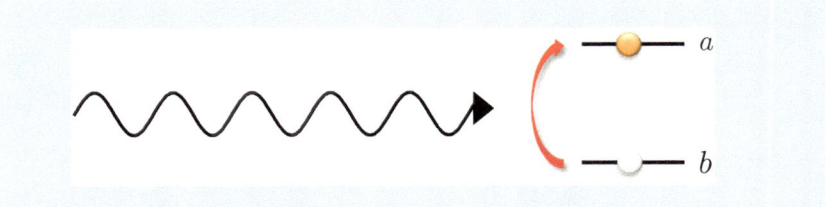

Fig. 15.6 *Absorption:* When an atom absorbs a photon it changes its quantum state from a low-energy state to a high-energy one

$$E = E_0 \sin\left(\omega t - 2\pi x/\lambda\right) = E_0 \sin\left(\phi\right), \tag{15.1}$$

where we have introduced the phase ϕ. The propagation velocity of such a wave can be found by asking: What is the position change Δx in a time Δt for a fixed value of the phase ϕ? One finds: $c = \Delta x/\Delta t = \omega\lambda/2\pi$, which is called the phase velocity.

Light carries energy, which, as figured out first by Max Planck in 1900, comes in quantized units. So a beam of light is composed of particles called photons. The amount of energy E contained in each of these photons is proportional to the oscillation frequency ω, i.e. it depends on the color, $E = \hbar\omega$, where the constant $\hbar$ entering here is the famous Planck constant. High-frequency photons, such as those of ultra-violet light or even X-rays, are very energetic, while low-frequency photons such as infrared light or microwaves, which we cannot see with our eyes, do contain much less energy per photon.

Matter, on the other hand, consists of atoms, which according to the laws of quantum mechanics have a number of states characterized by discrete energies. Very often it is sufficient to consider only two or three most relevant states. Atoms are also small quantum oscillators which can 'vibrate' at different frequencies corresponding to the energy differences between quantum states $\omega_{ab} = (E_a - E_b)/\hbar$. Many (but not all) of these 'vibration' modes are associated with an oscillating electric dipole. In this way an atom can absorb or emit radiation just like an antenna of a mobile phone. As we shall see later on, photons play the role of the fast racing cars described in the introductory section, whereas properly prepared atoms absorbing the photons play the role of the heavy trailers. When an atom absorbs a photon it changes its quantum state from the low-energy state to the high-energy state (see ◘ Fig. 15.6) and vice versa if it emits a photon.

There are actually two types of emission of an excited atom. The most common is spontaneous emission, where a photon is emitted in a random direction leading to the loss of information on the state, the propagation direction and the polarization of the photon that excited the atom in the first place, see ◘ Fig. 15.7a. The other one is stimulated emission which takes place in the presence of other identical photons and is pointed into the direction determined by these photons, see ◘ Fig. 15.7b. In addition to spontaneous emission there are a number of other relaxation processes for excited states in atoms. As a consequence of these processes and due to spontaneous emission, excited atomic states decay with some rate γ. Thus when light shines on a cloud of atoms or atoms arranged in a crystal, it can be absorbed by exciting some of the atoms into high-energy states which subsequently decay. Clearly how much a medium absorbs depends on the density of atoms, which in a gas is much less than, e.g., in a solid.

Still, why is it that some solids like diamond are transparent to visible light and others like coal are pitch black? Both are just slightly different forms of carbon and their density does not differ significantly. The reason is simple: In order for a photon to be efficiently absorbed, its frequency has to be close to the frequency of the atomic oscillator, i.e. the frequency should correspond more or less to the

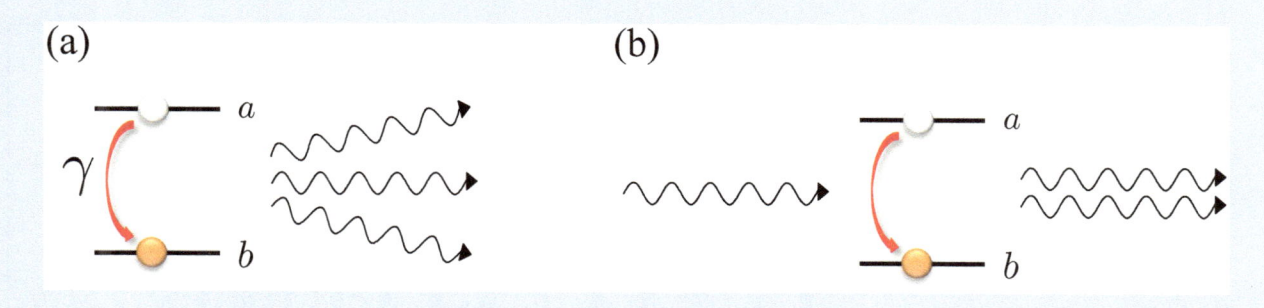

Fig. 15.7 *Spontaneous and stimulated emission:* An excited atom can loose its excitation energy either by spontaneously emitting a photon in an arbitrary direction (**a**) or stimulated by an incoming photon (**b**) in which case the emitted photon has the same direction than the incident one. In both cases the atom changes its quantum state from the high-energy to the low-energy one

energy difference between some lower and higher state $\omega_{ab} = (E_a - E_b)/\hbar$. When this is the case, one talks about resonance. If the photon frequency is very different from any of the vibration frequencies of the atomic oscillator, i.e. if the light is off-resonant, not much can happen. It is like if you are trying to make a bridge vibrate by jumping up and down but are doing it at the wrong pace. Only a tiny bit of the photon energy is transferred to the atom, stored there for a very little moment and then is reemitted into the stream of photons. In this process the atom is actually not completely transferred from the lower-energy state to the higher-energy state, as in ◻ Fig. 15.6, and the subsequent emission process is a bit different from the stimulated process shown in ◻ Fig. 15.7b, but in essence it is like this. A word of caution is needed here: This picture of absorption is a bit of an oversimplification if applied to solids rather than to sparse atomic gases. The quantum states and energies in a solid are not the same than those of isolated atoms as they are affected by atom–atom interactions. Also even off-resonant transitions can eventually lead to sizable absorption if there are very many of them.

As we have mentioned before, waves are characterized by a wavelength λ, which gives the spatial period of a wave and is directly related to the frequency. In vacuum the relation between the two is $\lambda_0 = 2\pi c_0/\omega$. Here c_0 is the vacuum speed of light, i.e. the fastest velocity allowed by the laws of nature. In a medium this relation is changed, however. The short moment for which the photon is stored in the atom causes a delay. The effect of the very many, tiny delays at every atom in the medium makes light appear to propagate with a modified phase velocity

$$c(\omega) = c_0/n(\omega). \tag{15.2}$$

Here $n(\omega)$ is called the refractive index. In vacuum the refractive index is unity. The name 'refractive index' stems from the fact that it characterizes the refraction of light beams at an interface between say air and a piece of glass, as illustrated in ◻ Fig. 15.8. Refraction comes about since along with the change of the phase velocity of a plane wave at frequency ω comes a change of the wavelength $\lambda = \lambda_0/n(\omega)$. This is because the frequency of the wave remains the same in the medium, giving $\omega = 2\pi c_0/\lambda_0 = 2\pi c/\lambda$.

The influence of a medium on the propagation of light is characterized by the susceptibility χ. In ◻ Fig. 15.9 we have plotted both the absorption strength (red line) represented by the imaginary part of the susceptibility $Im[\chi] = \chi''(\omega)$ together with its real part $Re[\chi] = \chi'(\omega)$ (blue line) as function of the frequency in the vicinity of an atomic resonance frequency ω_{ab}. The latter χ' describes the deviation of the index of refraction from unity, $n = 1 + \chi'/2$. One recognizes that the absorption peaks on resonance and falls off quickly with increasing frequency mismatch $\Delta = \omega - \omega_{ab}$, called detuning. The refractive index has a bit more

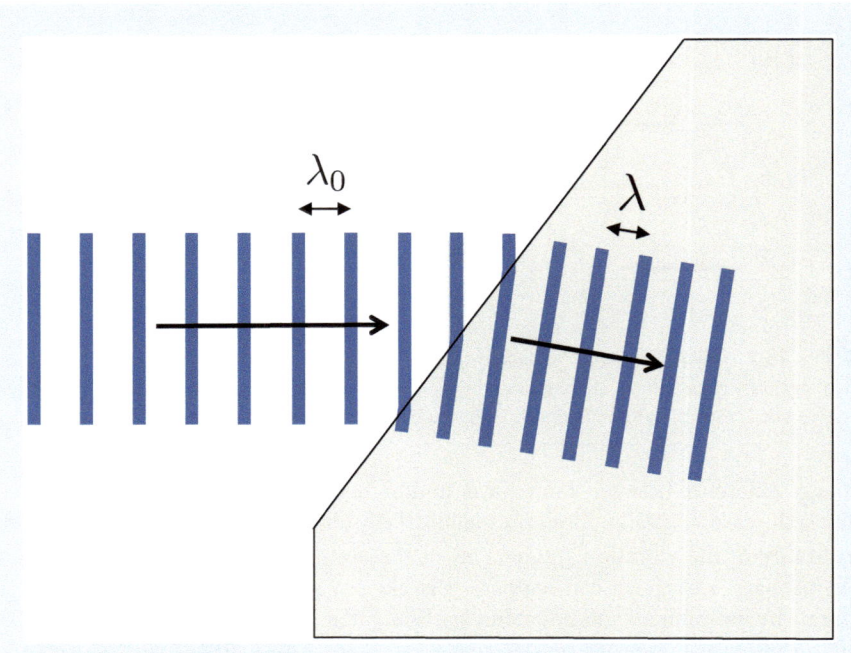

Fig. 15.8 *Refraction of a wave:* When a wave hits the surface of a medium with a different phase velocity, the wavelength has to change as the electric field oscillates in time always with the same frequency. This causes a change in the propagation direction of the wave

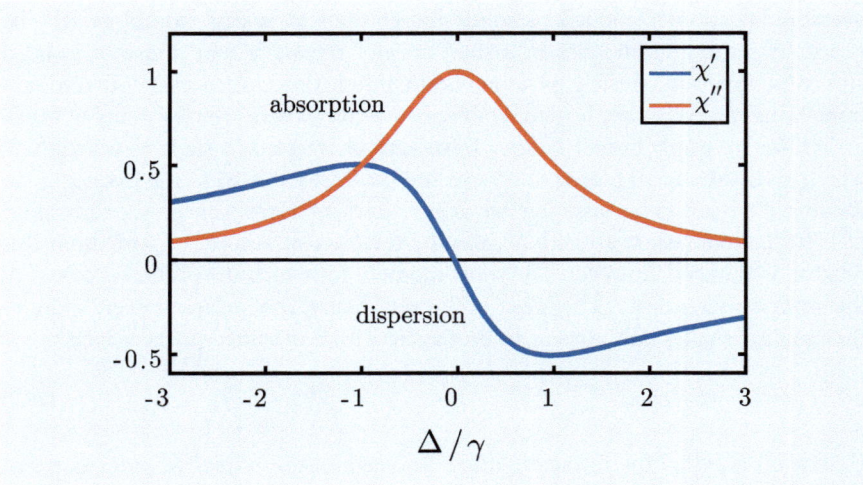

Fig. 15.9 *Absorption and dispersion of a two-level atom:* An atomic oscillator described by a two-level quantum systems leads to a strong absorption of light close to its resonance frequency. This is shown by the *red curve*, representing the imaginary part $\chi''(\omega)$ of the susceptibility as function of frequency ω. The refractive index $n(\omega) = 1 + \chi'(\omega)/2$, determined by the real part of the susceptibility $\chi'(\omega)$, is shown as the *blue curve*

complicated anti-symmetric shape. For frequencies above the resonance, $\omega > \omega_{ab}$, the medium leads to a reduction of the refractive index with respect to the background value, while below resonance, $\omega < \omega_{ab}$, the refractive index is enhanced. One notices the following from the figure: For large values of $|\Delta|$ the refractive index falls off much slower than the absorption, so for far off-resonant light only refractive effects of the medium matter. This is why even transparent media can still have a strong effect on the propagation of light. One of these effects

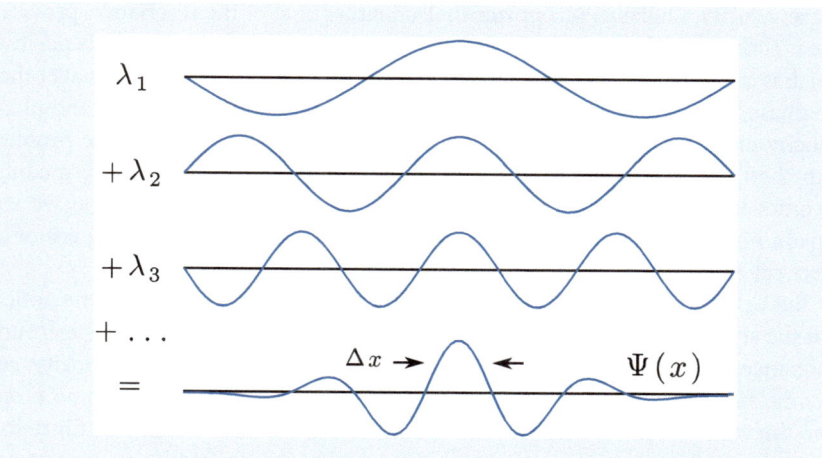

⬛ **Fig. 15.10** *Wavepackets:* In order to create pulses of light with a finite spatial length, one needs to superimpose plane waves with slightly different wavelength in a proper way. In a medium the phase velocity of these components can differ. As a consequence the effective speed of the wavepacket is not given by the phase velocity but by the group velocity defined in Eq. (15.3)

is the refraction of a light beam at an interface between two media with different refractive indices. Another one is the modification of the propagation velocity of pulses, discussed in the following subsection.

15.3.2 Group Velocity

We have seen that the dependence of the refractive index on the frequency leads to different wavelength of light in a transparent medium as compared to free space. This dependence has another equally important effect, it determines the effective propagation speed of photon *wavepackets*. As illustrated in ⬛ Fig. 15.10, one needs to superpose light waves with slightly different wavelength in order to create a wavepacket, i.e. a light pulse of finite length. In some sense we can envision photons as such wavepackets.

What is the propagation speed of such a wavepacket which consists of plane waves of different frequencies? In vacuum all frequency components propagate at the fundamental speed of light c_0, so wavepackets made of plane waves also propagate at this speed. But what about a medium, where each component has a different phase velocity $c(\omega) = c_0/n(\omega)$? It turns out that the slightly different phase velocities of each constituting plane wave cause the envelope of the pulse to move at the so-called group velocity v_{gr} which can be very different from the phase velocity $c = c_0/n(\omega)$. It is given by

$$v_{gr} = \frac{c_0}{n(\omega_0) + \left(\frac{\Delta n}{\Delta \omega}\omega_0\right)} \tag{15.3}$$

where ω_0 is the average frequency of the different components. The group velocity determines the effective speed of photons in a medium. When we talk about slow light, what we mean is light with a very small group velocity compared to c_0.

From Eq. (15.3) one recognizes that in addition to the refractive index itself, contained in the phase velocity $c = c_0/n(\omega)$, also the slope $\Delta n(\omega)/\Delta \omega$ enters at which the refractive index $n(\omega)$ changes by $\Delta n(\omega)$ when the frequency makes a small change $\Delta \omega$. As can be seen from ⬛ Fig. 15.9 this slope is typically small far off resonance and the second term in the denominator of Eq. (15.3) is irrelevant. Thus in this frequency range the group velocity is essentially equal to the average phase velocity. One also recognizes that on either side of the resonance, provided one is sufficiently far away from the resonance point, the slope of $n(\omega)$ is positive, which is called 'normal' dispersion. Here the group velocity is slightly smaller than the phase velocity. In order to see a dramatic difference between group and phase velocity one has to go closer to resonance. We immediately notice the problem with that: Whenever we are closer to resonance, the absorption of the medium becomes large and light gets quickly absorbed. In the following section we will explain how one can overcome this problem in an elegant way making use of an effect called EIT.

But before we proceed with this let's make a little side remark here: One notices that the situation is completely different in a very narrow frequency range around resonance: Here $\Delta n(\omega)/\Delta\omega$ is negative and large and the group velocity can become larger than the phase velocity. In principle it can even become larger than the vacuum speed of light c_0! But don't worry, this does not violate Einstein's principle of relativity as proven already by Arnold Sommerfeld [26]. One notices, for example, that in the same spectral region there is large absorption. As a consequence no signal can actually propagate faster than c_0.

15.4 Electromagnetically Induced Transparency

How can we get around the problem that strong effects on the group velocity of light seem to be always associated with large losses? The answer came from an effect known as EIT [2, 27, 28]. To understand what EIT is all about let us start with an analogy from mechanics [29]: Consider a mass m which can slide on a surface and is attached to a wall with a spring, as shown in ◘ Fig. 15.11a. This system forms an oscillator with frequency $\omega_0 = \sqrt{k/m}$, where k is the spring constant. Now assume that there is some friction, e.g. due to a rough surface on which the mass slides. If the oscillator is excited by a periodic force with frequency

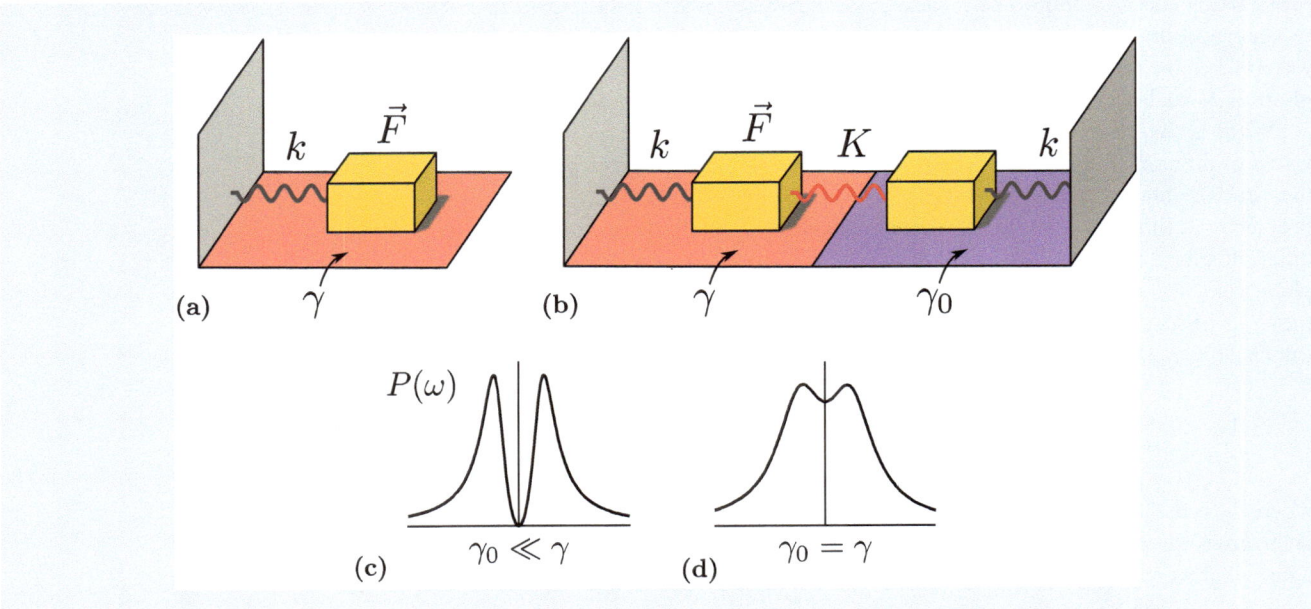

◘ **Fig. 15.11** *Coupled mechanical oscillators:* (**a**) A mechanical oscillator with resonance frequency $\omega_0 = \sqrt{k/m}$ driven by a periodic force **F** with frequency ω and subject to friction with energy loss rate γ generates a loss power spectrum similar to the absorption spectrum of a two-level system shown in Fig. 15.9. (**b**) If the mass is coupled to a second one with smaller friction (loss rate $\gamma_0 \ll \gamma$) a resonant periodic drive causes only the second mass to move and thus the power loss is dramatically reduced. (**c**) Loss power spectrum for $\gamma_0 \ll \gamma$. (**d**) If γ_0 equals γ, the total loss power spectrum is that of two independent absorption spectra slightly shifted in frequency (Adapted from [29])

ω close to the resonance frequency ω_0, energy is transferred to the oscillator and subsequently dissipated into heat due to the friction. The dissipated power $P(\omega)$ depends on the frequency mismatch between oscillator and drive frequency $\Delta = \omega - \omega_0$ and has a similar form as the absorption curve in ◘ Fig. 15.9.

Now suppose we couple this oscillator to another mass oscillating with the same frequency ω_0 using an additional spring with spring constant K (◘ Fig. 15.11b). Let us assume next that the second oscillator has little or no friction. If we now drive the first mass with a periodic force something interesting happens: Looking at ◘ Fig. 15.11c, where we have plotted the dissipated power again as function of frequency, one notices that if the driving frequency ω matches exactly the oscillator resonance frequency ω_0 little or no energy gets dissipated!

The reason is that the first mass, i.e. the one with friction, does not move at all. Only the second mass, the one with little or no friction, oscillates. It does this in such a way that it produces a force on the first mass exactly opposite to the external force F. The two forces compensate each other, and so the first mass stands still. One can say that the system of oscillators is driven into a dark mode, i.e. a mode without dissipation in which the lossy oscillator is not excited. Consequently the effect of friction is reduced considerably and no or little energy is dissipated.

The situation changes if the second mass also experiences a substantial friction. In particular, if the loss rates of both oscillators are the same, i.e. $\gamma_0 = \gamma$, the loss power spectrum is just the addition of two simple loss curves slightly shifted in frequency relative to each other, as shown in ◘ Fig. 15.11d. As long as γ_0 is not too large, there are two maxima corresponding to the two eigenfrequencies of the coupled oscillators. The splitting increases with $\sqrt{K}$, i.e. with the strength of the coupling. Most importantly if γ_0 vanishes or is very small, one can make the coupling very weak and still the dissipation essentially disappears when driving the first mass. This creates a situation where one can be close to resonance while there is almost no loss.

This principle can be translated to atomic oscillators. What is needed are two oscillators, one of them almost lossless, another one lossy, and the two oscillators need to be coupled by a 'spring'. This can be realized in a 3-level Λ-type system shown in ◘ Fig. 15.12. The atom-light coupling scheme is called Λ-type scheme because of the resemblance to the Greek letter Λ.

The first oscillator corresponds to the transition between the initially populated ground state g and the excited state e, as shown in ◘ Fig. 15.12a. This oscillator dissipates energy because of decay of the excited state e with rate γ, e.g. due to

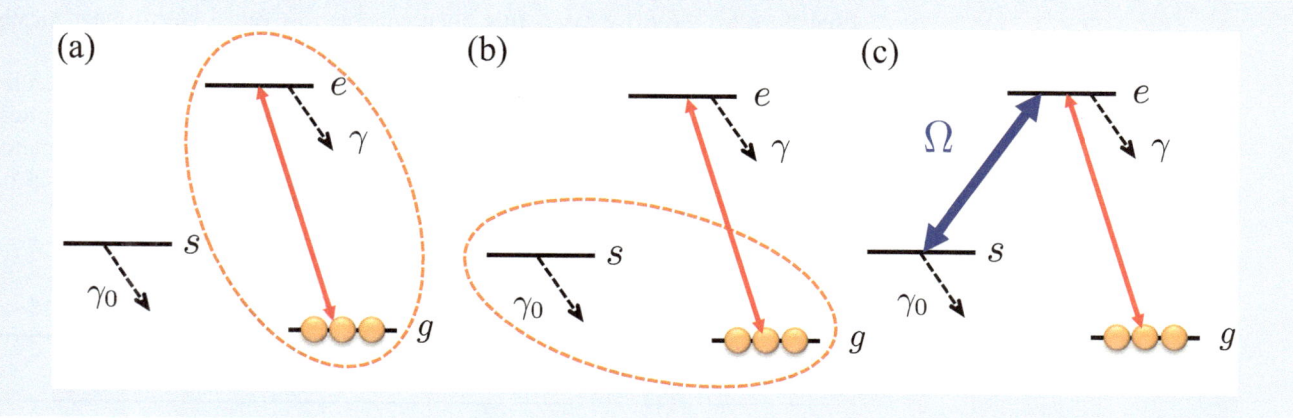

◘ **Fig. 15.12** *Principle of electromagnetically induced transparency:* (**a**) A lossy atomic oscillator consisting of the initially populated ground state g and an excited atomic state e is driven by a probe field (*red arrow*). (**b**) In a three-level Λ-type system there exists a second atomic oscillator between states g and s, which can be lossless or have very small losses, e.g. if s is a low-energy state. (**c**) Coupling the two oscillators by a control laser with a strength characterized by the Rabi frequency Ω produces a situation similar to that shown in ◘ Fig. 15.11b. Consequently the medium becomes (almost) transparent to the probe field

15

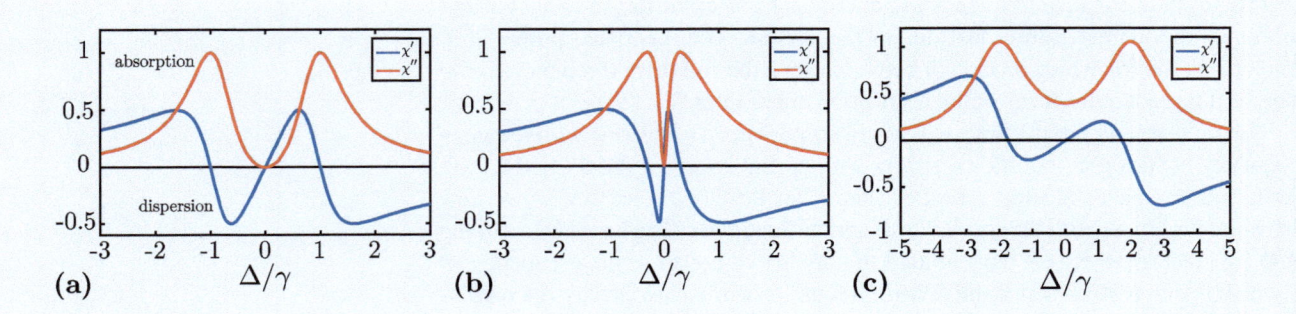

⬛ Fig. 15.13 *EIT versus two-level resonances:* (**a**) Real (χ') and imaginary (χ'') parts of the susceptibility of an EIT system characterizing the refraction and the absorption, respectively. Figure (**b**) shows the same with a smaller Rabi frequency of the drive field. For comparison we have shown in (**c**) the total susceptibility spectrum of two independent two-level systems with slightly shifted resonance frequencies. While χ', i.e. the index of refraction has a very similar shape in (**a**) and (**c**), there is an important difference in the absorption: In the EIT case it vanishes in between the two maxima, while for two two-level resonances it remains large

spontaneous emission. The oscillator is driven by the probe field (see ⬛ Fig. 15.12a) corresponding to the external driving force in the mechanical picture from above. The ground state g together with another metastable ground state s forms the second oscillator (see ⬛ Fig. 15.12b). The latter state s can be, e.g., a long-lived hyperfine spin state in the atomic ground state manifold, i.e. a low-energy state like g. Therefore the second oscillator is essentially lossless or has very small losses. Finally the role of the spring coupling the two oscillators is taken over by a coherent control laser field inducing transitions between the excited state e and state s (see ⬛ Fig. 15.12c). The strength of this coupling is directly proportional to the amplitude of the electric field of the control laser, and the resulting splitting of the absorption peak (shown in ⬛ Fig. 15.13) is denoted as Ω and is called Rabi frequency.

The absorption as a function of the probe field frequency ω relative to the resonance, expressed by the detuning $\Delta = \omega - \omega_0$ is shown in ⬛ Fig. 15.13a, b as red lines. It consists of two absorption peaks like the spectrum of two coupled mechanical oscillators in ⬛ Fig. 15.11c. Similar to the mechanical analog, the absorption shown in ⬛ Fig. 15.13a, b vanishes exactly on resonance for $\gamma_0 = 0$, or is insignificant for small γ_0. This is quite remarkable since this means that despite the fact that one is very close to the resonance frequencies of the coupled system, the absorption is vanishingly small! Since a non-absorbing medium is transparent and since this effect is induced by the coupling of the two atomic oscillators by the drive laser, this phenomenon was called electromagnetically induced transparency or in short EIT.

The phenomenon of EIT has a widespread application in atomic and molecular physics and in optics. It can be used, for example, to make nonlinear optical processes much more efficient as it allows to operate close to atomic resonance without suffering from absorption. Some of the interesting applications will be discussed in detail in the following section.

15.5 Slow Light, Stored Light and Dark-State Polaritons

15.5.1 Slow Light

As we have discussed in ⬛ Sect. 15.3 the absorption spectrum is associated with the imaginary part of the susceptibility. ⬛ Figure 15.13a, b show the absorption spectrum of the atomic medium at an EIT resonance. The spectrum consists of two lines separated by an amount proportional to the strength of the driving field (Ω)

and in between these two peaks the absorption goes to zero. Also shown is the real part of the susceptibility as a function of frequency, which is called dispersion. In ◘ Fig. 15.13c we have plotted the absorption and dispersion spectra of two uncoupled oscillators with slightly different frequencies. We notice that the dispersion curves look qualitatively very similar in ◘ Fig. 15.13a, c. In particular the real part of the susceptibility, i.e. the refractive index, has a positive slope around $\Delta = 0$. In the case of two uncoupled two-level systems this just results from superposing the below-resonance tail corresponding to one oscillator with the above-resonance tail of the other. The most important difference between the case of two uncoupled resonances and EIT is that in the former case the absorption does not vanish in between the two resonances.

The dispersion curve has a remarkable feature right on resonance. It has a linear slope that can become very steep. In fact the closer the two absorption peaks are, the steeper is the dispersion curve. From Eq. (15.3) we notice that a steep slope of the index of refraction leads to a very large denominator in the expression for the group velocity. This means close to resonance the medium is, on the one hand, transparent due to EIT and at the same time the group velocity can be extremely small. This is the origin of ultra-slow light in EIT.

The value of the group velocity in an EIT medium is determined by the general equation (15.3) with the second term in the denominator being much larger than the first one, giving

$$v_{gr} \approx \frac{c_0}{\omega_0 \frac{\Delta n}{\Delta \omega}} \sim \frac{\Omega^2}{\rho},$$ (15.4)

where Ω is the Rabi frequency of the drive laser, and ρ is the density of atoms. By turning down the intensity of the drive laser, i.e. by reducing Ω, or alternatively by increasing the atom density ρ, one can reach very small values of the group velocity. This can also be seen from ◘ Fig. 15.13a, b: Reducing Ω the separation between the absorption maxima decreases making the dispersion curve steeper in the center and hence the group velocity smaller.

The first experiments measuring the group velocity reduction in EIT where done by Harris et al. [3] in an atomic vapor cell reaching $v_{gr} = c_0/170$. The smallest group velocities achieved so far in experiments are obtained using very cold and dense clouds of atoms such as in a Bose Einstein Condensate and are on the order of 10 m/s, i.e. $v_{gr} = c_0/30,000,000$ [4].

When a light pulse enters a medium with a small group velocity it will be transmitted if its central frequency is close enough to the resonance and if its spectral width, i.e. the spread of frequencies associated with any pulse of finite duration, is much less than the distance between the two peaks in the absorption spectrum shown in ◘ Fig. 15.13. The very steep slope of the refractive index has also a profound effect on the spatial shape of the pulse, as illustrated in ◘ Fig. 15.14. When the pulse just enters the medium its front end will propagate with the group velocity v_{gr}, while its back end still propagates with the vacuum speed of light. As a consequence the pulse will be dramatically compressed in length inside the medium. The compression ratio is given by

$$l/l_0 = v_{gr}/c_0.$$ (15.5)

This resembles a situation where a number of vehicles moving fast on a highway suddenly approaches the beginning of an area with restricted speed. At this point the bunch of cars is compressed since when the first cars have already entered the area of restricted velocity, the ones at the back still drive at full speed. If the velocity of the vehicles is reduced by half, the distance between them becomes twice

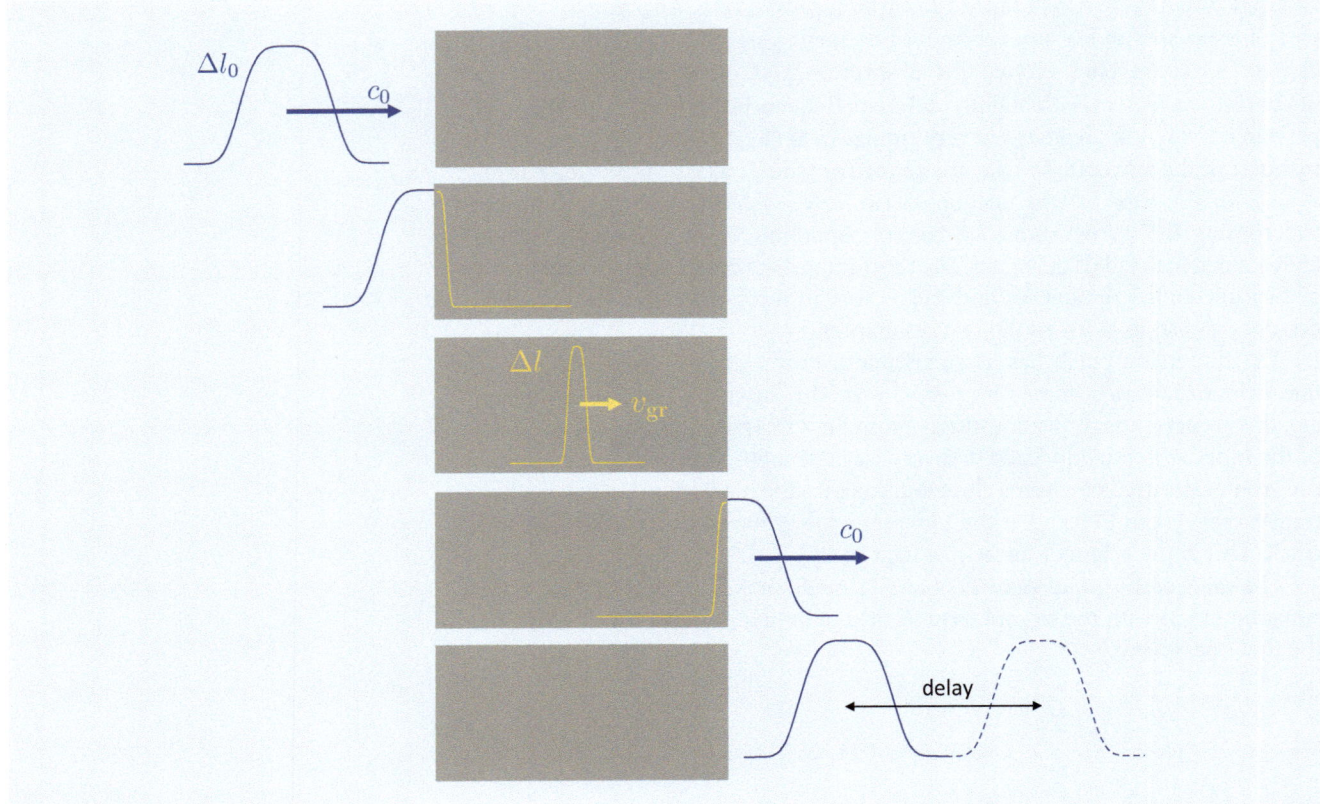

◻ Fig. 15.14 *Pulse compression:* When a light pulse enters a medium with a reduced group velocity it becomes spatially compressed by the ratio v_{gr}/c_0. When the front end is already in the medium it propagates with v_{gr}, while the back end still moves with the much larger speed c_0. This causes the pulse to shrink in space. The opposite is happening when the pulse leaves the medium

smaller, so the compression factor is 1/2. Since in the atomic media the light can be slowed down to such extremely small velocities as $v_{gr} \approx c_0/30,000,000 = 10$ m/s, the incoming pulse of fast light with original length l_0 of about 1 km will be compressed to a pulse of length l of about 30 µm (!). In this way even very long pulses of light can be made to fit into a small-sized material, such as an elongated (cigar shape) Bose Einstein Condensate of sodium atoms used in the 1999 experiment by the group of Hau [4]. When the pulse leaves the medium the opposite effect happens. The leading edge travels fast since it is in free space and the back end lags behind as it is still inside the medium. At the end the outgoing pulse has the same length as the incoming one, at least under ideal conditions. This is again like the spatial decompression of a bunch of cars when leaving the area of restricted speed on the highway.

15.5.2 Stopped Light and Quantum Memories for Photons

As can be seen from Eq. (15.4) the group velocity of slow light can be controlled by the strength of the coupling laser or the density of the medium. So what would happen if we turn the coupling laser off while the probe pulse propagates inside the EIT medium? The medium becomes immediately opaque for the probe light and thus we expect no probe field to survive. This is indeed the case. So does this mean the probe pulse is lost? Surprisingly this does not happen!

At the entrance of the medium most of the incoming photons are transferred to atomic excitations during the slowing down. In this process the pulse is also substantially compressed in space, so that it fits inside the medium. The atomic

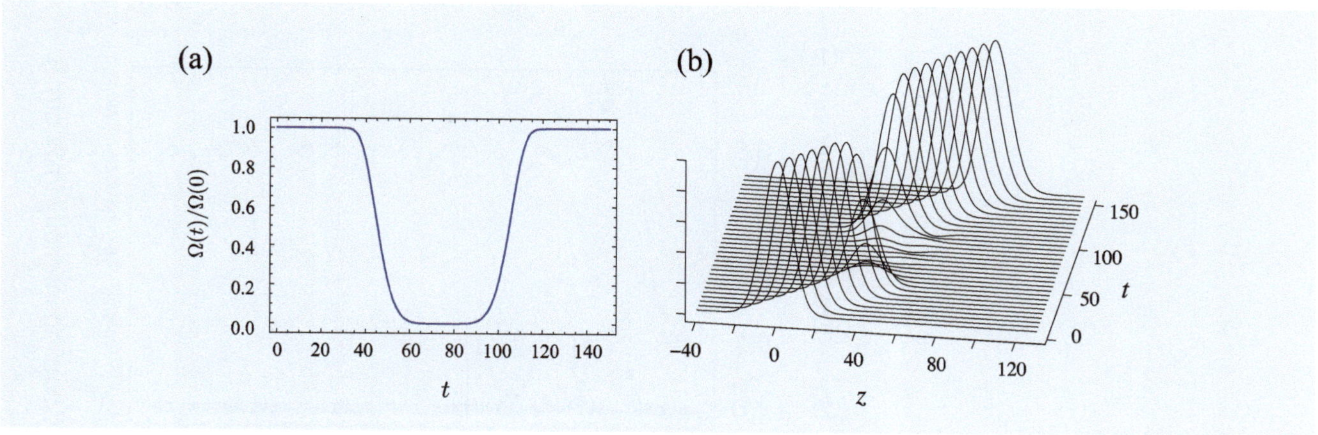

Fig. 15.15 *Light storage and retrieval:* When the strength of the control field is switched off smoothly while the slow-light pulse is in the medium (**a**) the pulse stops but also all photons disappear (**b**). If the control field is, however, switched on again at a later time, the pulse miraculously reappears and continues to propagate as a slow-light pulse in the medium

excitations, carrying information about the incoming pulse, travel together with the remaining photons. If the control field is switched off while the compressed pulse is still inside the medium the light disappears, i.e. no probe light survives. But if the control field is switched on again at a later instant of time, the pulse miraculously reappears! This is shown in the numerical simulation of Fig. 15.15. The right-hand side shows the propagation of the compressed light pulse inside the medium when the control laser is switched off and on again as illustrated on the left-hand side. So obviously we have somehow managed to stop (or more specifically to store) the light pulse for a while and sent it off its way a while later.

This remarkable phenomenon of light stopping (storing) was theoretically predicted in 2000 [12] and experimentally demonstrated in 2001 by two groups at Harvard University [13] and the Roland Institute of Science [14]. Figure 15.16 is a reproduction of the data obtained in one of these experiments from [14]. In these experiments a storage time of up to half a millisecond was reached. In 2009 the group of Immanuel Bloch at the Max Planck Institute for Quantum Optics in Garching, Germany in collaboration with colleagues from Israel has increased the storage time to 240 ms using ultracold atoms in a Mott insulating state in a three-dimensional optical lattice [30]. In the so-called Mott insulating phase atoms are particularly protected from perturbations such as collisions and diffusion, which leads to the prolonged storage duration. The current record for storage times is 1 min [31]. It has been obtained in doped glasses, where impurity atoms behave almost like free atoms in a vapor with the advantage that they do not move as in the Mott insulating state discussed above, and the atomic density is higher than in a gas.

15.5.3 Slow-Light Polaritons

We have seen in Sect. 15.3 that the microscopic picture of light propagation in a transparent medium is that each atomic oscillator absorbs a tiny little bit of an incoming photon, stores it for a short moment and releases it again with a small time delay as electromagnetic energy. The amount of the time delay is determined by the ratio of group velocity and vacuum speed of light. Furthermore the reduction in the group velocity also leads to a spatial compression of a photon pulse at the entrance to the atomic medium, as discussed in the previous subsection. If a light pulse is spatially compressed without increasing its amplitude, this

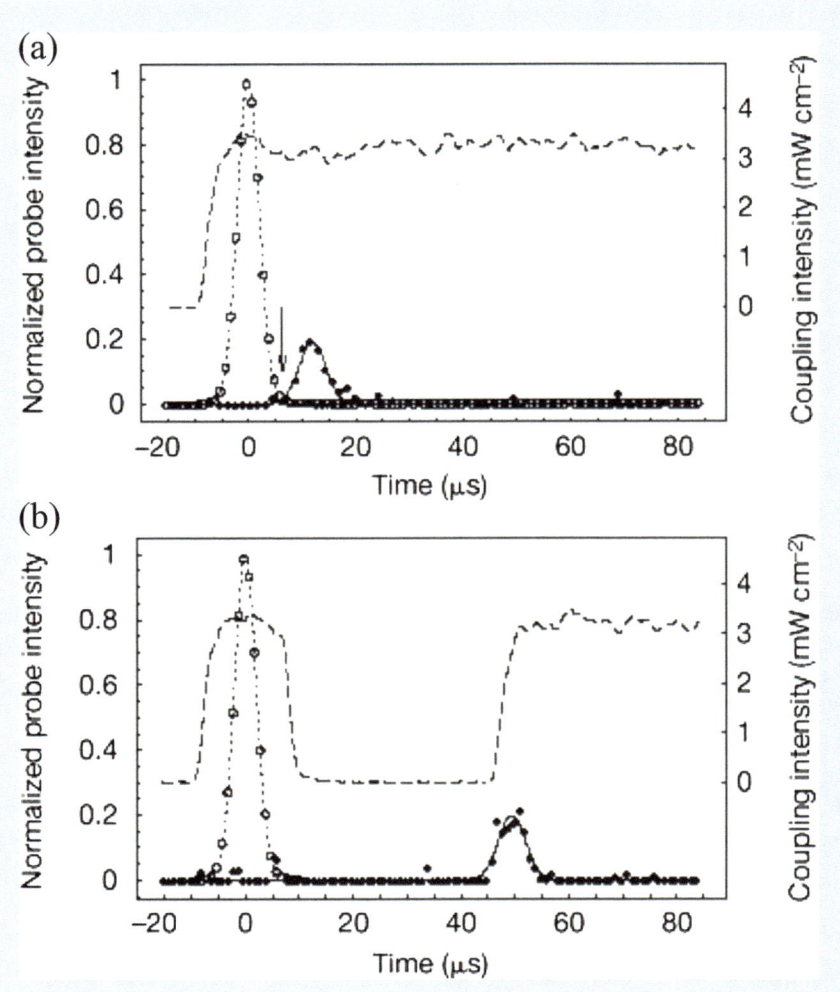

Fig. 15.16 *Light storage experiment:* Reproduction from one of the first experiments on light storage [14] (with permission of the journal Nature). Shown are the control field (*dashed*), the input probe pulse (*open circles and dotted line*) as well as the output probe pulse (*full circles and full line*). The *top curve* shows the pulse delay when the control field is on all the time, the *lower curve* shows the storage of the probe pulse when the control field is switched off and subsequently on again after some time

means that its content of photon energy decreases, i.e. the total number of photons contained in the pulse must be reduced according to the spatial compression. Where do these photons go if the medium is not absorbing? The answer is: They are temporarily stored in the form of atomic spin excitations.

In an usual transparent medium, such as glass, the ratio between the number of atomic excitations and photons is fixed and is very tiny. In an EIT medium this ratio can be large and it can be dynamically modified by tuning the strength of the control laser or by changing the atomic density. The best way to describe this is not to think in terms of photons and atoms separately but in terms of a combined quasiparticle, called polariton, containing a contribution due to both a photon and an atomic spin excitation, i.e. the excitation of the atom from the initially populated atomic ground states g to another ground state s [12, 32, 33]. The polariton picture has been introduced in [12] to describe storing and releasing of slow light following an earlier single-mode treatment [32] used to describe Raman adiabatic passage between the atomic ground states which did not include pulse propagation. We can visualize this polariton as a vector with two components, the

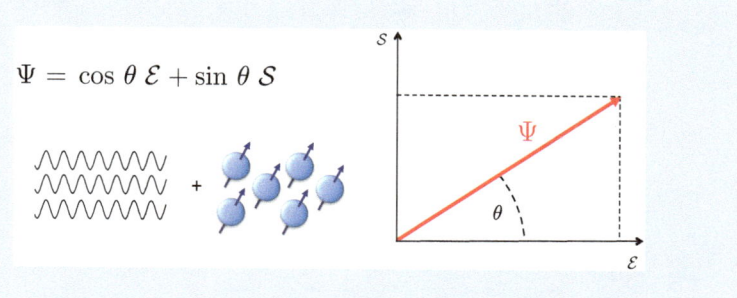

$$\Psi = \cos\theta\,\mathcal{E} + \sin\theta\,\mathcal{S}$$

◼ **Fig. 15.17** *Slow-light polariton:* Slow and stored light can most easily be understood in terms of quasiparticles called dark-state polaritons introduced in [12]. They are a superposition of the electric field $\mathcal{E}$ of the probe pulse and an atomic spin excitation S, like a vector in a two-dimensional plane. The mixing angle θ depends on the strength of the control laser and the atom density and thus can be changed. The angle θ also determines the properties of the polaritons, such as their velocity, see Eq. (15.6)

electric field $\mathcal{E}$ and an atomic excitation S indicated in ◼ Fig. 15.17, where the mixing angle θ determines the ratio between the photonic and atomic components making up the polariton. Since the polariton is only partially a photon and only the photons move, the propagation speed is determined by the fraction of photons comprising the polariton:

$$v_{gr}/c_0 = \cos^2\theta = \frac{\Omega^2}{\alpha\rho + \Omega^2}. \tag{15.6}$$

The group velocity v_{gr} is evidently less than that of pure photons. (Here α is some constant, which is not relevant for the present discussion.) When the probe pulse is outside the medium, where $\rho = 0$, it can be interpreted as a polariton with $\cos^2\theta = 1$ representing a pure photon without any atomic component. When it enters the medium, e.g. the cloud of ultracold atoms in the BEC experiment of Hau et al. [4], the density ρ increases smoothly in space. As a consequence the polariton turns smoothly into a mixed atomic-photonic excitation, with a large atomic component.

Since Ω, determining the group velocity in Eq. (15.6), is a tunable parameter, the composition of the slow-light polariton can be modified further while the pulse is propagating inside the medium. In the case of slow light, $\cos^2\theta$ is much less than unity already when the probe pulse has just entered the medium and most of the excitations which were originally photons propagate as an atomic excitation. By further reducing the strength of the control laser Ω from the initial value where $\cos^2\theta$ is finite (yet much smaller than unity) all the way to zero, the slow-light polariton looses its photon component altogether and reduces to a pure atomic excitation which does not move any more. By switching on the control laser again at a later time, $\cos^2\theta$ becomes finite again (yet much smaller than unity). The slow-light pulse resumes its motion inside the medium until reaching the end of the atomic cloud where it finally converts completely into a fast, purely photonic pulse. This explains the reappearance of the light pulse, when the control field is turned back on again. As shown in ◼ Fig. 15.18, illustrating the stopping and reacceleration of a slow-light pulse while inside the medium, the light storage and retrieval sequence becomes very clear in terms of the polariton picture. The polariton is there all the time. It only changes its character, first from fast light to slow light, and then to a frozen atomic spin excitation and finally back to a slow polariton and eventually to fast light again.

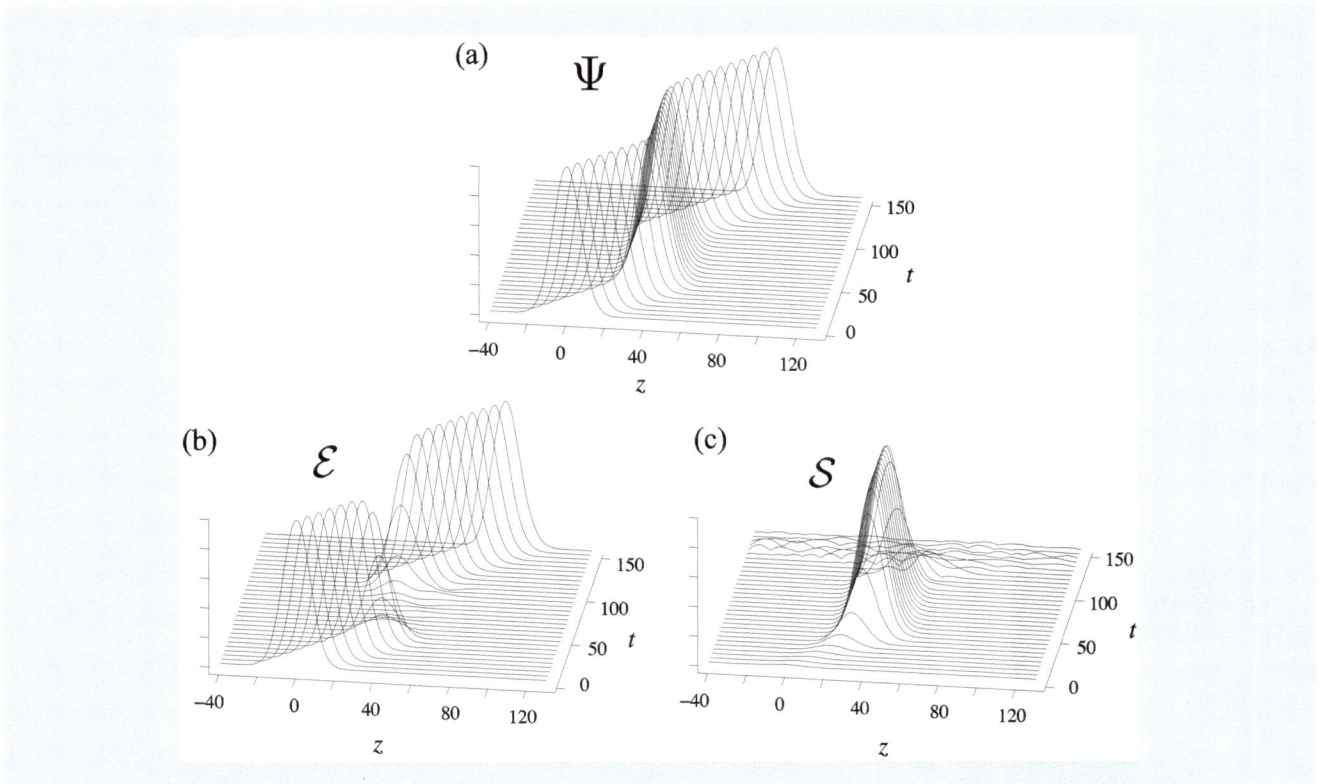

🔲 **Fig. 15.18** *Polariton picture of light storage:* Light storage and retrieval, as shown in 🔲 Fig. 15.15. This time also the propagation of the polariton (**a**) and the spin excitation (**c**) are shown. One recognizes that light storage is nothing else than a smooth conversion of the slow-light polariton from a polariton containing an electric-field component into a pure atomic excitation and back

15.6 Stationary Light

We have seen in the last section that a light pulse can be brought to a complete stop without loosing the information it contains by storing it in an atomic excitation. When the light pulse is at a halt, no photon is left in the medium anymore, so the polariton becomes entirely an atomic excitation. In the example of cars and trailers this corresponds to the case when all cars are converted into trailers. Thus there is no car left to pull and everything comes to a stop. There is, however, a way to keep the cars from driving without converting all of them to trailers. If two cars driving in opposite directions pull the same trailer, their forces can compensate and neither of the two can move forward. This is exactly what is happening in a situation called *stationary light*, which we will explain in the following.

A very interesting aspect of stationary light is that it mimics the behaviour of a massive quantum particle described by the Schrödinger equation for the amplitude of the stationary light polariton Ψ_{ss}:

$$ih\frac{d}{dt}\Psi_{ss} = -\frac{\hbar^2}{2m^*}\frac{d^2}{dx^2}\Psi_{ss}. \tag{15.7}$$

Unlike photons in free space, which always propagate at the speed of light c, massive particles can stand still, or more precisely, as we are talking about quantum particles, can have a zero average velocity. Importantly the effective mass m^* of the stationary light polaritons is not a fixed quantity such as the mass of an electron or a proton, but is a tunable parameter. It can be changed by the strength of the control laser fields. This property makes stationary light an

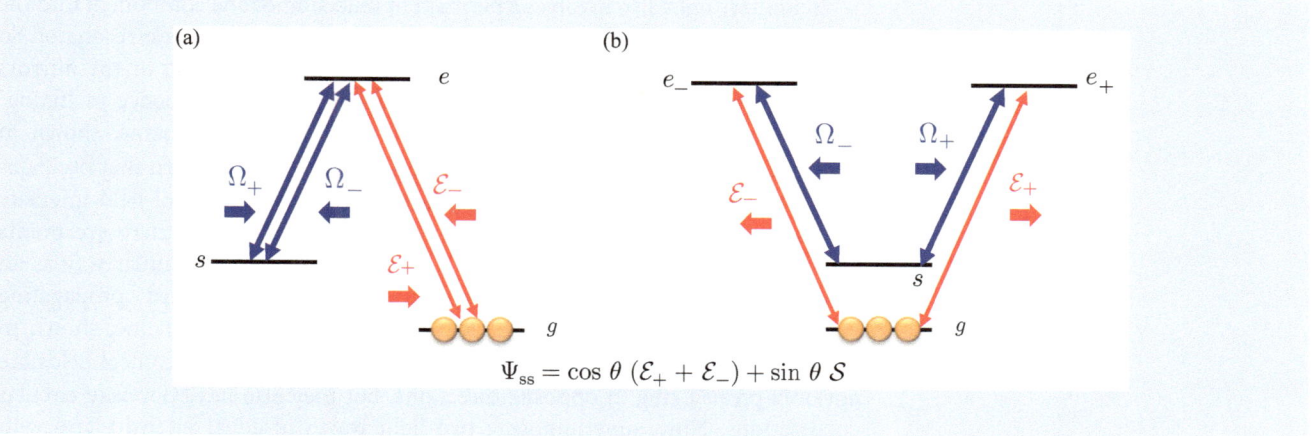

$$\Psi_{ss} = \cos\theta\,(\mathcal{E}_+ + \mathcal{E}_-) + \sin\theta\,\mathcal{S}$$

■ Fig. 15.19 *Stationary light:* If two counter-propagating drive fields of equal strength couple a Λ system of atomic levels (**a**), two counter-propagating probe field components of equal strength are formed. These fields interfere with each other and form a stationary-wave pattern. The same happens if the two drive fields have orthogonal polarizations and couple to two different transitions in a four-level (doubleΛ) scheme (**b**). In this case two counter-propagating probe fields are generated which also have orthogonal polarizations, but which nevertheless form a stationary-wave pattern. In these ways an excitation wavepacket is created which does not move and has still a non-vanishing electric-field component

interesting model system for analyzing fundamental properties of massive quantum particles.

What is the physics behind stationary light? Suppose there are two (rather than one) control laser beams of equal strength $\Omega_+ = \Omega_- = \Omega$ and two (rather than one) probe fields $\mathcal{E}_\pm$ inducing transitions in a three-level Λ-system or a four-level system, as shown in ■ Fig. 15.19.

The four-level system can be viewed as two Λ sub-systems, one for fields propagating in the forward direction (+), and another sub-system for the fields propagating in the backward direction (−). Since the two control fields have the same amplitude, so do the probe fields. In each of these Lambda systems the respective pairs of control and probe fields induce a transition from the ground atomic state g to the metastable state s (see ■ Fig. 15.19). In such a situation the two counter-propagating probe beams drive the same atomic transition $g \to s$. Since the amplitudes of both probe fields are the same, each photon propagating forward has its counterpart, a photon propagating backward, and a stationary pattern of light is formed, frozen in the medium. This is as if two racing cars driving in opposite directions try to pull the same trailer but are not able to move it since their forces compensate (see ■ Fig. 15.3).

For stationary light it is important that the counter-propagating probe fields are coupled to each other by the atomic medium. To see this let us draw an analogy with the string of a guitar: When a guitar player pulls the string at some place, *two* waves of equal frequency are created which propagate along the string in opposite directions. If two wavepackets of equal strength and opposite propagation directions are superimposed, a standing wave forms, but only for the short period of time for which they overlap. The two wavepackets would continue to propagate each in its own direction. Soon they would not overlap anymore and would be two spatially separated wavepackets. To prevent this another element is needed: At the points where the string is fixed to the body of the guitar, the wavepackets get reflected and the effect of this is a true standing wave that does not smear out. In a similar manner, one could produce a standing wave of light by confining the radiation in a resonator between parallel mirrors, so that the forward propagating light is permanently reflected to the backward propagating direction and vice versa. This principle is used, e.g., in a laser allowing the light to pass many times the lasing medium.

Stationary light also involves a permanent reflection of one component into the other but with no mirrors, and one can think of a kind of a mirrorless resonator. So what takes over the role of the fixing points of the guitar string or the mirrors reflecting the light? In fact we have here a whole periodic sequence of 'fixing' points, which causes reflection. In the case of a simple Λ-scheme, shown in ◙ Fig. 15.19a, the control lasers form a stationary *intensity* pattern that oscillates in space. Thus there is a periodic grating where the total control field intensity vanishes, which also means that in a periodic spatial pattern there are points without EIT for the probe light. This periodic array acts in a similar way as an absorption grating and reflects the forward and backward propagating components of the probe field. In the case of the four-level scheme, shown in ◙ Fig. 15.19b, the situation is somewhat different. Here the two control fields are not only propagating in opposite directions, but they also have opposite circular polarizations. Now, superimposing two light waves of equal intensity and with opposite circular polarization results in constant total field intensity with a linear polarization. Yet, since the two control beams propagate opposite to each other, the linear polarization rotates in space, forming a polarization grating. This polarization grating has the same effect as the intensity grating in the case of the simple Λ-scheme, it reflects forward- and backward propagating components into each other making the light stationary.

Stationary light has been first observed in 2003 by the group of Mikhail Lukin at Harvard University [17] using a Λ-type atom-light coupling which involves pairs of counter-propagating control (probe) beams with the same frequency, shown in ◙ Fig. 15.19a. One difficulty of these experiments is to make the 'non-moving light' visible. A trick used here is that the stationary light tends to excite also further off-resonant transitions to other excited states with a small probability. These excitations are then visible due to the spontaneous emission from these states.

Another form of stationary light, called bichromatic stationary light was observed in 2009 by the group of Ite Yu at the National Tsing Hua University in Taiwan [20] using a double Λ coupling scheme, as shown in ◙ Fig. 15.19b. Here the frequency (or color) of the two control fields and, respectively, the two probe fields were different, thus the name 'bichromatic'. Stationary light pulses maximize the interaction time and thus can provide a considerable interaction efficiency even at a single-photon level. Interaction of two stationary light pulses through the medium was experimentally demonstrated by the same group 3 years later [21].

15.7 Multi-Component Slow Light

We have seen that slow light can be turned into something that behaves like a massive quantum particle. It is known from quantum physics that certain particles can show up in different forms, i.e. they can have different internal states. Electrons, for example, possess two different spin states, spin-up and spin-down states. In a bit oversimplified picture the spin of a particle can be viewed as a tiny gyroscope resulting from rotation of the particle around its center. Such a rotation is often accompanied with a magnetic dipole, so an electron represents a little magnet pointing up or down depending on the spin state relative to the chosen axis. More exotic particles can have not only spin but also other internal degrees of freedom, such as isospin, colour or flavour. So an interesting question is: Can we give slow light internal properties such that it mimics massive quantum particles with, e.g., spin? The answer is yes, and this makes slow light an even more interesting object for quantum physicists. We note that in quantum mechanics the spin of, e.g., an electron is a relativistic effect, so slow light with spin can be used to investigate relativistic quantum physics.

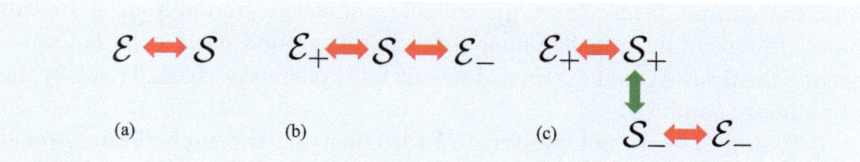

■ **Fig. 15.20** *Slow (a), stationary (b) and two-component (c) slow light:* In (**a**) a single probe field $\mathcal{E}$ is coupled to a single atomic coherence S. The radiation has to push the atomic coherence forwards and thus the light slows down. In (**b**) two counter-propagating probe beams $\mathcal{E}_\pm$ drive the same atomic coherence characterized by the amplitude S. One probe field pushes the atomic coherence forwards and the other backwards. The velocities of the probe photons compensate leading to stationary light. In (**c**) two counter-propagating probe beams $\mathcal{E}_\pm$ drive two different atomic coherences characterized by the amplitudes S_+ and S_-. If there is a coupling between these coherences indicated by the *green double arrow*, two-component stationary light is formed

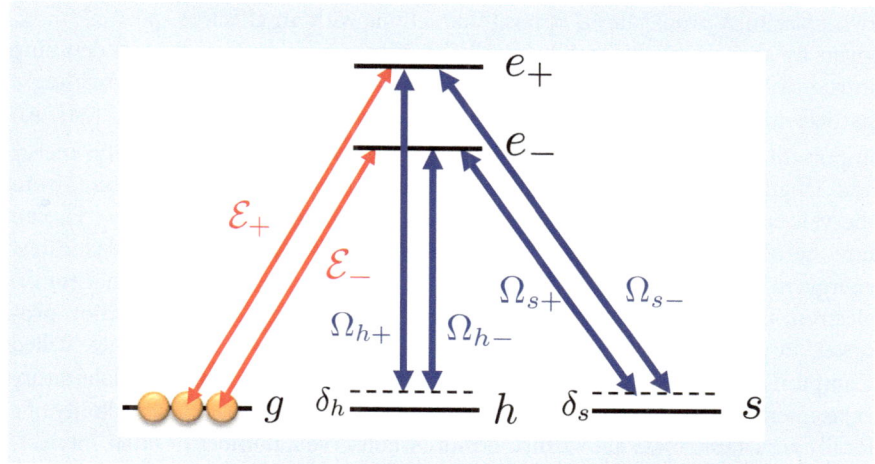

■ **Fig. 15.21** *Two-component slow light:* Atom-light coupling scheme of the double-tripod type for implementation of two-component stationary light adapted from [24]. The scheme involves three atomic ground states *g*, *s* and *h* coupled to two excited states $e_\pm$ by six fields: a pair of counter-propagating probe beams $\mathcal{E}_\pm$, as well as two pairs of counter-propagating control beams $\Omega_{s\pm}$ and $\Omega_{h\pm}$

Slow light as introduced in ◻ Sect. 15.5 and the stationary light discussed in ◻ Sect. 15.6 both involve only one spin component associated with a transition from the initially populated ground state *g* to one other ground state *s*, and described by the amplitude S, as illustrated in ◻ Fig. 15.20a, b. This represents a single normal mode of oscillations of the coupled atom-light system (a single polariton) even though there are two counter-propagating probe fields, as in the case of the stationary light depicted in ◻ Fig. 15.20b.

In order for stationary light to have two components, the counter-propagating probe fields $\mathcal{E}_\pm$ (together with a number of control beams) should drive *two different* spin coherences described by two amplitudes $S_\pm$. This is illustrated in ◻ Fig. 15.20c. Two-component stationary light can be implemented using a tripod [23] or a double-tripod [24] atom-light coupling scheme, the latter shown in ◻ Fig. 15.21. Here one has two pairs of counter-propagating control fields with Rabi frequencies $\Omega_{s\pm}$ and $\Omega_{h\pm}$ inducing the atomic transitions $s \rightarrow e_\pm$ and $h \rightarrow e_\pm$, respectively. Compared to the double Λ scheme used for stationary light (◻ Fig. 15.19b) now there is an extra pair of counter-propagating control laser beams $\Omega_{h\pm}$, as well as an extra atomic ground state *h*. This leads to EIT for a pair of counter-propagating probe fields $\mathcal{E}_\pm$ inducing transitions (together

15

with the control fields) from the initially populated ground state g to two superpositions of the initially unpopulated atomic ground states s and h. Consequently the fields $\mathcal{E}_+$ and $\mathcal{E}_-$ drive different spin coherences characterized by the amplitudes S_+ and S_-.

If S_+ and S_- were not coupled to each other, the two probe beams would propagate in opposite directions slowly and independently from each other. The coupling emerges though a two-photon detuning $\delta = \delta_s = -\delta_h$ shown in ◘ Fig. 15.21. The corresponding two types of polaritons behave like particles with positive and negative effective masses, i.e. like electrons and positrons representing particles and antiparticles in the relativistic Dirac theory. Thus the two-component (spinor) slow-light polaritons Ψ obey an effective one-dimensional Dirac equation

$$ih\frac{\partial}{\partial t}\Psi = \left(ihv_{\mathrm{gr}}\sigma_z\frac{\partial}{\partial z} + m^*c^{*2}\sigma_y\right)\Psi, \qquad \Psi = \begin{pmatrix}\Psi_1 \\ \Psi_2\end{pmatrix}, \tag{15.8}$$

where σ_z and σ_y are the 2×2 Pauli matrices. For zero two-photon detuning δ, the two polaritons propagate in opposite directions with an effective speed $c^* = v_{\mathrm{gr}}$ given by the slow-light group velocity. A non-vanishing two-photon detuning introduces a coupling between the counter-propagating polaritons, providing a particle–antiparticle type dispersion with a variable mass $m^* = h\delta/v_{\mathrm{gr}}^2$ [24]. An important feature of spinor slow light is that the relevant scales of velocity, energy and length, where relativistic effects start to matter, are very different from the values for say electrons. The effective 'vacuum speed of light' $c^* = v_{\mathrm{gr}}$ can now be a few meters per second instead of 300,000 km/s. The relativistic rest energy $m^*c^{*2} = h\delta$ can be many orders of magnitude smaller than that for an electron, making it possible to observe particle–antiparticle pair generation processes in a conventional laser lab. Finally the relativistic length scale, called Compton length $\lambda_C^* = \hbar/m^*c^*$, is now large enough to be resolved in laboratory experiments as opposed to the value of 10^{-12} m for an electron. The possibility of a locally adjustable mass allows furthermore to observe a number of other interesting phenomena. For instance, if the mass m^* of the Dirac particle suddenly changes at a certain point in space from the value $+|m|$ to $-|m|$, a localized, topological mid-gap (zero-energy) state is created. If m^* is a randomly varying function of space with a vanishing mean-value, there exist mid-gap states with unusual correlations [23, 34, 35].

Two-component slow light has been recently implemented in an experiment [25] using the double-tripod coupling scheme, like the one shown in ◘ Fig. 15.21 but with co-propagating rather than counter-propagating control and probe laser fields. Oscillations due to an effective interaction between the two components of the probe field have been observed revealing the two-component nature of the slow light. It was demonstrated that the double-tripod scheme enables precision measurements of frequency detunings. Furthermore a possible application of the double-tripod scheme as quantum memory/rotator for a two-colour qubit was experimentally demonstrated. This offers potential applications in quantum computation and quantum information processing.

15.8 Quo Vadis Slow Light?

Light is fascinating! Light has very many uses and modern life would be unthinkable without them. Thus there is plenty of reason for us to celebrate the Year of Light. We believe that the applications of slow light based on EIT and its generalizations, which we have discussed in this chapter of the book, are important additions to this list of reasons. We have seen that coupling light to atomic media, which are specially prepared by external laser fields, allows us to dramatically modify the property of photons. We can change their effective propagation velocity, can store them or more precisely their information content with important applications for quantum information networks based on light, and we can turn them into massive quantum particles with tunable mass. Finally we can even use them to model relativistic quantum particles with spin.

It is interesting to note that EIT and slow light are not restricted to light in the optical frequency spectrum coupled to atoms. EIT can also be generated in other type of coupled oscillators, such as meta-materials build up of periodic arrays of small metallic antennas [36]. This allows to access the microwave part of the electromagnetic spectrum. On the other hand, the storage and release of light can also be carried out beyond atomic systems. Recently the conversion of light pulses into mechanical excitations of a silica optomechanical resonator and the subsequent retrieval of radiation using a method closely related to the EIT was experimentally demonstrated [37].

All phenomena we have discussed so far in this chapter address the single-particle properties of slow light, i.e. properties of *individual* photons. Yet it is also highly desirable to make photons interact with each other sufficiently strongly. Strong and controlled interactions between individual photons would, e.g., allow to implement quantum logic operations in the so-called quantum gates, the second important ingredient next to a quantum memory for photon-based quantum information technology. Interactions are also crucial for most applications of slow light to fundamental science. Several ideas have been put forward here to exploit the properties of slow light for implementing strong interactions. For example, the possibility offered by EIT to operate close to atomic resonances without suffering from absorption can be exploited to enhance nonlinear optical processes in atomic media [21, 38–44]. Another very promising direction is to combine EIT with the so-called Rydberg atoms. Here the atomic state s populated during the propagation and storage of light is not a hyperfine (spin) ground state of an atom, but rather a Rydberg state corresponding to a very high atomic level close to the ionization threshold. Such a state is metastable and has a very long lifetime. Atoms in Rydberg states exhibit very strong and long-range dipole–dipole interactions. This property is carried over to slow-light polaritons, whose spin component contains the Rydberg state, thus making these Rydberg polaritons strongly interacting [45]. The strongly nonlinear and nonlocal interaction between Rydberg polaritons has been observed in a number of recent experiments [46–50]. This opens many more fascinating applications in fundamental science and in quantum technology, and we anticipate a bright future for slow light.

15.9 Conclusions

In this chapter we have explained what slow light is and what it is good for, how to understand the physics of it and how one can practically make light go so slow. To answer these questions, we used simple pictures, on the one hand, and supplemented them with a little bit of details, on the other hand, for those who want to go slightly deeper into the field. Subsequently we discussed recent generalizations of slow light, such as stationary and spinor slow light which are interesting model system and can be used to understand more complex quantum systems. The chapter also presents important applications of the slow light in photon-based quantum information technology.

References

1. Boller KJ, Imamoğlu A, Harris SE (1991) Observation of electromagnetically induced transparency. Phys Rev Lett 66:2593–2596
2. Harris SE (1997) Electromagnetically induced transparency. Phys Today 50(7):36–42
3. Harris SE, Field JE, Kasapi A (1992) Dispersive properties of electromagnetically induced transparency. Phys Rev A 46:R29–R32
4. Hau LV, Harris SE, Dutton Z, Behroozi CH (1999) Light speed reduction to 17 metres per second in an ultracold atomic gas. Nature 397:594–598
5. Kash MM, Sautenkov V A, Zibrov AS, Hollberg L, Welch GR, Lukin MD, Rostovtsev Y, Fry ES, Scully MO (1999) Ultraslow group velocity and enhanced nonlinear optical effects in a coherently driven hot atomic gas. Phys Rev Lett 82:5229–5232
6. Budker D, Kimball DF, Rochester SM, Yashchuk VV (1999) Nonlinear magneto-optics and reduced group velocity of light in atomic vapor with slow ground state relaxation. Phys Rev Lett 83:1767–1770
7. Boyd RW, Gauthier DJ (2002) "Slow" and "fast" light. In: Wolf E (ed) Progress in optics, vol 43. Elsevier, Amsterdam, pp 497–530
8. Lukin MD (2003) Colloquium: trapping and manipulating photon states in atomic ensembles. Rev Mod Phys 75:457–472
9. Fleischhauer M, Imamoğlu A, Marangos JP (2005) Electromagnetically induced transparency: optics in coherent media. Rev Mod Phys 77:633–673
10. Milonni PW (2005) Fast light, slow light and left-handed light. Taylor and Francis, New York
11. Firstenberg O, Shuker M, Ron A, Davidson N (2013) Colloquium: coherent diffusion of polaritons in atomic media. Rev Mod Phys 85:941–960
12. Fleischhauer M, Lukin MD (2000) Dark-state polaritons in electromagnetically induced transparency. Phys Rev Lett 84:5094–5097
13. Phillips DF, Fleischhauer A, Mair A, Walsworth RL, Lukin MD (2001) Storage of light in atomic vapor. Phys Rev Lett 86:783–786
14. Liu C, Dutton Z, Behroozi CH, Hau LV (2001) Observation of coherent optical information storage in an atomic medium using halted light pulses. Nature 409:490–493
15. Chaneliére T, Matsukevich DN, Jenkins SD, Lan SY, Kennedy TAB, Kuzmich A (2005) Storage and retrieval of single photons transmitted between remote quantum memories. Nature 438:833–836
16. Eisaman MD, André A, Massou F, Fleischhauer M, Zibrov AS, Lukin MD (2005) Electromagnetically induced transparency with tunable single-photon pulses. Nature 438:837–841
17. Bajcsy M, Zibrov AS, Lukin MD (2003) Stationary pulses of light in an atomic medium. Nature 426:638–641
18. André A, Lukin MD (2002) Manipulating light pulses via dynamically controlled photonic band gap. Phys Rev Lett 89:143602
19. André A, Bajcsy M, Zibrov AS, Lukin MD (2005) Nonlinear optics with stationary pulses of light. Phys Rev Lett 94:063902
20. Lin YW, Liao WT, Peters T, Chou HC, Wang JS, Cho HW, Kuan PC, Yu IA (2009) Stationary light pulses in cold atomic media and without Bragg gratings. Phys Rev Lett 102:213601
21. Chen YH, Lee MJ, Hung W, Chen YC, Chen YF, Yu IA (2012) Demonstration of the interaction between two stopped light pulses. Phys Rev Lett 108:173603
22. Otterbach J, Unanyan RG, Fleischhauer M (2009) Confining stationary light: Dirac dynamics and Klein tunneling. Phys Rev Lett 102:063602

23. Unanyan RG, Otterbach J, Fleischhauer M, Ruseckas J, Kudriašov V, Juzeliūnas G (2010) Spinor slow-light and Dirac particles with variable mass. Phys Rev Lett 105:173603

24. Ruseckas J, Kudriašov V, Juzeliūnas G, Unanyan RG, Otterbach J, Fleischhauer M (2011) Photonic-band-gap properties for two-component slow light. Phys Rev A 83:063811

25. Lee MJ, Ruseckas J, Lee CY, Kudriašov V, Chang KF, Cho HW, Juzeliūnas G, Yu IA (2014) Experimental demonstration of spinor slow light. Nat Commun 5:5542

26. Brillouin L (1960) Wave propagation and group velocity. Academic Press, New York

27. Arimondo E (1996) Coherent population trapping in laser spectroscopy. In: Wolf E (ed) Progress in optics, vol 35. Elsevier, Amsterdam, pp 257–354

28. Scully MO, Zubairy, MS (1997) Quantum optics. Cambridge University Press, Cambridge

29. Garrido Alzar CE, Martinez MAG, Nussenzveig P (2002) Classical analog of electromagnetically induced transparency. Am J Phys 70:37–41

30. Schnorrberger U, Thompson JD, Trotzky S, Pugatch R, Davidson N, Kuhr S, Bloch I (2009) Electromagnetically induced transparency and light storage in an atomic Mott insulator. Phys Rev Lett 103:033003

31. Heinze G, Hubrich C, Halfmann T (2013) Stopped light and image storage by electromagnetically induced transparency up to the regime of one minute. Phys Rev Lett 111:033601

32. Mazets IE, Matisov BG (1996) Adiabatic Raman polariton in a Bose condensate. JETP Lett 64:515–519

33. Juzeliūnas G, Carmichael HJ (2002) Systematic formulation of slow polaritons in atomic gases. Phys Rev A 65:021601(R)

34. Balents L, Fisher MPA (1997) Delocalization transition via supersymmetry in one dimension. Phys Rev B 56:12970–12991

35. Shelton DG, Tsvelik AM (1998) Effective theory for midgap states in doped spin-ladder and spin-Peierls systems: Liouville quantum mechanics. Phys Rev B 57:14242–14246

36. Liu N, Langguth L, Weiss T, Kästel J, Fleischhauer M, Pfau T, Giessen H (2009) Plasmonic analogue of electromagnetically induced transparency at the Drude damping limit. Nat Mater 8:758–762

37. Fiore V, Yang Y, Kuzyk MC, Barbour R, Tian L, Wang H (2011) Storing optical information as a mechanical excitation in a silica optomechanical resonator. Phys Rev Lett 107:133601

38. Schmidt H, Imamoğlu A (1996) Giant Kerr nonlinearities obtained by electromagnetically induced transparency. Opt Lett 21:1936–1938

39. Harris SE, Yamamoto Y (1998) Photon switching by quantum interference. Phys Rev Lett 81:3611–3614

40. Lukin MD, Imamoğlu A (2000) Nonlinear optics and quantum entanglement of ultraslow single photons. Phys Rev Lett 84:1419–1422

41. Wang ZB, Marzlin KP, Sanders BC (2006) Large cross-phase modulation between slow copropagating weak pulses in 87Rb. Phys Rev Lett 97:063901

42. Shiau BW, Wu MC, Lin CC, Chen YC (2011) Low-light-level cross-phase modulation with double slow light pulses. Phys Rev Lett 106:193006

43. Venkataraman V, Saha K, Gaeta AL (2013) Phase modulation at the few-photon level for weak-nonlinearity-based quantum computing. Nat Photon 7:138–141

44. Chen W, Beck KM, Bücker R, Gullans M, Lukin MD, Tanji-Suzuki H, Vuletic V (2013) All-optical switch and transistor gated by one stored photon. Science 341:768–770

45. Gorshkov AV, Otterbach J, Fleischhauer M, Pohl T, Lukin MD (2011) Photon-photon interactions via Rydberg blockade. Phys Rev Lett 107:133602

46. Mohapatra AK, Jackson TR, Adams CS (2007) Coherent optical detection of highly excited Rydberg states using electromagnetically induced transparency. Phys Rev Lett 98:113003

47. Peyronel T, Firstenberg O, Liang Q, Hofferberth S, Gorshkov AV, Pohl T, Lukin MD, Vuletić V (2012) Quantum nonlinear optics with single photons enabled by strongly interacting atoms. Nature 488:57–60

48. Hofmann CS, Günter G, Schempp H, Robert-de-Saint-Vincent M, Gärttner M, Evers S, Whitlock J, Weidemüller M (2013) Sub-Poissonian statistics of Rydberg-interacting dark-state polaritons. Phys Rev Lett 110:203601

49. Firstenberg O, Peyronel T, Liang QY, Gorshkov AV, Lukin MD, Vuletić V (2013) Attractive photons in a quantum nonlinear medium. Nature 502:71–76

50. Maxwell D, Szwer DJ, Paredes-Barato D, Busche H, Pritchard JD, Gauguet A, Weatherill KJ, Jones PA, Adams CS (2013) Storage and control of optical photons using Rydberg polaritons. Phys Rev Lett 110:103001

Optical Tests of Foundations of Quantum Theory

Yanhua H. Shih

Y.H. Shih (✉)
Department of Physics, University of Maryland, Baltimore, MD 21250, USA
e-mail: shih@umbc.edu

© The Author(s) 2016
M.D. Al-Amri et al. (eds.), *Optics in Our Time*, DOI 10.1007/978-3-319-31903-2_16

16.1 Introduction

Since the beginning of Quantum Theory scientists questioned its very basic concepts, such as locality, reality, and complementarity, because they are so different from classical theory and from our everyday experience.

16.1.1 Locality

Einstein posed his students a question [1]: suppose a photon with energy $h\nu$ is created from a point source, such as an atomic transition; how big is the photon after propagating 1 year? This question seems easy to answer. Since the photon is created from a point source, it would propagate in the form of a spherical wave and its wavefront must be a sphere with a diameter of 2 lightyears after 1 year propagating. Einstein then asked again: suppose a point-like photon counting detector located on the surface of the big sphere is triggered by that photon, how long does it take for the energy on the other side of the big sphere to arrive at the detector? Two years? For a fast photodetector, it takes only a few picoseconds to produce a photoelectron by annihilating a photon with energy $h\nu$. Does this mean something has happened faster than the speed of light? Bohr provided a famous answer to this question: the "wavefunction collapses" instantaneously! Why does the wavefunction need to "collapse"? Bohr did not explain. In quantum theory, perhaps, the wavefunction does not need to "collapse." A wavefunction is defined as the probability amplitude for a particle to be observed at a space-time coordinate $(\mathbf{r}, t)$.

Quantum theory, however, does allow nonlocal interference. Assuming Einstein continued his question: if two photons are created simultaneously from the point source, and we set up a measurement with two different yet indistinguishable alternative ways for the photon pair to produce a joint photodetection event between two distant point-like photon counting detectors, what is the chance to observe a joint photodetection event at $(\mathbf{r}_1, t_1)$ and $(\mathbf{r}_2, t_2)$? According to quantum theory, the probability is the result of the linear superposition between the two probability amplitudes,

$$P(\mathbf{r}_1, t_1; \mathbf{r}_2, t_2) = |\mathcal{A}_\mathrm{I}(\mathbf{r}_1, t_1; \mathbf{r}_2, t_2) + \mathcal{A}_\mathrm{II}(\mathbf{r}_1, t_1; \mathbf{r}_2, t_2)|^2 \qquad (16.1)$$

despite the distance between the two photodetection events, even if the two detectors are placed on the opposite sides of the big sphere.[1] How much time for this superposition to complete? Two years? Again, the two-photon interference must be completed within the "coincidence" time window which can be a few picoseconds. Furthermore, it is not necessary to use a hardware coincidence counter to count the coincidences. Two independent "event timers," which record the registration times of the two photodetection events of the two photon counting detectors, respectively, and PC software are able to calculate the joint photodetection probability. In some experiments, one detector-event timer package is placed on a satellite and the other one is placed in a ground laboratory. The recorded history of photodetection events are later brought together at the ground laboratory and analyzed by a PC. We found that the two-photon interferences are observable only when the time axis of the two event timers is correctly synchronized within the response time of the photodetectors which could be a few picoseconds.

---

1 This kind superposition has been named two-photon interference: a pair of photon interferes with the pair itself at distance.

16.1.2 Reality

Now, we ask a different kind of question: Does Einstein's photon have a defined momentum and position over the course of its propagation? On one hand, the photodetection event of a point-like photon counting detector tells us that the annihilated photon at $(\mathbf{r}, t)$ must carry momentum $\mathbf{p} = (\hbar\omega/c)\hat{\mathbf{n}}$, where $\hat{\mathbf{n}}$ is the unit vector normal to the sphere; on the other hand, Einstein's spherical wavefunction means the momentum of that photon cannot be a constant vector when it is created at the point source of $(\mathbf{r}_0, t_0)$, otherwise it would not propagate to all 4π directions, yet, the uncertainty principle prevents a point source from producing a photon with $\Delta\mathbf{p} = 0$. Does it mean the photon has no momentum in the course of its propagation until its annihilation? To Einstein, the statement from Copenhagen "no phenomenon is a phenomenon until it is a registered phenomenon" [1] was unacceptable! Einstein believed a photon must be created and propagated with a defined momentum, the same as that observed from its annihilation. In Einstein's opinion, momentum and position must be physical realities accompany with a photon, otherwise, we may have to accept that some kinds of phenomena happen faster than the speed of light, such as "wavefunction collapse," or we may have to accept that a photon can be divided into parts, or that part of $h\nu$ is able to excite a photoelectron.

In 1935, Einstein, Podolsky, and Rosen (EPR) published an article to defend their opinion on physical reality [2]. In that article, EPR proposed a *gedankenexperiment* and introduced an entangled two-particle system based on the superposition of two-particle wavefunctions. The EPR system is composed of two distant interaction-free particles which are characterized by the following wavefunction:

$$\Psi(x_1, x_2) = \frac{1}{2\pi\hbar} \int dp_1 dp_2 \; \delta(p_1 + p_2) \; e^{ip_1(x_1-x_0)/\hbar} e^{ip_2 x_2/\hbar}$$
$$= \delta(x_1 - x_2 - x_0) \tag{16.2}$$

where $e^{ip_1(x_1-x_0)/\hbar}$ and $e^{ip_2 x_2/\hbar}$ are the eigenfunctions, with eigenvalues $p_1 = p$ and $p_2 = -p$, respectively, of the momentum operators $\hat{p}_1$ and $\hat{p}_2$ associated with particles 1 and 2; x_1 and x_2 are the coordinate variables to describe the positions of particles 1 and 2, respectively; and x_0 is a constant. The EPR state is very peculiar. Although there is no interaction between the two distant particles, the two-particle superposition cannot be factorized into a product of two individual superpositions of two particles. Quantum theory does not prevent such states.

What can we learn from the EPR state of Eq. (16.2)?

(1) In the coordinate representation, the wavefunction is a delta function: $\delta(x_1 - x_2 - x_0)$. The two particles are always separated in space with a constant value of $x_1 - x_2 = x_0$, although the coordinates x_1 and x_2 of the two particles are both unspecified.

(2) The delta wavefunction $\delta(x_1 - x_2 - x_0)$ is the result of the superposition of the plane wavefunctions of free particle one, $e^{ip_1(x_1-x_0)/\hbar}$, and free particle two, $e^{ip_2 x_2/\hbar}$, with a particular distribution $\delta(p_1 + p_2)$. It is $\delta(p_1 + p_2)$ that made the superposition special: although the momentum of particle one and particle two may take on any values, the delta function restricts the superposition with only these terms in which the total momentum of the system takes a constant value of zero.

Now, we transfer the wavefunction from coordinate representation to momentum representation:

$$
\begin{aligned}
\Psi(p_1, p_2) &= \frac{1}{2\pi\hbar} \int dx_1 dx_2 \; \delta(x_1 - x_2 - x_0) \; e^{-ip_1(x_1-x_0)/\hbar} e^{-ip_2 x_2/\hbar} \\
&= \delta(p_1 + p_2).
\end{aligned}
\tag{16.3}
$$

What can we learn from the EPR state of Eq. (16.3)?

(1) In the momentum representation, the wavefunction is a delta function: $\delta(p_1 + p_2)$. The total momentum of the two-particle system takes a constant value of $p_1 + p_2 = 0$, although the momenta p_1 and p_2 are both unspecified.

(2) The delta wavefunction $\delta(p_1 + p_2)$ is the result of the superposition of the plane wavefunctions of free particle one, $e^{-ip_1(x_1-x_0)/\hbar}$, and free particle two, $e^{-ip_2 x_2/\hbar}$, with a particular distribution $\delta(x_1 - x_2 - x_0)$. It is $\delta(x_1 - x_2 - x_0)$ that made the superposition special: although the coordinates of particle one and particle two may take on any values, the delta function restricts the superposition with only these terms in which $x_1 - x_2$ is a constant value of x_0.

In an EPR system, *the value of the momentum (position) is not determined for either single subsystem. However, if one of the subsystems is measured to be at a certain momentum (position), the other one is determined to have a unique corresponding value despite the distance between them.* An idealized EPR state of a two-particle system is therefore characterized by $\Delta(p_1 + p_2) = 0$ and $\Delta(x_1 - x_2) = 0$ simultaneously, even if the momentum and position of each individual free particle are completely undefined, i.e., $\Delta p_j \sim \infty$ and $\Delta x_j \sim \infty$, $j = 1, 2$. In other words, each of the subsystems may have completely random values or all possible values of momentum and position in the course of their motion, but the correlations of the two subsystems are determined with certainty whenever a joint measurement is performed.[2]

According to EPR's criteria:

Locality - There is no action-at-a-distance;

Reality - If, without in any way disturbing a system, we can predict with certainty the value of a physical quantity, then there exists an element of physical reality corresponding to this quantity;

Completeness - Every element of the physical reality must have a counterpart in the complete theory;

momentum and position must be physical realities associated with particle one and two. This led to the title of their 1935 article: "Can Quantum-Mechanical Description of Physical Reality Be Considered Complete?" [2]

In early 1950s, Bohm simplified EPR's entangled two-particle state of continuous space-time variables to discrete spin variables [3]. Bohm suggested the singlet state of two spin 1/2 particles:

$$
|\Psi\rangle = \frac{1}{\sqrt{2}} \left[|\uparrow\rangle_1 |\downarrow\rangle_2 - |\downarrow\rangle_1 |\uparrow\rangle_2 \right]
\tag{16.4}
$$

---

2   There have been arguments considering $\Delta(p_1 + p_2)\Delta(x_1 - x_2) = 0$ a violation of the uncertainty principle. This argument is false. It is easy to find that $p_1 + p_2$ and $x_1 - x_2$ are not conjugate variables. As we know, non-conjugate variables correspond to commuting operators in quantum mechanics, if the corresponding operators exist. To have $\Delta(p_1 + p_2) = 0$ and $\Delta(x_1 - x_2) = 0$ simultaneously, or to have $\Delta(p_1 + p_2)\Delta(x_1 - x_2) = 0$ is not a violation of the uncertainty principle.

where the kets $|\!\uparrow\rangle$ and $|\!\downarrow\rangle$ represent the states of spin "up" and spin "down," respectively, along an *arbitrary* direction. Again, for this state, *the spin of neither particle is determined; however, if one particle is measured to be spin up along a certain direction, the other one must be spin down along that direction, despite the distance between the two spin 1/2 particles.* Similar to the original EPR state, Eq. (16.4) is independent of the choice of the spin directions and the eigenstates of the associated non-commuting spin operators.

The most widely used entangled two-particle states might have been the "Bell states" (or EPR-Bohm-Bell states) [4]. Bell states are a set of polarization states for a pair of entangled photons. The four Bell states which form a complete orthonormal basis of two-photon states are usually represented as

$$|\Phi_{12}^{(\pm)}\rangle = \frac{1}{\sqrt{2}}[\,|\,0_1\ 0_2\,\rangle \pm |\,1_1\ 1_2\,\rangle],$$

$$|\Psi_{12}^{(\pm)}\rangle = \frac{1}{\sqrt{2}}[\,|\,0_1\ 1_2\,\rangle \pm |\,1_1\ 0_2\,\rangle] \tag{16.5}$$

where $|0\rangle$ and $|1\rangle$ represent two arbitrary orthogonal polarization bases, for example, $|0\rangle = |H\rangle$ (horizontal linear polarization) and $|1\rangle = |V\rangle$ (vertical linear polarization); or $|0\rangle = |R\rangle$ (right-hand circular polarization) and $|1\rangle = |L\rangle$ (left-hand circular polarization). We will have a detailed discussion of Bell states based on two types of experiments: (1) EPR-Bohm-Bell correlation measurement; and (2) Bell's inequality testing.

16.1.3 Complementarity

Wave-particle duality, which Feynman called the basic mystery of quantum mechanics [5], says that there is always a trade-off between the knowledge of the particle-like and wave-like behavior of a quantum system. In slightly different words, Bohr suggested a complementarity principle in 1927: one can never measure the precise position and momentum of a quantum simultaneously [6]. Since then, complementarity has often been superficially identified with the "wave-particle duality of matter." How quantum mechanics enforces complementarity may vary from one experimental situation to another (◻ Fig. 16.1).

In a single-photon Young's double-slit experiment, is the photon going to pass "both slits" like a wave or will it choose "which slit" to pass like a particle? This question has been asked since the early days of quantum mechanics [8]. Among most physicists, the common "understanding" is that the position-momentum uncertainty relation makes it impossible to determine which slit a photon or wavepacket passes through without at the same time disturbing the photon or wavepacket enough to destroy the interference pattern. However, it has been shown that under certain circumstances this common "understanding" may not be true. In 1982, Scully and Drühl showed that a "quantum eraser" may erase the which-path information [9]. The "random delayed choice quantum eraser" has been experimentally demonstrated with interesting results: the which-path information is truly erasable even after the annihilation of the quantum itself [10, 11].

Popper's thought experiment evaluated the same fundamental problem from a slightly different position [12]. Popper proposed a coincidence measurement on a pair of entangled particles. If the position of particle one is learned within Δy through the joint measurement of its twin, particle two, do we expect an uncertainty relation on particle one $\Delta y \Delta p_y \geq \hbar$? Namely, if we place an array of detectors at a distance at which particle one is restricted within Δy, do we expect a diffraction pattern with a minimum width that is determined by $\Delta y \Delta p_y \geq \hbar$?

🔲 **Fig. 16.1** On one hand, a photon can never be divided into parts; on the other hand, we never lose interference at the single photon level. In fact, according to quantum theory, Young's double-slit interference is a single-photon phenomenon. In Diracts terms: "... photon ... only interferes with itself" [7]

Popper predicted a negative answer: particle one would not be diffracted unless a real slit of Δy is inserted. Similar to Einstein, Popper was a believer of realism. In his opinion, a particle must have defined momentum and position over the course of its propagation. We will review two experimental realizations of Popper's thought experiment. It is interesting to see that both experiments, one based on the measurement of entangled photon pairs, another based on the measurement of randomly paired photons in a thermal state, produced a similar result that $\Delta y \Delta p_y < \hbar$, agreeing with Popper's prediction [13, 14]. Is this result a violation of the uncertainty principle?

In the following, we will focus on three types of "optical tests of foundations of quantum theory": (1) EPR-Bohm-Bell correlation and Bell's inequality; (2) Quantum eraser; (3) Popper's experiment.

16.2 EPR-Bohm-Bell Correlation and Bell's Inequality

An important step to perform an optical test of the EPR-Bohm-Bell correlation and Bell's inequality is to prepare an entangled photon pairs in Bell state. Historically, the most popular entangled photon sources have been: (1) annihilation of positronium; (2) atomic cascade decay; (3) spontaneous parametric down-conversion (SPDC). Both atomic cascade decay and SPDC have been experimentally tested. Most of the early EPR-Bohm-Bell experiments demonstrated in the 1970s and early 1980s used atomic cascade decay [15–20]. Since Alley and Shih introduced SPDC to the preparation of entangled states in the middle of 1980s [21], the signal-idler photon pair of SPDC has played an important role, especially in the tests of Bell's inequality. Using SPDC now-a-days one could easily observe a violation of Bell's inequality with hundreds of standard deviations [22]. The photon pair produced from SPDC has received an interesting name: *biphoton* [23].

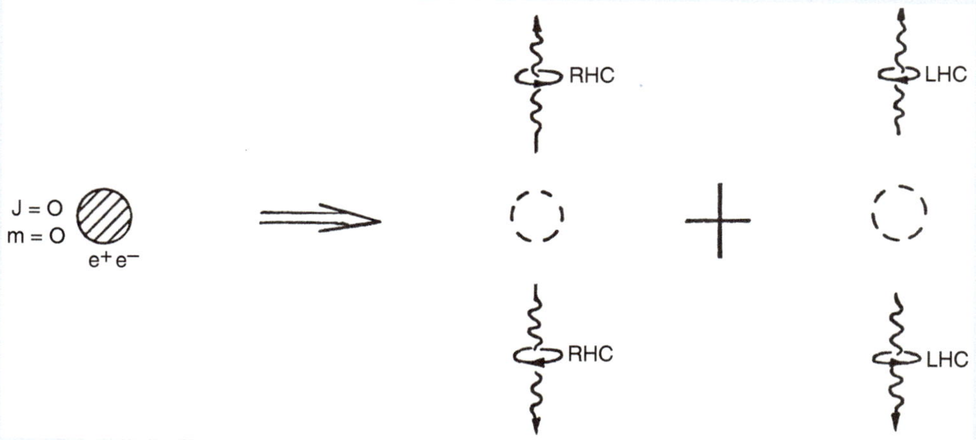

Fig. 16.2 Annihilation of Positronium. Due to the conservation of angular momentum, if photon 1 is right-hand circular (RHC) polarized, photon 2 must be right-hand circular polarized. If photon 1 is left-hand circular (LHC) polarized, then photon 2 has to be left-hand circular polarized

Figure 16.2 schematically illustrates the pair-creation mechanism of positronium annihilation [5]. Initially, we have a positron and an electron in the spin-zero state with antiparallel spins. The positronium cannot exist very long: it disintegrates into two γ-ray photons within $\sim 10^{-10}$ s of its lifetime. The spin zero state is symmetric under all rotations. Therefore, the photon pair may disintegrate into any direction in space with equal probability. The conservation of linear momentum, however, guarantees that if one of the photons is observed in a certain direction, its twin must be found in the opposite direction (with finite uncertainty $\Delta(\mathbf{p}_1 + \mathbf{p}_1) \neq 0$). The conservation of angular momentum will decide the polarization state of the photon pair. As shown in Fig. 16.2, in order to keep spin-zero, if photon 1 is right-hand circular polarized (RHC), photon 2 must also be right-hand circular polarized. The same argument shows that if photon 1 is left-hand circular (LHC) polarized, then photon 2 has to be left-hand circular polarized too. Therefore, the positronium may decay into two RHC photons or two LHC photons with equal probability.

Furthermore, the law of parity conservation must be satisfied in the disintegration: the spin-zero ground state of positronium holds an odd parity. Thus, the state of the photon pair must keep its parity odd:

$$|\Psi\rangle = \frac{1}{\sqrt{2}}[|R_1\rangle|R_2\rangle - |L_1\rangle|L_2\rangle], \qquad (16.6)$$

which is a non-factorizable pure state of a special superposition between the RHC and LHC states specified with a relative phase of π. Mathematically, "non-factorizable" means that the state cannot be written as a product state of photon 1 and photon 2. Physically, it means that photon 1 and photon 2 are not independent despite the distance between them. The two γ-ray photons are in an entangled polarization state, or spin state. The high energy γ-ray photon pair disintegrated from the annihilation of positronium is a good example to explore the physics of the EPR-Bohm state, however, the γ-ray photon pairs are difficult to handle experimentally: (1) There are no effective polarization analyzers available for the high energy γ-rays; (2) The uncertainty in momentum correlation, $\Delta(\mathbf{p}_1 + \mathbf{p}_2)$, has considerable large value, resulting in a "pair collection efficiency loophole" in Bell's inequality measurements, i.e., one may never have $\sim 100\%$ chance to "collect" a pair for joint photo-detection measurement [15]. Fortunately, the two-photon state of Eq. (16.6) is also observed in atomic cascade decay with visible-ultraviolet wavelengths and we have plenty of high efficiency polarization analyzers available in that wavelengths. Thus, most of the early EPR-Bohm-Bell experiments demonstrated in the 1970s and early 1980s used two-photon source of atomic cascade decay [15]. These experiments, unfortunately, still experienced the difficulties in the momentum uncertainty. The "pair collection" efficiency is as low as that of the annihilation of positronium. It was in the middle of 1980s, Alley and Shih introduced the nonlinear optical spontaneous parametric down-conversion to the preparation of entangled states [21]. The entangled signal-idler photon pair can be easily prepared in visible-infrared wavelengths, and very importantly, the uncertainty in momentum correlation was improved significantly. The "pair collection efficiency loophole" was finally removed.

16.2.1 Biphoton and Bell State Preparation

The state of a signal-idler photon pair created in SPDC is a typical EPR state [24]. Roughly speaking, the process of SPDC involves sending a pump laser beam into a nonlinear material, such as a non-centrasymmetric crystal. Occasionally, the nonlinear interaction leads to the annihilation of a high frequency pump photon and the simultaneous creation of a pair of lower frequency signal-idler photons into an entangled two-photon state:

$$|\Psi\rangle = \Psi_0 \sum_{s,i} \delta(\omega_s + \omega_i - \omega_p)\delta(\mathbf{k}_s + \mathbf{k}_i - \mathbf{k}_p) a_s^\dagger(\mathbf{k}_s)\, a_i^\dagger(\mathbf{k}_i)\, |\,0\rangle \qquad \text{(16.7)}$$

where $\omega_j, \mathbf{k}_j (j = s, i, p)$ are the frequency and wavevector of the signal (s), idler (i), and pump (p), $a_s^\dagger$ and $a_i^\dagger$ are creation operators for the signal and the idler photon, respectively, and Ψ_0 is the normalization constant. We have assumed a CW monochromatic laser pump, i.e., ω_p and $\mathbf{k}_p$ are considered as constants. The two delta functions in Eq. (16.7) are technically named as phase matching condition:

$$\omega_p = \omega_s + \omega_i, \qquad\qquad \mathbf{k}_p = \mathbf{k}_s + \mathbf{k}_i. \qquad \text{(16.8)}$$

The names *signal* and *idler* are historical leftovers. The names probably came about due to the fact that in the early days of SPDC, most of the experiments were done with non-degenerate processes. One radiation was in the visible range (and thus easily detected, the signal), and the other was in IR range (usually not detected, the idler). We will see in the following discussions that the role of the idler is not any less than that of the signal. The SPDC process is referred to as type-I if the signal and idler photons have identical polarizations, and type-II if they have orthogonal polarizations. The process is said to be *degenerate* if the SPDC photon pair have the same free space wavelength (e.g., $\lambda_i = \lambda_s = 2\lambda_p$), and *nondegenerate* otherwise. In general, the pair exit the crystal *non-collinearly*, that is, propagate to different directions defined by the second equation in Eq. (16.8) and the Snell's law. Of course, the pair may also exit *collinearly*, in the same direction, together with the pump.

The state of the signal-idler pair can be derived, quantum mechanically, by the first order perturbation theory with the help of the nonlinear interaction Hamiltonian. The SPDC interaction arises in a nonlinear crystal driven by a pump laser beam. The polarization, i.e., the dipole moment per unit volume, is given by

$$P_i = \chi_{i,j}^{(1)} E_j + \chi_{i,j,k}^{(2)} E_j E_k + \chi_{i,j,k,l}^{(3)} E_j E_k E_l + \dots \qquad \text{(16.9)}$$

where $\chi^{(m)}$ is the mth order electrical susceptibility tensor. In SPDC, it is the second order nonlinear susceptibility $\chi^{(2)}$ that plays the role. The second order nonlinear interaction Hamiltonian can be written as

$$H = \varepsilon_0 \int_V d\mathbf{r}\, \chi_{ijk}^{(2)} E_i E_j E_k \tag{16.10}$$

where the integral is taken over the interaction volume V.

It is convenient to use the Fourier representation for the electrical fields in Eq. (16.10):

$$\mathbf{E}(\mathbf{r},\, t) = \int d\mathbf{k}\,[\, \mathbf{E}^{(-)}(\mathbf{k})e^{-i(\omega(\mathbf{k})t - \mathbf{k}\cdot\mathbf{r})} + \mathbf{E}^{(+)}(\mathbf{k})e^{i(\omega(\mathbf{k})t - \mathbf{k}\cdot\mathbf{r})}\,]. \tag{16.11}$$

Substituting Eq. (16.11) into Eq. (16.10) and keeping only the terms of interest, we obtain the SPDC Hamiltonian in the interaction representation:

$$H_{\text{int}}(t) = \varepsilon_0 \int_V d\mathbf{r} \int d\mathbf{k}_s\, d\mathbf{k}_i\, \chi_{lmn}^{(2)} E_{p\,l}^{(+)} e^{i(\omega_p t - \mathbf{k}_p\cdot\mathbf{r})}$$

$$E_{s\,m}^{(-)} e^{-i(\omega_s(\mathbf{k}_s)t - \mathbf{k}_s\cdot\mathbf{r})} E_{i\,n}^{(-)} e^{-i(\omega_i(\mathbf{k}_i)t - \mathbf{k}_i\cdot\mathbf{r})} + h.c., \tag{16.12}$$

where $h.c.$ stands for Hermitian conjugate. To simplify the calculation, we have also assumed the pump field to be plane and monochromatic with wave vector $\mathbf{k}_p$ and frequency ω_p.

It is easily noticeable that in Eq. (16.12), the volume integration can be done for some simplified cases. At this point, we assume that V is infinitely large. Later, we will see that the finite size of V in longitudinal and/or transversal directions may have to be taken into account. For an infinite volume V, the interaction Hamiltonian Eq. (16.12) is written as

$$H_{\text{int}}(t) = \varepsilon_0 \int d\mathbf{k}_s\, d\mathbf{k}_i\, \chi_{lmn}^{(2)}\, E_{p\,l}^{(+)} E_{s\,m}^{(-)} E_{i\,n}^{(-)}$$

$$\times \delta(\mathbf{k}_p - \mathbf{k}_s - \mathbf{k}_i) e^{i(\omega_p - \omega_s(\mathbf{k}_s) - \omega_i(\mathbf{k}_i))t} + h.c. \tag{16.13}$$

It is reasonable to consider the pump field classical, which is usually a laser beam, and quantize the signal and idler fields, which are both in single-photon level:

$$E^{(-)}(\mathbf{k}) = i\sqrt{\frac{2\pi\hbar\omega}{V}} a^\dagger(\mathbf{k}),$$

$$E^{(+)}(\mathbf{k}) = i\sqrt{\frac{2\pi\hbar\omega}{V}} a(\mathbf{k}), \tag{16.14}$$

where $a^\dagger(\mathbf{k})$ and $a(\mathbf{k})$ are photon creation and annihilation operators, respectively. The state of the emitted photon pair can be calculated by applying the first order perturbation

$$|\Psi\rangle = -\frac{i}{\hbar} \int dt\, H_{\text{int}}(t)\, |0\rangle. \tag{16.15}$$

By using vacuum $|0\rangle$ for the initial state in Eq. (16.15), we assume that there is no input radiation in any signal and idler modes, that is, we have a spontaneous parametric down conversion (SPDC) process.

Further assuming an infinite interaction time, evaluating the time integral in Eq. (16.15) and omitting altogether the constants and slow (square root) functions of ω, we obtain the *entangled* two-photon state of Eq. (16.7) in the form of integral:

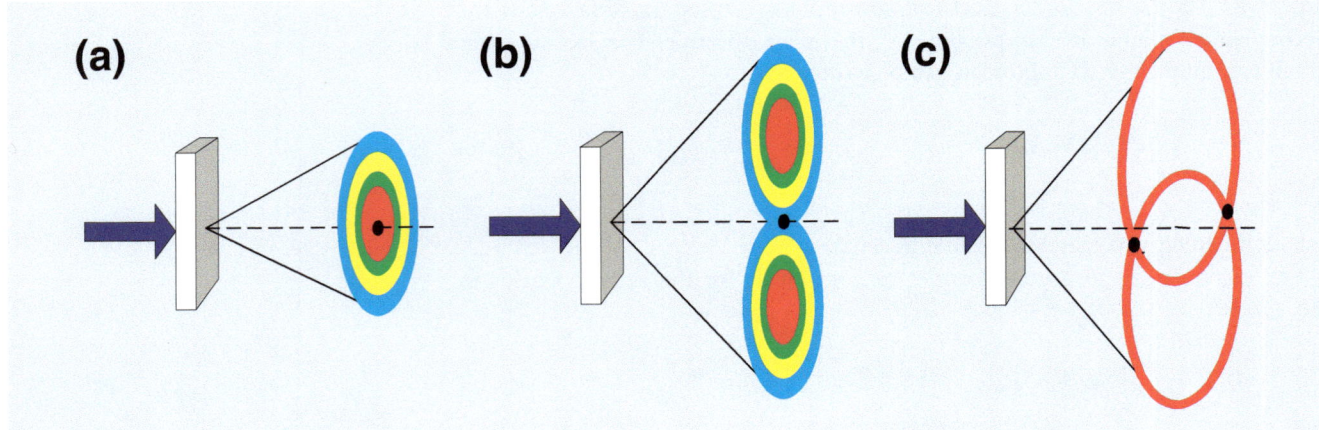

Fig. 16.3 Three widely used SPDC. (**a**) Type-I SPDC. (**b**) Collinear degenerate type-II SPDC. Two rings overlap at one region. (**c**) Non-collinear degenerate type-II SPDC. For clarity, only two degenerate rings, one for *e*-polarization and the other for *o*-polarization, are shown

$$
\begin{aligned}
|\Psi\rangle = \Psi_0 \int d\mathbf{k}_s d\mathbf{k}_i \delta[\omega_p - \omega_s(\mathbf{k}_s) - \omega_i(\mathbf{k}_i)] \\
\times \delta(\mathbf{k}_p - \mathbf{k}_s - \mathbf{k}_i) a_s^\dagger(\mathbf{k}_s) a_i^\dagger(\mathbf{k}_i)|0\rangle
\end{aligned}
\tag{16.16}
$$

where Ψ_0 is a normalization constant which has absorbed all omitted constants.

The way of achieving phase matching, i.e., the way of achieving the delta functions in Eq. (16.16) basically determines how the signal-idler pair "looks." For example, in a negative uniaxial crystal, one can use a linearly polarized pump laser beam as an extraordinary ray of the crystal to generate a signal-idler pair both polarized as the ordinary rays of the crystal, which is defined as type-I phase matching. One can alternatively generate a signal-idler pair with one ordinary polarized and another extraordinary polarized, which is defined as type II phase matching. **Figure 16.3** shows three examples of SPDC two-photon source. All three schemes have been widely used for different experimental purposes. Technical details can be found from textbooks and research references in nonlinear optics.

The two-photon state in the forms of Eq. (16.7) or Eq. (16.16) is a pure state, which describes the behavior of a signal-idler photon pair mathematically. Does the signal or the idler photon in the EPR state of Eq. (16.7) or Eq. (16.16) have a defined energy and momentum regardless of whether we measure it or not? Quantum mechanics answers: No! However, if one of the subsystems is measured with a certain energy and momentum, the other one is determined with certainty, despite the distance between them.

In the above calculation of the two-photon state we have approximated an infinite large volume of nonlinear interaction. For a finite volume of nonlinear interaction, we may write the state of the signal-idler photon pair in a more general form:

$$
|\Psi\rangle = \int dk_s \, dk_i \, F(\mathbf{k}_s, \mathbf{k}_i) \, a_i^\dagger(\mathbf{k}_s) \, a_s^\dagger(\mathbf{k}_i)|0\rangle
\tag{16.17}
$$

where

$$F(\mathbf{k}_s, \mathbf{k}_i) = \varepsilon \, \delta(\omega_p - \omega_s - \omega_i) \, f(\Delta_z L) \, h_{tr}(\vec{\kappa}_1 + \vec{\kappa}_2)$$

$$f(\Delta_z L) = \int_L dz \, e^{-i(k_p - k_{sz} - k_{iz})z}$$

$$h_{tr}(\vec{\kappa}_1 + \vec{\kappa}_2) = \int_A d\vec{\rho} \, \tilde{h}_{tr}(\vec{\rho}) \, e^{-i(\vec{\kappa}_s + \vec{\kappa}_i) \cdot \vec{\rho}}$$

$$\Delta_z = k_p - k_{sz} - k_{iz}$$

(16.18)

where ε is named as parametric gain, ε is proportional to the second order electric susceptibility $\chi^{(2)}$, and is usually treated as a constant; L is the length of the nonlinear interaction; the integral in $\vec{\kappa}$ is evaluated over the cross section A of the nonlinear material illuminated by the pump, $\vec{\rho}$ is the transverse coordinate vector, $\vec{\kappa}_j$ (with $j = s, i$) is the transverse wavevector of the signal and idler, and $f(|\vec{\rho}|)$ is the transverse profile of the pump, which can be treated as a Gaussion in most of the experimental conditions. The functions $f(\Delta_z L)$ and $h_{tr}(\vec{\kappa}_1 + \vec{\kappa}_2)$ can be approximated as δ-functions for an infinitely long ($L \sim \infty$) and wide ($A \sim \infty$) nonlinear interaction region. The reason we have chosen the form of Eq. (16.18) is to separate the "longitudinal" and the "transverse" correlations. We will show that $\delta(\omega_p - \omega_s - \omega_i)$ and $f(\Delta_z L)$ together can be rewritten as a function of $\omega_s - \omega_i$. To simplify the mathematics, we assume near co-linearly SPDC. In this situation, $|\vec{\kappa}_{s,i}| \ll |\mathbf{k}_{s,i}|$.

Basically, function $f(\Delta_z L)$ determines the "longitudinal" space-time correlation. Finding the solution of the integral is straightforward:

$$f(\Delta_z L) = \int_0^L dz \, e^{-i(k_p - k_{sz} - k_{iz})z} = e^{-i\Delta_z L/2} \, \text{sinc}(\Delta_z L/2).$$

(16.19)

where sinc $(x) = \sin(x)/x$.

Now, we consider $f(\Delta_z L)$ with $\delta(\omega_p - \omega_s - \omega_i)$ together, and taking advantage of the δ-function in frequencies by introducing a detuning frequency ν to evaluate function $f(\Delta_z L)$:

$$\omega_s = \omega_s^0 + \nu$$

$$\omega_i = \omega_i^0 - \nu$$

$$\omega_p = \omega_s + \omega_i = \omega_s^0 + \omega_i^0.$$

(16.20)

The dispersion relation $k(\omega)$ allows us to express the wave numbers through the detuning frequency ν:

$$k_s \approx k(\omega_s^0) + \nu \left. \frac{dk}{d\omega} \right|_{\omega_s^0} = k(\omega_s^0) + \frac{\nu}{u_s},$$

$$k_i \approx k(\omega_i^0) - \nu \left. \frac{dk}{d\omega} \right|_{\omega_i^0} = k(\omega_i^0) - \frac{\nu}{u_i}$$

(16.21)

where u_s and u_i are group velocities for the signal and the idler, respectively. Now, we connect Δ_z with the detuning frequency ν:

$$
\begin{aligned}
\Delta_z &= k_p - k_{sz} - k_{iz} \\
&= k_p - \sqrt{(k_s)^2 - (\vec{\kappa}_s)^2} - \sqrt{(k_i)^2 - (\vec{\kappa}_i)^2} \\
&\cong k_p - k_s - k_i + \frac{(\vec{\kappa}_s)^2}{2k_s} + \frac{(\vec{\kappa}_i)^2}{2k_i} \\
&\cong k_p - k(\omega_s^0) - k(\omega_i^0) + \frac{\nu}{u_s} - \frac{\nu}{u_i} + \frac{(\vec{\kappa}_s)^2}{2k_s} + \frac{(\vec{\kappa}_i)^2}{2k_i} \\
&\cong D\nu
\end{aligned}
\tag{16.22}
$$

where $D \equiv 1/u_s - 1/u_i$. We have also applied $k_p - k(\omega_s^0) - k(\omega_i^0) = 0$ and $|\vec{\kappa}_{s,i}| \ll |\mathbf{k}_{s,i}|$. The "longitudinal" wavevector correlation function is rewritten as a function of the detuning frequency ν: $f(\Delta_z L) \cong f(\nu DL)$. In addition to the above approximations, we have inexplicitly assumed the angular independence of the wavevector $k = n(\theta)\omega/c$. For type II SPDC, the refraction index of the extraordinary-ray depends on the angle between the wavevector and the optical axis and an additional term appears in the expansion. Making the approximation valid, we have restricted our calculation to near-collinear process. Thus, for a good approximation, in the near-collinear experimental setup:

$$
\Delta_z L \cong \nu DL = (\omega_s - \omega_i)DL/2.
\tag{16.23}
$$

Type-I degenerate SPDC is a special case. Due to the fact that $u_s = u_i$, and hence, $D = 0$, the expansion of $k(\omega)$ should be carried out up to the second order. Instead of (16.23), we have

$$
\Delta_z L \cong -\nu^2 D'L = -(\omega_s - \omega_i)^2 D'L/4
\tag{16.24}
$$

where

$$
D' \equiv \frac{d}{d\omega}\left(\frac{1}{u}\right)\Big|_{\omega^0}.
$$

The two-photon state of the signal-idler pair is then approximated as

$$
|\Psi\rangle = \int d\nu \, d\vec{\kappa}_s \, d\vec{\kappa}_i f(\nu) \, h_{tr}(\vec{\kappa}_s + \vec{\kappa}_i) \, a_s^\dagger(\omega_s^0 + \nu, \vec{\kappa}_s) \, a_i^\dagger(\omega_i^0 - \nu, \vec{\kappa}_i)|0\rangle
\tag{16.25}
$$

where the normalization constant has been absorbed into $f(\nu)$.

SPDC has been one of the most convenient two-photon sources for the preparation of Bell state. Although Bell state is for polarization (or spin), the space-time part of the state cannot be ignored. One important "preparation" is to make the two biphoton wavepackets, corresponding to the first and the second terms in the Bell state, completely "overlap" in space-time, or indistinguishable for the joint detection event. This is especially important for type-II SPDC.

A very interesting situation for type-II SPDC is that of "noncollinear phase matching." The signal-idler pair are emitted from an SPDC crystal, such as BBO, cut in type-II phase matching, into two cones, one ordinarily polarized, the other extraordinarily polarized, see ◘ Fig. 16.4. Along the intersection, where the cones overlap, two pinholes numbered 1 and 2 are used for defining the direction of the **k** vectors of the signal-idler pair. It is very reasonable to consider the polarization state of the signal-idler pair as

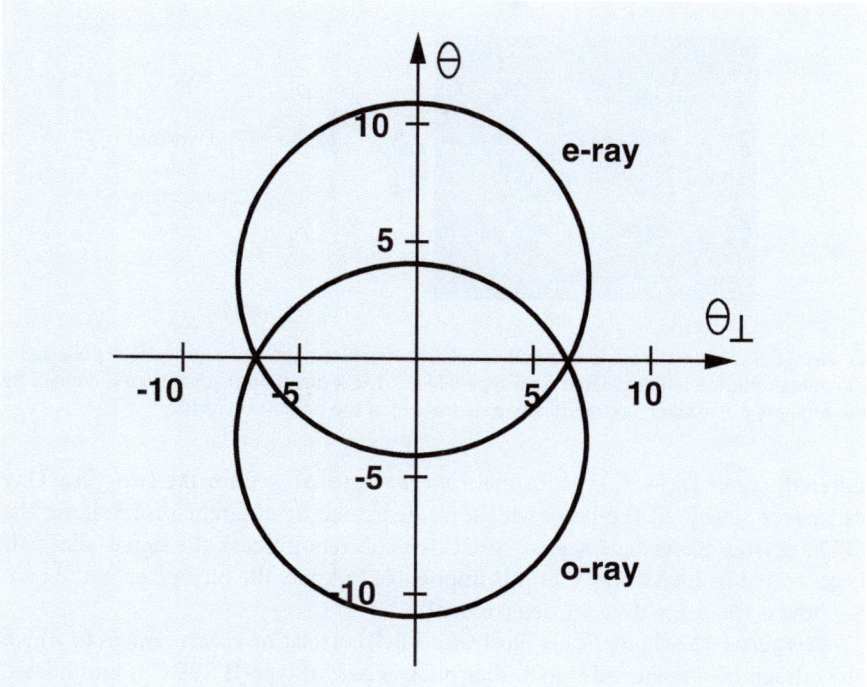

◘ Fig. 16.4 Type-II noncollinear phase matching: a cross section view of the degenerate 702.2 nm cones. The 351.1 nm pump beam is in the center. The numbers along the axes are in degrees

$$|\Psi\rangle = \frac{1}{\sqrt{2}}(|o_1 e_2\rangle + |e_1 o_2\rangle) = |\Psi^{(+)}\rangle \tag{16.26}$$

where o_j and e_j, $j = 1, 2$, are ordinarily and extraordinarily polarization, respectively. It seems straightforward to realize an EPR-Bohm-Bell measurement by simply setting up a polarization analyzer in series with a photon counting detector behind pinholes 1 and 2, respectively, and to expect observe the polarization correlation. This is, however, *incorrect*! One can never observe the EPR-Bohm-Bell polarization correlation unless a "compensator" is applied [4]. The "compensator" is a piece of birefringent material. For example, one may place another piece of nonlinear crystal behind the SPDC. It could be the same type of crystal as that of the SPDC, with the same cutting angle, except having half the length and a 90° rotation with respect to that of the SPDC crystal.

What is the role of the "compensator"? There have been naive explanations about the compensator. One suggestion was that the problem comes from the longitudinal "walk-off" of the type-II SPDC. For example, if one uses a type II BBO, which is a negative uni-axis crystal, the extraordinary-ray propagates faster than the ordinary-ray inside the BBO. *Suppose the $o - e \leftrightarrow e - o$ pair is generated in the middle of the crystal, the e-polarization will trigger the detector earlier than the o-polarization by a time $\Delta t = (n_o - n_e)L/2c$. This implies that D_2 would be fired first in $|o_1 e_2\rangle$ term; but D_1 would be fired first in $|e_1 o_2\rangle$ term. If Δt is greater than the coherence length of the signal-idler field, one would be able to distinguish which amplitude gave rise to the "click-click" coincidence event. One may compensate the "walk-off" by introducing an additional piece of birefringent material, like the compensator we have suggested above, to delay the e-ray relative to the o-ray by the same amount of time, Δt. If, however, the signal-idler pair is generated in the *front face* or the *back face* of the SPDC, the delay time would be very

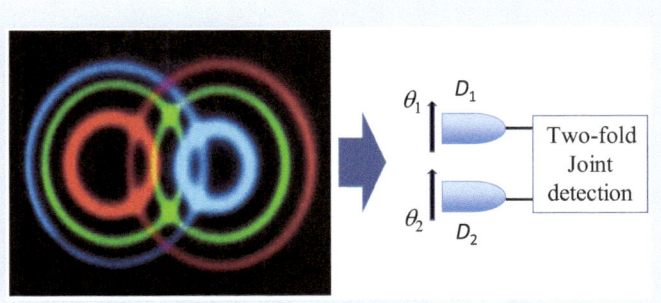

○ Fig. 16.5 Schematic setup of a Bell correlation measurement. The orthogonally polarized signal-idler photon pair is created from type-II SPDC. The X-direction (Y-direction) is defined by the ordinary polarization (extraordinary polarization) of the nonlinear crystal

different: $\Delta t = (n_o - n_e)L/c$ for the *front face* and $\Delta t = 0$ for the *back face*. One can never satisfy all the pairs which are generated at different places along the SPDC crystal. Nevertheless, since SPDC is a *coherent* process, the signal-idler pair is generated in such a way that it is impossible to know the birthplace of the pair. So, how is the delay time Δt determined?

○ Figure 16.5 schematically illustrates a Bell correlation measurement in which an orthogonally polarized signal-idler photon pair of type-II SPDC is annihilated at $(\mathbf{r}_1, t_1)$ and $(\mathbf{r}_2, t_2)$ jointly by two point-like photon counting detectors D_1 and D_2 with two polarization analyzers oriented at θ_1 and θ_2, respectively.

The coincidence counting rate of D_1 and D_2 measures the probability for a pair of photons to produce a joint photodetection event at D_1 and D_2. In this setup, the pair has two different yet indistinguishable ways to produce a coincidence count: (1) the X-polarized photon passes θ_1 triggering D_1, the Y-polarized photon passes θ_2 triggering D_2; (2) the Y-polarized photon passes θ_1 triggering D_1, the X-polarized photon passes θ_2 triggering D_2. If the above two alternatives are indistinguishable, quantum theory requires a superposition of the two probability amplitudes which results in an EPR-Bohm-Bell correlation:

$$R_c(\theta_1, \theta_2) \propto |\mathcal{A}_{\mathrm{I}}(\theta_1, \theta_2) + \mathcal{A}_{\mathrm{II}}(\theta_1, \theta_2)|^2 = \sin^2(\theta_1 + \theta_2). \tag{16.27}$$

To calculate the joint detection counting rate, we follow the Glauber formula [25]:

$$\begin{aligned} R_c &\propto \langle \Psi | E^{(-)}(\mathbf{r}_1, t_1) E^{(-)}(\mathbf{r}_2, t_2) E^{(+)}(\mathbf{r}_2, t_2) E^{(+)}(\mathbf{r}_1, t_1) | \Psi \rangle \\ &= \left| \langle 0 | E^{(+)}(\mathbf{r}_2, t_2) E^{(+)}(\mathbf{r}_1, t_1) | \Psi \rangle \right|^2. \end{aligned} \tag{16.28}$$

Adopting our earlier result, we may rewrite the state of the type-II signal-idler pair in the following form:

$$|\Psi\rangle = \int d\mathbf{k}_o d\mathbf{k}_e \delta(\omega_o + \omega_e - \omega_p) \Phi(\Delta_k L)\, \hat{\mathbf{o}}\, a_o^\dagger(\omega(\mathbf{k}_o)) \hat{\mathbf{e}}\, a_e^\dagger(\omega(\mathbf{k}_e)) |0\rangle \tag{16.29}$$

where $\hat{\mathbf{o}}$ and $\hat{\mathbf{e}}$ are unit vectors along the o-ray and the e-ray polarization direction of the SPDC crystal, and $\Delta_k = k_o + k_e - k_p$. In $\Phi(\Delta_k L)$, the finite length of the nonlinear crystal has been taken into account. Suppose the polarizers of the detectors D_1 and D_2 are set at angles θ_1 and θ_2, relative to the polarization direction of the o-ray of the SPDC crystal, respectively, the field operators can be written as

$$E_j^{(+)}(t_j, r_j) = \int d\omega\, \hat{\theta}_j\, a(\omega) e^{-i[\omega t_j - k(\omega)r_j]}$$

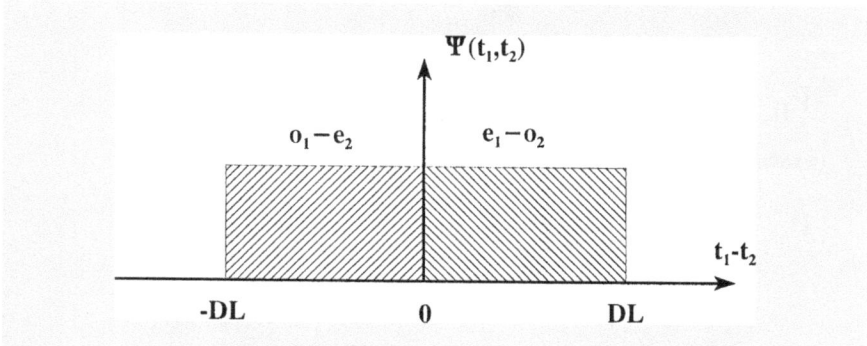

Fig. 16.6 Without "compensator," the two dimensional wavepackets of $\Psi(\tau_1^o, \tau_2^e)$ and $\Psi(\tau_1^e, \tau_2^o)$ do not overlap along $\tau_1 - \tau_2$ axis

where $j = 1, 2$, $\hat{\theta}_j$ is the unit vector along the orientation of the *i*th polarization analyzer. Substitute the field operator into Eq. (16.28),

$$
\begin{aligned}
R_c &\propto |(\hat{\theta}_1 \cdot \hat{\mathbf{o}})(\hat{\theta}_2 \cdot \hat{\mathbf{e}}) \, \Psi(\tau_1^o, \tau_2^e) + (\hat{\theta}_1 \cdot \hat{\mathbf{e}})(\hat{\theta}_2 \cdot \hat{\mathbf{o}}) \, \Psi(\tau_1^e, \tau_2^o)|^2 \\
&= |\mathcal{A}_1(\theta_1, \theta_2) + \mathcal{A}_2(\theta_1, \theta_2)|^2 \\
&= \cos^2\theta_1 \, \sin^2\theta_2 + \sin^2\theta_1 \, \cos^2\theta_2 \\
&\quad + \cos\theta_1 \, \sin\theta_2 \, \sin\theta_1 \, \cos\theta_2 \, \Psi^*(\tau_1^o, \tau_2^e) \, \Psi(\tau_1^e, \tau_2^o)
\end{aligned}
$$

(16.30)

where $\Psi(\tau_1^o, \tau_2^e)$ and $\Psi(\tau_1^e, \tau_2^o)$ are the effective two-photon wavefunctions, namely the biphoton wavepackets, and $\tau_j^o = t_j - r_j/u_o$, $\tau_j^e = t_j - r_j/u_e$. The third term of Eq. (16.30) determines the degree of two-photon coherence. Considering degenerate CW laser pumped SPDC, the biphoton wavepacket can be simplified as

$$
\Psi(\tau_1, \tau_2) = \Psi_0 \, e^{-i\omega_p(\tau_1+\tau_2)/2} \, \mathcal{F}_{\tau_-} \{f(\Omega)\}.
$$

The coefficient of $\cos\theta_1 \, \sin\theta_2 \, \sin\theta_1 \, \cos\theta_2$ in the third term of Eq. (16.30) is thus

$$
e^{-i\omega_p(\Delta\tau_1 - \Delta\tau_2)/2} \, \mathcal{F}_{\tau_1^o - \tau_2^e} \{f(\Omega)\} \otimes \mathcal{F}_{\tau_1^e - \tau_2^o} \{f(\Omega)\}.
$$

where $F_\tau - \{f(\Omega)\}$ labels a Fourier transform.

Therefore, two important factors will determine the result of the polarization correlation measurement: (1) the phase of $e^{-i\omega_p(\Delta\tau_1 - \Delta\tau_2)/2}$; and (2) the overlapping between the biphoton wavepackets $\Psi^*(\tau_1^o, \tau_2^e)$ and $\Psi(\tau_1^e, \tau_2^o)$, i.e., the chances for both $\Psi^*(\tau_1^o, \tau_2^e)$ and $\Psi(\tau_1^e, \tau_2^o)$ take nonzero values simultaneously at $\tau_1^o - \tau_2^e$.

Examining the two wavepackets associated with the $o_1 - e_2$ and $e_1 - o_2$ terms, we found the two dimensional biphoton wavepackets of type II SPDC do not overlap, due to the *asymmetrical* rectangular function of $\pi(\tau_1 - \tau_2)$ as indicated in Fig. 16.6. In order to make the two wavepackets overlap, we may either (1) move both wavepackets a distance of $DL/2$ (case I) or (2) move one of the wavepackets a distance of DL (case II). The use of "compensator" is for this purpose. After compensating the two asymmetrical function of $\pi(\tau_1 - \tau_2)$, we need to further manipulate the phase of $e^{-i\omega_p(\Delta\tau_1 - \Delta\tau_2)/2}$ to finalize the desired Bell states. This can be done by means of a retardation plate to introduce phase delay of 2π (+1) or π (−1) between the o-ray and the e-ray in either arm 1 or arm 2. The EPR-Bohm-Bell polarization correlation $R_c \propto \sin^2(\theta_1 \pm \theta_2)$ is expected only when the above two conditions are satisfied. We can simplify the polarization state of the signal-idler photon pair in the form of Bell states $|\Psi^{(\pm)}\rangle$ in this situation only.

In recent years, special attention has been paid to femtosecond laser pulse pumped SPDC due to its attractive applications in quantum information processing and communication. The biphoton wavepacket looks very different

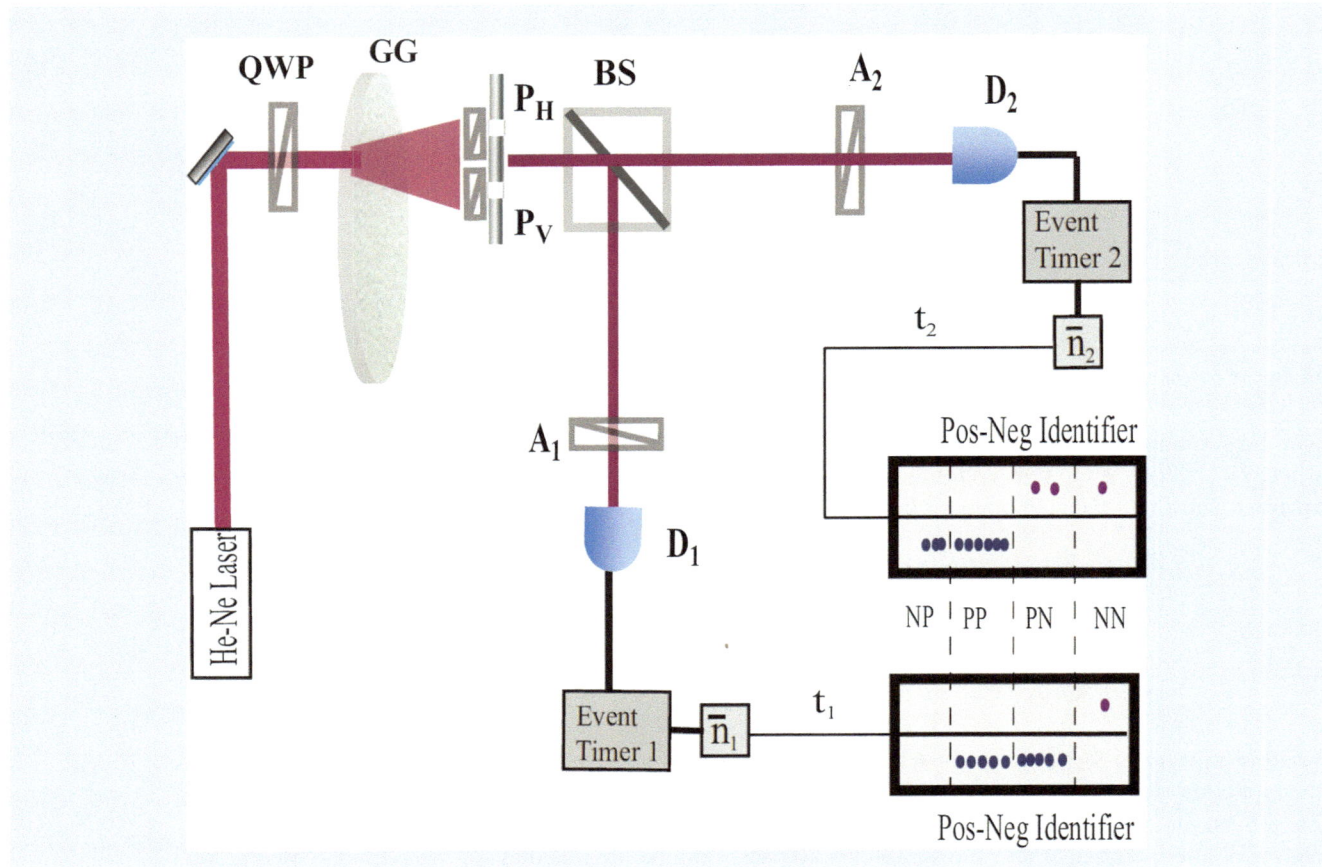

Fig. 16.7 Schematic setup of the experiment: polarization correlation measurement of thermal fields in photon-number fluctuations

in this case than that of the CW pump. One needs to exam the biphoton wavepackets carefully to be sure the superposed probability amplitudes overlap [26, 27].

From the above analysis, we may conclude the EPR-Bohm-Bell correlation is the result of a nonlocal interference: a pair of entangled photons interferes with the pair itself. This peculiar interference involves the superposition of two-photon wavepackets, or two-photon amplitudes, corresponding to different yet indistinguishable alternative ways for a pair of photons to produce a photodetection event at distant space-time coordinates.

16.2.2 Bell State Simulation of Thermal Light

Now we ask what would happen if we replace the entangled photons with a randomly paired photons, or wavepackets, in the thermal state? Can a randomly paired photons in a thermal state simulate the Bell state? The answer is positive. In the following we analyze a recent experiment of Peng et al. in which a Bell-type correlation was observed from the polarization measurement of thermal fields in photon-number fluctuations, indicating the successful simulation of Bell state [28]. Very importantly, the same mechanism can be easily extended to the simulation of a multi-photon GHZ state and N-qubits for $N \gg 2$.

🔲 Figure 16.7 schematically illustrates the experimental setup of Peng et al. A large number of circular polarized wavepackets at the single-photon level, such as the mth and the nth, come from a standard pseudo-thermal light source [29] consisting of a circularly polarized 633 nm CW laser beam and a rotating ground glass (GG). The diameter of the laser beam is ~ 2 mm. The size of the tiny diffusers on the GG is roughly a few micrometers. The randomly distributed wavepackets pass two pinholes P_H and P_V with two linear polarizers oriented at a horizontal polarization $\vec{H}$ ($\theta = 0°$) and vertical polarization $\vec{V}$ ($\theta = 90°$), respectively. The circular polarized wavepackets have 50 % chance to pass the upper pinhole P_H with horizontal polarization and 50 % chance to pass the lower pinhole P_V with vertical polarization. The separation between the two pinholes is much greater than the coherence length of the pseudo-thermal field. Therefore, (1) the $\vec{H}$ polarization and $\vec{V}$ polarization are first-order incoherent and the mixture of the two polarizations results in a unpolarized field; (2) the fluctuations of the $\vec{H}$ polarization and the $\vec{V}$ polarization are completely independent and random without any correlation. A 50–50 non-polarizing beamsplitter (BS) is used to divide the unpolarized thermal field, i.e., the 50–50 mixture of the two polarizations, into arms 1 and 2. Two polarization analyzers A_1, oriented at θ_1, and A_2, oriented at θ_2, followed by two photon counting detectors D_1 and D_2, are placed into arms 1 and 2 for the measurement of the polarization of the wavepackets. The registration time and the number of photodetection events of D_1 and D_2 at each jth time window are recorded, respectively, by two independent but synchronized event timers. The width of the time window, Δt_j, can be adjusted from nanoseconds to milliseconds. For each detector, D_β, $\beta = 1, 2$, at each chosen value of θ_β, the mean photon number, $\tilde{n}_\beta$, is calculated from $\tilde{n}_\beta = \left(\sum_{j=1}^{N} n_{\beta j} \right)/N$,

where N is the total number of time windows recorded for each data point in which θ_1 and θ_2 are set at certain chosen values. In our experiments, the total number and the width of the time window were $N \approx 4 \times 10^5$, and $\Delta t_j = 800\ \mu$ s. The mean photon number was chosen $\tilde{n}_1 \sim \tilde{n}_2 \sim 20$. In addition, the counting rate of D_1 and D_2 is monitored to be constants, independent of θ_1 and θ_2. The number fluctuation is then calculated for each time window, $\Delta n_{\beta j} = n_{\beta j} - \tilde{n}_\beta$ [30]:

$$\langle \Delta n_1(\theta_1) \Delta n_2(\theta_2) \rangle = \frac{1}{N} \left[\sum_{j=1}^{N} \Delta n_{1j}(\theta_1) \Delta n_{2j}(\theta_2) \right]. \tag{16.31}$$

Achieving the maximum space-time correlation in photon-number fluctuations, we place D_1 and D_2 at equal longitudinal and transverse coordinates, $z_1 = z_2$ and $\vec{\rho}_1 = \vec{\rho}_2$.

◻ Figure 16.8 reports a typical measurement of the polarization correlation in photon-number fluctuation correlation. In this measurement, we fixed $\theta_1 = 45°$ and rotated θ_2 to a set of different values. The black dots are experimental data, the red sinusoidal curve is the theoretical fitting of $\cos^2(\theta_1 - \theta_2)$ based on Eq. (16.44) with a ~ 92.5 % contrast. For other values of $\theta_1 \neq 45°$ we have observed the same sinusoidal correlation function. ◻ Figure 16.9 reports a measurement of $\langle \Delta n_1(\theta_1) \Delta n_2(\theta_2) \rangle$ by scanning the values of θ_1 and θ_2 (2-D scanning). Based on these measurements, we conclude that our observed polarization correlation is the same as that of the Bell state $|\Phi^{(+)}\rangle$. Apparently, the post-selection measurements of the reported experiment has "entangled" a product state of polarization into the Bell state $|\Phi^{(+)}\rangle$.

To explain the experimental observation, we start from the analysis of chaotic-thermal light. Chaotic-thermal light may come from a natural thermal light source, such as the sun, or from a pseudo-thermal light source, usually consisting of a laser beam, either CW or pulsed, and a fast rotating ground glass containing a large number of tiny scattering diffusers (usually on the order of a few micrometers). For a natural thermal light source, each radiating atom among a large number of

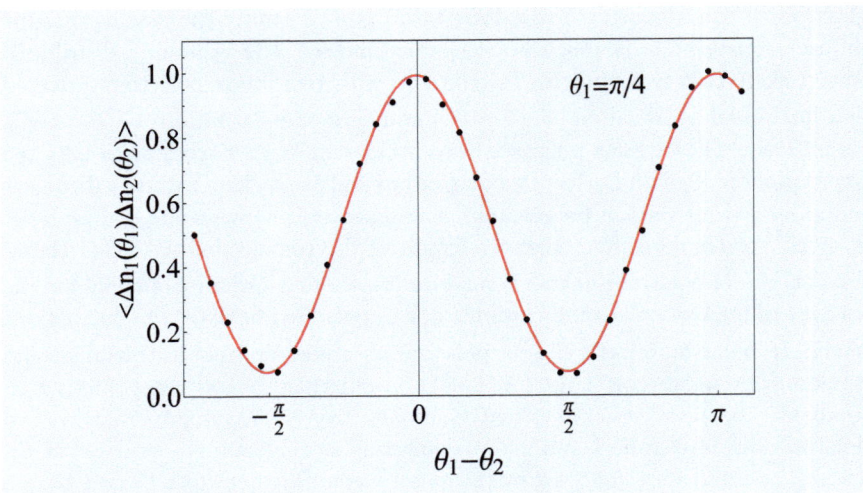

Fig. 16.8 Experimental observation of a Bell correlation with $\sim 92.5\%$ contrast. The *black dots* are experimental data and the *red sinusoidal curve* is a theoretical fitting. The *horizontal axis* labels $\varphi = \theta_1 - \theta_2$ while θ_1 was fixed at $45°$, the *vertical axis* reports the normalized photon-number fluctuation correlation $\langle \Delta n_1(\theta_1) \Delta n_2(\theta_2) \rangle$

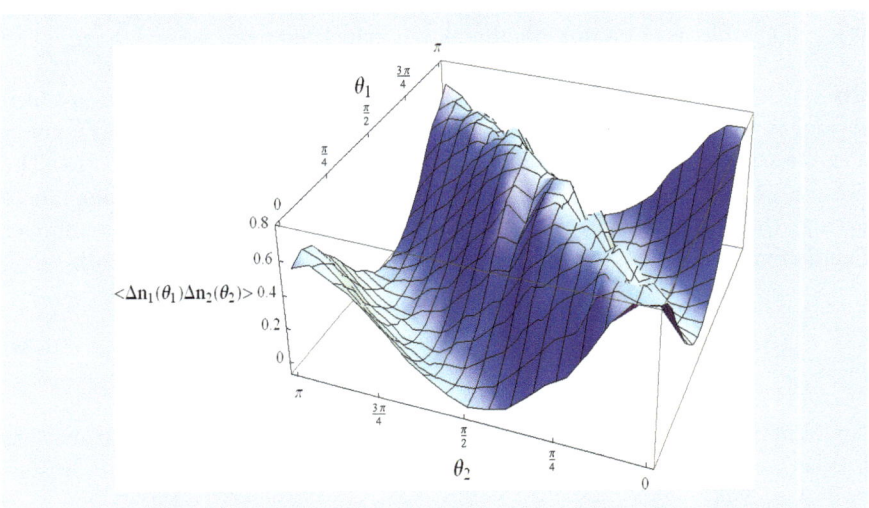

Fig. 16.9 A typical measurement of $\langle \Delta n_1(\theta_1) \Delta n_2(\theta_2) \rangle$ by a 2-D scanning of θ_1 and θ_2

randomly distributed and randomly radiated atomic transitions can be considered a sub-source. A photon may be created from an atomic transition, or sub-source, such as the mth atomic transition, or the mth sub-source, at space-time coordinate $(\mathbf{r}_{0m}, t_{0m})$, where $(\mathbf{r}_{0m})$ indicates the spatial coordinate of the mth atomic transition, and t_{0m} is the creation time of the photon. With a pseudo-thermal light source, each tiny scattering diffuser in the ground glass is a sub-source which scatters a wavepacket from the laser beam at space-time coordinate $(\vec{\rho}_{0m}, t_{0m})$ with random phase φ_{0m}, where $(\vec{\rho}_{0m})$ indicates the transverse spatial coordinates of the mth scattering diffuser of the fast rotating ground glass, and t_{0m} is the scattering time of the subfield. It is reasonable to model thermal light, either from a natural thermal light source or from a pseudo-thermal light source, in the coherence state representation [4, 31]:

$$|\Psi\rangle = \prod_m |\{\alpha_m\}\rangle = \prod_{m,\mathbf{k}} |\alpha_m(\mathbf{k})\rangle, \tag{16.32}$$

where m labels the mth photon that is created from the mth atomic transition of a natural thermal source, or the mth wavepacket that is scattered from the mth sub-source of the pseudo-thermal source, and $\mathbf{k}$ is a wavevector. $|\alpha_m(\mathbf{k})\rangle$ is an eigenstate of the annihilation operator with an eigenvalue $\alpha_m(\mathbf{k})$,

$$\hat{a}_m(\mathbf{k})|\alpha_m(\mathbf{k})\rangle = \alpha_m(\mathbf{k})|\alpha_m(\mathbf{k})\rangle. \tag{16.33}$$

Thus, we have

$$\hat{a}_m(\mathbf{k})|\Psi\rangle = \alpha_m(\mathbf{k})|\Psi\rangle. \tag{16.34}$$

The field operator corresponding to the mth subfield at the detector can be written in the following form:

$$\hat{E}_m^{(+)}(\mathbf{r},t) = \int d\omega \, \hat{a}_m(\omega) g_m(\omega;\mathbf{r},t) \tag{16.35}$$

with $g_m(\omega;\mathbf{r},t)$ the Green's function that propagates the ω mode of the mth subfield from the source to $(\mathbf{r},t)$. A point-like photon counting detector, behind a polarizer oriented at angle $\vec{\theta}$, at space-time coordinate $(\mathbf{r},t)$ counts the photon number that is polarized along $\vec{\theta}$, $n(\theta;\mathbf{r},t)$, which is usually written as the sum of mean photon-number $\langle n(\theta;\mathbf{r},t)\rangle$ and the photon-number fluctuation $\Delta n(\theta;\mathbf{r},t)$:

$$
\begin{aligned}
n(\theta;\mathbf{r},t) &= \sum_m \vec{p}_m \langle \alpha_m | \sum_p \hat{E}_p^{(-)}(\mathbf{r},t) \sum_q \hat{E}_q^{(+)}(\mathbf{r},t) \sum_n \vec{p}_n |\alpha_n\rangle \\
&= \sum_m (\vec{p}_m \cdot \vec{\theta})\Psi_m^*(\mathbf{r},t) \sum_n (\vec{p}_n \cdot \vec{\theta})\Psi_n(\mathbf{r},t) \\
&= \sum_m (\vec{p}_m \cdot \vec{\theta})\Psi_m^*(\mathbf{r},t)\,(\vec{p}_m \cdot \vec{\theta})\Psi_m(\mathbf{r},t) \\
&\quad + \sum_{m\neq n} (\vec{p}_m \cdot \vec{\theta})\Psi_m^*(\mathbf{r},t)\,(\vec{p}_n \cdot \vec{\theta})\Psi_n(\mathbf{r},t) \\
&\equiv \sum_m \Psi_m^*(\theta;\mathbf{r},t)\Psi_m(\theta;\mathbf{r},t) + \sum_{m\neq n} \Psi_m^*(\theta;\mathbf{r},t)\Psi_n(\theta;\mathbf{r},t) \\
&= \langle n(\theta;\mathbf{r},t)\rangle + \Delta n(\theta;\mathbf{r},t),
\end{aligned}
\tag{16.36}
$$

where $\vec{p}_m$ is the polarization of the mth wavepacket, $|\alpha_m\rangle$ is the state of the mth photon or the mth group of identical photons in the thermal state. In Eq. (16.36) we have introduced the effective wavefunction of a photon or a group of identical photons:

$$\Psi_m(\mathbf{r},t) = \langle \alpha_m | \hat{E}_m^{(+)}(\mathbf{r},t) | \alpha_m \rangle = \int d\omega \, a_m(\omega) g_m(\omega;\mathbf{r},t). \tag{16.37}$$

An effective wavefunction or a wavepacket, corresponding to the classical concept of an electromagnetic subfield $\mathbf{E}_m(\mathbf{r},t)$ however, represents a very different physical reality. The effective wavefunction represents the "probability amplitude" for a photon or a group of identical photons to produce a photoelectron event at space-time coordinate $(\mathbf{r},t)$. From Eq. (16.36), we find that the mean photon-number $\langle n(\theta;\mathbf{r},t)\rangle = \sum_m \Psi_m^*(\theta;\mathbf{r},t)\Psi_m(\theta;\mathbf{r},t)$ involves the effective wavefunction of a photon or a wavepacket while the photon-number fluctuation $\Delta n(\theta;\mathbf{r},t) = \sum_{m\neq n} \Psi_m^*(\theta;\mathbf{r},t)\Psi_n(\theta;\mathbf{r},t)$ involves the effective wave functions of two different photons, or a random pair of wavepackets, $m \neq n$. The measurement of mean photon-number gives the self-coherence of a photon or a group of identical photons while the measurement of photon-number fluctuation gives the mutual-coherence between different photons or different groups of identical photons. In the polarization-based photon counting measurement, the above equation can be used to calculate either the polarization correlation or the space-time correlation.

In general, a Bell type experiment measures the statistical correlation between $n(\theta_1; \mathbf{r}_1, t_1)$ and $n(\theta_2; \mathbf{r}_2, t_2)$. For thermal light, the photon-number correlation can be written as the sum of two contributions [4]:

$$\langle n_1(\theta_1) n_2(\theta_2) \rangle = \langle n_1(\theta_1) \rangle \langle n_2(\theta_2) \rangle + \langle \Delta n_1(\theta_1) \Delta n_2(\theta_2) \rangle$$
$$= \sum_m \Psi_{m1}^* \Psi_{m1} \sum_n \Psi_{n2}^* \Psi_{n2} + \sum_{m \neq n} \Psi_{m1}^* \Psi_{n1} \Psi_{n2}^* \Psi_{m2}. \qquad (16.38)$$

Here, we have shortened the notations of the effective wavefunction: the subindex β, $\beta = 1, 2$, indicates θ_β and $(\mathbf{r}_\beta, t_\beta)$. The first contribution is the result of two independent mean photon-number measurements, the statistics and coherence involves the measurement of single photons only while the second contribution is the result of photon-number fluctuation correlation, the statistics and coherence involves randomly paired photons. A Bell type experiment studies the polarization correlation of a pair of photons, obviously, we need to measure the photon-number fluctuation correlation $\langle \Delta n_1(\theta_1) \Delta n_2(\theta_2) \rangle$. The measurement of $\langle \Delta n_1(\theta_1) \Delta n_2(\theta_2) \rangle$ gives both polarization correlation and space-time correlation between two randomly paired photons. In the Bell type measurements, we usually manage to achieve a maximum correlation in space-time, then test the polarization correlation $\langle \Delta n_1(\theta_1) \Delta n_2(\theta_2) \rangle$ as a function of $\varphi = \theta_1 - \theta_2$ by varying θ_1 and θ_2 to all possible different values.

The effective wavefunctions: Ψ_{m1}^*, Ψ_{n1}, Ψ_{n2}^*, and Ψ_{m2} are calculated in the flowing. In general, each operator of the subfield is identified to be

$$\hat{E}_{m\beta}^{(+)} = \int d\omega \, \hat{a}_m(\omega) g_m(\omega; \mathbf{r}_\beta, t_\beta). \qquad (16.39)$$

Examine the experiment detail, we find

$$g_m(\omega; \mathbf{r}_\beta, t_\beta) = \frac{1}{\sqrt{2}} \Big[(\vec{H} \cdot \vec{\theta}_\beta) g_m(\omega; \mathbf{r}_H, t_H) g_H(\omega; \mathbf{r}_\beta, t_\beta) $$
$$+ (\vec{V} \cdot \vec{\theta}_\beta) g_m(\omega; \mathbf{r}_V, t_V) g_V(\omega; \mathbf{r}_\beta, t_\beta) \Big], \qquad (16.40)$$

where $g_m(\omega; \mathbf{r}_H, t_H)$ and $g_H(\omega; \mathbf{r}_\beta, t_\beta)$ are the Green's functions that propagate the ω mode of the mth subfield from the source to the upper pinhole P_H and from P_H to D_β, respectively. The effective wavefunction $\Psi_{m\beta}$ is thus

$$\Psi_{m\beta} = \int d\omega \, a_m \frac{1}{\sqrt{2}} \Big[(\vec{H} \cdot \vec{\theta}_\beta) g_m(\omega; \mathbf{r}_H, t_H) g_{P_H}(\omega; \mathbf{r}_\beta, t_\beta) $$
$$+ (\vec{V} \cdot \vec{\theta}_\beta) g_m(\omega; \mathbf{r}_V, t_V) g_{P_V}(\omega; \mathbf{r}_\beta, t_\beta) \Big] \qquad (16.41)$$
$$= \frac{1}{\sqrt{2}} \big[\Psi_{mH\beta} + \Psi_{mV\beta} \big],$$

where $\Psi_{mH\beta}$ is the mth wavepacket passes P_H, θ_β, triggers D_β, and $\Psi_{mV\beta}$ the mth wavepacket passes P_V, θ_β, triggers D_β. The normalized photon-number fluctuation correlation is thus

$$\langle \Delta n_1(\theta_1) \Delta n_2(\theta_2) \rangle = \sum_{m,n} \Psi_{m1}^* \Psi_{n1} \Psi_{m2} \Psi_{n2}^*$$

$$= \sum_{m,n} \frac{1}{\sqrt{2}} [\Psi_{mH1}^* + \Psi_{mV1}^*] \frac{1}{\sqrt{2}} [\Psi_{nH1} + \Psi_{nV1}] \quad \textbf{(16.42)}$$

$$\times \frac{1}{\sqrt{2}} [\Psi_{mH2} + \Psi_{mV2}] \frac{1}{\sqrt{2}} [\Psi_{nH2}^* + \Psi_{nV2}^*]$$

where, for example, Ψ_{mH1} is the mth wavepacket passing through the upper pinhole with $\vec{H}$ polarization contributing to the photodetection event of D_1 at space-time $(\mathbf{r}_1, t_1)$. In this experiment, we have separated the pinholes P_H and P_V beyond the transverse coherence length of the thermal field. Therefore, only four of the above sixteen terms survive from the sum of m and n,

$$\langle \Delta n_1(\theta_1) \Delta n_2(\theta_2) \rangle$$
$$\propto \sum_{m,n} [\Psi_{mH1}^* \Psi_{mH2} \Psi_{nH1} \Psi_{nH2}^* + \Psi_{mH1}^* \Psi_{mH2} \Psi_{nV1} \Psi_{nV2}^* \quad \textbf{(16.43)}$$
$$+ \Psi_{mV1}^* \Psi_{mV2} \Psi_{nH1} \Psi_{nH2}^* + \Psi_{mV1}^* \Psi_{mV2} \Psi_{nV1} \Psi_{nV2}^*]$$

corresponding to an interference effect which involves the joint detection of two wavepackets at two independent photodetectors located a distance apart. Adding the four cross terms that involve the random pair, i.e., the mth and the nth wavepackets, which is observable in the photon-number fluctuations, we obtain

$$\langle \Delta n_1(\theta_1) \Delta n_2(\theta_2) \rangle$$
$$\propto [\cos\theta_1 \cos\theta_2 \cos\theta_1 \cos\theta_2 + \cos\theta_1 \cos\theta_2 \sin\theta_1 \sin\theta_2$$
$$+ \sin\theta_1 \sin\theta_2 \cos\theta_1 \cos\theta_2 + \sin\theta_1 \sin\theta_2 \sin\theta_1 \sin\theta_2] \quad \textbf{(16.44)}$$
$$= |\cos\theta_1 \cos\theta_2 + \sin\theta_1 \sin\theta_2|^2$$
$$= \cos^2(\theta_1 - \theta_2),$$

which is the same correlation as that of the Bell state $|\Phi^{(+)}\rangle$.

Under the experimental condition of equal transverse and longitudinal (temporal) coordinates of D_1 and D_2, the observed correlation is the result of four groups of nonlocal superposition between different yet indistinguishable probability amplitudes of a randomly paired photons, or wavepackets. ◘ Figure 16.10 illustrates the four groups of such superposition. Group $\mathcal{A}$: (I) the mth wavepacket passes P_H, θ_1, triggers D_1 and the nth wavepacket passes P_V, θ_2, triggers D_2; (II) the mth wavepacket passes P_H, θ_2, triggers D_2 and the nth wavepacket passes P_V, θ_1, triggers D_1. Group $\mathcal{B}$: (I) the mth wavepacket passes P_V, θ_1, triggers D_1 and the nth wavepacket passes P_H, θ_2, triggers D_2; (II) the mth wavepacket passes P_V, θ_2, triggers D_2 and the nth wavepacket passes P_H, θ_1, triggers D_1. Group $\mathcal{C}$: (I) the mth wavepacket passes P_H, θ_1, triggers D_1 and the nth wavepacket passes P_H, θ_2, triggers D_2; (II) the mth wavepacket passes P_H, θ_2, triggers D_2 and the nth wavepacket passes P_H, θ_1, triggers D_1. Group $\mathcal{A}$: (I) the mth wavepacket passes P_V, θ_1, triggers D_1 and the nth wavepacket passes P_V, θ_2, triggers D_2; (II) the mth wavepacket passes P_V, θ_2, triggers D_2 and the nth wavepacket passes P_V, θ_1, triggers D_1. Calculating the four superpositions,

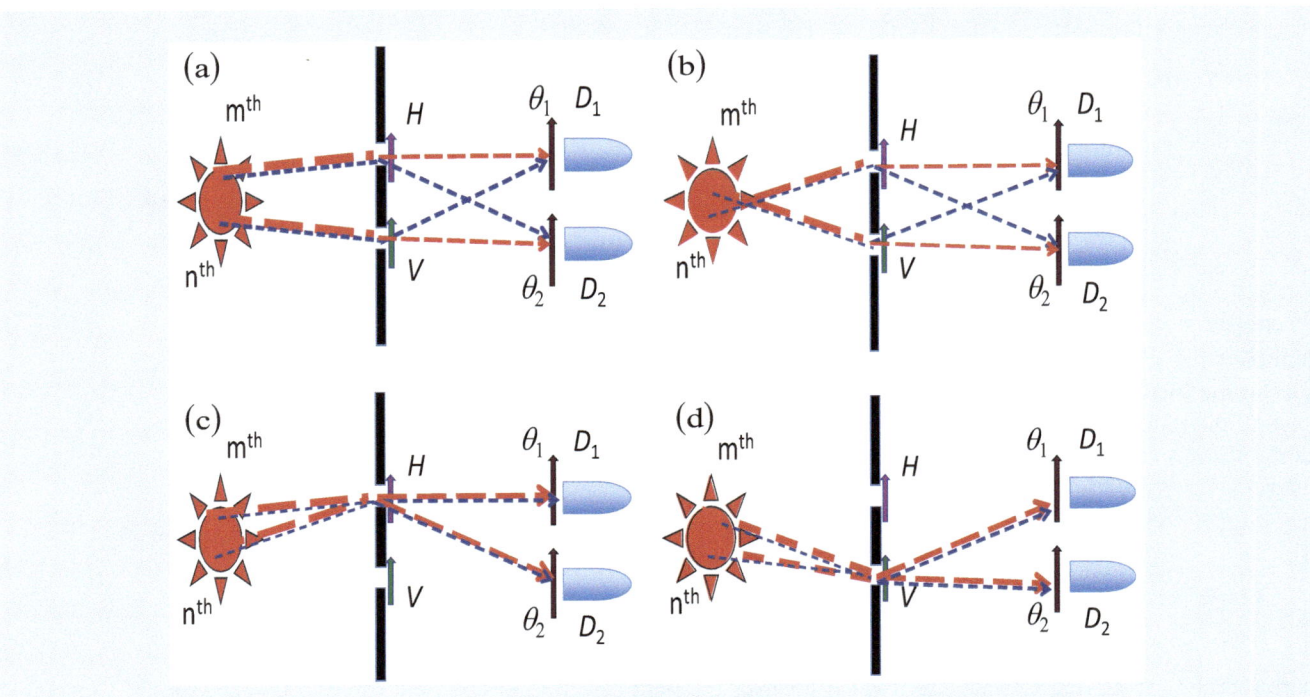

Fig. 16.10 There exist four groups of different yet indistinguishable amplitudes of the *m*th and the *n*th wavepacket to produce a joint photodetector event

$$\left|\mathcal{A}_I(\theta_1,\theta_2) + \mathcal{A}_{II}(\theta_1,\theta_2)\right|^2 = \left|\Psi_{mH1}\Psi_{nV2} + \Psi_{mH2}\Psi_{nV1}\right|^2$$
$$\left|\mathcal{B}_I(\theta_1,\theta_2) + \mathcal{B}_{II}(\theta_1,\theta_2)\right|^2 = \left|\Psi_{mV1}\Psi_{nH2} + \Psi_{mV2}\Psi_{nH1}\right|^2$$
$$\left|\mathcal{C}_I(\theta_1,\theta_2) + \mathcal{C}_{II}(\theta_1,\theta_2)\right|^2 = \left|\Psi_{mH1}\Psi_{nH2} + \Psi_{mH2}\Psi_{nH1}\right|^2 \tag{16.45}$$
$$\left|\mathcal{D}_I(\theta_1,\theta_2) + \mathcal{D}_{II}(\theta_1,\theta_2)\right|^2 = \left|\Psi_{mV1}\Psi_{nV2} + \Psi_{mV2}\Psi_{nV1}\right|^2.$$

Adding the four cross terms that involve the random pair, i.e., the *m*th and the *n*th wavepackets, which is observable in the photon-number fluctuations, we obtain Eq. (16.44).

16.2.3 Bell's Inequality

In 1964, Bell derived an inequality to distinguish quantum mechanics from local realistic probability theory of hidden variable [32]. In his pioneer work Bell introduced a "more complete specification effected by means of parameter λ" with probability distribution $\rho(\lambda)$ for the classical statistical estimation of the expectation value of the spin correlation measurement $\langle \Psi | (\vec{\sigma}_1 \cdot \hat{\mathbf{a}}) \, (\vec{\sigma}_2 \cdot \hat{\mathbf{b}}) | \Psi \rangle$ of particle-1 and particle-2, such as the spin-1/2 particle pair of Bohm, in the directions $\hat{\mathbf{a}}$ and $\hat{\mathbf{b}}$, simultaneously and respectively. The quantum mechanical result of this measurement gives

$$E_{ab} = \langle \Psi | (\vec{\sigma}_1 \cdot \hat{\mathbf{a}}) (\vec{\sigma}_2 \cdot \hat{\mathbf{b}}) | \Psi \rangle = -\hat{\mathbf{a}} \cdot \hat{\mathbf{b}}. \tag{16.46}$$

A special case of this result contains the determinism implicit in this idealized system When the Stern–Gerlach analyzers (SGA) are parallel, we have

$$E_{ab} = \langle \Psi | (\vec{\sigma}_1 \cdot \hat{\mathbf{a}}) (\vec{\sigma}_2 \cdot \hat{\mathbf{a}}) | \Psi \rangle = -1 \tag{16.47}$$

for all λ and all $\hat{\mathbf{a}}$. Thus, we can predict with certainty the result B by obtaining the result of A. Since $|\Psi\rangle$ does not determine the result of an individual measurement, this fact (via EPR's argument) suggests that there exits a more complete specification of the state by a single symbol λ it may have many dimensions, discrete and/or continuous parts, and different parts of it interacting with either apparatus, etc. Let Λ be the space of λ for an ensemble composed of a very large number of the particle systems. Bell represented the distribution function for the state λ on the space Λ by the symbol $\rho(\lambda)$ and take $\rho(\lambda)$ to be normalized

$$\int_\Lambda \rho(\lambda) \, d\lambda = 1. \tag{16.48}$$

In a deterministic hidden variable theory the observable $[A(\hat{\mathbf{a}})B(\hat{\mathbf{b}})]$ has a defined value $[A(\hat{\mathbf{a}})B(\hat{\mathbf{b}})](\lambda)$ for the state λ.

The locality is defined as follows: a deterministic hidden variable theory is local if for all $\hat{\mathbf{a}}$ and $\hat{\mathbf{b}}$ and all $\lambda \in \Lambda$

$$[A(\hat{\mathbf{a}})B(\hat{\mathbf{b}})](\lambda) = A(\hat{\mathbf{a}}, \lambda) \, B(\hat{\mathbf{b}}, \lambda). \tag{16.49}$$

This is, once λ is specified and the particle has separated, measurements of A can depend only upon λ and $\hat{\mathbf{a}}$ but not $\hat{\mathbf{b}}$. Likewise measurements of B depend only upon λ and $\hat{\mathbf{b}}$. Any reasonable physical theory that is realistic and deterministic and that denies action-at-a-distance is local in this sense. For such theories the expectation value of $[A(\hat{\mathbf{a}})B(\hat{\mathbf{b}})]$ is given by

$$\begin{aligned} E(\hat{\mathbf{a}}, \hat{\mathbf{b}}) &= \int_\Lambda d\lambda \, \rho(\lambda) \, [A(\hat{\mathbf{a}})B(\hat{\mathbf{b}})](\lambda) \\ &= \int_\Lambda d\lambda \, \rho(\lambda) \, A(\hat{\mathbf{a}}, \lambda) \, B(\hat{\mathbf{b}}, \lambda), \end{aligned} \tag{16.50}$$

where $E(\hat{\mathbf{a}}, \hat{\mathbf{b}}) \equiv E_{ab}$, corresponding to our previous notation. It is clear that Eq. (16.47) can hold if only if

$$A(\hat{\mathbf{a}}, \lambda) = -B(\hat{\mathbf{b}}, \lambda) \tag{16.51}$$

hold for all $\lambda \in \Lambda$.

Using Eq. (16.51) we calculate the following expectation values, which involves three different orientations of the SGA analyzers:

$$
E(\hat{\mathbf{a}}, \hat{\mathbf{b}}) - E(\hat{\mathbf{a}}, \hat{\mathbf{c}})
$$
$$
= \int_{\Lambda} d\lambda\, \rho(\lambda)\, \left[A(\hat{\mathbf{a}}, \lambda)\, B(\hat{\mathbf{b}}, \lambda) - A(\hat{\mathbf{a}}, \lambda)\, B(\hat{\mathbf{c}}) \right]
$$
$$
= -\int_{\Lambda} d\lambda\, \rho(\lambda)\, \left[A(\hat{\mathbf{a}}, \lambda)\, A(\hat{\mathbf{b}}, \lambda) - A(\hat{\mathbf{a}}, \lambda)\, A(\hat{\mathbf{c}}, \lambda) \right] \qquad \textbf{(16.52)}
$$
$$
= -\int_{\Lambda} d\lambda\, \rho(\lambda)\, A(\hat{\mathbf{a}}, \lambda)\, A(\hat{\mathbf{b}}, \lambda) \left[1 - A(\hat{\mathbf{b}}, \lambda)\, A(\hat{\mathbf{c}}, \lambda) \right].
$$

Since $A(\hat{\mathbf{a}}, \lambda) = \pm 1$, $A(\hat{\mathbf{b}}, \lambda) = \pm 1$, this expression can be written as

$$
\left| E(\hat{\mathbf{a}}, \hat{\mathbf{b}}) - E(\hat{\mathbf{a}}, \hat{\mathbf{c}}) \right| \leq \int_{\Lambda} d\lambda\, \rho(\lambda) \left[1 - A(\hat{\mathbf{b}}, \lambda)\, A(\hat{\mathbf{c}}, \lambda) \right], \qquad \textbf{(16.53)}
$$

and consequently,

$$
\left| E(\hat{\mathbf{a}}, \hat{\mathbf{b}}) - E(\hat{\mathbf{a}}, \hat{\mathbf{c}}) \right| \leq 1 + E(\hat{\mathbf{b}}, \hat{\mathbf{c}}). \qquad \textbf{(16.54)}
$$

This inequality is the first of a family of inequalities which are collectively called "Bell's inequalities."

It is easy to find a disagreement between the quantum mechanics prediction of Eq. (16.46) and the inequality of Eq. (16.54). When we choose $\hat{\mathbf{a}}$, $\hat{\mathbf{b}}$, and $\hat{\mathbf{c}}$ to be coplanar with $\hat{\mathbf{c}}$ making an angle of $2\pi/3$ with $\hat{\mathbf{a}}$, and $\hat{\mathbf{b}}$ making an angle of $\pi/3$ with both $\hat{\mathbf{a}}$ and $\hat{\mathbf{c}}$, the quantum prediction gives

$$
\left| \left[E(\hat{\mathbf{a}}, \hat{\mathbf{b}}) - E(\hat{\mathbf{a}}, \hat{\mathbf{c}}) \right]_{QM} \right| = 1, \qquad \textbf{(16.55)}
$$

while

$$
1 + \left[E(\hat{\mathbf{b}}, \hat{\mathbf{c}}) \right]_{QM} = \frac{1}{2}. \qquad \textbf{(16.56)}
$$

It does not satisfy inequality of Eq. (16.54).

It was soon realized that the Bell's inequality of Eq. (16.54) cannot be tested in a real experiment. Because Eq. (16.47) cannot be realized exactly in a realistic measurement. Any real detector cannot have a perfect quantum efficiency of 100 %, and any real analyzer cannot have a perfect distinguish ratio between orthogonal channels. In 1971, Bell proved a new inequality [33] which includes these concerns by assuming the outcomes of measurement A or B may take one of the following possible results:

$$
A(\hat{\mathbf{a}}, \lambda) \text{ or } B(\hat{\mathbf{b}}, \lambda) = \begin{cases} +1 & \text{"spin} - \text{up"} \\ -1 & \text{"spin} - \text{down"} \\ 0 & \text{particle not detected} \end{cases} \qquad \textbf{(16.57)}
$$

For a given state λ, we define the measured values for these quantities by the symbols $\bar{A}(\hat{\mathbf{a}}, \lambda)$ and $\tilde{B}(\hat{\mathbf{b}}, \lambda)$, which satisfy

$$
\left| \overline{A}(\hat{\mathbf{a}}, \lambda) \right| \leq 1 \qquad \text{and} \qquad \left| \overline{B}(\hat{\mathbf{b}}, \lambda) \right| \leq 1. \qquad \textbf{(16.58)}
$$

Following the same definition of locality, the expectation value of $A(\hat{\mathbf{a}})B(\hat{\mathbf{b}})$ is calculated as

$$E(\hat{\mathbf{a}}, \hat{\mathbf{b}}) = \int_{\Lambda} d\lambda \, \rho(\lambda) \, \overline{A}(\hat{\mathbf{a}}, \lambda) \, \overline{B}(\hat{\mathbf{b}}, \lambda). \tag{16.59}$$

Consider a measurement which involves $E(\hat{\mathbf{a}}, \hat{\mathbf{b}})$ and $E(\hat{\mathbf{a}}, \hat{\mathbf{b}}')$

$$E(\hat{\mathbf{a}}, \hat{\mathbf{b}}) - E(\hat{\mathbf{a}}, \hat{\mathbf{b}}')$$
$$= \int_{\Lambda} d\lambda \, \rho(\lambda) \, \left[\overline{A}(\hat{\mathbf{a}}, \lambda) \, \overline{B}(\hat{\mathbf{b}}, \lambda) - \overline{A}(\hat{\mathbf{a}}, \lambda) \, \overline{B}(\hat{\mathbf{b}}', \lambda) \right], \tag{16.60}$$

which can be written in the following form:

$$E(\hat{\mathbf{a}}, \hat{\mathbf{b}}) - E(\hat{\mathbf{a}}, \hat{\mathbf{b}}')$$
$$= \int_{\Lambda} d\lambda \, \rho(\lambda) \, \overline{A}(\hat{\mathbf{a}}, \lambda) \, \overline{B}(\hat{\mathbf{b}}, \lambda) \left[1 \pm \overline{A}(\hat{\mathbf{a}}', \lambda) \, \overline{B}(\hat{\mathbf{b}}', \lambda) \right]$$
$$- \int_{\Lambda} d\lambda \, \rho(\lambda) \, \overline{A}(\hat{\mathbf{a}}, \lambda) \, \overline{B}(\hat{\mathbf{b}}', \lambda) \left[1 \pm \overline{A}(\hat{\mathbf{a}}', \lambda) \, \overline{B}(\hat{\mathbf{b}}, \lambda) \right]. \tag{16.61}$$

Applying the triangle theorem, and considering $\rho(\lambda)[1 \pm \overline{A}(\hat{\mathbf{a}}', \lambda)\overline{B}(\hat{\mathbf{b}}', \lambda)]$ and $\rho(\lambda)[1 \pm \overline{A}(\hat{\mathbf{a}}', \lambda) \, \overline{B}(\hat{\mathbf{b}}, \lambda)]$ cannot take negative values, then using inequality in Eq. (16.58), we obtain

$$\left| E(\hat{\mathbf{a}}, \hat{\mathbf{b}}) - E(\hat{\mathbf{a}}, \hat{\mathbf{b}}') \right| \leq \int_{\Lambda} d\lambda \, \rho(\lambda) \, \left[1 \pm \overline{A}(\hat{\mathbf{a}}', \lambda) \, \overline{B}(\hat{\mathbf{b}}', \lambda) \right]$$
$$+ \int_{\Lambda} d\lambda \, \rho(\lambda) \, \left[1 \pm \overline{A}(\hat{\mathbf{a}}', \lambda) \, \overline{B}(\hat{\mathbf{b}}, \lambda) \right], \tag{16.62}$$

or

$$\left| E(\hat{\mathbf{a}}, \hat{\mathbf{b}}) - E(\hat{\mathbf{a}}, \hat{\mathbf{b}}') \right| \leq \pm \left[E(\hat{\mathbf{a}}', \hat{\mathbf{b}}') + E(\hat{\mathbf{a}}', \hat{\mathbf{b}}) \right] + 2 \int_{\Lambda} d\lambda \, \rho(\lambda). \tag{16.63}$$

We thus derive a measurable inequality

$$-2 \leq E(\hat{\mathbf{a}}, \hat{\mathbf{b}}) - E(\hat{\mathbf{a}}, \hat{\mathbf{b}}') + E(\hat{\mathbf{a}}', \hat{\mathbf{b}}) + E(\hat{\mathbf{a}}', \hat{\mathbf{b}}') \leq 2. \tag{16.64}$$

The quantum mechanical prediction of the EPR-Bhom state in a realistic measurement with imperfect detectors, analyzers etc., can be written as

$$\left[E(\hat{\mathbf{a}}, \hat{\mathbf{b}}) \right]_{QM} = C \, \hat{\mathbf{a}} \cdot \hat{\mathbf{b}} \tag{16.65}$$

where $| C | \leq 1$. Suppose we take $\hat{\mathbf{a}}, \hat{\mathbf{a}}', \hat{\mathbf{b}}, \hat{\mathbf{b}}'$ to be coplanar with $\phi = \pi/4$, we can easily find a disagreement between the quantum mechanics prediction and the inequality of Eq. (16.64):

$$\left[E(\hat{\mathbf{a}}, \hat{\mathbf{b}}) - E(\hat{\mathbf{a}}, \hat{\mathbf{b}}') + E(\hat{\mathbf{a}}', \hat{\mathbf{b}}) + E(\hat{\mathbf{a}}', \hat{\mathbf{b}}') \right]_{QM} = 2\sqrt{2} \, C. \tag{16.66}$$

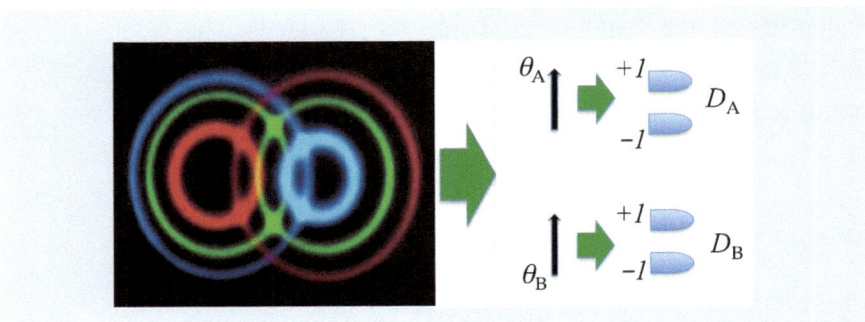

Fig. 16.11 Schematic setup of a Bell's inequality measurement. The expectation value is calculated from Eq. (16.67) which involves the measurement of four joint photodetections of D_A^+ &D_B^+, D_A^- &D_B^+, D_A^+&D_B^-, and D_A^-&D_B^-

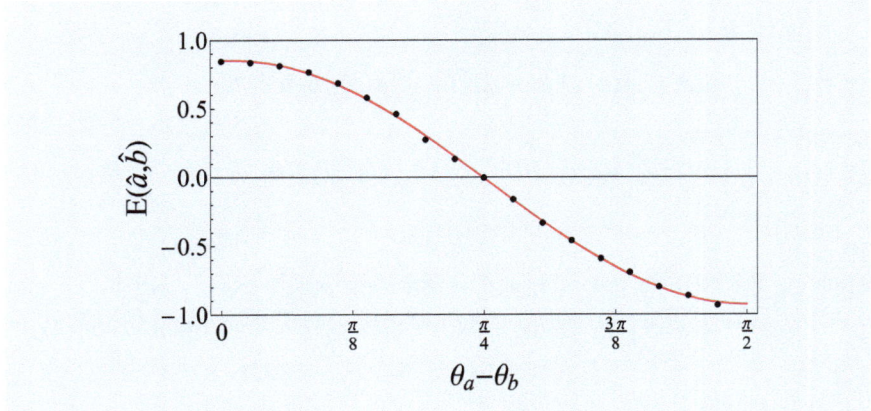

Fig. 16.12 Experimental observation of $E(\theta_A, \theta_B)$ from a typical Bell's inequality measurement. The *black dots* are experimental data and the *red sinusoidal curve* is a theoretical fitting

Although Bell derived his inequalities based on the measurement of spin-1/2 particle pairs, Eqs. (16.54) and (16.64) are not restricted to the measurement of spin-1/2 particle pairs. In fact, most of the historical experimental testing have been the polarization measurements of photon pairs. The photon pairs are prepared in similar states which have been called EPR–Bohm–Bell states, or Bell states in short. Most of the experimental observations violated Bell's inequalities which may have different forms and have their violation occur at different orientations of the polarization analyzers. However, the physics behind the violations is all similar to that of Bell's theorem.

■ Figure 16.11 is a schematic experimental setup for a Bell's inequality measurement. Since the space of Λ in this measurement is spanned into four regions with classical probabilities P_{ab}, P_{-ab}, P_{a-b}, P_{-a-b} in which A and B have values ± 1, the expectation value evaluation of Eq. (16.50) can be explicitly calculated as

$$E_{ab} = (+1)(+1)P_{ab} + (-1)(+1)P_{-ab} + (+1)(-1)P_{a-b}$$
$$+ (-1)(-1)P_{-a-b} = P_{ab} - P_{-ab} - P_{a-b} + P_{-a-b},$$
(16.67)

P_{ab}, P_{-ab}, P_{a-b} and P_{-a-b}, respectively, are measurable quantities by means of the joint photodetections of D_A^+&D_B^+, D_A^-&D_B^+, D_A^+&D_B^-, and D_A^-&D_B^-. A typical experimental observation of $E(\theta_A, \theta_B)$ from a Bell state is shown in ■ Fig. 16.12. Bell's inequality violation is expected from this measurement.

16.3 Scully's Quantum Eraser

Quantum eraser, proposed by Scully and Drühl in 1982 [9], is another thought experiment challenge the "basic mystery" of quantum mechanics: wave-particle duality. So far, several quantum eraser experiments have been demonstrated with interesting results supporting the ideas of Scully and Drühl [10, 11, 34, 35].

A double-slit type quantum eraser experiment, closing to the original Scully–Drühl thought experiment of 1982, is illustrated in ◘ Fig. 16.13. A pair of entangled photons, photon 1 and photon 2, is excited by a weak laser pulse either from atom A, which is located in slit A, or from atom B, which is located in slit B. Photon 1, propagates to the right, is registered by detector D_0, which can be scanned by a step motor along its x_0-axis for the examination of interference fringes. Photon 2, propagating to the left, is injected into a beamsplitter. If the pair is generated in atom A, photon 2 will follow the A path meeting BSA with 50 % chance of being reflected or transmitted. If the pair is generated in atom B, photon 2 will follow the B path meeting BSB with 50 % chance of being reflected or transmitted. In view of the 50 % chance of being transmitted by either BSA or BSB, photon 2 is detected by either detector D_3 or D_4. The registration of D_3 or D_4 provides which-path information (path A or path B) on photon 2 and in turn provides which-path information for photon 1 because of the entanglement nature of the two-photon state generated by atomic cascade decay. Given a reflection at either BSA or BSB photon 2 will continue to follow its A or B path to meet another 50–50 beamsplitter BS and then be detected by either detectors D_1 or D_2.

The experimental condition was arranged in such a way that no interference is observable in the single counting rate of D_0., i.e., the distance between A and B is large enough to be "distinguishable" for D_0 to learn which-path information of photon 1. However, the "clicks" at D_1 or D_2 will erase the which-path information of photon 1 and help to restore the interference. On the other hand, the "clicks" at D_3 or D_4 record which-path information. Thus, no observable interference is expected with the help of these "clicks." It is interesting to note that both the "erasure" and "recording" of the which-path information can be made as a "delayed choice": the experiment is designed in such a way that L_0, the optical distance between atoms A, B and detector D_0, is much shorter than L_A (L_B), which is the optical distance between atoms A, B and the beamsplitter BSA (BSB) where the "which-path" or "both-paths" "choice" is made randomly by photon 2. Thus, after the annihilation of photon 1 at D_0, photon 2 is still on its way to BSA (BSB), i.e., "which-path" or "both-path" choice is "delayed" compared to the detection of

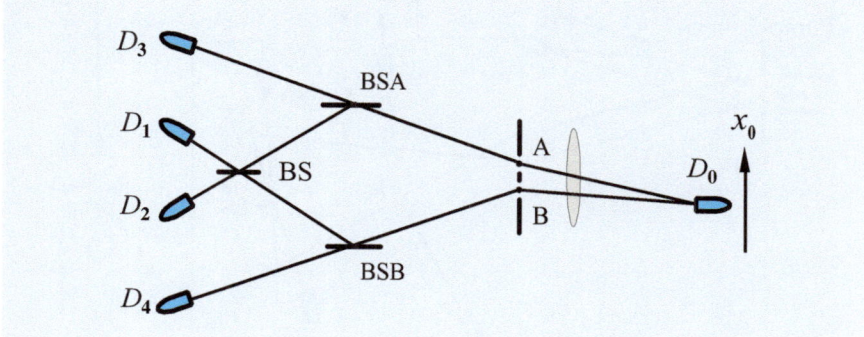

◘ **Fig. 16.13** Quantum erasure: a thought experiment of Scully–Drühl. A pair of entangled photons is emitted from either atom A or atom B by atomic cascade decay. The experimental condition guarantees no interference fringes is observable in the single detector counting rate of D_0. The "clicks" at D_1 or D_2 erase the which-path information, thus helping to restore the interference even after the "click" of D_0. On the other hand, the "clicks" at D_3 or D_4 record which-slit information. Thus, no observable interference is expected with the help of these "clicks"

photon 1. After the annihilation of photon 1, we look at these "delayed" detection events of D_1, D_2, D_3, and D_4 which have constant time delays, $\tau_i \simeq (L_i - L_0)/c$, relative to the triggering time of D_0. L_i is the optical distance between atoms A, B and detectors D_1, D_2, D_3, and D_4, respectively. It was predicted that the "joint-detection" counting rate R_{01} (joint-detection rate between D_0 and D_1) and R_{02} will show an interference pattern as a function of the position of D_0 on its x-axis. This reflects the wave nature (both-path) of photon 1. However, no interference fringes will be observable in the joint detection counting events R_{03} and R_{04} when scanning detector D_0 along its x-axis. This is as would be expected because we have now inferred the particle (which-path) property of photon 1. It is important to emphasize that all four joint detection rates R_{01}, R_{02}, R_{03}, and R_{04} are recorded at the same time during one scanning of D_0. That is, in the present experiment, we "see" both wave (interference) and which-path (particle-like) with the same measurement apparatus.

It should be mentioned that (1) the "choice" in this experiment is not actively switched by the experimentalist during the measurement. The "delayed choice" associated with either the wave or particle behavior of photon 1 is "randomly" made by photon 2. The experimentalist simply looks at which detector D_1, D_2, D_3 or D_4 is triggered by photon 2 to determine either wave or particle properties of photon 1 after the annihilation of photon 1; (2) the photo-detection event of photon 1 at D_0 and the delayed choice event of photon 2 at BSA (BSB) are space-like separated events. The "coincidence" time window is chosen to be much shorter than the distance between D_0 and BSA (BSB). Within the joint-detection time window, it is impossible to have the two events "communicating."

16.3.1 Random Delayed Choice Quantum Eraser One

Kim et al. realized the above random delayed choice quantum eraser in 2000 [10]. The schematic diagram of the experimental setup of Kim et al. is shown in ▫ Fig. 16.14. Instead of atomic cascade decay, SPDC is used to prepare the entangled two-photon state.

In the experiment, a 351. 1 nm Argon ion pump laser beam is divided by a double-slit and directed onto a type-II phase matching nonlinear crystal BBO at regions A and B. A pair of 702. 2 nm orthogonally polarized signal-idler photon is

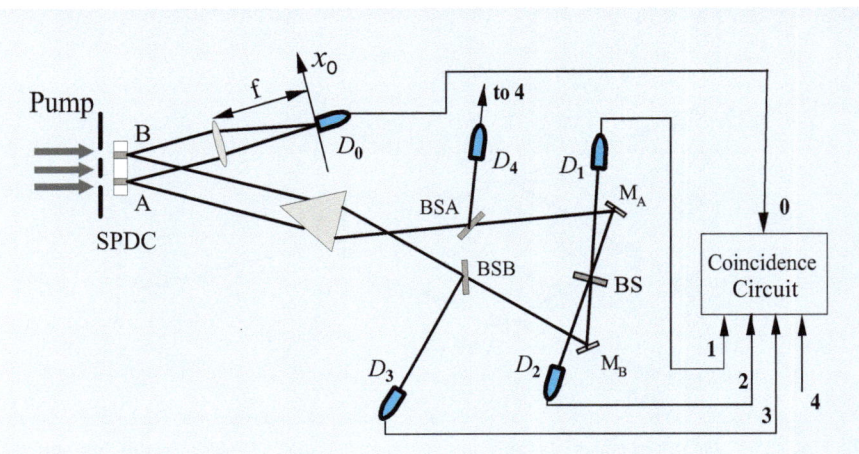

▫ **Fig. 16.14** Delayed choice quantum eraser: Schematic of an actual experimental setup of Kim et al. Pump laser beam is divided by a double-slit and makes two regions A and B inside the SPDC crystal. A pair of signal-idler photons is generated either from the A or B region. The "delayed choice" to observe either wave or particle behavior of the signal photon is made randomly by the idler photon about 7. 7 ns after the detection of the signal photon

generated either from region A or region B. The width of the region is about $a = 0.3$ mm and the distance between the center of A and B is about $d = 0.7$ mm. A Glen-Thompson prism is used to split the orthogonally polarized signal and idler. The signal photon (photon 1, coming either from A or B) propagates through lens LS to detector D_0, which is placed on the Fourier transform plane of the lens. The use of lens LS is to achieve the "far field" condition, but still keep a short distance between the slit and the detector D_0. Detector D_0 can be scanned along its x-axis by a step motor for the observation of interference fringes. The idler photon (photon 2) is sent to an interferometer with equal-path optical arms. The interferometer includes a prism PS, two 50–50 beamsplitters BSA, BSB, two reflecting mirrors M_A, M_B, and a 50–50 beamsplitter BS. Detectors D_1 and D_2 are placed at the two output ports of the BS, respectively, for erasing the which-path information. The triggering of detectors D_3 and D_4 provides which-path information for the idler (photon 2) and, in turn, which-path information for the signal (photon 1). The detectors are fast avalanche photodiodes with less than 1 ns rise time and about 100 ps jitter. A constant fractional discriminator is used with each of the detectors to register a single photon whenever the leading edge of the detector output pulse is above the threshold. Coincidences between D_0 and D_j ($j = 1, 2, 3, 4$) are recorded, yielding the joint detection counting rates R_{01}, R_{02}, R_{03}, and R_{04}.

In the experiment, the optical delay $(L_{A,\,B} - L_0)$ is chosen to be $\simeq 2.3\ m$, where L_0 is the optical distance between the output surface of BBO and detector D_0, and L_A (L_B) is the optical distance between the output surface of the BBO and the beamsplitter BSA (BSB). This means that any information (which-path or both-path) one can infer from photon 2 must be at least 7.7 ns later than the registration of photon 1. Compared to the 1 ns response time of the detectors, 2.3 m delay is thus enough for "delayed erasure." Although there is an arbitrariness about when a photon is detected, it is safe to say that the "choice" of photon 2 is delayed with respect to the detection of photon 1 at D_0 since the entangled photon pair is created simultaneously.

◘ Figure 16.15 reports the joint detection rates R_{01} and R_{02}, indicating the regaining of standard Young's double-slit interference pattern. An expected π

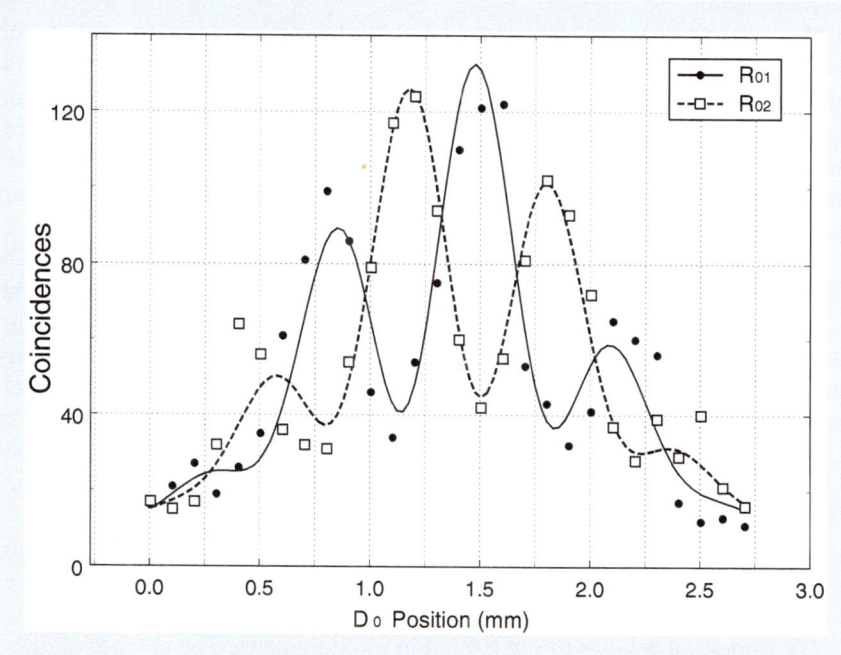

◘ **Fig. 16.15** Joint detection rates R_{01} and R_{02} against the x coordinates of detector D_0. Standard Young's double-slit interference patterns are observed. Note the π phase shift between R_{01} and R_{02}. The *solid line* and the *dashed line* are theoretical fits to the data

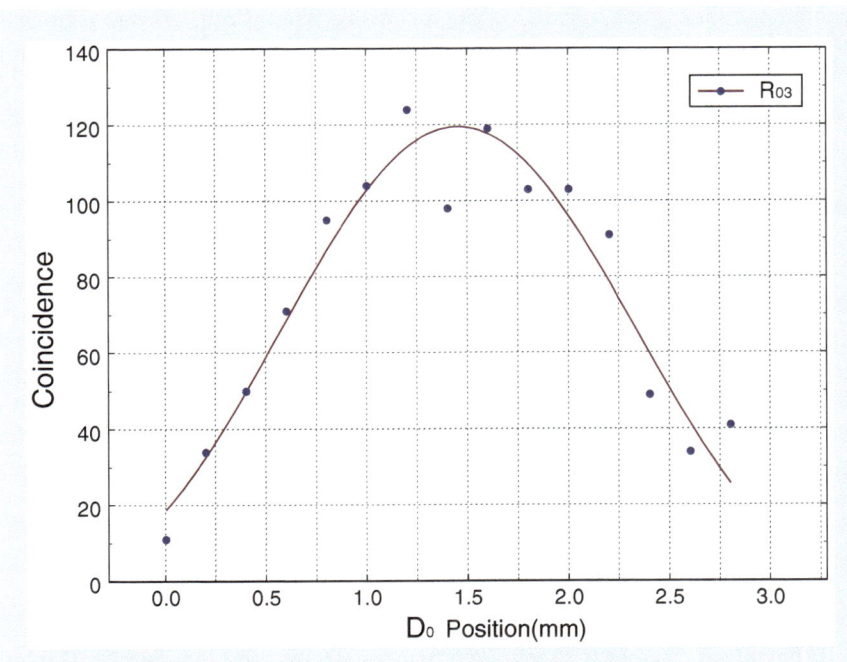

■ **Fig. 16.16** Joint detection counting rate of R_{03}. Absence of interference is clearly demonstrated. The *solid line* is a sinc-function fit

phase shift between the two interference patterns is clearly shown in the measurement. The single detector counting rates of D_0 and D_1 are recorded simultaneously. Although interference is observed in the joint detection counting rate, there is no significant modulation in any of the single detector counting rate during the scanning of D_0. R_0 is a constant during the scanning of D_0. The absence of interference in the single detector counting rate of D_0 is simply because the separation between slits A and slit B is much greater than the coherence length of the single field.

■ Figure 16.16 reports a typical R_{03} (R_{04}), joint detection counting rate between D_0 and "which-path detector" D_3 (D_4). An absence of interference is clearly demonstrated. The fitting curve of the experimental data indicates a sinc-function like envelope of the standard Young's double slit interference-diffraction pattern. Two features should bring to our attention that (1) there is no observable interference modulation as expected, and (2) the curve is different from the constant single detector counting rate of D_0.

The experimental result is surprising from a classical point of view. The result, however, is easily explained in the contents of quantum theory. In this experiment, there are two kinds of very different interference phenomena: single-photon interference and two-photon interference. As we have discussed earlier, single-photon interference is the result of the superposition between single-photon amplitudes, and two-photon interference is the results of the superposition between two-photon amplitudes. Quantum mechanically, single-photon amplitude and two-photon amplitude represent very different measurements and, thus, very different physics.

In this regard, we analyze the experiment by answering the following questions:

(1) Why is there no observable interference in the single-detector counting rate of D_0?

This question belongs to single-photon interferometry. The absence of interference in single-detector counting rate of D_0 is very simple: the separation between slit A and slit B is much greater than the coherence length of the signal filed.

(2) Why is there observable interference in the joint detection counting rate of D_{01} and D_{02}?

This question belongs to two-photon interferometry. Two-photon interference is very different from single-photon interference. Two-photon interference involves the addition of different yet indistinguishable two-photon amplitudes. The coincidence counting rate R_{01}, again, is proportional to the probability P_{01} of joint detecting the signal-idler pair by detectors D_0 and D_1,

$$R_{01} \propto P_{01} = \langle \Psi | E_0^{(-)} E_1^{(-)} E_1^{(+)} E_0^{(+)} | \Psi \rangle = \left| \langle 0 | \; E_1^{(+)} E_0^{(+)} | \Psi \rangle \right|^2. \tag{16.68}$$

To simplify the mathematics, we use the following "two-mode" expression for the state, bearing in mind that the transverse momentum δ-function will be taken into account.

$$|\Psi\rangle = \varepsilon \left[a_s^\dagger a_i^\dagger e^{i\varphi_A} + b_s^\dagger b_i^\dagger \quad e^{i\varphi_B} \right] |0\rangle$$

where ε is a normalization constant that is proportional to the pump field and the nonlinearity of the SPDC crystal, φ_A and φ_B are the phases of the pump field at A and B, and $a_j^\dagger$ $(b_j^\dagger)$, $j = s, \, i$, are the photon creation operators for the lower (upper) mode in Fig. 16.14.

In Eq. (16.68), the fields at the detectors D_0 and D_1 are given by

$$\begin{aligned} E_0^{(+)} &= a_s \; e^{ikr_{A0}} + b_s \; e^{ikr_{B0}} \\ E_1^{(+)} &= a_i \; e^{ikr_{A1}} + b_i \; e^{ikr_{B1}} \end{aligned} \tag{16.69}$$

where r_{Aj} (r_{Bj}), $j = 0, \, 1$ are the optical path lengths from region A (B) to the jth detector. Substituting the biphoton state and the field operators into Eq. (16.68),

$$\begin{aligned} R_{01} &\propto \left| \; e^{i(kr_A + \varphi_A)} + e^{i(kr_B + \varphi_B)} \right|^2 = |\Psi_A + \Psi_B|^2 \\ &= 1 + \cos\left[k(r_A - r_B) \right] \simeq \cos^2\left(x_0 \pi d / \lambda z_0 \right) \end{aligned} \tag{16.70}$$

where $r_A = r_{A0} + r_{A1}$, $r_B = r_{B0} + r_{B1}$; Ψ_A and Ψ_B are the two-photon effective wave functions of path A and path B, representing the two different yet indistinguishable probability amplitudes to produce a joint photodetection event of D_0 and D_1, indicating a two-photon interference. In Dirac's language: a signal-idler photon pair interferes with the pair itself.

To calculate the diffraction effect of a single-slit, again, we need an integral of the effective two-photon wavefunction over the slit width (the superposition of infinite number of probability amplitudes results in a click-click joint detection event):

$$R_{01} \propto \left| \int_{-a/2}^{a/2} dx_{AB} \; e^{-ik \; r(x_0, \; x_{AB})} \right|^2 \cong \mathrm{sinc}^2(x_0 \pi a / \lambda z_0) \tag{16.71}$$

where $r(x_0, \, x_{AB})$ is the distance between points x_0 and x_{AB}, x_{AB} belongs to the slit's plane, and the far-field condition is applied.

Repeating the above calculations, the combined interference-diffraction joint detection counting rate for the double-slit case is given by

$$R_{01} \propto \text{sinc}^2(x_0\pi a/\lambda z_0)\cos^2(x_0\pi d/\lambda z_0).$$ (16.72)

If the finite size of the detectors is taken into account, the interference visibility will be reduced.

(3) Why is there no observable interference in the joint detection counting rate of R_{03} and R_{04}?

This question belongs to two-photon interferometry. From the view of two-photon physics, the absence of interference in the joint detection counting rate of R_{03} and R_{04} is obvious: only one two-photon amplitude contributes to the joint detection events.

16.3.2 Random Delayed Choice Quantum Eraser Two

Now we ask, again, what would happen if we replace the entangled photons with a randomly paired photons, or wavepackets, in thermal state? Can a randomly paired photons in thermal state erase the which-path information? The answer is, again, positive. A random delayed choice quantum eraser using randomly paired photons in thermal state has been demonstrated by Peng et al. recently [11].

The experiment setup of Peng et al. is schematically illustrated in ◘ Fig. 16.17. The experimental setup is almost the same as that of the experiment of Kim et al. of 2000, except the photon source and the coincidence measurement: the randomly paired photons, or wavepackets are created from a standard pseudo-thermal source and the purse this quantum eraser measures the photon-number fluctuation-correlation of thermal light. Thermal light has a peculiar "spatial coherence" property: the fluctuation of the measured photon-numbers, or intensities, correlated within its spatial coherence area only. When the measurements are beyond its coherence area, the photon-numbers fluctuation correlation vanish:

$$\langle \Delta n_A \Delta n_B \rangle \propto \left| G^{(1)}(\vec{\rho}_A, \vec{\rho}_B) \right|^2,$$ (16.73)

where $\Delta n_j, j = A, B$, is the photon-number fluctuation at $(\vec{\rho}_j, t_j)$ of the double-slit plane, $G^{(1)}(\vec{\rho}_A, \vec{\rho}_B)$ is the first-order spatial coherence function of the thermal field. The spatial coherence of thermal light guarantees the photon-numbers fluctuate correlatively only when $|\vec{\rho}_A - \vec{\rho}_B| < l_c$, where l_c is the spatial coherence length. In this experiment, we choose $|\vec{\rho}_A - \vec{\rho}_B| \gg l_c$. Under this condition, we have achieved $\langle \Delta n_A \Delta n_{A'} \rangle \neq 0$, $\langle \Delta n_B \Delta n_{B'} \rangle \neq 0$ but $\langle \Delta n_A \Delta n_B \rangle = 0$. Note, again, here $(\vec{\rho}_j)$ is on the double-slit plane, see ◘ Fig. 16.17. This peculiar property of thermal light together with the photon-number fluctuation-correlation measurement between D_0 and D_3 (or D_4) provides the which-path information. It is interesting, the which-slit information is erasable in the fluctuation-correlation measurement between D_0 and D_1 (or D_2).

The experimental setup in ◘ Fig. 16.17 can be divided into four parts: a thermal light source, a Young's double-slit interferometer, a Mach–Zehnder-like interferometer, and a photon-number fluctuation-correlation measurement circuit. (1) The light source is a standard pseudo-thermal source [29] which consists of a He-Ne laser beam ($\sim$ 2 mm diameter) and a rotating ground glass (GG). Within the $\sim$ 2 mm diameter spot, the ground glass contains millions of tiny diffusers. A large number of randomly distributed sub-fields, or wavepacket, are scattered from millions of randomly distributed tiny diffusers with random phases. The pseudothermal field then passes a double-slit which is about 25 cm away from

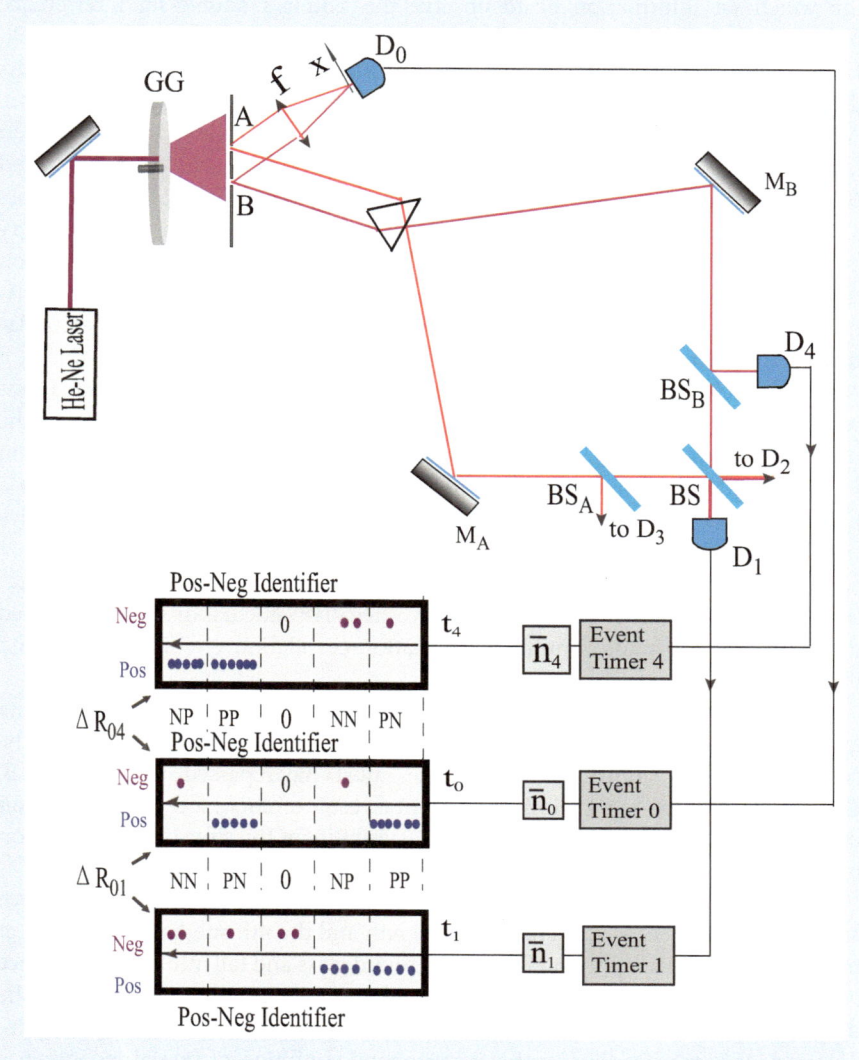

● **Fig. 16.17** Schematic of a random delayed choice quantum eraser. The He-Ne laser beam spot on the rotating ground glass has a diameter of ∼ 2 mm. A double-slit, with slit-width 150 μm and slit-separation 0. 7 mm, is placed ∼ 25 cm away from the GG. The spatial coherence length of the pseudo-thermal field on the double-slit plane is calculated from Eq. (16.73), $l_c = \lambda/\Delta\theta \sim 160\ \mu$, which guarantees the two fields E_A and E_B are spatially incoherent. Under this experimental condition, the photon number fluctuates correlatively only within slit A or slit B. All beamsplitters are non-polarizing and 50/50. The two fields from the two slits may propagate to detector D_0 which is transversely scanned on the focal plan of lens f for observing the interference pattern of the double-slit interferometer; and may also pass a long a Mach–Zehnder-like interferometer and finally reach at D_1 or D_4 (D_2 or D_3). A positive-negative fluctuation-correlation protocol is followed to evaluate the photon-number fluctuation-correlations from the coincidences between D_0-D_1 and D_0-D_4 (or D_0-D_2 and D_0-D_3)

the GG. The spatial coherence length of the pseudo-thermal field on the double slit plane is calculated from Eq. (16.73), $l_c = \lambda/\Delta\theta \sim 160\ \mu$, which guarantees the spatial incoherence of the two fields E_A and E_B that passing through slit A and slit B, respectively. Under this experimental condition, the photon-numbers fluctuate correlatively only within slit-A or slit-B. We therefore learn the which-slit information in a photon number-fluctuation correlation measurement. (2) The double-slit has a slit-width 150 μm, and a slit-separation 0. 7 mm (distance between the center of two slits). A lens, f, is placed following the double-slit. On the focal-plane of the lens a scannable point-like photodetector D_0 is used to learn

16

the which-slit information or to observe the Young's double-slit interference pattern. (3) The Mach–Zehnder-like interferometer and the photodetectors D_1, D_2 are used to "erase" the which-slit information. Simultaneously, the joint-detection between D_0 and D_3 or D_4 is used to "learn" the which-slit information. All five photodetectors are photon-counting detectors working at single-photon level. The Mach–Zehnder-like interferometer has three beamsplitters, BS, BS_A, and BS_B, all of them are 50/50 non-polarizing beamsplitters. Moreover, the detectors are fast avalanche photodiodes with rise time less than 1 ns, and the path delay between BS_A or BS_B, and D_0 is $\approx 1.\,5$ m which ensure that, at each joint-detection measurement, when a photon chooses to be reflected (read which-way) or transmitted (erase which-way) at BS_A or BS_B, it is already 5 ns later than the annihilation of its partner at D_0. Comparing the 1 ns rise time, we are sure this is a "delayed choice" made by that photon. (4) The photon-number fluctuation-correlation circuit consists of five synchronized "event-timers" which record the registration times of D_0, D_1, D_2, D_3, and D_4. A positive-negative fluctuation identifier follows each event-timer to distinguish "positive-fluctuation" Δn^+, from "negative-fluctuation" Δn^-, for each photodetector within each coincidence time window. The photon-number fluctuation-correlations of D_0-D_1: $\Delta R_{01} = \langle \Delta n_0 \Delta n_1 \rangle$ and D_0-D_4: $\Delta R_{04} = \langle \Delta n_0 \Delta n_4 \rangle$ are calculated, accordingly and respectively, based on their measured positive-negative fluctuations. The detailed description of the photon-number fluctuation-correlation circuit can be found in [30].

The experimental observation of ΔR_{04} is reported in ◘ Fig. 16.18. The data excludes any possible existing interferences. This measurement means the coincidences that contributed to ΔR_{04} must have passed through slit B. ◘ Figure 16.19 reports a typical experimental result of ΔR_{01}: a typical double-slit interference-diffraction pattern. The 100 % visibility of the sinusoidal modulation indicates complete erasure of the which-slit information.

Assuming a random pair of sub-fields at single-photon level, such as the mth and nth wavepackets, is scattered from the mth and the nth sub-sources located at transverse coordinates $\vec{\rho}_{0m}$ and $\vec{\rho}_{0n}$ of the ground glass and fall into the coincidence time windows of D_0-D_1 and D_0-D_4, the mth wavepacket may propagate to the double-slit interferometer and the nth wavepacket may pass through the Mach–Zehnder, or vice versa. Under the experimental condition of spatial incoherence between E_A and E_B, the which-slit information is learned from the photon-number

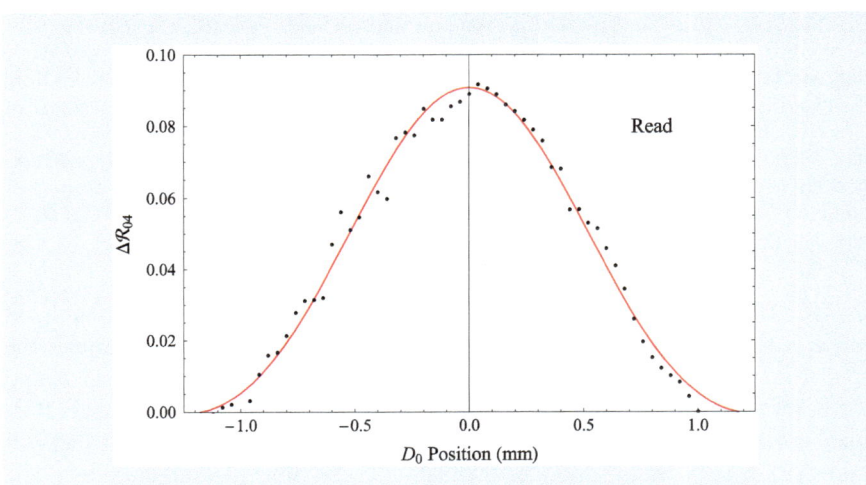

◘ Fig. 16.18 The measured ΔR_{04} by scanning D_0 on the observation plane of the Young's double-slit interferometer. The *black dots* are experimental data, the *red line* is the theoretical fitting with Eq. (16.81)

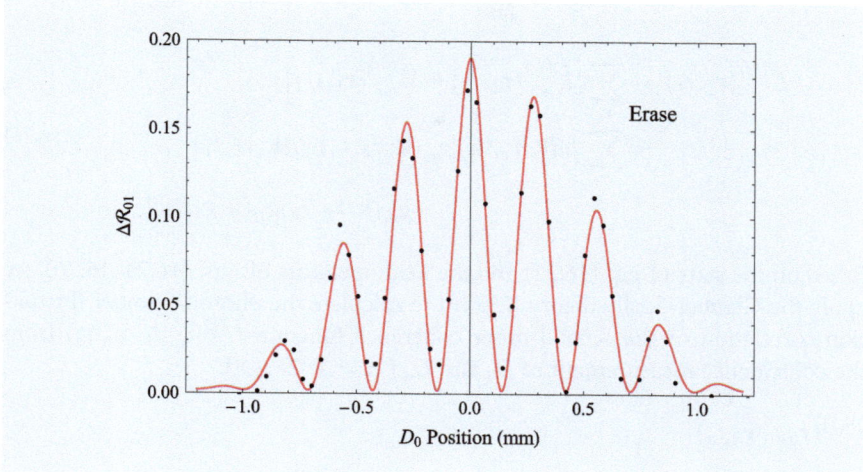

Fig. 16.19 The measured ΔR_{01} as a function of the transverse coordinate of D_0. The *black dots* are experimental data, the *red line* is the theoretical fitting with Eq. (16.82)

fluctuation-correlation measurements $\Delta R_{04} = \langle \Delta n_0 \Delta n_4 \rangle = \langle \Delta n_{B0} \Delta n_{B4} \rangle$ of D_0-D_4, and no interference is observable by scanning D_0. It is interesting that the which-slit information are erasable in the photon-number fluctuation-correlation measurements of $\Delta R_{01} = \langle \Delta n_0 \Delta n_1 \rangle$ of D_0-D_1, resulting in a reappeared interference pattern as a function of the scanning coordinate of D_0.

The field operator at detector D_0 can be written in the following form in terms of the subfields:

$$
\begin{aligned}
\hat{E}^{(+)}(\mathbf{r}_0, t_0) &= \hat{E}_A^{(+)}(\mathbf{r}_0, t_0) + \hat{E}_B^{(+)}(\mathbf{r}_0, t_0) \\
&= \sum_m \left[\hat{E}_{mA}^{(+)}(\mathbf{r}_0, t_0) + \hat{E}_{mB}^{(+)}(\mathbf{r}_0, t_0) \right] \\
&= \sum_m \int d\mathbf{k}\, \hat{a}_m(\mathbf{k}) [g_m(\mathbf{k}; \mathbf{r}_A, t_A) g_A(\mathbf{k}; \mathbf{r}_0, t_0) \\
&\qquad\qquad + g_m(\mathbf{k}; \mathbf{r}_B, t_B) g_B(\mathbf{k}; \mathbf{r}_0, t_0)].
\end{aligned}
\tag{16.74}
$$

where $g_m(\mathbf{k}; \mathbf{r}_s, t_s)$ is a Green's function which propagates the mth subfield from the mth sub-source to the sth slit ($s = A, B$). $g_s(\mathbf{k}; \mathbf{r}_0, t_0)$ is another Green's function that propagates the field from the sth slit to detector D_0. It is easy to notice that, although there are two ways a photon can be detected at D_0, due to the first order incoherence of E_A and E_B, there should be no interference at the detection plane.

D_4 (D_3) in the experiment can only receive photons from slit B (slit A), so the field operator is then:

$$
\begin{aligned}
\hat{E}^{(+)}(\mathbf{r}_4, t_4) &= \sum_m \hat{E}_{mB}^{(+)}(\mathbf{r}_4, t_4) \\
&= \sum_m \int d\mathbf{k}\, \hat{a}_m(\mathbf{k}) g_m(\mathbf{k}; \mathbf{r}_B, t_B) g_B(\mathbf{k}; \mathbf{r}_4, t_4).
\end{aligned}
\tag{16.75}
$$

The detector D_1 (D_3), however, can receive photons from both slit A and slit B through the Mach–Zehnder-like interferometer, so the field operator has two terms:

$$\hat{E}^{(+)}(\mathbf{r}_1, t_1) = \sum_m \left[\hat{E}_{mA}^{(+)}(\mathbf{r}_1, t_1) + \hat{E}_{mB}^{(+)}(\mathbf{r}_1, t_1)\right]$$

$$= \sum_m \int d\mathbf{k}\, \hat{a}_m(\mathbf{k})[g_m(\mathbf{k}; \mathbf{r}_A, t_A)g_A(\mathbf{k}; \mathbf{r}_1, t_1) \tag{16.76}$$

$$+ g_m(\mathbf{k}; \mathbf{r}_B, t_B)g_B(\mathbf{k}; \mathbf{r}_1, t_1)].$$

Based on the state of Eq. (16.32) and the field operators of Eqs. (16.74–16.76), we apply the Glauber-Scully theory [25, 36] to calculate the photon-number fluctuation-correlation or the second-order coherence function $G^{(2)}(\mathbf{r}_0, t_0; \mathbf{r}_\alpha, t_\alpha)$ from the coincidence measurement of D_0 and D_α,($\alpha = 1, 2, 3, 4$):

$$G^{(2)}(\mathbf{r}_0, t_0; \mathbf{r}_\alpha, t_\alpha)$$

$$= \langle\langle\Psi|E^{(-)}(\mathbf{r}_0, t_0)E^{(-)}(\mathbf{r}_\alpha, t_\alpha)E^{(+)}(\mathbf{r}_\alpha, t_\alpha)E^{(+)}(\mathbf{r}_0, t_0)|\Psi\rangle\rangle_{Es}$$

$$= \left\langle\left\langle \Psi\left|\sum_m E_m^{(-)}(\mathbf{r}_0, t_0)\sum_n E_n^{(-)}(\mathbf{r}_\alpha, t_\alpha)\sum_q E_q^{(+)}(\mathbf{r}_\alpha, t_\alpha)\sum_p E_p^{(+)}(\mathbf{r}_0, t_0)\right|\Psi\right\rangle\right\rangle_{Es}$$

$$\tag{16.77}$$

$$= \sum_m \psi_m^*(\mathbf{r}_0, t_0)\psi_m(\mathbf{r}_0, t_0)\sum_n \psi_n^*(\mathbf{r}_\alpha, t_\alpha)\psi_n(\mathbf{r}_\alpha, t_\alpha)$$

$$+ \sum_{m,n} \psi_m^*(\mathbf{r}_0, t_0)\psi_n(\mathbf{r}_0, t_0)\psi_n^*(\mathbf{r}_\alpha, t_\alpha)\psi_m(\mathbf{r}_\alpha, t_\alpha)$$

$$= \langle n_0\rangle\langle n_\alpha\rangle + \langle\Delta n_0 \Delta n_\alpha\rangle.$$

Here $\psi_m(\mathbf{r}_\alpha, t_\alpha)$ is the effective wavefunction of the mth subfield at $(\mathbf{r}_\alpha, t_\alpha)$. In the case of $\alpha = 1, 2$

$$\psi_m(\mathbf{r}_\alpha, t_\alpha) = \psi_{mA\alpha} + \psi_{mB\alpha}$$

$$= \int d\mathbf{k}\alpha_m(\mathbf{k})[g_m(\mathbf{k}; \mathbf{r}_A, t_A)g_A(\mathbf{k}; \mathbf{r}_\alpha, t_\alpha) + g_m(\mathbf{k}; \mathbf{r}_B, t_B)g_B(\mathbf{k}; \mathbf{r}_\alpha, t_\alpha)]. \tag{16.78}$$

This shows that the measured effective wavefunction $\psi_m(\mathbf{r}_\alpha, t_\alpha)$ is the result of a superposition between two alternative amplitudes in terms of path-A and path-B, $\psi_{m\,\alpha} = \psi_{mA\,\alpha} + \psi_{mB\,\alpha}$. When $\alpha = 4$ (or $\alpha = 3$), the effective wavefunction has only one amplitude

$$\psi_m(\mathbf{r}_4, t_4) = \psi_{mB4} = \int d\mathbf{k}\alpha_m(\mathbf{k})g_m(\mathbf{k}; \mathbf{r}_B, t_B)g_B(\mathbf{k}; \mathbf{r}_4, t_4). \tag{16.79}$$

From Eq. (16.77) and the measurement circuit in ◻ Fig. 16.17, it is easy to find that what we measure in this experiment is the photon-number fluctuation-correlation:

$$\langle\Delta n_0 \Delta n_\alpha\rangle = \sum_{m,n} \psi_m^*(\mathbf{r}_0, t_0)\psi_n(\mathbf{r}_0, t_0)\psi_n^*(\mathbf{r}_\alpha, t_\alpha)\psi_m(\mathbf{r}_\alpha, t_\alpha). \tag{16.80}$$

We thus obtain

$$\Delta R_{04} \propto \langle\Delta n_0 \Delta n_4\rangle = \sum_{n\neq m} \psi_{mB0}^* \psi_{nB0}\psi_{nB4}^*\psi_{mB4} \propto \mathrm{sinc}^2(x\pi a/\lambda f), \tag{16.81}$$

indicating a diffraction pattern which agrees with the experimental observation of ◻ Fig. 16.18.

In the case of $\alpha = 1$, 2, we obtain

$$
\begin{aligned}
\Delta R_{01} \quad &\propto \quad \langle \Delta n_0 \Delta n_1 \rangle \\
&\propto \quad \sum_{n \neq m} \left[\psi_{mA0}^* \psi_{nA0} \psi_{nA1}^* \psi_{mA1} + \psi_{mB0}^* \psi_{nB0} \psi_{nB1}^* \psi_{mB1} \right. \\
&\qquad \left. + \psi_{mA0}^* \psi_{nB0} \psi_{nB1}^* \psi_{mA1} + \psi_{mB0}^* \psi_{nA0} \psi_{nA1}^* \psi_{mB1} \right] \\
&\propto \quad \mathrm{sinc}^2 (x\pi a / \lambda f) \cos^2 (x\pi d / \lambda f),
\end{aligned}
\tag{16.82}
$$

which agrees with the experimental observation in ◘ Fig. 16.19.

16.4 Popper's Experiment

Popper's original thought experiment is schematically shown in ◘ Fig. 16.20 [12]. A point source S, positronium as Popper suggested, is placed at the center of the experimental arrangement from which entangled pair of particle 1 and particle 2 are emitted in opposite directions along the respective positive and negative x-axes towards two screens A and B. There are slits on both screens parallel to the y-axis and the slits may be adjusted by varying their widths Δy. Beyond the slits on each side stand an array of Geiger counters for the joint measurement of the particle pairs as shown in the figure. The entangled pair could be emitted to any direction in 4π solid angles from the point source. However, if particle 1 is detected in a certain direction, particle 2 is then known to be in the opposite direction due to the momentum conservation of the quanta pair.

First, let us imagine the case in which slits A and B are both adjusted very narrowly. In this circumstance, particle 1 and particle 2 experience diffraction at slit A and slit B, respectively, and exhibit greater Δp_y for smaller Δy of the slits.

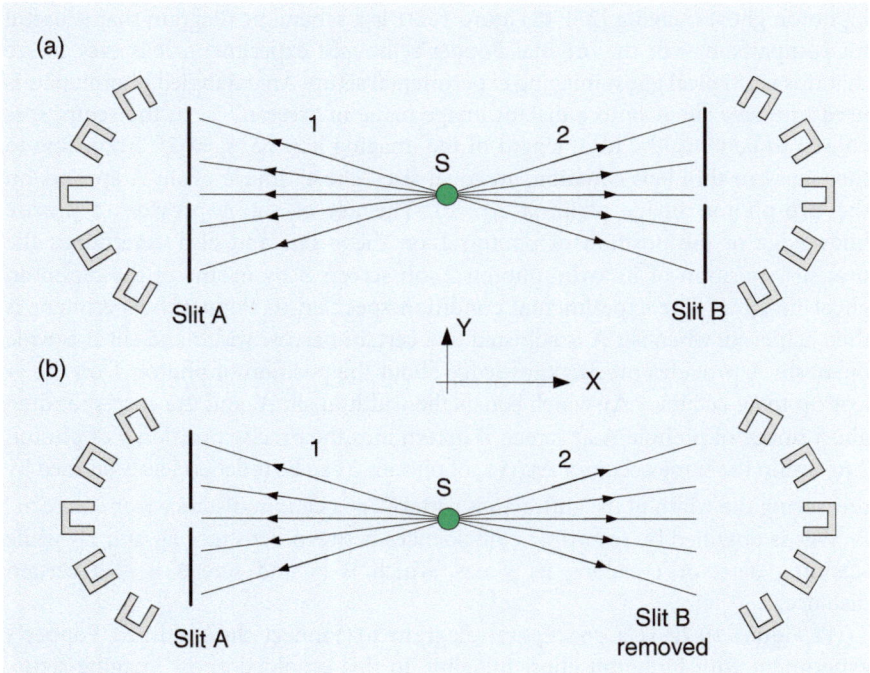

◘ **Fig. 16.20** Popper's thought experiment. An entangled pair of particles are emitted from a point source with momentum conservation. A narrow slit on screen A is placed in the path of particle 1 to provide the precise knowledge of its position on the y-axis and this also determines the precise y-position of its twin, particle 2, on screen B. (**a**) Slits A and B are both adjusted very narrowly. (**b**) Slit A is kept very narrow and slit B is left wide open

There seems to be no disagreement in this situation between Copenhagen and Popper.

Next, suppose we keep slit A very narrow and leave slit B wide open. The main purpose of the narrow slit A is to provide the precise knowledge of the position y of particle 1 and this subsequently determines the precise position of its twin (particle 2) on side B through quantum entanglement. Now, Popper asks, in the absence of the physical interaction with an actual slit, does particle 2 experience a greater uncertainty in Δp_y due to the precise knowledge of its position? Based on his beliefs, Popper provides a straightforward prediction: *particle 2 must not experience a greater Δp_y unless a real physical narrow slit B is applied*. However, if Popper's conjecture is correct, this would imply the product of Δy and Δp_y of particle 2 could be smaller than h ($\Delta y \, \Delta p_y < h$). This may pose a serious difficulty for Copenhagen and perhaps for many of us. On the other hand, if particle 2 going to the right does scatter like its twin, which has passed though slit A, while slit B is wide open, we are then confronted with an apparent *action-at-a-distance*!

The use of a *point source* in Popper's proposal has been criticized historically as the fundamental error Popper made. It is true that a point source can never produce a pair of entangled particles which preserves EPR correlation in momentum as Popper expected. However, notice that a *point source* is *not* a necessary requirement for Popper's experiment. What is required is a precise position-position EPR correlation: if the position of particle 1 is precisely known, the position of particle 2 is 100 % determined. Ghost imaging is a perfect tool to achieve this.

16.4.1 Popper's Experiment One

In 1999, Popper's experiment was realized by Y.H. Kim et al. [13] with the help of biphoton ghost imaging [37]. ■ Figure 16.21 is a schematic diagram that is useful for comparison with the original Popper's thought experiment. It is easy to see that this is a typical ghost imaging experimental setup. An entangled photon pair is used to image slit A onto a distant image plane of "screen" B. In the setup, s_o is chosen to be twice the focal length of the imaging lens LS, $s_o = 2f$. According to the Gaussian thin lens equation, an equal size "ghost" image of slit A appears on the two-photon image plane at $s_i = 2f$. The use of slit A provides a precise knowledge of the position of photon 1 on the y-axis and also determines the precise y-position of its twin, photon 2, on screen B by means of the biphoton ghost imaging. The experimental condition specified in Popper's experiment is then achieved: when slit A is adjusted to a certain narrow width and slit B is wide open, slit A provides precise knowledge about the position of photon 1 on the y-axis up to an accuracy Δy which equals the width of slit A, and the corresponding ghost image of pinhole A at screen B determines the precise position y of photon 2 to within the same accuracy Δy. Δp_y of photon 2 can be independently studied by measuring the width of its "diffraction pattern" at a certain distance from "screen" B. This is obtained by recording coincidences between detectors D_1 and D_2 while scanning detector D_2 along its y-axis, which is behind screen B at a certain distance.

■ Figure 16.22 is a conceptual diagram to connect the modified Popper's experiment with biphoton ghost imaging. In this unfolded ghost imaging setup, we assume the entangled signal-idler photon pair holds a perfect EPR correlation in momentum with $\delta(\mathbf{k}_s + \mathbf{k}_i) \sim 0$, which can be easily realized in a large transverse sized SPDC. In this experiment, we have chosen $s_o = s_i = 2f$. Thus, an equal size ghost image of slit A is expected to appear on the image plane of screen B.

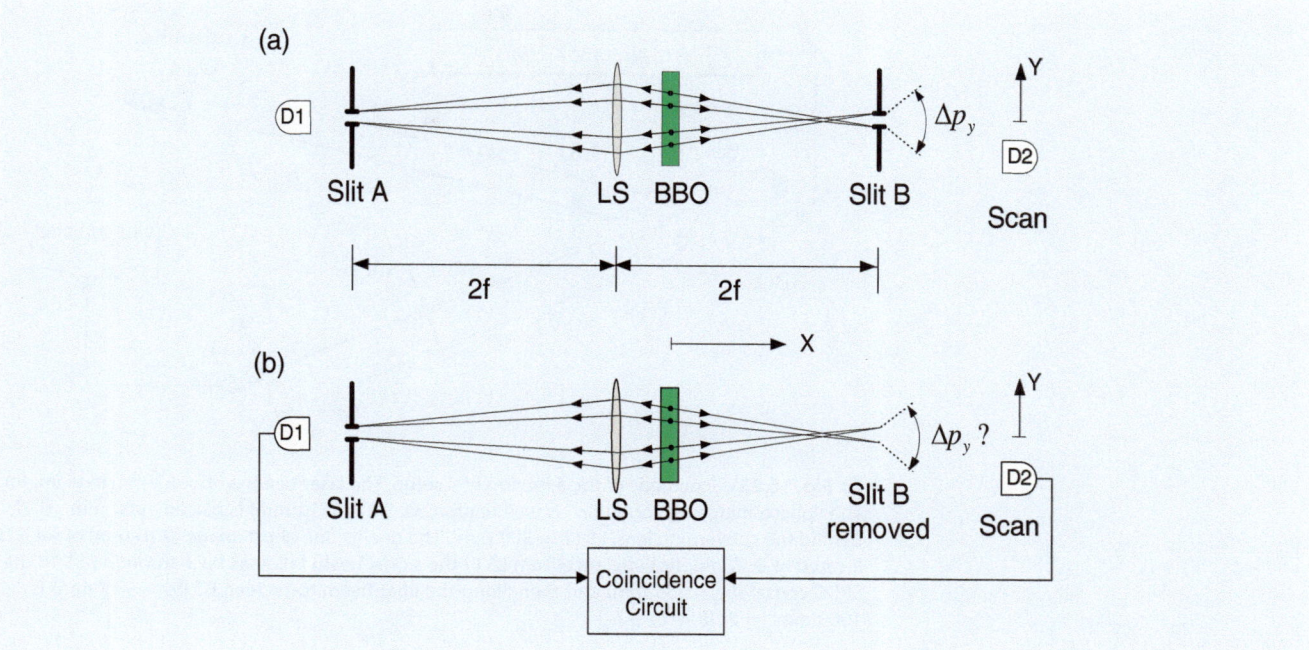

Fig. 16.21 Modified version of Popper's experiment. An entangled photon pair is generated by SPDC. A lens and a narrow slit A are placed in the path of photon 1 to provide the precise knowledge of its position on the y-axis and also to determine the precise y-position of its twin, photon 2, on screen B by means of biphoton ghost imaging. Photon counting detectors D_1 and D_2 are used to scan in y-directions for joint detections. (**a**) Slits A and B are both adjusted very narrowly. (**b**) Slit A is kept very narrow and slit B is left wide open

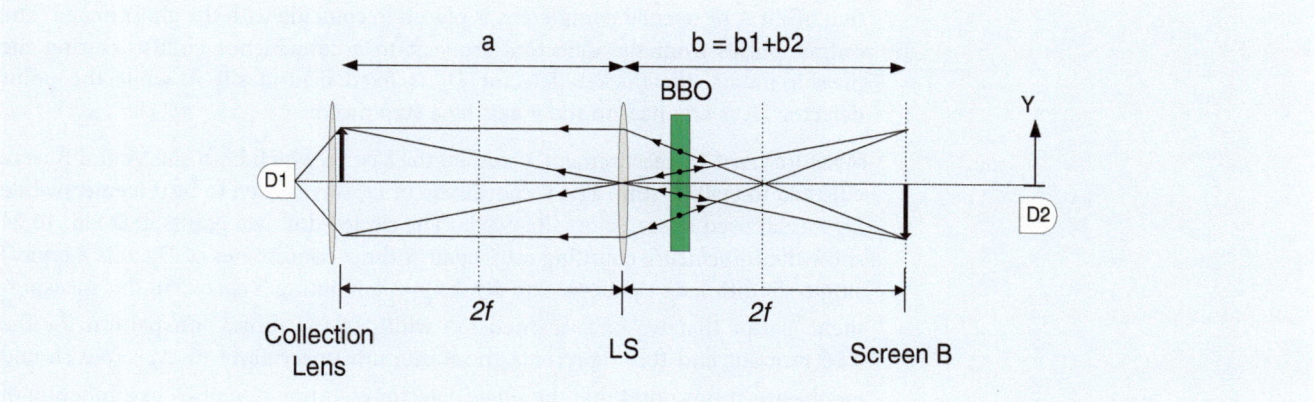

Fig. 16.22 An unfolded schematic of ghost imaging. We assume the entangled signal-idler photon pair holds a perfect momentum correlation $\delta(\mathbf{k}_s + \mathbf{k}_i) \sim 0$. The locations of the slit A, the imaging lens LS, and the ghost image must be governed by the Gaussian thin lens equation. In this experiment, we have chosen $s_o = s_i = 2f$. Thus, the ghost image of slit A is expected to be the same size as that of slit A

The detailed experimental setup is shown in ◘ Fig. 16.23 with indications of the various distances. A CW Argon ion laser line of $\lambda_p = 351.1$ nm is used to pump a 3 mm long beta barium borate (BBO) crystal for type-II SPDC to generate an orthogonally polarized signal-idler photon pair. The laser beam is about 3 mm in diameter with a diffraction limited divergence. It is important not to focus the pump beam so that the phase-matching condition, $\mathbf{k}_s + \mathbf{k}_i = \mathbf{k}_p$, is well reinforced in the SPDC process, where $\mathbf{k}_j$ ($j = s, i, p$) is the wavevectors of the signal (s), idler (i), and pump (p) respectively. The collinear signal-idler beams, with $\lambda_s = \lambda_i = 702$ .2 $nm = 2\lambda_p$ are separated from the pump beam by a fused quartz dispersion prism, and then split by a polarization beam splitter PBS. The signal beam

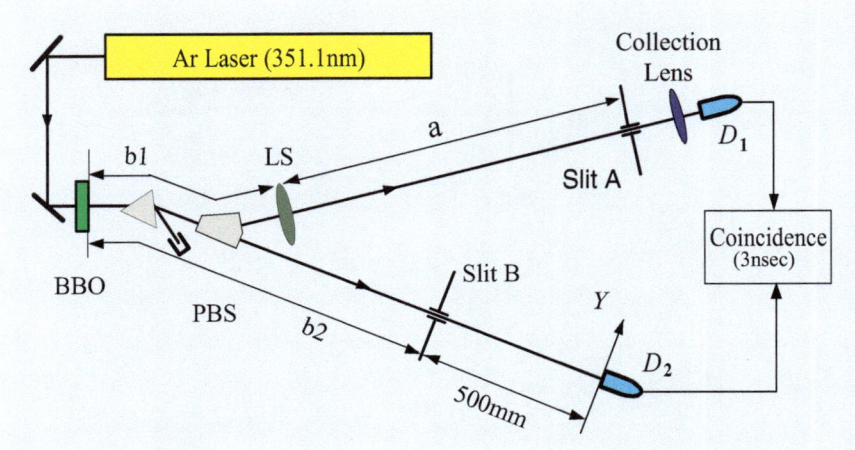

🔲 **Fig. 16.23** Schematic of the experimental setup. The laser beam is about 3 mm in diameter. The "phase-matching condition" is well reinforced. Slit A (0.16 mm) is placed 1000 mm = 2f behind the converging lens, LS (f = 500 mm). The one-to-one ghost image (0.16 mm) of slit A is located at B. The optical distance from LS in the signal beam taken as back through PBS to the SPDC crystal (b_1 = 255 mm) and then along the idler beam to "screen B" (b_2 = 745 mm) is 1000 mm = 2f (b = b_1 + b_2)

(photon 1) passes through the converging lens LS with a 500 mm focal length and a 25 mm diameter. A 0.16 mm slit is placed at location A which is 1000 mm (s_o = 2f) behind the lens LS. A short focal length lens is used with D_1 for collecting all the signal beam that passes through slit A. The point-like photon counting detector D_2 is located 500 mm behind "screen B." "Screen B" is the image plane defined by the Gaussian thin equation. Slit B, either adjusted as the same size as that of slit A or opened completely, is placed to coincide with the ghost image. The output pulses from the detectors are sent to a coincidence circuit. During the measurements, the bucket detector D_1 is fixed behind slit A while the point detector D_2 is scanned on the y-axis by a step motor.

Measurement 1 Measurement 1 studied the case in which both slits A and B were adjusted to be 0.16 mm. The y-coordinate of D_1 was chosen to be 0 (center) while D_2 was allowed to scan along its y-axis. The circled dot data points in 🔲 Fig. 16.24 show the *coincidence* counting rates against the y-coordinates of D_2. It is a typical single-slit diffraction pattern with $\Delta y \, \Delta p_y = h$. Nothing is special in this measurement except that we have learned the width of the diffraction pattern for the 0.16 mm slit and this represents the minimum uncertainty of Δp_y. We should emphasize at this point that the *single* detector counting rate of D_2 as a function of its position y is basically the same as that of the coincidence counts except for a higher counting rate.

Measurement 2 The same experimental conditions were maintained except that slit B was left wide open. This measurement is a test of Popper's prediction. The y-coordinate of D_1 was chosen to be 0 (center) while D_2 was allowed to scan along its y-axis. Due to the entangled nature of the signal-idler photon pair and the use of coincidence measurement circuit, only those twins which have passed through slit A and the "ghost image" of slit A at screen B with an uncertainty of $\Delta y = 0.16 \, mm$ (which is the same width as the real slit B we have used in measurement 1) would contribute to the coincidence counts through the joint detection of D_1 and D_2. The diamond dot data points in 🔲 Fig. 16.24 report the measured coincidence counting rates against the y coordinates of D_2. The measured width of the pattern is narrower than that of the diffraction pattern shown in measurement 1. It is also interesting to notice that the single detector counting rate of D_2 keeps constant in

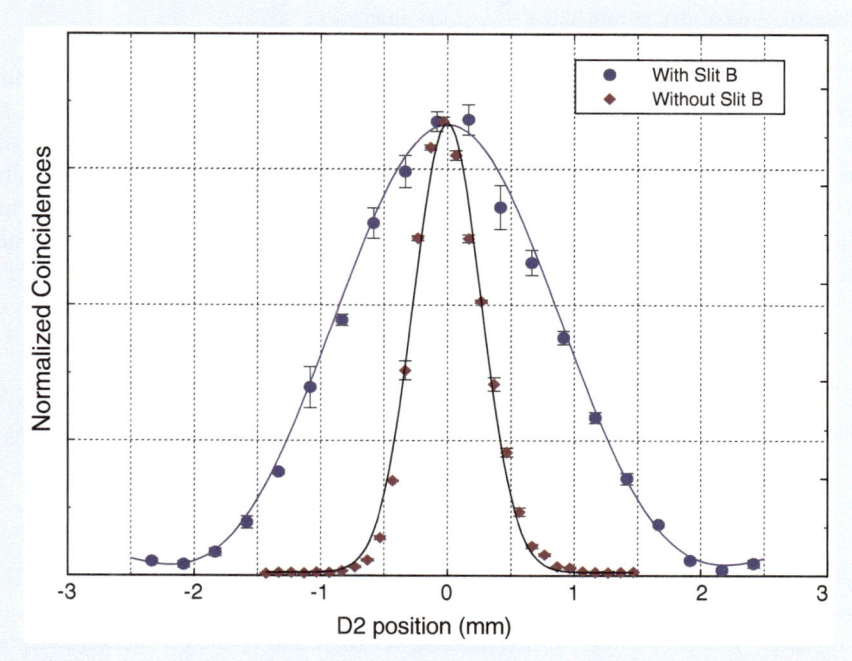

□ **Fig. 16.24** The observed coincidence patterns. The y-coordinate of D_1 was chosen to be 0 (center) while D_2 was allowed to scan along its y-axis. *Circled dot points: Slit A = Slit B = 0.16 mm. Diamond dot points: Slit A = 0.16 mm , Slit B wide open.* The width of the sinc-function curve fitted by the *circled dot points* is a measure of the minimum Δp_y determined by a 0.16 mm slit. The fitting curve for the *diamond dots* is numerical result of Eq. (16.83), indicating a *blurred* ghost image of silt A

the entire scanning range, which is very different from that in measurement 1. The experimental data has provided a clear indication of $\Delta y\,\Delta p_y < h$ in the joint measurements of the entangled photon pairs.

Given that $\Delta y\,\Delta p_y < h$, is this a violation of the uncertainty principle? Does quantum mechanics agree with this peculiar experimental result? If quantum mechanics does provide a solution with $\Delta y\,\Delta p_y < h$ for photon 2, we would indeed be forced to face a paradox as EPR had pointed out in 1935.

Quantum mechanics does provide a solution that agrees with the experimental result. However, it is not the solution for photon 2. Instead, it is for a joint measurement of the entangled photon pair.

We now examine the experimental results with the quantum mechanical calculation by adopting the formalisms from the ghost image experiment with two modifications:

Case (I): - slits A = 0. 16 mm, slit B = 0. 16 mm.

This is the experimental condition for measurement one: slit B is adjusted to be the same as slit A. There is nothing surprise for this measurement. The measurement simply provides us the knowledge of Δp of photon 2 after the diffraction coursed by slit B of $\Delta y = 0.16$ mm. The experimental data shown in □ Fig. 16.24 agrees with the calculation. Notice that slit B is about 745 mm far away from the 3 mm two-photon source, the angular size of the light source is roughly the same as $\lambda/\Delta y$, $\Delta\theta \sim \lambda/\Delta y$, where $\lambda = 702$ nm is the wavelength and $\Delta y = 0.16$ mm is the width of the slit. The calculated diffraction pattern is very close to that of the "far-field" Fraunhofer diffraction of a 0.16 mm single-slit.

16

Case (II): - slit A = 0. 16 mm, slits $B \sim \infty$ (wide open).

Now we remove slit B from the ghost image plane. The calculation of the transverse effective two-photon wavefunction and the second-order correlation is the same as that of the ghost image except the observation plane of D_2 is moved from the image plane a distance of 500 mm behind. The two-photon image of slit A is located at a distance $s_i = 2f = 1000$ mm $(b_1 + b_2)$ from the imaging lens, in this measurement D_2 is placed at $d = 1500$ mm from the imaging lens. The measured pattern is simply a "blurred" two-photon image of slit A. The "blurred" two-photon image can be calculated from Eq. (16.83)

$$\Psi(\vec{\rho}_o, \vec{\rho}_2) \propto \int_{lens} d\vec{\rho}_l \, G\left(|\vec{\rho}_2 - \vec{\rho}_l|, \frac{\omega}{cd}\right) G\left(|\vec{\rho}_l|, \frac{\omega}{cf}\right) G\left(|\vec{\rho}_l - \vec{\rho}_o|, \frac{\omega}{cs_o}\right)$$

$$\propto \int_{lens} d\vec{\rho}_l \, G\left(|\vec{\rho}_l|, \frac{\omega}{c}[\frac{1}{s_o} + \frac{1}{d} - \frac{1}{f}]\right) e^{-i\frac{\omega}{c}(\frac{\vec{\rho}_o}{s_o} + \frac{\vec{\rho}_i}{d})\cdot\vec{\rho}_l}$$

(16.83)

where d is the distance between the imaging lens and D_2. In this measurement, D_2 was placed 500 mm behind the image plane, i.e., $d = s_i + 500$ mm. The numerical calculated "blurred" image, which is narrower than that of the diffraction pattern of the 0.16 mm slit B, agrees with the measured result of ◘ Fig. 16.24 within experimental error.

The measurement does show a result of $\Delta y \, \Delta p_y < h$. The measurement, however, has nothing to do with the uncertainty relation that governs the behavior of photon 2 (the idler). Popper and EPR were correct in the prediction of the outcomes of their experiments. Popper and EPR, on the other hand, made the same error by applying the results of two-particle physics to the explanation of the behavior of an individual particle.

In both the Popper and EPR experiments, the measurements are *joint detection* between two detectors applied to entangled states. Quantum mechanically, an entangled two-particle state only provides *the precise knowledge of the correlations of the pair*. The behavior of *photon 2* observed in the joint measurement is conditioned upon the measurement of its twin. A quantum must obey the uncertainty principle but the *conditional behavior* of a quantum in an entangled biparticle system is different in principle. We believe paradoxes are unavoidable if one insists the *conditional behavior* of a particle is the *behavior* of the particle. This is the central problem in the rationale behind both Popper and EPR. $\Delta y \, \Delta p_y \geq h$ is not applicable to the conditional behavior of either *photon 1* or *photon 2* in the experiments of Popper and EPR.

The behavior of photon 2 being conditioned upon the measurement of photon 1 is well represented by the two-photon amplitudes. Each of the *straight lines* in ◘ Fig. 16.22 corresponds to a two-photon amplitude. Quantum mechanically, the superposition of these two-photon amplitudes is responsible for a "click-click" measurement of the entangled pair. A "click-click" joint measurement of the two-particle entangled state projects out certain two-particle amplitudes, and only these two-particle amplitudes are featured in the quantum formalism. In the above analysis we never consider photon 1 or photon 2 *individually*. Popper's question about the momentum uncertainty of photon 2 is then inappropriate. The correct question to ask in these measurements should be: what is the uncertainty of Δp_y for the signal-idler *pair* which are localized within $\Delta y = 0.16$ mm at "screen" A with and without slit B? This is indeed the central point for Popper's experiment.

Once again, the demonstration of Popper's experiment calls our attention to the important message: the physics of the entangled two-particle system must inherently be very different from that of individual particles.

16.4.2 Popper's Experiment Two

In fact, the nonfactorizable, point-to-point image-forming correlation is not only the property of entangled photon pairs; it can also be realized in the joint-detection of a randomly paired photons in thermal state. In 2005, 10 years after the first ghost imaging experiment, a near-field lensless ghost imaging experiment that uses chaotic-thermal radiation source was demonstrated by Valencia et al. [38]. This experiment opened a door for the realization of Popper's thought experiment through the joint measurement of randomly paired photons in thermal state.

With the help of a novel joint detection scheme, namely the photon-number fluctuation correlation (PNFC) circuit [30], which distinguishes the positive and negative photon-number fluctuations measured by two single-photon counting detectors, and calculates the correlation between them, we were able to produce the ghost image of an object at a distance with 100 % visibility. By modifying the 1999 Kim-Shih experiment with a different light source and a lensless configuration, Peng and Shih realized Popper's thought experiment again in 2015 [14]. ◘ Figure 16.25 is an unfolded schematic, in which a large enough angular sized thermal source produces an equal-sized ghost image of slit-A at the plane $d_B = d_A$. The ghost image of slit-A can be verified by scanning the point-like photodetector D_B in the plane of slit-B. This ghost image provides the value of Δy through the

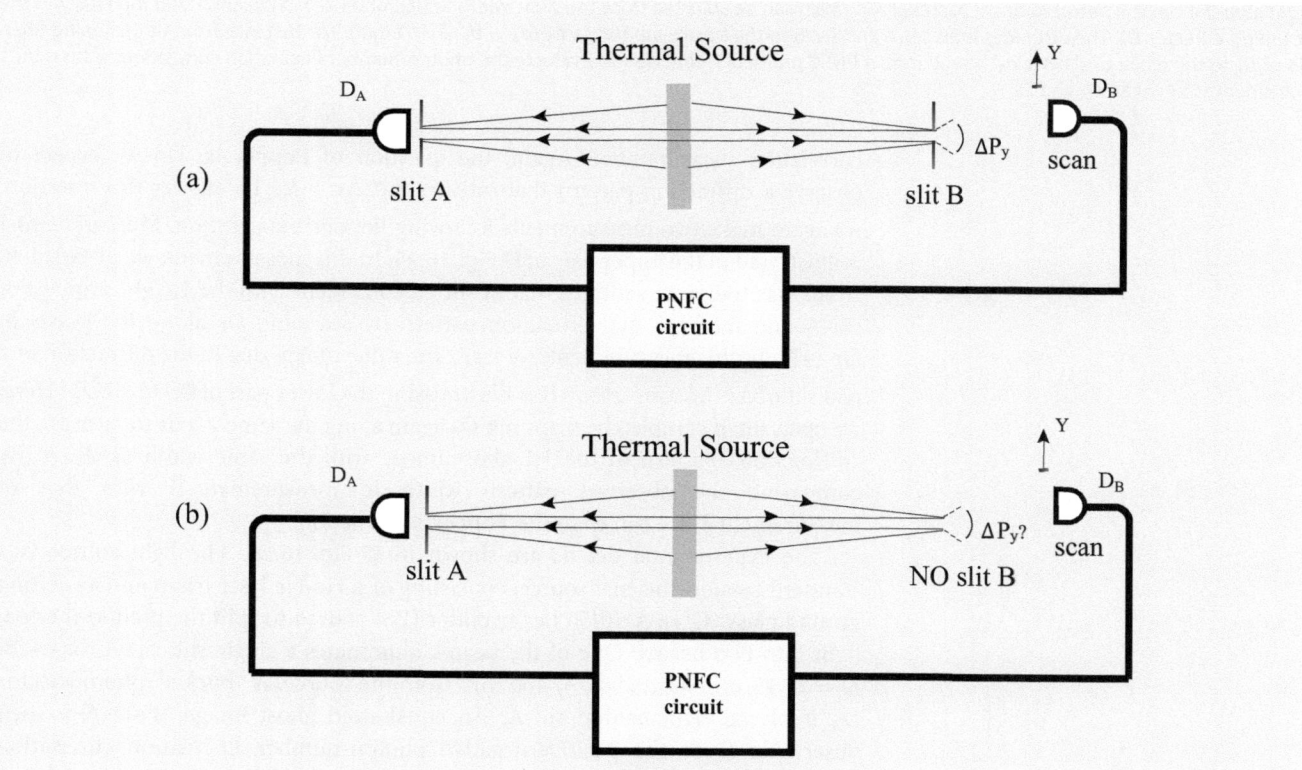

◘ **Fig. 16.25** Unfolded schematic of Popper's experiment with thermal light. The lensless ghost imaging setup with PNFC protocol produces an equal sized 100 % visibility ghost image of slit-A at the position of slit-B. Detector D_B is scanning transversely in the y direction to measure the photon-number fluctuation correlation with D_A when (**a**) Slit-A and slit-B are adjusted both very narrowly, and (**b**) Slit-A is kept very narrow and slit-B is left wide open

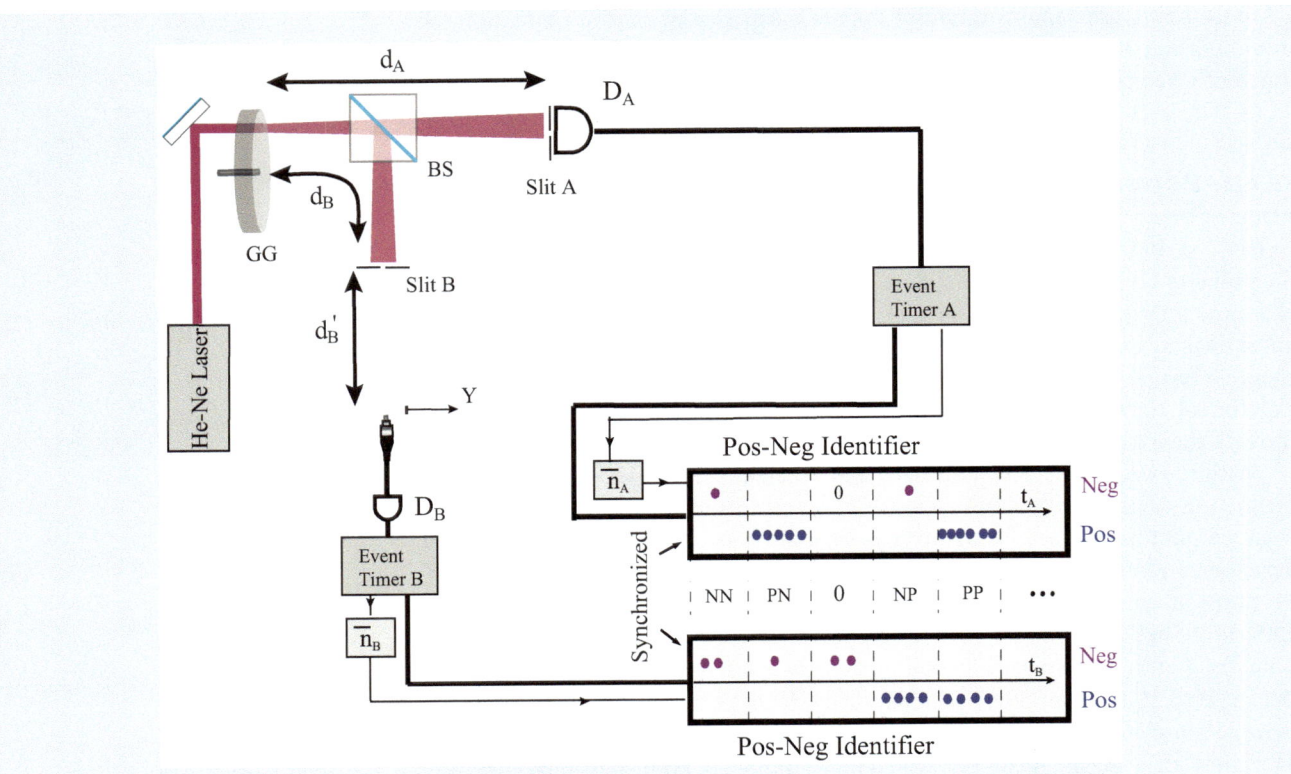

Fig. 16.26 Schematic of the experimental setup. A rotating ground glass (GG) is employed to produce pseudo-thermal light. BS is a 50/50 non-polarizing beam splitter. After BS, the transmitted beam passes through slit-A (0.15 mm) and collected by a "bucket" detector D_A which is put right after the slit. The reflected beam passes slit-B, which can be adjusted to be the same width as that of slit-A or wide open, and then reaches the scanning detector D_B. The distances from slit-A and slit-B to the source are the same ($d_A = d_B = 400$ mm). The distance from the scanning fiber tip of D_B to the plane of slit-B is $d_B' = 900$ mm. A PNFC protocol is followed to evaluate the photon-number fluctuation correlations from the coincidences between D_A and D_B

correlation measurement. Again, the question of Popper is: Do we expect to observe a diffraction pattern that satisfies $\Delta p_y \Delta y > h$? To answer this question, we again make two measurements following Popper's suggestion. Measurement-I is illustrated in the upper part of ◘ Fig. 16.25. In this measurement, we place slit-B, which has the same width as that of slit-A, coincident with the 1:1 ghost image of slit-A and measure the diffraction pattern by scanning D_B along the y-axis in far-field. In this measurement, we learn the value of Δp_y due to the diffraction of a real slit of Δy. Measurement-II is illustrated in the lower part of ◘ Fig. 16.25. Here, we open slit-B completely, scanning D_B again along the same y-axis to measure the "diffraction" pattern of the 1:1 ghost image with the same width as slit-A. By comparing the observed pattern width in measurement-II with that of measurement-I, we can examine Popper's prediction.

The experimental details are shown in ◘ Fig. 16.26. The light source is a standard pseudo-thermal source, consisting of a He-Ne laser beam and a rotating ground glass (GG). A 50/50 beamsplitter (BS) is used to split the pseudo-thermal light into two beams. One of the beams illuminates a single slit, slit-A, of width $D = 0.15$ mm located $d_A \sim 400$ mm from the source. A "bucket" photodetector D_A is placed right behind slit-A. An equal-sized ghost image of slit-A is then observable from the positive-negative photon-number fluctuation correlation measurement between the "bucket" detector D_A and the transversely scanning point-like photodetector D_B, if D_B is scanned on the ghost image plane located at $d_B = d_A = 400$ mm. In this experiment, however, D_B is scanned on a plane that is located $d_B' \sim 900$ mm behind the ghost image plane, to measure the "diffraction"

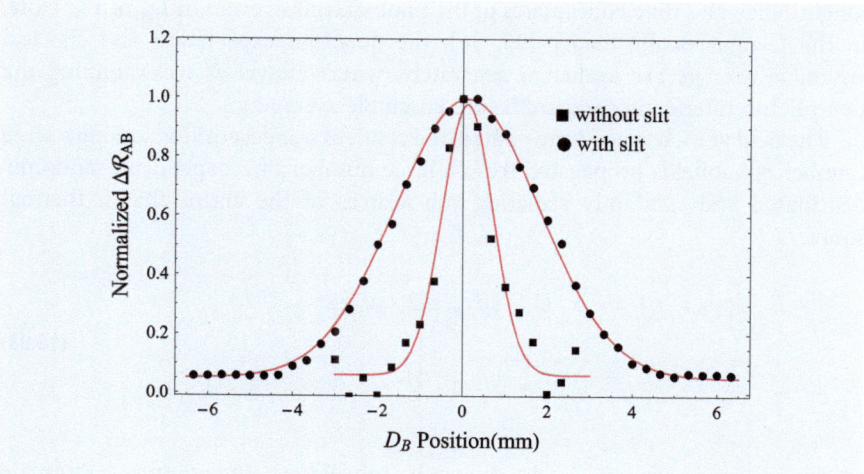

Fig. 16.27 The observed diffraction patterns. *Circles*: slit-A and slit-B are both adjusted for 0. 15 mm. *Squares*: slit-A is 0.15 mm, slit-B is wide open. The width of the curve without the slit is almost three times narrower than that of the curve with slit, agreeing well with the theoretical predictions from Eqs. (16.89) and (16.91)

pattern of the ghost image. The output pulses from the two single-photon counting detectors are then sent to a PNFC circuit, which starts from two Pos-Neg identifiers follow two event-timers distinguish the "positive-fluctuation" Δn^{+}, from the "negative-fluctuation" Δn^{-}, measured by D_A and D_B, respectively, within each coincidence time window. The photon-number fluctuation-correlations of D_A-D_B: $\Delta R_{AB} = \langle \Delta n_A \Delta n_B \rangle$ is calculated, accordingly and respectively, based on their measured positive-negative fluctuations. The detailed description of the PNFC circuit can be found in [30].

The experiment was performed in two steps after confirming the 1:1 ghost image of slit-A. In measurement-I, we place slit-B ($D = 0. 15$ mm) coincident with the ghost image and move D_B to a plane at $d_B' \sim 900$ mm to measure the diffraction pattern of slit-B. In measurement-II, we keep the same experimental condition as that of measurement-I, except slit-B is set wide open.

■ Figure 16.27 reports the experimental results. The circles show the normalized photon-number fluctuation correlation from the PNFC protocol against the position of D_B along the y-axis for Popper's measurement-I. As expected, we observed a typical single-slit diffraction pattern giving us the uncertainty in momentum, Δp_y^{real}. The squares show the experimental observation from the PNFC for Popper's measurement-II, when slit-B is wide open. The measured curves agree well with our theoretical fittings. We found the width of the curve representing no physical slit is much narrower than that of the real diffraction pattern, which agrees with Popper's prediction.

Similar to our early analysis in Bell state and in quantum eraser, we chose the coherent state representation for the calculation of the joint photodetection counting rate of D_A and D_B which is proportional to the second-order coherence function $G_{AB}^{(2)}$:

$$G_{AB}^{(2)} = \left\langle \langle \hat{E}^{(-)}(\vec{\rho}_A, z_A, t_A) \hat{E}^{(-)}(\vec{\rho}_B, z_B, t_B) \right.$$
$$\left. \times \hat{E}^{(+)}(\vec{\rho}_A, z_B, t_B) \hat{E}^{(+)}(\vec{\rho}_A, z_A, t_A) \rangle_{\text{QM}} \right\rangle_{\text{Es}}, \qquad (16.84)$$

where $E^{(+)}(\vec{\rho}_j, z_j, t_j)$ $(E^{(-)}(\vec{\rho}_j, z_j, t_j))$ is the positive (negative) field operator at space-time coordinate $(\vec{\rho}_j, z_j, t_j)$, $j = A, B$, with $(\vec{\rho}_j, z_j, t_j)$ the transverse, longitudinal, and time coordinates of the photodetection event of D_A or D_B. Note, in the Glauber-Scully theory [25, 36], the quantum expectation and classical ensemble average are evaluated separately, which allows us to examining the two-photon interference picture before ensemble averaging.

The field at each space-time point is the result of a superposition among a large number of subfields propagated from a large number of independent, randomly distributed and randomly radiating sub-sources of the entire chaotic-thermal source,

$$
\begin{aligned}
\hat{E}^{(\pm)}(\vec{\rho}_j, z_j, t_j) &= \sum_m \hat{E}^{(\pm)}(\vec{\rho}_{0m}, z_{0m}, t_{0m}) g_m(\vec{\rho}_j, z_j, t_j) \\
&\equiv \sum_m \hat{E}_m^{(\pm)}(\vec{\rho}_j, z_j, t_j),
\end{aligned}
\tag{16.85}
$$

where $\hat{E}^{(\pm)}(\vec{\rho}_{0m}, z_{0m}, t_{0m})$ is the mth subfield at the source coordinate $(\vec{\rho}_{0m}, z_{0m}, t_{0m})$, and $g_m(\vec{\rho}_j, z_j, t_j)$ is the optical transfer function that propagates the mth subfield from coordinate $(\vec{\rho}_{0m}, z_{0m}, t_{0m})$ to $(\vec{\rho}_j, z_j, t_j)$. We can write the field operators in terms of the annihilation and creation operators:

$$
\hat{E}_m^{(+)}(\vec{\rho}_j, z_j, t_j) = C \int d\mathbf{k}\, \hat{a}_m(\mathbf{k})\, g_m(\mathbf{k}; \vec{\rho}_j, z_j, t_j),
\tag{16.86}
$$

C is a normalization constant, $g_m(\mathbf{k}; \vec{\rho}_j, z_j, t_j)$, $j = A, B$, is the optical transfer function for mode $\mathbf{k}$ of the mth subfield propagated from the mth sub-source to the jth detector, and $\hat{a}_m(\mathbf{k})$ is the annihilation operator for the mode $\mathbf{k}$ of the mth subfield.

Substituting the field operators and the state, in the multi-mode coherent representation, into Eq. (16.84), we then write $G_{AB}^{(2)}$ in terms of the superposition of a large number of effective wavefunctions, or wavepackets:

$$
\begin{aligned}
&G^{(2)}(\vec{\rho}_B, z_B, t_B; \vec{\rho}_B, z_B, t_B) \\
&= \left\langle \sum_{m,n,p,q} \psi_m^*(\vec{\rho}_A, z_A, t_A)\psi_n^*(\vec{\rho}_B, z_B, t_B)\psi_p(\vec{\rho}_B, z_B, t_B)\psi_q(\vec{\rho}_A, z_A, t_A) \right\rangle_{Es} \\
&= \left\langle \sum_{m,n} \left| \psi_m(\vec{\rho}_A, z_A, t_A)\psi_n(\vec{\rho}_B, z_B, t_B) + \psi_n(\vec{\rho}_A, z_A, t_A)\psi_m(\vec{\rho}_B, z_B, t_B) \right|^2 \right\rangle_{Es} \\
&= \left\langle \sum_m \left| \psi_m(\vec{\rho}_A, z_A, t_A) \right|^2 \sum_n \left| \psi_n(\vec{\rho}_B, z_B, t_B) \right|^2 \right. \\
&\quad + \left. \sum_{m \neq n} [\psi_m^*(\vec{\rho}_A, z_A, t_A)\psi_m(\vec{\rho}_B, z_B, t_B)\psi_n(\vec{\rho}_A, z_A, t_A)\psi_n^*(\vec{\rho}_B, z_B, t_B)] \right\rangle_{Es} \\
&\equiv \langle n_A \rangle \langle n_B \rangle + \langle \Delta n_A \Delta n_B \rangle.
\end{aligned}
\tag{16.87}
$$

with

$$
\psi_s(\vec{\rho}_j, z_j, t_j) = \int d\mathbf{k}\, \alpha_s(\mathbf{k}) e^{i\varphi_{0s}} g_s(\mathbf{k}; \vec{\rho}_j, z_j, t_j),
$$

where $s = m, n, p, q$, $j = A, B$, and the phase factor $e^{i\varphi_{0s}}$ represents the random initial phase of the mth subfield. In Eq. (16.87), we have completed the ensemble average in terms of the random phases of the subfields, i.e. φ_{0s}, and kept the nonzero terms only. Equation (16.87) indicates the second-order coherence function is the result of a sum of a large number of subinterference patterns, each subpattern indicates an interference in which a random pair of wavepackets interfering with the pair itself. For example, the mth and the nth wave packets have two different yet indistinguishable alternative ways to produce a joint photodetection event, or a coincidence count, at different space-time coordinates: (1) the mth wavepacket is annihilated at D_A and the nth wavepacket is annihilated at D_B; (2) the mth wavepacket is annihilated at D_B and the nth wavepacket is annihilated at D_A. In quantum mechanics, the joint detection probability of D_A and D_B is proportional to the normal square of the superposition of the above two probability amplitudes. We name this kind of superposition "nonlocal interference." The superposition of the two amplitudes for each random pair results in an interference pattern, and the addition of these large number of interference patterns yields the nontrivial correlation of the chaotic-thermal light.

The cross interference term in Eq. (16.87) indicates the photon-number fluctuation correlation $\langle \Delta n_A \Delta n_B \rangle$:

$$\langle \Delta n_A(\vec{\rho}_A, z_A, t_A) \Delta n_B(\vec{\rho}_B, z_B, t_B) \rangle_{Es}$$

$$= \left\langle \sum_{m \neq n} \left[\psi_m^*(\vec{\rho}_A, z_A, t_A) \psi_n(\vec{\rho}_A, z_A, t_A) \right] \left[\psi_m(\vec{\rho}_B, z_B, t_B) \psi_n^*(\vec{\rho}_B, z_B, t_B) \right] \right\rangle_{Es} \quad \text{(16.88)}$$

$$\simeq \left\langle \sum_m \psi_m^*(\vec{\rho}_A, z_A, t_A) \psi_m(\vec{\rho}_B, z_B, t_B) \sum_n \psi_n(\vec{\rho}_A, z_A, t_A) \psi_n^*(\vec{\rho}_B, z_B, t_B) \right\rangle_{Es}.$$

In measurement-I, the optical transfer functions that propagate the fields from the source to D_A and D_B are

$$g_m(\vec{\kappa}, \omega; \vec{\rho}_A, z_A = d_A) = \frac{-i\omega e^{i(\omega/c)z_A}}{2\pi c d_A} \int d\vec{\rho}_s f(\vec{\rho}_s) e^{i\vec{\kappa} \cdot \vec{\rho}_s} G(|\vec{\rho}_s - \vec{\rho}_o|)_{[\omega/(cd_A)]},$$

and

$$g_n(\vec{\kappa}, \omega; \vec{\rho}_B, z_B = d_B + d'_B)$$
$$= \frac{-\omega^2 e^{i(\omega/c)z_B}}{(2\pi c)^2 d_B d'_B} \int d\vec{\rho}_s \int d\vec{\rho}_i f(\vec{\rho}_s) e^{i\vec{\kappa} \cdot \vec{\rho}_s} G(|\vec{\rho}_s - \vec{\rho}_i|)_{[\omega/(cd_B)]}$$
$$t(\vec{\rho}_i) G(|\vec{\rho}_i - \vec{\rho}_B|)_{[\omega/(cd'_B)]},$$

where $\vec{\rho}_s$ is defined on the output plane of the source and $f(\vec{\rho}_s)$ denotes the aperture function of the source. We also assumed a perfect "bucket" detector D_A, which is placed at the object plane of slit-A ($\vec{\rho}_A = \vec{\rho}_o$), in the following calculation. $\vec{\rho}_i$ is defined on the ghost image plane, which coincides with the plane of slit-B, and $\vec{\rho}_B$ is defined on the detection plane of D_B, $t(\vec{\rho}_i)$ is the aperture function of slit-B. The function $G(|\alpha|)_{[\beta]}$ is the Gaussian function $G(|\alpha|)_{[\beta]} = e^{\frac{i\beta}{2}|\alpha|^2}$. The measured fluctuation correlation can be calculated from Eq. (16.88)

$$\Delta R_{AB} = \int d\vec{\rho}_o |t(\vec{\rho}_o)|^2 \text{sinc}^2[\omega_0 D \frac{\vec{\rho}_B}{2cd'_B}] \equiv C' \times \text{sinc}^2[\omega_0 D \frac{\vec{\rho}_B}{2cd'_B}], \quad \text{(16.89)}$$

where $t(\vec{\rho}_o)$ is the aperture function of slit-A. The above calculation indicates a product between a constant C', which is from the integral on the "bucket" detector D_A, and a first order diffraction pattern of slit-B. With our experimental setup, the width of the diffraction pattern is estimated to be ~ 4 mm, which agrees well with the experimental observation, as shown in ◻ Fig. 16.27.

16

In measurement-II, with slit-B wide open, the field at D_B becomes

$$g_n(\vec{\kappa}, \omega; \vec{\rho}_B, z_B) = \frac{-i\omega e^{i(\omega/c)z_B}}{2\pi c z_B} \int d\vec{\rho}_s f(\vec{\rho}_s) e^{i\vec{\kappa} \cdot \vec{\rho}_s} G(|\vec{\rho}_s - \vec{\rho}_B|)_{[\omega/(cz_B)]}.$$

We first check if a ghost image of slit-A is present when scanning D_B in the ghost image plane of $d_B = d_A$. The photon-number fluctuation correlation is calculated to be

$$\Delta R_{AB} = \int d\vec{\rho}_o |t(\vec{\rho}_o)|^2 \mathrm{sin}\, c^2 \left[\frac{\omega_0 a}{c d_A} |\vec{\rho}_o - \vec{\rho}_B| \right]$$
$$= |t(\vec{\rho}_o)|^2 \otimes \mathrm{sin}\, c^2 \left[\frac{\omega_0 a}{c d_A} |\vec{\rho}_o - \vec{\rho}_B| \right] \approx |t(\vec{\rho}_B)|^2. \tag{16.90}$$

Note, we have placed D_A right behind slit-A and thus $\vec{\rho}_A = \vec{\rho}_o$. This suggests an equal-sized 100 % visibility ghost image on the plane of $d_B = d_A$.

When we move D_B away from the ghost image plane to the far-field plane of $d_B + d_{B'}$, the photon-number fluctuation correlation becomes:

$$\Delta R_{AB} = \int d\vec{\rho}_o |t(\vec{\rho}_o)|^2 \tilde{\mathcal{F}}_s^{\,2} (m\vec{\rho}_o - \vec{\rho}_{B'} = |t(\vec{\rho}_o)|^2 \otimes \tilde{\mathcal{F}}_s^{\,2} (m\vec{\rho}_o - \vec{\rho}_{B'}), \tag{16.91}$$

where $\tilde{\mathcal{F}}_s$ is the Fourier transform of the defocused pupil function $\mathcal{F}_s = f(\vec{\rho}_s) e^{-i(\omega_0/2c\mu)\vec{\rho}_s^2}$ and μ, m are defined as $1/\mu = 1/d_A - 1/(d_B + d_{B'})$, $m = (d_B + d_{B'})/d_A$, respectively. The measured result of measurement-II is thus a convolution between the aperture function of slit-A, $t(\vec{\rho}_o)$, and the correlation function $\tilde{\mathcal{F}}_s (m\vec{\rho}_o - \vec{\rho}_{B'}$, resulting in a "blurred" image of slit-A. With our experimental setup, the width of the "diffraction" pattern is estimated to be ~ 1.4 mm, which is almost three times narrower than the diffraction pattern of measurement-I and agrees well with the experimental observation, as shown in ◘ Fig. 16.27. Compared with the Kim-Shih experimental result [13], we can see that although the number varies due to different experimental parameters, we have obtained a very similar result: the measured width of the "diffraction pattern" in measurement-II is much narrower than that of the diffraction pattern in measurement-I.

The above analysis indicates that the experimental observations are reasonable from the viewpoint of the coherence theory of light. The important physics we need to understand is to distinguish the first-order coherent effect and the second-order coherent effect, even if the measurement is for thermal light. In Popper's measurement-I, the fluctuation correlation is the result of first-order coherence. The joint measurement can be "factorized" into a product of two first-order diffraction patterns. After the integral of the "bucket" detector, which turns the diffraction pattern of slit-A into a constant, the joint measurement between D_A and D_B is a product between a constant and the standard first-order diffraction pattern of slit-B. There is no question the measured width of the diffraction pattern satisfies $\Delta p_y \Delta y \geq h$. In Popper's measurement-II when slit-B is wide open or removed, the measurement can no longer be written as a product of single-photon detections but as a non-separable function, i.e., a convolution between the object aperture function and the photon-number fluctuation correlation function of randomly paired photons, or the second-order coherence function of the thermal field. We thus consider the observation of $\Delta p_y \Delta y < h$ the result of the second-order coherence of thermal field which is caused from nonlocal interference: a randomly paired photon interferes with the pair itself at a distance by means of a joint photodetection event between D_A and D_B. The result of nonlocal two-photon interference does not contradict the uncertainty principle that governs the behavior of single photons. Again, the observation of this experiment is not a violation of the uncertainty principle. The observation of $\Delta p_y \Delta y < h$ from thermal light, however, may reveal a concern about nonlocal interference.

16.5 Conclusion

This chapter reviewed three types of optical tests of the foundations of quantum theory: (1) EPR-Bohm-Bell correlation and Bell's inequality; (2) Scully's quantum eraser; (3) Popper's experiment. The results of these experiments are very interesting. On one hand, the experimental observations confirm the predictions of EPR-Bell, Scully, and Popper. On the other hand, the calculations from quantum theory perfectly agree with the experimental data. Moreover, apparently, the experimental observations do not lead to any "violations" of the principles of quantum mechanics. One important conclusion we may draw from these optical tests is that all the observations are the results of multi-photon interference: a group of photons interferes with the group itself at distance. The nonlocal multi-photon interference phenomena may never be understood in classical theory, however, it is legitimate in quantum mechanics. The superposition principle of quantum theory supports the superposition of multi-photon amplitudes, whether the photons are entangled or randomly grouped and despite the distances between these individual photodetection events. Perhaps we must accept the probabilistic nature of the "wavefunction" associated with a quantum or a group of quanta. Although a photon does not have a "wavefunction," we have developed the concept of an effective wavefunction for a photon and for a group of photons which have similar physical meanings as that of the wavefunction of a particle or the wavefunction of a group of particles. In terms of the superposition, although the effective wavefunction plays the same role as that of the electromagnetic wave, apparently, the effective wavefunction is different from the electromagnetic field in nature. Any efforts attempting to physically equal the two concepts would trap us in the question posed by Einstein: how long does it take for the energy on the other side of the 2-lightyear diameter sphere to arrive at the detector? Is it possible god of the quantum world does play dice?

References

1. Wheeler JA (1982) A delayed choice experiment. Maryland lectures collection
2. Einstein A, Podolsky B, Rosen N (1935) Can quantum-mechanical description of physical reality be considered complete? Phys Rev 47:777
3. Bohm D (1951) Quantum theory. Prentice-Hall, Yew York
4. Shih YH (2011) An introduction to quantum optics: photon and biphoton physics. CRC press, Taylor & Francis, London

5. Feynman RF, Leighton RB, Sands ML (1965) Lectures on physics. Addison-Wesley, Reading
6. Bohr N (1928) Das Quantenpostulat und die neuere Entwicklung der Atomistik. Naturwissenschaften 16:245
7. Dirac P (1930) The principle of quantum mecanics. Oxford University Press, Oxford
8. Wheeler JA, Zurek WH (1983) Quantum theory and measurement. Princeton University Press, Princeton
9. Scully MO, Druhl H (1982) Quantum eraser: a proposed photon correlation Experiment concerning observation and "delayed choice" in quantum mechanics. Phys Rev A 25:2208
10. Kim YH, Yu SP, Kulik SP, Shih YH, Scully MO (2000) A delayed "choice" quantum eraser. Phys Rev Lett 84:1
11. Peng T, Chen H, Shih YH, Scully MO (2014) Delayed-choice quantum eraser with thermal light. Phys Rev Lett 112:180401
12. Popper K (1934) Zur Kritik der Ungenauigkeitsrelationen. Naturwissenschaften 22:807
13. Kim YH, Shih YH (1999) Experimental realization of Popper's experiment: violation of the uncertainty principle? Found Phys 29:1849
14. Peng T, Simon J, Chen H, French R, Shih YH (2015) Popper's experiment with randomly paired photons in thermal state. Euro Phys Lett 109:14003
15. Clauser JF, Shimony A (1978) Bell's theorem. Experimental tests and implications. Rep Prog Phys 41:1881. An excellent review on Bell measurement before the introduction of SPDC
16. Freedman SJ, Clauser JF (1972) Experimental test of local hidden-variable theories. Phys Rev Lett 28:938
17. Clauser JF (1976) Experimental investigation of a polarization correlation anomaly. Phys Rev Lett 36:1223
18. Fry ES, Thompson PC (1976) Experimental test of local hidden-variable theories. Phys Rev Lett 37:465
19. Aspect A, Grangier P, Roger G (1981) Experimental tests of realistic local theories via Bell's theorem. Phys Rev Lett 47:460
20. Aspect A, Grangier P, Roger G (1982) Experimental realization of Einstein-Podolsky-Rosen-Bohm gedankenexperiment: a new violation of Bell's inequalities. Phys Rev Lett 49:91
21. Alley CO, Shih YH (1986) In: Namiki M et al (ed) Foundations of quantum mechanics in the light of new technology. Physical Society of Japan, Tokyo; Shih YH, Alley CO (1988) New type of Einstein-Podolsky-Rosen-Bohm experiment using pairs of light quanta produced by optical parametric down conversion. Phys Rev Lett 61:2921
22. Kwiat PG, Mattle K, Weinfurter H, Zeilinger A, Sergienko AV, Shih YH (1995) New high-intensity source of polarization-entangled photon pairs. Phys Rev Lett 75:4337
23. Klyshko DN (1988) Photons and nonlinear optics. Gorden and Breach, New York
24. Rubin MH, Klyshko DN, Shih YH, Sergienko AV (1994) The theory of two-photon entanglement in type-II optical parametric down conversion. Phys Rev A 50:5122
25. Glauber RJ (1963) The quantum theory of optical coherence. Phys Rev 130:2529
26. Kim YH, Berardi V, Chekhova MV, Shih YH (2001) Anti-correlation effect in femtosecond-pulse pumped type-II spontaneous parametric down-conversion. Phys Rev A 64:R011801
27. Kim YH, Kulk SP, Shih YH (2001) Bell state preparation using pulsed non-degenerate two-photon entanglement. Phys Rev A 63:R060301
28. Peng T, Shih YH (2015) Bell correlation of thermal fields in photon-number fluctuations. Europhys Lett 112: 60006
29. Martienssen W, Spiller E (1964) Coherence and fluctuations in light beams. Am J Phys 32:919
30. Chen H, Peng T, Shih YH (2013) 100% correlation of chaotic thermal light. Phys Rev A 88:023808
31. Glauber RJ (1963) Coherent and incoherent states of the radiation field. Phys Rev 131:2766
32. Bell JS (1964) On the Einstein Podolsky Rosen paradox. Physics 1:195
33. Bell JS (1987) Speakable and unspeakable in quantum mechanics. Cambridge University Press, Cambridge
34. Herzog TJ, Kwiat PG, Weinfurter H, Zeilinger A (1995) Complementarity and the quantum eraser. Phys Rev Lett 75:3034
35. Walborn SP, Terra Cunha MO, Padua S, Monken CH (2002) Double-slit quantum erasure. Phys Rev A 65:033818
36. Scully MO, Zubairy MS (1997) Quantum optics. Cambridge University Press, Cambridge
37. Pittman TB, Shih YH, Strekalov DV, Sergienko AV (1995) Optical imaging by means of two-photon entanglement. Phys Rev A 52:R3429
38. Valencia A, Scarcelli G, D'Angelo M, Shih YH (2005) Two-photon imaging with thermal light. Phys Rev Lett 94:063601

Quantum Mechanical Properties of Light Fields Carrying Orbital Angular Momentum

Robert W. Boyd and Miles J. Padgett

R.W. Boyd (✉)
Department of Physics and School of Electrical Engineering and Computer Science, University of Ottawa,
Ottawa, ON, Canada

Max Planck Centre for Extreme and Quantum Photonics, University of Ottawa, Ottawa, ON, Canada

Department of Physics and Astronomy, The Institute of Optics, University of Rochester, Rochester, NY, USA

School of Physics and Astronomy, University of Glasgow, Glasgow G12 8QQ, UK
e-mail: boydrw@mac.com

M.J. Padgett
School of Physics and Astronomy, University of Glasgow, Glasgow G12 8QQ, UK
e-mail: Miles.Padgett@glasgow.ac.uk

© The Author(s) 2016
M.D. Al-Amri et al. (eds.), *Optics in Our Time*, DOI 10.1007/978-3-319-31903-2_17

17.1 Introduction

There is a growing appreciation of the importance of quantum mechanics and in particular of quantum information science both for understanding the nature of the world in which we live and in the development of new technologies for communication and imaging. It is in this spirit that this chapter is written. The chapter deals with structured light fields, especially fields that carry orbital angular momentum (OAM), and their application to communication systems.

The topic of structured light fields in the quantum domain is intimately related to the topic of quantum imaging [1]. Quantum imaging is a discipline that studies quantum aspects of image formation and that uses quantum properties of light to produce images. Imaging techniques inherently require the ability to encode massive amounts of information in a light field. The quantum aspects of optical images can therefore be a key resource for quantum information and communication systems.

In this chapter we will review several examples of the quantum properties of structured light field. In broad concept, there are two sorts of quantum states that appear in our examples. One sort is a "single-photon" state, a state in which it is known that there is one and only one photon present in the field of interest. Such a state shows strong quantum properties because, for example, if you send such a beam onto a beamsplitter, the photon will emerge in one of the two output ports but not half and half in both. The other sort of quantum state of interest in this chapter is an entangled state of two photons. In fact, the concept of entanglement is one of the great mysteries of quantum mechanics. The term entanglement and the first explicit description of this phenomenon were introduced by Schrödinger in 1935 [2]. Here is quote from his paper:

"When two systems, of which we know the states by their respective representatives, enter into temporary physical interaction due to known forces between them, and when after a time of mutual influence the systems separate again, then they can no longer be described in the same way as before, viz. by endowing each of them with a representative of its own. I would not call that one but rather the characteristic trait of quantum mechanics, the one that enforces its entire departure from classical lines of thought."

This concept of entanglement leads to what today is often called "quantum weirdness," a term that arises from the seemingly paradoxical effects that can occur in an entangled system. Entanglement has, for instance, played a key role in compelling laboratory demonstrations [3, 4] of the nonlocality of quantum phenomena. However, this weirdness has in fact turned into an asset. This entanglement weirdness leads to effects such as quantum teleportation and certain forms of quantum communication with guaranteed security. One of the easiest ways to produce quantum entanglement uses nonlinear optical methods, and it is these methods that therefore give the field of photonics a special and elevated position in the arena of quantum technologies.

The concept of entanglement generation as applied to light fields can be visualized using the drawing shown in ◼ Fig. 17.1. Here a laser beam at frequency ω_p excites a second-order nonlinear optical crystal, whose nonlinear response can be characterized in terms of its second-order susceptibility $\chi^{(2)}$. Occasionally, a pump photon can be absorbed and generates two lower-frequency photons of frequencies ω_s and ω_i, a process known as spontaneous parametric downconversion (SPDC) [5–7]. It can be shown that the rate at which photon pairs are created is proportional to the product $[\chi^{(2)}]^2 L^2 I_p$, where L is the length of the nonlinear crystal and I_p is the intensity of the pump laser.

By conservation of energy, the condition $\omega_p = \omega_s + \omega_i$ must be satisfied, as illustrated in part (b) of the figure, furthermore these two new photons are

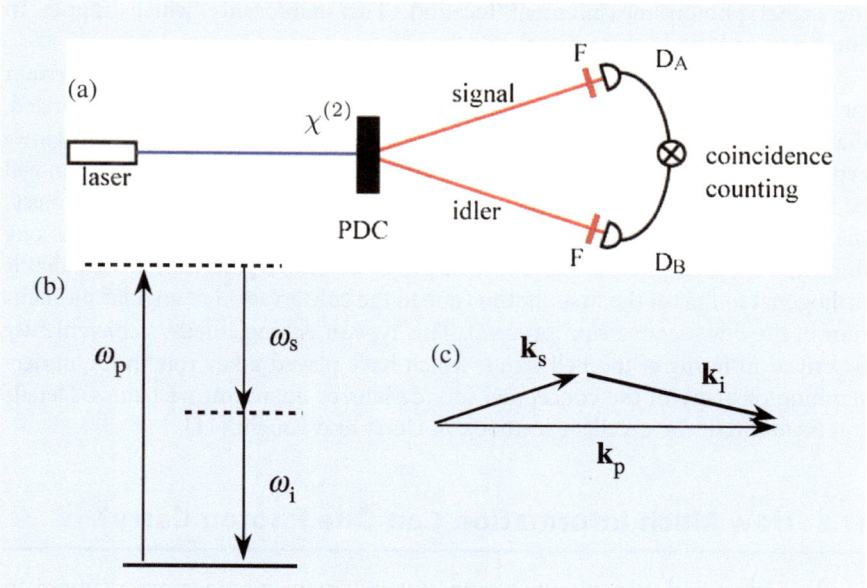

● **Fig. 17.1** (**a**) Schematic illustration of the process of spontaneous parametric downconversion (SPDC). A laser beam excites a second-order ($\chi^{(2)}$) nonlinear optical crystal, leading to the generation of pairs of photons conventionally known as signal and idler photons. This process must obey the conservation of both energy (**b**) and momentum (**c**). These conditions lead to quantum correlations known as entanglement between the signal and idler photons, as discussed further in the text

generated at the same position (i.e., position correlated). However, photon momentum must also be conserved in this generation process, as illustrated in part (c) of the figure, and hence the two new photons are generated with opposite transverse momentum components (i.e., momentum anti correlated). It is these simultaneous conditions on position and momentum that lead to the paradox of Einstein et al. [8] and the concept of quantum entanglement [2].

The photons created by SPDC form entangled pairs, and in fact these photons can be entangled simultaneously in more than one pair of degrees of freedom. The possible types of entanglement that are often studied are

- position and transverse momentum
- angular position and orbital angular momentum
- time and energy
- polarization in different measurement bases.

Examples of the first two types of entanglement will be presented later in this chapter. Here we present a brief discussion of the other two types of entanglement.

By time-energy entanglement, one means that if one measures, for example, the energy of the signal photon, one is able to predict with certainty that the energy of the idler photon will be given by $\hbar\omega_i = \hbar\omega_p - \hbar\omega_s$. However, if one instead chooses to measure the moment of time at which the signal photon is emitted, one will always find that the idler photon is emitted at exactly the same moment. It seems that the product of uncertainty in tightness in the correlation of energies multiplied the uncertainty in the correlation of times can be arbitrarily small and certainly smaller than the value $\frac{1}{2}\hbar$ that one might have envisaged from the naive application of uncertainty relations [9, 10]. The situation is the essence of entanglement: the resolution of this seeming paradox is that a measurement that one performs on the signal photon results in a restriction of our ability to predict the properties of the idler photon, even if that idler photon is arbitrarily distant from the signal photon measurement location. This nonlocality which applies to entangled systems leads to the phrase "spooky action at a distance."

Polarization entanglement can be similarly described. Under certain circumstances [6], each of the photons emitted by SPDC will be unpolarized, that is a complete statistical mixture of two orthogonal polarization states. However, for any one particular measurement the polarization of the signal photon will be found to have a defined value; one says that the measurement process projects the polarization state unto one of the polarization eigenstates. Furthermore, one finds that the idler photon will always be projected onto a polarization state that is orthogonal to that of the first photon (due to the conservation of angular momentum in the down conversion process). This type of entanglement is conveniently described in terms of the Bell states, which have played a key role in the understanding of many of the conceptual foundations of quantum mechanics. Details can be found in the excellent textbook of Gerry and Knight [11].

17.2 How Much Information Can One Photon Carry?

In classical optical telecommunication systems, many photons are required to transmit one bit of information. But it can be interesting to turn this question around and ask it differently: How much information can be carried by a single photon? Perhaps surprising to some is that research conducted over the last decade shows that there is no fundamental limit to the amount of information that can be carried by a single photon.

We start this section by giving a specific example of the ability to transfer many bits of information for each photon; this example will be developed in greater depth in the following sections. Laboratory procedures now exist for switching between single-photon states in any one of the Laguerre–Gaussian modes (Eq. (17.3)) of light using, for example, liquid crystal-based spatial light modulators [12] or a digital micromirror devices (DMD) [13]. It is crucial to recall that the Laguerre–Gaussian modes constitute an infinite set of basis functions. Thus, to the extent that one can perform OAM encoding and decoding with high efficiency, there is no limit to the amount of information that can be carried by a single photon.

The ability to encode more than one bit per photon is, of course, not restricted to the Laguerre–Gaussian light beams. More generally, the transverse degree of freedom of the light field offers a means to carry and manipulate quantum information. An example of multi-bit information transfer relating to imaging is provided in an experiment performed by Broadbent et al. [14]. A schematic of this experiment is shown in ◘ Fig. 17.2. Part (a) of the figure shows a multiplexed hologram of objects A and B. By saying that the hologram is multiplexed, we mean that different write-beam directions are used to form the interference fringes for each object. Part (b) of the figure shows the read-out stage. It makes use of entangled photons created by parametric downconversion in a BiBO crystal. One of these photons falls onto the trigger detector, which heralds the presence of the photon in the other arm. This photon falls onto an object in its arm, which could be either object A or B. This photon is diffracted from the hologram into the path of either detector A or B, depending on which object is placed in this arm. In this manner, one can determine with high reliability which object is located in this arm, even though only one photon is used to make this determination. Quantitative results are presented in the paper. It is shown that the likelihood of a misidentification (that is, for example, that the photon is detected by detector B when in fact object A is present) is less than 1 %.

The experiment just described shows that one can discriminate between two objects using single-photon illumination. A subsequent experiment [15]

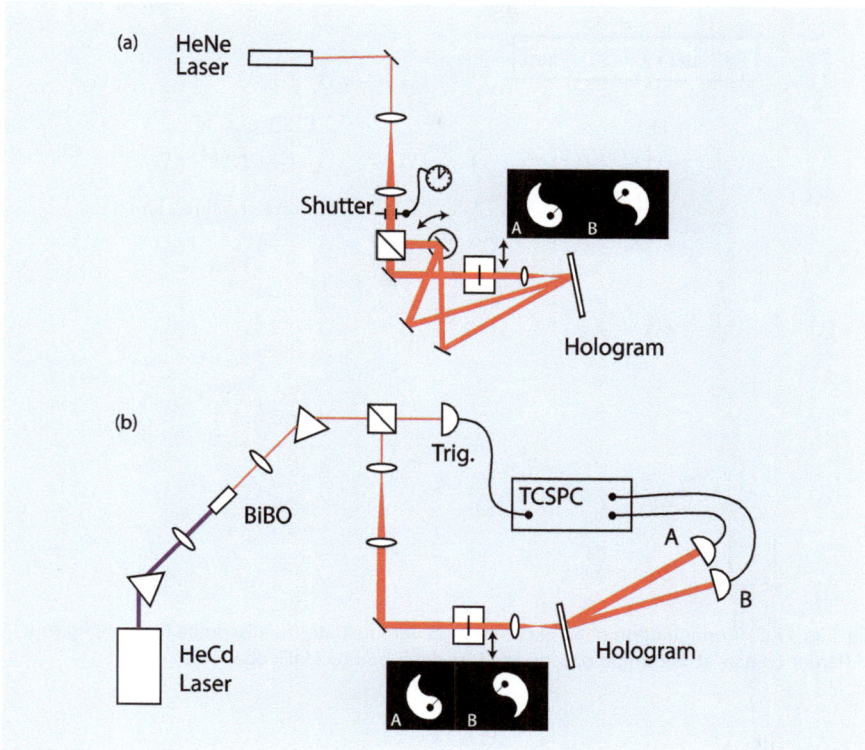

Fig. 17.2 Configuration of the "single-photon imaging" experiment of Broadbent et al. [14] described in the text. Part (**a**) shows the procedure for writing a multiplexed hologram, and part (**b**) shows the read-out stage, which operates at the single-photon level. The TCPSC is a time-correlated single-photon counter

demonstrated the ability to discriminate among four objects, again using only single-photon illumination. For this experiment a "quantum ghost-imaging" protocol [16] is used. The setup is shown schematically in ▪ Fig. 17.3. Spatially entangled photons are again created by the process of parametric downconversion. One of these photons illuminates one of the four test objects (only two are displayed in the diagram to avoid clutter) and the other falls onto a multiplexed hologram, where it is diffracted into one of four output ports. Coincidence events between the reference detector R and one of the detectors A, B, etc. are recorded. In this figure, DM denotes a dichroic mirror for blocking the pump laser and IF is an interference filter with a 10 nm bandwidth, centered at 727.6 nm.

The results of this experiment are shown in ▪ Fig. 17.4. Note that input object a produces counts predominantly in detector A, and similarly for the other three object-detector combinations. The data are displayed using two different normalization conventions. In part (a), data for each object-detector combination are normalized by the maximum coincidence count for the corresponding object. In part (b), the T/A ratio is calculated by dividing the total coincidences by the accidental coincidences for each object-detector combination. Part (c) of the figure shows the four test objects. These results show that one can reliably discriminate among four objects even when they are illuminated with weak light at a single photon level. However, in this experiment the detection efficiency was low, and thus more than one photon needed to illuminate the object in order to make an unambiguous determination. In fact, for the sort of simple, multiplexed hologram used in this experiment the maximum detection efficiency (that is, assuming lossless optical elements and unit quantum efficiency detectors) is equal to $1/N$ (where N is the number of objects). However, there seems to be no reason in principle [17] why a hologram could not be designed to give a maximum detection efficiency of unity.

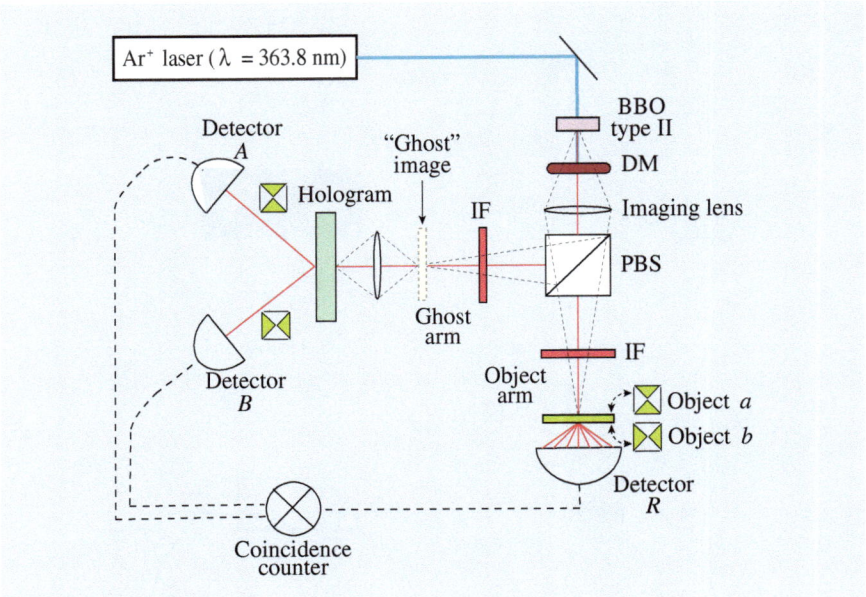

Fig. 17.3 Configuration of an experiment to demonstrate the discrimination among four different objects at the single-photon level, as described by Malik et al. [15]

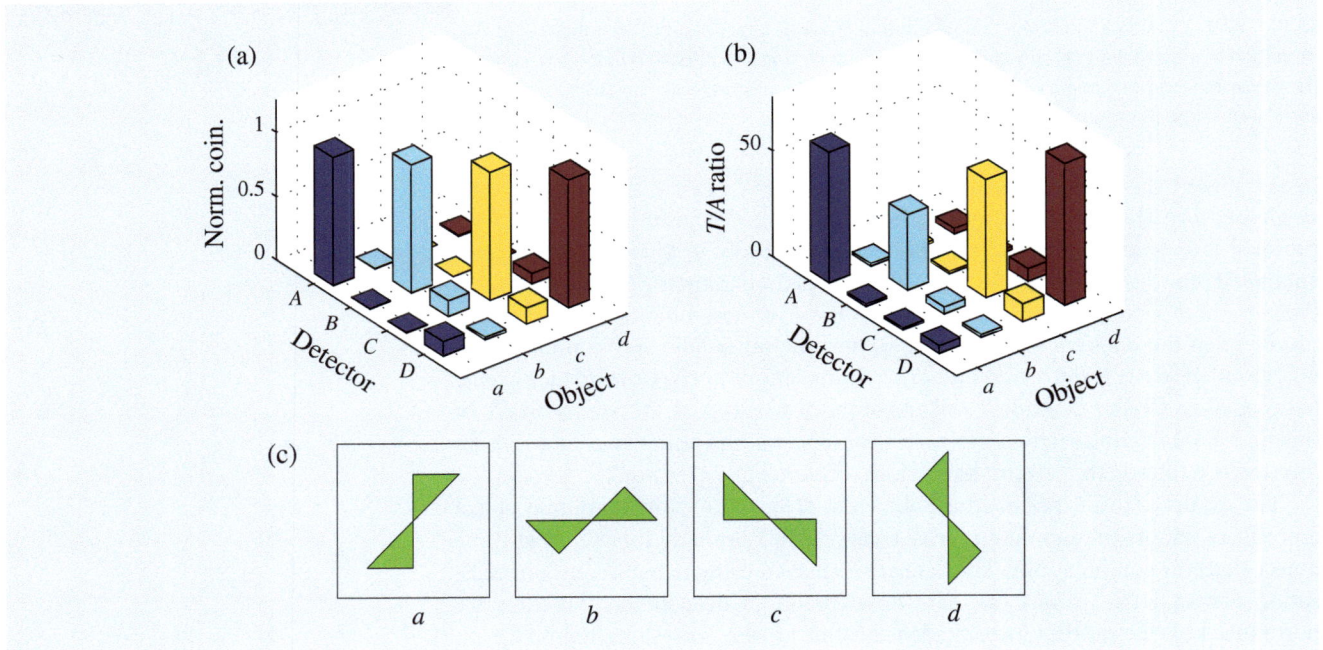

Fig. 17.4 Results of the single-photon ghost imaging experiment described in Fig. 17.3, which can distinguish between four different (non-overlapping) objects [15]

We note that this sort of ghost imaging relies upon correlations between photon pairs. If detector R registers a photon, we know with certainly that this photon possessed the transverse mode structure given by the transmission function of the object in its path. Since this detected photon is entangled with the photon in the other arm, this detected photon must therefore acquire the same conjugate mode structure, and thus be diffracted by the hologram into a specific output port.

17.3 Light Beams that Carry Orbital Angular Momentum

We turn now to another example of a structured light field, namely one carrying orbital angular momentum, which displays interesting quantum properties that can lead to important applications. First, we consider a light field of the form

$$E(r, t) = u(x, y)e^{i\ell\phi}e^{i(kz-\omega t)} \tag{17.1}$$

Here $u(x, y)$ is some function of the transverse coordinates x and y, ℓ is a positive or negative integer, $k = \omega/c$ is the propagation constant, z is the longitudinal coordinate, and ω is the angular frequency. We assume propagation through vacuum. It is well known that such a field carries angular momentum of amount $\ell\hbar$ per photon [18]. For this reason, ℓ is often referred to as the OAM quantum number or OAM mode index. This contribution to the angular momentum is referred to as orbital angular momentum (OAM), distinguishing it from spin angular momentum, which is associated with circular polarization of a light field. These two contributions are additive, and in the paraxial limit considered here independent of each other.

We can understand why the field given by Eq. (17.1) carries angular momentum with the help of the sketch in part (a) of ☐ Fig. 17.5. We see that such a field possesses a wavefront structure in the form of a helix, and that the phase at each point advances in the azimuthal direction at a rate proportional to the value of l. One might well imagine that a small particle placed in such a beam would experience a radiation pressure in the direction of phasefront normal and hence a force with an azimuthal component that induces the object to begin to rotate around the beam axis, and in fact this is just what has been observed experimentally [19, 20].

Equation (17.1) shows that the a light field will carry OAM for any transverse mode function $u(x, y)$. However, some specific mode functions are especially important in the utilization of structured light fields. One such example is that of the Laguerre–Gaussian modes, and we will now briefly explore their properties. The paraxial approximation to the wave equation $(\nabla^2 - \partial^2/\partial t^2)E(x, y, z) = 0$ gives us the paraxial wave equation, which is written in the cartesian coordinate system as

$$\left(\frac{\partial^2}{\partial x^2} + \frac{\partial^2}{\partial y^2} + 2ik\frac{\partial}{\partial z}\right)E(x, y, z) = 0. \tag{17.2}$$

The paraxial wave equation is satisfied by the Laguerre–Gaussian modes, a family of orthogonal modes that have a well-defined orbital angular momentum. The field amplitude, in cylindrical coordinates, $LG_p{}^l(\rho, \phi, z)$ of a normalized Laguerre–Gaussian mode is given by

$$G_p^\ell(\rho, \phi, z) = \sqrt{\frac{2p!}{\pi(|\ell| + p)!}}\frac{1}{w(z)}\left[\frac{\sqrt{2}\rho}{w(z)}\right]^{|\ell|}L_p^\ell\left[\frac{2\rho^2}{w^2(z)}\right]\exp\left[-\frac{\rho^2}{w^2(z)}\right]$$
$$\times\exp\left[-\frac{ik^2\rho^2z}{2(z^2 + z_R^2)}\right]\exp\left[i(2p + |\ell| + 1)tan^{-1}\left(\frac{z}{z_R}\right)\right]e^{-i\ell\phi}, \tag{17.3}$$

where k is the wave-vector magnitude of the field, z_R the Rayleigh range, $w(z)$ the radius of the beam at z, ℓ is the azimuthal quantum number, and p is the radial quantum number. $L_p{}^l$ is the associated Laguerre polynomial.

17

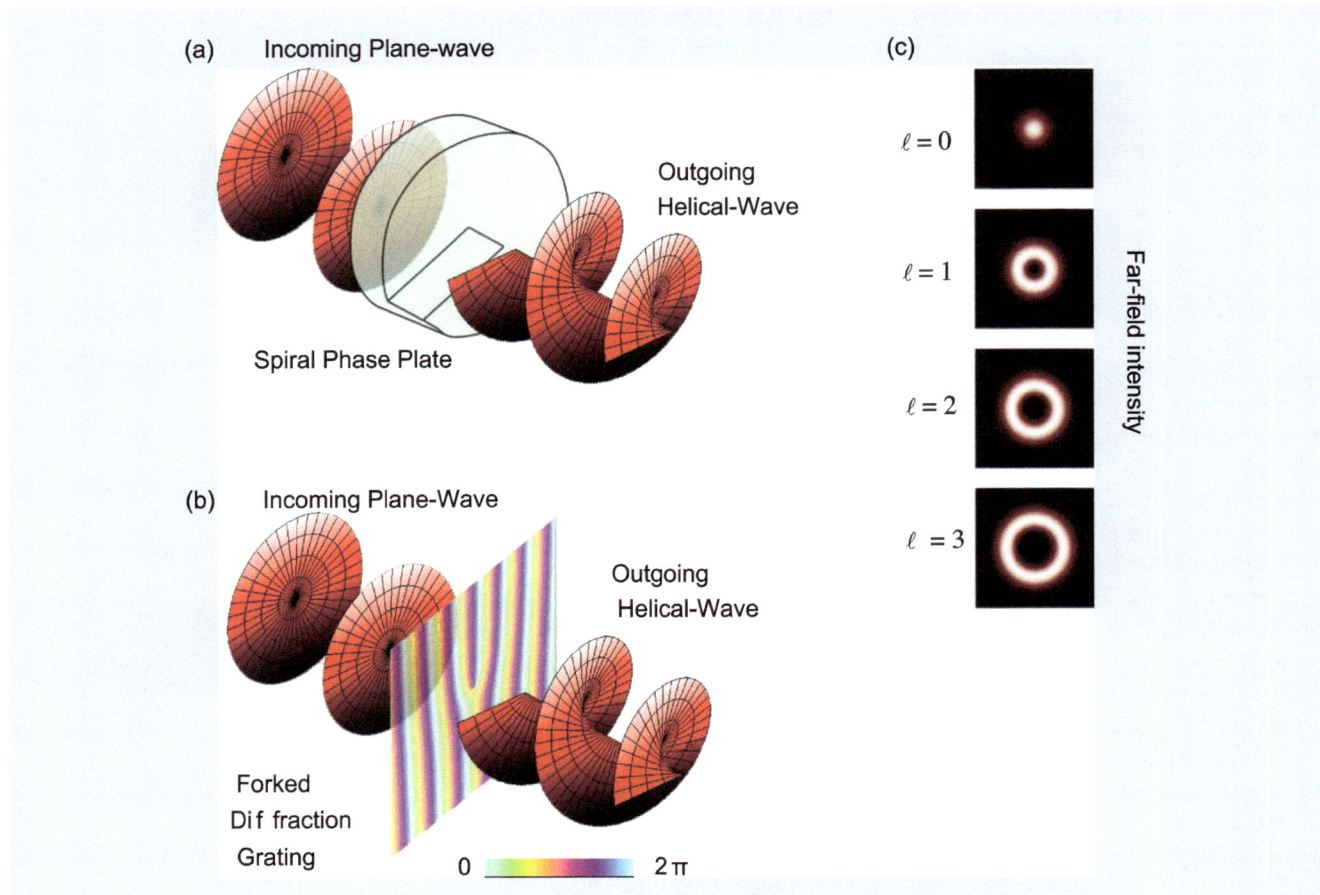

Fig. 17.5 Two methods of producing a light beam that carries orbital angular momentum. (**a**) The conceptually simplest way to form a beam carrying OAM is to pass a plane wave beam through a spiral phase plate, an optical element whose thickness increases linearly with the azimuthal angle. After transmission through such an element, a incident plane wave is transformed into a light beam with helical phasefronts. The height of the phase step controls the azimuthal index ℓ of the transmitted beam. (**b**) Alternatively, one can replace the phase plate with its holographic equivalent with a phase or amplitude structure in the form of a pitchfork as shown. The first-order diffracted beam will have helical phasefronts with an azimuthal index given by the number of dislocations in the pitchfork. These holograms are conveniently created by using a spatial light modulator. (**c**) Examples of some OAM beams produced by these approaches

Some methods for the production of beams that carry OAM are sketched in Fig. 17.5. The use of a spiral phase plate is shown in part (a) [19]. Another important method for generating OAM light beams (part b) is to impress a specially designed computer-generated hologram (CGH) taking the form of a forked diffraction grating [12, 20–23] or a digital micromirror device (DMD) [13]. If a beam with nearly plane wave fronts, such as a Gaussian laser beam, is made to fall onto such a CGH, the diffracted light will acquire the desired form of a beam carrying OAM. Another means to form beams carrying OAM is through the use of a device known as a q-plate [24, 25]. This device is a birefringent phase plate in which the orientation of the birefringent axes varies uniformly as a function of azimuthal position around the axis of the plate. Such a device acts as a spin angular momentum to OAM converter, that is, the OAM carried by the output beam depends on the polarization state of the input beam. A q-plate can thus serve as a quantum interface between polarization-encoded quantum light states and OAM-encoded quantum light states.

17.4 Fundamental Quantum Studies of Structured Light Beams

In ◘ Sect. 17.1 of this chapter we noted that the process of spontaneous parametric downconversion can lead to entanglement in several different degrees of freedom, including position-momentum [26], time-energy [9, 10], polarization [3, 4], and superpositions of OAM modes [27] or explicitly angle-OAM [28]. In this section we provide a brief account of work aimed at studying these various types of entanglement.

In Einstein, Podolski, and Rosen's (EPR's) classic paper [8], they argued against the completeness of quantum mechanics. Their argument was based on the situation of two particles that were strongly correlated both in position and momentum. Later, David Bohm [29] restated this argument in terms of two particles entangled in their spin (or polarization), and it was this spin-version of the EPR paradox that was treated by John Bell in devising his celebrated Bell inequalities. In the ensuing decades, most subsequent work [3, 4] has concentrated on the polarization of the EPR paradox. In 1990 Rarity and Tapster [30] extended the Bell violation to one based on measurement of phase and momentum. Howell et al. [26] later performed an experimental investigation in which they studied the original (i.e., Einstein et al. [8]) position-momentum version of the EPR paradox. Some of their results are summarized in ◘ Fig. 17.6. Their experimental procedure is as follows. Photons entangled in position and momentum were created by type-II parametric downconversion in a BBO nonlinear crystal, and the two photons were separated by a polarizing beamsplitter (PBS) and traveled over separate paths.

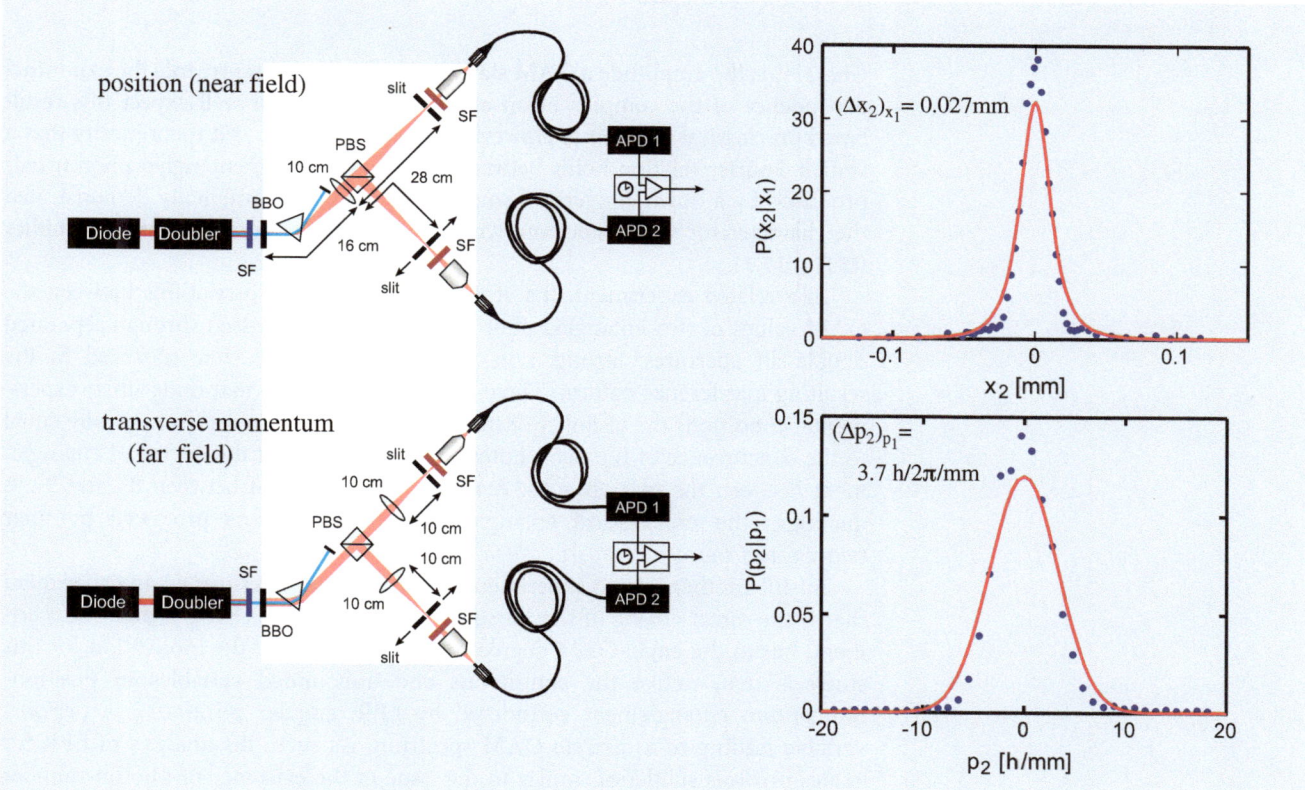

◘ **Fig. 17.6** Laboratory setups (*left*) and measured conditional count rates (*right*) for a laboratory demonstration of the EPR effect for position and transverse momentum variables. The notation $P(x_2 \mid x_1)$ means that probability of measuring one photon at position x_1 conditioned on the other photon being detected at position x_1, and analogously for $P(p_2 \mid p_1)$. The measured conditional uncertainty product is $0.1\hbar$, which violates the Heisenberg uncertainty relation for independent particles [26]

Either the birthplace (i.e., position) of each photon or its transverse momentum could be determined by placing a slit followed by an area detector either in a focal plane of the crystal or in its far field, respectively. Coincidence counts between the two detectors were measured, and the conditional count rates associated with position and momentum are shown in the graphs on the right-hand side of the figure. The measured conditional uncertainty product is found to be $(\Delta x_2)_{x_1}(\Delta p_2)_{p_1} = 0.1\hbar$, which is $\approx$ five times smaller than which might be expected for the uncertainty principle as applied to independent particles.

There has also been considerable interest in studies of time-energy entanglement. For example, Ali-Khan et al. [10] have developed a protocol for quantum key distribution (QKD) that can encode as much as ten bits of information onto a single photon. In a separate study, Jha et al. [10] have studied time-energy entanglement controlled by a geometrical (Berry) phase on the Poincare sphere instead of by using a dynamical phase. The ability to manipulate entanglement by means of a geometrical phase could have important consequences for quantum information technology, because polarization controllers can be much more stable than translation stages needed to actively control optical path lengths.

We next turn to a description of angle-OAM entanglement. We first note that angle and OAM form a Fourier transform pair [22, 31]:

$$A_\ell = \frac{1}{2\pi} \int_{-\pi}^{\pi} \psi(\phi)\exp(-i\ell\phi)d\phi \tag{17.4}$$

$$\psi(\phi) = \sum_{\ell=-\infty}^{\ell=\infty} A_\ell \exp(i\ell\phi) \tag{17.5}$$

where A_ℓ is the amplitude a OAM state ℓ and where $\psi(\phi)$ represents the azimuthal dependence of the complex beam amplitude. One might well expect this result based on classical reasoning. However, Jha et al. [32] showed theoretically that a similar Fourier relation holds between the photons of an entangled photon pair produced by a down-conversion source. They also experimentally demonstrated the characteristic OAM sideband structure that this Fourier relationship implies (◘ Fig. 17.7).

In a related experiment, Jha et al. [33] studied the correlations between the OAM values of two entangled photons after each had passed through separated double-slit apertures. Strong, non-classical, correlations were observed in the resulting interference pattern. These authors also showed that under their experimental conditions the visibility of this interference pattern was numerically equal to the concurrence of the two-photon state, a measure of the degree of entanglement between the two photons. A measured visibility of between 85 and 92 % quantifies the nonclassical entanglement of the photons produced by their two-photon source.

In still another related experiment, Leach et al. [28] performed an experiment that is the direct analog of the Einstein–Podolsky–Rosen (EPR) gedankenexperiment, but in the angle-OAM degrees of freedom. Part of the motivation for this study is that, unlike the continuous and unbounded variables in position-momentum entanglement considered by EPR, angular position is a periodic variable leading to a discrete OAM spectrum. As such, the analysis of EPR for angles involves subtleties similar to the issue of the existence of photon-number photon-phase uncertainty relation [34]. The details of this experiment are presented in ◘ Fig. 17.8.

There has also been great recent interest in harnessing the radial modes of Laguerre–Gaussian (LG) beams in addition to the azimuthal modes that we have

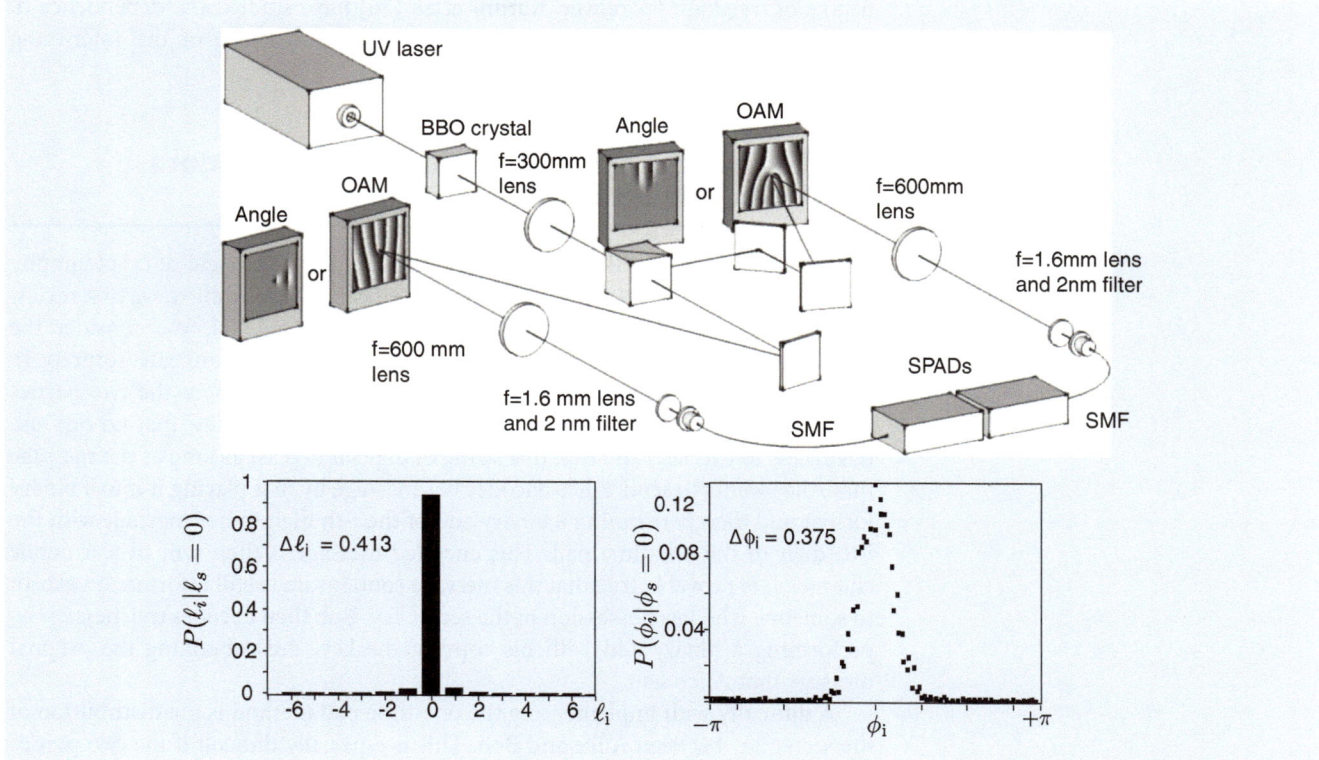

Fig. 17.7 Laboratory setup (*top*) and conditional probability of detection (*bottom*) for a laboratory demonstration of the EPR effect for angular position and orbital angular momentum variables. The conditional uncertainty product is 0.024, which violates the Heisenberg uncertainty relation for independent particles [28]

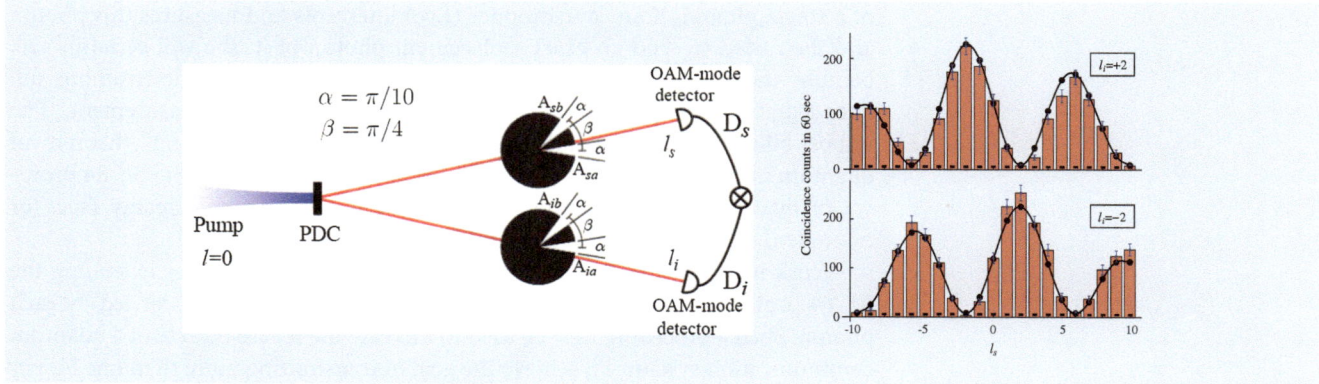

Fig. 17.8 (*left*) Laboratory setup to study angular two-photon interference and angular two-qubit states. In this experiment, an entangled two-photon state is created by parametric downconversion, and each photon falls onto a different aperture having the form of a double angular slit. The OAM content of each photon is then measured, and correlations between the two outputs are calculated. (*right*) Some of the results of this experiment. Here the coincident count rate is plotted as a function of the OAM of the signal photon for two different values of the OAM of the idler photon. The high visibility of the interference fringes is an indication of the high level of entanglement between the two photons [33]

primarily discussed up to now. One reason for this interest is to increase the information capacity of a light beam of a given restricted diameter. We note that the LG modes of Eq. (17.3) depend on two indices, the azimuthal index ℓ and the radial index p. But there are also further subtleties involved in exploiting the radial distribution, related to the fact that the radial coordinate ρ ranges from 0 to ∞, unlike the azimuthal coordinate ϕ, which ranges from 0 to 2π. Recently, Karimi et al. [35] presented a theoretical analysis of the operator nature of the radial degree of freedom. Moreover, Karimi et al. [36] have studied the dependence of Hong–Ou–Mandel interference on the transverse structure of the interfering photons.

17.5 Secure Quantum Communication with More than One Bit Per Photon

We now turn to an application for the OAM of light in the field of cryptography and secure communication. To put this application topic in context, we first review the use of a one-time pad in cryptography (Shannon [37]). We consider the situation in which one party, A (or Alice), wants to communicate securely to another party, B (or Bob). We assume that by pre-arrangement the two parties share the same string of random binary digits known as the key, that no one else has access to this key, and that this string of digits is at least as long as the message that Alice wants to send. Alice encodes her message by first placing it into a binary format and then performing a binary add of the i-th digit in her message with the i-th digit in the one-time pad. This encoded message is then sent over a public channel. It is provably true that this message contains no useful information except to someone who has possession of the secret key. Bob then decodes the message by performing a binary add with his copy of the key, thus obtaining the original message that Alice sent.

A difficulty with implementing the one-time pad method is the distribution of the secret key between Alice and Bob. This is especially difficult if the two parties are not and cannot be in the same place, where the key can simply be handed from one to the other. When not in the same place, a procedure proposed by Bennet and Brassard in 1984 (know as the BB-84 protocol) can be used to distribute the key in an entirely secure manner. In brief (some of the details are provided below), Alice sends the key one element at a time, and each digit is encoded in the quantum state of a single photon. If an eavesdropper (Eve) intercepts and measures this photon and then tried to send an exact replacement photon of it, she will certainly fail, because the laws of quantum mechanics prohibit her from determining full knowledge of the quantum state of a photon in a single measurement. The impossibility of doing so results from the celebrated "no-cloning" theorem of quantum mechanics [38]. Secure communication through use of the BB-84 protocol of quantum key distribution (QKD) is now a commercial reality (see, for example the website ▶ http://www.idquantique.com).

Work in which the present authors have participated involves extending the BB-84 protocol so that more than one bit of information can be carried by each photon. Such a procedure may be used to increase the secure bit rate of a quantum communication system. To achieve the goal of transmitting more than one bit per photon, we encode information in the transverse degree of freedom of the light field. For the transverse degree of freedom one can choose any complete set of orthonormal modes. In keeping with the context of this chapter we consider encoding in OAM modes such as Laguerre–Gaussian (LG) modes. In the original QKD proposal of Bennett and Brassard, information is encoded in the polarization degree of freedom of an individual photon. As a result, only one bit of information could be impressed onto each photon. In contrast, when using OAM, there is no limit to how many bits of information can be impressed onto a single photon, as the LG modes span an infinite-dimensional state space. As mentioned above, one motivation for doing this is that rate of data transmission is thereby increased. Another more subtle motivation is that the security of the protocol can be increased by encoding information within a higher-dimensional state space.

The system that we envisage is illustrated in broad scope in ◻ Fig. 17.9. It consists of a sender, Alice, and a receiver, Bob. Alice impresses information onto

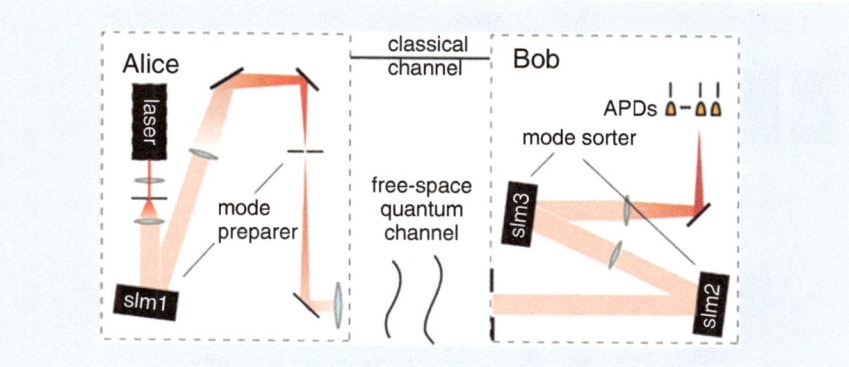

■ **Fig. 17.9** System schematic of the baseline QKD protocol of Mirhosseini et al. [41]. A sender (A or Alice) impresses information onto an individual photon through use of a spatial light modulator (SLM). This photon is then sent to the receiver (B or Bob) through a free-space link, where it may experience degradation by means of atmospheric turbulence. The receiver then determines the quantum state of this photon

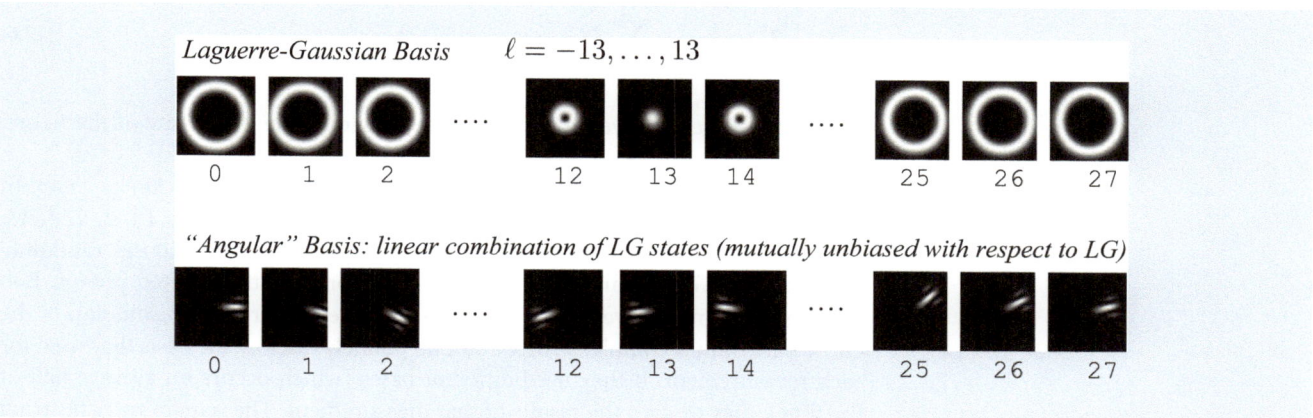

■ **Fig. 17.10** The LG basis (*top*) and a linear combination of the LG states (*bottom*) that constitutes the angular basis (AB). The information is encoded by launching individual photons that have been prepared in one of these modes [40]

the transverse degree of freedom of individual photons through the use of a spatial light modulator (SLM). Bob then randomly guesses which basis (OAM or angle) Alice might be using and makes a measurement of the quantum state of the received photon in this basis. The procedure for ensuring the security of the transmission is a generalization of that of the BB84 protocol and is described in the review of Gisin and Thew [39]. In the remainder of the present section we describe in more detail our laboratory procedure and present some laboratory data.

The BB84 QKD protocol entails Alice sending each photon in a randomly chosen basis. At least two mutually unbiased bases (MUBs) must be used. Certain advantages accrue from using more than two MUBs. It is known that the maximum number B_{max} of MUBs is related to the dimension D of the state space by $B_{max} = D + 1$. In our laboratory investigations we use the minimum number of MUBs, $B = 2$. We choose this value for convenience and to maximize our data transmission rate. Our two basis sets are illustrated in ■ Fig. 17.10. One basis is comprised of the LG states themselves. The other basis is composed of a linear combination of the LG states of the form

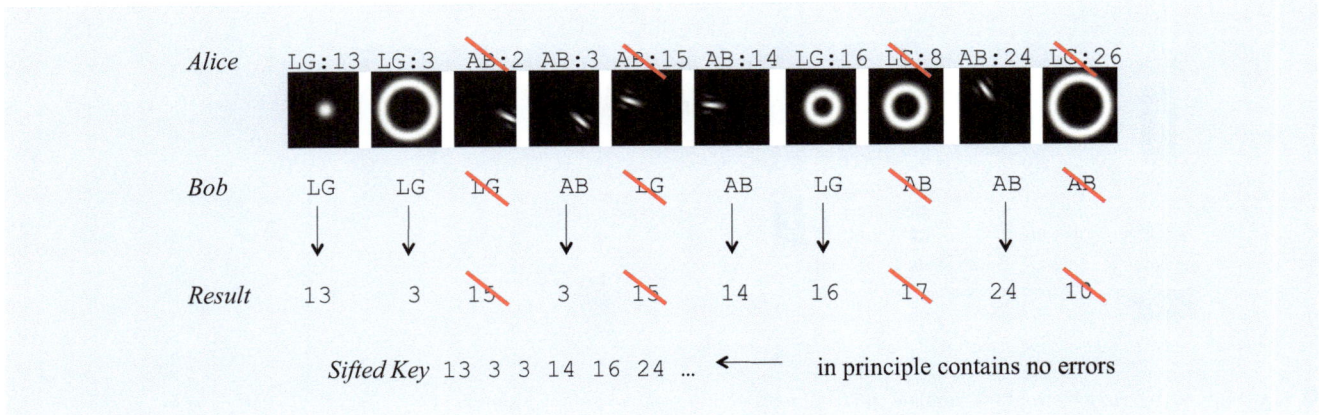

Fig. 17.11 Example of a proposed implementation of a generalized BB84 protocol in a high-dimensional (27-dimensions as illustrated) state space [40]

$$\Psi^N_{AB} = \frac{1}{\sqrt{27}} \sum_{\ell=-13}^{13} LG_{\ell,0}\exp(2\pi i\ell/27). \qquad (17.6)$$

From this expression we obtain the states shown in the lower row of the figure, which is referred to as the angular basis (AB).

An example of the implementation of this protocol is shown in ◘ Fig. 17.11. In this example, Alice is attempting to send the string of numbers 13, 3, 2, 3, 15, 14, 16, 8, 24, 26 to Bob. For each transmitted photon, Alice chooses randomly between the LG basis and the AB basis. Also, for each transmitted photon Bob chooses randomly between the OAM and AB bases. After the transmission of the entire data train is complete, Alice and Bob publicly disclose the basis they used for each measurement. If they used different bases (which occurs on average half of the time), they discard the results of that measurement. The remaining data string is known as the sifted data, and this data should contain no errors. Any error in this data string could be the result of measurement errors or to the presence of an eavesdropper. For reasons of extreme caution, one must ascribe all errors to the presence of an eavesdropper. To test for errors, Alice and Bob sacrifice some fraction of their data for public comparison. If errors are detected, they conclude that an eavesdropper is present and take appropriate corrective measures.

We have implemented this BB84-type protocol in our laboratory. ◘ Figure 17.11 shows how Alice forms each of the basis states. Basically, she programs a spatial light modulator (SLM) to convert an individual photon in a plane-wave state into one of the desired LG or AB modes [41]. The upper row shows the LG basis and the lower row shows the angular AB basis. The panel on the left shows representative examples of the pattern displayed on the SLM. The panels on the right show examples of the field distribution written onto the light field. These frames show actual laboratory results, although read out with intense classical light, not with single photons (◘ Fig. 17.12).

Special considerations apply to the configuration of the receiver, or Bob. He is presented with a single photon and needs to determine its quantum state. Thus, he is allowed to perform only one measurement to determine in which of a large number of quantum states the received photon resides. This sorting task has eluded the scientific community until very recently, when Berkhout et al. [40] demonstrated a means for performing this task. Their approach is illustrated in ◘ Fig. 17.13. The key element of this approach is the ability to map the azimuthal phase distribution of an incident mode onto a linear phase distribution at the

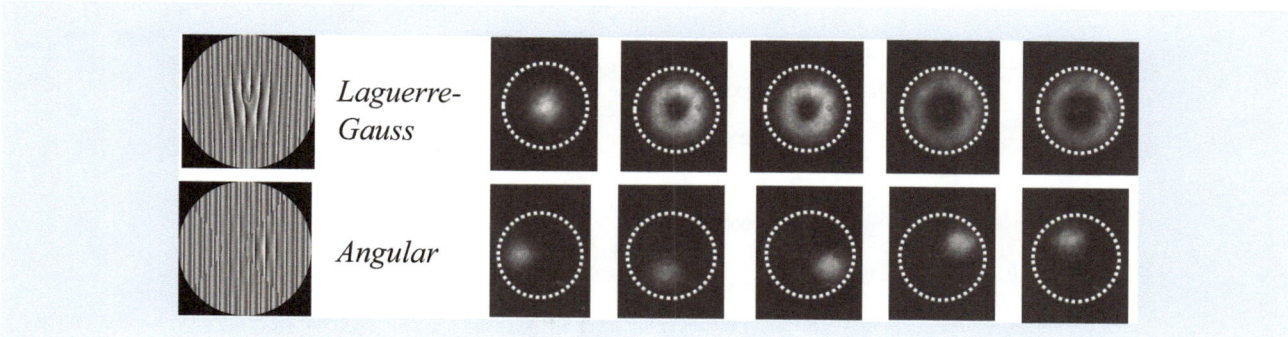

◘ **Fig. 17.12** Illustration of the procedure for producing light fields in one of the Laguerre–Gaussian or angular basis states, shown for the case of a five-state bases (D = 5). The *dotted circles* in the panels on the right denote the aperture of the transmitting optics [Unpublished laboratory results of M.N. O'Sullivan]

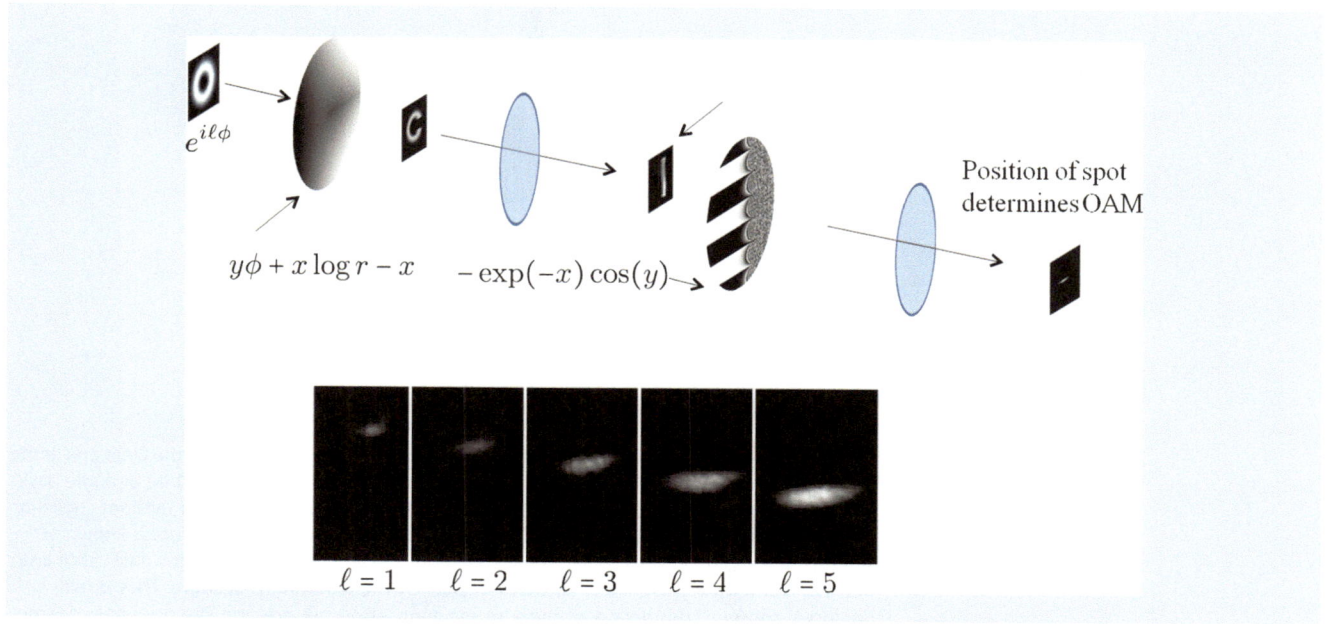

◘ **Fig. 17.13** The angular-to-linear reformatter (Glasgow mode-sorter). (*top*) Physical layout of the reformatter. (*bottom*) Some results showing the performance of the reformatter used as a sorter. Note that the vertical position of the light beam at the output of the sorter depends on the OAM value ℓ of the beam (Unpublished data from the Boyd laboratory)

output of the device. Of course, a linear phase ramp in one cartesian dimension is simply a wavefront tilt, and leads to a shift in the position of the beam in the far field. It turns out that one can determine analytically the form of the phase function that needs to be applied to a light field to perform this mapping. In their original implementation of this sorting procedure, Berkhout et al. [42] applied this phase mapping through the use of an SLM. In a more recent work they have fabricated refractive elements that perform this same function but with much higher conversion efficiency than those based on diffraction from an SLM.

Some laboratory results validating the performance of this sorter are shown in ◘ Fig. 17.14. These results demonstrate our ability to discriminate among various quantum states in either the LG or angular basis. In each basis we include only four basis states. This limitation is due to the number of photodetectors (APDs) available to us. We see no fundamental limit to our ability to distinguish among all of the states in our protocol, 27 in this particular situation. We see that there is a small amount of crosstalk among the various channels.

17

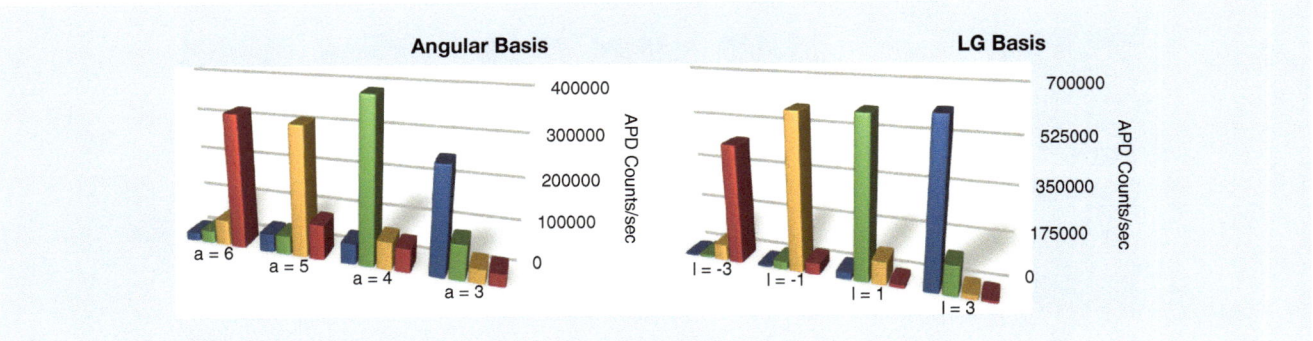

Fig. 17.14 Laboratory data demonstrating Bob's ability to discriminate among various quantum states in either the LG or angular basis through the use of the Glasgow mode sorter. Note that discrimination is good but not perfect; there is cross-talk among the channels (Unpublished data from the Boyd laboratory)

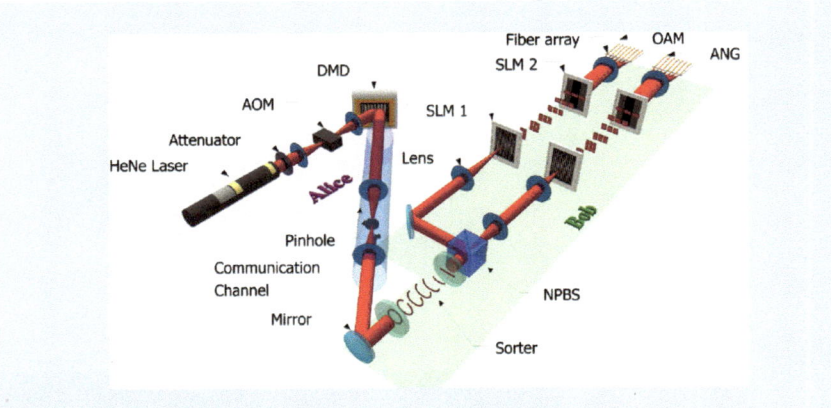

Fig. 17.15 Experimental setup of Mirhosseini et al. [43]. This system uses the OAM and angle bases to implement a QKD system. Alice encodes information in either the OAM or angle basis (chosen randomly), and Bob performs a measurement after making a further random choice of basis. Data obtained when they use different bases is later discarded, in a process known as sifting. If there is no eavesdropper, there should in concept be no errors in this data. Alice and Bob can test for the presence of an eavesdropper by the following procedure. They openly disclose a subset of this data, and check to see if any errors are present. The presence of errors suggests the presence of an eavesdropper

Using the procedures described here, we recently performed a realistic demonstration of quantum key distribution based on OAM encoding [43]. Our experimental setup is shown in Fig. 17.15 and is composed of the various components described above. Alice prepares state to be sent by first carving out pulses from a highly attenuated He-Ne laser through the use of an AOM. Then spatial mode information is impressed on these pulses with a digital micromirror device (DMD). Bob's mode sorter and fan-out elements map the OAM modes and the ANG modes onto separated spots that are collected by an array of fibers and sent to individual APD detectors.

Some of the results of this demonstration are shown in Fig. 17.16. The top row (left) shows the string of numbers sent by Alice and the top row right shows the string of numbers received by Bob. Note that the strings are not identical; due to various transmission and detection errors, some of the symbols are not detected as transmitted. In the figure errors are marked in red and are underlined. As a test of their system, Alice and Bob could publicly disclose these results to determine the fractional error rate. However, in an operating system, Alice and Bob would want

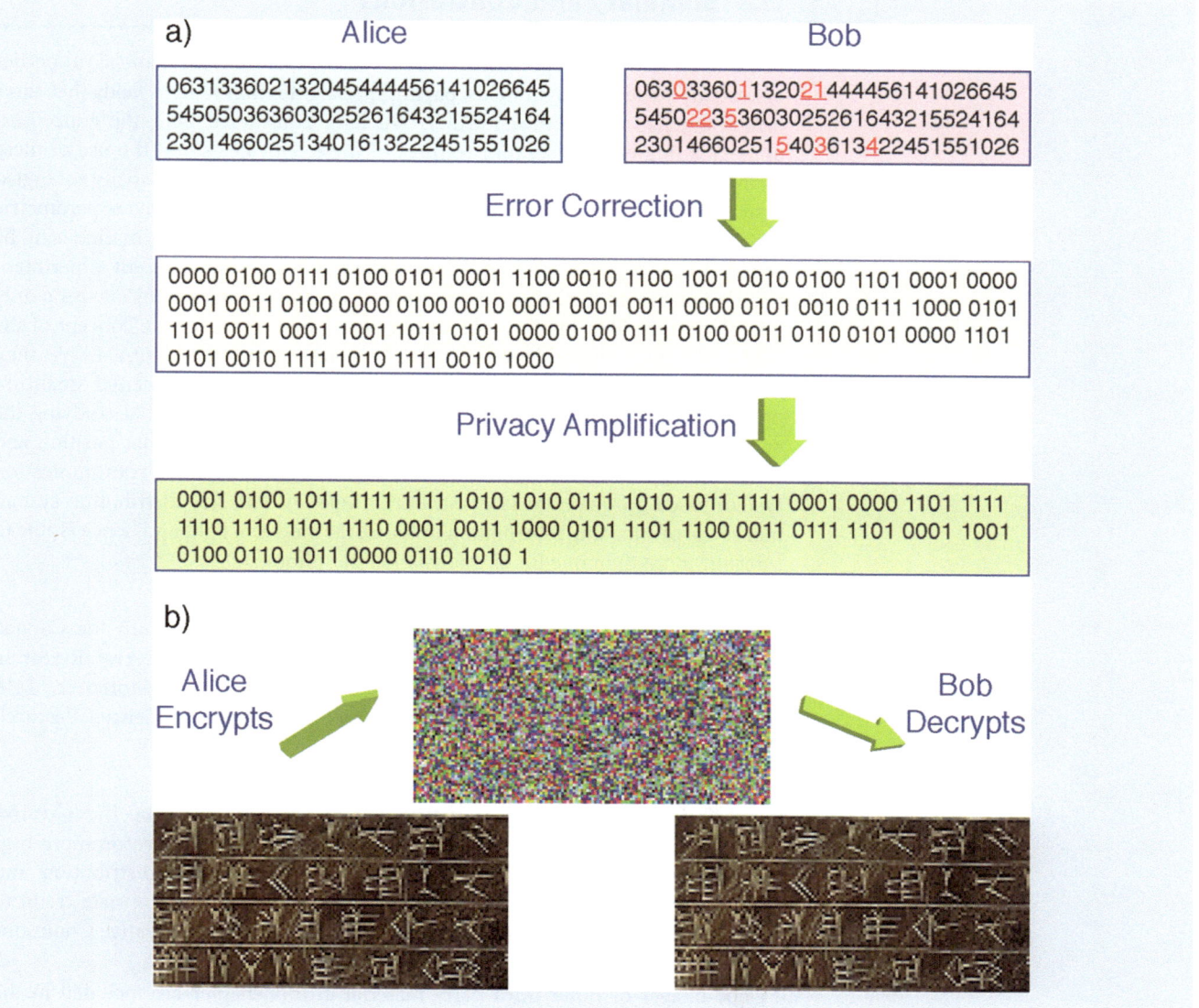

Fig. 17.16 Experimental results from the study of Mirhosseini et al. [43] (**a**) Example of a random sifted key from the experiment. The spatial modes are mapped to numbers between 0 and 6, and errors are marked in *red* and are *underlined*. Each symbol is converted into a three digit binary number first and the binary key is randomized before the error-correction. Privacy amplification minimizes Eve's information by shortening the key length. (**b**) Alice encrypts the secret message (in this case an image of an ancient Persian tablet) using the shared secure key and Bob subsequently decrypts it

to sacrifice only a small fraction of their data to test system security. Alice and Bob therefore employ an error correction algorithm and a method known as privacy amplification (which shortens the length of their shared string) to decrease the number of errors in the shared strings. They end up with a shared key that contains essentially no errors. (For this reason we show the string constituting the shared key only once.) As a graphic demonstration of the use of this procedure, we also show how it could be used for the secure transmission of an image. The image is separated into pixels which are then digitized and transmitted using the secret key shared by Alice and Bob. An eavesdropper who intercepted the signal would see only the noisy pattern that is also displayed.

17.6 Summary and Conclusions

In this chapter we have presented a review of the quantum mechanical properties of spatially structured light fields, paying special attention to light fields that carry orbital angular momentum (OAM). We have considered both the conceptual understanding of the quantum features of these light fields and the use of these quantum features for applications. We describe how to produce spatially entangled light fields by means of the nonlinear optical process of spontaneous parametric downconversion. We address the question of how much information can be encoded onto a single photon. As an example, we review a recent experiment that demonstrated the ability to discriminate among four target objects using only one photon for illumination. We also present a description of the concept of the OAM of light, and we describe means to generate and detect OAM. We then present a brief survey of some recent studies of the fundamental quantum properties of structured light beams. Much of this work is aimed at studying the nature of entanglement for the complementary variables of angular position and OAM. Finally, as a real-world application, we describe a secure communication system based on quantum key distribution (QKD). This key distribution system makes use of encoding information in the OAM modes of light and hence is able to transmit more than one bit of information per photon.

Acknowledgements One of us (RWB) acknowledges support from the Canada Excellence Research Chair program, from the US Office of Naval Research, DARPA, and the Air Force Office of Scientific Research. Moreover, MJP acknowledges support from the Engineering and Physical Sciences Research Council of the UK and from the European Research Council.

References

1. Kolobov MI (ed) (2006) Quantum imaging. Springer, New York
2. Schrödinger E (1935) Discussion of probability between separated systems. Proc Camb Phys Soc 31:555
3. Freedman SJ, Clauser JF (1972) Experimental test of local hidden-variable theories. Phys Rev Lett 28(938):938–941
4. Aspect A, Grangier P, Roger G (1981) Experimental tests of realistic local theories via Bell's theorem. Phys Rev Lett 47(7):460–463

5. Burnham DC, Weinberg DL (1970) Observation of simultaneity in parametric production of optical photon pairs. Phys Rev Lett 25:84

6. Kwiat PG, Mattle K, Weinfurter H, Zeilinger A, Sergienko AV, Shih Y (1995) New high-intensity source of polarization-entangled photon Pairs. Phys Rev Lett 75:4337

7. Ling A, Lamas-Linares A, Kurtsiefer C (2008) Absolute emission rates of spontaneous parametric down-conversion into single transverse Gaussian modes. Phys Rev A 77:043834

8. Einstein A, Podolsky B, Rosen N (1935) Can quantum-mechanical description of physical reality be considered complete? Phys Rev 47:777

9. Ali-Khan I, Broadbent CJ, Howell JC (2007) Large-alphabet quantum key distribution using energy-time entangled bipartite States. Phys Rev Lett 98:060503

10. Jha AK, Malik M, Boyd RW (2008) Exploring energy-time entanglement using geometric phases. Phys Rev Lett 101:180405

11. Gerry C, Knight PL (2005) Introductory quantum optics. Cambridge University Press, Cambridge

12. Curtis JE, Koss BA, Grier DG (2002) Dynamic holographic optical tweezers. Opt Commun 207:169–175

13. Mirhosseini M, Magaña-Loaiza OS, Chen C, Rodenburg B, Malik M, Boyd RW (2013) Rapid generation of light beams carrying orbital angular momentum. Opt Exp 21:30204

14. Broadbent CJ, Zerom P, Shin H, Howell JC, Boyd RW (2009) Discriminating orthogonal single-photon images. Phys Rev A 79:033802

15. Malik M, Shin H, O'Sullivan M, Zerom P, Boyd RW (2010) Quantum ghost image discrimination with correlated photon Pairs. Phys Rev Lett 104:163602

16. Strekalov DV, Sergienko AV, Klyshko DN, Shih YH (1995) Observation of two-photon "Ghost" interference and diffraction. Phys Rev Lett 74:3600

17. Miller DAB (2013) Self-configuring universal linear optical component. Photon Res 1(1):1. ► http://doi.org/10.1364/PRJ.1.000001

18. Allen L, Beijersbergen MW, Spreeuw RJC, Woerdman JP (1992) Orbital angular-momentum of light and the transformation of Laguerre-Gaussian laser modes. Phys Rev A, 45(11):8185–8189

19. He H, Friese M, Heckenberg N, Rubinsztein-Dunlop H (1995) Direct observation of transfer of angular momentum to absorptive particles from a laser beam with a phase singularity. Phys Rev Lett 75:826–829

20. O'Neil AT, MacVicar I, Allen L, Padgett MJ (2002) Intrinsic and extrinsic nature of the orbital angular momentum of a light beam. Phys Rev Lett 88(5):053601

21. Leach J, Dennis MR, Courtial J, Padgett MJ (2005) Vortex knots in light. New J Phys 7:55. ► http://doi.org/10.1088/1367-2630/7/1/055

22. Yao E, Franke-Arnold ES, Courtial J, Barnett SM, Padgett MJ (2006) Fourier relationship between angular position and optical orbital angular momentum. Opt Exp14:9071

23. Bolduc E, Bent N, Santamato E, Karimi E, Boyd RW (2013) Exact solution to simultaneous intensity and phase masking with a single phase-only hologram. Opt Lett 38:3546

24. Marrucci L, Manzo C, Paparo D (2006) Optical spin-to-orbital angular momentum conversion in inhomogeneous anisotropic media. Phys Rev Lett 96:1605

25. Karimi E, Piccirillo B, Nagali E, Marrucci L, Santamato E (2009) Efficient generation and sorting of orbital angular momentum eigenmodes of light by thermally tuned q-plates. Appl Phys Lett 94:1124

26. Howell JC, Bennink RS, Bentley SJ, Boyd RW (2004) Realization of the Einstein-Podolsky-Rosen paradox using momentum and position-entangled photons from spontaneous parametric down conversion. Phys Rev Lett 92:210403

27. Mair A, Vaziri A, Weihs G, Zeilinger A (2001) Entanglement of the orbital angular momentum states of photons. Nature 412(6844):313–316

28. Leach J, Jack B, Romero J, Jha AK, Yao AM, Franke-Arnold S, Ireland DG, Boyd RW, Barnett SM, Padgett MJ (2010) Quantum correlations in optical angle-orbital angular momentum variables. Science 329:662

29. Bohm D (1951). Quantum theory. Prentice-Hall, Englewood Cliffs, p 29, and Chapter 5 Section 3, and Chapter 22 Section 19

30. Rarity JG, Tapster PR (1990) Experimental violation of Bell?s inequality based on phase and momentum. Phys Rev Lett 64(21):2495–2498

31. Pors JB, Aiello A, Oemrawsingh SSR, van Exter MP, Eliel ER, Woerdman JP (2008) Angular phase-plate analyzers for measuring the dimensionality of multimode fields. Phys Rev A 77:033845

32. Jha AK, Jack B, Yao E, Leach J, Boyd RW, Buller GS, Barnett SM, Franke-Arnold S, Padgett MJ (2008) Fourier relationship between the angle and angular momentum of entangled photons. Phys Rev A 78:043810

33. Jha AK, Leach J, Jack B, Franke-Arnold S, Barnett SM, Boyd RW, Padgett MJ (2010) Angular two-photon interference and angular two-qubit states. Phys Rev Lett 104:010501

34. Pegg DT, Barnett SM (1997) Tutorial review - quantum optical phase. J Mod Opt 44:225–264

35. Karimi E, Boyd RW, de la Hoz P, de Guise H, Rehacek J, Hradil Z, Aiello A, Leuchs G, Sánchez-Soto LL (2014) Radial quantum number of Laguerre-Gaussian modes. Phys Rev A89:063813

36. Karimi E, Giovannini D, Bolduc E, Bent N, Miatto FM, Padgett MJ, Boyd RW (2014) Exploring the quantum nature of the radial degree of freedom of a photon via Hong-Ou Mandel interference. Phys Rev A 89:013829

37. Shannon C (1949) Communication theory of secrecy systems. Bell Syst Tech J 28 (4):656–715. As is often the case, the idea of the one-time pad appears to have been invented independently several times in the past. Shannon presents a good review of communication security based on the one-time pad

38. Wootters WK, Zurek WH (1982) A single quantum cannot be cloned. Nature 299:802–803

39. Gisin N, Thew R (2007) Quantum communications. Nat Photon 1:165

40. Boyd RW, Jha A, Malik M, O'Sullivan C, Rodenburg B, Gauthier DJ (2011) Quantum key distribution in a high-dimensional state space: exploiting the transverse degree of freedom of the photon. In: Proceedings of the SPIE 7948, 79480L

41. Gruneisen MT, Miller WA, Dymale RC, Sweiti AM (2008) Holographic generation of complex fields with spatial light modulators: application to quantum key distribution. Appl Opt 47:A32

42. Berkhout GCG, Lavery MPJ, Courtial J, Beijersbergen MW, Padgett MJ (2010) Efficient sorting of orbital angular momentum states of light. Phys Rev Lett 105:153601

43. Mirhosseini M, Magaña-Loaiza OS, O'Sullivan MN, Rodenburg B, Malik M, Lavery MPJ, Padgett MJ, Gauthier DJ, Boyd RW (2015) High-dimensional quantum cryptography with twisted light. New J Phys 17:033033

17

Quantum Communication with Photons

Mario Krenn, Mehul Malik, Thomas Scheidl, Rupert Ursin, and Anton Zeilinger

M. Krenn (✉) • M. Malik • T. Scheidl • A. Zeilinger (✉)
Vienna Center for Quantum Science and Technology (VCQ), Faculty of Physics, University of Vienna,
Boltzmanngasse 5, A-1090 Vienna, Austria

Institute for Quantum Optics and Quantum Information (IQOQI), Austrian Academy of Sciences (ÖAW),
Boltzmanngasse 3, A-1090 Vienna, Austria
e-mail: mario.krenn@univie.ac.at; anton.zeilinger@univie.ac.at

R. Ursin
Institute for Quantum Optics and Quantum Information (IQOQI), Austrian Academy of Sciences (ÖAW),
Boltzmanngasse 3, A-1090 Vienna, Austria

© The Author(s) 2016
M.D. Al-Amri et al. (eds.) *Optics in Our Time*, DOI 10.1007/978-3-319-31903-2_18

18.1   Introduction

Ever since its inception, quantum physics has changed our understanding of the fundamental principles of nature. Apart from their impact on all fields of academic research, these insights have merged together with the field of information science to create the novel field of quantum information. Quantum information science provides qualitatively new concepts for communication, computation, and information processing, which are much more powerful than their classical counterparts. Quantum information is an intriguing example where purely fundamental and even philosophical research can lead to new technologies. The developments in this young field recently experienced a worldwide boom—as is evidenced by the increasing number of quantum information centers being founded in countries all over the world. Although its long-term industrial applications cannot be clearly anticipated, it is clear that quantum information science entails a huge potential economic impact. For reasons of space we limit ourselves to polarization and orbital angular momentum (OAM) as information carrying degrees of freedom.

18.1.1   The Quantum Bit

In classical information and computation science, information is encoded in the most fundamental entity, the bit. Its two possible values **0** and **1** are physically realized in many ways, be it simply by mechanical means (as a switch), in solids by magnetic or ferroelectric domains (hard drives), or by light pulses (optical digital media). All of these methods have one thing in common—one state of the device mutually excludes the simultaneous presence of the other—the switch is either **on** or **off** (◘ Fig. 18.1).

The superposition principle entails one of the most fundamental aspects of quantum physics, namely to allow the description of a physical system as being in a probabilistic combination of its alternative states. This so-called *superposition of states* not only provides all predictions for the outcome of a physical measurement, but it also has drastic consequences for the nature of the physical state that we ascribe to a system. Its most important direct implication is the so-called

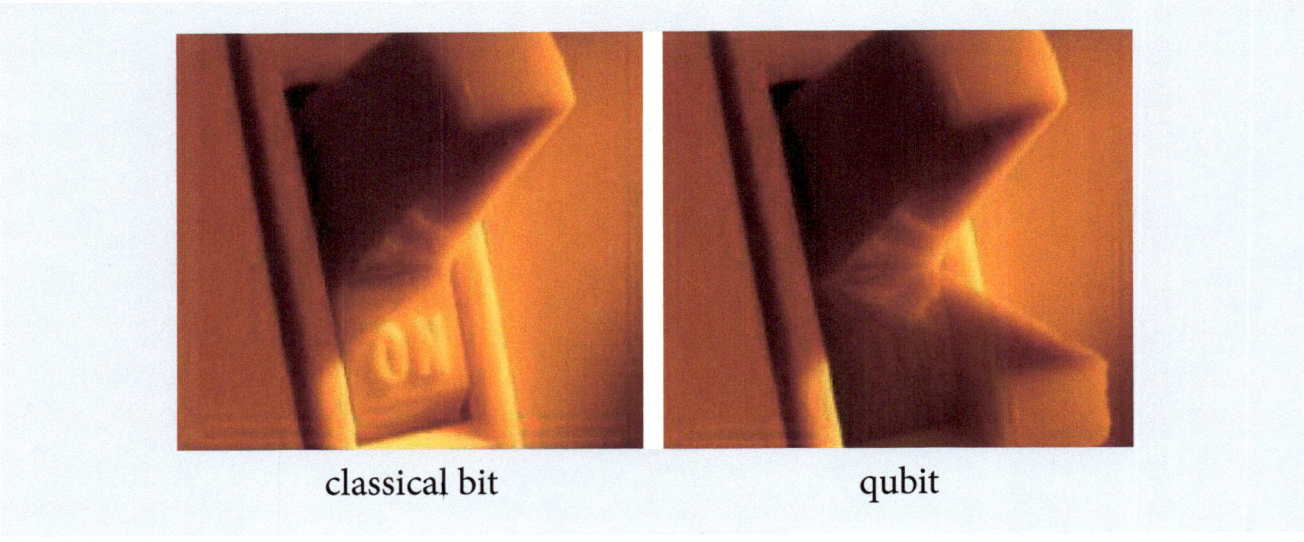

classical bit   qubit

◘ **Fig. 18.1** An illustration of the difference between a classical bit and a qubit. The classical bit is always in a well-defined state while the qubit can also exist in a superposition of orthogonal states (copyright University of Vienna)

no-cloning theorem, which states that it is impossible to obtain a perfect copy of a qubit in an unknown state without destroying the information content of the original. The no-cloning theorem is the basis for the security of all quantum communication schemes described in the following sections, and will be explained later in more detail.

A qubit can be realized in many different physical systems such as atoms, ions, and super-conducting circuits. The most prominent physical realization of a qubit in view of a potential global-scale quantum communication network is with photons. Using photons, the two values of a bit, **0** and **1**, can be encoded in many different ways. One possibility is to use two orthogonal polarization states of a single photon, referred to as a polarization qubit. In the latter case, one can ascribe the horizontal polarization state of the photon with the logical value **0** and the vertical polarization state with the value of **1**. Any arbitrary polarization state can be obtained via a superposition of the horizontal and vertical state. The advantage of using photonic polarization qubits is that they can be easily generated, controlled, and manipulated with rather simple linear optical devices like wave plates. Furthermore, since photons rarely exhibit interaction with the environment they are the best candidates for long-distance free-space transmission as would be required in a future network involving ground-to-space links.

To fully understand a qubit, it is important to distinguish between a coherent superposition and a mixture of possible states. For its use in quantum communication, it is important that a photon exists in a coherent superposition of its possible states. For example, a polarization qubit being in a coherent superposition of horizontal and vertical polarizations (with a certain phase relation) can be understood as a photon polarized diagonally at + 45°. A polarizer set at this angle will always transmit such a photon with 100 % probability (and zero probability when set to − 45°). However, a photon in a mixture (incoherent superposition) of horizontal and vertical polarization states will be transmitted with 50 % probability.

Quantum superpositions, however, are not limited to just two possible states. The information carried by a photon is potentially enormous. While polarization is necessarily a two-level (qubit) property, other degrees of freedom of a photon such as its spatial or temporal structure can have many orthogonal levels. For example, a photon can exist in a coherent superposition of different paths coming out of a multi-port beam splitter. These types of superpositions are referred to as "high-dimensional" by virtue of their ability to encode large amounts of information. Consider a photon that is carrying a complicated image, such as that shown in ◘ Fig. 18.2. This image can be decomposed in terms of any orthonormal basis of spatial modes. The number of modes required for a complete description of this image dictates the number of levels, or dimensionality of this photon. One such basis is the set of Laguerre–Gaussian modes, which are described by a photon carrying a twisted wavefront. The phase structure of such a photon winds from

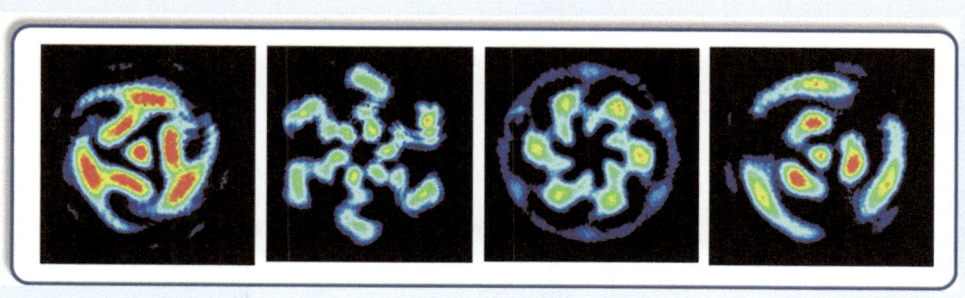

◘ **Fig. 18.2** Some types of higher-order spatial modes, which can carry more information than one bit per photon (Image by Mario Krenn, copyright University of Vienna)

0 to 2π azimuthally around the optical axis, with the number of twists dictating the photon state dimensionality. Using such high-dimensional degrees of freedom of a photon for encoding surely increases the amount of information one can send per photon. However, a more subtle advantage of doing this is found in quantum communication—not only can one vastly increase the information capacity of quantum communication systems, one can also increase their security. This point is discussed in detail later in this chapter.

18.1.2 Entanglement

The principle of superposition also holds for states containing several qubits. This allows for multi-qubit systems, which can only be described by joint properties. Such states are called *entangled*, describing the fact that none of the particles involved can be described by an individual quantum state [6, 19, 70]. This is equivalent to the astonishing property of entangled quantum systems, that all of their information content is completely entailed in the correlations between the individual subsystems and none of the subsystems carry any information on their own. For example, when performing measurements on only one of two entangled qubits, the outcome will be perfectly random, i.e., it is impossible to obtain information about the entangled system. However, since the entangled state consists of two qubits, the correlations shared between them must consist of two bits of classical information. As a consequence, these two bits of information can only be obtained when the outcomes of the individual measurements on the separate subsystems are compared (see ◘ Fig. 18.3).

Another intriguing feature of entangled states is that a measurement on one of the entangled qubits instantaneously projects the other one onto the corresponding perfectly correlated state, thereby destroying the entanglement. Since these perfect correlations between entangled qubits are in theory independent of the distance between them, the entanglement is in conflict with the fundamental concepts of classical physics—locality (i.e., distant events cannot interact faster than the speed of light) and realism (i.e., each physical quantity

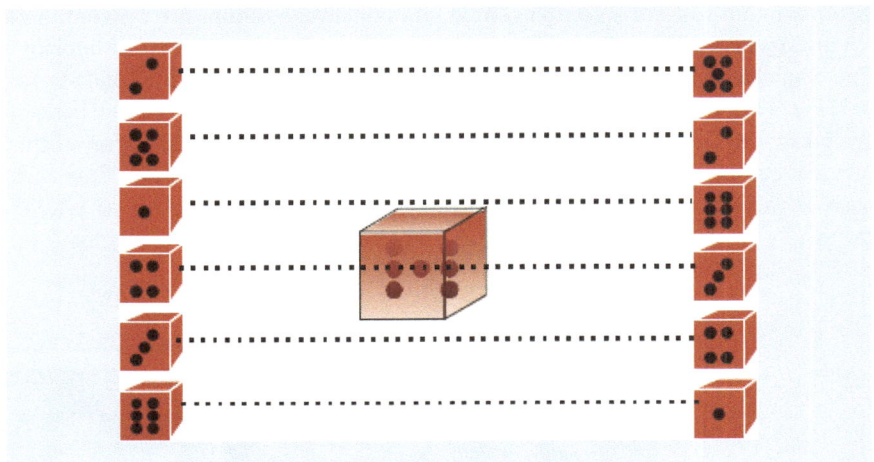

◘ **Fig. 18.3** If one could entangle a pair of dices with respect to their numbers, one can encode the message **7** by using their entanglement. None of the dices would carry this information on its own and a local measurement of the dice will result in a completely random result (without revealing the information). However, the results are perfectly correlated to add up to **7** for every joint measurement on the two dices. Note that a rolling dice corresponds to a six-dimensional quDit (where D stands for Dimension), which was prepared in a way unknown to us, and which is about to be measured in one out of six orthogonal bases (copyright University of Vienna)

that can be predicted with certainty corresponds to an ontological entity, a so-called element of reality) [6]. This has led to various philosophical debates about whether quantum mechanics can serve as a complete description of reality. However, there have been many experiments performed addressing this issue, and to date each of them has confirmed the predictions of quantum mechanics [4, 25, 66, 68, 84]. One should note that while here we focus on polarization and orbital-angular momentum entanglement, light can be entangled in its other degrees of freedom as well, such as time-frequency [24, 33] and position-momentum [11, 37, 45].

18.1.3 Mutually Unbiased Bases

One fascinating concept in quantum mechanics is the possibility to encode quantum information in different ways. In the simple example of the polarization of light, there are three bases in which one can encode one bit of information (see ◘ Fig. 18.4). These are the horizontal and vertical (H/V) basis, the diagonal and anti-diagonal (D/A) basis, and the left- and right-circular (L/R) basis. One can encode a bit in the H/V basis by considering 0 to be horizontal polarization and 1 to be vertical polarization. If a photon encoded in either H or V polarization is measured in any of the other two bases, its information cannot be extracted. For example, in the case of measurements made in the D/A basis, in 50 % of the cases, a diagonally polarized photon will be observed; in the other cases, the photon will be measured as anti-diagonally polarized. This property is the main ingredient for quantum cryptography, as we will see later. Furthermore, in higher-dimensional systems, fundamental properties of mutually unbiased bases are still open questions that are significant for quantum communication.

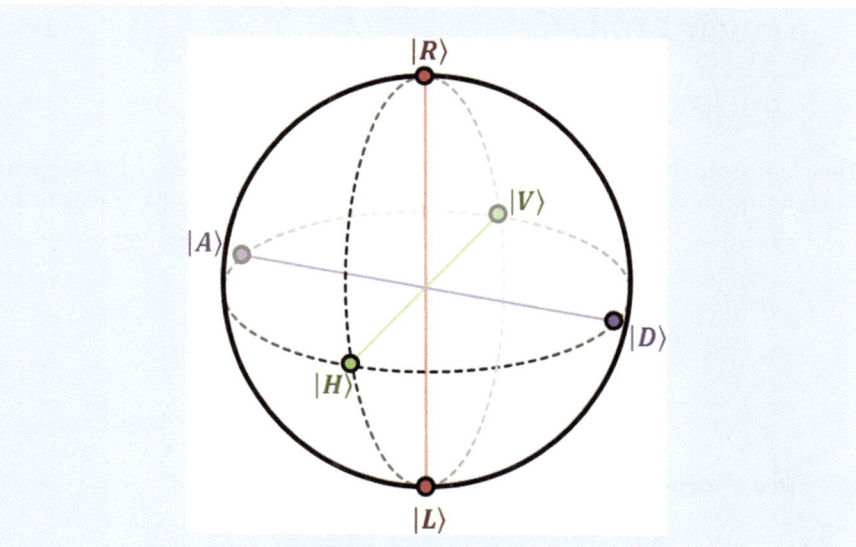

◘ **Fig. 18.4** The Bloch-sphere: Graphical representation of a two-dimensional qubit. There are three mutually unbiased bases—three ways of encoding information in different ways. In the case of polarization, they correspond to horizontal and vertical (*violet*), diagonal and anti-diagonal (*green*) and right- and left-circular (*red*) polarization (Image by Mario Krenn, copyright University of Vienna)

18.1.4 Faster-than-Light Communication and the No-Cloning Theorem

As discussed above, two entangled photons are connected even though they can be spatially separated by hundreds of kilometers. The measurement of the first photon immediately defines the state of the second photon. Can one use that to transmit information faster than the speed of light? If Alice and Bob share an entangled state and measure their respective photon in the same mutually unbiased basis (for instance, in the horizontal/vertical basis), they will always find the same result. However, whether they detect a horizontal or vertical photon is intrinsically random—there is no way that Alice could influence the outcome of Bob. Regardless, there could exist a workaround, as shown in ◻ Fig. 18.5. Alice could use her choice of measurement basis to convey information: either horizontal/vertical (H/V) if she wants to transmit 0 or diagonal/antidiagonal (D/A) if she wants to send 1. When she does this, Bob's photon is immediately defined in that specific basis. If Bob could now clone his photon, he could make several measurements in both bases and find out in which of the two bases his photon is well defined: If Alice measured in the H/V basis and finds a H outcome, all of Bob's measurements in the H/V basis will be H. However, his measurements in the D/A basis will show 50 % diagonal and 50 % antidiagonal. Thus, he knows that Alice has chosen the H/V basis, and thereby transmitted the bit value 0.

Unfortunately, there is one problem with that protocol: It cannot exist. In 1982, Wootters and Zurek found that quantum mechanics forbids one to perfectly clone a quantum state [86]. This profound result originates from a simple property of quantum mechanics, namely the linear superposition principle. We can inspect what a potential cloning-operation $\hat{C}$ would do. We use an input quantum state, and an undefined second photon $|X\rangle$. After the cloning operation, the second photon should have the polarization property of the first photon. This is how our cloning machine would act on states in the H/V-basis:

$$\hat{C}\left(|H\rangle|X\rangle\right) = |H\rangle|H\rangle \tag{18.1}$$

$$\hat{C}\left(|V\rangle|X\rangle\right) = |V\rangle|V\rangle \tag{18.2}$$

The cloning-machine should work in every basis, thus we inspect what happens when we try to clone a diagonally polarized photon $|D\rangle$. Note that a diagonally

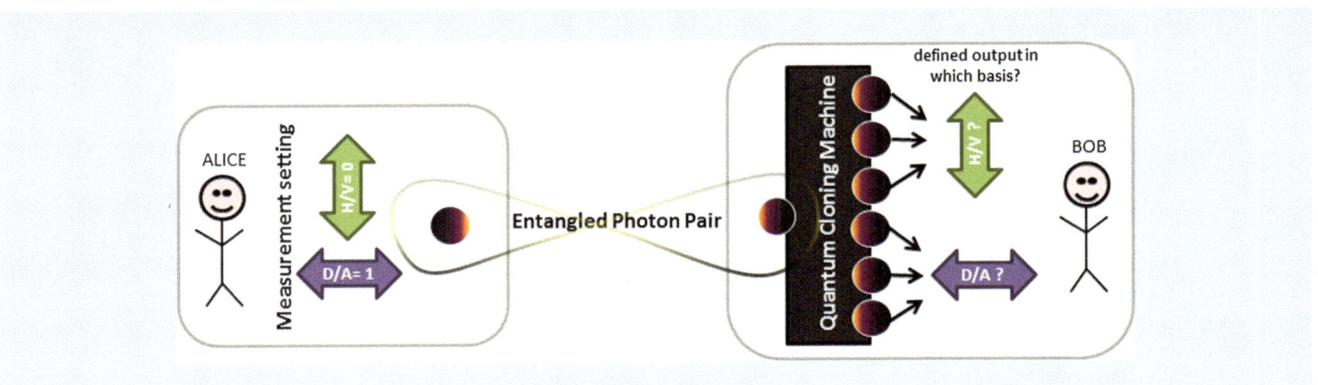

◻ **Fig. 18.5** Visualization of a faster-than-light quantum communication protocol, if (!) quantum states could be cloned: Alice and Bob share an entangled photon pair. By choosing the measurement basis between horizontal/vertical or diagonal/antidiagonal polarization, Alice projects the whole state into an eigenstate of that basis. This means that Bob's state is also defined in that basis. To find the basis chosen by Alice, Bob would need to measure more than one photon. If he could perfectly clone his photon, he could find the basis, and receive the information faster than light. Unfortunately, this is prohibited by the no-cloning theorem, a fundamental rule in quantum mechanics (Image by Mario Krenn, copyright University of Vienna)

polarized photon can be expressed in the H/V basis as a coherent superposition of a horizontal and a vertical part $|D\rangle = \frac{1}{\sqrt{2}}(|H\rangle + |V\rangle)$. The quantum cloning machine acts as

$$
\begin{aligned}
\hat{C}(|D\rangle|X\rangle) &= \hat{C}\left(\frac{1}{\sqrt{2}}(|H\rangle + |V\rangle)|X\rangle\right) \\
&= \frac{1}{\sqrt{2}}(\hat{C}|H\rangle|X\rangle + \hat{C}|V\rangle|X\rangle) \\
&= \frac{1}{\sqrt{2}}(|H\rangle|H\rangle + |V\rangle|V\rangle)
\end{aligned}
\tag{18.3}
$$

The last line in Eq. (18.3) was obtained by using Eqs. (18.1) and (18.2) for the cloning operator $\hat{C}$. The result is an entangled state that cannot be factorized into $|D\rangle\,|D\rangle$. If one were to measure either of the entangled photons individually, the result would be random, and certainly not $|D\rangle$. From this simple example it is clear that quantum cloning is not possible. This property prohibits faster-than-light communication, but it opens the door to many different quantum secret sharing protocols, such as quantum cryptography.

18.1.5 Quantum Communication Schemes

The counterintuitive quantum principles of superposition and entanglement are not only the basis of acquiring a deeper understanding of nature, but also enable new technologies that allow one to perform tasks which are not possible by classical means. When speaking about such "quantum technologies," we refer to technologies that make explicit use of these kinds of quantum properties that do not have a classical analog. Quantum information science and quantum communication are important ingredients in future quantum information processing technologies. They enable the transfer of a quantum state from one location to another. All quantum communication schemes have in common that two or more parties are connected via both a classical communication channel and a quantum channel (i.e., a channel over which quantum systems are transmitted). Typically, measurements are performed on the individual quantum (sub-) systems and the measurement bases used for every measurement are communicated via the classical channel. Here, we focus on quantum communication with discrete variables. However, we should mention that there exists a parallel branch of quantum communication that is based on continuous variables, where extensive theoretical and experimental work has been performed. More information on this field can be found in [83] and references therein.

18.1.6 Quantum Key Distribution

If two parties want to share a secret message, they have two options: the first possibility is to share a random key that is the size of the message that needs to be encrypted with it (shown in ◘ Fig. 18.6). The sender, let's call her Alice, performs a simple logical operator (an *exclusive or*, XOR) of the message with the key, and gets the cipher. The cipher can only be read if the key is known. The receiver of the encrypted text, whom we will call Bob, can use the key to undo Alice's operation, which gives him the original message. The challenge lies in Alice and Bob having to share the entire secret key.

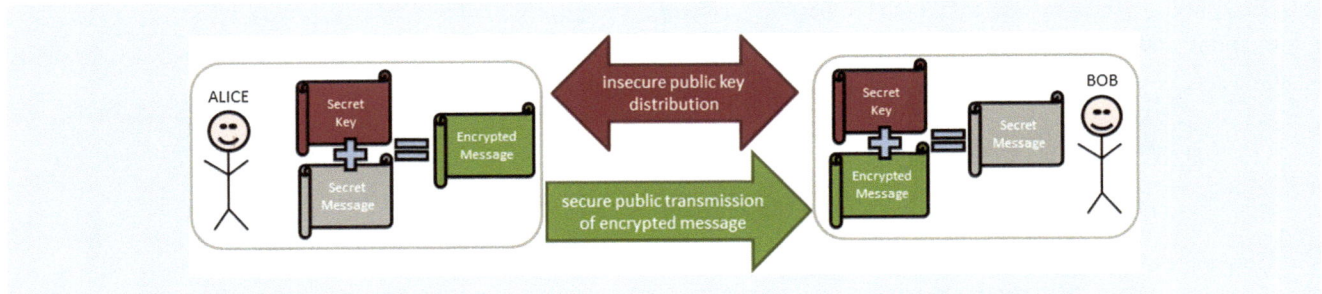

Fig. 18.6 Scheme of a classical symmetric cryptographical system. Alice wants to send a secret message to Bob. In order to do so, Alice and Bob have to share a secret key. With this key, they can distribute messages securely. The bottleneck is the distribution of the key. This problem is solved by quantum cryptography (Image by Mario Krenn, copyright University of Vienna)

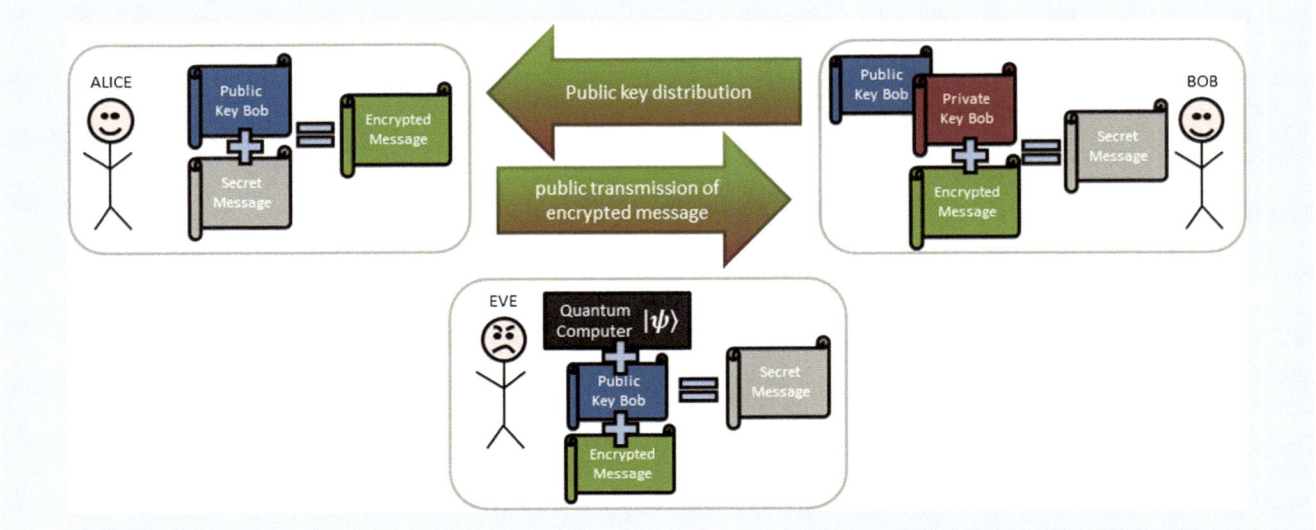

Fig. 18.7 Scheme of a classical asymmetric cryptographical system. Alice wants to send a secret message to Bob. In order to do so, Bob prepares a public and private key. Alice can then prepare an encrypted message for Bob with his public key. Usually, the message can only be decrypted by Bob with his private key. However, a powerful enough eavesdropper (for example, one with a quantum computer!) can infer Bob's private key from the public key, and can thus break the encryption protocol (Image by Mario Krenn, copyright University of Vienna)

The alternative is a public–private key cryptography. This method, invented in the 1970s, is based on the computational complexity of finding the prime factors of large numbers. Again, Alice wants to send a secret message to Bob. Now Bob creates a pair of keys, a private and a public one. Everybody who has Bob's public key can encrypt messages for him. However, only Bob can decrypt those messages with his private key. However, it has been discovered by Peter Shor in 1994 that a quantum computer could factor prime numbers significantly faster than classical computers [71]. It would allow an eavesdropper to read the secret message with only the information that is distributed publicly (see ◘ Fig. 18.7). One possible way to circumvent this problem is quantum key distribution.

Quantum key distribution (QKD) allows two authorized parties to establish a secret key at a distance. The generation of this secret key is based on the same quantum physical principles that a quantum computer relies on. In contrast to classical cryptography, QKD does not simply rely on the difficulty of solving a mathematical problem (such as finding the prime powers of a large number). Therefore, even a quantum computer could not break the key. QKD consists of two phases (see ◘ Fig. 18.8). In the first phase the two communicating parties, usually called Alice and Bob, exchange quantum signals over the quantum channel

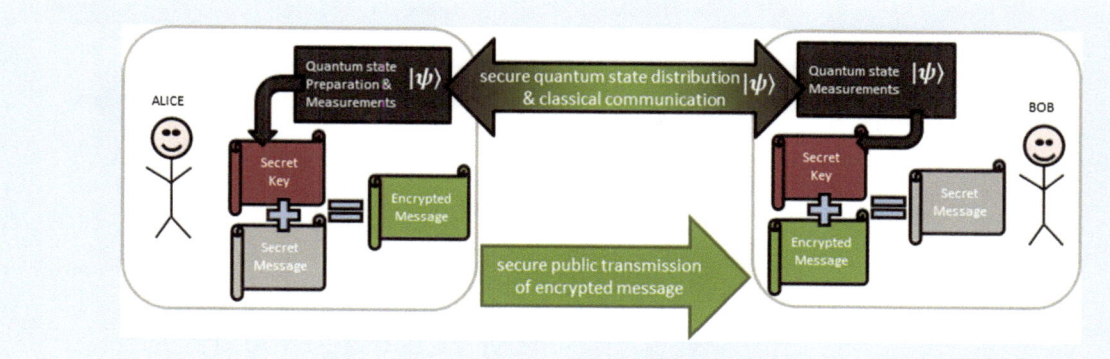

Fig. 18.8 Scheme of a quantum cryptographical system. Alice wants to send a secret message to Bob. In order to do so, a secret key is established over public (quantum) channels. Alice prepares a quantum state and transmits it to Bob. By making appropriate measurements, Alice and Bob can obtain a shared secret key. Alice then encrypts the message with this key and sends it to Bob; Bob can decrypt it with his copy of the key. Eavesdropping attempts during the key transmission appear as errors in the measurement results, allowing the presence of an eavesdropper to be detected (Image by Mario Krenn, copyright University of Vienna)

and perform measurements, obtaining a raw key (i.e., two strongly correlated but nonidentical and only partly secret strings). In the second phase, Alice and Bob use the classical channel to perform an interactive post-processing protocol, which allows them to distill two identical and completely secret (known only to themselves) strings, which are two identical copies of the generated secret key. The classical channel in this protocol needs to be authenticated: this means that Alice and Bob identify themselves; a third person can listen to the conversation but cannot participate in it. The quantum channel, however, is open to any possible manipulation from a third person. Specifically, the task of Alice and Bob is to guarantee security against an adversarial eavesdropper, usually called Eve, tapping on the quantum channel and listening to the exchanges on the classical channel.

In this context security explicitly means that a nonsecret key is never used: either the authorized parties can indeed create a secret key, or they abort the protocol. Therefore, after the transmission of the quantum signals, Alice and Bob must estimate how much information about raw keys has leaked out to Eve. Such an estimate is obviously impossible in classical communication: if someone is tapping on a telephone line, or when Eve listens to the exchanges on the classical channel, the communication goes on unmodified. This is where quantum physics plays a crucial role: in a quantum channel, leakage of information is quantitatively related to a degradation of the communication. The origin of security of QKD can be traced back to the fundamental quantum physical principles of superposition and no-cloning. If Eve wants to extract some information from the quantum states, this is a generalized form of measurement, which will usually modify the state of the system. Alternatively, if Eve's goal is to have a perfect copy of the state that Alice sends to Bob, she will fail due to the no-cloning theorem, which states that one cannot duplicate an unknown quantum state while keeping the original intact. In summary, the fact that security can be based on general principles of physics allows for unconditional security, i.e. the possibility of guaranteeing security without imposing any restriction on the power of the eavesdropper.

The first quantum cryptography scheme was published by Bennett and Brassard in 1984 [8] and is known today as the BB84 protocol. It requires four different qubit states that form two complementary bases (i.e., if the result of a measurement can be predicted with certainty in one of the two bases, it is completely undetermined in the other). These states are usually realized with four linear polarization states of a photon forming two complementary bases, for, e.g., horizontal (H), vertical (V), diagonal (D), and anti-diagonal (A). As illustrated in ▣ Fig. 18.9, Alice sends single photons to Bob, which were prepared randomly

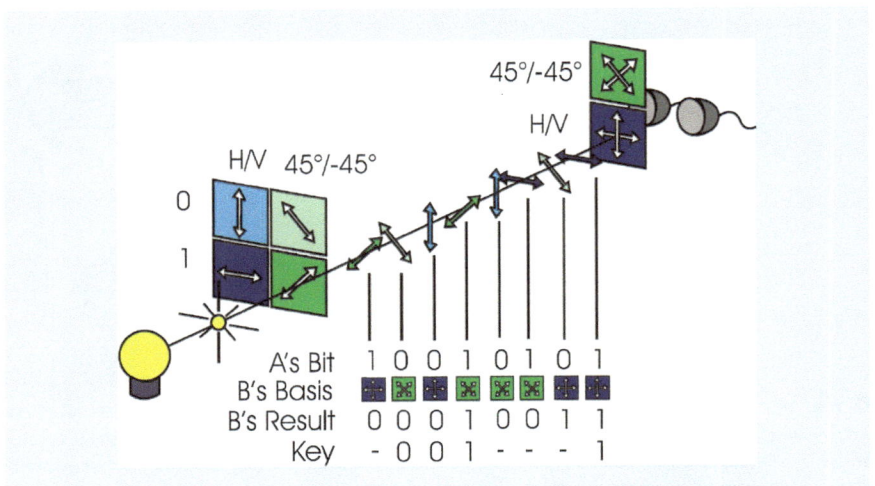

A's Bit	1	0	0	1	0	1	0	1
B's Basis								
B's Result	0	0	0	1	0	0	1	1
Key	-	0	0	1	-	-	-	1

Fig. 18.9 An illustration of the coherent state BB84 protocol. Alice sends polarized single photons, prepared randomly in either of two complementary bases. Bob measures them, again randomly in one of the two bases. After publicly announcing their choice of bases, they obtain the sifted key from their data (Copyright Univ. of Vienna)

in any of the four polarization states and records the state of any sent photon. Bob receives and analyzes them with a two-channel analyzer, again randomly in one of the two complementary bases H/V or D/A. He records his measurement results together with the corresponding measurement basis. After enough photons have been transmitted, Bob communicates publicly with Alice and tells her which photons actually arrived and in which basis it was measured, but does not reveal the measurement result. In return, Alice tells Bob when she has used the same bases to prepare them, because only in these cases Bob obtains the correct result. Assigning the binary value **0** to H and D and the value **1** to V and A leaves Alice and Bob with an identical set of **0** s and **1** s. This set is called the sifted key.

The security of the key distribution is based on the fact that a measurement of an unknown quantum system will (in most cases) disturb the system: If Alice's and Bob's sifted keys are perfectly correlated (which can be proven by comparing a small subset of the whole sifted key via classical communication), no eavesdropper tried to listen to the transmission and the key can be used for encoding a confidential message using the one-time pad (i.e., a specific key is exactly as long as the message to be encrypted and this key is only used once). In practical systems, however, there will always be some inherent noise due to dark counts in the detectors and transmission errors. As it cannot be distinguished whether the errors in the sifted key come from noise in the quantum channel or from eavesdropping activity, they all must be attributed to an eavesdropping attack. If the error is below a certain threshold, Alice and Bob can still distill a final secret key using classical protocols for error correction and privacy amplification. If the error is above the threshold, the key is discarded and a new distribution has to be started.

In contrast to the *single-photon* protocols described above, entanglement based QKD uses entangled photon pairs to establish the secure key [20]. Let's assume that Alice and Bob share a polarization entangled two-photon state. Due to the perfect polarization correlations between entangled photons, Alice and Bob will always obtain the same result, when they measure the polarization state of their photon in the same measurement basis. Since both measure randomly in one of the two complementary bases (just as in the BB84 protocol), they have to publicly communicate after they have finished their measurements, which photons they actually detected and in which basis it was measured. Again, they discard those results in which they disagreed in the measurement basis and finally end up with an identical set of **0** s and **1** s—the sifted key. Just as in the BB84 protocol, Alice and Bob authenticate their keys by openly comparing (via classical communication) a small subset of their keys and evaluating the bit error rate.

There are two big advantages in using entangled photons for implementing the QKD protocol. First, the randomness of the individual measurement results is intrinsic to the entangled state and therefore the randomness of the final key is ascertained. Second, an eavesdropper cannot mimic an entangled state by sending single photons in correlated polarization states simultaneously to Alice and Bob. Hence, when using a subset of the transmitted photon pairs to examine the entanglement between them, secure communication is possible even though the operator of the entangled photon source might not be trustworthy.

18.1.7 Quantum Teleportation

Quantum teleportation is a process by which the state of a quantum system is transferred onto another distant quantum system without ever existing at any location in between [10]. In contrast to what is often wrongly stated, this does not even in principle allow for faster-than-light communication or transport of matter. This becomes clearer when considering the entire three-step protocol of quantum teleportation (an illustration is shown in ◘ Fig. 18.10).

First, it is necessary that Alice (the sender) and Bob (the receiver) share a pair of entangled qubits (qubits 2 and 3 in the figure). Next, Alice is provided with a third qubit (qubit 1), the state of which she wants to teleport and which is unknown to her. In the last step, Alice destroys any information about the state of qubit 1 by performing a so-called Bell-state measurement (BSM) between qubits 1 and 2. As a consequence of this measurement and due to the initial entanglement between qubit 2 and 3, qubit 3 is instantaneously projected onto the same state as qubit 1. However, the teleportation protocol only works in cases, where the BSM resulted in exactly one out of four possible random outcomes. As a consequence, Bob needs to be notified by Alice about the outcome of the BSM in order to being able to identify the successful teleportation events. This requires classical communication between Alice and Bob and essentially limits the speed of information transfer within the teleportation protocol to the speed of the classical communication channel.

Quantum teleportation is an essential prerequisite for a so-called quantum repeater. A quantum repeater will be an important building block in a future

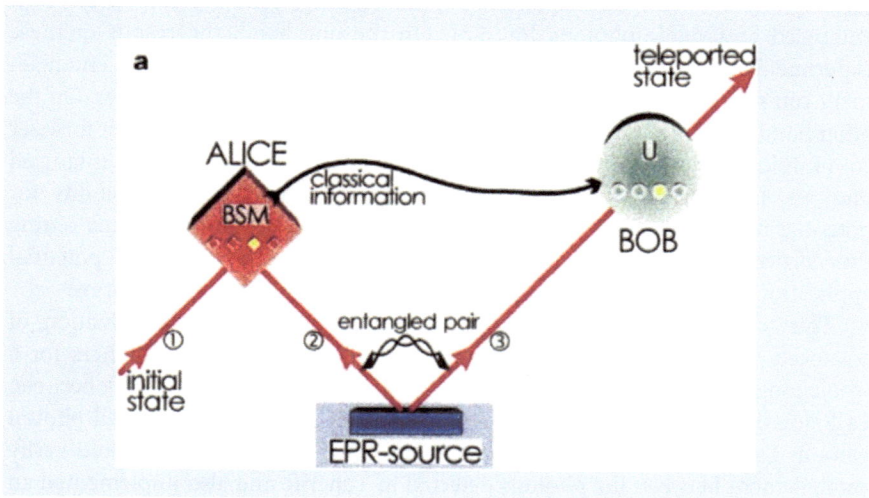

◘ **Fig. 18.10** Quantum state teleportation scheme. Picture taken from [14]

18

network, since it allows to interconnect different network nodes. In a quantum repeater, two particles of independent entangled pairs are combined within a BSM, such that the entanglement is relayed onto the remaining two particles. This process is called entanglement swapping and will eventually allow to overcome any distance limitations in a global-scale network. However, in order to efficiently execute entanglement swapping, it has to be supplemented with an entanglement purification step requiring quantum memories.

18.2   Long Distance Quantum Communication

18.2.1   Ground-Based Long-Distance Experiments

Quantum physics was invented to describe nature at the microscopic level of atoms and light. It remains an open question to what extent these laws are applicable in the macroscopic domain. In this respect, numerous ongoing research efforts pursue the goal of extending the distance between entangled quantum systems. They aim at investigating whether there are any possible fundamental limitations to quantum entanglement and if it is feasible to establish a global-scale quantum communication network in the future. In the past years, several free-space quantum communication experiments have been performed by several groups over various distances [5, 58, 63, 69, 76, 90], studying the feasibility of different quantum communication protocols over large distances. Starting with fairly short free-space links in the order of a few kilometers, the range was quickly extended up to today's world-record distance of 144 km, held by some of the authors of this article.

One of the first experiments using a 144 km free-space link between the Canary Islands of La Palma and Tenerife was performed by Ursin et al. in 2007 [76]. In this experiment (see ◘ Fig. 18.11), a source of entangled photon pairs was installed in La Palma at the top of the volcano mountain Roque de los Muchachos at an altitude of 2400 m.

One of the photons of an entangled pair was detected locally, while the other photon was sent to Tenerife. There, the optical ground station (OGS) of the European Space Agency (ESA), located at the Observatory del Teide at an altitude of 2400 m, was used as the receiving telescope for the photons coming from La Palma. After analyzing the polarization correlations between the associated photons on both islands, the scientists could verify that the photons are still entangled even though they have been separated by 144 km. Additionally, the same group implemented quantum key distribution protocols based on both entangled and single photons [69, 76]. On the one hand, the results of these experiments addressed a question of fundamental physical interest that entanglement can survive global-scale separations between the entangled particles. On the other hand, it verified that the OGS in Tenerife, which was originally built for laser communication with satellites, is also suitable to faithfully receive entangled photons. In combination, these results demonstrate the general feasibility for potential future space-based quantum communication experiments, thus setting the cornerstone for fundamental physical research as well as for potential applications of quantum mechanical principles in future network scenarios.

The achievements of these experiments were based on a combination of advanced techniques, laying the cornerstone for the Austrian researchers for a whole range of continuative activities employing the same free-space link between La Palma and Tenerife. In 2008, Fedrizzi et al. [21] generated entangled photon pairs in La Palma and sent both photons to Tenerife. The authors could verify entanglement between the photons detected in Tenerife and also implemented an entanglement based QKD protocol. This experiment was an important step

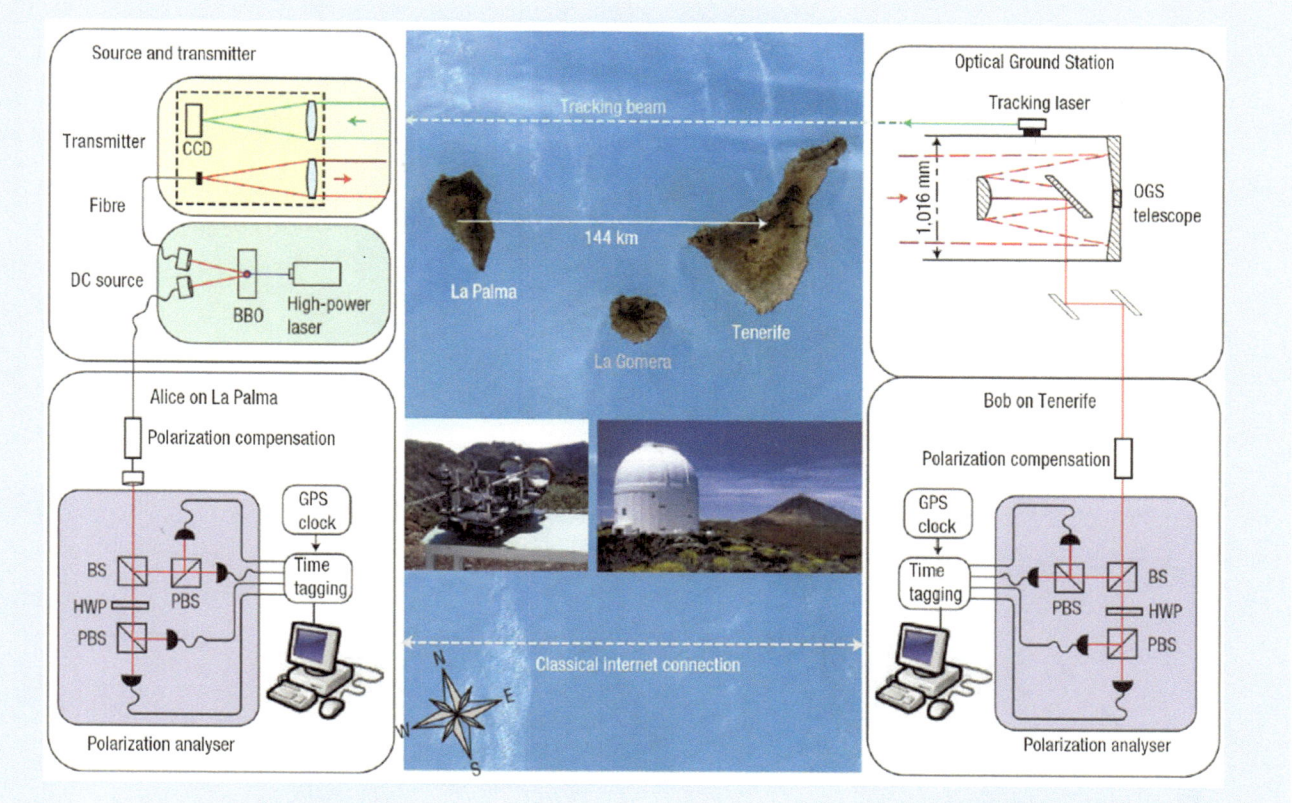

Fig. 18.11 An illustration of the experimental setup in the inter-island experiment from Ursin et al., distributing entangled photons over 144 km between La Palma and Tenerife. Figure taken from [76]

towards a potential future quantum communication network, because with respect to the transmission loss, their experimental configuration was equivalent to a basic future network scenario, where entangled pairs are transmitted from a satellite to two separate receiving stations on ground.

The long-distance experiments of our group so far involved only two photons. However, quantum communication protocols like teleportation or entanglement swapping, as described earlier, require more than two photons and will be of utmost importance in a future network. Its experimental implementation, however, is substantially more complex than the two-photon protocols, necessitating a step back regarding the communication distance (◘ Fig. 18.12).

In 2010, a group of Chinese researchers were the first to report on a long-distance free-space quantum teleportation experiment [34], demonstrating this protocol outside the shielded laboratory environment. They implemented a variant of the teleportation scheme described earlier and teleported the quantum states of photons over a distance of 16 km. This achievement triggered a race between the Austrian and Chinese groups to push the distance record for teleportation even further. It lasted until 2012 that the Chinese group reported on a successful demonstration of quantum teleportation over a 97 km free-space link across the Qinghai lake [90]. But it was only 8 days later that also the Austrian group with the results of their work on long-distance quantum teleportation between La Palma and Tenerife, reporting a new distance record of 143 km [47].

The communication distances spanned in these experiments were in fact more challenging than expected for a satellite-to-ground link and thus the results of both groups proved the feasibility of quantum repeaters in a future space- and ground-based worldwide quantum internet. Together with a reliable quantum memory, these results set the benchmark for an efficient quantum repeater at the heart of a global quantum-communication network (◘ Fig. 18.13).

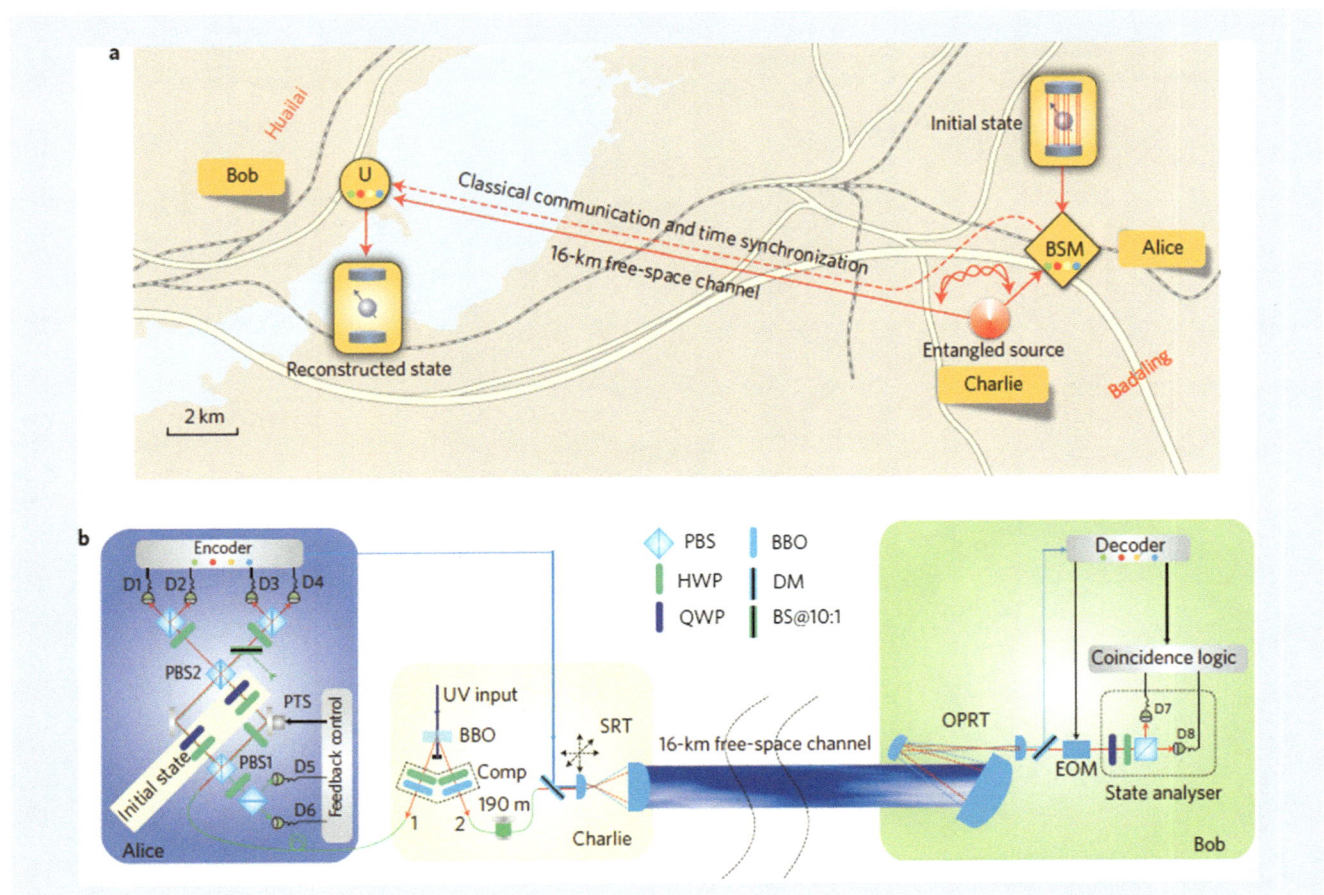

◘ Fig. 18.12 Free-space quantum teleportation experiment over 16 km in 2010 in China. (**a**) A bird's-eye view of the experiment. (**b**) The details of the quantum optical experiment at Alice's and Bob's station. Figure taken from [34], where [34] is Experimental free-space quantum teleportation

18.2.2 Space-Based Quantum Communication

The experiments described above represent the state of the art of long-distance quantum communication. Significantly longer distances are no longer possible on ground, since the curvature of the earth will then prevent direct line of sight links. The logical next step is to bring quantum technology into space and several international research initiatives in Europe, Singapore, China, USA, and the Canada are currently pursuing related projects.

It is a clear vision of the science community to establish a worldwide quantum communication network with all the advantages over its classical counterpart described above. That requires significantly expanding the distances for distributing quantum systems beyond the capabilities of terrestrial experiments and can only be realized by tackling the additional challenge of bringing the concepts and technologies of quantum physics to a space environment. Long-distance quantum communication experiments have been underway for some time sending single photons through long optical fibers. The first scientific demonstration, still in the shielded laboratory, was conducted in the late 1990s. The question to be answered at that time was if the peculiar and fragile laboratory experiments can also be executed facing harsh real-world environmental conditions as are present in optical telecommunication networks.

There are limitations for high-speed quantum communication in optical fibers. For example, the maximum speed of generating, preparing, and detecting single photons is on the order of a few Mbit per second using state-of-the-art high speed

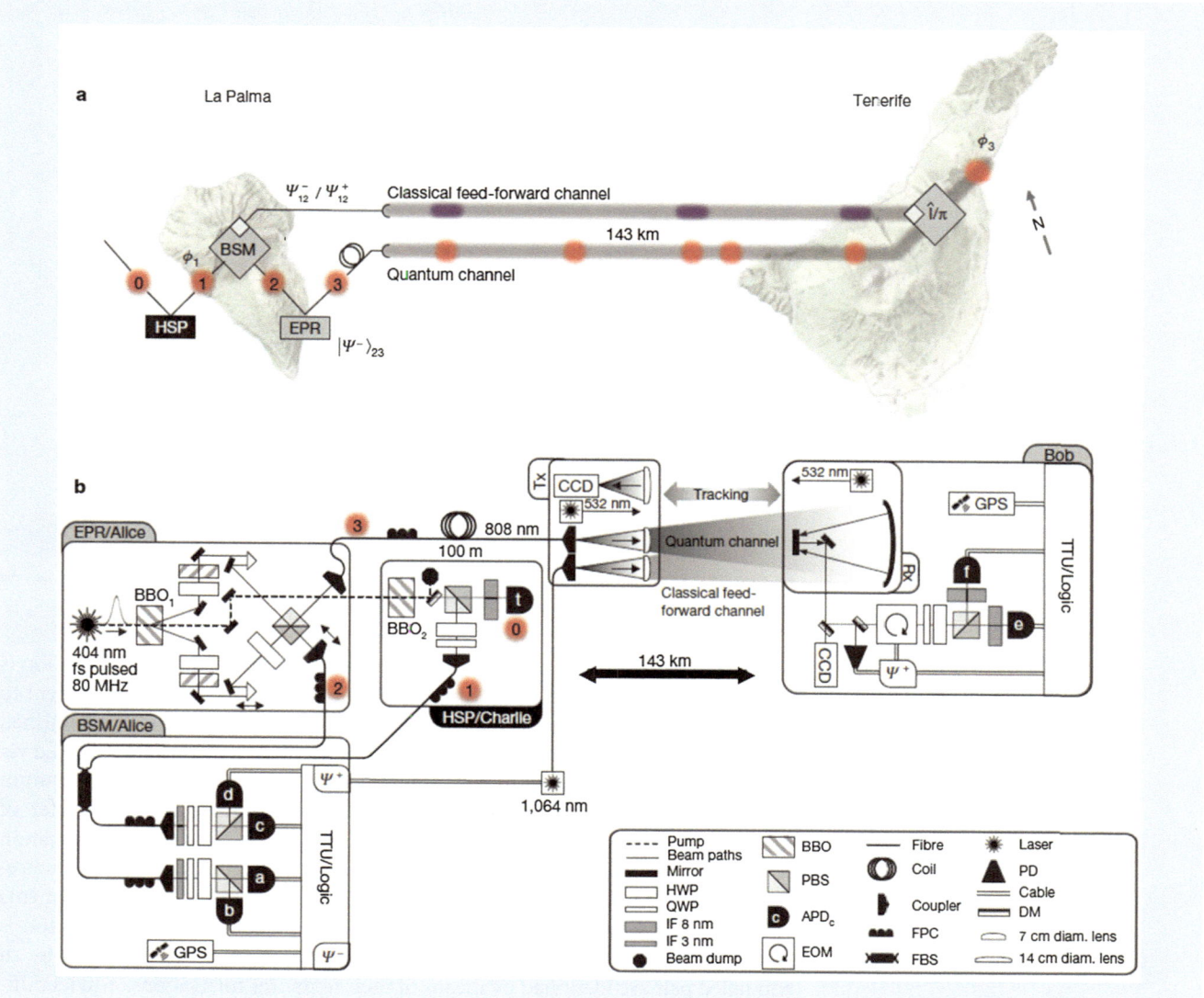

Fig. 18.13 Free-space quantum teleportation experiment over 143 km in 2012 by the Austrian group, conducted at the Canary Islands. (**a**) A bird's-eye view of the experiment. (**b**) The details of the quantum optical experiment at the stations in La Palma and Tenerife. Figure taken from [47], where [47] is Quantum teleportation over 143 kilometres using active freed-forward

electronics. Due to the combination of noise in real detector-devices and transmission loss in the optical fiber, the distance over which quantum information can be communicated is restricted to a few 100 km [79]. Hence, for bridging distances on a global scale using optical fiber networks, the implementation of the so-called quantum repeaters is paramount. Quantum repeaters are the quantum analog to classical optical amplifiers making global fiber communication as of today yet feasible. Quantum repeaters are a theoretical concept proposed in 1998 [16] and require as basic building blocks the concepts of quantum teleportation and quantum memories. Specifically, the combination of both is highly complex from a technological point of view, such that the development of a quantum repeater is yet in the early stages. The second solution to bridge distances on a global scale is to use satellite-to-earth and inter-satellite optical free-space connections [76].

Figure 18.14 depicts a typical space-mission scenario for the distribution of entanglement from a transmitter terminal to two receiver stations (Alice and Bob). The quantum source installed on the transmitter emits pairs of photons in a desired entangled state. The photon pairs exhibit strong correlations in time,

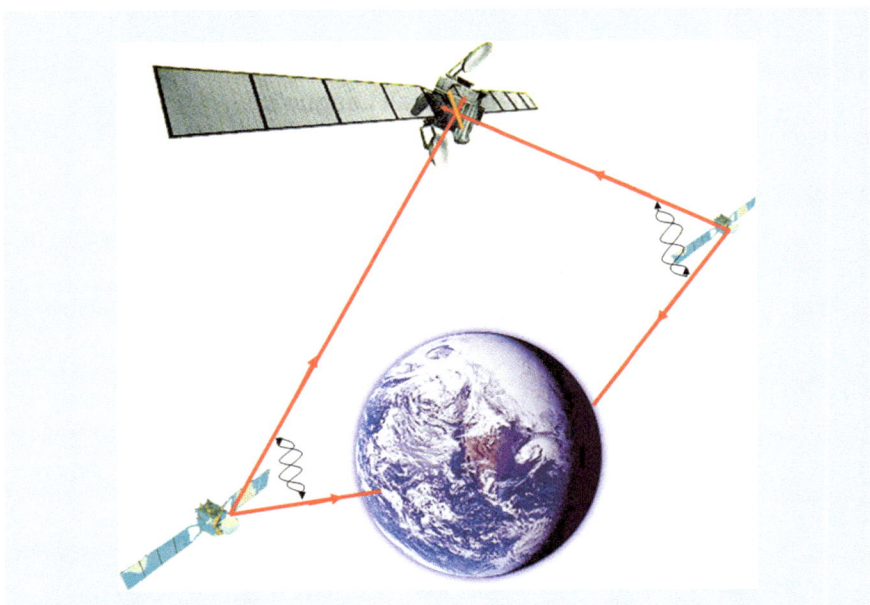

Fig. 18.14 A vision: Global Quantum Communication via satellites connecting any point on ground requiring optical ground station (taken from [77])

and entanglement in the degree of freedom in which the quantum information is encoded. The single photons comprising each of these entangled pairs are sent to Alice and Bob via free-space communication links (quantum links) established between the satellites and an optical ground station. The photons are collected via telescopes at the receiver terminals, where Alice and Bob each perform quantum measurements on their respective photons. Before initializing the transfer of information, the transmitter must establish a separate standard communication channel with Alice and Bob. This classical communications channel is subsequently used to send information about which basis state the measurements were performed on a given pair. The detection time of every arriving photon is recorded using fast single-photon detectors, and detection events that comprise an entangled pair are identified by means of their temporal correlations. The identification of photon pairs by their detection times requires the transmitter and receiver modules to establish and maintain a synchronized time basis, which can be achieved using an external reference, or autonomously via the classical communication link. Once the pair-detection events have been identified, Alice and Bob can reveal their stronger-than-classical correlations by communicating the bases of the quantum measurements performed on each photon pair via the classical communication channel.

Distributing entangled photon pairs over long-distance links and revealing their quantum correlations is an immensely challenging task from a technological point of view, in particular due to the fact that, as a result of unavoidable losses in the quantum link, only a fraction of the photons emitted by the transmitter actually arrive at the receiver modules. The main sources contributing to losses along the optical transmission channel are atmospheric absorption and scattering, on the one hand, and diffraction, telescope pointing errors, and atmospheric turbulence, which all lead to beam broadening and thus limit the fraction of photons collected by the receiver aperture, on the other. Typical losses in such scenarios are in the order of -30 to -40 dB.

Nevertheless, in order to achieve feasible pair-detection rates at such huge link losses requires a very bright source of entangled photon pairs as well as minimizing losses in the transmission channel and the receivers. Note that since correlated photon pairs are identified by their arrival times, there is an upper limit to how effective the photon production rate can mitigate against link loss. Once the time between two successive pair emissions at the source decreases below the timing jitter of the detectors, these two successive photons can no longer be distinguished from each other, such that as a result the quantum bit error ratio (QBER) will be increased.

The pairs detected by the two terminals will ultimately comprise of photons steaming not only from the entangled photons source (the signal) but also from unavoidable sources of uncorrelated background photons (the noise). The background is from stray light the detector might see and the intrinsic dark counts of the photon avalanche detectors in use. The background can be mitigated to a certain extent by using very narrow-band filters, allowing only those photons to be guided to the detector, who are at the wavelength of the quantum source in use. Also the common timing of the entangled photons is useful to mitigate noise pair counts.

Entangled photon sources maintaining both their high brightness and the quality of the emitted quantum state will have to be manufactured in a very reliable and stable manner to survive the launch of the satellite as well as the harsh space environment (radiation). The first research and development projects funded by the European Space Agency were dedicated to the nonlinear periodically poled potassium titanyl phosphate (ppKTP) crystal, which is used in state-of-the-art entangled photon sources. Additionally, the implementation of the rather complex structure of lenses and beam-splitters is addressed in these studies and radiation effects on single photon detectors have already been investigated in detail [35]. These first attempts do show that a quantum mission based on state-of-the-art technology is feasible and requires the integration into commercially available space-laser terminals as a next step.

As outlined above, quantum communication provides a novel way of information transfer. Even though it is still under development, it has the potential to become our future technology for communication and computation. The first proposed experiments in space will serve as a very good platform to test these concepts and could pave the way for follow-up industrial systems. From a very long-term perspective it is highly interesting to test quantum mechanics at distances on the order of millions of km, and even beyond. Furthermore, an ultimate experiment regarding the role of randomness and humans free-will could be performed by two individuals, separated by at least one light second, who each measures entangled particles and separately chooses the setting of their analyzer. To extend the scale of quantum mechanical states over astronomical distances might provide us with a suitable insight on the link between gravitation, quantum mechanics, and even more. Clearly, these experiments require advances in technology not even foreseeable today. Nevertheless, the proposed experiments are a major step in investigating these fundamental questions as well as enhancing the technology for the society's benefit.

18.3 Higher Dimensions

So far, we have focused only on qubits, which are quantum mechanical two-level systems. This is a natural choice, as all of our classical data storage, transmission, and processing are based on classical two-level systems that encode zeros and ones. There are only a very few exotic exceptions, such as the *Setun* computer built in Soviet union in the late 1950s, which used trinary logic.

However, if one were to look at nature's way of encoding and processing information, one would be surprised to find that it uses a higher-level system: DNA (deoxyribonucleic acid) uses four types of nucleobase (adenine, guanine, cytosine, and thymine) to encode information. Three nucleobases together encode one amino acid, the basis of biological life. If nature—optimized over hundreds of million of years through evolution—uses a higher-level system for encoding information, we see no reason why one shouldn't investigate its use in quantum information as well!

There are two types of high-dimensional systems that depend on whether one considers discrete or continuous parameters. An example of a continuous degree-of-freedom (DoF) is the position (or likewise, the momentum) of a photon. Quantum correlations in this DoF have been used for interesting new types of imaging schemes such as quantum ghost imaging, where the image of the object can only be seen in the correlations of the photons [50, 60, 73]. A different, even more counterintuitive quantum imaging procedure was recently demonstrated where an object was imaged without ever detecting the photons which were in contact with the imaged object [46].

In some scenarios, a discrete basis is more advantageous. In classical communications or data storage, for example, information is encoded either as a 0 or a 1; fractional numbers in between are not used. The same is true for quantum communication or quantum computation, even with larger alphabets. A natural basis that uses a discrete DoF of a photon is its orbital angular momentum, which is presented in the next section. Other possible bases can be constructed by the discretization of continuous parameters such as position or wavelength.

18.3.1 Twisted Photons

If one investigates the spatial profile of a laser beam with a camera, one usually finds that it has a Gaussian shape. However, that is only a special case of a much more complex family of fundamental spatial structures or modes. One very convenient set of modes are the so-called Laguerre–Gaussian modes [3, 57, 88]. In ◘ Fig. 18.15, the intensity and phase structure of a Gaussian mode ($\ell = 0$) compared to Laguerre–Gaussian modes ($|\ell| > 0$) are shown.

In contrast to its polarization, which is a property related to its spin angular momentum, a photon with a Laguerre–Gaussian mode structure can also carry orbital angular momentum (OAM). The spin and orbital-angular momenta have

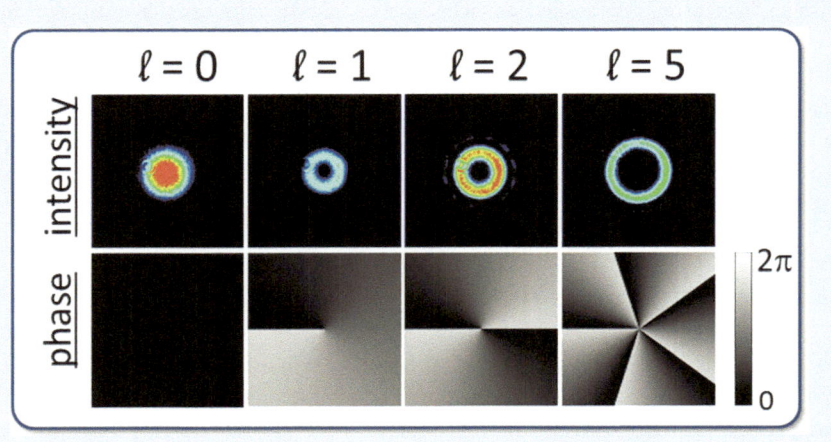

◘ **Fig. 18.15** Intensities and phase information of orbital angular momentum beams. The intensity is collected with a camera. The OAM = 0 mode is the well-known Gaussian distribution. OAM larger than 0 show a ring, or doughnut structure. The *lower line* shows that these structures have a twisted phase-front, with $2\pi\,\ell$ phase-change in a ring. In the center, they have a phase singularity—also known as Vortex. The vortex is the reason why there is no intensity in the center (Image by Mario Krenn, copyright University of Vienna)

distinct physical properties: if a laser beam with circular polarization illuminates a small particle, the particle will start to rotate around its own axis. However, if a beam with orbital angular momentum shines on a particle, it starts to rotate around the external orbit defined by the laser beam [29]. Surprisingly, the OAM of photons and its connection to Laguerre–Gauss modes was identified only recently in 1992 [2].

Interestingly, the OAM quantum number of a photon can theoretically take on any integer number between $-\infty$ and ∞. This allows one to encode a huge amount of data onto a single photon [17, 26]. In classical communications, this can improve the data rates enormously. Recent experiments have demonstrated data transmission of 100 Tbit/s by using the OAM of light together with other DoFs [30, 82]. In quantum communication, secret sharing protocols have been developed that use OAM modes as an alphabet for encoding [28, 48, 54, 56, 80]. Not only do such protocols offer an increased data rate, they also provide an improved level of security against eavesdropping attacks [31, 81].

18.3.2 High-Dimensional Entanglement

Earlier in this chapter, entanglement was explained in the context of photon polarization, which is a two-level system. In such systems, the separated photon pair can share one bit of information in a nonlocal manner, referred to as an entangled bit or "ebit" (◘ Fig. 18.16).

However, if we consider larger dimensional systems such as the OAM of photons, one can easily imagine that a pair of photons entangled in their OAM could share much more information than photons entangled in their polarization. Such modes get bigger in size as the OAM quantum number ℓ is increased. Thus, the amount of information carried by them is only limited by the size of the optical devices used, or more generally, by the size of the universe itself! A natural question that arises is whether there exists a limit to the amount of information that can be non-locally shared between two entangled photon pairs. This question

◘ **Fig. 18.16** Two classical 100-sided dice. If one were to roll them, it is very unlikely that they would both show the same number. However, were they high-dimensionally entangled, they would both always show the same number. Note: such a metaphor for quantum entanglement is limited in that one cannot visualize the results of correlated measurement outcomes in superposition bases. This is key for distinguishing entanglement from classical correlations (Image by Mario Krenn, copyright University of Vienna)

is being investigated in several laboratories around the world [1, 13, 18, 27, 39, 52, 55, 61, 65, 67, 75, 78]. These efforts have confirmed that two distant photons can be entangled in hundred and more dimensions of their spatial mode structure. This means that by measuring the first photon of the entangled pair, one will observe one definite result out of the hundred possible outcomes. This immediately tells us the outcome of a similar measurement on the second, distant photon. However, the strangeness lies in the fact that the two photons did not have a definite value before they were measured. Only when the first photon is observed does the common state become a reality, and the second photon gets a defined value.

Photons entangled in their orbital angular momentum also enable the possibility to explore more complex types of entanglement that is not possible with two-dimensional entangled states. Recent state-of-the-art experiments have shown the entanglement of eight photons [89], nine superconducting circuits [36], and fourteen ions [42]. However, these experiments have singularly focused on increasing the number of particles entangled, while remaining in a two-dimensional space for each particle. The OAM of light was recently used to create the first entangled state where both the number of particles and the number of dimensions were greater than two [51]. This state involved three photons asymmetrically entangled in their OAM: two photons resided in a three-dimensional space, while one photon lived in two dimensions. These experiments have been designed by a computer algorithm [41]. Interestingly, this asymmetric structure only appears when one considers multi-particle entanglement in dimensions greater than two [32]. Such states also enable a novel "layered" quantum communication protocol. For example, if three parties were to share the state described above, all three would have access to one bit of secure information, allowing them to generate a secure random key for sharing information. However, part of the time, two of the parties would have access to another bit of secure information. This would allow them to share an additional layer of information unknown to the third party in the communication scheme. This protocol can be generalized to include multiple layers of information shared asymmetrically amongst many different parties.

18.3.3 Mutually Unbiased Bases in High Dimensions

Earlier in this chapter we have learned that for 2-dimensional systems, three unbiased bases exist. For larger dimensions, one finds more of these unbiased bases: in 3 dimensions there are 4 bases, in 4 dimensions there are 5 bases. In fact, it is known that for every prime-power dimension (with $d = p^n$), the number of MUBs is $(d+1)$. That means, in dimension d, there are $(d+1)$ different ways to encode information. Now there is one very surprising fact: If the dimension of the space is not a prime-power, it is not known how many MUBs there are. The first of those cases is dimension $2 \cdot 3 = 6$ [7, 85]. Numerical search has only found 3 MUBs, and it is a conjecture that there are only 3 MUBs. It is fascinating because it means that in 5 dimensions, there are more ways to encode information in different ways than in 6 dimensions, even though intuitively one might think that a larger space allows for more ways to embed information in different ways. This is crucial for quantum communication, because the number of MUBs is directly connected to the robustness (against noise and eavesdropping-attacks) of the protocol. The more different ways of encoding the information, the more secure the system is.

18.3.4 High-Dimensional Quantum Key Distribution

Quantum cryptography based on photons carrying OAM is similar to the schemes developed for polarization that are explained earlier in this chapter. High-dimensional analogs to the BB84 and Ekert QKD protocols have been developed that use OAM for encoding [49]. Similar to polarization-based QKD, OAM-based QKD requires measurements to be performed in mutually unbiased bases to guarantee security against eavesdropping. The earliest such protocol was demonstrated with photons entangled in three dimensions of their OAM ($\ell = 0$, $+1$, and -1) [28]. The high-dimensionally entangled photon pairs were produced in a BBO crystal and sent to two separate stations, where basis transformations were randomly performed by two holograms mounted on moving motorized stages at each station. The photons were then probabilistically split into three paths where their OAM content was measured by three additional holograms. In this manner, a three-dimensional key was generated with an error rate of 10 %. Security was verified by testing for the presence of entanglement via a high-dimensional Bell inequality.

One of the challenges in using OAM modes for quantum communication is the ability to sort single photons carrying OAM. The QKD scheme described above used beam splitters and holograms to projectively measure the OAM content of the single photons. This resulted in a scheme that was photon-inefficient, i.e. only one out of every nine photons was actually used for communication. While techniques for efficiently sorting the OAM of single photons existed, they relied on N cascaded Mach–Zehnder interferometers for sorting $N + 1$ OAM modes [44]. Thus, the use of such a device in a quantum communication scheme was impractical due to issues of complexity and stability. However, in 2010, the group of Miles Padgett developed a refractive device that could sort the OAM of a single photon [12]. This device "unwrapped" the helical wavefront of an OAM mode, transforming it into a plane wave with a tilted wavefront. The amount of tilt was proportional to the OAM quantum number ℓ, allowing these modes to be separated by a simple lens. This device provided a diffraction-limited sorting efficiency of 75 %, which was improved to 93 % by the addition of two additional holographic transformations [53].

The development of this device allowed photon-efficient OAM-based quantum communication schemes to be realized in the laboratory. Recently, a BB84 protocol using a seven-dimensional OAM alphabet was performed which made heavy use of the OAM sorter discussed above [54]. Additionally, a digital micro-mirror device (DMD) was used to generate OAM modes at a rate of 4 kHz, which is much faster than the rates attainable with spatial light modulators. The key was encoded in the OAM basis as well as the mutually unbiased of the so-called angular modes (ANG), as shown in ◘ Fig. 18.17b. Using this scheme, Alice and Bob were able to communicate securely at a rate of 2.05 bits per sifted photon. Their generated key had an error rate of approximately 10 %, which was below the bounds for security against coherent attacks in a seven-dimensional QKD link. This experiment served as a proof-of-principle demonstration of OAM-based QKD. Several technological improvements (discussed in Ref. [54]) will be required to take such a scheme into the real world.

18.3.5 Large Quantum Number Entanglement

Twisted photons not only allow access to a very large state space, but also give access to very high quantum numbers. Photons can carry $\ell\hbar$ of angular momentum, and ℓ can be arbitrarily large. Usually, quantum phenomena are only

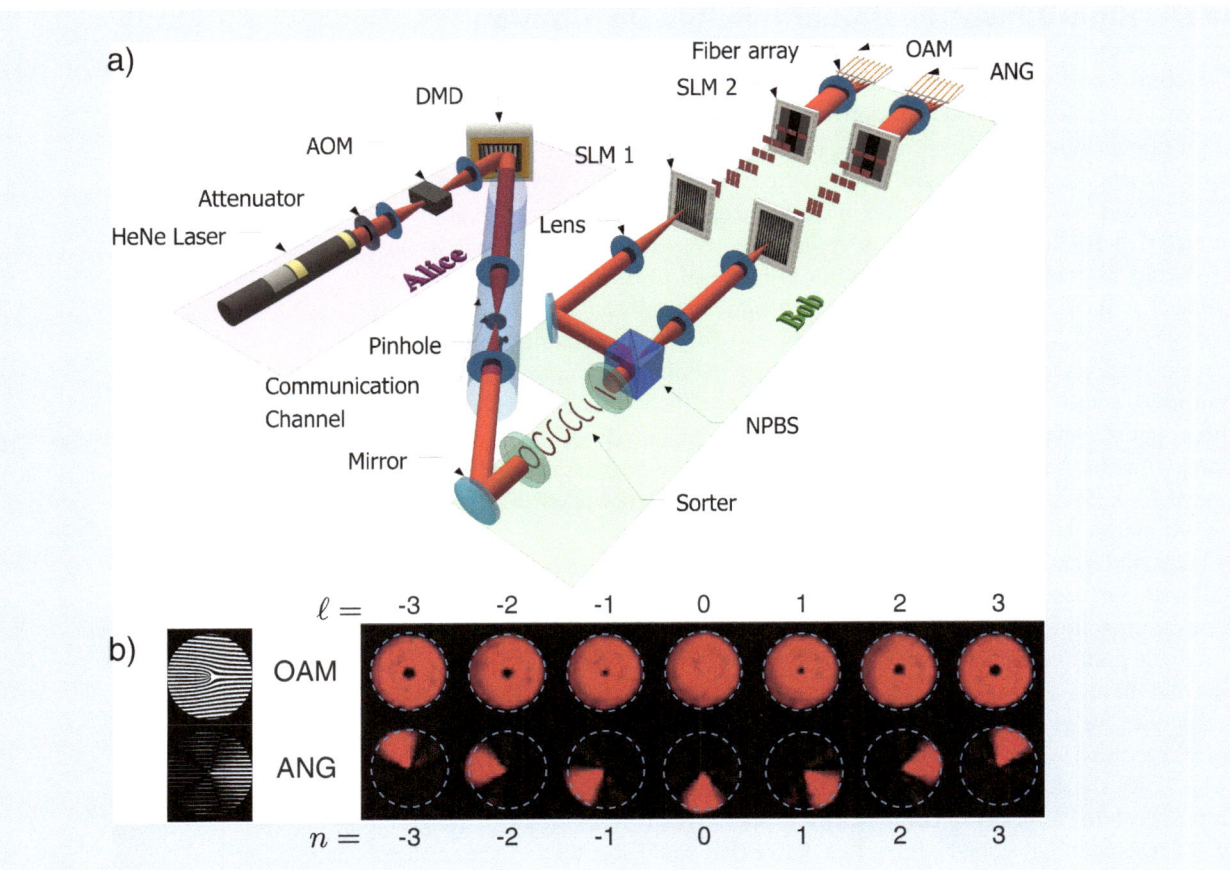

Fig. 18.17 (**a**) An OAM-based BB84 scheme for quantum key distribution. Alice encodes a random key in a seven-dimensional alphabet consisting of OAM modes using a high-speed digital micro-mirror device (DMD). Bob sorts these modes using an OAM sorter and four additional holograms implemented on spatial light modulators (SLMs). Using this scheme, Alice and Bob are able to communicate with a channel capacity of 2.05 bits per sifted photon. (**b**) CCD images showing the intensity profiles of the seven-dimensional alphabet in the OAM basis, as well as the mutually unbiased basis of angular (ANG) modes. Examples of binary holograms for generating these modes are shown on the *left* (figure adapted from Ref. [54])

observed in the microscopic world. Here, however, with twisted photons it is possible to create entanglement between photons that differ by a very large amount of angular momentum. Theoretically, there is no upper limit of the number of angular momentum, which would give rise to the possibility of entanglement of macroscopic values of angular momentum.

With this method, it was possible to show that two photons with a difference of $600\hbar$ can be entangled [22]. If the first photon carries $300\hbar$ of angular momentum, the second carried $-300\hbar$, and vice versa. While being entangled in a two-dimensional subspace, it was the largest quantum number difference achieved. In that experiment, a spatial light modulator has been used, which can be seen in Fig. 18.18. Recently, using novel methods to encode very large angular momentum at single photons, it was able to show entanglement of photons with a quantum number difference of $10,000\hbar$ [23].

An important question that needs to be answered is the definition of *macroscopic angular momentum*, and which phenomena might arise from that. For example, there are predictions that photons close to a black hole change their angular momentum [74]. As black holes are purely general relativistic objects, and entanglement is a purely quantum mechanical phenomenon, a deeper investigation into these effects will be exciting.

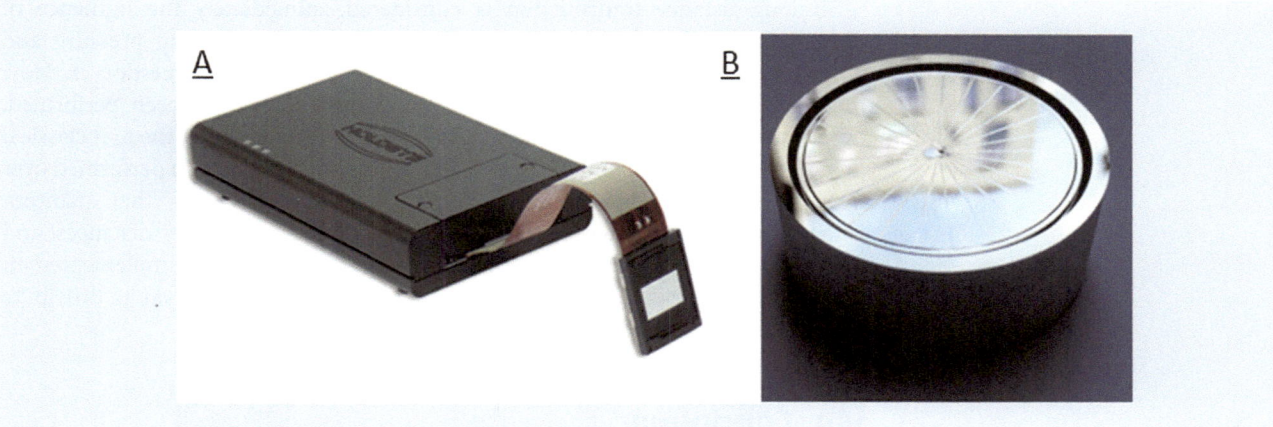

Fig. 18.18 Different ways to create photons with large angular momentum. (**a**): A spatial light modulator consists of a liquid crystal display. The display consists of roughly 1000 × 1000 pixels, which performs phase shifts from zero to 2π. The flexibility allows to create arbitrary phase structure, thus arbitrary structures of the modulated light. However, due to their finite resolution, there is an upper limit of roughly $300\hbar$. (**b**): A different method that can create angular momentum of up to $10.000\hbar$ are fixed phase holograms built out of aluminium. In compensation for the lower flexibility, the holograms can be produced very precise, which is responsible for the much larger possible angular momentum. (Image by Robert Fickler, copyright University of Vienna)

Fig. 18.19 Receiver at the Hedy Lamarr quantum communication telescope for the first free-space long-distance entanglement distribution experiment with a high-dimensional degree of freedom (Image by Mario Krenn, copyright University of Vienna)

18.3.6 Long-Distance Transmission of Twisted Photons

In a quantum communication scenario, the encoded information needs to be distributed between two parties. Usually one would think that optical fibers are the ideal solutions. Unfortunately, the information in twisted photons is not conserved in propagation through conventional fibers: Different modes mix in fibers, therefore the output is different than the input. Although recent advances show that special fibers can be used to transmit the first higher-order OAM modes for more than 1 km [15], and reach a classical communication rate in the order of Terabit, this technology is still in its infancy. Specifically, it hasn't been used in the realm of quantum physics yet. An alternative method is the transmission through free-space. In the case of earth-to-satellite quantum communication, this is the only possibility in any case (Fig. 18.19).

If long-distance transmission is considered, immediately the influence of atmospheric turbulence has to be taken into account. Varying pressure and temperature influence the structure of twisted photons. The question is: How much? While many mathematical and lab-scale studies have been performed, experimental investigation of that question is rare. Only recently, the first classical [38, 43] and quantum communication [40] experiments have been performed over free-space intra-city link of 3 km distance. Those results show that quantum entanglement with twisted photons can be distributed over larger distances, and the quality can be improved with technology that is already implemented in lab-scale experiments [62, 64, 87]. As such, it could be a reliable way to distribute high-dimensional entanglement in a future quantum network.

18.4 Conclusion

The possibility to share secret messages is of utmost importance for our society. From simple things like sending emails which can't be read by an eavesdropper to the transmission of highly sensitive information between governments that needs to be secure for decades—cryptography plays a key role in ensuring privacy, economic stability, and stable relations between countries worldwide.

As we have seen, classical cryptographic systems are vulnerable to various types of eavesdropping attacks. The problem is that either the secret key needs to be transmitted over insecure channels or (in a public–private cryptography system) the security relies on mathematical conjectures that specific properties are difficult to calculate. Furthermore, quantum computing algorithms can significantly reduce the required time to find solutions for such problems (finding prime factors of large numbers, or calculating a discrete logarithm). On top of all this, back-doors can be implemented into these algorithms such that they perform as expected, but the creator of the algorithm obtains additional information. Such attacks have been widely discussed in connection with a weak generator for pseudo random numbers certified by NIST [59, 72].

The need for overcoming these problems posed by classical asymmetric cryptographic systems has led to the development of a field called post-quantum-cryptography. There, problems which are believed to be more difficult than factoring large numbers are used to prepare a public and private key. Such methods are not practically used yet because of performance issues and unclear results on their security. While there are no classical or quantum algorithms to solve such problems yet, it is only conjectured that they are difficult to solve—a breakthrough in (quantum) complexity theory or novel kind of computations might only shift the problem into the future.

The only unconditionally secure encryption requires a random key with the same size as the message, a so-called one-time pad. The question is, how can such a key be distributed securely? Quantum key distribution provides a solution to that question, by exploiting quantum mechanical properties of individual particles. Several newly founded companies already provide small-scale quantum key distribution systems, such as ID Quantique in Switzerland, MagiQ Technologies in the USA, QuintessenceLabs in Australia, or SeQureNet in France.

As shown in this chapter, fundamental investigations test the feasibility of global quantum networks, on the order of 100 km on the Earth's surface, as well as between ground and space. A second path of research focuses on more complex quantum states, to improve data-rates and robustness against noise and eavesdropping attacks. The experiments discussed in this chapter form only a small subset of experimental efforts currently in progress around the world. It is clear that we are perched on the edge of a quantum communication revolution that will change information security and how we understand privacy for years to come.

Acknowledgements We acknowledge cooperation with Jian-Wei Pan and the Chinese Academy of Sciences. This work was supported by the European Space Agency, the European Research Council (ERC Advanced Grant No. 227844 "QIT4QAD" and SIQS Grant No. 600645 EU-FP7-ICT), the European Commission (Marie Curie grant "OAMGHZ"), the Austrian Science Fund (FWF), the Austrian Academy of Sciences (ÖAW), and the Austrian Research Promotion Agency (FFG) within the ASAP program from the Federal Ministry of Science and Research (BMWF), as well as the John Templeton Foundation.

References

1. Agnew M, Leach J, McLaren M, Roux FS, Boyd RW (2011) Tomography of the quantum state of photons entangled in high dimensions. Phys Rev A 84(6):062,101
2. Allen L, Beijersbergen M, Spreeuw R, Woerdman JP (1992) Orbital angular momentum of light and the transformation of Laguerre-Gaussian laser modes. Phys Rev A 45 (11):8185–8189
3. Allen L, Padgett MJ, Babiker M (1999) The orbital angular momentum of light. Prog Opt (39):291–372
4. Aspect A, Dalibard J, Roger G (1982) Experimental test of bell's inequalities using time-varying analyzers. Phys Rev Lett 49(25):1804
5. Aspelmeyer M, Böhm HR, Gyatso T, Jennewein T, Kaltenbaek R, Lindenthal M, Molina-Terriza G, Poppe A, Resch K, Taraba M et al (2003) Long-distance free-space distribution of quantum entanglement. Science 301(5633):621–623
6. Bell J (1964) On the einstein-podolsky-rosen paradox. Physics 1(3):195–200
7. Bengtsson I (2007) Three ways to look at mutually unbiased bases. In: AIP Conf Proc 889, 40. ▶ http://dx.doi.org/10.1063/1.2713445
8. Bennett C, Brassard G (1984) Quantum cryptography: public key distribution and coin tossing. In: Proceedings of IEEE International Conference on Computers, Systems, and Signal Processing. Bangalore, p 175
9. Bennett C, Wiesner S (1992) Communication via one- and two-particle operators on Einstein-Podolsky-Rosen states. Phys Rev Lett 69(20):2881–2884
10. Bennett CH, Brassard G, Crépeau C, Jozsa R, Peres A, Wootters WK (1993) Teleporting an unknown quantum state via dual classical and Einstein-Podolsky-Rosen channels. Phys Rev Lett 70(13):1895–1899
11. Bennink R, Bentley S, Boyd RW, Howell J (2004) Quantum and classical coincidence imaging. Phys Rev Lett 92(3):033,601
12. Berkhout GCG, Lavery MPJ, Courtial J, Beijersbergen MW, Padgett MJ (2010) Efficient sorting of orbital angular momentum states of light. Phys Rev Lett 105(15):153,601
13. Bolduc E, Gariepy G, Leach J (2016) Direct measurement of large-scale quantum states via expectation values of non-Hermitian matrices. Nat Commun 7
14. Bouwmeester D, Pan JW, Mattle K, Eibl M, Weinfurter H, Zeilinger A (1997) Experimental quantum teleportation. Nature 390(6660):575–579
15. Bozinovic N, Yue Y, Ren Y, Tur M, Kristensen P, Huang H, Willner AE, Ramachandran S (2013) Terabit-scale orbital angular momentum mode division multiplexing in fibers. Science 340(6140):1545–1548

16. Briegel HJ, Dür W, Cirac JI, Zoller P (1998) Quantum repeaters: the role of imperfect local operations in quantum communication. Phys Rev Lett 81(26):5932
17. Čelechovský R, Bouchal Z (2007) Optical implementation of the vortex information channel. New Journal of Physics 9(9):328
18. Dada AC, Leach J, Buller GS, Padgett MJ, Andersson E (2011) Experimental high-dimensional two-photon entanglement and violations of generalized Bell inequalities SI. Nat Phys 7(9):677–680
19. Einstein A, Podolsky B, Rosen N (1935) Can quantum-mechanical description of physical reality be considered complete? Phys Rev 47(10):777–780
20. Ekert AK (1991) Quantum cryptography based on Bell's theorem. Phys Rev Lett 67(6):661–663
21. Fedrizzi A, Ursin R, Herbst T, Nespoli M, Prevedel R, Scheidl T, Tiefenbacher F, Jennewein T, Zeilinger A (2009) High-fidelity transmission of entanglement over a high-loss free-space channel. Nat Phys 5(6):389–392
22. Fickler R, Lapkiewicz R, Plick WN, Krenn M, Schaeff C, Ramelow S, Zeilinger A (2012) Quantum entanglement of high angular momenta. Science 338(6107):640–643
23. Fickler R, Campbell GT, Buchler BC, Lam PK, Zeilinger A (2016) Quantum entanglement of angular momentum states with quantum numbers up to 10010. arXiv preprint arXiv:1607.00922.
24. Franson J (1989) Bell inequality for position and time. Phys Rev Lett 62(19):2205–2208
25. Freedman SJ, Clauser JF (1972) Experimental test of local hidden-variable theories. Phys Rev Lett 28(14):938
26. Gibson G, Courtial J, Padgett MJ, Vasnetsov M, Pas'ko V, Barnett SM, Franke-Arnold S (2004) Free-space information transfer using light beams carrying orbital angular momentum. Opt Express 12(22):5448–5456
27. Giovannini D, Romero J, Leach J, Dudley A, Forbes A, Padgett MJ (2013) Characterization of high-dimensional entangled systems via mutually unbiased measurements. Physical Rev Lett 110(14):143,601
28. Groblacher S, Jennewein T, Vaziri A, Weihs G, Zeilinger A (2006) Experimental quantum cryptography with qutrits. New J Phys 8:75
29. He H, Friese M, Heckenberg N, Rubinsztein-Dunlop H (1995) Direct observation of transfer of angular momentum to absorptive particles from a laser beam with a phase singularity. Phys Rev Lett 75(5):826
30. Huang H, Xie G, Yan Y, Ahmed N, Ren Y, Yue Y, Rogawski D, Tur M, Erkmen B, Birnbaum K et al (2013) 100 tbit/s free-space data link using orbital angular momentum mode division multiplexing combined with wavelength division multiplexing. In: Optical Fiber Communication Conference, Optical Society of America, paper OTh4G.5
31. Huber M, Pawłowski M (2013) Weak randomness in device-independent quantum key distribution and the advantage of using high-dimensional entanglement. Phys Rev A 88(3):032,309
32. Huber M, de Vicente J (2013) Structure of multidimensional entanglement in multipartite systems. Phys Rev Lett 110(3):030,501
33. Jha AK, Malik M, Boyd RW (2008) Exploring energy-time entanglement using geometric phase. Phys Rev Lett 101(18):180,405
34. Jin XM, Ren JG, Yang B, Yi ZH, Zhou F, Xu XF, Wang SK, Yang D, Hu YF, Jiang S, et al (2010) Experimental free-space quantum teleportation. Nat Photonics 4(6):376–381
35. Kaiser KH, Aulenbacher K, Chubarov O, Dehn M, Euteneuer H, Hagenbuck F, Herr R, Jankowiak A, Jennewein P, Kreidel HJ et al (2008) The 1.5 gev harmonic double-sided microtron at mainz university. Nucl Instrum Methods Phys Res Sect A Accelerators Spectrom Detect Assoc Equip 593(3):159–170
36. Kelly J, Barends R, Fowler AG, Megrant A, Jeffrey E, White TC, Sank D, Mutus JY, Campbell B, Chen Y, Chen Z, Chiaro B, Dunsworth A, Hoi IC, Neill C, O'Malley PJJ, Quintana C, Roushan P, Vainsencher A, Wenner J, Cleland AN, Martinis JM (2015) State preservation by repetitive error detection in a superconducting quantum circuit. Nature 519(7541):66–69
37. Klyshko DN (1988) A simple method of preparing pure states of an optical field, of implementing the Einstein–Podolsky–Rosen experiment, and of demonstrating the complementarity principle. Sov Phys Usp 31(1):74–85
38. Krenn M, Fickler R, Fink M, Handsteiner J, Malik M, Scheidl T, Ursin R, Zeilinger A (2014) Communication with spatially modulated light through turbulent air across Vienna. New J Phys 16(11):113,028
39. Krenn M, Huber M, Fickler R, Lapkiewicz R, Ramelow S, Zeilinger A (2014) Generation and confirmation of a (100 x 100)-dimensional entangled quantum system. Proc Natl Acad Sci 111:6243
40. Krenn M, Handsteiner J, Fink M, Fickler R, Zeilinger A (2015) Twisted photon entanglement through turbulent air across vienna. Proc Natl Acad Sci 112(46):14197–14201

41. Krenn M, Malik M, Fickler R, Lapkiewicz R, Zeilinger A (2016) Automated search for new quantum experiments. Phys Rev Lett 116(9):090405

42. Lanyon BP, Zwerger M, Jurcevic P, Hempel C, Dür W, Briegel HJ, Blatt R, Roos CF (2014) Experimental violation of multipartite bell inequalities with trapped ions. Phys Rev Lett 112(10):100,403

43. Lavery MP, Heim B, Peuntinger C, Karimi E, Magaña-Loaiza OS, Bauer T, Marquardt C, Boyd RW, Padgett M, Leuchs G et al (2015) Study of turbulence induced orbital angular momentum channel crosstalk in a 1.6 km free-space optical link. In: CLEO: Science and Innovations, Optical Society of America, paper STu1L.4

44. Leach J, Padgett MJ, Barnett SM, Franke-Arnold S (2002) Measuring the orbital angular momentum of a single photon. Phys Rev Lett 88:257,901

45. Leach J, Warburton RE, Ireland DG, Izdebski F, Barnett SM, Yao AM, Buller GS, Padgett MJ (2012) Quantum correlations in position, momentum, and intermediate bases for a full optical field of view. Phys Rev A 85(1):013,827

46. Lemos GB, Borish V, Cole GD, Ramelow S, Lapkiewicz R, Zeilinger A (2014) Quantum imaging with undetected photons. Nature 512(7515):409–412

47. Ma XS, Herbst T, Scheidl T, Wang D, Kropatschek S, Naylor W, Wittmann B, Mech A, Kofler J, Anisimova E et al (2012) Quantum teleportation over 143 kilometres using active feed-forward. Nature 489(7415):269–273

48. Mafu M, Dudley A, Goyal S, Giovannini D, McLaren M, Padgett MJ, Konrad T, Petruccione F, Lütkenhaus N, Forbes A (2013) Higher-dimensional orbital-angular-momentum-based quantum key distribution with mutually unbiased bases. Phys Rev A 88(3):032,305

49. Malik M, Boyd RW (2014) Quantum imaging technologies. Riv Nuovo Cimento 37:273

50. Malik M, Shin H, O'Sullivan MN, Zerom P, Boyd RW (2010) Quantum ghost image identification with correlated photon pairs. Phys Rev Lett 104(16):163,602

51. Malik M, Erhard M, Huber M, Krenn M, Fickler R, Zeilinger A (2016) Multi-photon entanglement in high dimensions. Nat Photonics 10(4):248–252

52. McLaren M, Agnew M, Leach J, Roux FS, Padgett MJ, Boyd RW, Forbes A (2012) Entangled bessel-gaussian beams. Opt Express 20(21):23,589–23,597

53. Mirhosseini M, Malik M, Shi Z, Boyd RW (2013) Efficient separation of the orbital angular momentum eigenstates of light. Nat Commun 4:2781

54. Mirhosseini M, Magaña-Loaiza OS, O'Sullivan MN, Rodenburg B, Malik M, Lavery MPJ, Padgett MJ, Gauthier DJ, Boyd RW (2015) High-dimensional quantum cryptography with twisted light. New J Phys 17(3):033,033

55. Molina-Terriza G, Vaziri A, Řeháček J, Hradil Z, Zeilinger A (2004) Triggered qutrits for quantum communication protocols. Phys Rev Lett 92(16):167,903

56. Molina-Terriza G, Vaziri A, Ursin R, Zeilinger A (2005) Experimental quantum coin tossing. Phys Rev Lett 94(4):40,501

57. Molina-Terriza G, Torres JP, Torner L (2007) Twisted photons. Nat Phys 3(5):305–310

58. Peng CZ, Yang T, Bao XH, Zhang J, Jin XM, Feng FY, Yang B, Yang J, Yin J, Zhang Q et al (2005) Experimental free-space distribution of entangled photon pairs over 13 km: towards satellite-based global quantum communication. Phys Rev Lett 94(15):150,501

59. Perlroth N (2013) Government announces steps to restore confidence on encryption standards. New York Times. Available at: ▶ bits.blogs.nytimes.com/2013/09/10/government-announces-steps-to-restore-confidence-on-encryption-standards/

60. Pittman T, Shih Y, Strekalov D, Sergienko A (1995) Optical imaging by means of two-photon quantum entanglement. Phys Rev A 52(5):R3429–R3432

61. Pors J, Oemrawsingh S, Aiello A, Van Exter M, Eliel E, Woerdman J et al (2008) Shannon dimensionality of quantum channels and its application to photon entanglement. Physical review letters 101(12):120,502

62. Ren Y, Xie G, Huang H, Ahmed N, Yan Y, Li L, Bao C, Lavery MP, Tur M, Neifeld MA et al (2014) Adaptive-optics-based simultaneous pre-and post-turbulence compensation of multiple orbital-angular-momentum beams in a bidirectional free-space optical link. Optica 1(6):376–382

63. Resch K, Lindenthal M, Blauensteiner B, Böhm H, Fedrizzi A, Kurtsiefer C, Poppe A, Schmitt-Manderbach T, Taraba M, Ursin R et al (2005) Distributing entanglement and single photons through an intra-city, free-space quantum channel. Opt Express 13(1):202–209

64. Rodenburg B, Mirhosseini M, Malik M, Magaña-Loaiza OS, Yanakas M, Maher L, Steinhoff NK, Tyler GA, Boyd RW (2014) Simulating thick atmospheric turbulence in the lab with application to orbital angular momentum communication. J Phys 16(3):033,020

65. Romero J, Giovannini D, Franke-Arnold S, Barnett S, Padgett M (2012) Increasing the dimension in high-dimensional two-photon orbital angular momentum entanglement. Phys Rev A 86(1):012,334

66. Rowe MA, Kielpinski D, Meyer V, Sackett CA, Itano WM, Monroe C, Wineland DJ (2001) Experimental violation of a bell's inequality with efficient detection. Nature 409 (6822):791–794

67. Salakhutdinov V, Eliel E, Löffler W (2012) Full-field quantum correlations of spatially entangled photons. Phys Rev Lett 108(17):173,604

68. Scheidl T, Ursin R, Kofler J, Ramelow S, Ma XS, Herbst T, Ratschbacher L, Fedrizzi A, Langford NK, Jennewein T et al (2010) Violation of local realism with freedom of choice. Proc Natl Acad Sci 107(46):19,708–19,713

69. Schmitt-Manderbach T, Weier H, Fürst M, Ursin R, Tiefenbacher F, Scheidl T, Perdigues J, Sodnik Z, Kurtsiefer C, Rarity JG et al (2007) Experimental demonstration of free-space decoy-state quantum key distribution over 144 km. Phys Rev Lett 98(1):010,504

70. Schrödinger E (1935) Die gegenwärtige situation in der quantenmechanik. Naturwissenschaften 23(49):823–828

71. Shor PW (1994, November) Algorithms for quantum computation: Discrete logarithms and factoring. In Foundations of Computer Science, 1994 Proceedings., 35th Annual Symposium on (pp. 124-134). IEEE

72. Shumow D, Ferguson N (2007) On the possibility of a back door in the nist sp800–90 dual ec prng. In: Proc. Crypto, vol 7

73. Strekalov D, Sergienko A, Klyshko D (1995) Observation of two-photon "Ghost" interference and diffraction. Phys Rev Lett 74:3600

74. Tamburini F, Thidé B, Molina-Terriza G, Anzolin G (2011) Twisting of light around rotating black holes. Nat Phys 7(3):195–197

75. Torres JP, Deyanova Y, Torner L, Molina-Terriza G (2003) Preparation of engineered two-photon entangled states for multidimensional quantum information. Phys Rev A 67(5):052,313

76. Ursin R, Tiefenbacher F, Schmitt-Manderbach T, Weier H, Scheidl T, Lindenthal M, Blauensteiner B, Jennewein T, Perdigues J, Trojek P et al (2007) Entanglement-based quantum communication over 144 km. Nat Phys 3(7):481–486

77. Ursin R, Jennewein T, Kofler J, Perdigues JM, Cacciapuoti L, de Matos CJ, Aspelmeyer M, Valencia A, Scheidl T, Acin A et al (2009) Space-quest, experiments with quantum entanglement in space. Europhys. News 40(3):26–29

78. Vaziri A, Weihs G, Zeilinger A (2002) Experimental two-photon, three-dimensional entanglement for quantum communication. Phys Rev Lett 89(24):240,401

79. Waks E, Zeevi A, Yamamoto Y (2002) Security of quantum key distribution with entangled photons against individual attacks. Phys Rev A 65(5):052,310

80. Walborn S, Lemelle D, Almeida M, Ribeiro P (2006) Quantum Key Distribution with Higher-Order Alphabets Using Spatially Encoded Qudits. Phys Rev Lett 96(9):090,501

81. Wang C, Deng F, Li Y, Liu X, Long G (2005) Quantum secure direct communication with high-dimension quantum superdense coding. Phys Rev A 71(4):–

82. Wang J, Yang JY, Fazal IM, Ahmed N, Yan Y, Huang H, Ren Y, Yue Y, Dolinar S, Tur M, Willner AE (2012) Terabit free-space data transmission employing orbital angular momentum multiplexing. Nat Phot 6(7):488–496

83. Weedbrook C, Pirandola S, García-Patrón R, Cerf NJ, Ralph TC, Shapiro JH, Lloyd S (2012) Gaussian quantum information. Rev Mod Phys 84(2):621–669

84. Weihs G, Jennewein T, Simon C, Weinfurter H, Zeilinger A (1998) Violation of bell's inequality under strict einstein locality conditions. Phys Rev Lett 81(23):5039

85. Wieśniak M, Paterek T, Zeilinger A (2011) Entanglement in mutually unbiased bases. J Phys 13(5):053,047

86. Wootters WK, Zurek WH (1982) A single quantum cannot be cloned. Nature 299 (5886):802–803

87. Xie G, Ren Y, Huang H, Lavery MP, Ahmed N, Yan Y, Bao C, Li L, Zhao Z, Cao Y, et al (2015) Phase correction for a distorted orbital angular momentum beam using a zernike polynomials-based stochastic-parallel-gradient-descent algorithm. Opt Lett 40 (7):1197–1200

88. Yao A, Padgett MJ (2011) Orbital angular momentum: origins, behavior and applications. Adv Opt Photon 3(2):161–204

89. Yao XC, Wang TX, Xu P, Lu H, Pan GS, Bao XH, Peng CZ, Lu CY, Chen YA, Pan JW (2012) Observation of eight-photon entanglement. Nat Phot 6(4):225–228

90. Yin J, Ren JG, Lu H, Cao Y, Yong HL, Wu YP, Liu C, Liao SK, Zhou F, Jiang Y, et al (2012) Quantum teleportation and entanglement distribution over 100-kilometre free-space channels. Nature 488(7410):185–188

Wave-Particle Dualism in Action

Wolfgang P. Schleich

W.P. Schleich (✉)
Institut für Quantenphysik and Center for Integrated Quantum Science and Technology (IQST), Universität Ulm, D-89069 Ulm, Germany

Institute for Quantum Science and Engineering (IQSE), Texas A&M University Institute for Advanced Study (TIAS), College Station 77843-4242, TX, USA

Department of Physics and Astronomy, Texas A&M University, College Station 77843-4242, TX, USA

© The Author(s) 2016
M.D. Al-Amri et al. (eds.), *Optics in Our Time*, DOI 10.1007/978-3-319-31903-2_19

19.1 Introduction

The wave-particle dualism, that is the wave nature of particles and the particle nature of light together with the uncertainty relation of Werner Heisenberg and the principle of complementarity formulated by Niels Bohr represent pillars of quantum theory. We provide an introduction into these fascinating yet strange aspects of the microscopic world and summarize key experiments confirming these concepts so alien to our daily life.

> » "It looks strange and it looks strange and it looks very strange; and then suddenly it doesn't look strange at all and you can't understand what made it look strange in the first place."
>
> Gertrude Stein

The opening quote refers to modern art but might as well refer to the light quantum, that is the photon. Indeed, in his lecture entitled "Delayed choice experiment and the Bohr-Einstein-Dialogue" on June 5, 1980 in a joint session of the American Philosophical Society and the Royal Society John Archibald Wheeler notes:

> » "The quantum, the most revolutionary principle in all of science and the strangest continues today to unfold its wonders and raise every deeper questions about the relation between man and the universe."

The year of the light constitutes an excellent opportunity to review the progress in our understanding of the light quantum and its idiosyncrasies made possible only recently thanks to novel experimental techniques of addressing and manipulating single particles.

19.1.1 The Strange Photon

Although we have learned a lot we still lack the full picture. In particular, there is still no unique answer to the long-standing question: "What is a photon?"

In the present essay we of course do not answer this deep question either but illuminate one important aspect of the photon that on first sight looks very strange that is the wave-particle dualism. Indeed, according to the quantum theory of radiation the photon is a wave *and* a particle at the same time and their respective distinct features manifest themselves in countless phenomena. The double-slit experiment with individual photons is one of them.

The ultimate goal of our article is to discuss a rather special double-slit experiment based on two entangled photons which seems to show simultaneously the wave and the particle nature of light. Such a behavior which is strictly forbidden by quantum theory and, in particular, by the principle of complementarity makes the photon even stranger. However, a closer look at the details of the light generation reveals that there is no violation of quantum mechanics, and in the words of G. Stein: ". . .suddenly it doesn't look strange at all."

19.1.2 Overview

In order to lay the foundations for our study we first recall important concepts of quantum mechanics such as the wave nature of matter, the uncertainty principle, complementarity, and the quantum eraser. We then focus on a brief description and an elementary analysis of this experiment.

Our article is organized as follows: In ▣ Sect. 19.2 we focus on *the* trademark of quantum mechanics, that is, discrete events and yes/no answers arising from measurements of single particles. Closely associated with this notion is the wave nature of particles discussed in more detail in ◘ Sect. 19.3. Here we consider not only matter waves but also light waves.

We then dedicate ◘ Sect. 19.4 to a historical overview starting with the Heisenberg uncertainty principle and arriving via the formulation of the principle of complementarity at the delayed-choice experiment and the game of twenty questions in its surprise version. In ◘ Sect. 19.5 we turn to the Bohr–Einstein dialogue on the recoiling double-slit and the quantum eraser.

Finally, ◘ Sect. 19.6 is devoted to the discussion of the double-slit experiment using two entangled photons suggesting "which-path" information while observing at the same time interference. We explain these rather counter-intuitive results by considering an elementary model. In particular, we demonstrate that the mutually exclusive scattering arrangements involve different atoms. Therefore, there is no contradiction to the principle of complementarity. We conclude in ◘ Sect. 19.7 by summarizing our results and by providing ideas for further research.

19.2   From the Macro- to the Microcosmos

The transition from the macroscopic to the microscopic world, that is from our daily life to that of an electron orbiting a nucleus, is not as smooth as the limit of classical mechanics of a particle moving with a large velocity to that with a small velocity, or vice versa. In the present section we provide an elementary introduction into some peculiarities of the quantum world, in particular the importance of single events. This fact which emerges from an elementary gedanken experiment suggests that trajectories of particles do not exist in the microscopic world.

19.2.1   Atommechanik

Newtonian mechanics is extremely successful and describes correctly the motion of macroscopic bodies, such as cars, trains, planes, and even planets. Indeed, the description of the motion of the earth around the sun on a Kepler orbit has been a great triumph of classical mechanics.

Of course there are deviations from Newtonian mechanics, for example, due to special relativity when the velocity of the moving object approaches the speed c of light, or due to general relativity, when the curvature of spacetime is no more negligible. An example for the latter is the perihelion shift of mercury.

Why not apply the Newtonian concept of planetary motion which has worked so beautifully for the macro-cosmos to problems of the microscopic world, such as a hydrogen atom. In complete analogy to the earth–sun system we now consider the motion of a single electron around the proton. The resulting Rutherford model of hydrogen supplemented by the appropriate quantization conditions of the actions as proposed by Niels Bohr, Arnold Sommerfeld, and William Wilson gives us a first glimpse of the inner workings of the atom.

However, the early success of "Atommechanik" as this field was called quickly faded. There were too many features of the atom this theory could not explain. Only quantum mechanics developed by W. Heisenberg, Erwin Schrödinger, and Paul Adrien Maurice Dirac could provide a complete and consistent picture.

What is the crucial element not included in Atommechanik? What is the unique feature distinguishing the macro- from the micro-world? Where is the borderline between them, as asked in ◘ Fig. 19.1?

Fig. 19.1 The interface of the classical and the quantum world depicted as the border between two countries with either well-defined structures such as the right part of the guard house, or fuzzy ones represented, for example, by the Schrödinger cat sitting alive in the tree *and* laying dead in the grass. [Taken from Zurek WH (1991) Decoherence and the transition from quantum to classical. Physics Today 44:36–44]

19.2.2 Single Events and Probabilities

In order to provide at least some partial answers to these questions we now consider gedanken experiments which are extremely popular and very helpful in quantum theory. Gedanken experiments whose outcome is predicted by quantum mechanics are constructed to emphasize certain alien aspects of the underlying theory and can be performed in our brain without ever really going to a laboratory.

We illustrate the concept of a gedanken experiment using a specific example. How to determine the motion of an electron in an atom around the nucleus?

Since the electron has an average separation from the nucleus which is of the order of a few Bohr radii we cannot just simply take a camera and take pictures of the electron, or look at it with a microscope. The only way to gain more information is to send a probe into the atom.

In the discussion of what defines the borderline between the microscopic and the macroscopic world one quantity stands out most clearly and allows us to make such a decisive cut: it is Planck's constant $\hbar$. For a given object we can compare its angular momentum $\mathbf{J}$ to $\hbar$. When $\mathbf{J}$ is of the order of $\hbar$ we certainly deal with a problem from the microscopic world.

In classical mechanics, that is in the mechanics of macroscopic bodies we observe trajectories. At every instance of time we can determine uniquely the position of the body. The positions at different times form a world line in spacetime.

However, there are no continuous trajectories for a quantum particle. This feature originates from the discreteness of the particles and reflects the fact that we are trying to learn something about the properties of the microscopic world. Since we do not have a microscope with a resolving power large enough to observe the electron in the atom we have to send a probe from the outside into the atom. By measuring the change of that probe induced by the interaction with the electron we learn something about the electron.

When we use a single particle as a probe we get one bit of information from the detection of the scattered probe. In order to obtain more information we have to repeat this experiment many times. In this way these scattering events, each obtained from single quantum probes, provide us with information about the inside of the atom.

This analysis brings out most clearly that we do not see the electron in the atom move around the nucleus as suggested by the Bohr–Sommerfeld–Wilson atom model, but rather find *probabilities* that the electron had been at a certain position. Obviously, the scattering events do not tell us with certainty the locations where the electron was at a given time.

19.2.3 Single Clicks Reconstruct the Microcosmos

In summary, the microscopic world is only accessible through probes which have to be of the same size as the elements of the microscopic world that we want to investigate. Hence, we probe quantum objects by single microscopic particles.

From every scattering event we gain one bit of information and complete our picture of the microscopic world by recording a multitude of single events, that is, single clicks. Once more we are reminded of a quote by J. A. Wheeler who summarized this situation in his poetic style:

» "Do we not do better to recognize that what we call existence consists of countably many iron posts of observations between which we fill in by an elaborate papier-mâché construction of imagination and theory?"

19.3 Double-Slit Experiments with Light and Matter

Wave-like aspects of light have already been observed around 1660. For example, Francesco Grimaldi noticed that when light passes a narrow slit in a wall the edges of the narrow band of brightness are slightly blurred suggesting that light diffracts.

However, it was only in the beginning of the twentieth century that a similar revolution took place for matter. Up to that moment electrons, atoms, or molecules were considered particles. However, the experiments of Clinton Joseph Davisson and Lester Halbert Germer in 1926 who scattered electrons from a nickel crystal brought out most clearly that also matter displays wave features as proposed earlier by Louis-Victor Pierre Raymond de Broglie.

In the present section we first recall the transition from the corpuscular theory of light due to Isaac Newton to the wave interpretation of Thomas Young. We then briefly review various double-slit experiments with matter waves and conclude by emphasizing subtleties associated with this arrangement.

19.3.1 Light: Corpuscle Versus Wave

More than 200 years ago, Th. Young demonstrated the wave nature of light. However, he did not use slits in an opaque screen as widely believed but rather pinholes. Despite this fact this famous experiment carries the name *double-slit experiment*.

Before his impressive demonstration the dominance of the corpuscular theory of light proposed by I. Newton had suppressed any wave theory. The following quote from Th. Young's article may illustrate this strong influence of I. Newton even almost 100 years later:

» "In making some experiments on the fringes of colors accompanying shadows, I have found so simple and so demonstrative a proof of the general law of the interference of two portions of light, which I have already endeavored to establish, that I think it right to lay before the Royal Society, a short statement of the facts which appear to me so decisive. The proposition on which I mean to insist at the present, is simply this, that fringes of colors are produced by the interference of two portions of light; and I think it will not be denied by the most prejudiced, that the assertion is proved by the experiments I am about to relate, which may be repeated with great ease, whenever the sun shines, and without any other apparatus than is at hand to every one."

He continues his critique of the corpuscular theory by stating:

» "Those who are attached to the Newton theory of light, or to hypotheses of modern opticians, founded on views still less enlarged, would do well to endeavor to imagine anything like an explanation of these experiments, derived from their own doctrines; and, if they fail in the attempt, to refrain at least from idle declamation against a system which is founded on the accuracy of its application to all these facts, and to a thousand others of a similar nature."

It is amusing that the quantum theory of radiation brings back the particle aspect of light in the form of the photon, that is, the quantized excitation of a mode of the radiation field. In this way I. Newton and his corpuscular theory were vindicated after all.

19.3.2 Matter: Particle Versus Wave

Next we turn to the wave nature of matter which under appropriate conditions can also manifest itself in interference fringes in the far field of a double-slit. Here we discuss "slits in space" as well as "slits in time."

Slits in Space

In ◘ Fig. 19.2 we show the essential ingredients of a double-slit experiment for matter waves consisting of a particle source, an opaque screen with two slits, and a detector in the far field. We assume that the source emits one particle at a time and there is a long delay between two successive emissions. In this case there is only a single particle between the source and the detector at a time.

Each particle can only go either through the upper or the lower slit. After many particles have passed the slits, we should observe a double-hump distribution where the two maxima correspond to the two slits. However, numerous experiments clearly show that under appropriate conditions this by classical notions motivated picture is incorrect.

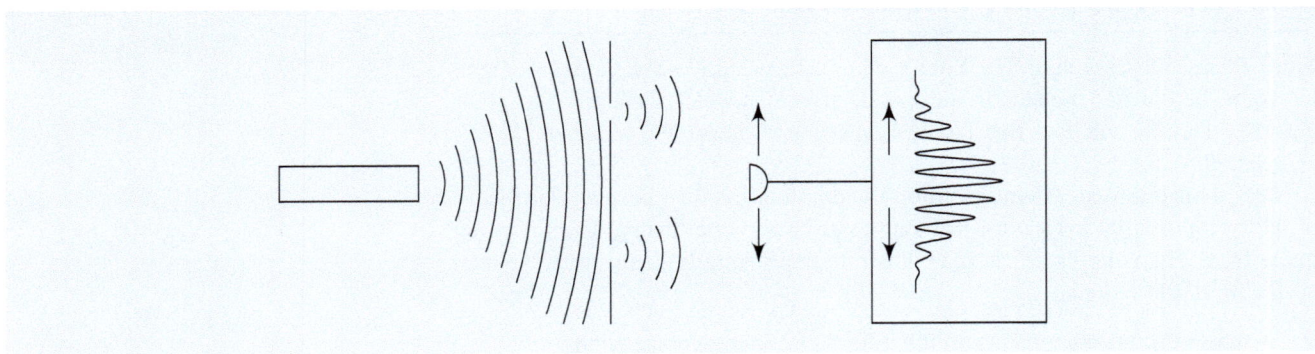

◘ **Fig. 19.2** Elementary building blocks of a double-slit experiment for particles involving a source of particles (*left*), a screen with two slits (*middle*), and a detector in the far field (*right*). The particles to be scattered could be electrons, neutrons, atoms, or rather large molecules. We observe an oscillatory count rate (*far right*) as the detector moves along an axis parallel to the screen demonstrating the existence of matter waves

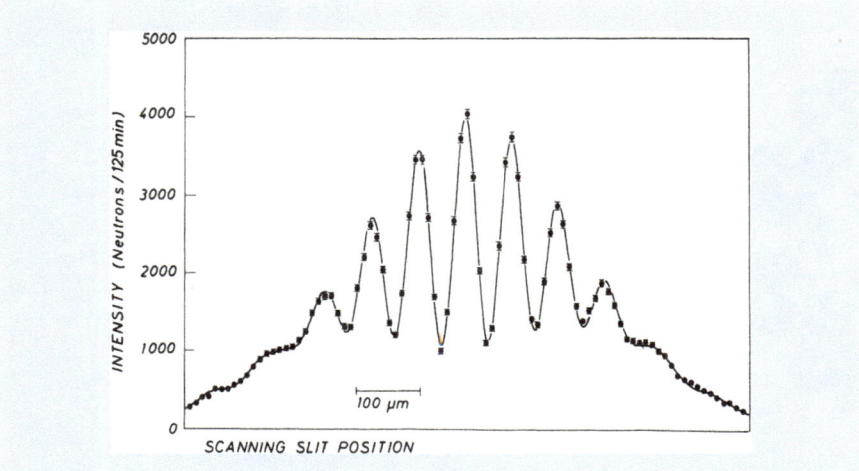

Fig. 19.3 Interference pattern of a double-slit experiment with cold neutrons. [Taken from Zeilinger A et al. (1988) Single and double-slit diffraction of neutrons. Rev Mod Phys 60:1067–1073]

Fig. 19.4 Distribution of electrons scattered from two slits in a screen. [Taken from Jönsson C (1961) Elektroneninterferenzen an mehreren künstlich hergestellten Feinspalten. Z f Phys 161:454–474, for an English translation see: Jönsson C (1974) Electron Diffraction at Multiple Slits. Am J Phys 42(1), 4–11]

◘ Figure 19.3 depicts the intensity pattern of neutrons in the far field of a mechanical double-slit which displays interference fringes. This effect is quite remarkable when we recognize the count rates on the vertical axis. At a maximum of the fringe we find approximately 4000 counts per 125 min. This rate corresponds to two neutrons going through the apparatus per second. Since the velocity of the neutrons was 200 m/sec there was never more than one neutron in the apparatus.

Similar experiments have been performed earlier by Claus Jönsson in the group of Gottfried Möllenstedt at the Universität Tübingen with electrons. ◘ Figure 19.4

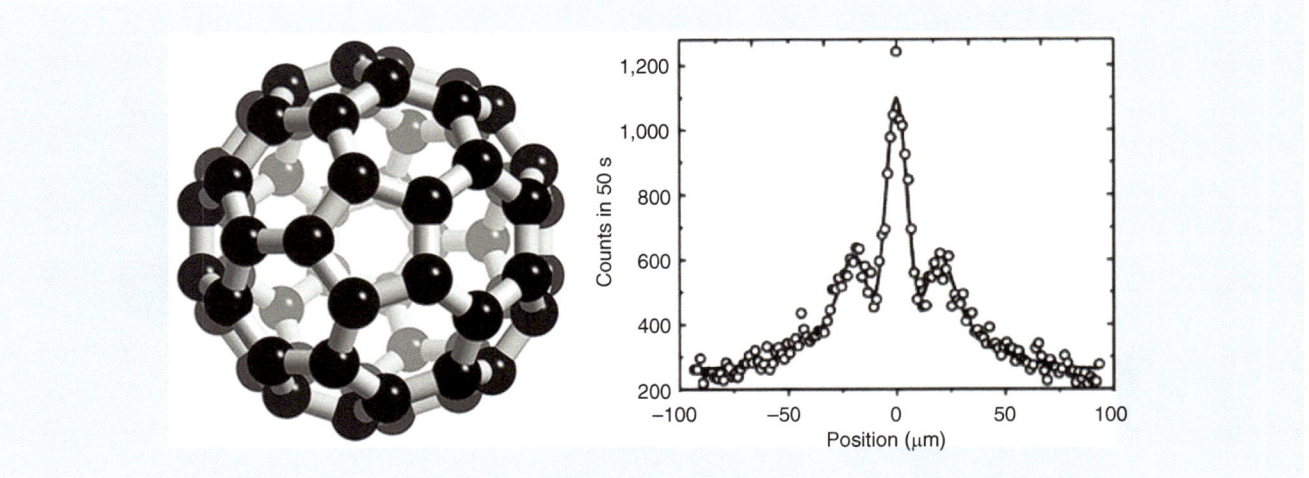

Fig. 19.5 Double-slit interference pattern (*right*) of fullerene molecules (*left*) which are regular structures of 60 carbon atoms in the shape of a soccer ball. For this reason the fullerene molecule is sometimes jokingly referred to as soccerballium. When individual molecules are sent one at a time through a double-slit the pattern found on a screen in the far field shows clear interference fringes. [Taken from Arndt M et al. (1999) Wave particle duality of C_{60} molecules. Nature 401:680–682]

shows the count rate of a double-slit experiment with electrons and the interference fringes are clearly visible.

It is interesting to note that in September 2002 the journal *Physics World* reported a poll concerning the top ten most beautiful experiments in physics. The Jönsson experiment was the number one.

This phenomenon of matter–wave interference is not limited to neutrons or electrons. Even bigger objects such as the fullerene molecule C_{60} exhibit an interference pattern, as shown in ◘ Fig. 19.5.

Slits in Time

A rather intriguing version of a double-slit experiment with electrons was carried out by the group of the late Herbert Walther using ultra-short laser pulses. Here the interference appears in the time rather than the space domain.

Light pulses in the femto-second regime consist of a few optical cycles and can ionize single electrons in atoms, as shown in ◘ Fig. 19.6. However, this process only occurs when the associated electric field is above a threshold. Since the pulses are short the intensity necessary for ionization exists only during one or two time periods with an extension of an atto-second. By shifting the envelope of the pulse relative to the oscillation we can control the time window of ionization and create in this way a single- or a double-slit type of excitation of the atom.

Indeed, the double-slit situation appears when the sub-cycle pulse contains *two* narrow time windows in which the atom can be ionized. Therefore, the electron which reaches the detector with a well-defined momentum results from field ionization either in the first, or in the second time window. As long as we cannot decide in principle in which one the electron was born the two ionization paths in time must interfere. Since only a single electron is ionized in each event the interference takes place on the level of individual particles. Moreover, it manifests itself in the energy spectrum of the electron shown in the bottom of ◘ Fig. 19.6. If there is only one time window for ionization representing a single-slit situation there is no interference.

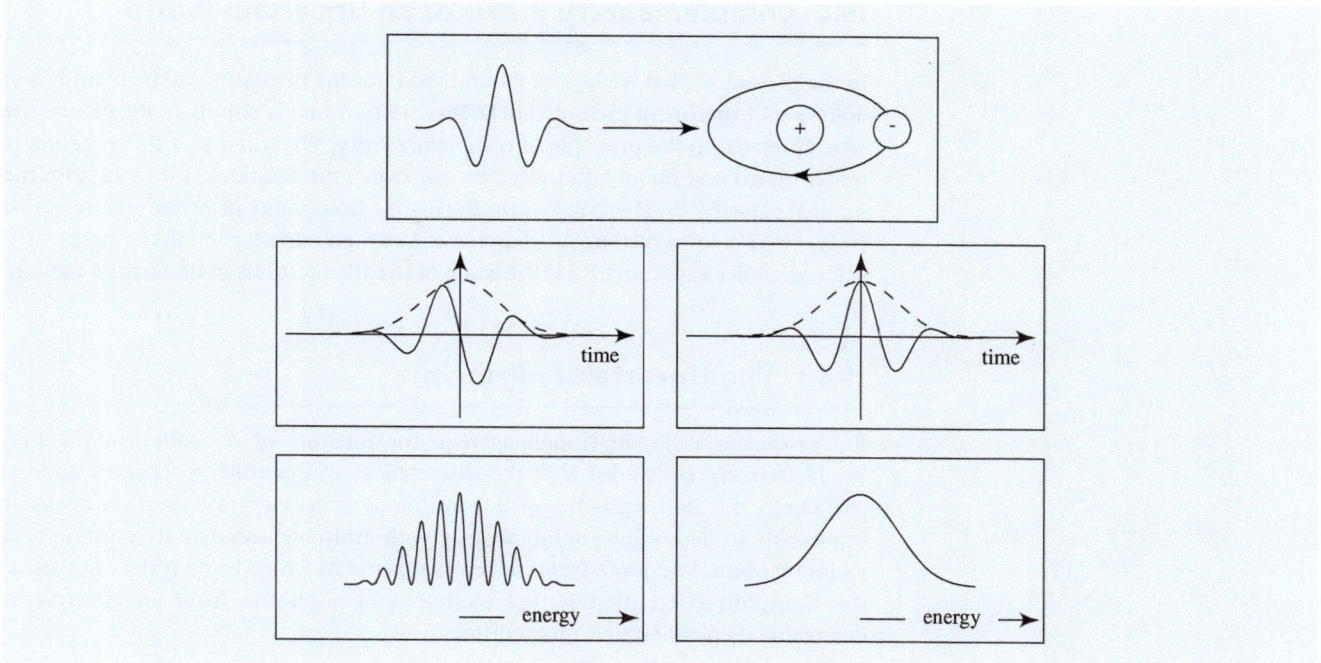

Fig. 19.6 Double-slit experiment in time. A sub-cycle laser pulse ionizes an electron in an atom (*top*). Depending on the phase of the oscillation relative to the envelope (*middle*) we either have two maxima (*left*) or a single dominant maximum (*right*) that can ionize the electron. In case of two maxima we cannot identify which one ionized the electron. As a result, the ionization current (*bottom*) displays oscillations (*left*) as a function of energy due to the indistinguishable excitation paths. In contrast, no interference arises for a single intensity maximum (*right*). [After Lindner F et al. (2005) Attosecond double-slit experiment. Phys Rev Lett 95:040401]

19.3.3 The Mystery of the Double-Slit Experiment

The physics of the double-slit experiment has occupied physicists since the early days of quantum mechanics. The rather paradoxical phenomenon of an interference pattern for particles can be expressed most vividly by the following situation borrowed from a wild-west movie.

Imagine a person shooting bullets towards a screen with two slits. Due to the wave nature of the bullets the interference pattern in the far field of the slits enjoys positions where no bullet will ever hit. Hence, a person standing behind the screen at one of these zeros of the fringes is safe. However, if one of the holes gets closed the interference and, hence, the zeros cease to exist. As a consequence, the next bullet might kill the person.

This discussion of the double-slit experiment also brings out most clearly the importance of the single event emphasized already in ■ Sect. 19.2.3 as a building block of our conception of the microscopic world. Indeed, we send individual particles through the apparatus, one at a time. Each particle is detected after it has passed the slits and will either hit the screen at this position or at another position. After we have sent many such particles through the apparatus we have a histogram, that is, a number of counts at every position on the screen which will not be uniform but will show oscillations. From individual counts we have built up a continuous distribution.

19.4 Complementary Views of an Uncertain World

In the present section we lay the ground work for the discussion of the two-photon double-slit experiment presented in ◘ Sect. 19.6 by briefly summarizing pioneering articles related to the principle of complementarity. We admit that this selection is rather biased and mainly motivated by our own considerations. We start from the seminal paper by W. Heisenberg introducing the uncertainty principle and then turn to N. Bohr's introduction of complementarity culminating in the concept of a delayed-choice experiment and the game of twenty questions in its surprise version.

19.4.1 The Uncertainty Principle

It was during a stay at Copenhagen in the institute of N. Bohr in 1927 that W. Heisenberg concluded that the discreteness of quantum mechanics and, in particular, the non-existence of a continuous trajectory of a particle make it impossible to determine simultaneously with arbitrary accuracy its position and its momentum. We quote from his seminal paper in which he considers the use of the Compton effect, that is, the scattering of a photon from an electron to determine its position and momentum:

> "At the instant at which the position of the electron is known, its momentum therefore can be known up to magnitudes which correspond to that discontinuous change. Thus, the more precisely the position is determined the less precisely the momentum is known, and conversely. In this circumstance we see a direct physical interpretation of the equation $\mathbf{pq} - \mathbf{qp} = -i\hbar$. Let q_1 be the precision with which the value q is known (q_1 is, say, the mean error of q), therefore here the wavelength of the light. Let p_1 be the precision with which the value p is determinable: that is here the discontinuous change of p in the Compton effect. Then, according to the elementary laws of the Compton effect p_1 and q_1 stand in the relation
>
> $$p_1 q_1 \sim h. \tag{19.1}$$
>
> That this relation (19.1) is a straight-forward mathematical consequence of the rule $pq - qp = -i\hbar$ will be shown below.[1]"

It is interesting to note that in this article W. Heisenberg does not use the notation Δq and Δp for the uncertainties but rather q_1 and p_1. Moreover, it is also amusing that he applies Schrödinger wave functions rather than matrices to

---

1   "In dem Moment, in dem der Ort des Elektrons bekannt ist, kann daher sein Impuls nur bis auf Größen, die jener unstetigen Änderung entsprechen, bekannt sein; also je genauer der Ort bestimmt ist, desto ungenauer ist der Impuls bekannt und umgekehrt; hierin erblicken wir eine direkte anschauliche Erläuterung der Relation $\mathbf{pq} - \mathbf{qp} = \frac{h}{2\pi i}$. Sei q_1 die Genauigkeit, mit der der Wert q bekannt ist (q_1 ist etwa der mittlere Fehler von q), also hier die Wellenlänge des Lichtes, p_1 die Genauigkeit, mit der der Wert p bestimmbar ist, also hier die unstetige Änderung von p beim Comptoneffekt, so stehen nach elementaren Formeln des Comptoneffekts p_1 und q_1 in der Beziehung

$$p_1 q_1 \sim h. \tag{19.1}$$

Daß diese Beziehung (19.1) in direkter mathematischer Verbindung mit der Vertauschungsrelation $pq - qp = \frac{h}{2\pi i}$ steht, wird später gezeigt werden."

illustrate the consequences of the uncertainty principle. The only remnants of matrix mechanics are the non-commuting operators **q** and **p**.

Heisenberg submitted his manuscript during a skiing vacation of N. Bohr who upon his return pointed out various mistakes and brought to light a deeper concept. As a result, W. Heisenberg felt obliged to include the following note added in proof:

» "After the conclusion of the foregoing paper, more recent investigations of Bohr have led to a point of view which permits an essential deepening and sharpening of the analysis of quantum-mechanical correlations attempted in this work. In this connection Bohr has brought to my attention that I have overlooked essential points in the course of several discussions in this paper. Above all, the uncertainty in our observation does not arise exclusively from the occurrence of discontinuities, but is tied directly to the demand that we ascribe equal validity to the quite different experiments which show up in the corpuscular theory on one hand, and in the wave theory on the other hand.[2]"

According to N. Bohr the uncertainty principle does not arise from the discontinuities but from the choice of the wave versus particle description demanded by the specific experimental setup. Hence, he supports the idea of an uncertainty relation but identifies a different origin of it.

19.4.2 The Birth of Complementarity

The article by N. Bohr summarizing his point of view appeared a year later, that is in 1928, with the title "The quantum postulate and the recent development of atomic theory." It was based on a lecture he gave on September 16, 1927 in Como at the International Congress of Physics in commemoration of the centenary of the death of Alessandro Volta. The reason for this delay originated from an unusual twist of events associated with his original manuscript on complementarity, his passport, and his train to Como.[3]

In his typical style N. Bohr draws attention to the fundamental difference between the classical and the quantum world when he states:

» "The very nature of the quantum theory thus forces us to regard the space-time coordination and the claim of causality, the union of which characterises the classical theories, as complementary but exclusive features of the description, symbolising the idealisation of observation and definition respectively."

Here the words "complementary" and "exclusive" enter the stage of physics for the first time.

---

2   "Nach Abschluß der vorliegenden Arbeit haben neuere Untersuchungen von Bohr zu Gesichtspunkten geführt, die eine wesentliche Vertiefung und Verfeinerung der in dieser Arbeit versuchten Analyse der quantenmechanischen Zusammenhänge zulassen. In diesem Zusammenhang hat mich Bohr darauf aufmerksam gemacht, daß ich in einigen Diskussionen dieser Arbeit wesentliche Punkte übersehen hatte. Vor allem beruht die Unsicherheit in der Beobachtung nicht ausschließlich auf dem Vorkommen von Diskontinuitäten, sondern hängt direkt zusammen mit der Forderung, den verschiedenen Erfahrungen gleichzeitig gerecht zu werden, die in der Korpuskulartheorie einerseits, der Wellentheorie andererseits zum Ausdruck kommen."

3   For the details of this amusing story see the commentary by Leon Rosenfeld on page 85 of the book by J. A. Wheeler and W. H. Zurek listed in Further Reading.

Moreover, N. Bohr has a clear picture how the observer changes the microscopic world by the intrusion necessary for his measurement. Indeed, he writes:

» "...the measurement of the positional coordinates of a particle is accompanied not only by a finite change in the dynamical variables, but also the fixation of its position means a complete rupture in the causal description of its dynamical behaviour, while the determination of its momentum always implies a gap in the knowledge of its spatial propagation. Just this situation brings out most strikingly the complementary character of the description of atomic phenomena which appears as an inevitable consequence of the contrast between the quantum postulate and the distinction between object and agency of measurement, inherent in our very idea of observation."

Obviously N. Bohr associates with the act of the measurement physical effects on the system to be measured. We shall return to this aspect in ◘ Sect. 19.5.

19.4.3 A Mechanical Model of Complementarity?

In 1939 at the world exhibition in New York the University of Copenhagen presented a mechanical model illustrating the principle of complementarity. ◘ Figure 19.7 shows a sketch of this device originating from J. A. Wheeler.

An unusual filing cabinet contains a drawer which can be pulled out in the front as well as in the back and which is divided into two compartments each of which contains a die. The task is to read the number shown on the top of *both* dice.

However, there is a slight problem. We cannot observe both dice simultaneously. When we pull the drawer to the front we can see only the die in the front compartment. When we push the drawer through to the back we can observe only the one in the back.

Of course, we could first copy down the number on the top of the first die and then pull the drawer out on the other side to have a look at the other die. However, a devilish device hidden underneath the floor of the drawer, such as a little hammer, is set into action as the drawer slides through the cabinet. Due to the kick imparted onto the floor the die whose number has just been recorded gets knocked over.

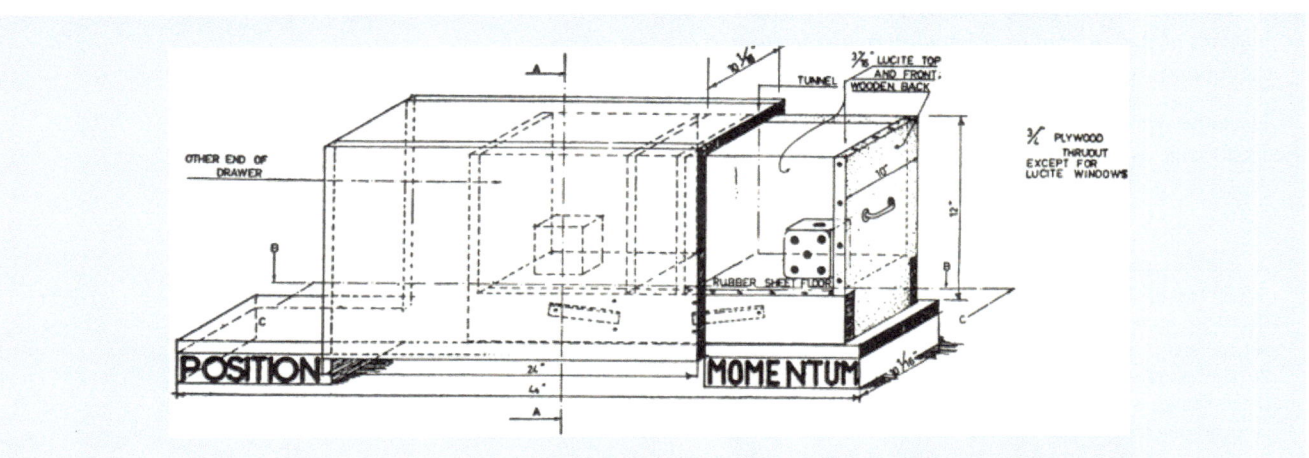

◘ **Fig. 19.7** Mechanical model of the principle of complementarity designed by the University of Copenhagen. Reading the numbers on the top faces of the dice stored in the two different compartments of a drawer in a filing cabinet corresponds to measuring two conjugate variables such as position and momentum, or path and interference. Unfortunately, this mechanical model misses the central lesson of quantum mechanics: There is no number on the dice until we make an observation. [Taken from Wheeler JA (1994) At Home in the Universe. AIP Press]

Although we can now record the number shown on the die confined to the back compartment it does not even make sense to do so. In our attempt to obtain information about the back die we have lost the information about the front one.

Unfortunately, we face the same dilemma if we start from the back die and move the drawer forward. Again the hammer is set in action and makes the knowledge we have just obtained redundant.

Obviously, the top faces of the dice play the role of two complementary quantities, such as position and momentum, and on first sight this model seems to illustrate in an impressive way the principle of complementarity. However, it lacks the fundamental ingredient of quantum mechanics summarized by J. A. Wheeler in the pregnant phrase:

» "No elementary quantum phenomenon is a phenomenon until it is a recorded phenomenon, brought to a close by an irreversible act of amplification."

According to quantum mechanics it is not a meaningful question to ask: What would the numbers on the dice have been, if we measured them. They do not exist until they are observed.

19.4.4 No Existence Without Measurement

The unusual property of quantum mechanical observables such as position or momentum, or components of angular momentum to take on a definite value only after observation comes out most clearly in the delayed-choice experiment. When we inject a single particle, one at a time, into the upper entrance of the Mach–Zehnder interferometer shown in ◘ Fig. 19.8 we expect the particle to either go on

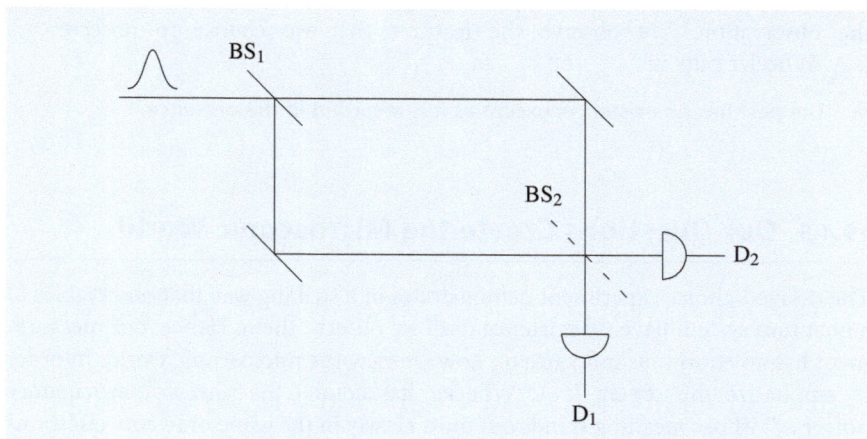

◘ **Fig. 19.8** Mach–Zehnder interferometer in a delayed-choice mode. A particle entering the interferometer in one of the two ports at the upper left beam splitter BS$_1$ either takes the high or the low road. When the beam splitter BS$_2$ at the lower right corner is inserted the two paths interfere at the two detectors D$_1$ and D$_2$. However, when we remove BS$_2$ we obtain "which-path" information. We can delay our decision concerning BS$_2$ till the particle hits it. Loosely speaking, at the moment of our decision we force the particle to retrace its tracks and change its history. Indeed, when BS$_2$ is part of the interferometer as the particle passes BS$_1$ and we remove BS$_2$ later we force the particle to alter its nature. Before our decision it was supposed to display interference properties, but afterwards it needs to provide us with path information. In this arrangement the particle was supposed to move along a *single* path rather than on *both* paths. This paradoxical situation can only be resolved by the assumption that the particle does not have any path whatsoever until we observe it, nor can it display any interference fringes until we measure them. In the language of J. A. Wheeler: "The particle is a great smoky dragon that is only sharp where it enters the interferometer and where it leaves the interferometer biting the detector"

the upper or the lower road. When we leave out the beam splitter at the lower right corner we can detect the path of the particle, because only one of the two detectors will respond.

However, when we do insert the beam splitter and have many particles, one at a time, pass through the interferometer we will find all particles in one of the exit ports provided we adjust the arm length appropriately. This behavior is a consequence of the wave nature of matter. Indeed, due to destructive interference no particles are in the other port.

So far we have only discussed another manifestation of the complementarity principle. However, we can now go one step further, and use our knowledge of the time at which the particle entered the interferometer. Moreover, if we know the initial velocity of the particle we can predict the time at which it will impinge on the second beam splitter. Within the period defined by the entrance of the particle into and its exit from the Mach–Zehnder interferometer we can now decide if we want to insert the second beam splitter or not. In this way we make a *delayed* choice between our ultimate observation of interference or "which-path."

We start from a situation where the second beam splitter is present when the particle enters the interferometer. In the language of the macroscopic world the particle has to display its interference nature and has to move on both paths.

This interpretation runs into problems when in the last moment before the particle hits the second beam splitter, we take it out. Now, we are asking for path information and the particle had to go on one path only. However, by that time the particle has almost reached the beam splitter. It can therefore not go back and retrace its tracks. Our procrastination in making a decision, that is, the delayed choice of interference versus "which-path" highlights the idea that in the microscopic world the properties of particles are not well-defined until they are observed.

We emphasize that many delayed-choice experiments with light and matter waves have been performed. They clearly show that the delay has no influence on the observation. We observe the features that we choose to observe. As J. A. Wheeler puts it:

» "The past has no existence, except as it is recorded in the presence."

19.4.5 Our Questions Create the Microscopic World

The delayed-choice experiment demonstrates in a striking way that observables of a quantum system have no existence until we observe them. Hence, our measurement has an enormous influence on how we view the microscopic world. In order to emphasize this aspect J. A. Wheeler has coined the phrase "participatory universe," whose meaning stands out most clearly in the game of twenty questions in its surprise version.

A group of friends sends one victim out of the room while his/her remaining colleagues agree on a word to be guessed. After the person has returned he/she is allowed to ask twenty questions. The answers must be given truthfully with "yes" or "no" and after this question-answer-period the person is confronted with the challenge to produce the word.

However, when once the turn came to J. A. Wheeler to be the victim he found upon reentering the room his friends with a grin on their faces. He knew something was up.

He started by asking: "Is it a cloud?" A quick response came: "No!" The second question: "Is it a car?" Now the answer took a little bit longer. His friends had to think about it and finally they answered: "No!" The more questions he asked the longer it took them to answer.

This hard work on their part was difficult to understand because a word had been agreed on and all that had to be done was to see if his guess was correct. Nevertheless, it took even longer as they approached the final trial answers.

Finally J. A. Wheeler had to make a decision. Challenging one of his opponents he put forward one final question: "Is it a bear?"

Again the challenged had to think for a long time before he eventually admitted: "Yes, you are right!" – Laughter broke out in the room.

How come his friends had to think at least as hard as he? The answer to this question originates from the fact that when he had left the room they had decided not to agree on a word at all. However, their individual answers would have to be consistent with each other and he would only win if his guess was consistent with the chain of their answers.

As a result of these new rules the game was as difficult for them as it was for him. No word existed in the room until it was challenged by the observer who became the "creator" of the word.

This game of twenty questions in its surprise version encapsulates the crucial point of quantum mechanics: The microscopic world does not exist until we observe it.

19.5 Physical Disturbance Versus Correlations

In our discussion of the principle of complementarity in ◩ Sect. 19.4.2 the second quote of N. Bohr shows that he associated with a measurement of the microscopic world a physical disturbance. His point of view stands out most clearly in the Bohr–Einstein discussion of the recoiling double-slit designed by Albert Einstein to obtain "which-path" information together with interference fringes. This dialogue started at the Solvay meeting of 1927 and continued for almost 30 years.

We dedicate this section to a brief introduction of this gedanken experiment which later has been analyzed by William Wootters and Wojciech Hubert Zurek using the formalism of modern quantum mechanics and, in particular, of joint measurements. We conclude by highlighting the key ingredients of the quantum eraser developed in various forms by Marlan Orvil Scully and coworkers.

19.5.1 Recoiling Double-Slit

With his friend the philosopher Harald Høffding, N. Bohr frequently discussed the double-slit experiment and Høffding asked: "Where can the particle be said to be?" Bohr answered in the familiar Hamlet way: "To be? To be? What does it mean 'to be'?"

What does it mean to talk about a particle going through the upper or lower slit, or through both slits if we do not make a measurement to prove our claim? But how can we make such a measurement?

The proposal of A. Einstein for such a measurement of "which-path" information *and* interference involves a *movable* rather than a *fixed* screen. By measuring the momentum transfer of the scattering particle on the slit, and the interference fringes in the far field A. Einstein argued that in principle we can observe simultaneously position and momentum with arbitrary accuracy. However, N. Bohr showed that this claim is not correct since the momentum transfer of the scattering particle wipes out the fringes. Measurements of this type are still limited by the uncertainty principle.

In 1979 W. Wootters and W. H. Zurek revisited this arrangement of a recoiling double-slit and demonstrated that the interference pattern is surprisingly sharp even when the trajectories have been determined with a fairly high accuracy. In their analysis the entanglement between the center-of-mass motions of the scattering particle and the slit plays a crucial role.

19.5.2 Quantum Eraser

We emphasize that the mechanism for the destruction of the interference fringes discussed in the preceding subsection relies heavily on random phase disturbances. Indeed, the key argument is always the physical transfer of momentum which leads to an uncontrollable phase disturbance and wipes out the fringes. We now analyze two situations which show that this notion is not correct.

Double-Slit with Two Atoms

A new era in the analysis of the double-slit experiment started in 1982 with the proposal of the quantum eraser illustrated in ◘ Fig. 19.9. Here the two mechanical slits are replaced by two identical atoms excited by the incident radiation. Each atom decays with the emission of a photon which is detected in the far field. When

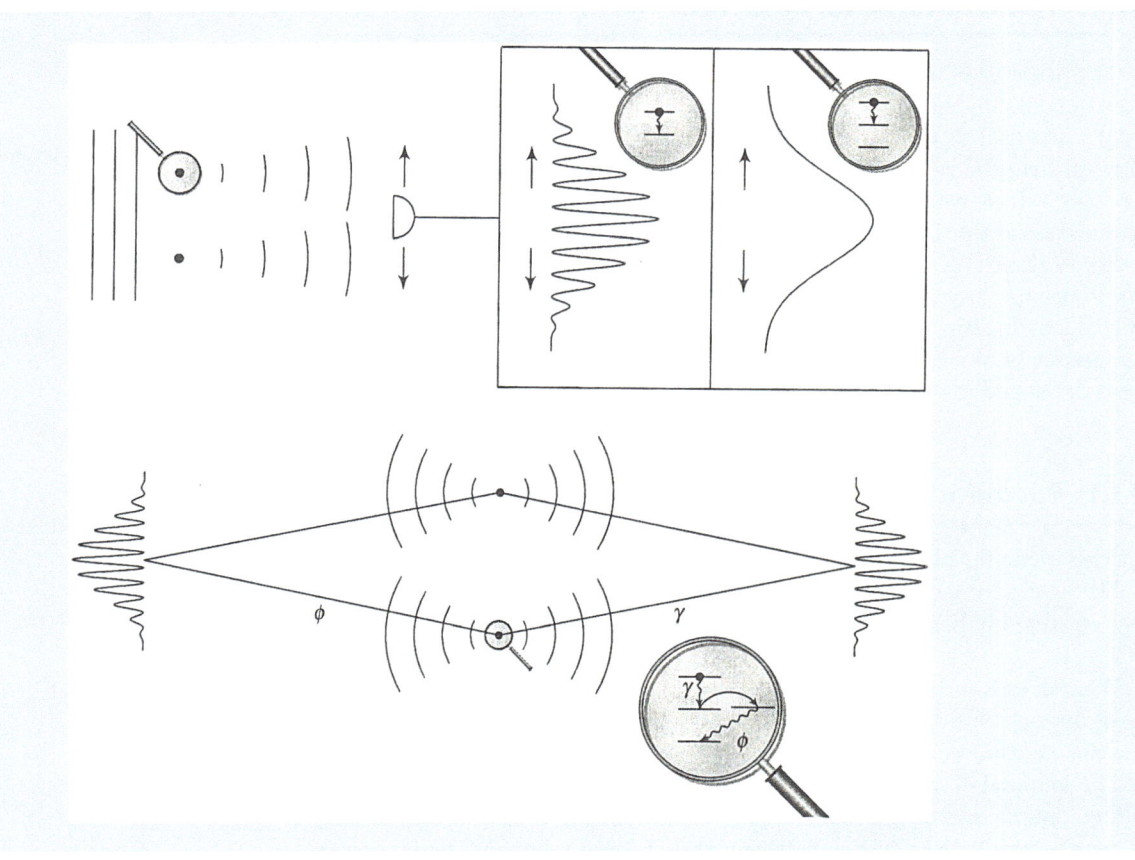

◘ **Fig. 19.9** Realization of a double-slit experiment based on the scattering of light from two atoms (*top*) and the concept of the quantum eraser (*bottom*). In this arrangement we replace the two slits in the screen by two atoms in the absence of any screen and scatter one light quantum from both atoms which are initially in the ground state. After the scattering event both atoms are again in their respective ground states provided we deal with *two*-level atoms. Since in this case it is impossible to tell which atom scattered the quantum we observe interference in the far field (*top-left*). For two *three*-level atoms one will always remain in the long-living intermediate state which provides us with "which-path" information and no interference occurs (*top right*). However, the fringes reemerge when instead we use two *four*-level atoms (*bottom*) together with a joint measurement between the two emitted quanta γ and ϕ. In this case we have erased the "which-path" information of the scattering since both atoms have again returned to their ground states. It is this process of erasing the "which-path" information which recreates the interference fringes

the radiation is so weak that only one of the two atoms gets excited at a time the paths of excitation can either interfere giving rise to fringes in the far field, or leave "which-path" information in the atom creating a smooth intensity pattern.

This decisive difference is dictated by the internal structure of the atoms. In order to bring this fact out most clearly we first consider two *two*-level atoms which are initially in their ground states. After the excitation by and subsequent emission of the photon both atoms are again in their ground states. As a result, we cannot tell from the final arrangement which atom has scattered the light and interference fringes occur.

However, when we use two *three*-level atoms with a long-living intermediate state the scattering path can be reconstructed from the final internal state. Indeed, the atom that has been excited and has reemitted the photon will be left in the middle state, whereas the atom that did not participate in the scattering process is still in the ground state. Due to the availability of "which-path" information no fringes appear in this situation.

We finally consider a *four*-level atom. After the emission of the photon γ and the decay to the intermediate state we pump into the fourth level which decays rapidly to the ground state by emitting a second photon ϕ. The intensity pattern of γ-photons measured in the far field now conditioned on the detection of ϕ-photons displays oscillations since we have erased the information about the path of the excitation.

The quantum eraser brings out clearly that the disappearance of the interference fringes is not related to uncontrollable disturbances of the phases of the atoms, or the field, but rather originates from the correlations established between the internal states of the atom and the field.

Double-Slit with Two Cavities

This shift of paradigm is also emphasized by the gedanken experiment shown in
▣ Fig. 19.10 combining the wave nature of matter, that is, atom optics with cavity

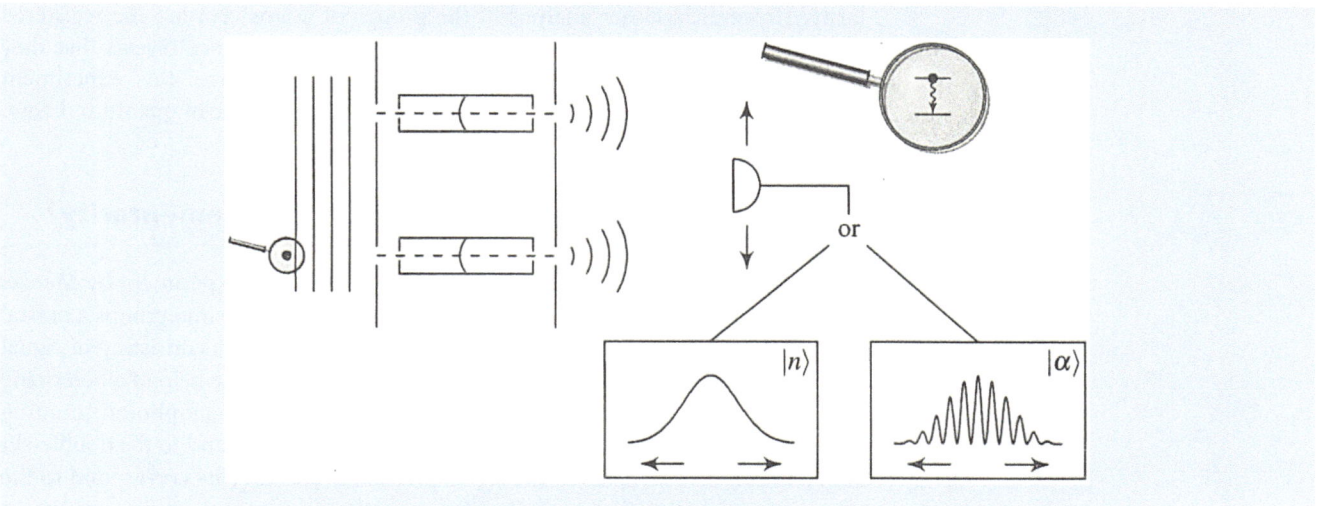

▣ **Fig. 19.10** "Which-path" information encoded in the quantum states of two *cavity fields*. A single two-level atom initially in its excited state and with a center-of-mass motion in a coherent superposition of passing the upper and the lower cavities interacts with the single-mode fields indicated here by their lowest modes in such a way as to deposit with certainty its internal excitation. When both fields are in a number state $|n\rangle$ one of them must change by one quantum, that is go from $|n\rangle$ to $|n+1\rangle$ and in this way create "which-path" information. As a consequence, no interference occurs in the center-of-mass motion in the far field of the second double-slit. However, when both fields are in a coherent state $|\alpha\rangle$ of large amplitude one quantum does not make a difference. In this case the path of the atom cannot be reconstructed and fringes emerge

quantum electrodynamics. Here a single two-level atom whose transverse center-of-mass motion is in a superposition of two locations, prepared, for example, by a double-slit, passes two high Q-cavities whose mode maxima are aligned with these slits. The atom is initially in the excited state and the experimental parameters are chosen such that the atom must deposit its internal excitation in one of the two cavities. Hence, the manifestation of interference in the transverse center-of-mass motion depends on the initial state of the cavities.

Indeed, when both are in a number state $|n\rangle$ the photon placed by the atom in one of the cavities increases the corresponding photon number n by one unit. We can reconstruct by this change the path the atom has taken, and since number states corresponding to different photon numbers such as $|n\rangle$ and $|n+1\rangle$ are orthogonal no interference fringes occur.

However, for a coherent state $|\alpha\rangle$ of large average number $\overline{n}$ of photons, that is $1 \ll \overline{n} \equiv |\alpha|^2$, this change by one photon is negligible and the two coherent states corresponding to $\overline{n}$ and $\overline{n}+1$ are *not* orthogonal. As a consequence, in this case the path is unknown to us and interference fringes emerge.

Orthogonality of field states instead of uncontrollable phase changes as the eraser of interferences—this statement serves as the one-sentence-summary of this version of "which-path" detectors. It is interesting to note that this insight has led to a lively controversy. Indeed, the group of the late Daniel F. Walls has repeatedly emphasized that there is still room for an interpretation in terms of phase disturbances. However, the seminal experiment based on atom interferometry by the group of Gerhard Rempe has tilted the scale towards the notion of orthogonality.

19.6 A Two-Photon Double-Slit Experiment

We now briefly highlight the key features of a recent experiment on wave-particle dualism using entangled photons performed in the group of Ralf Menzel. On first sight their results seem to indicate a break-down of the principle of complementarity. However, a closer analysis of the groups of atoms creating the registered photons leading to "which-path" information and interference reveals that they are different in the two arrangements. As a consequence, this experiment constitutes another impressive verification of this corner stone of quantum theory.

19.6.1 A Violation of the Principle of Complementarity?

◘ Figure 19.11 highlights the essential components of the experiment by Menzel et al. Here, a laser in a mode which displays two distinct maxima pumps a crystal and two entangled photons are born. The resulting two pairs consisting of signal and idler photons at the exit of the crystal are imaged with the help of a polarizing beam splitter onto the two slits of a double-slit and onto a single-photon counting detector D_1. Since the distances from the beam splitter to D_1 and to the double-slit are identical, the two spots of the signal photon in the two slits correspond to the two spots of the idler photon on D_1. Moreover, the low intensity of the pump beam ensures that only one photon pair is created at a time.

When we now observe on D_1 the idler photon in the left (right) intensity spot the signal photon is measured by the detector D_2 behind the upper (lower) slit. In this way we employ the entanglement between the signal and the idler photon to obtain "which-path" information about the signal photon without ever touching it.

However, we can also observe interference fringes while at the same time we gain "which-path" information when we scan D_2 in the *far-field* region of the

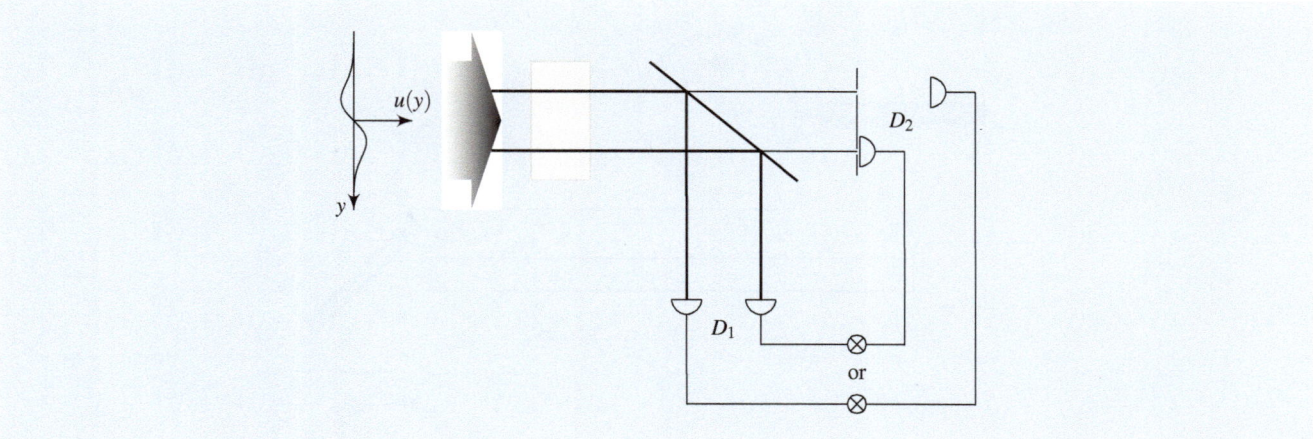

■ **Fig. 19.11** Essential ingredients of the Menzel et al. experiment aimed at observing interference while obtaining "which-path" information in a double-slit arrangement. We pump a nonlinear crystal (*gray area*) with a light beam of transverse mode function $u = u(y)$ with two distinct intensity maxima and a node between them. The correlated photon pair consisting of the signal and the idler photon and emerging from the crystal in two distinct spots is divided by a polarizing beam splitter. The signal photon passes a double-slit and is detected on detector D_2 which is either in the near field, or the far field. The widths and separation of the two slits are adjusted to match the two intensity spots on the end of the crystal. The idler photon is measured on the detector D_1 which is arranged in a way as to ensure that the distances from the beam splitter to D_1 are identical to the ones to the two slits. In this way the idler photon which is entangled with the signal photon allows us to obtain "which-path" information about the latter without ever touching it. Indeed, D_1 can be positioned on the left or right spot as indicated in the figure. We always perform measurements on the signal photon in coincidence with the idler photon

double-slit along the vertical direction and measure in coincidence with the idler photons detected on D_1 in the left spot. In this case we still find interference fringes.

The appearance of fringes in coincidence is surprising since the idler photon provides us with "which-path"-information about the signal photon which still displays interference. In contrast, the principle of complementarity seems to suggest no interference.

19.6.2 Different Atoms Yield "Which-Path" or Interference

We now discuss these rather counter-intuitive results of the Menzel et al. experiment using the model summarized in ■ Fig. 19.12. In particular, we explain the appearance of an interference pattern in the far field in coincidence with the "which-path" information contained in the correlations of the near field. However, we emphasize that there is no violation of the principle of complementarity.

In order to understand the origin of the observed near-field coincidence measurements, and in particular, the perfect correlations between the signal and the idler photon created by the subsequent emissions from a three-level atom we assume for the sake of simplicity that both detectors are in the plane parallel to the screen and consider a vanishing time delay between the clicks on D_1 and D_2. Hence, the atoms emitting the two photons must be in the plane located between the two detectors. When both detectors are in the same intensity maximum of the double-hump structure this plane is at the center of the maximum as well. As a result, we find simultaneous clicks in the two detectors.

However, when they are in different maxima the only atoms satisfying the coincidence condition are half-in-between the detectors, that is, at the node of the mode as indicated in the top of ■ Fig. 19.12. Therefore, the only atoms that could cause the appropriate clicks are not excited. From these perfect correlations between idler and signal photon on the left detector-upper slit or right detector-

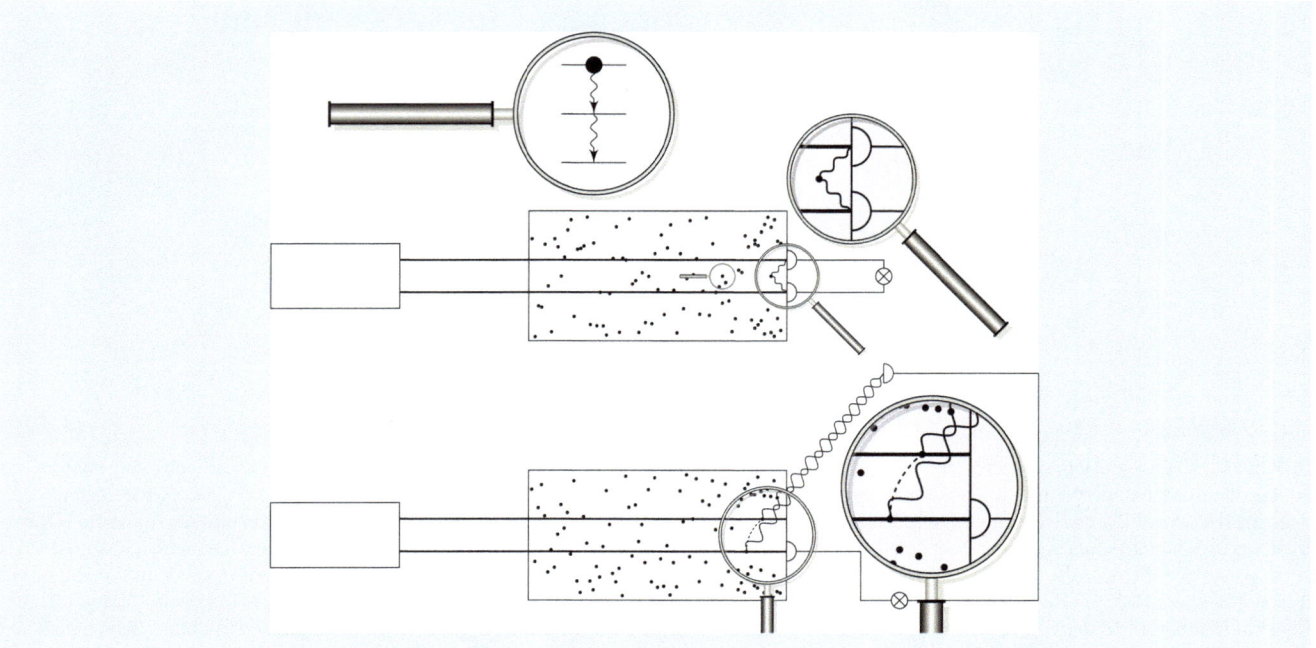

Fig. 19.12 Elementary model to explain the correlations observed in the Menzel et al. experiment. We consider a frozen gas of three-level atoms which are excited by a single photon in a mode consisting of a coherent superposition of two light pencils. When the two detectors are both located at the end of the cell (*near field*) two clicks at the same time registering the signal and idler photon can only result from atoms that are on the symmetry line half-way between the two pencils (*top*). Due to the special mode function these atoms are not excited and the probability for this event to occur vanishes. However, when one detector is in the near field and the other in the far field (*bottom*) the idler photon triggers the detector in the near field. Due to the double-hump structure of the mode function, this click could have come from any atom excited by the two pencils with identical separations from the detector (*dashed line*). As a result, we observe interference fringes in the far field, in complete accordance with the scattering situation of two-level atoms discussed in Fig. 19.9

lower slit we deduce our "which-path" information. We get this knowledge about the signal photon from the idler photon.

Next we consider the measurements which involve the detector D_1 in the near field and D_2 in the far field. Here the idler photon which triggers D_1 was emitted by one atom anywhere in the light pencils. However, we cannot distinguish between two atoms with identical separations from D_1 as indicated by the bottom of Fig. 19.12. As a result, we find the interference of the signal photon emitted from two indistinguishable atoms very much in the spirit of Fig. 19.9.

19.6.3 No Contradiction But Confirmation

We are now in a position to summarize our main results. Three features were crucial for obtaining the on first-sight surprising results of the Menzel et al. experiment: (1) A special mode function of the electromagnetic field consisting of a coherent superposition of two maxima, (2) the entanglement between two photons, and (3) a joint measurement of both of them.

When both detectors are in the near field there are no counts at the same time since the only atoms that could have caused such a result sit at a node of the field and cannot be excited. When one detector is in the near field and one in the far field two indistinguishable atoms lead to an interference signal in the joint count statistics. Since the two experiments correspond to two different arrangements and different atoms are involved there is no contradiction to the principle of complementarity.

19.7   More Questions than Ever

At the center of our essay was the double-slit experiment symbolizing the wave-particle dualism of quantum theory. We have provided a historical perspective and at the end have addressed a rather counter-intuitive experiment. Although there exists a straight-forward explanation many questions associated with this outcome and avenues for further research offer themselves. Here we only allude to one.

Is it possible to perform the Menzel et al. experiment with atoms and would we find the same result? On first sight the answer is "yes" since we can create entangled atoms in complete analogy to photons. But is the measurement process the same?

This question is closely related to the discomfort of W. E. Lamb with the concept of the photon. Whereas in the case of the atoms a description based on single-particle quantum mechanics suffices the photon experiment requires quantum field theory. In the latter the measurement does not reduce the mode function but annihilates the quantum in the mode and leaves the mode intact. However, in the case of the atom the measurement leads to a localization of the particle due to the reduction of the wave packet.

We have started our article by a quote from J. A. Wheeler. Therefore, it is appropriate to close it with the following summary from the same lecture which emphasizes again the influence of the measurement and the role of the observer:

» "Are billions upon billions of acts of observer—participancy the foundations of everything? We are about as far as we can be today from knowing enough about the deeper machinery of the universe to answer this question. Increasing knowledge about detail has brought an increasing ignorance about plan. The very fact that we can ask such a strange question shows how uncertain we are about the deeper foundations of the quantum and its ultimate implications."

Although we have made impressive progress in our understanding of the photon since Wheeler's lecture we are still far away from being able to say that we have discovered "'the plan" or with the words of G. Stein that we "can't understand what made it look strange in the first place."

Acknowledgements We thank W. Becker, R. W. Boyd, K. Dechoum, W. Demtröder, H. Carmichael, M. Efremov, M. Freyberger, R. J. Glauber, L. Happ, C. Henkel, A. Heuer, M. Hillery, J. Leach, M. Komma, G. Leuchs, R. Menzel, H. Paul, D. Puhlmann, E. M. Rasel, M. O. Scully, M. J. A. Spähn, M. Wilkens, S.Y. Zhu, and M. S. Zubairy for many fruitful and stimulating discussions. The author is grateful to Texas A&M University for a Texas A&M University Institute for Advanced Study (TIAS) Faculty Fellowship which made this work possible.

Further Reading

1. For the historical aspects mentioned in this article we refer to Crease RP (2003) The prism and the pendulum: The ten most beautiful experiments in science. Random House, New York

Wheeler JA, Zurek WH (1984) Quantum Theory and Measurement. Princeton University Press, Princeton

Scully RJ, Scully MO (2010) The Demon and the Quantum. Wiley, Berlin

2. The quotes of J. A. Wheeler can be found in Wheeler JA (1994) At Home in the Universe. AIP Press

3. A discussion of the concept of the photon with many historical remarks is presented in Mack H, Schleich WP (2003) A photon viewed from Wigner phase space. Optics and Photonics News Trends 3:29–35

4. An excellent introduction to quantum optics emphasizing the measurement process is provided by Scully MO, Zubairy MS (1997) Quantum Optics. Cambridge University Press, Cambridge

5. For an in-depth discussion of the fascinating facettes of the quantum eraser proposed by Scully MO, Drühl K (1982) Quantum eraser: A proposed photon correlation experiment concerning observation and "delayed choice" in quantum mechanics. Phys Rev A 25:2208–2213

and further developed by Scully MO, Englert BG, Walther H (1991) Quantum optical tests of complementarity. Nature 351:111–116

see for example Aharonov Y, Zubairy MS (2005) Time and the quantum: erasing the past and impacting the future. Science 307(5711):875–879

6. An up-to-date review on delayed-choice experiments can be found in Ma X, Kofler J, Zeilinger, A (2016) Delayed-choice gedanken experiments and their realizations. Rev Mod Phys 88,015005

7. The two-photon version of the double-slit experiment is discussed in Menzel R, Puhlmann D, Heuer A, Schleich WP (2012) Wave-particle dualism and complementarity unraveled by a different mode. Proc Natl Acad Sci 109(24):9314–9319; Bolduc E, Leach J, Miatto FM, Leuchs G, Boyd RW (2014) Fair sampling perspective on an apparent violation of duality. Proc Natl Acad Sci 111(34):12337–12341